Advances in Intelligent Systems and Computing

Volume 993

The series "Advances in Intelligent Systems and Computing" contains publications on theory, applications, and design methods of Intelligent Systems and Intelligent Computing. Virtually all disciplines such as engineering, natural sciences, computer and information science, ICT, economics, business, e-commerce, environment, healthcare, life science are covered. The list of topics spans all the areas of modern intelligent systems and computing such as: computational intelligence, soft computing including neural networks, fuzzy systems, evolutionary computing and the fusion of these paradigms, social intelligence, ambient intelligence, computational neuroscience, artificial life, virtual worlds and society, cognitive science and systems, Perception and Vision, DNA and immune based systems, self-organizing and adaptive systems, e-Learning and teaching, human-centered and human-centric computing, recommender systems, intelligent control, robotics and mechatronics including human-machine teaming, knowledge-based paradigms, learning paradigms, machine ethics, intelligent data analysis, knowledge management, intelligent agents, intelligent decision making and support, intelligent network security, trust management, interactive entertainment, Web intelligence and multimedia.

The publications within "Advances in Intelligent Systems and Computing" are primarily proceedings of important conferences, symposia and congresses. They cover significant recent developments in the field, both of a foundational and applicable character. An important characteristic feature of the series is the short publication time and world-wide distribution. This permits a rapid and broad dissemination of research results.

** Indexing: The books of this series are submitted to ISI Proceedings, EI-Compendex, DBLP, SCOPUS, Google Scholar and Springerlink **

More information about this series at http://www.springer.com/series/11156

Leonard Barolli · Farookh Khadeer Hussain ·
Makoto Ikeda
Editors

Complex, Intelligent, and Software Intensive Systems

Proceedings of the 13th International
Conference on Complex, Intelligent,
and Software Intensive Systems (CISIS-2019)

 Springer

Editors
Leonard Barolli
Department of Information
and Communication Engineering
Fukuoka Institute of Technology,
Faculty of Information Engineering
Fukuoka, Japan

Makoto Ikeda
Department of Information
and Communication Engineering
Fukuoka Institute of Technology,
Faculty of Information Engineering
Fukuoka, Japan

Farookh Khadeer Hussain
School of Software
University of Technology Sydney (UTS)
Ultimo, NSW, Australia

ISSN 2194-5357   ISSN 2194-5365   (electronic)
Advances in Intelligent Systems and Computing
ISBN 978-3-030-22353-3   ISBN 978-3-030-22354-0   (eBook)
https://doi.org/10.1007/978-3-030-22354-0

This Springer imprint is published by the registered company Springer Nature Switzerland AG
The registered company address is: Gewerbestrasse 11, 6330 Cham, Switzerland

Welcome Message of CISIS-2019 International Conference Organizers

Welcome to the 13th International Conference on Complex, Intelligent and Software Intensive Systems (CISIS-2019), which will be held from July 3 to July 5, 2019, at University of Technology Sydney (UTS), Sydney, Australia in conjunction with the 13th International Conference on Innovative Mobile and Internet Services in Ubiquitous Computing (IMIS-2019).

The aim of the conference is to deliver a platform of scientific interaction between the three interwoven challenging areas of research and development of future ICT-enabled applications: software intensive systems, complex systems, and intelligent Systems.

Software intensive systems are systems, which heavily interact with other systems, sensors, actuators, devices, other software systems, and users. More and more domains are involved with software intensive systems, e.g., automotive, telecommunication systems, embedded systems in general, industrial automation systems, and business applications. Moreover, the outcome of Web services delivers a new platform for enabling software intensive systems. The conference is thus focused on tools, practically relevant and theoretical foundations for engineering software intensive systems.

Complex systems research is focused on the overall understanding of systems rather than its components. Complex systems are very much characterized by the changing environments in which they act by their multiple internal and external interactions. They evolve and adapt through internal and external dynamic interactions.

The development of intelligent systems and agents which is each time more characterized by the use of ontologies and their logical foundations build a fruitful impulse for both software intensive systems and complex systems. Recent researches in the field of intelligent systems, robotics, neuroscience, artificial intelligence, and cognitive sciences are very important factor for the future development and innovation of software intensive and complex systems.

CISIS-2019 is aiming at delivering a forum for in-depth scientific discussions among the three communities. The papers included in the proceedings cover all aspects of theory, design, and application of complex systems, intelligent systems, and software intensive systems. The conference received 166 papers and accepted 45 papers (about 27% acceptance rate), which were selected after a careful review process.

We are very proud and honored to have two distinguished keynote talks by Prof. Wanlei Zhou, University of Technology Sydney, Australia, and Dr. Nadeem Javaid, COMSATS University Islamabad, Pakistan, who will present their recent work and will give new insights and ideas to the conference participants.

The organization of an international conference requires the support and help of many people. A lot of people have helped and worked hard to produce a successful CISIS-2019 technical program and conference proceedings. First, we would like to thank all the authors for submitting their papers, the Program Committee Members, and the reviewers who carried out the most difficult work by carefully evaluating the submitted papers. We are grateful to Honorary Co-Chairs Prof. Makoto Takizawa, Hosei University, Japan, and Prof. Jie Lu, University of Technology Sydney, Australia, for their guidance and advice.

This year in conjunction with CISIS-2019 we have seven international workshops that complemented CISIS-2019 program with contributions for specific topics. We would like to thank the Workshops Co-Chairs and all Workshops Organizers for organizing these workshops.

Finally, we would like to thank Web Administrator Co-Chairs and Local Arrangement Co-Chairs for their excellent and timely work.

We hope you will enjoy the conference and have a great time in Sydney, Australia.

Leonard Barolli
Farookh Khadeer Hussain
CISIS-2019 General Co-Chairs

Omar Hussain
Hiroaki Nishino
Kin Fun Li
CISIS-2019 Program Committee Co-chairs

Welcome Message from CISIS-2019 Workshops Co-chairs

Welcome to the Workshops of the 13th International Conference on Complex, Intelligent and Software Intensive Systems (CISIS-2019), which will be held from July 3 to July 5, 2019, at University of Technology Sydney, Sydney, Australia.

We are pleased that for this edition of CISIS International Conference we have seven international workshops. Some of these workshops are in 9th, 10th, 11th, 12th, and 13th editions. The objective was to complement as much as possible the main theme of CISIS-2019 with specific topics of different workshops in order to cover topics from the three challenging areas of ICT-enabled applications: software intensive systems, complex systems, and intelligent systems.

The list of workshops is as follows:

1. The 13th International Workshop on Engineering Complex Distributed Systems (ECDS-2019)
2. The 12th International Workshop on Intelligent Informatics and Natural Inspired Computing (IINIC-2019)
3. The 10th International Workshop on Frontiers in Complex, Intelligent and Software Intensive Systems (FCISIS-2019)
4. The 10th International Workshop on Virtual Environment and Network-Oriented Applications (VENOA-2019)
5. The 9th Semantic Web/Cloud Information and Services Discovery and Management (SWISM-2019)
6. The 6th International Workshop on Hybrid/Cloud Computing Infrastructure for E-Science Application (HCCIEA-2019)
7. The 1st International Workshop on Knowledge Creation and Innovation in Digital World (IKIDW-2019)

These workshops bring to the researchers conducting research in specific themes the opportunity to learn from this rich multidisciplinary experience. The Workshop Co-Chairs would like to thank CISIS-2019 International Conference Organizers for their help and support. We are grateful to the Workshops Organizers for their great efforts and hard work in proposing the workshops, selecting the papers, the interesting programs, and the arrangements of the workshops during the conference days. We are grateful to Web Administrator Co-Chairs for their excellent work and support.

We hope you enjoy the workshops program and proceedings.

Mohammad Alshehri
Tomoya Enokido
Beniamino Di Martino
Workshops Co-chairs of CISIS-2019 International Conference

CISIS-2019 Organizing Committee

Honorary Co-chairs

Makoto Takizawa	Hosei University, Japan
Jie Lu	University of Technology Sydney (UTS), Australia

General Co-chairs

Leonard Barolli	Fukuoka Institute of Technology, Japan
Farookh Khadeer Hussain	University of Technology Sydney (UTS), Australia

Program Committee Co-chairs

Omar Hussain	University of New South Wales, Canberra, Australia
Hiroaki Nishino	Oita University, Japan
Kin Fun Li	University of Victoria, Canada

Workshops Co-chairs

Mohammad Alshehri	Taif University, Saudi Arabia
Tomoya Enokido	Rissho University, Japan
Beniamino Di Martino	University of Campania Luigi Vanvitelli, Italy

International Advisory Board

Yoshitaka Shibata	Iwate Prefectural University, Japan
David Taniar	Monash University, Australia
Minoru Uehara	Toyo University, Japan
Arjan Durresi	IUPUI, USA

Award Co-chairs

Wenny Rahayu	La Trobe University, Australia
Nadeem Javaid	COMSATS University Islamabad, Pakistan
Olivier Terzo	LINKS Foundation, Italy
Hiroshi Shigeno	Keio University, Japan

International Liaison Co-chairs

Akio Koyama	Yamagata University, Japan
Asma Alkalbani	College of Applied Sciences, Oman
Hui-Huang Hsu	Tamkang University, Taiwan
Flora Amato	University of Naples Federico II, Italy

Publicity Co-chairs

Yahya AlHadhrami	University of Technology Sydney (UTS), Australia
Fumiaki Sato	Toho University, Japan
Markus Aleksy	ABB Corporate Research Center, Germany

Finance Chair

Makoto Ikeda	Fukuoka Institute of Technology, Japan

Local Arrangement Co-chairs

Alka Vishwa	University of Technology Sydney (UTS), Australia
Ebtesam Almansour	University of Technology Sydney (UTS), Australia

Web Administrator Co-chairs

Donald Elmazi	Fukuoka Institute of Technology, Japan
Miralda Cuka	Fukuoka Institute of Technology, Japan
Kevin Bylykbashi	Fukuoka Institute of Technology, Japan

Track Areas and PC Members

1. Database and Data Mining Applications

Track Co-chairs

Kin Fun Li	University of Victoria, Canada
Pavel Krömer	Technical University of Ostrava, Czech Republic

PC Members

Antonio Attanasio	LINKS Foundation, Italy
Tibebe Beshah	Addis Ababa University, Ethiopia
Jana Heckenbergerova	University of Pardubice, Czech Republic
Konrad Jackowski	Wroclaw University of Technology, Poland
Petr Musílek	University of Alberta, Canada
Aleš Zamuda	University of Maribor, Slovenia
Genoveva Vargas-Solar	French Council of Scientific Research, LIG-LAFMIA, France
Xiaolan Sha	Sky, UK
Deepali Arora	University of Victoria, Canada
Kosuke Takano	Kanagawa Institute of Technology, Japan
Masahiro Ito	Toshiba Laboratories, Japan
Watheq ElKharashi	Ain Shams University, Egypt
Martine Wedlake	IBM, USA

2. Artificial Intelligence and Bio-inspired Computing

Track Co-chairs

Hai Dong	Royal Melbourne Institute of Technology, Australia
Salvatore Vitabile	University of Palermo, Italy
Chin-Teng Lin	University of Technology Sydney, Australia

PC Members

Kit Yan Chan	Curtin University, Australia
Shang-Pin Ma	National Taiwan Ocean University, Taiwan
Pengcheng Zhang	Hohai University, China
Le Sun	Nanjing University of Information Science and Technology, China
Sajib Mistry	University of Sydney, Australia
Klodiana Goga	LINKS Foundation, Italy
Vincenzo Conti	Kore University of Enna, Italy

Minoru Uehara   Toyo University, Japan
Philip Moore   Lanzhou University, China
Mauro Migliardi   University of Padua, Italy
Dario Bonino   LINKS Foundation, Italy
Andrea Tettamanzi   University of Nice, France
Cornelius Weber   Hamburg University, Germany
Tim Niesen   German Research Center for Artificial
                               Intelligence (DFKI), Germany
Rocco Raso   German Research Center for Artificial
                               Intelligence (DFKI), Germany
Fulvio Corno   Politecnico di Torino, Italy

3. Multimedia Systems and Virtual Reality

Track Co-chairs

Yoshinari Nomura   Okayama University, Japan
Christy (Jie) Liang   University of Technology Sydney, Australia

PC Members

Shunsuke Mihara   Lockon Inc., Japan
Shunsuke Oshima   Kumamoto National College of Technology,
                               Japan
Yuuichi Teranishi   NICT, Japan
Kazunori Ueda   Kochi University of Technology, Japan
Hideaki Yanagisawa   Tokuyama National College of Technology,
                               Japan
Kaoru Sugita   Fukuoka Institute of Technology, Japan
Keita Matsuo   Fukuoka Institute of Technology, Japan
Santi Caballé   Open University of Catalonia, Spain

4. Next-Generation Wireless Networks

Track Co-chairs

Yunfei Chen   University of Warwick, UK
Sriram Chellappan   University of South Florida, USA

PC Members

Elis Kulla   Okayama University of Science, Japan
Santi Caballé   Open University of Catalonia, Spain
Admir Barolli   Aleksander Moisiu University of Durres, Albania
Makoto Ikeda   Fukuoka Institute of Technology, Japan

Keita Matsuo	Fukuoka Institute of Technology, Japan
Shinji Sakamoto	Seikei University, Japan
Omer Wagar	University of Engineering & Technology, Poland
Zhibin Xie	Jiangsu University of Science and Technology, China
Jun Wang	Nanjing University of Posts and Telecommunication, China
Vamsi Paruchuri	University of Central Arkansas, USA
Arjan Durresi	IUPUI, USA
Bhed Bista	Iwate Prefectural University, Japan

5. Semantic Web, Web Services, and Data Integration

Track Co-chairs

| Antonio Messina | Italian National Research Center (CNR), Italy |
| Natalia Kryvinska | Comenius University in Bratislava, Slovakia |

PC Members

Alba Amato	Italian National Research Center (CNR), Italy
Nik Bessis	Edge Hill University, UK
Robert Bestak	Czech Technical University in Prague, Czech Republic
Ivan Demydov	Lviv Polytechnic National University, Ukraine
Marouane El Mabrouk	Abdelmalek Essaadi University, Morocco
Corinna Engelhardt-Nowitzki	University of Applied Sciences, Austria
Michal Gregus	Comenius University in Bratislava, Slovakia
Jozef Juhar	Technical University of Košice, Slovakia
Nikolay Kazantsev	National Research University Higher School of Economics, Russia
Manuele Kirsch Pinheiro	Université Paris 1 Panthéon-Sorbonne, France
Cristian Lai	CRS4 Center for Advanced Studies, Research and Development in Sardinia, Italy
Michele Melchiori	University of Brescia, Italy
Giovanni Merlino	University of Messina, Italy
Kamal Bashah Nor Shahniza	Universiti Teknologi MARA, Malaysia
Eric Pardede	La Trobe University, Australia
Aneta Poniszewska-Maranda	Lodz University of Technology, Poland
Pethuru Raj	IBM Global Cloud Center of Excellence, India
Jose Luis Vazquez Avila	University of Quintana Roo, México
Salvatore Venticinque	University of Campania Luigi Vanvitelli, Italy

6. Security and Trusted Computing

Track Co-chairs

Hiroaki Kikuchi	Meiji University, Japan
Omar Khadeer Hussain	University of New South Wales, Canberra, Australia
Rajat Saxena	Indian Institute of Technology Indore, India

PC Members

Saqib Ali	Sultan Qaboos University, Oman
Zia Rehman	COMSATS University Islamabad, Pakistan
Morteza Saberi	UNSW Canberra, Australia
Sazia Parvin	UNSW Canberra, Australia
Farookh Hussain	University of Technology Sydney, Australia
Walayat Hussain	University of Technology Sydney, Australia
Sabu Thampi	Indian Institute of Information Technology and Management Kerala (IIITM-K), Technopark Campus, India
Sun Jingtao	National Institute of Informatics, Japan
Anitta Patience Namanya	University of Bradford, UK
Smita Rai	Uttarakhand Board of Technical Education, Roorkee, India
Abhishek Saxena	American Tower Corporation Limited, India

7. Cloud Computing Services and Orchestration Tools

Track Co-chairs

Olivier Terzo	LINKS Foundation, Italy
Jan Martinovič	IT4Innovations National Supercomputing Center, VSB Technical University of Ostrava, Czech Republic

PC Members

Alberto Scionti	LINKS Foundation, Italy
Antonio Attanasio	LINKS Foundation, Italy
Jan Platos	VŠB-Technical University of Ostrava, Czech Republic
Rustem Dautov	Kazan Federal University, Russia
Giovanni Merlino	University of Messina, Italy
Francesco Longo	University of Messina, Italy
Dario Bruneo	University of Messina, Italy

Nik Bessis	Edge Hill University, UK
Ming-Xue Wang	Ericsson, Ireland
Luciano Gaido	Istituto Nazionale di Fisica Nucleare (INFN), Italy
Giacinto Donvito	Istituto Nazionale di Fisica Nucleare (INFN), Italy
Andrea Tosatto	Open-Xchange, Germany

8. Parallel, Distributed, and Multicore Computing

Track Co-chairs

Eduardo Alchieri	University of Brasilia, Brazil
Maruf Ahmed	University of Technology Sydney, Australia
Valentina Casola	University of Naples Federico II, Italy

PC Members

Aldelir Luiz	Catarinense Federal Institute, Brazil
Edson Tavares	Federal University of Technology - Parana, Brazil
Fernando Dotti	Pontificia Universidade Catolica do Rio Grande do Sul, Brazil
Hylson Neto	Catarinense Federal Institute, Brazil
Jacir Bordim	University of Brasilia, Brazil
Lasaro Camargos	Federal University of Uberlandia, Brazil
Luiz Rodrigues	Western Parana State University, Brazil
Marcos Caetano	University of Brasilia, Brazil

9. Systems for Biological and Medical Applications

Track Co-chairs

Parag Chatterjee	National Technological University, Argentina
Ricardo Armentano	University of the Republic, Uruguay

PC Members

Leandro Cymberknop	National Technological University, Argentina
Rajiv Pandey	Amity University, India
Robin Singh Bhadoria	Indian Institutes of Information Technology, India
Walter Legnani	National Technological University, Argentina
Eduardo Jaime Quel	National Technological University, Argentina
Andrea Malizia	University of Rome Tor Vergata, Italy

Pablo Ristori	Institute of Scientific and Technical Research for Defense, Argentina
Fernanda Beatríz Martínez Micakoski	National Technological University, Argentina
Asoke Nath	St. Xavier's College, University of Calcutta, India

10. E-learning and Groupware Systems

Track Co-chairs

| Philip Moore | Lanzhou University, China |
| Santi Caballé | Open University of Catalonia, Spain |

PC Members

Nicola Capuano	University of Salerno, Italy
Jordi Conesa	Open University of Catalonia, Spain
Farzin Asadi	Kocaeli University, Kocaeli, Turkey
David Gañan	Open University of Catalonia, Spain
Le Hoang Son	Vietnam National University, Vietnam
Jorge Miguel	Grupo San Valero, Spain
David Newell	Bournemouth University, UK
Antonio Sarasa	Universidad Complutense de Madrid, Spain
Mak Sharma	Birmingham City University, UK
Hai Van Pham	Hanoi University of Science and Technology, Vietnam
Franz Wotawa	Graz University of Technology, Austria
Zhili Zhao	Lanzhou University, China
Fang Zheng	Lanzhou University, China

11. Energy-Aware Computing and Systems

Track Co-chairs

| Muzammil Behzad | University of Oulu, Finland |
| Zahoor Ali Khan | Higher Colleges of Technology, United Arab Emirates |

PC Members

Naveed Ilyas	Gwangju Institute of Science and Technology, South Korea
Muhammad Sharjeel Javaid	University of Hafr Al Batin, Saudi Arabia
Muhammad Talal Hassan	COMSATS University Islamabad, Pakistan

Waseem Raza	University of Lahore, Pakistan
Ayesha Hussain	COMSATS University Islamabad, Pakistan
Umar Qasim	University of Alberta, Canada
Nadeem Javaid	COMSATS University Islamabad, Pakistan
Yasir Javed	Higher Colleges of Technology, UAE
Kashif Saleem	King Saud University, Saudi Arabia
Hai Wang	Saint Mary's University, Canada

12. Complex Systems and Software Modeling and Analytics

Track Co-chairs

Yogesh Beeharry	University of Mauritius, Mauritius
Nabin Sharma	University of Technology Sydney, Australia
Lidia Fotia	Università Mediterranea di Reggio Calabria, Italy

PC Members

Nalini Sooknanan Pillay	Eskom Research Testing & Development, Sustainability & Risk, South Africa
Maleika Heenaye-Mamode Khan	University of Mauritius, Mauritius
T. P. Fowdur	University of Mauritius, Mauritius
Robert T. F. Ah King	University of Mauritius, Mauritius
Anshu Prakash Murdan	University of Mauritius, Mauritius

13. Multi-agent Systems, SLA Cloud, and Social Computing

Track Co-chairs

Douglas Macedo	Federal University of Santa Catarina, Brazil
Giuseppe Sarnè	Mediterranea University of Reggio Calabria, Italy
Takahiro Uchiya	Nagoya Institute of Technology, Japan

PC Members

Mario Dantas	Federal University of Juiz de Fora, Brazil
Luiz Bona	Federal University of Parana, Brazil
Márcio Castro	Federal University of Santa Catarina, Brazil
Fabrizio Messina	University of Catania, Italy
Hideyuki Takahashi	Tohoku University, Japan
Kazuto Sasai	Ibaraki University, Japan
Satoru Izumi	Tohoku University, Japan
Domenico Rosaci	Mediterranea University of Reggio Calabria, Italy
Lidia Fotia	Mediterranea University of Reggio Calabria, Italy

14. Smart Environments and Assistive Technologies

Track Co-chairs

Seyed Shahrestani	Western Sydney University, Australia
Mahmoud Elkhodr	Central Queensland University, Australia

PC Members

Andreas Pitsillides	University of Cyprus, Cyprus
Chun Ruan	Western Sydney University, Australia
Hao Wang	Shandong Normal University, China
Friedbert Kohler	South Western Sydney Local Health District, Australia
Farnaz Farid	University of Sydney, Australia
Kumudu Munasinghe	University of Canberra, Australia
Mehregan Mahdavi	API College, Sydney, Australia
Ergun	Central Queensland University, Australia
Omid Ameri	Victoria University, Sydney, Australia
Nizar Ben Neji	University of Carthage, Tunisia

15. IoT, Semantics, and Adaptive M2M/HCI Interfaces

Track Co-chairs

Farhaan Mirza	Auckland University of Technology, New Zealand
Alessandra De Benedictis	University of Naples Federico II, Italy
Yahya Alhadrami	University of Technology Sydney, Australia

PC Members

Hoa Nguyen	Auckland University of Technology, New Zealand
Muhammad Asif Naeem	Auckland University of Technology, New Zealand
Mirza Baig	Auckland University of Technology, New Zealand
David Sundaram	University of Auckland, New Zealand
Hamid Gholam Hosseini	Auckland University of Technology, New Zealand
Zahoor Ali Khan	Higher Colleges of Technology, UAE
Umar Qasim	University of Alberta, Canada
Farookh Hussain	University of Technology Sydney, Australia
Elis Kulla	Okayama University of Science, Japan
Keita Matsuo	Fukuoka Institute of Technology, Japan

16. FPGA Heterogeneous Architecture

Track Co-chairs

Yuiciiro Shibata	Nagasaki University, Japan
Fujio Kurokawa	Nagasaki Institute of Applied Science, Japan

PC Members

Hidenori Maruta	Nagasaki University, Japan
Haruhi Eto	Nagasaki Institute of Applied Science, Japan
Leonard Barolli	Fukuoka Institute of Technology, Japan
Makoto Takizawa	Hosei University, Japan
Tomoya Enokido	Rissho University, Japan
Olivier Terzo	LINKS Foundation, Italy
Fatos Xhafa	Technical University of Catalonia, Spain

17. Internet of Everything and Machine Learning Applications

Track Co-chairs

Omid Ameri Sianaki	Victoria University, Sydney, Australia
Khandakar Ahmed	Victoria University, Sydney, Australia

PC Members

Farhad Daneshgar	Victoria University, Sydney, Australia
M. Reza Hoseiny F.	University of Sydney, Australia
Kamanashis Biswas (KB)	Australian Catholic University, Australia
Khaled Kourouche	Victoria University, Sydney, Australia
Huai Liu	Victoria University, Sydney, Australia
Mark A Gregory	RMIT University, Australia
Nazmus Nafi	Victoria Institute of Technology, Australia
Mashud Rana	CSIRO, Australia
Farshid Hajati	Victoria University, Sydney, Australia
Ashkan Yousefi	Victoria University, Sydney, Australia
Nedal Ababneh	Abu Dhabi Polytechnic, UAE

CISIS-2019 Reviewers

Ali Khan Zahoor
Alfarraj Osama
Alhussain Thamer
Amato Alba
Amato Flora
Barolli Admir
Barolli Leonard
Bista Bhed
Caballé Santi
Chellappan Sriram
Chen Hsing-Chung
Chen Xiaofeng
Conti Vincenzo
Cui Baojiang
De Benedictis Alessandra
Di Martino Beniamino
Dong Hai
Durresi Arjan
Enokido Tomoya
Esposito Antonio
Ficco Massimo
Fotia Lidia
Fun Li Kin
Gotoh Yusuke
Hussain Farookh
Hussain Omar
Javaid Nadeem
Ikeda Makoto
Ishida Tomoyuki
Kikuchi Hiroaki
Koyama Akio
Kryvinska Natalia

Kulla Elis
Lee Kyungroul
Matsuo Keita
Moore Philip
Nishino Hiroaki
Ogiela Lidia
Ogiela Marek
Palmieri Francesco
Paruchuri Vamsi Krishna
Platos Jan
Rahayu Wenny
Rawat Danda
Saberi Morteza
Saito Takamichi
Sakamoto Shinji
Sato Fumiaki
Scionti Alberto
Sianaki Omid Ameri
Spaho Evjola
Sugita Kaoru
Takizawa Makoto
Taniar David
Terzo Olivier
Uchida Noriki
Uehara Minoru
Venticinque Salvatore
Vitabile Salvatore
Wang Xu An
Woungang Isaac
Xhafa Fatos
Yim Kangbin
Yoshihisa Tomoki

Welcome Message from ECDS-2019 International Workshop Co-chairs

It is our great pleasure to welcome you to the 13th International Workshop on Engineering Complex Distributed Systems (ECDS-2019), which will be held in conjunction with the 13th International Conference on Complex, Intelligent and Software Intensive Systems (CISIS-2019) from July 3rd to July 5th, 2019, at University of Technology Sydney (UTS), Sydney, Australia.

In the past, this field included technology concerns related to middleware solutions, dealing with the heterogeneity of the miscellaneous hardware and software environments and computing infrastructure. These technologies have been used to address the integration of existing legacy applications and improve the interoperability between applications across enterprises. The advances in wireless communication and pervasive computing extend this traditional wired area of distributed systems and make the new advanced application possible. The complexity of today's applications requires additional approaches to be able to realize an enterprise application time- and cost-saving. This includes the ability to model business processes, business policies, and event-oriented aspects of large systems and express these models through design solutions to address the complexity of enterprise applications and ease software design efforts. In addition, the engineering of complex distributed systems also requires a good understanding of the problem areas of concern for information systems and business administration, such as process management, supply chain management, security issues, and electronic business. These topics need to be addressed in order to deal with the complexity of today's increasingly dynamic, mobile, cross-organizational, and cross-jurisdictional systems.

In this workshop, various aspects of the design and implementation of distributed systems will be discussed. The scope of the presented papers ranges from engineering approaches and techniques to applications.

This workshop would not have been possible without the help of many people. First of all, we would like to thank all the authors for submitting their papers to our workshop. We also like to thank the Program Committee Chair, Program

Committee Members, and additional reviewers, who carefully evaluated the submitted papers.

We hope that you find the ECDS-2019 program inspiring and that the workshop provides you with the opportunity to interact, share ideas with, and learn from other distributed systems researchers from around the world. We also encourage you to continue to participate in future ECDS workshops, to increase its visibility, and to interest others in contributing to this growing community.

Leonard Barolli
Makoto Takizawa
ECDS-2019 Workshop Co-chairs

ECDS-2019 Organizing Committee

Workshop Co-chairs

Leonard Barolli	Fukuoka Institute of Technology (FIT), Japan
Makoto Takizawa	Hosei University, Japan

Program Committee Members

Markus Aleksy	ABB Corporate Research, Germany
Irfan Awan	University of Bradford, UK
Bhed Bahadur Bista	Iwate Prefectural University, Japan
Arjan Durresi	Indiana University–Purdue University at Indianapolis, USA
Tomoya Enokido	Rissho University, Japan
Akio Koyama	Yamagata University, Japan
Takahiro Uchiya	Nagoya Institute of Technology, Japan
Takuo Suganuma	Tohoku University, Japan
Kaoru Sugita	Fukuoka Institute of Technology, Japan
David Taniar	Monash University, Australia
Minoru Uehara	Toyo University, Japan
Marten van Sinderen	University of Twente, The Netherlands
Fatos Xhafa	Technical University of Catalonia, Spain
Muhammad Younas	Oxford Brookes University, UK
Maciej Zygmunt	ABB Corporate Research, Poland
Stefan Kuhlins	Heilbronn University, Germany

Web Administrators

Donald Elmazi	Fukuoka Institute of Technology (FIT), Japan
Miralda Cuka	Fukuoka Institute of Technology (FIT), Japan

Message from IINIC-2019 International Workshop Organizers

Advanced information processing technologies have the potential to significantly accelerate research in different fields. In particular, techniques from artificial intelligence, machine learning, and data mining can assist researchers in the discovery of new knowledge for next-generation applications. This workshop aims to attract state-of-the-art solutions and novel attempts in this direction.

The 12th International Workshop on Intelligent Informatics and Natural Inspired Computing (IINIC-2019) will provide a platform for researchers to meet and exchange their thoughts. IINIC-2019 will be held in conjunction with the 13th International Conference on Complex, Intelligent and Software Intensive Systems (CISIS-2019) from July 3rd to July 5th, 2019, at University of Technology Sydney (UTS), Sydney, Australia.

Many people contributed to the success of IINIC-2019. We wish to thank the Program Committee Members for their great effort. We also would like to express our gratitude to the main organizers of CISIS-2019 for their excellent work in organizing the conference. Last but not least, we would like to thank and congratulate all the contributing authors for their support to the workshop.

Takahiro Uchiya
Leonard Barolli
IINIC-2019 Workshop Co-chairs

IINIC-2019 Organizing Committee

Workshop Co-chairs

Takahiro Uchiya   Nagoya Institute of Technology, Japan
Leonard Barolli   Fukuoka Institute of Technology, Japan

Program Committee Members

Tun-Wen Pai	National Taiwan Ocean University, Taiwan
Oliver Ray	University of Bristol, UK
Tetsuya Oda	Okayama University of Science, Japan
Elis Kulla	Okayama University of Science, Japan
Salvatore Vitabile	University of Palermo, Italy
Omar Hussain	University of New South Wales, Canberra, Australia
Takuo Suganuma	Tohoku University, Japan
Makoto Ikeda	Fukuoka Institute of Technology, Japan
Fatos Xhafa	Technical University of Catalonia, Spain
Santi Caballé	Open University of Catalonia, Spain
Farookh Hussain	University of Technology Sydney, Australia

Message from FCISIS-2019 International Workshop Organizers

It is our great pleasure to welcome you for the 10th International Workshop on Frontiers in Complex, Intelligent and Software Intensive Systems (FCISIS-2019). The workshop will be held in conjunction with the 13th International Conference on Complex, Intelligent and Software Intensive Systems (CISIS-2019) from July 3rd to July 5th, 2019, at University of Technology Sydney (UTS), Sydney, Australia.

The objective of FCISIS Workshop is to foster the discussion in a rich inter-disciplinary context of the three challenging areas of ICT-enabled applications: software intensive systems, complex systems, and intelligent systems. FCISIS-2019 is conceived in terms of special papers, which were also carefully selected, from the organizers.

We would like to thank all participants of the workshop for submitting their research works and for their participation and look forward to meet you again in forthcoming editions of the workshop.

Leonard Barolli
FCISIS-2019 Workshop Chair

FCISIS-2019 Organizing Committee

Workshop Chair

Leonard Barolli   Fukuoka Institute of Technology, Japan

Program Committee Members

Makoto Ikeda   Fukuoka Institute of Technology, Japan
Tomoya Enokido   Rissho University, Japan

Farookh Hussain   University of Technology Sydney, Australia
Hiroaki Kikuchi   Meiji University, Japan
Akio Koyama   Yamagata University, Japan
Keita Matsuo   Fukuoka Institute of Technology, Japan
Hiroaki Nishino   Oita University, Japan
Tetsuya Shigeyasu   Prefectural University of Hiroshima, Japan
Makoto Takizawa   Hosei University, Japan
Salvatore Vitabile   University of Palermo, Italy
Admir Barolli   Aleksander Moisiu University of Durres, Albania
Elis Kulla   Okayama University of Science, Japan
Evjola Spaho   Polytechnic University of Tirana, Albania
Noriki Uchida   Fukuoka Institute of Technology, Japan
Hiroshi Maeda   Fukuoka Institute of Technology, Japan

Message from VENOA-2019 International Workshop Organizers

Welcome to the 10th International Workshop on Virtual Environment and Network-Oriented Applications (VENOA-2019), which will be held in conjunction with the 13th International Conference on Complex, Intelligent and Software Intensive Systems (CISIS-2019) from July 3rd to July 5th, 2019, at University of Technology Sydney (UTS), Sydney, Australia.

The past eight workshops were very successful, and many high-quality papers were presented and published in these workshops. We are pleased to announce the continuation of this workshop for serving as a forum for the exchange of information and ideas in the field of 3D computer graphics, virtual reality (VR), augmented reality (AR), mobile communications, IoT, and Web and network applications. We again received many unique and high-quality paper submissions in this workshop. We strictly followed the CISIS review procedures and finally selected excellent papers for publication and presentation. The program shows a variety of research activities with high relevance to the scope of the workshop.

This workshop cannot be organized without hard and excellent work of CISIS-2019 conference organizers. We would like to express our sincere appreciation to VENOA-2019 Program Committee Members and reviewers for their cooperation in completing their efforts under a very tight schedule. We also give our special thanks to all authors for their valuable contributions. We hope that these papers will have significant impacts and stimulate future research activities.

Yong-Moo Kwon
Hiroaki Nishino
VENOA-2019 Workshop Co-chairs

VENOA-2019 Organizing Committee

Workshop Co-chairs

Yong-Moo Kwon   Korea Institute of Science and Technology, Korea
Hiroaki Nishino   Oita University, Japan

Program Committee Members

Minoru Ikebe   Oita University, Japan
Eiji Aoki   Institute for Hypernetwork Society, Japan
Byungrae Cha   Gwangju Institute of Science and Technology, Korea
Makoto Fujimura   Nagasaki University, Japan
Nobuo Funabiki   Okayama University, Japan
Ken'ichi Furuya   Oita University, Japan
Nobukazu Iguchi   Kinki University, Japan
Tsuneo Kagawa   Oita University, Japan
Laehyun Kim   Korea Institute of Science and Technology, Korea
JongWon Kim   Gwangju Institute of Science and Technology, Korea
Byung-Gook Lee   Dongseo University, Korea
Jong Weon Lee   Sejong University, Korea
Yukikazu Murakami   Kagawa National College of Technology, Japan
Makoto Nakashima   Oita University, Japan
Dahlan Nariman   Ritsumeikan Asia Pacific University, Japan
Satoshi Ohtake   Oita University, Japan
Yoshihiro Okada   Kyushu University, Japan
Yoshitaka Sakurai   Meiji University, Japan
Shinji Sugawara   Chiba Institute of Technology, Japan
Shigeto Tajima   Osaka University, Japan
Kenzi Watanabe   Hiroshima University, Japan
Kazuyuki Yoshida   Oita University, Japan

Message from SWISM-2019 International Workshop Organizers

Welcome to the 9th International Workshop on Semantic Web/Cloud Information and Services Discovery and Management (SWISM-2019), which is held in conjunction with the 13th International Conference on Complex, Intelligent, and Software Intensive Systems (CISIS-2019) from July 3rd to July 5th, 2019, at University of Technology Sydney (UTS), Sydney, Australia.

SWISM-2019 will bring together scientists, engineers, computer users, and students to exchange and share their experiences, new ideas, and research results about all aspects (theory, applications, and tools) of intelligent and semantic methods applied to Web- and Cloud-based systems, and to discuss the practical challenges encountered and the solutions adopted.

The program of SWISM-2019 includes papers related to information retrieval, ontologies, intelligent agents, intelligent techniques for management and programming of Cloud services and business processes. The program for the conference is the result of the excellent work of reviewers and Program Committee Members. We hope you will find the final program enriching and stimulating.

We believe that all the papers and topics will provide novel ideas, new theoretical and experimental results, and will stimulate the future research activities in this area.

The papers collected in this international workshop were carefully reviewed by reviewers. According to the review results, the Program Committee Members selected high-quality papers to be presented in this workshop.

We would like to express our sincere appreciation to all Program Committee Members for their cooperation. We are thankful to Honorary Co-Chairs, General Co-Chairs, Program Committee Co-Chairs, and Workshops Co-Chairs of CISIS-2019 for excellent conference organization. It was a great pleasure in working with them.

Last but not least, we are grateful to all authors for their valuable contributions and attendees who contributed to the success of the program with their papers and speeches on their research results, and with their participation in the conference.

We hope you will enjoy the workshop and conference and have a great time in Sydney, Australia.

Beniamino Di Martino
Salvatore Venticinque
Antonio Esposito
SWISM-2019 Workshop Co-chairs

SWISM-2019 Organizing Committee

Workshop Co-chairs

Beniamino Di Martino	University of Campania Luigi Vanvitelli, Italy
Salvatore Venticinque	University of Campania Luigi Vanvitelli, Italy
Antonio Esposito	University of Campania Luigi Vanvitelli, Italy

Program Committee Members

Omer Rana	University of Cardiff, UK
Siegfred Benkner	University of Vienna, Austria
Marios Dikaiakos	University of Cyprus, Cyprus
Dieter Kranzlmueller	University Ludwig Maximilian of Munich, Germany
Antonino Mazzeo	University of Naples Federico II, Italy
Domenico Talia	University of Calabria, Italy
Rocco Aversa	University of Campania Luigi Vanvitelli, Italy
Thomas Fahringer	University of Innsbruck, Austria
Vincenzo Loia	University of Salerno, Italy

Welcome Message from HCCIEA-2019 International Workshop Chair

On behalf of the Organizing Committee, we would like to welcome you to the 6th International Workshop on Hybrid/Cloud Computing Infrastructure for E-Science Application (HCCIEA-2019) which will be held in conjunction with the 13th International Conference on Complex, Intelligent, and Software Intensive Systems (CISIS-2019) from July 3rd to July 5th, 2019, at University of Technology Sydney (UTS), Australia.

The workshop aims to promote research and development activities focused on E-science applications using distributed computing infrastructure, such as grid, Cloud computing, and hybrid system. With the rapid emergence of software systems and their applicability, the amount of data is growing exponentially. Existing computing infrastructure, software system designs, and use cases must take into account the enormity in volume of requests, size of data, and computing load. A complementary goal is to identify the open issues and the challenges to fix them, especially on security, flexibility, reliability, and privacy aspects.

Cloud computing has become a scalable services consumption and delivery platform in the field of services computing. Cloud is a platform or infrastructure that allows execution of code in a managed and elastic way. We want to put the emphasis of scientific and technologies progress on Cloud solutions and infrastructures, in particular concerning research activities on scalability and adaptability using effective scheduling for the virtualization.

All people involved in this workshop (authors and PC members) are researchers with high expertise, working on related research areas and projects. We are really grateful for their support, and we thank them for contributing their knowledge toward a successful event.

We would like to thank CISIS organizers for giving us the opportunity to organize HCCIEA Workshop series. We hope that the results of this event will advance the related research in multifold ways.

Olivier Terzo
HCCIEA-2019 chairs

HCCIEA-2019 Organizing Committee

Workshop Chair

Olivier Terzo   LINKS Foundation, Italy

Program Committee Members

Alexander Jungmann   University of Paderborn, Germany
Antonio Attanasio   LINKS Foundation, Italy
Antonio Parodi   CIMA Foundation, Italy
Fatos Xhafa   Technical University of Catalonia, Spain
Giuseppe Caragnano   LINKS Foundation, Italy
Alberto Scionti   LINKS Foundation, Italy
Klodiana Goga   LINKS Foundation, Italy
Leonard Barolli   Fukuoka Institute of Technology, Japan
Vincenzo Romano   INGV, Italy

Welcome Message from IKIDW-2019 International Workshop Co-chairs

Welcome to the 1st International Workshop on Knowledge Creation and Innovation in Digital World (IKIDW-2019). The workshop will be held in conjunction with the 13th International Conference on Complex, Intelligent and Software Intensive Systems (CISIS-2019) at University of Technology Sydney, Australia, from July 3rd to July 5th, 2019.

The value of most organizations today greatly exceeds their net tangible assets. The IKIDW-2019 workshop aims to address contemporary issues in managing knowledge, intellectual capital, and other intangible assets in the digital world with the help of IT application. The digital era contributes to the amount of knowledge available in various qualities. This is a challenge for business people in strategic decision making. IT application is expected to reduce knowledge ambiguity so that it will improve the quality of organizational decisions. Beginning with a view that knowledge becomes strategic assets, the workshop will discuss the fundamentals of managing knowledge and intellectual capital, understanding some of the measurement issues, processes, and cycles involved in their management and the specific issues in managing knowledge, especially with the availability of big data and with the help of IT application.

We would like to express our sincere gratitude to the members of the Program Committee for their efforts. We thank the 13th International Conference on Complex, Intelligent, and Software Intensive Systems (CISIS-2019) for co-hosting IKIDW-2019. Most importantly, we thank all the authors for their submission and contribution to the workshop.

We hope all of you will enjoy IKIDW-2019 and find this a productive opportunity to exchange ideas with many researchers.

Olivia Fachrunnisa
Ardian Adhiatma
IKIDW-2019 Workshop Co-chairs

IKIDW-2019 Organizing Committee

Workshop Co-chairs

Olivia Fachrunnisa	UNISSULA, Indonesia
Ardian Adhiatma	UNISSULA, Indonesia

Program Committee Members

Ahmed A. Al-Absi	Kyungdong University, Korea
Baharom Abdul Rahman	Universiti Teknologi MARA (UiTM) Terengganu, Indonesia
Chih-Peng Chu	National Dong Hwa University, Taiwan
Shu-Ling Chen	National Dong Hwa University, Taiwan
Farookh Hussain	University of Technology Sydney (UTS), Australia
Omar Hussain	University of New South Wales, Canberra, Australia

CISIS-2019 Keynote Talks

Trust, Security and Privacy in Low-Cost RFID Systems

Wanlei Zhou

University of Technology Sydney, Sydney, Australia

Abstract. Radio Frequency Identification (RFID) enables the automatic identification of objects using radio waves without the need for physical contact with the objects. RFID has been widely used in various fields such as logistics, manufacturing, pharmaceutical, supply chain management, healthcare, defense, aerospace and many other areas, apart from touching our everyday lives through RFID enabled car keys, ePassports, clothing, electronic items and others. However, the wide adoptions of RFID technologies also introduce serious security and privacy risks as the information stored in RFID tags can easily be retrieved by any malicious party with a compatible reader. In this talk, we will introduce some trust, security and privacy challenges in RFID technologies, and based on our research, we will outline a number of schemes for authentication, ownership transfer, secure search and grouping proof in Low-cost RFID systems.

Intelligent Context Awareness in Internet of Agricultural Things

Nadeem Javaid

COMSATS University Islamabad, Islamabad, Pakistan

Abstract. Variability in climate and recession in water reservoirs, diminishing the agrarian sector ecosystem production day by day. There is an imperative requirement to restore robustness and ensure high production rate with the use of smart communication infrastructure. Moreover, the farmers will be able to make resource efficient decisions with the availability of modern monitoring systems like Internet of agricultural things (IoAT). However, the data generated through IoAT devices is disparate which needs to be handled intelligently to bring artificial intelligence (AI), machine learning (ML) and data analytic (DA) techniques into play. In this talk, we will recommend the intensive use of coordination between AI, ML and DA at middleware to optimize the performance of IoAT system along with context awareness. Additionally, it will enable horizontal functionality for diverse services to mitigate the problem of inter-operability. An analysis is carried out using TOWS matrix to consider the effects of internal and external factors on the performance of automation techniques collaboration. This analysis points out various opportunities to innovate the livelihood of agrarian society around the globe.

Contents

The 10th International Workshop on Virtual Environment and Network-Oriented Applications (VENOA-2019)

The 9th Semantic Web/Cloud Information and Services Discovery
and Management (SWISM-2019)

The 6th International Workshop on Hybrid/Cloud Computing
Infrastructure for E-Science Application (HCCIEA-2019)

The 1st International Workshop on Knowledge Creation and Innovation in Digital World (IKIDW-2019)

l Contents

The 13th International Conference on Complex, Intelligent and Software Intensive Systems (CISIS-2019)

Implementation of a Fuzzy-Based Simulation System and a Testbed for Improving Driving Conditions in VANETs

Kevin Bylykbashi[1(✉)], Donald Elmazi[2], Keita Matsuo[2], Makoto Ikeda[2], and Leonard Barolli[2]

[1] Graduate School of Engineering, Fukuoka Institute of Technology (FIT),
3-30-1 Wajiro-Higashi, Higashi-Ku, Fukuoka 811–0295, Japan
`bylykbashi.kevin@gmail.com`
[2] Department of Information and Communication Engineering,
Fukuoka Institute of Technology (FIT),
3-30-1 Wajiro-Higashi, Higashi-Ku, Fukuoka 811–0295, Japan
`donald.elmazi@gmail.com`, {`kt-matsuo,barolli`}`@fit.ac.jp`,
`makoto.ikd@acm.org`

Abstract. Vehicular Ad Hoc Networks (VANETs) have gained a great attention due to the rapid development of mobile internet and Internet of Things (IoT) applications. With the evolution of technology, it is expected that VANETs will be massively deployed in upcoming vehicles. In addition, ambitious efforts are being done to incorporate Ambient Intelligence (AmI) technology in the vehicles, as it will be an important factor for VANET to accomplish one of its main goals, the road safety. In this paper, we propose an intelligent system for improving driving condition using fuzzy logic. The proposed system considers in-car environment data and driver's vital signs data to make the decision. Then uses the smart-box to inform the driver and to provide a better assistance. We aim to realize a new system to support the driver for safe driving. We evaluated the performance of proposed system by computer simulations and experiments. From the evaluation results, we conclude that the driver's heart rate, noise level and vehicle's inside temperature have different effects to the driver's condition.

1 Introduction

Traffic accidents, road congestion and environmental pollution are persistent problems faced by both developed and developing countries, which have made people live in difficult situations. Among these, the traffic incidents are the most serious ones because they result in huge loss of life and property. For decades, we have seen governments and car manufacturers struggle for safer roads and car accident prevention. The development in wireless communications has allowed companies, researchers and institutions to design communication systems that

© Springer Nature Switzerland AG 2020
L. Barolli et al. (Eds.): CISIS 2019, AISC 993, pp. 3–12, 2020.
https://doi.org/10.1007/978-3-030-22354-0_1

provide new solutions for these issues. Therefore, new types of networks, such as Vehicular Ad Hoc Networks (VANETs) have been created. VANET consists of a network of vehicles in which vehicles are capable of communicating among themselves in order to deliver valuable information such as safety warnings and traffic information.

Nowadays, every car is likely to be equipped with various forms of smart sensors, wireless communication modules, storage and computational resources. The sensors will gather information about the road and environment conditions and share it with neighboring vehicles and adjacent road side units (RSU) via vehicle-to-vehicle (V2V) or vehicle-to-infrastructure (V2I) communication. However, the difficulty lies on how to understand the sensed data and how to make intelligent decisions based on the provided information.

As a result, Ambient Intelligence (AmI) becomes a significant factor for VANETs. Various intelligent systems and applications are now being deployed and they are going to change the way manufacturers design vehicles. These systems include many intelligence computational technologies such as fuzzy logic, neural networks, machine learning, adaptive computing, voice recognition, and so on, and they are already announced or deployed [1]. The goal is to improve both vehicle safety and performance by realizing a series of automatic driving technologies based on the situation recognition. The car control relies on the measurement and recognition of the outside environment and their reflection on driving operation.

On the other hand, we are focused on the in-car information and driver's vital information to detect the danger or risk situation and inform the driver about the risk or change his mood. Thus, our goal is to prevent the accidents by supporting the drivers. In order to realize the proposed system, we will use some Internet of Things (IoT) devices equipped with various sensors for in-car monitoring.

In this paper, we propose a fuzzy-based system for improving driving conditions considering three parameters: Heart Rate (HR), Noise Level (NL) and Vehicle Inside Temperature (VIT) to decide the Driver's Situation Condition (DSC).

The structure of the paper is as follows. In Sect. 2, we present an overview of VANETs. In Sect. 3, we present a short description of AmI. In Sect. 4, we describe the proposed fuzzy-based system and its implementation. In Sect. 5, we discuss the simulation and experimental results. Finally, conclusions and future work are given in Sect. 6.

2   Vehicular Ad Hoc Networks (VANETs)

VANETs are a type of wireless networks that have emerged thanks to advances in wireless communication technologies and the automotive industry. VANETs are considered to have an enormous potential in enhancing road traffic safety and traffic efficiency. Therefore, various governments have launched programs dedicated to the development and consolidation of vehicular communications and networking and both industrial and academic researchers are addressing many related challenges, including socio-economic ones, which are among the most important [2].

The VANET technology uses moving vehicle as nodes to form a wireless mobile network. It aims to provide fast and cost-efficient data transfer for the advantage of passenger safety and comfort. To improve road safety and travel comfort of voyagers and drivers, Intelligent Transport Systems (ITS) are developed. The ITS manages the vehicle traffic, support drivers with safety and other information, and provide some services such as automated toll collection and driver assist systems [3].

The VANETs provide new prospects to improve advanced solutions for making reliable communication between vehicles. VANETs can be defined as a part of ITS which aims to make transportation systems faster and smarter, in which vehicles are equipped with some short-range and medium-range wireless communication [4]. In a VANET, wireless vehicles are able to communicate directly with each other (i.e., emergency vehicle warning, stationary vehicle warning) and also served various services (i.e., video streaming, internet) from access points (i.e., 3G or 4G) through roadside units.

3 Ambient Intelligence (AmI)

The AmI is the vision that technology will become invisible, embedded in our natural surroundings, present whenever we need it, enabled by simple and effortless interactions, attuned to all our senses, adaptive to users and context and autonomously acting [5]. High quality information and content must be available to any user, anywhere, at any time, and on any device.

In order that AmI becomes a reality, it should completely envelope humans, without constraining them. Distributed embedded systems for AmI are going to change the way we design embedded systems, in general, as well as the way we think about such systems. But, more importantly, they will have a great impact on the way we live. Applications ranging from safe driving systems, smart buildings and home security, smart fabrics or e-textiles, to manufacturing systems and rescue and recovery operations in hostile environments, are poised to become part of society and human lives.

The AmI deals with a new world of ubiquitous computing devices, where physical environments interact intelligently and unobtrusively with people. AmI environments can be diverse, such as homes, offices, meeting rooms, hospitals, control centers, vehicles, tourist attractions, stores, sports facilities, and music devices.

In the future, small devices will monitor the health status in a continuous manner, diagnose any possible health conditions, have conversation with people to persuade them to change the lifestyle for maintaining better health, and communicates with the doctor, if needed [6]. The device might even be embedded into the regular clothing fibers in the form of very tiny sensors and it might

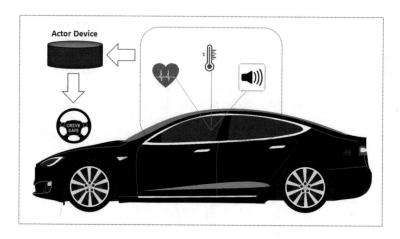

Fig. 1. Proposed system architecture.

communicate with other devices including the variety of sensors embedded into the home to monitor the lifestyle. For example, people might be alarmed about the lack of a healthy diet based on the items present in the fridge and based on what they are eating outside regularly.

The AmI paradigm represents the future vision of intelligent computing where environments support the people inhabiting them [7–9]. In this new computing paradigm, the conventional input and output media no longer exist, rather the sensors and processors will be integrated into everyday objects, working together in harmony in order to support the inhabitants [10]. By relying on various artificial intelligence techniques, AmI promises the successful interpretation of the wealth of contextual information obtained from such embedded sensors and will adapt the environment to the user needs in a transparent and anticipatory manner.

4   Proposed System

In this work, we use fuzzy logic to implement the proposed system. Fuzzy sets and fuzzy logic have been developed to manage vagueness and uncertainty in a reasoning process of an intelligent system such as a knowledge based system, an expert system or a logic control system [11–16]. In Fig. 1 we show the architecture of our proposed system.

4.1   Proposed Fuzzy-Based Simulation System

The proposed system called Fuzzy-based System for improving Driver's Situation Condition (FSDSC) is shown in Fig. 2. For the implementation of our system, we consider three input parameters: Heart Rate (HR), Noise Level (NL) and Vehicle Inside Temperature (VIT) to determine the Driver's Situation Condition. These

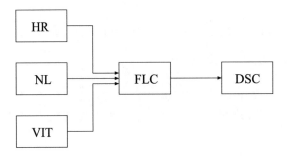

Fig. 2. Proposed system structure.

three input parameters are not correlated with each other, for this reason we use fuzzy system. The input parameters are fuzzified using the membership functions showed in Fig. 3(a), (b) and (c). In Fig. 3(d) are shown the membership functions used for the output parameter. We use triangular and trapezoidal membership functions because they are suitable for real-time operation. The term sets for each linguistic parameter are shown in Table 1. We decided the number of term sets by carrying out many simulations. In Table 2, we show the Fuzzy Rule Base (FRB) of FSDSC, which consists of 27 rules. The control rules have the form: IF "conditions" THEN "control action". For instance, for Rule 1: "IF HR is S, NL is Q and VIT is L, THEN DSC is B" or for Rule 11: "IF HR is No, NL is Q and VIT is M, THEN DSC is VG".

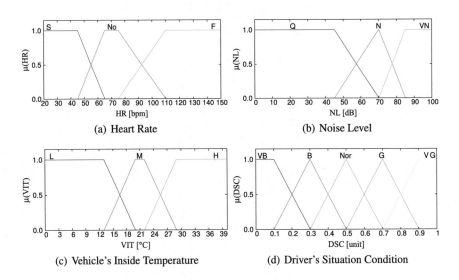

Fig. 3. Membership functions.

Table 1. Parameters and their term sets for FSDSC.

Parameters	Term sets
Heart Rate (HR)	Slow (S), Normal (No), Fast (F)
Noise Level (NL)	Quiet (Q), Noisy (N), Very Noisy (VN)
Vehicle Inside Temperature (VIT)	Low (L), Medium (M), High (H)
Driver's Situation Condition (DSC)	Very Bad (VB), Bad (B), Normal (Nor), Good (G), Very Good (VG)

Table 2. The fuzzy rule base of FSDSC.

Rule	HR	NL	VIT	DSC
1	S	Q	L	B
2	S	Q	M	Nor
3	S	Q	H	B
4	S	N	L	VB
5	S	N	M	B
6	S	N	H	VB
7	S	VN	L	VB
8	S	VN	M	VB
9	S	VN	H	VB
10	No	Q	L	G
11	No	Q	M	VG
12	No	Q	H	G
13	No	N	L	Nor
14	No	N	M	G
15	No	N	H	Nor
16	No	VN	L	B
17	No	VN	M	Nor
18	No	VN	H	B
19	F	Q	L	B
20	F	Q	M	Nor
21	F	Q	H	B
22	F	N	L	VB
23	F	N	M	B
24	F	N	H	VB
25	F	VN	L	VB
26	F	VN	M	VB
27	F	VN	H	VB

4.2   Testbed Description

In order to evaluate the proposed system, we implemented a testbed and carried out experiments in a real scenario. A snapshot of testbed is shown in Fig. 4. The testbed is composed of sensing and processing components. The sensing system is implemented in the Arduino Platform. We set-up sensors on Arduino Uno to measure the environment temperature and noise, and the driver's heart rate. Then, we implemented a processing device to get the sensed data and to run our fuzzy system. The processing device is connected to Arduino via USB cable. It consists on a Raspberry Pi 3 model B running on Linux Raspbian. The Raspbian is an operating system based on Debian optimized for the Raspberry Pi hardware. We used Arduino IDE and Processing language to get the sensed data. Then, we use FuzzyC [17] to fuzzify these data and to determine the Driver's Situation Condition which is the output of our proposed system. Based on the DSC an appropriate task can be performed.

Fig. 4. Snapshot of testbed.

5   Proposed System Evaluation

5.1   Simulation Results

In this subsection, we present the simulation results for our proposed system. The simulation results are presented in Fig. 5. We consider the HR as a constant parameter. The NL values considered for simulations are from 40 to 85 dB. We show the relation between DSC and VIT for different NL values. We vary the VIT parameter from 0 to 40 °C. From the results, we see that for VIT values under 13 and over 29 °C, the DSC value is very low. Also, we can see that when NL is increased, the DSC is decreased. Better DSC values are achieved for VIT 20–22 °C and NL 40 dB.

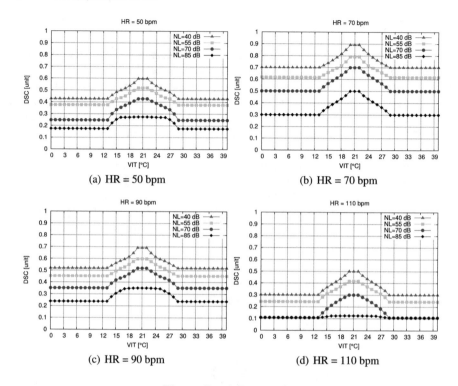

Fig. 5. Simulation results.

In Fig. 5(a), we consider the HR value 50 bpm. A normal situation for the driver with his heart beating 50 times per minute is when there is not any annoying noise and also the ambient temperature is between 17 and 25 °C.

In Fig. 5(b), we present the simulation results for HR 70 bpm. The DSC values are better than the first case. This is because of the heart rate value which lies within the normal range, so the driver's situation is much better, and he can endure situations when a noise can be present or the temperatures are not the best ones.

In Fig. 5(c), we show the simulation results for HR 90 bpm. This rate is not a big concern as it is not a very fast rate but it can be seen that any noise or temperature other than 15 and 27 °C range could affect very much the driver.

In Fig. 5(d), we increase the value of HR to 110 bpm. We can see that there is not a situation that can be decided to be either good or very good by our system.

In the cases where the driver's situation is decided as bad of very bad continuously for relatively long time, the system can perform a certain action. For example, the system may limit the vehicle's maximal speed, suggest him to have a rest, or to call the doctor if his heart beats at very low/high rates.

5.2 Experimental Results

The experimental results are presented in Fig. 6. In Fig. 6(a) are shown the results of DSC when HR is Low. As we can see there are just a few DSC values decided as normal situation by the system. All other are values are bad and very bad.

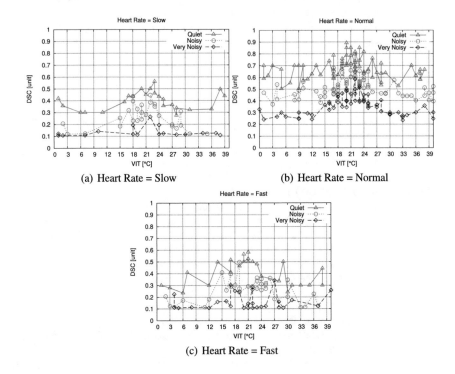

(a) Heart Rate = Slow

(b) Heart Rate = Normal

(c) Heart Rate = Fast

Fig. 6. Experimental results.

The results of DSC when HR is Normal are presented in Fig. 6(b). Here the driver is in better conditions and many values are decided as good or very good. These values are achieved especially when the ambient is quiet and the temperature is within 18–24 °C range. As we explained in the simulation results, Fig. 5(b), we get better DSC values because the heart rate is within the normal range.

In Fig. 6(c) are shown the results of DSC when HR is High. The results are almost the same with that of Fig. 5(d) where any situation was decided as bad and very bad. However, in the experimental results we can see some values that are decided as Normal by our system. This is due to the fact that the heart rate values that were considered in the simulations were constant (110 bpm), while here are shown the DSC values for HR values that might have happen to be from 95 to 150 bpm. The heart rate values near 95 bpm are those which increased the DSC.

6 Conclusions

In this paper, we proposed a fuzzy-based system to decide the DSC. We took into consideration three parameters: HR, NL and VIT. We evaluated the performance of proposed system by simulations and experiments. From the evaluation results, we conclude that the driver's heart rate, noise level and vehicle's inside temperature have different effects to the driver's condition.

In the future, we would like to make extensive simulations and experiments to evaluate the proposed systems and compare the performance with other systems.

References

1. Gusikhin, O., Filev, D., Rychtyckyj, N.: Intelligent vehicle systems: applications and new trends. In: Informatics in Control Automation and Robotics, pp. 3–14. Springer, Heidelberg (2008)
2. Santi, P.: Mobility Models for Next Generation Wireless Networks: Ad Hoc, Vehicular and Mesh Networks. Wiley, Hoboken (2012)
3. Hartenstein, H., Laberteaux, L.: A tutorial survey on vehicular ad hoc networks. IEEE Commun. Mag. **46**(6), 164–171 (2008)
4. Karagiannis, G., Altintas, O., Ekici, E., Heijenk, G., Jarupan, B., Lin, K., Weil, T.: Vehicular networking: a survey and tutorial on requirements, architectures, challenges, standards and solutions. IEEE Commun. Surv. Tutor. **13**(4), 584–616 (2011)
5. Lindwer, M., Marculescu, D., Basten, T., Zimmennann, R., Marculescu, R., Jung, S., Cantatore, E.: Ambient intelligence visions and achievements: linking abstract ideas to real-world concepts. In: 2003 Design, Automation and Test in Europe Conference and Exhibition, pp. 10–15, March 2003
6. Acampora, G., Cook, D.J., Rashidi, P., Vasilakos, A.V.: A survey on ambient intelligence in healthcare. Proc. IEEE **101**(12), 2470–2494 (2013)
7. Aarts, E., Wichert, R.: Ambient intelligence. In: Technology Guide, pp. 244–249. Springer, Heidelberg (2009)
8. Aarts, E., De Ruyter, B.: New research perspectives on ambient intelligence. J. Ambient Intell. Smart Environ. **1**(1), 5–14 (2009)
9. Vasilakos, A., Pedrycz, W.: Ambient Intelligence, Wireless Networking, and Ubiquitous Computing. Artech House Inc., Norwood (2006)
10. Sadri, F.: Ambient intelligence: a survey. ACM Comput. Surv. (CSUR) **43**(4), 36 (2011)
11. Kandel, A.: Fuzzy Expert Systems. CRC Press, Boca Raton (1991)
12. Zimmermann, H.-J.: Fuzzy Set Theory and Its Applications. Springer, Heidelberg (1991)
13. McNeill, F.M., Thro, E.: Fuzzy Logic: A Practical Approach. Academic Press, Boston (1994)
14. Zadeh, L.A., Kacprzyk, J.: Fuzzy Logic for the Management of Uncertainty. Wiley, Hoboken (1992)
15. Klir, G.J., Folger, T.A.: Fuzzy Sets, Uncertainty, and Information (1988)
16. Munakata, T., Jani, Y.: Fuzzy systems: an overview. Commun. ACM **37**(3), 69–77 (1994)
17. Inaba, T., Sakamoto, S., Oda, T., Barolli, L., Takizawa, M.: A new FACS for cellular wireless networks considering QoS: a comparison study of FuzzyC with MATLAB. In: Proceedings of the 18th International Conference on Network-Based Information Systems (NBiS-2015), pp. 338–344 (2015)

Performance Analysis of WMNs by WMN-PSOHC-DGA Simulation System Considering Random Inertia Weight and Linearly Decreasing Vmax Router Replacement Methods

Admir Barolli[1], Shinji Sakamoto[2(✉)], Seiji Ohara[3], Leonard Barolli[4], and Makoto Takizawa[5]

[1] Department of Information Technology, Aleksander Moisiu University of Durres, L.1, Rruga e Currilave, Durres, Albania
admir.barolli@gmail.com

[2] Department of Computer and Information Science, Seikei University, 3-3-1 Kichijoji-Kitamachi, Musashino-shi, Tokyo 180-8633, Japan
shinji.sakamoto@ieee.org

[3] Graduate School of Engineering, Fukuoka Institute of Technology, 3-30-1 Wajiro-Higashi, Higashi-Ku, Fukuoka 811-0295, Japan
seiji.ohara.19@gmail.com

[4] Department of Information and Communication Engineering, Fukuoka Institute of Technology, 3-30-1 Wajiro-Higashi, Higashi-Ku, Fukuoka 811-0295, Japan
barolli@fit.ac.jp

[5] Department of Advanced Sciences, Faculty of Science and Engineering, Hosei University, Kajino-Machi, Koganei-Shi, Tokyo 184-8584, Japan
makoto.takizawa@computer.org

Abstract. The Wireless Mesh Networks (WMNs) are becoming an important networking infrastructure because they have many advantages such as low cost and increased high speed wireless Internet connectivity. In our previous work, we implemented a Particle Swarm Optimization (PSO) and Hill Climbing (HC) based hybrid simulation system, called WMN-PSOHC, and a simulation system based on Genetic Algorithm (GA), called WMN-GA, for solving node placement problem in WMNs. Then, we implemented a hybrid simulation system based on PSOHC and distributed GA (DGA), called WMN-PSOHC-DGA. In this paper, we analyze the performance of WMNs using WMN-PSOHC-DGA simulation system considering Random Inertia Weight Method (RIWM) and Linearly Decreasing Vmax Method (LDVM). Simulation results show that a good performance is achived for RIWM compared with the case of LDVM.

© Springer Nature Switzerland AG 2020
L. Barolli et al. (Eds.): CISIS 2019, AISC 993, pp. 13–21, 2020.
https://doi.org/10.1007/978-3-030-22354-0_2

1 Introduction

The wireless networks and devices are becoming increasingly popular and they provide users access to information and communication anytime and anywhere [2,6–8,10,14,20,26–28]. Wireless Mesh Networks (WMNs) are gaining a lot of attention because of their low cost nature that makes them attractive for providing wireless Internet connectivity. A WMN is dynamically self-organized and self-configured, with the nodes in the network automatically establishing and maintaining mesh connectivity among them-selves (creating, in effect, an ad hoc network). This feature brings many advantages to WMNs such as low up-front cost, easy network maintenance, robustness and reliable service coverage [1]. Moreover, such infrastructure can be used to deploy community networks, metropolitan area networks, municipal and corporative networks, and to support applications for urban areas, medical, transport and surveillance systems.

Mesh node placement in WMN can be seen as a family of problems, which are shown (through graph theoretic approaches or placement problems, e.g. [4,11]) to be computationally hard to solve for most of the formulations [31]. We consider the version of the mesh router nodes placement problem in which we are given a grid area where to deploy a number of mesh router nodes and a number of mesh client nodes of fixed positions (of an arbitrary distribution) in the grid area. The objective is to find a location assignment for the mesh routers to the cells of the grid area that maximizes the network connectivity and client coverage. Node placement problems are known to be computationally hard to solve [9,32]. In some previous works, intelligent algorithms have been recently investigated [3,5,12,15–18,23,24].

In [27], we implemented a Particle Swarm Optimization (PSO) and Hill Climbing (HC) based simulation system, called WMN-PSOHC. Also, we implemented another simulation system based on Genetic Algorithm (GA), called WMN-GA [3,13], for solving node placement problem in WMNs. Then, we designed a Hybrid Intelligent System Based on PSO, HC and DGA, called WMN-PSOHC-DGA [25].

In this paper, we evaluate the performance of WMNs using WMN-PSOHC-DGA simulation system considering Random Inertia Weight Method (RIWM) and Linearly Decreasing Vmax Method (LDVM).

The rest of the paper is organized as follows. The mesh router nodes placement problem is defined in Sect. 2. We present our designed and implemented hybrid simulation system in Sect. 3. The simulation results are given in Sect. 4. Finally, we give conclusions and future work in Sect. 5.

2 Node Placement Problem in WMNs

For this problem, we have a grid area arranged in cells we want to find where to distribute a number of mesh router nodes and a number of mesh client nodes of fixed positions (of an arbitrary distribution) in the considered area. The objective is to find a location assignment for the mesh routers to the area that maximizes the network connectivity and client coverage. Network connectivity is measured by Size of Giant Component (SGC) of the resulting WMN graph, while the user coverage is simply the number of mesh client nodes that fall within the radio coverage of at least one mesh router node and is measured by Number of Covered Mesh Clients (NCMC).

An instance of the problem consists as follows.

- N mesh router nodes, each having its own radio coverage, defining thus a vector of routers.
- An area $W \times H$ where to distribute N mesh routers. Positions of mesh routers are not pre-determined and are to be computed.
- M client mesh nodes located in arbitrary points of the considered area, defining a matrix of clients.

It should be noted that network connectivity and user coverage are among most important metrics in WMNs and directly affect the network performance.

In this work, we have considered a bi-objective optimization in which we first maximize the network connectivity of the WMN (through the maximization of the SGC) and then, the maximization of the NCMC.

In fact, we can formalize an instance of the problem by constructing an adjacency matrix of the WMN graph, whose nodes are router nodes and client nodes and whose edges are links between nodes in the mesh network. Each mesh node in the graph is a triple $v = <x, y, r>$ representing the 2D location point and r is the radius of the transmission range. There is an arc between two nodes u and v, if v is within the transmission circular area of u.

3   Proposed and Implemented Simulation System

3.1   WMN-PSOHC-DGA Hybrid Simulation System

Distributed Genetic Algorithm (DGA) has been focused from various fields of science. DGA has shown their usefulness for the resolution of many computationally hard combinatorial optimization problems. Also, Particle Swarm Optimization (PSO) has been investigated for solving NP-hard problem.

PSOHC part
WMN-PSOHC-DGA decide the velocity of particles by a random process considering the area size. For instance, when the area size is $W \times H$, the velocity is decided randomly from $-\sqrt{W^2 + H^2}$ to $\sqrt{W^2 + H^2}$. Each particle's velocities are updated by simple rule [19].

For HC mechanism, next positions of each particle are used for neighbor solution s'. The fitness function f gives points to the current solution s. If $f(s')$ is better than $f(s)$, the s is updated to s'. However, if $f(s')$ is not better than $f(s)$, the s is not updated. It should be noted that the positions are not updated but the velocities are updated even if the $f(s)$ is better than $f(s')$.

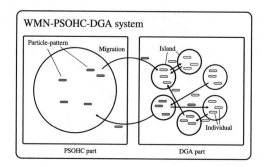

Fig. 1. Model of WMN-PSOHC-DGA migration.

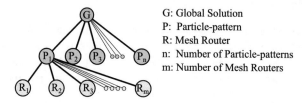

G: Global Solution
P: Particle-pattern
R: Mesh Router
n: Number of Particle-patterns
m: Number of Mesh Routers

Fig. 2. Relationship among global solution, particle-patterns and mesh routers in PSOHC part.

Routers Replacement Method for PSO Part

A mesh router has x, y positions and velocity. Mesh routers are moved based on velocities. There are many moving methods in PSO field, such as:

Random Inertia Weight Method (RIWM)
  In RIWM, the ω parameter is changing randomly from 0.5 to 1.0. The C_1 and C_2 are kept 2.0. The ω can be estimated by the week stable region. The average of ω is 0.75 [21,30].

Linearly Decreasing Vmax Method (LDVM)
  In LDVM, PSO parameters are set to unstable region ($\omega = 0.9$, $C_1 = C_2 = 2.0$). A value of V_{max} which is maximum velocity of particles is considered. With increasing of iteration of computations, the V_{max} is kept decreasing linearly [22,29].

DGA part

Population of individuals: Unlike local search techniques that construct a path in the solution space jumping from one solution to another one through local perturbations, DGA use a population of individuals giving thus the search a larger scope and chances to find better solutions. This feature is also known as "exploration" process in difference to "exploitation" process of local search methods.

Selection: The selection of individuals to be crossed is another important aspect in DGA as it impacts on the convergence of the algorithm. Several selection schemes have been proposed in the literature for selection operators trying to cope with premature convergence of DGA. There are many selection methods in GA. In our system, we implement 2 selection methods: Random method and Roulette wheel method.

Crossover operators: Use of crossover operators is one of the most important characteristics. Crossover operator is the means of DGA to transmit best genetic features of parents to offsprings during generations of the evolution process. Many methods for crossover operators have been proposed such as Blend Crossover (BLX-α), Unimodal Normal Distribution Crossover (UNDX), Simplex Crossover (SPX).

Mutation operators: These operators intend to improve the individuals of a population by small local perturbations. They aim to provide a component of randomness in the neighborhood of the individuals of the population. In our system, we implemented two mutation methods: uniformly random mutation and boundary mutation.

Escaping from local optima: GA itself has the ability to avoid falling prematurely into local optima and can eventually escape from them during the search process. DGA has one more mechanism to escape from local optima by considering some islands. Each island computes GA for optimizing and they migrate its gene to provide the ability to avoid from local optima.

Convergence: The convergence of the algorithm is the mechanism of DGA to reach to good solutions. A premature convergence of the algorithm would cause that all individuals of the population be similar in their genetic features and thus the search would result ineffective and the algorithm getting stuck

Table 1. WMN-PSOHC-DGA parameters.

Parameters	Values
Clients distribution	Normal distribution
Area size	32.0×32.0
Number of mesh routers	16
Number of mesh clients	48
Number of migrations	200
Evolution steps	9
Number of GA islands	16
Radius of a mesh router	2.0
Selection method	Roulette wheel method
Crossover method	SPX
Mutation method	Boundary mutation
Crossover rate	0.8
Mutation rate	0.2
Replacement method	RIWM, LDVM

into local optima. Maintaining the diversity of the population is therefore very important to this family of evolutionary algorithms.

In following, we present our proposed and implemented simulation system called WMN-PSOHC-DGA. We show the fitness function, migration function, particle-pattern, gene coding and client distributions.

Fitness Function
The determination of an appropriate fitness function, together with the chromosome encoding are crucial to the performance. Therefore, one of most important thing is to decide the determination of an appropriate objective function and its encoding. In our case, each particle-pattern and gene has an own fitness value which is comparable and compares it with other fitness value in order to share information of global solution. The fitness function follows a hierarchical approach in which the main objective is to maximize the SGC in WMN. Thus, the fitness function of this scenario is defined as

$$\text{Fitness} = 0.7 \times \text{SGC}(\boldsymbol{x}_{ij}, \boldsymbol{y}_{ij}) + 0.3 \times \text{NCMC}(\boldsymbol{x}_{ij}, \boldsymbol{y}_{ij}).$$

Migration Function
Our implemented simulation system uses Migration function as shown in Fig. 1. The Migration function swaps solutions between PSOHC part and DGA part.

Particle-Pattern and Gene Coding
In order to swap solutions, we design particle-patterns and gene coding carefully. A particle is a mesh router. Each particle has position in the considered area and velocities. A fitness value of a particle-pattern is computed by combination of mesh routers and mesh clients positions. In other words, each particle-pattern is a solution as shown is Fig. 2.

A gene describes a WMN. Each individual has its own combination of mesh nodes. In other words, each individual has a fitness value. Therefore, the combination of mesh nodes is a solution.

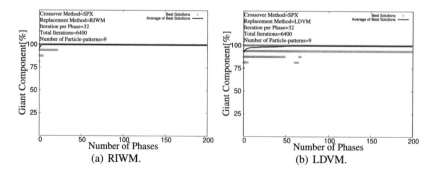

(a) RIWM.   (b) LDVM.

Fig. 3. Simulation results of WMN-PSOHC-DGA for SGC.

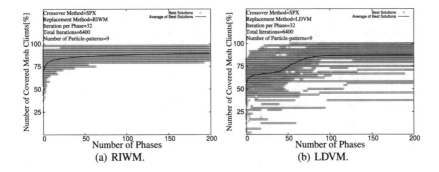

Fig. 4. Simulation results of WMN-PSOHC-DGA for NCMC.

4 Simulation Results

In this section, we show simulation results using WMN-PSOHC-DGA system. In this work, we analyse the performance of WMNs considering RIWM and LDVM router replacement methods. The number of mesh routers is considered 16 and the number of mesh clients 48. We conducted simulations 100 times, in order to avoid the effect of randomness and create a general view of results. We show the parameter setting for WMN-PSOHC-DGA in Table 1.

We show simulation results in Figs. 3 and 4. We see that for both SGC and NCMC, the performance of RIWM is better than LDVM.

5 Conclusions

In this work, we evaluated the performance of WMNs using a hybrid simulation system based on PSOHC and DGA (called WMN-PSOHC-DGA) considering RIWM and LDVM router replacement methods. Simulation results show that the performance is better for RIWM compared with the case of LDVM.

In our future work, we would like to evaluate the performance of the proposed system for different parameters and patterns.

References

1. Akyildiz, I.F., Wang, X., Wang, W.: Wireless mesh networks: a survey. Comput. Netw. **47**(4), 445–487 (2005)
2. Barolli, A., Sakamoto, S., Barolli, L., Takizawa, M.: Performance analysis of simulation system based on particle swarm optimization and distributed genetic algorithm for WMNs considering different distributions of mesh clients. In: International Conference on Innovative Mobile and Internet Services in Ubiquitous Computing, pp. 32–45. Springer, Heidelberg (2018)
3. Barolli, A., Sakamoto, S., Ozera, K., Barolli, L., Kulla, E., Takizawa, M.: Design and implementation of a hybrid intelligent system based on particle swarm optimization and distributed genetic algorithm. In: International Conference on Emerging Internetworking, Data & Web Technologies, pp. 79–93. Springer, Heidelberg (2018)

4. Franklin. A.A., Murthy, C.S.R.: Node placement algorithm for deployment of two-tier wireless mesh networks. In: Proceedings of Global Telecommunications Conference, pp. 4823–4827 (2007)
5. Girgis, M.R., Mahmoud, T.M., Abdullatif, B.A., Rabie, A.M.: Solving the wireless mesh network design problem using genetic algorithm and simulated annealing optimization methods. Int. J. Comput. Appl. **96**(11), 1–10 (2014)
6. Inaba, T., Elmazi, D., Sakamoto, S., Oda, T., Ikeda, M., Barolli, L.: A secure-aware call admission control scheme for wireless cellular networks using fuzzy logic and its performance evaluation. J. Mob. Multimedia **11**(3&4), 213–222 (2015)
7. Inaba, T., Obukata, R., Sakamoto, S., Oda, T., Ikeda, M., Barolli, L.: Performance evaluation of a QoS-aware fuzzy-based CAC for LAN access. Int. J. Space-Based Situated Comput. **6**(4), 228–238 (2016)
8. Inaba, T., Sakamoto, S., Oda, T., Ikeda, M., Barolli, L.: A testbed for admission control in WLAN: a fuzzy approach and its performance evaluation. In: International Conference on Broadband and Wireless Computing, Communication and Applications, pp. 559–571. Springer, Heidelberg (2016)
9. Maolin, T., et al.: Gateways placement in backbone wireless mesh networks. Int. J. Commun. Netw. Syst. Sci. **2**(1), 44 (2009)
10. Matsuo, K., Sakamoto, S., Oda, T., Barolli, A., Ikeda, M., Barolli, L.: Performance analysis of WMNs by WMN-GA simulation system for two WMN architectures and different TCP congestion-avoidance algorithms and client distributions. Int. J. Commun. Netw. Distrib. Syst. **20**(3), 335–351 (2018)
11. Muthaiah, S.N., Rosenberg, C.P.: Single gateway placement in wireless mesh networks. In: Proceedings of 8th International IEEE Symposium on Computer Networks, pp 4754–4759 (2008)
12. Naka, S., Genji, T., Yura, T., Fukuyama, Y.: A hybrid particle swarm optimization for distribution state estimation. IEEE Trans. Power Syst. **18**(1), 60–68 (2003)
13. Sakamoto, S., Kulla, E., Oda, T., Ikeda, M., Barolli, L., Xhafa, F.: A comparison study of hill climbing, simulated annealing and genetic algorithm for node placement problem in WMNs. J. High Speed Netw. **20**(1), 55–66 (2014)
14. Sakamoto, S., Kulla, E., Oda, T., Ikeda, M., Barolli, L., Xhafa, F.: A simulation system for WMN based on SA: performance evaluation for different instances and starting temperature values. Int. J. Space-Based Situated Comput. **4**(3–4), 209–216 (2014)
15. Sakamoto, S., Kulla, E., Oda, T., Ikeda, M., Barolli, L., Xhafa, F.: Performance evaluation considering iterations per phase and SA temperature in WMN-SA system. Mob. Inf. Syst. **10**(3), 321–330 (2014)
16. Sakamoto, S., Lala, A., Oda, T., Kolici, V., Barolli, L., Xhafa, F.: Application of WMN-SA simulation system for node placement in wireless mesh networks: a case study for a realistic scenario. Int. J. Mob. Comput. Multimedia Commun. (IJMCMC) **6**(2), 13–21 (2014)
17. Sakamoto, S., Oda, T., Ikeda, M., Barolli, L., Xhafa, F.: An integrated simulation system considering WMN-PSO simulation system and network simulator 3. In: International Conference on Broadband and Wireless Computing, Communication and Applications, pp. 187–198. Springer, Heidelberg (2016)
18. Sakamoto, S., Oda, T., Ikeda, M., Barolli, L., Xhafa, F.: Implementation and evaluation of a simulation system based on particle swarm optimisation for node placement problem in wireless mesh networks. Int. J. Commun. Netw. Distrib. Syst. **17**(1), 1–13 (2016)

19. Sakamoto, S., Oda, T., Ikeda, M., Barolli, L., Xhafa, F.: Implementation of a new replacement method in WMN-PSO simulation system and its performance evaluation. In: The 30th IEEE International Conference on Advanced Information Networking and Applications (AINA-2016), pp. 206–211 (2016). https://doi.org/10.1109/AINA.2016.42

20. Sakamoto, S., Obukata, R., Oda, T., Barolli, L., Ikeda, M., Barolli, A.: Performance analysis of two wireless mesh network architectures by WMN-SA and WMN-TS simulation systems. J. High Speed Netw. **23**(4), 311–322 (2017)

21. Sakamoto, S., Ozera, K., Barolli, A., Ikeda, M., Barolli, L., Takizawa, M.: Performance evaluation of WMNs by WMN-PSOSA simulation system considering random inertia weight method and linearly decreasing Vmax method. In: International Conference on Broadband and Wireless Computing, Communication and Applications (BWCCA-2017), pp 114–124. Springer, Heidelberg (2017)

22. Sakamoto, S., Ozera, K., Ikeda, M., Barolli, L.: Performance evaluation of WMNs by WMN-PSOSA simulation system considering constriction and linearly decreasing inertia weight methods. In: International Conference on Network-Based Information Systems, pp. 3–13. Springer, Heidelberg (2017)

23. Sakamoto, S., Ozera, K., Oda, T., Ikeda, M., Barolli, L.: Performance evaluation of intelligent hybrid systems for node placement in wireless mesh networks: a comparison study of WMN-PSOHC and WMN-PSOSA. In: International Conference on Innovative Mobile and Internet Services in Ubiquitous Computing, pp. 16–26. Springer, Heidelberg (2017)

24. Sakamoto, S., Ozera, K., Oda, T., Ikeda, M., Barolli, L.: Performance evaluation of WMN-PSOHC and WMN-PSO simulation systems for node placement in wireless mesh networks: a comparison study. In: International Conference on Emerging Internetworking, Data & Web Technologies, pp. 64–74. Springer, Heidelberg (2017)

25. Sakamoto, S., Barolli, A., Barolli, L., Takizawa, M.: Design and implementation of a hybrid intelligent system based on particle swarm optimization, hill climbing and distributed genetic algorithm for node placement problem in WMNs: a comparison study. In: The 32nd IEEE International Conference on Advanced Information Networking and Applications (AINA-2018), pp. 678–685. IEEE (2018)

26. Sakamoto, S., Ozera, K., Barolli, A., Barolli, L., Kolici, V., Takizawa, M.: Performance evaluation of WMN-PSOSA considering four different replacement methods. In: International Conference on Emerging Internetworking, Data & Web Technologies, pp. 51–64. Springer, Heidelberg (2018)

27. Sakamoto, S., Ozera, K., Ikeda, M., Barolli, L.: Implementation of intelligent hybrid systems for node placement problem in wmns considering particle swarm optimization, hill climbing and simulated annealing. Mob. Netw. Appl. **23**(1), 27–33 (2018)

28. Sakamoto, S., Ozera, K., Barolli, A., Ikeda, M., Barolli, L., Takizawa, M.: Implementation of an intelligent hybrid simulation systems for WMNs based on particle swarm optimization and simulated annealing: performance evaluation for different replacement methods. Soft Comput. **23**(9), 3029–3035 (2019)

29. Schutte, J.F., Groenwold, A.A.: A study of global optimization using particle swarms. J. Glob. Optim. **31**(1), 93–108 (2005)

30. Shi, Y.: Particle swarm optimization. IEEE Connections **2**(1), 8–13 (2004)

31. Vanhatupa, T., Hannikainen, M., Hamalainen, T.: Genetic algorithm to optimize node placement and configuration for WLAN planning. In: Proceedings of the 4th IEEE International Symposium on Wireless Communication Systems, pp. 612–616 (2007)

32. Wang, J., Xie, B., Cai, K., Agrawal, D.P.: Efficient mesh router placement in wireless mesh networks. In: Proceedings of IEEE International Conference on Mobile Adhoc and Sensor Systems (MASS-2007), pp. 1–9 (2007)

IoT Node Selection and Placement:
A New Approach Based on Fuzzy Logic
and Genetic Algorithm

Miralda Cuka[1(✉)], Donald Elmazi[2], Makoto Ikeda[2], Keita Matsuo[2],
and Leonard Barolli[2]

[1] Graduate School of Engineering, Fukuoka Institute of Technology (FIT),
3-30-1 Wajiro-Higashi, Higashi-Ku, Fukuoka 811-0295, Japan
`mcuka91@gmail.com`
[2] Department of Information and Communication Engineering,
Fukuoka Institute of Technology (FIT), 3-30-1 Wajiro-Higashi,
Higashi-Ku, Fukuoka 811-0295, Japan
`donald.elmazi@gmail.com`, `makoto.ikd@acm.org`,
`{kt-matsuo,barolli}@fit.ac.jp`

Abstract. The enormous growth of devices having access to the Internet, along the vast evolution of the Internet and the connectivity of objects and devices, has evolved as Internet of Things (IoT). There are different issues for these networks. One of them is the selection and placement of IoT nodes. In this work, we propose a simulating system based on Fuzzy Logic and Genetic Algorithm for IoT node selection and placement. We consider three input parameters for our Fuzzy-based selection system: IoT Node Density (IND), IoT Node's Remaining Energy (INRE) and IoT Node's Distance to Event (INDE). We also present a simulation system based on Genetic Algorithm which is implemented in Rust, for IoT node placement. We consider different aspects of an IoT network, considering coordination, connectivity and coverage. We describe the implementation and show the interface of simulation system. We evaluated the performance of the proposed system by a simulation scenario. For the IoT node fuzzy-based selection system, we show that the system makes a proper selection of IoT nodes. The simulation results of GA-based system show that the constructed network, can cover both events.

1 Introduction

The Internet of Things (IoT) can seamlessly connect the real world and cyberspace via physical objects embedded with various types of intelligent sensors. A large number of Internet-connected machines will generate and exchange an enormous amount of data that make daily life more convenient, help to make a tough decision and provide beneficial services. The IoT probably becomes one of the most popular networking concepts that has the potential to bring out many benefits [1,2].

© Springer Nature Switzerland AG 2020
L. Barolli et al. (Eds.): CISIS 2019, AISC 993, pp. 22–35, 2020.
https://doi.org/10.1007/978-3-030-22354-0_3

There are still many open challenges and research trends in IoT, such as standardization, security (e.g., data privacy and confidentiality), reliability, scalability, performance, data mining, data storage, green IoT, and smart objects using AI. Security, data mining, and data storage are recurrently stated as three of the top research challenges in IoT [3].

The Fuzzy Logic (FL) is a unique approach that is able to simultaneously handle numerical data and linguistic knowledge. The FL works on the levels of possibilities of input to achieve the definite output. Fuzzy set theory and FL establish the specifics of the nonlinear mapping.

When it comes to IoT node placement, we have random deployment, where nodes are randomly scattered within the field; and deterministic placement, where nodes are placed at desired locations. In the latter case, the fundamental research question is how to find the least number of best locations to place nodes while guaranteeing the deployment quality indicators like field coverage and network connectivity [4]. Genetic Algorithms (GAs) are a family of computational models inspired by evolution. These algorithms encode a potential solution to a specific problem on a simple chromosome-like data structure and apply recombination operators to these structures so as to preserve critical information. The GAs are often viewed as function optimizers, although the range of problems to which GAs have been applied is quite broad. An implementation of a GA begins with a population of (typically random) chromosomes. Then evaluates these structures and allocates reproductive opportunities in such a way that those chromosomes which represent a better solution to the target problem, are given more chances to reproduce than those chromosomes which are poorer solutions [5,16,17].

In this paper, we propose and implement a simulation system for Internet of Things (IoT) node selection and placement. The selection system is based on Fuzzy logic, while the placement system is based on GA. We describe the implementation of proposed system and show its interface. Our Fuzzy-based system used three input parameters: IoT Node Density (IND), IoT Node's Remaining Energy (INRE) and IoT Node's Distance to Event (INDE), to make a proper selection of an IoT node for a certain task. For the GA parameter configuration, we use: number of independent runs, population size, crossover probability, mutation probability, initial placement and selection methods.

The remainder of the paper is organized as follows. In the Sect. 2, we present the basics of IoT. In Sect. 3, we describe IoT Node selection and placement problems. In Sect. 4, we present the Fuzzy logic and Genetic Algorithm. In Sect. 5, we show the design and implementation of our simulation system. Simulation results are shown in Sect. 6. Finally, conclusions and future work are given in Sect. 7.

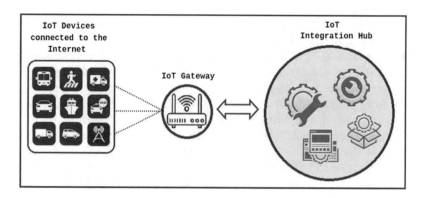

Fig. 1. An Iot network architecture.

2   IoT

2.1   IoT

IoT allows the integration of physical and virtual objects. Virtual reality, which was recently available only on the monitor screens, now integrates with the real world, providing users with completely new opportunities: interact with objects on the other side of the world and receive the necessary services that became real due the wide interaction [6]. The IoT will support substantially higher number of end users and devices. In Fig. 1, we present an example of an IoT network architecture. The IoT network is a combination of IoT nodes which are connected with different mediums using IoT Gateway to the Internet. The data transmitted through the gateway is stored, processed securely within cloud server. These new connected things will trigger increasing demands for new IoT applications that are not only for users. The current solutions for IoT application development generally rely on integrated service-oriented programming platforms. In particular, resources (e.g., sensory data, computing resource, and control information) are modeled as services and deployed in the cloud or at the edge. It is difficult to achieve rapid deployment and flexible resource management at network edges, in addition, an IoT system's scalability will be restricted by the capability of the edge nodes [7].

3   IoT Node Selection and Placement Problems

Due to high diversity, an IoT network consists of different nodes with different resource capabilities. When multiple IoT nodes are deployed densely, there is a possibility that a node may reside in the coverage area of multiple different nodes. When a specific task request requires an IoT node to complete it, it is challenging to determine which is the best one for that specific request. First, the IoT networks are heterogeneous rather than homogeneous, which consists of many diverse IoT nodes which have largely different demands on data traffic and data processing [8]. To maintain network quality and to have better resource allocation, IoT nodes are selected based on different parameters or based on event coverage. IoT node selection proves useful in mitigating common IoT-related issues like resource allocation, network lifetime, and the confidence in the collected data, by having the right IoT nodes active at a given time. IoT node selection helps in saving and better managing resources by choosing the right subset of nodes to be active depending on the task requirements [9].

IoT node placement problems have been long investigated in the optimization field due to numerous applications in location science (facility location, logistics, services, etc.) and classification (clustering). In such problems, we are given a number of potential facilities to serve to costumers connected to facilities aiming to find locations such that the cost of serving to all customers is minimized [10]. In traditional versions of the problem, facilities could be hospitals, polling centers, fire stations serving to a number of clients and aiming to minimize some distance function in a metric space between clients and such facilities. One classical version of the problem is that of p-median problem, defined as follows.

The problem, which is known for its intractability, has many applications not only in location science but also in communication networks, where facilities could be servers, routers, etc., offering connectivity services to clients. In an IoT network, nodes provide network connectivity services to events. The good performance and operability of an IoT network largely depends on placement of nodes in the geographical deployment area to achieve network connectivity, stability and user coverage. The objective is to find an optimal and robust topology of the network nodes to support connectivity services to events.

Facility location problems are thus showing their usefulness to communication networks. In a general setting, location models in the literature have been defined as follows. We are given:

(a) a universe $\mathcal{U}$, from which a set $\mathcal{E}$ of event input positions is selected;
(b) an integer, $\mathcal{N} \geq 1$, denoting the number of facilities to be deployed;
(c) one or more metrics of the type d: $\mathcal{U} \times \mathcal{U} \to \mathcal{R}_+$, which measure the quality of the location; and,
(d) an optimization model.

The optimization model takes in input the universe where facilities are to be deployed, a set of client positions and returns a set of positions for facilities that optimize the considered metrics. It should be noted that different models can be established depending on whether the universe is considered: (a) continuous (universe is a region, where clients and facilities may be placed anywhere within the continuum leading to an uncountably infinite number of possible locations); (b) discrete (universe is a discrete set of predefined positions); and, (c) network (universe is given by an undirected weighted graph; in the graph, client positions are given by the vertices and facilities may be located anywhere on the graph) [11].

For most IoT formulations, node placement problems are shown to be computationally hard to solve to optimality and therefore heuristic and meta-heuristic approaches are useful approaches to solve the problem for practical purposes.

4   Fuzzy Logic and Genetic Algorithm

4.1   Fuzzy Logic

One of the problems that frequently arises when designing a decision making system, is to represent the vagueness and uncertainty that typically affects information which cannot be handled with traditional (crisp) mathematical models. The proposed approach takes into account such vagueness and uncertainty by means of fuzzy sets and emulates the decision process of a human expert by means of a rule-based inference engine. A basic fuzzy logic controller is constituted of four components: a rules set, a fuzzifier, an inference engine and a defuzzifier. The core of a fuzzy intelligent system is its knowledge base, which is expressed in terms of fuzzy rules and allows for approximate reasoning [12].

(a) *Rule Set:* The inference rules, expressed in the form of IF-THEN rules, provide the necessary connection between the controller input and output fuzzy sets.
(b) *Fuzzifier:* The fuzzifier maps crisp numbers into fuzzy sets. It is needed in order to activate rules which are in terms of linguistic variables, which have fuzzy sets associated with them.
(c) *Inference Engine:* The inference engine maps a given input set to a fuzzy set. It handles the way in which rules are combined just as humans use many different types of inferential procedures to understand things or to make decisions. There are many different FL inferential procedures [13].
(d) *Deffuzifier:* Defuzzification is the process of transforming the result of the inference engine into a crisp output for further processing.

4.2   Genetic Algorithm

Genetic Algorithm (GA) is one of the most powerful heuristics for solving optimization problems that is based on natural selection, the process that drives biological evolution. The GA repeatedly modifies a population of individual solutions as shown in Fig. 2. At each step, the genetic algorithm selects individuals at random from the current population to be parents and uses them to produce the children for the next generation. Over successive generations, the population "evolves" towards an optimal solution [14]. In Algorithm 1 is shown the pseudocode for a GA.

(a) *Selection:* As selection operator, we use roulette-wheel selection [15–17]. In roulette-wheel selection, each individual in the population is assigned a roulette wheel slot sized in proportion to its fitness. That is, in the biased roulette wheel, good solutions have a larger slot size than the less fit solutions. The roulette wheel can obtain a reproduction candidate.
(b) *Crossover:* The crossover operators are the most important ingredient of GAs. Indeed, by selecting individuals from the parental generation and interchanging their genes, new individuals (descendants) are obtained. The aim

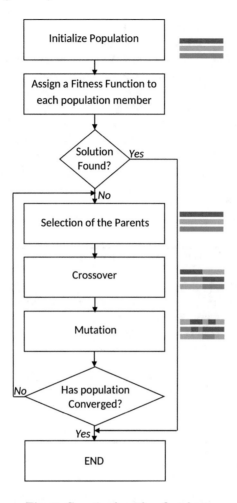

Fig. 2. Genetic algorithm flowchart.

is to obtain descendants of better quality that will feed the next generation and enable the search to explore new regions of solution space not explored yet [18]. There exist many types of crossover operators explored in the evolutionary computing literature. It is very important to stress that crossover operators depend on the chromosome representation. This observation is especially important for an IoT network scenario problem, since in our case, instead of having strings we have an area of IoT nodes located in certain positions. The crossover operator should thus take into account the specifics of IoT nodes encoding. We have considered the following crossover operator, called intersection operators (denoted CrossRegion, hereafter), which take in input two individuals and produce in output two new individuals.

(c) *Mutation:* Mutation operator is one of the GA ingredients. Unlike crossover operators, which achieve to transmit genetic information from parents to offsprings, mutation operators usually make some small local perturbation of the individuals, having thus less impact on newly generated individuals. Crossover is "a must" operator in GA and is usually applied with high probability, while mutation operators when implemented are applied with small probability. The rationale is that a large mutation rate would make the GA search to resemble a random search. Due to this, mutation operator is usually considered as a secondary operator. In the case of IoT node placement, the matrix representation is chosen for the individuals of the population, in order to keep the information on IoT nodes positions, events positions, links among nodes and links among nodes and events [19]. The definition of the mutation operators is therefore specific to a matrix based encoding of the individuals of the population.

Algorithm 1. Genetic Algorithm Pseudocode

---

t ← **0**;
Initialize population $(P(0))$;
Assign a fitness function to each population member $(P(0))$ of size Θ
while max_nr_of_generations_reached **do**
    Select the parent pool $(P_p(t))$ of size Φ;
    Crossover pairs from parents pool $P_p(t)$ with probability p_c; $P_c(t) = Crossover(P_p(t))$;
    Mutate individuals in $P_c(t)$ with probability p_m; $P_m(t) = Mutate(P_c(t))$;
    Create new population with individuals from crossover and mutation;
    $P(t+1) = Individuals(P_c(t) \bigcup P_m(t))$;
    $t \leftarrow t + 1$
end while

---

5 Design and Implementation of IoT Node Selection and Placement System

5.1 System Parameters

In this section, we present the design and implementation of a simulation system based on FL and GA for IoT node selection and placement. The simulation system structure is shown in Fig. 3. It consists of two main parts, a selection system based on fuzzy logic and a placement system based on GA. The fuzzy-based system makes selection decisions based on three input parameters IoT Node Density (IND), IoT Node Remaining Energy (INRE), IoT Node's Distance to Event (INDE).

IoT Node Density (IND): IoT nodes can be randomly redistributed, or clustered in a specific area. When IoT node density is sparse and nodes are more uniformly distributed in the network.

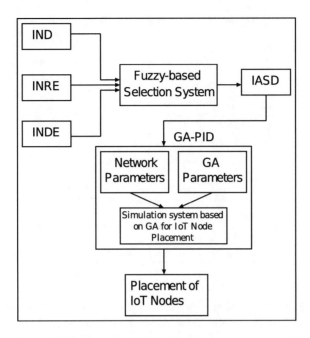

Fig. 3. Simulation system structure.

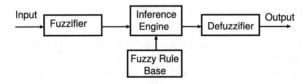

Fig. 4. FLC structure.

Table 1. Parameters and their term sets for Fuzzy based selection system.

Parameters	Term sets
IoT Node Density (IND)	Low (Lo), Medium (Med), High (Hgh)
IoT Node's Remaining Energy (INRE)	Low (Lw), Medium (Mdm), High (Hi)
IoT Node's Distance to Event (INDE)	Near (Ne), Medium (Me), Far (Fr)
IoT Actor Selection Decision (IASD)	Very Low Selection Possibility (VLSP), Low Selection Possibility (LSP), Medium Selection Possibility (MSP), High Selection Possibility (HSP), Very High Selection Possibility (VHSP)

Table 2. GA parameter configuration.

Parameters	Values
Number of IoT nodes	15
Number of IoT actors	5
Grid width	32 units
Grid height	32 units
Independent runs	10
Number of generations	50, 100
Population size	8, 16, 32
Crossover rate	65%, 78%, 80%
Mutate method	Single
Mutate rate	5%, 15%, 20%, 35%
Distribution	Normal

IoT Node's Remaining Energy (INRE): The IoT nodes are active and can perform tasks and exchange data in different ways from each other. Consequently, some IoT nodes may have a lot of remaining power and other may have very little, when an event occurs.

IoT Node's Distance to Event (INDE): Our system makes decisions based on the availability of the IoT node when it is called for action near an event, the distance of the node from the event varies for different scenarios.

In a fuzzy based system, each fuzzy set corresponds to a linguistic concept. In Table 1 is shown the mapping of the quantitative parameters to the term set of the linguistic variables. The structure of the fuzzy based selection system is shown in Fig. 4. It consists of one Fuzzy Logic Controller (FLC), which is the main part of our system and its basic elements. They are the fuzzifier, inference engine, Fuzzy Rule Base (FRB) and defuzzifier. Each point of the input space is mapped to a membership value between 0 and 1 as shown in Fig. 5. Curved or nonlinear membership functions increase the computation time, so we have used linear trapezoidal and triangular functions.

Our placement system based on GA, can generate instances of the problem using different distributions of events and IoT actor nodes. In Table 2 are shown the GA parameters. For the network configuration, we use: distribution of events, number of events, number of IoT nodes, area size and radius of communication range. For the GA parameter configuration, we use: number of independent runs, GA evolution steps, population size, crossover probability, mutation probability, initial placement methods, selection methods. We consider SingleMutate mutation operator which is a move-based operator. It selects an IoT node in the problem area and moves it to another cell of the problem area.

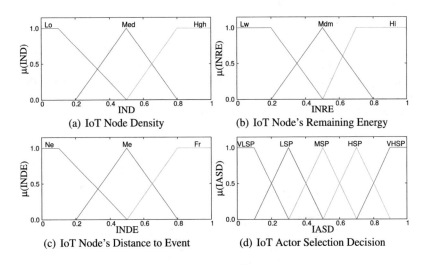

Fig. 5. Fuzzy membership functions.

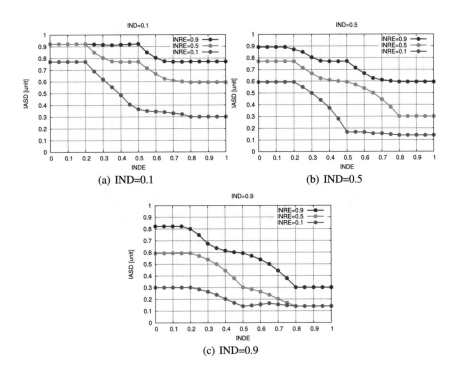

Fig. 6. Results for different values of IND, $INRE$, and $INDE$.

6   Simulation Results

The simulation results for our integrated system are shown in Figs. 6 and 7 for the fuzzy-based system and GA, respectively. For our fuzzy-based system, we see how different values of parameters affect the possibility of an IoT node to be selected for a specific task. IoT nodes that compose the network, have different properties and values, due to the diverse nature of an IoT network. By using fuzzy logic we are able to give insight on which IoT nodes are preferred over others.

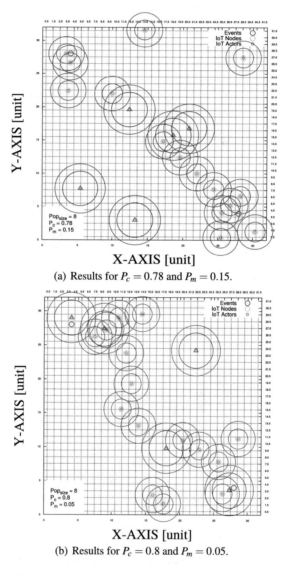

(a) Results for $P_c = 0.78$ and $P_m = 0.15$.

(b) Results for $P_c = 0.8$ and $P_m = 0.05$.

Fig. 7. Simulation results for population size 8 number of generations 50.

In Fig. 6 are shown the simulation results of the IoT node selection possibility. It shows the relation between IoT Actor Selection Decision and IoT Node's Remaining Energy, IoT Node's Distance to Event and IoT Node Density. In a network of nodes with different resources, some are limited by having different energy levels than the others. Energy affects the lifetime of an IoT node in the network, so higher energy means the IoT node will be available for a longer time. In Fig. 6(a), we see that for higher levels of remaining energy, IoT Node has a higher possibility to be selected.

In case of an event, some IoT nodes are further away than others, these nodes will take a longer time to reach the event and drain their battery life. In Fig. 6(b), we see that the possibility for an IoT actor to be selected decreases, because the distance between the event and the IoT node increases.

Another important factor that affects IoT Actor Selection Decision, is density. Comparing Fig. 6(a) with 6(b), and Fig. 6(b) with 6(c), we see that IoT actor selection possibility decreases. Node density is an important factor as it affects the connectivity among nodes in the network. However, if beyond a certain boundary, network performance will degrade, because in case of a limited number of nodes, IoT nodes on the network edges will not connect. When density is sparse, IoT nodes tend to be more uniformly distributed in the network.

In Fig. 7 are shown the simulation results of our GA-based placement system. We carried out many simulations to evaluate the coverage of two events. After each generation, the fitness of each solution is calculated. The next generation is created from selection, crossover and mutation. We used a pool of 20 individuals and two static events and did repeated runs by varying the population size 8, 16, 32 and the number of generations 50 and 100. However, the computational time increased significantly for higher number of population size and generation, so for our simulations we have used a population size of 8 and number of generations 50. We carried many simulations for different crossover P_c and mutation P_m rates. In Fig. 7(a), are shown a crossover rate $P_c = 0.78$ and mutation rate $P_m = 0.15$. We see that the two events are covered but the network of IoT nodes is not connected. We increased the crossover rate to $P_c = 0.8$ and decreased the mutation rate to $P_m = 0.05$. In this case both events are covered and the network is connected.

7   Conclusions and Future Work

In this paper, we proposed and implemented an integrated system for IoT node selection and placement based on FL and GA. We evaluated the proposed system by computer simulations.

The simulation results for the fuzzy-based selection system show that for low IoT node density, the IoT nodes are better distributed and are more likely to be selected for carrying out a job. We can see that by decreasing INDE and increasing INRE, the IASD is also increased.

The simulation results for IoT node placement, show that for $P_c = 0.8$ and $P_m = 0.05$ both events are covered and the network is well connected.

In the future work, we will also consider other parameters for IoT node selection, evaluate different scenarios for IoT node placement and make extensive simulations to evaluate the proposed system.

References

1. Kraijak, S., Tuwanut, P.: A survey on internet of things architecture, protocols, possible applications, security, privacy, real-world implementation and future trends. In: 16th International Conference on Communication Technology (ICCT), pp. 26–31. IEEE (2015)
2. Arridha, R., Sukaridhoto, S., Pramadihanto, D., Funabiki, N.: Classification extension based on iot-big data analytic for smart environment monitoring and analytic in real-time system. Int. J. Space-Based Situated Comput. **7**(2), 82–93 (2017)
3. Braulio, L.D.C., Moreno, E.D., de Macedo, D.D.J., Kreutz, D., Dantas, M.A.R.: Towards a hybrid storage architecture for IoT. In: 2018 IEEE Symposium on Computers and Communications (ISCC), pp. 00470–00473, June 2018
4. Lu, D., Bang, W.: Sensor placement based on an improved genetic algorithm for connected confident information coverage in an area with obstacles. In: 2017 IEEE 42nd Conference on Local Computer Networks (LCN), pp. 595–598. IEEE (2018)
5. Holland, J.H., et al.: Adaptation in Natural and Artificial Systems: An Introductory Analysis with Applications to Biology, Control, and Artificial Intelligence. MIT Press, Cambridge (1992)
6. Popereshnyak, S., Suprun, O., Suprun, O., Wieckowski, T.: IoT application testing features based on the modelling network. In: 2018 XIV-th International Conference on Perspective Technologies and Methods in MEMS Design (MEMSTECH), pp. 127–131 (2018)
7. Chen, N., Yang, Y., Li, J., Zhang, T.: A fog-based service enablement architecture for cross-domain IoT applications. In: 2017 IEEE Fog World Congress (FWC), pp. 1–6. IEEE (2017)
8. Zhao, Z., Min, G., Gao, W., Wu, Y., Duan, H., Ni, Q.: Deploying edge computing nodes for large-scale IoT: a diversity aware approach. IEEE Internet Things J. **5**(5), 3606–3614 (2018)
9. Alagha, A., Singh, S., Mizouni, R., Ouali, A., Otrok, H.: Data-driven dynamic active node selection for event localization in IoT applications - a case study of radiation localization. IEEE Access **7**, 16168–16183 (2019)
10. Oda, T., Barolli, A., Xhafa, F., Barolli, L., Ikeda, M., Takizawa, M.: WMN-GA: a simulation system for wmns and its evaluation considering selection operators. J. Ambient Intell. Human. Comput. **4**(3), 323–330 (2013)
11. Xhafa, F., Sánchez, C., Barolli, L.: Genetic algorithms for efficient placement of router nodes in wireless mesh networks. In: 2010 24th IEEE International Conference on Advanced Information Networking and Applications, pp. 465–472. IEEE (2010)
12. Aiello, G., Certa, A., Enea, M.: A fuzzy inference expert system to support the decision of deploying a military naval unit to a mission. In: International Workshop on Fuzzy Logic and Applications, pp. 320–327. Springer, Heidelberg (2009)
13. Mendel, J.M.: Fuzzy logic systems for engineering: a tutorial. Proc. IEEE **83**(3), 345–377 (1995)

14. Bhondekar, A.P., Vig, R., Singla, M.L., Ghanshyam, C., Kapur, P.: Genetic algorithm based node placement methodology for wireless sensor networks. In: Proceedings of the International Multiconference of Engineers and Computer Scientists, vol. 1, pp. 18–20 (2009)
15. Holland, J.: Adaptation in Natural and Artificial Systems. University of Michigan Press, Ann Arbor (1975)
16. Goldberg, D.E., Holland, J.H.: Genetic algorithms and machine learning. Mach. Learn. 3(2), 95–99 (1988)
17. Sastry, K., Goldberg, D., Kendall, G.: Genetic algorithms. In: Search Methodologies, pp. 97–125. Springer, Heidelberg (2005)
18. Xhafa, F., Sánchez, C., Barolli, L., Spaho, E.: Evaluation of genetic algorithms for mesh router nodes placement in wireless mesh networks. J. Ambient Intell. Human. Comput. 1(4), 271–282 (2010). Springer
19. Barolli, A., Sakamoto, S., Oda, T., Spaho, E., Barolli, L., Xhafa, F.: Performance evaluation of WMN-GA system for different settings of population size and number of generations. Hum.-Centric Comput. Inf. Sci. 4(1), 5–19 (2013)

Day Ahead Electric Load Forecasting by an Intelligent Hybrid Model Based on Deep Learning for Smart Grid

Ghulam Hafeez[1,2], Nadeem Javaid[1(✉)], Muhammad Riaz[3], Ammar Ali[1],
Khalid Umar[4], and Zafar Iqbal[5]

[1] COMSATS University Islamabad, Islamabad 44000, Pakistan
nadeemjavaidqau@gmail.com
[2] University of Engineering and Technology, Mardan 23200, Pakistan
[3] Wah Engineering College University of Wah, Wah Cantt 47070, Pakistan
[4] Bahria University Islambad, Islamabad 44000, Pakistan
[5] PMAS Agriculture University, Rawalpindi 46000, Pakistan
http://www.njavaid.com

Abstract. Electrical load forecasting is a challenging problem due to random and non-linear behavior of the consumers. With the emergence of the smart grid (SG) and advanced metering infrastructure (AMI), people are capable to record, monitor, and analyze such a complicated non-linear behavior. Electric load forecasting models are indispensable in the decision making, planning, and contract evaluation of the power system. In this regard, various load forecasting models are proposed in the literature, which exhibit trade-off between forecast accuracy and execution time (convergence rate). In this article, a fast and accurate short-term load forecasting model is proposed. The abstractive features from the historical data are extracted using modified mutual information (MMI) technique. The factored conditional restricted boltzmann machine (FCRBM) is empowered via learning to predict the electric load. Eventually, the proposed genetic wind driven optimization (GWDO) algorithm is used to optimize the performance. The remarkable advantages of the proposed framework are the improved forecast accuracy and convergence rate. The forecast accuracy is improved through the use of MMI technique and FCRBM model. On the other side, convergence rate is enhanced by GWDO algorithm. Simulation results illustrate that the proposed fast and accurate model outperforms existing models i.e., Bi-level, MI-artificial neural network (MI-ANN), and accurate fast converging short-term load forecast (AFC-STLF) in terms of forecast accuracy and convergence rate.

1 Introduction

Smart grid (SG) emerged as a smart power grid that has recently achieved lot of popularity [1,2]. In the SG, one vault problem i.e., electrical load forecasting is

© Springer Nature Switzerland AG 2020
L. Barolli et al. (Eds.): CISIS 2019, AISC 993, pp. 36–49, 2020.
https://doi.org/10.1007/978-3-030-22354-0_4

very challenging due to non-linear behavior of consumers. An accurate load forecasting is of great importance for both electric utility companies and consumers due to its application in the decision making and operation of power system [3].

In recent years, both classical (time-series methods) and computational intelligence methods are applied for electrical load forecasting [4]. The classical methods are blamed for their limited ability to handle non-linear data. The computational intelligence methods are criticized for the problems like handcrafted features, limited learning capacity, and impotent learning. Although, there are some existing machine learning models applied for electric load forecasting, which partially resolve the problems and have improved performance due to the use of ingenious design [5].

Thus, a suitable mechanism is required to solve the aforementioned problems because low forecast accuracy results in prominent economic loss and slow convergence rate results in user frustration. Therefore, electric utility companies are trying to develop a fast, accurate, and simple short-term electric load forecasting model. Boroojeni *et al.* proposed a generalized method to model offline data that have different seasonal cycles (e.g., daily, weekly, quarterly, and annually). Both seasonal and non-seasonal load cycles are modeled individual with the help of auto-regressive (AR) and moving-average (MA) components [6]. Xiaomin Xu *et al.* investigated ensemble subsampled support vector regression (SVR) for forecasting and estimation of load [7]. A deep belief network i.e., restricted boltzmann machine (RBM) is used for electric load forecasting. The network reduced the forecast error with affordable execution time [8]. Hong *et al.* forecasts electric load of southeast China with the help of hybrid model using seasonal recurrent SVR model and chaotic artificial bee colony (CABC) algorithm. This hybrid model outperforms the existing AR integrated MA (ARIMA) model [9]. Inspired by the above literature, a fast and accurate short-term electric load forecasting model is proposed. The main contributions of this paper are demonstrated as follows:

1. A fast and accurate short-term electric load forecasting model is proposed based on FCRBM to forecast the future electric load. The superiority of the proposed model is validated by comparing with the existing models like, Bi-level, MI-artificial neural network (MI-ANN), and accurate fast converging-STLF (AFC-STLF).
2. Based on the existing mutual information (MI) technique [10–12], a new MMI technique for feature selection is proposed (Sect. 3). The proposed technique includes filtering approach, which ranks the candidate inputs according to their information value, select abstractive features by maximizing the relevancy and minimizing redundancy.
3. Finally, the GWDO algorithm is proposed, which is a hybrid of genetic algorithm (GA) and wind driven optimization (WDO) algorithm for performance optimization.

The remaining organization of the paper is as: Recent literature review is presented in Sect. 2. In Sect. 3, the proposed system model is introduced. Section 4 includes simulation results and discussions, and Sect. 5 concludes the paper.

2   Recent Literature Review

Short-term electric load forecasting normally covers hours to week prediction horizon and is crucial in the decision making of power system. In literature, both statical models and machine learning models are commonly used for short-term load forecasting. Let us discuss some of these models adopted for forecasting, in recent years.

Authors proposed distributed methods based on ARIMA and grey models in [13] to forecast the future load using weather information. Authors in [14] introduced combined bluetooth home energy management system (HEMS) with ANN to forecast the load. Deep recurrent neural network (DRNN) based model is proposed to forecast the household load [15]. This method overcome the problems of overfitting created by classical deep learning methods. In [16], long term short-term memory RNN (LSTM-RNN) based forecasting framework is proposed to forecast the future residential load. The accuracy of the proposed framework is enhanced by embedding appliance consumption sequences in the training data. The proposed framework is validated on the real world dataset. The proposed framework performance analysis is carried out using historical load data of steel powder manufacturing.

To optimally harvest the potential of solar energy, forecasting of solar power energy is indispensable. Thus, least absolute shrinkage and selection operator model is proposed for forecasting solar energy generation [17]. The proposed model is trained using historical weather data aiming not only to reduce prediction error. Authors in [18], presented probabilistic forecasting model to forecast the solar power, electrical energy consumption, and net load across the seasonal variations and scalability. The hybrid model is proposed in [19] for short-term load prediction. This model is based on improved empirical mode decomposition, ARIMA, and wavelet neural network (WNN) optimized by fruit-fly optimization algorithm. In [20], a deep learning based electric load prediction model is proposed to forecast the future load. The proposed model extract abstracted features using stacked denoising auto-encoders technique.

3   Proposed System Model

In order to forecast the future load, prediction models must have the ability to learn the non-linear input/output mapping in most efficient way. In machine learning, ANN is one of the techniques mostly used to forecast non-linear load due to easy and flexible implementation. However, the performance of ANN is compromised for large datasize. The learning algorithms for training neural network such as gradient decent, multivariate AR, and back propagation may suffer from premature convergence and overfitting [10]. To cure the aforementioned problems, hybrid forecast strategies in literature have been proposed. However, hybrid forecast strategies have improved modeling capabilities as compared to individual methods. Still there is a problem of slow convergence and high execution time due to their molding complexity. In [11], the authors used

Bi-level strategy, which is based on ANN and DEA for electric load forecasting. An accurate fast convergence strategy based on ANN and MEDEA is proposed to forecast the future load [12]. However, the performance of the aforesaid strategies are satisfactory for small data size and their performance is compromised as the size of the data increases. There is no mechanism proposed to handle the large data (big data) and in real life the data size is increasing dramatically.

In this paper, a hybrid model based on MMI, deep learning (FCRBM), and GWDO algorithm is proposed for fast and accurate short-term load forecasting, as shown in Fig. 1. The proposed model comprises of three modules as illustrated in Fig. 1: (a) data pre-processing and feature selection module based on MMI, (b) FCRBM based training and forecasting module, and (c) the proposed GWDO algorithm based optimization module. The detailed demonstration of the proposed system model is as follows:

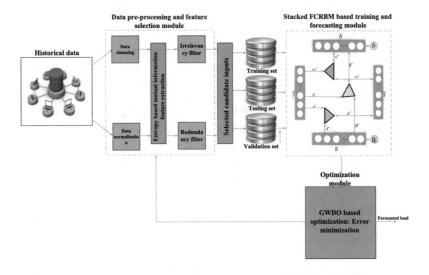

Fig. 1. Proposed system model

3.1 Data Pre-processing and Feature Selection Module

Let, E is the historical electric load data, which is represented in the matrix form. This historical data is fed into the data pre-processing and feature selection module.

$$E = \begin{bmatrix} E\,(1,1) & E\,(2,1) & E\,(3,1) & E\,(4,1) & \cdots & E\,(x,1) \\ E\,(1,2) & E\,(2,2) & E\,(3,2) & E\,(4,2) & \cdots & E\,(x,2) \\ E\,(1,3) & E\,(2,3) & E\,(3,3) & E\,(4,3) & \cdots & E\,(x,3) \\ E\,(1,4) & E\,(2,4) & E\,(3,4) & E\,(4,4) & \cdots & E\,(x,4) \\ \cdot & \cdot & \cdot & \cdot & \cdot & \cdot \\ \cdot & \cdot & \cdot & \cdot & \cdot & \cdot \\ \cdot & \cdot & \cdot & \cdot & \cdot & \cdot \\ E\,(1,y) & E\,(2,y) & E\,(3,y) & E\,(4,y) & \cdots & E\,(x,y) \end{bmatrix} \tag{1}$$

where $E(1,1)$ is the electric load of first day first hour, $E(2,1)$ is the electric load of second day first hour, and such that $E(x,y)$ is the electric load of xth day and y^{th} hour. The rows show the number of hours and columns show the number of days. The value of x is linked with tuning of FCRBM training, larger the value of x performs fine tuning and vice versa. There is a performance tradeoff between fine tuning and convergence rate. This input data is first passed through the data cleansing phase, where defective and missing values are replaced by the average value of preceding days. The cleansed data is passed through normalization phase because the data have outliers and weight matrix is extremely small, to make the overall weighted sum within the limit of activation function. In machine learning, feature extraction/selection is a process of selecting a subset of abstracted features from a given dataset to avoid curse of dimensionality. In this regard, MI features selection technique is developed and used by [10] and [12] for features selection. In this work, the MI technique is improved by modification (MMI) subjected to accuracy and convergence rate. The MMI based feature extraction technique rank the inputs according to the information importance. The ranked inputs are filtered using the irrelevancy and redundancy filters in order to remove irrelevant and redundant information. The subset of selected features contain best and more relevant information which contributes highly to the accuracy and convergence. The detailed demonstration of MMI feature extraction technique is as:

For two discrete random variables the joint entropy based MI is defined as, the information obtained while observing both discrete random variables at the same time. The mathematical description is as follow:

$$H\left(E, E^t\right) = -\sum_i \sum_j p\left(E_i, E_j^t\right) \log_2 \left(p\left(E_i, E_j^t\right)\right) \; \forall i, j \in \{1,\, 2\} \tag{2}$$

where $p\left(E_i, E_j^t\right)$ is the two discrete random variables joint probability, E_i is the input discrete random variables, and E_j^t is the target value. In feature selection the information which is common among both variables are indispensable, which is formulated as in [10]:

$$MI\left(E, E^t\right) = \sum_i \sum_j p\left(E_i, E_j^t\right) \log_2 \left(\frac{p\left(E_i, E_j^t\right)}{p\left(E_i\right) p\left(E_j^t\right)}\right) \tag{3}$$

where $MI\left(E_i, E_j^t\right)$ is the common information among the two variables known as mutual information. In this case, the candidate inputs are ranked by MI technique between input and target value. From entropy based MI technique, the following three reasonings can be drawn:

- If $MI\left(E_i, E_j^t\right) = 0$, it indicates that the discrete random variables E_i and E_j^t are irrelevant.
- If $MI\left(E_i, E_j^t\right)$ has some larger value, it indicates that discrete random variables E_i and E_j^t highly relevant.
- If $MI\left(E_i, E_j^t\right)$ has smaller value, it indicates that discrete variables E_i and E_j^t are lightly related.

In [10], among the training data samples last value of every hour of the day is chosen as the target value. The target value or last sample is very close to next day with respect to time. However, it will cause problems while forecasting. In [12], the authors used average value in addition to the target value because both average and target values are of equal importance. They modified Eq. 3 for three variables as:

$$MI\left(E, E^t, E^n\right) = \sum_i \sum_j \sum_k p\left(E_i, E_j^t, E_k^n\right) \times \log_2 \left(\frac{p\left(E_i, E_j^t, E_k^n\right)}{p(E_i)p\left(E_j^t\right)p\left(E_k^n\right)}\right) \quad (4)$$

where E_k^n is the average value. However, the average value is very low, if some values in the selected features are very small. The addition of average with other two parameters are not sufficient because it will cause prediction problems. The Eq. 3 is modified as for four variables:

$$MI\left(E, E^t, E^m, E^n\right) = \sum_i \sum_j \sum_k \sum_l p\left(E_i, E_j^t, E_k^n, E_l^m\right) \times \log_2 \left(\frac{p\left(E_i, E_j^t, E_k^n, E_l^m\right)}{p(E_i)p\left(E_j^t\right)p\left(E_k^n\right)p\left(E_l^m\right)}\right)$$
$$(5)$$

where the third target value E_l^m is mode.

The Eq. 5 is the MMI technique equation, which is used to find the mutual information between the four variables such as E_i, E^t, E^n, E^m. The candidate inputs are ranked on the basis of these mutual information to remove the irrelevant and redundant information. The MMI feature selection technique provides two fold benefits: (a) selection of suitable and relevant features minimize the forecast error, and (b) selection of subset of features improves the convergence rate. Before to fed the training and forecasting module the selected features are split into training, testing, and validation data samples for training and validation of the FCRBM.

3.2 FCRBM Based Training and Forecasting Module

In literature, a wide variety of short-term load forecasting strategies using ANN are proposed in [21]. These forecasting strategies are capable to handle the non-linear behavior of electric load. However, the performance of these strategies are compromised with the increase in the datasize. In this regard, FCRBM among the deep learning models is selected to forecast the future load because it has more layers and capable to improve the forecast accuracy.

The training and forecasting module is based on FCRBM. Therefore, at first, the architecture of FCRBM model is determined. The model has four layers such as the of number hidden layers, number of visible layers, number of style layers, and number of history layers. Then, FCRBM training process is described as: first, train the FCRBM with the training data samples using rectified linear unit (ReLU) activation function as shown in Eq. 6 because it tackles the problems of overfitting and vanishing gradient encountered in other activation functions. Thus, FCRBM technique is empowered via training and learning to predict the load. The output of this module is fed into the GWDO based optimization module to further improve forecast accuracy with affordable convergence rate.

$$f(x) = \max (0, \ x)$$
$$\Delta f(x) \begin{cases} 1 & \text{if } x \geq 0 \\ 0 & \text{otherwise} \end{cases} \tag{6}$$

3.3   GWDO Based Optimization Module

The objective of this module is to minimize the forecast error with affordable convergence rate. The authors used DEA [10] and MEDEA [12] to optimize the performance of the forecasting model. Both of these algorithms have slow convergence rate and low precision [22]. Furthermore, aforesaid algorithms trapped into local optimum [22]. To remedy the aforementioned problems, GWDO algorithm is proposed, which is a hybrid of WDO and GA algorithms. The proposed algorithm takes benefit from the features of both algorithms (GA and WDO). The GA algorithm enables the diversity of population and WDO has faster convergence. The GWDO based module receives the forecasted load with some error that is minimum as per the ability of FCRBM. This forecast error can be minimized with the proposed GWDO optimization technique. The sole objective of GWDO based optimization module is to improve forecast accuracy with affordable convergence rate and mathematically modeled as:

$$\min \ MAPE(i) \tag{7}$$

The GWDO based optimization module sets the thresholds for MMI to select the abstracted features from the given data. The integration of optimization module to forecasting module increased the execution time, which disturb the convergence rate because tradeoff between execution time and convergence rate. However, the proposed fast and accurate short-term load forecasting outperforms existing models i.e., MI-ANN [10], Bi-level [11], and AFC-STLF [12]. It is due to the fact that as the size of the data increases the ANN based model performance degraded because of their shallow layout. The FCRBM have improved performance with the large datasize due to their deeper layers layout.

4   Simulation Results and Discussions

For performance evaluation of the proposed fast and accurate short-term load forecasting model, simulations are conducted in MATLAB. In simulations, the proposed model is compared with three literature short-term load forecasting models i.e., MI-ANN [10], Bi-level [11], and AFC-STLF [12]. Historical electric load data is taken from publicly available PJM market [23] for performance evaluation of the proposed model. The data is the monthly electric load data of three USA grids (FE, EKPC, and Daytown) of years 2014–2017. The first three years of data are used to train the network and last one year of data is used to test the network. The aforementioned models are selected as benchmark models due their closer architectural similarities with the proposed model. Two performance metrics i.e., accuracy and convergence rate are used for performance evaluation. Accuracy is defined as: accuracy = 100-MAPE and is measured in %. Execution time is defined as: the time spent by the forecast strategy during execution and is measured in seconds. The detailed demonstration is as follows.

4.1   Learning Curve Analysis in Terms of Resolve Overfitting

Learning curve is a graphical representation that compares the performance of models on training and testing data samples across a varying number of epochs. The analysis in Fig. 2 is to verify whether the chosen model is learning or memorizing the data. When there is high variance and bias, the learning curve is bad, and the model is memorizing not learning. Due to high bias, the training and testing error rate is high and the convergence rate is fast. In contrast, the high variance occurs, when the gap between training and testing errors is large. In both cases the model is not good and leads to poor generalization. Overfitting occurs when the test error at certain point starts to increase and training error decrease. This shows that the model memorizing training data and prediction is inaccurate. Thus, such model leads to bad generalization. The overfitting problem is prevented using dropout method and early stoping. However, the adopted deep learning model i.e., FCRBM, it is observe that the testing error gradually decrease as the training error does for FE, Daytown, and EKPC grids of USA as illustrated in Fig. 2. Thus, the FCRBM model is relearning and resolving the problem of overfitting. Moreover, the gap between training error and testing error is small and there is no bias and variance as clearly depicted in Fig. 2 for FE, Daytown, and EKPC grids of USA.

4.2   Predicted Load Analysis in Terms of Forecast Accuracy

The hourly forecasted future electric load profile of the proposed and existing models (Bi-level, MI-ANN, and AFC-STLF) for three girds of USA (FE, Daytown, and EKPC) is illustrated in Fig. 3. It is obvious from this graphical illustration that all models are capable to capture the complex behavior of load. It is also clear that all existing models (Bi-level, MI-ANN, and AFC-STLF) use sigmoidal activation function, levenburg-marquardt, and multi-variate AR

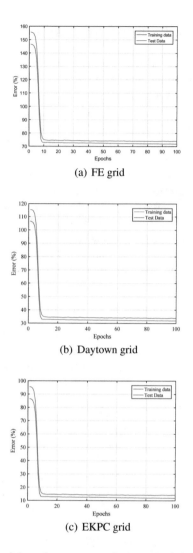

(a) FE grid

(b) Daytown grid

(c) EKPC grid

Fig. 2. FCRBM model performance analysis on training and testing data

algorithms to train the network. The adopted FCRBM network is trained using ReLU and stochastic gradient descent due to having small execution time. It is verified from the Fig. 3 that the proposed model closely follows the target load as compared to existing models for all three USA grids: FE, Daytown, and EKPC. The MAPE error of proposed FRCBM based model is 0.1231%, MAPE of Bi-level model is 2.5186%, MAPE of MI-ANN model is 4.3371%, and MAPE of AFC-STLF model is 0.7741%. The MAPE of the proposed model is lower as compared to existing models, lower MAPE results in better accuracy. The forecasted load of AFC-STLF is more suitable than Bi-level, and Bi-level is more suitable than MI-ANN in terms of accuracy. The reason is that AFC-

STLF model used MEDEA for optimization and Bi-level model used DEA for optimization, which improves the forecast accuracy by minimizing the error. However, this accuracy is improved at the cost of large execution time. The proposed FCRBM based model outperforms Bi-level, MI-ANN, and AFC-STLF models due to the integration of MMI technique and GWDO based optimization module. The MAPE of the proposed model, Bi-level, MI-ANN, and AFC-STLF are 0.1180%, 2.6220%, 4.4988%, and 0.7633%, respectively. Furthermore, the future forecasted load based on FCRBM is more accurate as compared to the existing models (Bi-level, MI-ANN, and AFC-STLF) due to the use of MMI technique, deep neural network, and GWDO for the proposed model. The MAPE

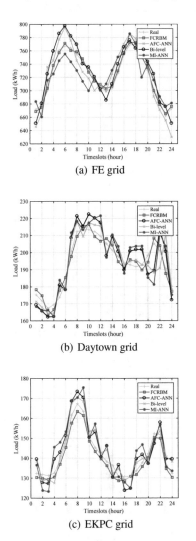

(a) FE grid

(b) Daytown grid

(c) EKPC grid

Fig. 3. Comparatively evaluation of predicted load in terms forecast accuracy

the proposed model is 0.1325%, Bi-level is 2.4202%, MI-ANN is 4.3280, and AFC-STLF is 0.7530%. Thus, the proposed model is better as compared to the existing models (Bi-level, MI-ANN, and AFC-STLF) in terms of forecast accuracy. From this discussion, it is concluded that the proposed FCRBM based model outperforms existing models, with the FCRBM based model the MAPE for FE grid 0.1231%, 0.1180% for Daytown, and 0.1325% for grids.

4.3 Predicted Load Analysis in Terms of Convergence Rate

The convergence rate analysis of the proposed model and existing models for three grids of USA (FE, Daytown, and EKPC) is illustrated in Fig. 4. There

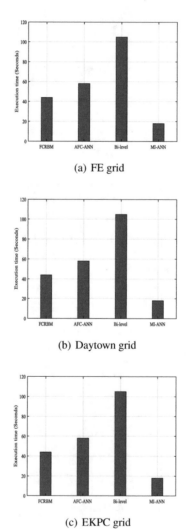

(a) FE grid

(b) Daytown grid

(c) EKPC grid

Fig. 4. Comparative forecasting analysis in terms of convergence rate

is a trade-off between forecast accuracy and convergence rate. The accuracy of the Bi-level strategy is improved as compared to MI-ANN model at the cost of more execution time due to the integration of DEA based optimization module with Bi-level strategy. It is obvious from the Fig. 4 the execution time is increased from 6.52 s to 102 s as the optimization module is integrated with the forecasting module. The proposed model have reduced execution time due to the following reasons: (i) highly abstractive features are given as an input to training and forecasting module which reduces the network training time, (ii) it replaces the sigmoidal activation function by ReLU, and (iii) it uses GWDO instead of MEDEA algorithm due to relatively faster convergence rate. In the proposed fast and accurate short-term load forecasting model leads to decrease the execution time (from 103 s to 43 s) due to aforesaid modifications in the existing models. On the other side, the proposed model have excellent performance as compared to the existing models which is depicted in Fig. 4.

5   Conclusion

Load forecasting is an indispensable part of the SG for effective operation and disciplined planning of the power systems. Therefore, it is a significant goal for scholars and industry to develop a forecasting model, which provides fast and accurate load forecasting. However, the performance of single module based forecasting models is not satisfactory due to the inherent limitations. On the other hand, hybrid models fully utilize the advantages of individual techniques and have improved performance. In this paper, a fast and accurate short-term load forecasting model is proposed. The model have three modules data preprocessing and feature extraction module based on MMI, training and forecasting module based on FCRBM, and optimization module based on GWDO. Based on simulation results, the following conclusions can be drawn. First, the proposed MMI technique improves the forecast accuracy by extracting abstractive features from the given dataset and then input these abstracted features into the training module based on FCRBM to reduce the training time. Secondly, the adopted deep learning model i.e., FCRBM is empowered via learning to forecast the load. Thirdly, the proposed GWDO algorithm is used in the optimization module due to its ability to fine tune the control parameters with affordable execution time. Finally, the proposed model is validated by comparing with AFC-STLF, Bi-level, and MI-ANN models in terms of accuracy and convergence rate. In short, the proposed fast and accurate short-term load forecasting model has MAPE of 0.0424%, AFC-STLF model has MAPE of 0.8108%, Bi-level model has MAPE of 0.0718%, and MI-ANN model has MAPE of 0.6978%. Furthermore, the proposed model has execution time of 43 s, AFC-STLF model has execution time of 59 s, Bi-level model has execution time of 102 s, and MI-ANN has execution time of 19 s. It is concluded that the proposed model outperforms existing models in terms forecast accuracy and convergence rate.

References

1. Javaid, N., Hafeez, G., Iqbal, S., Alrajeh, N., Alabed, M.S., Guizani, M.: Energy efficient integration of renewable energy sources in the smart grid for demand side management. IEEE Access **6**, 77077–77096 (2018)
2. Hafeez, G., Javaid, N., Iqbal, S., Khan, F.: Optimal residential load scheduling under utility and rooftop photovoltaic units. Energies **11**(3), 611 (2018). Xiao, L., Shao, W., Wang, C., Zhang, K., Lu, H.: Research and application of a hybrid model based on multi-objective optimization for electrical load forecasting. Appl. Energy **180**, 213–233 (2016)
3. Metaxiotis, K., Kagiannas, A., Askounis, D., Psarras, J.: Artificial intelligence in short term electric load forecasting: a state-of-the-art survey for the researcher. Energy Convers. Manag. **44**(9), 1525–1534 (2003)
4. Hernandez, L., Baladron, C., Aguiar, J.M., Carro, B., Sanchez-Esguevillas, A.J., Lloret, J., Massana, J.: A survey on electric power demand forecasting: future trends in smart grids, microgrids and smart buildings. IEEE Commun. Surv. Tutor. **16**(3), 1460–1495 (2014)
5. Rahman, A., Srikumar, V., Smith, A.D.: Predicting electricity consumption for commercial and residential buildings using deep recurrent neural networks. Appl. Energy **212**, 372–385 (2018)
6. Boroojeni, K.G., Amini, M.H., Bahrami, S., Iyengar, S.S., Sarwat, A.I., Karabasoglu, O.: A novel multi-time-scale modeling for electric power demand forecasting: from short-term to medium-term horizon. Electr. Power Syst. Res. **142**, 58–73 (2017)
7. Xu, X., Niu, D., Wang, Q., Wang, P., Wu, D.D.: Intelligent forecasting model for regional power grid with distributed generation. IEEE Syst. J. **11**(3), 1836–1845 (2017)
8. Dedinec, A., Filiposka, S., Dedinec, A., Kocarev, L.: Deep belief network based electricity load forecasting: an analysis of Macedonian case. Energy **115**, 1688–1700 (2016)
9. Hong, W.-C.: Electric load forecasting by seasonal recurrent SVR (support vector regression) with chaotic artificial bee colony algorithm. Energy **36**(9), 5568–5578 (2011)
10. Amjady, N., Keynia, F., Zareipour, H.: Short-term load forecast of microgrids by a new bilevel prediction strategy. IEEE Trans. Smart Grid **1**(3), 286–294 (2010)
11. Amjady, N., Keynia, F.: Day-ahead price forecasting of electricity markets by mutual information technique and cascaded neuro-evolutionary algorithm. IEEE Trans. Power Syst. **24**(1), 306–318 (2009)
12. Ahmad, A., Javaid, N., Guizani, M., Alrajeh, N., Khan, Z.A.: An accurate and fast converging short-term load forecasting model for industrial applications in a smart grid. IEEE Trans. Ind. Inform. **13**(5), 2587–2596 (2017)
13. Liu, D., Zeng, L., Li, C., Ma, K., Chen, Y., Cao, Y.: A distributed short-term load forecasting method based on local weather information. IEEE Syst. J. **12**(1), 208–215 (2018)
14. Collotta, M., Pau, G.: An innovative approach for forecasting of energy requirements to improve a smart home management system based on BLE. IEEE Trans. Green Commun. Netw. **1**(1), 112–120 (2017)
15. Shi, H., Minghao, X., Li, R.: Deep learning for household load forecasting-a novel pooling deep RNN. IEEE Trans. Smart Grid **9**(5), 5271–5280 (2018)

16. Kong, W., Dong, Z.Y., Hill, D.J., Luo, F., Xu, Y.: Short-term residential load forecasting based on resident behaviour learning. IEEE Trans. Power Syst. **33**(1), 1087–1088 (2018)

17. Tang, N., Mao, S., Wang, Y., Nelms, R.M.: Solar power generation forecasting with a LASSO-based approach. IEEE Internet Things J. **5**, 1090–1099 (2018)

18. van der Meer, D.W., Munkhammar, J., Widén, J.: Probabilistic forecasting of solar power, electricity consumption and net load: investigating the effect of seasons, aggregation and penetration on prediction intervals. Solar Energy **171**, 397–413 (2018)

19. Zhang, J., Wei, Y.-M., Li, D., Tan, Z., Zhou, J.: Short term electricity load forecasting using a hybrid model. Energy **158**, 774–781 (2018)

20. Tong, C., Li, J., Lang, C., Kong, F., Niu, J., Rodrigues, J.J.P.C.: An efficient deep model for day-ahead electricity load forecasting with stacked denoising auto-encoders. J. Parallel Distrib. Comput. **117**, 267–273 (2018)

21. Abedinia, O., Amjady, N., Zareipour, H.: A new feature selection technique for load and price forecast of electrical power systems. IEEE Trans. Power Syst. **32**(1), 62–74 (2017)

22. Bao, Z., Zhou, Y., Li, L., Ma, M.: A hybrid global optimization algorithm based on wind driven optimization and differential evolution. In: Mathematical Problems in Engineering 2015 (2015)

23. https://www.pjm.com/. Accessed 8 Mar 2018

Subprocess Transmission Strategies for Recovering from Faults in the Tree-Based Fog Computing (TBFC) Model

Ryuji Oma[1(✉)], Shigenari Nakamura[1], Dilawaer Duolikun[1], Tomoya Enokido[2], and Makoto Takizawa[1]

[1] Hosei University, Tokyo, Japan
ryuji.oma.6r@stu.hosei.ac.jp, nakamura.shigenari@gmail.com,
dilewerdolkun@gmail.com, makoto.takizawa@computer.org
[2] Rissho University, Tokyo, Japan
eno@ris.ac.jp

Abstract. In order to increase the performance in the IoT (Internet of Things), the fog computing model is proposed. Here, subprocesses to handle sensor data are performed on fog nodes in addition to servers. Output data processed by a subprocess of a fog node is sent to a succeeding node. If a fog node is faulty, the preceding nodes are disconnected, i.e. no output data of preceding nodes can be delivered to servers. Another operational node which supports the same subprocess as the faulty node is an alternate node. In our previous studies, the FTBFC (Fault-tolerant TBFC) and MFTBFC (Modified FTBFC) models are proposed where disconnected nodes send the output data to alternate nodes. In this paper, we newly propose another strategy where a subprocess of a faulty node is transmitted to surrogate nodes which hold data to be processed. In this paper, we propose an SMSGD (Selecting Multiple Surrogates in Grandparent and Disconnected nodes) algorithm to select surrogate nodes. In the evaluation, we show the energy consumption and execution time of each surrogate node can be reduced compared with the data transmission strategy.

Keywords: Fault-tolerant TBFC (FTBFC) model ·
Process transmission strategy · Surrogate nodes · SMSGD algorithm

1 Introduction

In the Internet of Things (IoT) [4], not only computers like servers and clients but also sensors and actuators are interconnected in networks. In the cloud computing model [1], data collected by sensors is transmitted to servers in a cloud. Networks are congested to transmit huge volume of sensor data and servers are also overloaded to process the sensor data. The fog computing (FC) model [11]

© Springer Nature Switzerland AG 2020
L. Barolli et al. (Eds.): CISIS 2019, AISC 993, pp. 50–61, 2020.
https://doi.org/10.1007/978-3-030-22354-0_5

is proposed to reduce the communication and processing traffic. Here, subprocesses to handle sensor data are performed on fog nodes in addition to servers. Sensor data is processed and the output data is sent to another fog node. On receipt of output data from fog nodes, a fog node further processes the data and sends the output data to fog nodes. Thus, data processed by fog nodes is finally delivered to servers.

It is critical to reduce the electric energy consumed by fog nodes and servers since the IoT includes a huge number of nodes. In order to reduce the energy consumption and execution time of fog nodes and servers, the TBFC (Tree-Based Fog Computing) model [5,6,10] is proposed. Here, fog nodes are hierarchically structured in a height-balanced tree. A root node shows a cluster of servers. Fog nodes at the bottom level are edge nodes which communicate with sensors and actuators. Sensors first send data to edge nodes. Each edge node generates output data by processing input data and sends the output data to a parent node [5]. An application process is assumed to be a sequence of subprocesses. Every node at each level is equipped with a same subprocess. Each node receives input data from child nodes and sends a parent node output data obtained by processing the input data by the subprocess.

If a node gets faulty, child nodes of the faulty node are disconnected. Disconnected nodes cannot deliver output data to nodes of higher levels. An operational node at the same level as a faulty node is an *alternate* node which supports a same subprocess as the faulty node. In the FTBFC (Fault-tolerant TBFC) [7,9] and MFTBFC (Modified FTBFC) models [8], one alternate node is selected as a new parent node of each disconnected node. Data processed by each disconnected node is transmitted to the selected alternate node.

In this paper, we newly propose another strategy where a subprocess of a faulty node is transmitted to a node where data processed by disconnected nodes can be processed. A node to which a subprocess of a faulty node is transmitted is a *surrogate* node. We newly propose an SMSGD (Selecting Multiple Surrogates in Grandparent and Disconnected nodes) algorithm to select surrogate nodes for disconnected nodes so that the energy to be consumed by the surrogate nodes can be reduced.

We evaluate the subprocess transmission strategy with the SMSGD algorithm compared with the data transmission strategy [7–9] in terms of the energy consumption and execution time of surrogate nodes and alternate nodes. In the evaluation, we show the energy consumption and execution time of each new parent node can be reduced in the SMSGD algorithm.

In Sect. 2, we present the TBFC model. In Sect. 3, we propose the subprocess transmission strategy and the SMSGD algorithm. In Sect. 4, we evaluate the SMSGD algorithm.

2   Tree-Based Fog Computing (TBFC) Model

2.1   Tree Structure of Fog Nodes

Sensor data is sent to servers and application processes to handle the data are performed on servers in the cloud computing (CC) model [1]. Networks are congested and servers are overloaded due to heavy traffic from sensors. The fog computing (FC) model [11] is composed of sensors and actuators, fog nodes, and clouds of servers. Sensors and actuators are connected to edge nodes. Some subprocesses of an application process are performed on a fog node to process a collection of input data sent by sensors or other fog nodes. Then, output data obtained by processing the input data is sent to another node. Servers finally receive data processed by fog nodes. Actions decided by servers are sent to edge nodes and the edge nodes issue the actions to actuator nodes.

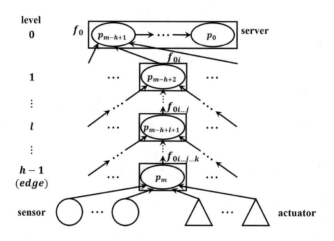

Fig. 1. TBFC model.

In the tree-based fog computing (TBFC) model [5–10], fog nodes are hierarchically structured in a height-balanced tree. Here, the root node f_0 is a cloud of servers. Each node f_R has l_R child nodes f_{R0}, ..., f_{R,l_R-1} ($l_R \geq 1$). Here, f_{Ri} stands for the $(i+1)$th child node of the node f_R. $ch(f_R)$ is a set $\{f_{R0}, ..., f_{R,l_R-1}\}$ of child nodes of a node f_R and $pt(f_{Ri})$ is a parent node f_R of a node f_{Ri}. Let $as(f_R)$ and $ds(f_R)$ be sets of ancestor and descendant nodes of a f_R, respectively. An edge node f_R receives data from child sensors s_{R0}, ..., s_{R,sl_R-1} ($sl_R \geq 1$) and sends actions to child actuators a_{R0}, ..., a_{R,al_R-1} ($al_R \geq 1$).

A node f_R takes a collection D_R of input data d_{R0}, ..., d_{R,l_R-1} from child nodes f_{R0}, ..., f_{R,l_R-1}, respectively. A subprocess $p(f_R)$ of the node f_R generates output data d_R by processing input data D_R. Then, the node f_R sends the output data d_R to a parent node $pt(f_R)$. A notation $|d|$ shows the size [Byte] of data d. The ratio $|d_R|/|D_R|$ is the *output ratio* ρ_R of a node f_R.

In this paper, we assume an application process p is realized as a sequence of subprocesses p_0, p_1, ..., p_m. The subprocess p_m takes sensor data d_{m+1} from sensors, generates output data d_m by doing computation on the input data d_{m+1}, and sends the output data d_m to the subprocess p_{m-1}. Thus, each subprocess p_i receives input data d_{i+1} from a preceding subprocess p_{i+1} and outputs data

d_i to a succeeding subprocess p_{i-1}. In the CC model, all the subprocesses p_0, ..., p_m are performed on a server. In the TBFC model, the subprocess p_m is performed on edge nodes of level h - 1 as shown in Fig. 1. The subprocess p_m on each edge node f_R receives sensor data d_{Ri} from each descendant sensor s_{Ri} and sends output data d_R to the succeeding subprocess p_{m-1} of a parent node $pt(f_R)$. A same subprocess $p_{m-h+l+1}$ is performed on each node of each level l (l = 0, 1, ..., $h - 1$). The subprocess p_{m-h+2} is performed on nodes f_{00}, ..., f_{0,l_0-1} of level 1, i.e. child nodes of the root node. Subprocesses p_{m-h+1}, ..., p_0 ($m \geq h$) are performed on the root node f_0, i.e. server in the cluster.

2.2   Energy Consumption of Fog Nodes

A node f_R takes input data D_R of size i_R ($= |D_R|$) from child nodes and sends output data d_R of size o_R ($= |d_R|$) [5,6,10] to a parent node $pt(f_R)$, where o_R = $\rho_R \cdot i_R$ for the output ratio ρ_R. $EI_R(x)$, $EC_R(x)$, $ES_R(x)$, and $EO_R(x)$ show the energy consumption [J] of input, computation, storage, and output modules [5] to process data of size x, respectively. $TI_R(x)$, $TC_R(x)$, $TS_R(x)$, and $TO_R(x)$ are execution time [sec] of input, computation, storage, and output modules for data of size x, respectively. $TC_R(x)$ depends on the computation complexity of a subprocess $p(f_R)$ of the node f_R. In this paper, $TC_R(x)$ is assumed to be x or x^2. $TI_R(x)$, $TS_R(x)$, and $TO_R(x)$ are proportional to the data size x. It takes $TF_R(x)$ [sec] to process data of size x in a node f_R:

$$TF_R(x) = TI_R(x) + TC_R(x) + TS_R((1 + \rho_R) \cdot x) + \delta_R \cdot TO_R(\rho_R \cdot x). \quad (1)$$

Here, if f_R is a root, $\delta_R = 0$, else $\delta_R = 1$.

In this paper, each node f_R follows the SPC (simple power consumption) model [2,3]. The power consumption of a node f_R to perform the computation module C_R is $maxE_R$ [W]. The energy consumption $EC_R(x)$ [J] to process input data of size x (> 0) is $EC_R(x) = maxE_R \cdot TC_R(x)$.

A pair of the power consumption PI_R and PO_R [W] of the input I_R and output O_R modules are proportional to the receiving and transmission rates of a node f_R, respectively [2,3]. Hence, the energy consumption $EI_R(x)$ and $EO_R(x)$ [J] to receive and send data of size x (> 0) are $EI_R(x) = PI_R \cdot TI_R(x)$ and $EO_R(x) = PO_R \cdot TO_R(x)$. PS_R [W] shows the power consumption of a node f_R to store data in a database which depends on the access rate a_R [bps]. Hence, the energy consumption $ES_R(x)$ of a fog node f_R to store data of size x ($>$ 0) is $ES_R(x) = PS_R \cdot TS_R(x)$. Each node f_R consumes the energy $EF_R(x)$ to process input data D_R of size x:

$$EF_R(x) = EI_R(x) + EC_R(x) + ES_R((1 + \rho_R) \cdot x) + \delta_R \cdot EO_R(\rho_R \cdot x). \quad (2)$$

3 Fault-Tolerant Strategies

3.1 Data Transmission Strategy

If a fog node f_R stops by fault, every child node f_{Ri} is *disconnected*. Disconnected nodes cannot deliver output data to nodes of higher level even if the nodes receive input data from child nodes. A node at the same level as a node f_R is an *alternate* node of the node f_R which supports the subprocess $p(f_R)$. An alternate node of a faulty node f_R can be a new parent node of disconnected nodes of f_R. Let $an(f_R)$ be a set of alternate nodes of a node f_R. We assume each node f_{Ri} knows every alternate node of the parent node f_R in $an(f_R)$. In the FTBFC model [7], one alternate node f_U is selected as a new parent node of all the disconnected nodes $f_{R0}, ..., f_{R,l_R-1}$ of a faulty node f_R. A new parent node f_U has to process input data D_R of size i_R from the disconnected nodes in addition to input data D_U of size i_U from its own child nodes $f_{U0}, ..., f_{U,l_U-1}$. Hence, the energy consumption $EF_U(i_U)$ [J] and execution time $TF_U(i_U)$ [sec] of the new parent node f_U increase to $EF_U(i_U + i_R)$ and $TF_U(i_U + i_R)$, respectively. In a parent node f_T of the faulty node f_R, the energy consumption EF_T and execution time TF_T decrease to $NEF_T = EF_T(i_T - \rho_R \cdot i_R)$ and $NTF_T = TF_T(i_T - \rho_R \cdot i_R)$, respectively.

In the ME (Minimum Energy) algorithm [9], an alternate node f_U whose energy consumption $EF_U(i_U + i_R)$ is minimum is selected to be a new parent node of the disconnected nodes. Here, the total energy consumption and execution time of nodes can be more reduced. However, the energy consumption of the selected new parent node f_U increases while the reduction of total energy consumption of the tree is small [7]. In order to reduce the energy consumption and execution time of a new parent node, the SMPR (Selecting Multiple Parents for Recovery) algorithm [8] is proposed. Here, an alternate node f_U is selected for each disconnected node f_{Ri} as shown in Fig. 2. Since the output data of disconnected nodes is distributed to multiple new parent nodes, the energy consumption and execution time of each new parent node can be smaller than the ME algorithm.

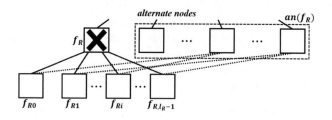

Fig. 2. SMPR algorithm.

3.2   Subprocess Transmission Strategy

In this paper, we newly propose a strategy to transfer a subprocess $p(f_R)$ of a faulty node f_R to a node f_S where the output data of disconnected nodes can be processed. Here, the node f_S is a *surrogate* node of f_R. A node which supports a subprocess $p(f_R)$ is a *source* node of $p(f_R)$. A source node f_U sends a subprocess $p(f_R)$ to a surrogate node f_S. A subprocess $p(f_R)$ of f_R of level l is $p_{m-h+l+1}$ as shown in Fig. 1.

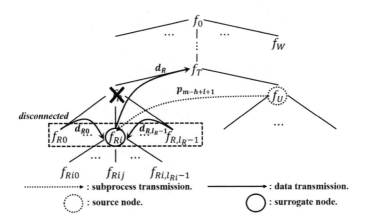

Fig. 3. Subprocess transmission strategy.

We consider the following surrogate nodes:

1. A disconnected node f_{Ri} of the faulty node f_R is a surrogate node of f_R.
2. A parent node f_T of the faulty node f_R is a surrogate node of f_R.

Suppose a node f_R is faulty and a node f_S is a surrogate node of the faulty node f_R. $|p(f_R)|$ shows the size of the subprocess $p(f_R)$. In Fig. 3, a node f_R is faulty and an alternate node f_U is a source node of the subprocess $p(f_R)$. A disconnected node f_{Ri} is a surrogate node. The source node f_U sends the subprocess $p(f_R)$ to f_{Ri}. The other disconnected nodes send output data to f_{Ri}. Then, the surrogate node f_{Ri} sends the output data to a grandparent node f_T of disconnected nodes. Thus, each disconnected node f_{Ri} sends output data to the surrogate node f_S. In the surrogate node f_S, the subprocess $p(f_R)$ is performed on input data from disconnected nodes in addition to its own subprocess $p(f_S)$. A surrogate node f_S is selected and receives the subprocess $p(f_R)$, e.g. from an alternate node f_U as shown in Fig. 3. Then, the other disconnected nodes send output data to the node f_{Ri}. The node f_{Ri} processes not only output data by itself but also output data received from the other disconnected nodes. Then, the node f_{Ri} sends the output data to its grandparent node f_T. If a parent node f_S of f_R is a surrogate node, the energy consumption PEF_S and execution time

PET_S of the surrogate node f_S for input data D_S of size i_S from its child nodes and input data of size i from disconnected nodes are given as follows:

$$
\begin{aligned}
PEF_S(i_S, i) = \; & EI_S(i) + EC_S(i) + ES_S((1 + \rho_R) \cdot i) + \\
& EI_S(i_S) + EC_S(i_S + \rho_R \cdot i) + ES_S((1 + \rho_S) \cdot (i_S + \rho_R \cdot i)) + \\
& \delta_S \cdot EO_S(\rho_S \cdot (i_S + \rho_R \cdot i)) + EI_S(|p(f_R)|). \quad (3)
\end{aligned}
$$

$$
\begin{aligned}
PTF_S(i_S, i) = \; & TI_S(i) + TC_S(i) + TS_S((1 + \rho_R) \cdot i) + \\
& TI_S(i_S) + TC_S(i_S + \rho_R \cdot i) + TS_S((1 + \rho_S) \cdot (i_S + \rho_R \cdot i)) + \\
& \delta_S \cdot TO_S(\rho_S \cdot (i_S + \rho_R \cdot i)) + TI_S(|p(f_R)|). \quad (4)
\end{aligned}
$$

If the disconnected node f_S ($\in ch(f_R)$) is a surrogate node, the energy consumption DEF_S and execution time DTF_S are given as follows:

$$
\begin{aligned}
DEF_S(i_S, i) = \; & EI_S(i_S) + EC_S(i_S) + ES_S((1 + \rho_S) \cdot i_S) + \\
& EI_S(i) + EC_S(i + \rho_S \cdot i_S) + ES_S((1 + \rho_R) \cdot (i + \rho_S \cdot i_S)) + \\
& \delta_S \cdot EO_S(\rho_R \cdot (i + \rho_S \cdot i_S)) + EI_S(|p(f_R)|). \quad (5)
\end{aligned}
$$

$$
\begin{aligned}
DTF_S(i_S, i) = \; & TI_S(i_S) + TC_S(i_S) + TS_S((1 + \rho_S) \cdot i_S) + \\
& TI_S(i) + TC_S(i + \rho_S \cdot i_S) + TS_S((1 + \rho_R) \cdot (i + \rho_S \cdot i_S)) + \\
& \delta_S \cdot TO_S(\rho_R \cdot (i + \rho_S \cdot i_S)) + TI_S(|p(f_R)|). \quad (6)
\end{aligned}
$$

In this paper, we assume every fog node is homogeneous, i.e. $DEF(x, y) = DEF_S(x, y) = DEF_T(x, y)$ and $DTF(x, y) = DTF_S(x, y) = DTF_T(x, y)$ for every pair of nodes f_S and f_T. It takes times to receive the subprocess $p(f_R)$. We assume the energy consumption $EI_S(x)$ and execution time $TI_S(x)$ depends on only the size x of the subprocess $p(f_R)$. If a disconnected node f_{Ri} is a surrogate node, the surrogate node f_{Ri} sends the output data to the grandparent node f_T, which is the same as the output data d_R to be sent by the faulty node f_R. If a parent node f_T of the faulty node f_R is a surrogate node, disconnected nodes send output data to the node f_T.

A disconnected node f_{Ri} is *cheaper* than another disconnected node f_{Rj} (f_{Rj} is *richer* than f_{Ri}) iff f_{Ri} consumes smaller energy than f_{Rj}. If a node f_R gets faulty, surrogate nodes and disconnected nodes to be their child nodes are selected in an SMSGD (Selecting Multiple Surrogates in Grandparent and Disconnected nodes) algorithm which is shown in Algorithm 1. Here, multiple surrogate nodes are taken in the grandparent and disconnected nodes of the faulty node f_R. First, a parent node f_T of the node f_R is taken as a surrogate. Then, a richest disconnected node f_X whose energy consumption $DEF_X(i_X, 0)$ is the largest is selected as a child node of f_T. Each time a disconnected node f_X is connected to f_T, ci_T is incremented by o_X since the output data of f_X is processed by f_T. Here, f_X is removed from the set F. If $PEF_T(i_T, ci_T) \le EF_S(i_R/|ch(f_R)|)$, a next richest node f_X is taken as a child node of f_T. Otherwise, a cheapest disconnected node f_M in F is selected as a surrogate node. A largest disconnected node is selected to be a child of the surrogate node f_M.

Algorithm 1. SMSGD algorithm

Input   : f_R = faulty node;
Output: SN = a set of surrogate nodes;
1 $F = ch(f_R)$; /* set of disconnected nodes. */ $SN = \phi$; /* set of surrogate nodes */
2 $f_G = pt(f_R)$; /* parent node of faulty node */ $ci_G = 0$;
3 $E = 0$; /* energy consumption of minimum node in F */
4 $AE = EF_G(i_R/|F|)$; /* energy consumption of average size of input data of f_R */
5 **select** a node f_M in F where $DEF_M(i_M, 0)$ is minimum; $ci_M = 0$;
6 **while** $F \neq \phi$ **do**
7   **select** a node f_X in F whose energy consumption $DEF_X(i_X, 0)$ is maximum;
8   **if** $PEF_G(i_G, ci_G) \leq AE$ **then**
9   **if** $p(f_R) \notin p(f_G)$ **then**
10   f_G receives $p(f_R)$ from a source node; $p(f_G) = p(f_G) \cup \{p(f_R)\}$;
11              $SN = SN \cup \{f_G\}$; /* f_G is a surrogate */
12          $ch(f_G) = ch(f_G) \cup \{f_X\}$; /* f_X connects to f_G */ $ci_G = ci_G + o_X$;
13   **else**
14   **if** $p(f_R) \notin p(f_M)$ **then**
15   f_M receives $p(f_R)$ from a source node; $p(f_M) = p(f_M) \cup \{p(f_R)\}$;
16              $F = F - \{f_M\}$; $SN = SN \cup \{f_M\}$; /* f_M is a surrogate */
17              $ch(f_M) = ch(f_G) \cup \{f_M\}$; /* f_M connects to f_G */ $E = DEF_M(i_M, 0)$;
18   f_X connects to f_M; $ch(f_M) = ch(f_M) \cup \{f_X\}$;
19          $ci_M = ci_M + o_X$;
20   **if** $2 \cdot E \leq DEF_M(i_M, ci_M)$ **then**
21   **select** a node f_M in F where $DEF_M(i_M, 0)$ is minimum; $ci_M = 0$;
22      $F = F - \{f_X\}$;

---

4   Evaluation

We evaluate the SMSGD algorithm to select a surrogate node for each disconnected node. We consider a height-balanced four-ary tree with height h (≥ 1)
of the TBFC model, where each non-edge node f_R has four child nodes f_{R0}, ...,
$f_{R,k-1}$ ($k = 4$) and every edge node is at level h - 1. There are totally 4^{h-1} edge
nodes. Let SD be the total size [bit] of sensor data sent by all the sensors, i.e.
the sensors totally send sensor data SD [bit] to 4^{h-1} edge nodes. In this paper,
we assume SD is 1 [MB]. The size of sensor data which each edge node receives
is randomly decided.

In this evaluation, an application process p is a sequence of subprocesses p_0,
p_1, ..., p_m ($m \geq h$ - 1) where p_m is an edge subprocess. As presented in this
paper, a subprocess $p_{m-h+1+l}$ is deployed on fog nodes of each level l ($0 \leq l$
$< h$). Subprocesses p_{m-h+1}, p_{m-h}, ..., p_0 are performed on a root node. We
consider a pair of types of subprocesses which receive input data of size x for
the computation complexity $O(x)$ or $O(x^2)$.

In the evaluation, we assume one node f_R is randomly selected to be faulty
for each level l ($1 < l < h$ - 1). We assume neither a root node nor every edge
node is faulty for simplicity and a surrogate node is selected in fog node. That

is, nodes at levels 0, 1, and $h - 1$ are not faulty. If a node f_R is faulty, k child nodes $f_{R0}, ..., f_{R,k-1}$ of the faulty node f_R are disconnected.

We consider a RD (Random), ME [9], SMPR [7], RS (Random Surrogate), and SMSGD algorithms. In the RD, ME, and SMPR algorithms, disconnected nodes are selected among the alternate nodes for disconnected nodes in the data transmission strategy. In the RD algorithm, an alternate node is randomly selected to be a new parent node of all the disconnected nodes. In the ME algorithm [9], an alternate node is selected as a new parent node, whose energy consumption is minimum in the alternate nodes. In the SMPR algorithm [8], an alternate node whose energy consumption is minimum is selected as a new parent node for each disconnected node. In the RS and SMSGD algorithms, surrogate nodes are selected for disconnected nodes in the subprocess transmission strategy. In the RS algorithm, a surrogate node is randomly selected in the grandparent node and disconnected nodes for disconnected nodes. In the SMSGD algorithm, multiple surrogate nodes are selected for disconnected nodes as discussed in this paper.

Figures 4 and 5 show the average energy consumed by new parent nodes in the tree for the computation complexity $O(x)$ and $O(x^2)$ of subprocesses, respectively. Figures 4 and 5 show the ratios of the energy consumption of the new parent node selected by the SMSGD, RS, SMPR, ME, and RD algorithms. The energy consumption of a new parent node selected by the SMSGD algorithm decreases by about 33%, 21%, and 40% for computation complexity $O(x)$ and 27%, 33%, and 62% for $O(x^2)$ in the RS, ME, and RD algorithms, respectively. However, the energy consumption of a new parent node in the SMSGD algorithm increases by 30% for $O(x)$ and 83% in the SMPR algorithm.

Figures 6 and 7 show the execution time of a new parent node selected in the SMSGD, RS, SMPR, ME, and RD algorithms. The execution time of a new parent fog node selected in the SMPR algorithm is about 34% and 25% for computation complexity $O(x)$ and 57% and 44% for $O(x^2)$ shorter than the RD

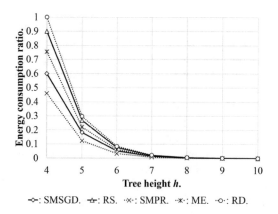

-◇-: SMSGD.   -▲-: RS.   ⋯×⋯: SMPR.   ⋯✱⋯: ME.   ⋯○⋯: RD.

Fig. 4. Energy consumption of a new parent node of computation complexity $O(x)$ for height h.

and ME algorithms, respectively. The execution time of a new parent fog node selected in the SMSGD algorithm decreases by about 33%, 21%, and 40% for computation complexity $O(x)$ and 27%, 33%, and 62% for $O(x^2)$ smaller than the RS, ME, and RD algorithms, respectively. However, the execution time of a new parent node selected in the SMSGD algorithm increases by 30% for $O(x)$ and 83%.

Figure 8 shows matching ratios (≤ 1) of paths from disconnected nodes to the root node f_0 in the SMSGD, RS, SMPR, ME, and RD algorithms compared with non-fault case. Matching ratios of paths from disconnected nodes to root node increases in the SMSGD and RS algorithms in the subprocess transmission strategy while decreasing in the SMPR, ME, and RD algorithms in the data transmission strategy. This means, even if a node gets faulty, each sensor data is more processed by same subprocesses as no node is faulty.

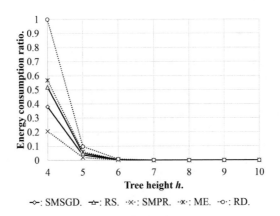

Fig. 5. Energy consumption of a new parent node of computation complexity $O(x^2)$ for height h.

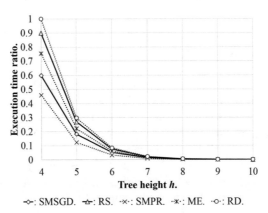

Fig. 6. Execution time of a new parent node of computation complexity $O(x)$ for height h.

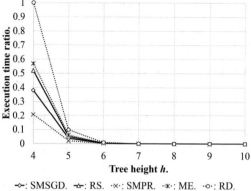

Fig. 7. Execution time of a new parent node of computation complexity $O(x^2)$ for height h.

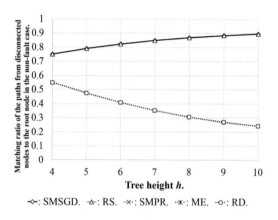

Fig. 8. Matching ratio of paths from disconnected nodes to root node in non-fault case.

5   Concluding Remarks

In the FTBFC and MFTBFC models, if a node gets faulty, alternate nodes of the faulty node are selected and disconnected nodes send output data to the alternate nodes. In this paper, we newly proposed the subprocess transmission strategy where a subprocess of a faulty node is transmitted to a node where output data obtained by disconnected nodes are to be processed. The node is a surrogate node. A subprocess of a faulty node is first transmitted to a surrogate node and disconnected nodes send output data to the surrogate node. We also proposed the SMSGD algorithm to select surrogate nodes for disconnected nodes. In the evaluation, we showed the energy consumption and execution time of a surrogate node can be reduced in the SMSGD algorithm compared with the RS, ME, and RD algorithms.

References

1. Creeger, M.: Cloud computing: an overview. Queue **7**(5), 3–4 (2009)
2. Enokido, T., Ailixier, A., Takizawa, M.: Process allocation algorithms for saving power consumption in peer-to-peer systems. IEEE Trans. Ind. Electron. **58**(6), 2097–2105 (2011)
3. Enokido, T., Ailixier, A., Takizawa, M.: An extended simple power consumption model for selecting a server to perform computation type processes in digital ecosystems. IEEE Trans. Ind. Inform. **10**, 1627–1636 (2014)
4. Hanes, D., Salgueiro, G., Grossetete, P., Barton, R., Henry, J.: IoT Fundamentals: Networking Technologies, Protocols, and Use Cases for the Internet of Things. Cisco Press (2018)
5. Oma, R., Nakamura, S., Duolikun, D., Enokido, T., Takizawa, M.: An energy-efficient model for fog computing in the internet of things (IoT). Internet Things **1–2**, 14–26 (2018). https://doi.org/10.1016/j.iot2018.08.003
6. Oma, R., Nakamura, S., Duolikun, D., Enokido, T., Takizawa, M.: Evaluation of an energy-efficient tree-based model of fog computing. In: Proceedings of the 21st International Conference on Network-Based Information Systems (NBiS-2018), pp. 99–109 (2018)
7. Oma, R., Nakamura, S., Duolikun, D., Enokido, T., Takizawa, M.: Fault-tolerant fog computing models in the IoT. In: Proceedings of the 13th International Conference on P2P, Parallel, Grid, Cloud and Internet Computing (3PGCIC-2018), pp. 14–25 (2018)
8. Oma, R., Nakamura, S., Duolikun, D., Enokido, T., Takizawa, M.: Energy-efficient recovery algorithm in the fault-tolerant tree-based fog computing (FTBFC) model. In: Proceedings of the 33rd International Conference on Advanced Information Networking and Applications (AINA-2019), pp. 132–143 (2019)
9. Oma, R., Nakamura, S., Duolikun, D., Enokido, T., Takizawa, M.: A fault-tolerant tree-based fog computing model (accepted). Int. J. Web Grid Serv. (IJWGS) (2019)
10. Oma, R., Nakamura, S., Enokido, T., Takizawa, M.: A tree-based model of energy-efficient fog computing systems in IoT. In: Proceedings of the 12th International Conference on Complex, Intelligent, and Software Intensive Systems (CISIS-2018), pp. 991–1001 (2018)
11. Rahmani, A.M., Liljeberg, P., Preden, J.S., Jantsch, A.: Fog Computing in the Internet of Things. Springer, Heidelberg (2018)

Message Ordering Based on the Object-Based-Causally (OBC) Precedent Relation

Takumi Saito[1]($\boxtimes$), Shigenari Nakamura[1], Tomoya Enokido[2],
and Makoto Takizawa[1]

[1] Hosei University, Tokyo, Japan
takumi.saito.3j@stu.hosei.ac.jp, nakamura.shigenari@gmail.com,
makoto.takizawa@computer.org
[2] Rissho University, Tokyo, Japan
eno@ris.ac.jp

Abstract. The P2PPS model is a peer-to-peer (P2P) type of publish/subscribe (PS) model where peer processes (peers) are cooperating by publishing and subscribing messages with no centralized coordinator. Messages carry objects and are only received by target peers which are interested in the objects. Messages carrying a common object are related. After receiving messages, only related messages have to be causally delivered to target peers. In topic-based PS systems, objects are denoted by topics. Messages are characterized by publication topics which denote objects in the messages. Each peer specifies subscription topics and receives messages whose publication topics include some subscription topic. A pair of messages which have a common publication topic are considered to be related with respect to topics. Here, it is noted a pair of messages m_i and m_j may not be related in terms of objects even if m_i and m_j are related in terms of topics. In this paper, we propose an MTBC (Modified Topic-Based Causally ordering) protocol which uses topic vectors. In the evaluation, we show clear how many number of messages are unnecessarily ordered in the MTBC protocol.

Keywords: Topic-based Publish/Subscribe system · P2P model ·
OBC-precedent relation · TBC-precedent relation · MTBC protocol ·
Unnecessarily ordered messages

1 Introduction

In the publish/subscribe (PS) model [2,4,15,16], processes either publish or receive messages. In this paper, we consider a peer-to-peer (P2P) model of a topic-based PS (P2PPS) model [7–9,14] where there is no centralized coordinator and every peer process (peer) can publish and subscribe messages. Each message carries objects as contents. In the topic-based PS model [14], contents of messages are denoted by publication topics. Subscriber processes specify interesting

© Springer Nature Switzerland AG 2020
L. Barolli et al. (Eds.): CISIS 2019, AISC 993, pp. 62–72, 2020.
https://doi.org/10.1007/978-3-030-22354-0_6

topics named subscription topics. A message published by a publisher process is delivered to only subscribers which are interested in the message, i.e. the publication topics and subscription topics include a common topic.

In the P2PPS model, each peer receives messages published by other peers. Here, some pair of peers may receive a pair of messages published by different peers in different orders. A message m_1 causally precedes a message $m_2(m_1 \rightarrow m_2)$ iff (if and only if) m_1 is published before m_2 [5]. We have to deliver messages to peers in the causal order [5]. An object is a unit of data in a system. A peer p_i publishes a message to carry objects to target peers. A pair of messages are defined to be *related* iff the messages carry a common object. Here, a message m_1 object-based-causally (OBC) precedes a message $m_2(m_1 \Rightarrow m_2)$ [5,13] iff m_1 and m_2 are related and $m_1 \rightarrow m_2$. Each target peer is required to deliver messages in the OBC-precedent order.

In the topic-based PS systems [2,15], objects are denoted by topics. A message m carrying objects is characterized by publication topics $m.PT$ of the objects. Each peer p_i specifies subscription topics $p_i.ST$. Each peer p_i only receives a message m where $m.P$ includes some subscription topic in $p_i.ST$. A pair of messages m_1 and m_2 which have common publication topics are *related* with respect to topics. Here, a message m_1 topic-based-causally (TBC) precedes a message $m_2(m_1 \Rightarrow m_2)$ if $m_1 \rightarrow m_2$ and $m_1.PT \cap m_2.PT \neq \phi$. Suppose a peer p_i receives a message m_1 and then publishes a message m_2. Here, the message m_1 causally precedes the message $m_2(m_1 \rightarrow m_2)$. However, if a pair of messages m_1 and m_2 are not related, the causally precedent relation $m_1 \rightarrow m_2$ is meaningless. The TBC protocol [11] is proposed for each peer to TBC-deliver messages by using vector clocks and linear clocks. In this paper, we propose an MTBC (Modified TBC) protocol which only uses the topic vector. A pair of messages m_1 and m_2 are defined to be unnecessarily ordered in a peer p_i iff p_i receives m_1 before m_2 and delivers m_2 before m_1, but $m_1 \not\rightarrow m_2$ [11,12]. Even if $m_1 \Rightarrow m_2, m_1 \Rightarrow m_2$ may not hold. Thus, a pair of messages m_1 and m_2 are unnecessarily ordered in a peer p_i with respect to topics and objects iff p_i receives m_1 before m_2 and delivers m_2 before m_1 but $m_1 \not\Rightarrow m_2$ and $m_1 \not\Rightarrow m_2$, respectively. In the evaluation, we make clear how many number of messages are unnecessarily ordered in the MTBC protocol.

In Sect. 2, we present a system model. In Sect. 3, we discuss the OBC-precedent relation among messages. In Sect. 4, we discuss the TBC precedent relation among messages. In Sect. 5, we newly propose the MTBC protocol. In Sect. 6, we evaluate the MTBC protocol and OBC precedent relation.

2 System Model

In the P2PPS (peer-to-peer publish/subscribe) model [1,8–10,14], peer processes (peers) $p_1, ..., p_n (n \geq 1)$ are cooperating with one another by exchanging messages in underlying networks. Here, there is no centralized coordinator [3] to deliver messages to peers. If a peer p_i publishes a message m_i after receiving another message m_j published by a peer p_j, the message m_i causally precedes the message $m_j (m_i \rightarrow m_j)$ since the publication event of the message m_i happens before the message m_j according to the causality theory [5]. Each peer p_i can publish messages while receiving messages from other peers. A message m is only received by a target peer which is interested in the objects of the message m. If a peer p_i publishes a message m_1 before another message m_2, every common target peer p_j receives the messages m_1 then m_2 in the publishing order. However, some pair of messages m_1 and m_2 published by different peers p_i and p_j may not be received by every pair of common destination peers p_k and p_l in the same order. For example, a peer p_k receives the message m_1 then the message m_2 but another peer p_l receives m_2 then m_1. Here, if m_1 causally precedes m_2 ($m_1 \rightarrow m_2$), every target peer is required to receive m_1 before m_2. Even if m_1 $\rightarrow m_2$, if a pair of messages m_1 and m_2 are not related, every common target peer can receive the messages m_1 and m_2 in any order.

We have to discuss whether or not messages are related. The contents of each message are composed of objects. An object is a unit of data resource. Thus, peers exchange objects with one another by publishing and receiving messages. Let O be a set of objects in a system. Let $m.O (\subseteq O)$ be a collection of objects carried by a message m. A peer which is interested in the objects $m.O$ of a message m is a *target* peer of the message m. A peer p_i supports a memory $p_i.M$ to receive and store objects. On receipt of a message m, a peer p_i stores objects $m.O$ carried by the message m in the memory $p_i.M$. In this paper, a pair of object collections O_1 and O_2 are defined to be related iff (if and only if) $O_1 \cap O_2 \neq \phi$, i.e. O_1 and O_2 include a common object. A transaction is a unit of work of each peer p_i. Each time, a peer p_i initiates a transaction, the storage $p_i.O$ is made empty.

A transaction is a sequence of publication and receiving events of messages. A peer p_i is modeled to be a sequence of transactions $T_{i1}, ..., T_{il_i}$. This means, objects are not persistent. Messages published and received in a pair of different transactions T_{ik} and T_{ih} are not related. That is, each time a transaction starts on a peer p_i, the memory $p_i.M$ is flushed.

We assume the underlying network supports a pair of peers p_i and p_j with the following communication service:

1. Each peer p_i receives only and all messages in whose objects the peer p_i is interested. Here, the peer p_i is a target peer of the message m.
2. Messages published by each peer can be received by every target peer in the publishing order with neither message loss nor duplication.
3. Every message published by a peer arrives at every peer.

In the topic-based PS model [14], messages and interest of each peer are characterized by topics. Each object o_i is characterized in terms of topics. Let $o_i.T (\subseteq T)$ be a set of topics of an object o_i. A peer p_i subscribes interesting topics. $p_i.ST (\subseteq T)$ is a set of subscription topics of the peer p_i. The publication $m.PT$ of a message m is a set of topics of objects in the message m, i.e. $m.PT = \cup_{o_i \in m.O} o_i.T$. A peer p_i receives a message m only if the message m

carries topics subscribed by the peer p_i, i.e. $m.PT \cap p_i.ST \neq \phi$. Here, p_i is a target peer of the message m. A peer p_i publishes a message m after including some objects in the memory $p_i.M$.

Let us consider a pair of messages m_1 and m_2. If a pair of publications $m_1.O$ and $m_2.O$ include a common object o, $m_1.PT$ and $m_2.PT$ include common topics $o.T$ of the common object o, i.e. $m_1.PT \cap m_2.PT \supseteq o.T(\neq \phi)$ and $o \in m_1.O \cap m_2.O$. Next, suppose a pair of the messages m_1 and m_2 only include objects o_1 and o_2. Suppose, $o_1.T = \{a, b\}$ and $o_2.T = \{b, c\}$. Here, $m_1.O(= \{o_1\}) \cap m_2.O(= \{o_2\}) = \phi$, but $o_1.T \cap o_2.T = \{b\}$. Here, $m_1.PT(= o_1.T) \cap m_2.PT(= o_2.T) \neq \phi$ but the messages m_1 and m_2 are not related in terms of objects. If a pair of messages m_1 and m_2 are related, $m_1.PT \cap m_2.PT \neq \phi$. However, it is noted, even if m_1 and m_2 are related with respect to topics, i.e. $m_1.PT \cap m_2.PT \neq \phi$, m_1 and m_2 may not be related.

3 An OBC (Object-Based-Causally) Precedent Relation of Messages

A message m_j carries a set $m_j.O(\subseteq O)$ of objects to a target peer p_i. On receipt of the message m_j, the peer p_i stores the objects $m_j.O$ in the memory $p_i.M$. After receiving the message m_j, the target peer p_i manipulates objects in the memory $p_i.M$. Then, the peer p_i publishes a message m_i including objects in the memory $p_i.M$. Suppose an object $o_j(\in m_j.O)$ in a message m_j is also carried by another message m_i. Here, the message m_i is related with the message m_j since $o_j \in m_i.O \cap m_j.O$. If no object in the message m_i is carried by the message m_j, the message m_i is not related with the message m_j. Thus, a message m_i is *related* with a message m_j iff $m_i.O \cap m_j.O \neq \phi$.

Next, an *object-based-causally (OBC) precedent* relation $\Rightarrow$ on messages is defined as follows [11,13]:

[**Definition**]. Let m_i and m_j be a pair of messages and A be a subset of objects $(A \subseteq m_i.O \cap m.j.O)$.

1. m_i *primarily OBC (object-based-causally)* precedes m_j with respect to $A(m_i \hookrightarrow_A m_j)$ iff (if and only if) a peer p_i publishes the message m_j after receiving the message m_i.
2. m_i *OBC (object-based-causally)* precedes m_j with respect to $A(m_i \Rightarrow_A m_j)$ iff one of the following conditions holds:
   - m_i primarily OBC-precedes m_j with respect to the subset $A(m_i \hookrightarrow_A m_j)$.
   - $m_i \hookrightarrow_A m_k$ and $m_k \Rightarrow_A m_j$ for some message m_k.
3. m_i and m_j are OBC-concurrent with respect to $A(m_i \parallel_A m_j)$ iff nether $m_i \Rightarrow_A m_j$ nor $m_j \Rightarrow_A m_i$.

A message m_i primarily OBC-precedes a message $m_j(m_i \hookrightarrow m_j)$ iff $m_i \hookrightarrow_A m_j$ for some subset $A(\subseteq m_i.O \cap m_j.O)$ of objects. A message m_i OBC-precedes a message $m_j(m_i \Rightarrow m_j)$ iff $m_i \Rightarrow_A m_j$ for some subset $A(\subseteq$

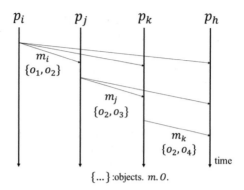

Fig. 1. OBC-precedency.

$m_i.O \cap m_j.O$) of objects. A pair of messages m_i and m_j are *OBC-concurrent* ($m_i \parallel m_j$) iff $m_i \parallel_A m_j$ for some subset A of objects.

Figure 1 shows four peers p_i, p_j, p_k, and p_h which are cooperating with one another. The peer p_i first publishes a message m_i with a pair of objects o_1 and o_2, i.e. $m_i.O = \{o_1, o_2\}$. Then, the peer p_j publishes a message m_j where $m_j.O = \{o_2, o_3\}$. After receiving the message m_j, the peer p_k publishes a message m_k with $m_k.O = \{o_2, o_4\}$. Here, the message m_i causally precedes the message $m_j(m_i \rightarrow m_j)$ and $m_j \rightarrow m_k$. Since $o_2 \in m_i.O \cap m_j.O$, the message m_i OBC-precedes the message m_j with respect to the object $o_2(m_i \Rightarrow_{\{o_2\}} m_j)$. Similarly, $m_j \Rightarrow_{\{o_2\}} m_k$. Since $m_i \Rightarrow_{\{o_2\}} m_j$ and $m_j \Rightarrow_{\{o_2\}} m_k$, $m_i \Rightarrow_{\{o_2\}} m_k$. All the messages m_i, m_j, and m_k carry the common object o_2, i.e. related with respect to the object o_2.

Next, suppose the message m_k carries objects o_3 and o_4, i.e. $m_k.O = \{o_3, o_4\}$. Here, $m_j \Rightarrow_{\{o_3\}} m_k$. However, $m_i \nRightarrow m_k$ because $m_i \nRightarrow_{\{o_3\}} m_j$. From the definitions, if $m_1 \Rightarrow_A m_2$ for some subset $A(\subseteq O)$ of objects, $m_1 \rightarrow m_2$. Even if $m_1 \rightarrow m_2$, $m_1 \Rightarrow_A m_2$ may not hold,

It is not easy for each peer p_i to keep in record every object carried by each message. It also takes time for each peer p_i to check if objects of each message are related with objects in the memory $p_i.M$. We consider a topic-based PS system [14] where each peer p_i receives messages only based on publication topics of the messages and the subscription topics of the peer p_i on behalf of objects. It is simpler to decide on which message causally precedes another message by using publication topics carried by the messages and subscription topics of peers.

It is noted, even if $m_1 \Rightarrow_B m_2$, $m_1 \Rightarrow_A m_2$ may not hold.

4   An MTBC (Modified Topic-Based Causally Ordering) Relation of Messages

In a topic-based P2PPS system, a message m with publication topics $m.PT$ is received by a target peer p_i if $m.PT \cap p_i.ST \neq \phi$. Suppose a pair of peers p_i and p_j publish messages m_i and m_j, respectively, and each of target peers p_h

and p_k receives both the messages m_i and m_j. Here, the peer p_h first receives the message m_i and then the message m_j. Another peer p_k first receives the message m_j and then the message m_i in an order different from the peer p_h. Thus, messages published by different peers may be received by different common target peers in different orders depending on network delay time because there is no centralized coordinator. If the message m_i OBC-precedes the message $m_j (m_i \Rightarrow m_j)$, each common target peer p_k is required to receive the message m_i before the message m_j. However, it is not easy for each peer to hold every object received and manipulate objects to check if the objects are related. The *TBC (topic-based-causally) precedent* relation on messages is defined [11]:

[**Definition**]. Let m_1 and m_2 be a pair of messages and B be a subset of topics $(B \subseteq m_1.PT \cap m_2.PT)$.

1. m_1 *primarily TBC-precedes* m_2 with respect to B $(m_1 \mapsto_B m_2)$ iff m_1 is received and then m_2 is published before m_2.
2. m_1 *TBC-precedes* m_2 with respect to $B(m_1 \Rightarrow_B m_2)$ iff one of the following conditions holds:
   - $m_1 \mapsto_B m_2$.
   - $m_1 \mapsto_B m_3$ and $m_3 \Rightarrow_B m_2$ for some message m_3.
3. m_1 is TBC-concurrent with m_2 with respect to $B(m_1 \; |||_B \; m_2)$ iff nether $m_1 \Rightarrow_B m_2$ nor $m_2 \Rightarrow_B m_1$.

$m_1 \Rightarrow_B m_k$ means that the topic sets B_1, ..., B_{k-1} include common topics B, i.e. $B \subseteq B_1 \cap ... \cap B_{k-1}$.

A message m_1 TBC-precedes a message m_2 $(m_1 \Rightarrow m_2)$ iff $m_1 \Rightarrow_B m_2$ for some subset B $(\subseteq T)$ of topics. A pair of messages m_1 and m_2 are TBC-concurrent $(m_1 \; ||| \; m_2)$ iff $m_1 \; |||_B \; m_2$ for some subset B $(\subseteq T)$.

Suppose there are four peers p_i, p_j, p_k, and p_h as shown in Fig. 2. The peer p_i first publishes a message m_i. The peer p_j publishes a message m_j after receiving the message m_i. Lastly, the peer p_k publishes a message m_k after receiving the message m_j. Suppose $m_i.PT = \{a, b\}$, $m_j.PT = \{a, c\}$, and $m_k.PT = \{a, d\}$. Here, the message m_i primarily TBC-precedes the message $m_j (m_i \mapsto_{\{a\}} m_j)$ since m_i causally precedes m_j $(m_i \rightarrow m_j)$ and $\{a\} \subseteq m_i.PT \cap m_j.PT$. Similarly, $m_j \mapsto_{\{a\}} m_k$. Hence, the message m_i TBC-precedes a pair of the messages m_j and m_k with respect to a topic a, i.e. $m_i \Rightarrow_{\{a\}} m_j$ and $m_i \Rightarrow_{\{a\}} m_k$. It is noted, each of the messages m_i, m_j, and m_k includes the same publication topic a.

Suppose, a message m_i OBC-precedes a message $m_j (m_i \Rightarrow m_j)$ and a pair of the messages m_i and m_j carry the object o_2 as shown in Fig. 1. A pair of the objects o_1 and o_2 carried by the message m_i are denoted by topics $o_1.T = \{b\}$ and $o_2.T = \{a\}$, respectively, and the object o_3 carried by the message m_j is denoted by a topic $o_3.T = \{c\}$. Here, $m_i.PT = \{a, b\}$ and $m_j.PT = \{a, c\}$. Here, $m_i \Rightarrow_{\{o_2\}} m_j$ and $m_i \Rightarrow_{\{a\}} m_j$.

Next, suppose the message m_j carries a pair of the objects o_3 and o_4 where $o_4.T = \{a\}$. Hence, $m_j.PT = \{a, c\}$. However, the messages m_i and m_j carry the objects o_2 and o_4 of the topic a, respectively, while the objects o_2 and o_4

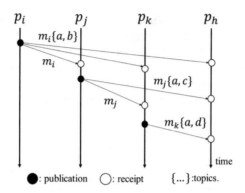

Fig. 2. TBC-precedent relation.

are different. This means, the messages m_i and m_j are not related while related with respect to topics. Thus, $m_i \Rightarrow_{\{a\}} m_j$ but $m_i \not\Rightarrow m_j$.

[**Property**]. For a pair of messages m_1 and m_2, $m_1 \Rightarrow_B m_2$ if $m_1 \Rightarrow_A m_2$ where $B = \cup_{O \in A} o.T$.

5   An MTBC Protocol

In this paper, we newly propose an MTBC (modified TBC) protocol to TBC-deliver messages in a P2PPS system composed of $n(\geq 1)$ peers $p_1, ..., p_n$. We assume every message is reliably broadcast to every peer in a system. That is, every message published by a peer arrives at every peer in a system in the publishing order without message loss and duplication.

In the TBC protocol [11], the linear time is used in addition to the topic vector. In the MTBC protocol, only the topic vector is used. Each message m arrives at every peer in an underlying network. Then, each peer p_i receives the message m if $m.PT \cap p.ST \neq \phi$. Publication $m.PT$ of each message m and subscription $p_i.ST$ of each peer p_i are realized in a bitmap $\langle b_1, ..., b_l \rangle$ a set T of for topics $t_1, ..., t_l$ where each bit b_j is 0 or 1. If a topic t_k is included in a publication or subscription, the kth bit b_k is 1 in the bitmap, otherwise $b_k = 0 (k = 1, ..., l)$.

In papers [7–10], the topic vector is proposed to TBC-deliver messages. Each peer p_i manipulates a topic vector $TV = \langle tv_1, ..., tv_l \rangle$ for the topics $t_1, ..., t_l$. Each element tv_k in the topic vector TV is manipulated for a topic t_k. Initially, each element tv_k is 0.

For a pair of topic vectors $A = \langle a_1, ..., a_l \rangle$ and $B = \langle b_1, ..., b_l \rangle$ and a subset $V(\subseteq T)$ of the topics, the following relations are defined:

1. $A >_V B$ iff $a_k > b_k$ for every topic t_k in V.
2. $A \equiv_V B$ iff $a_k = b_k$ for every topic t_k in V.
3. $A \geq_V B$ iff $A >_V B$ or $A \equiv_V B$.
4. $A \mid_V B$ (A and B are uncomparable) iff neither $A \geq_V B$ nor $A \leq_V B$.

Suppose a peer p_i would like to publish a message m. An element tv_k in the topic vector TV is incremented by one for every topic t_k the publication in $m.PT$. The message m carries the topic vector TV in a field $m.TV$, i.e. $m.TV = TV$. Then, the message m is published. On arrival of a message m, $tv_k = max(tv_k, m.tv_k)$ for every topic t_k in every peer p_i. The following properties hold for the topic vector:

[**Properties**]

1. A message m_1 causally precedes a message $m_2 (m_1 \rightarrow m_2)$ if $m_1.TV < m_2.TV$.
2. If a message m_1 TBC-precedes a message m_2 with respect to topics B in $m_1.PT \cap m_2.PT$ ($m_1 \Rrightarrow_B m_2$), $m_1.TV <_B m_2.TV$

Suppose a peer p_i publishes a message m with publication $m.PT$. A message m carries a topic vector $m.TV$. The message m is published by the peer p_i the MTBC protocol as follows:

[**Publication** at a peer p_i]

1. $m.O =$ objects which m carries ($\subseteq p_i.M$);
2. $m.PT =$ topics of objects in $m.O$;
3. $tv_k = tv_k + 1$ for each element tv_k in TV where a topic t_k is in the publication $m.PT$;
4. $m.TV = TV$;
5. **publish** m;

On arrival of a message m, a peer p_i manipulates the topic vector TV and receives the message m as follows:

[**Receipt** at a peer p_i]

1. A message m arrives at a peer p_i.
2. $tv_k = max(tv_k, m.tv_k)$ for each element tv_k in TV;
3. **receive** m in a receipt queue RQ if $m.T \cap p_i.ST \neq \phi$, else **neglect** m;
4. Objects in $m.O$ are stored in the memory $p_i.M$.

For a pair of messages m_1 and m_2 received by a peer p_i, m_1 precedes $m_2 (m_1 \prec m_2)$ in the peer p_i with the MTBC protocol iff $m_1.TV \leq_A m_2.TV$ for topics $A = m_1.PT \cap m_2.PT$. In the receipt queue RQ, message are ordered in the precedent relation $\prec$. Even if $m_1 \prec m_2$, "$m_1 \Rrightarrow_B m_2$" may not hold. Suppose a peer p_i receives a message m_2 before a message m_1. If $m_1 \prec m_2$, the peer p_i cannot deliver the message m_2 until receiving the message m_1. Unless $m_1 \Rrightarrow_B m_2$ for $B = m_1.PT \cap m_2.PT$, the peer p_i does not need to wait for the message m_1. Here, a pair of the messages m_1 and m_2 are unnecessarily ordered in the peer p_i.

[**Definition**]. A message m_1 *unnecessarily* precedes a message m_2 with respect to topics in a target peer p_i with the MTBC protocol ($m_1 \preceq^T m_2$) iff m_2 is received before m_1 and $m_1 \prec m_2$ in the peer p_i but $m_1 \not\Rrightarrow_B m_2$ for topics B in $m_1.PT \cap m_2.PT$.

[**Definition**]. A message m_1 unnecessarily precedes a message m_2 with respect to objects in a target peer p_i with the MTBC protocol ($m_1 \preceq^O m_2$) iff m_2 is received before m_1 and $m_1 \prec m_2$ in p_i but $m_1 \not\Rightarrow_A$ for objects A in $m_1.O \cap m_2.O$.

It is noted, $m_1 \preceq^O m_2$ if $m_1 \preceq^T m_2$ but even if $m_1 \preceq^O m_2$, $m_1 \preceq^T m_2$ may not hold. We have to reduce the number of messages unnecessarily ordered. The more number of messages are unnecessarily ordered, the larger messages are delayed to be delivered.

6 Evaluation

We evaluate the MTBC protocol in terms of the number of pairs of messages ordered in the OBC precedent relation and the MTBC protocol. A system includes peers $p_1, ..., p_n (n \geq 1)$. Let T be a set of topics $t_1, ...t_l$ ($l \geq 1$). Let O be a set of objects $o_1, ...o_m$ ($m \geq 1$).

Each peer p_i randomly publishes messages. That is, the publishing time $m.PBT$ of a message m published by a peer p_i is randomly decided from time 0 to time $maxT$ [time unit (tu)]. In the evaluation, $maxT$ is 1000 [tu]. At each time unit, a peer p_i is randomly taken in all the peers. Then, the peer p_i creates one message m which carries objects and publishes the message m. The receiving time $m.RVT_j$ of a message m of each peer p_j is $m.PBT + \delta_{ij}$. Here, the delay time δ_{ij} between a pair of peers p_i and p_j is randomly taken from 1 to 10 [tu]. Each message m carries the topic vector $m.TV$.

In the evaluation, np_i publication and ns_i subscription topics of each peer p_i are randomly taken from the topic set T. The numbers np_i and ns_i of topics are randomly take from 1 to 10. Each peer p_i creates five objects $o_{1i}, ..., o_{5i}$, $m.O = \{o_{1i}, ..., o_{5i}\}$ whose topics are subsets of subscription topics of the peer p_i, i.e. $o_{ki}.T \subseteq p_i.ST$. Publication topics $m.PT$ of a message m are topics of objects in $m.O$. In the evaluation, each peer p_i manipulates the vector time variable VT [6] and each message m carries the vector time $m.VT$ to check the causally precedent relation $\rightarrow$ among messages. Each peer p_i includes the vector time VT in a message m as $m.VT$. In the vector time, "$m_i.VT < m_j.VT$" iff a message m_i causally precedes a message $m_j(m_i \rightarrow m_j)$. Even if $m_i.TV < m_j.TV$ holds, m_i may not causally precede m_j. Here, if a message m_j is received before another message m_i, a pair of the messages m_i and m_j are unnecessarily ordered.

Suppose a peer p_i receives a message m_1 before a message m_2 but $m_2 \prec m_1$. In the MTBC protocol, the messages m_1 and m_2 are exchanged in the receipt queue RQ. Here, if $m_2.VT < m_1.VT$ does not hold, the messages m_1 and m_2 are unnecessarily ordered with respect to topics, i.e. $m_2 \preceq^T m_1$. In addition, m_2 unnecessarily precedes $m_1(m_2 \preceq^O m_2)$ if $m_1.O \cap m_2.O = \phi$. In the simulation, we obtain the number of pairs of messages unnecessarily ordered.

Figure 3 shows the numbers of pairs of messages which are ordered in the OBC precedent relation and the MTBC protocol. In the evaluation, there are ten peers ($n = 10$) and twenty topics ($l = 20$) and fifty objects ($m = 50$). The number of pairs of messages ordered in the OBC precedent relation is about 13% of the number of pairs of messages ordered in the MTBC protocol. For example,

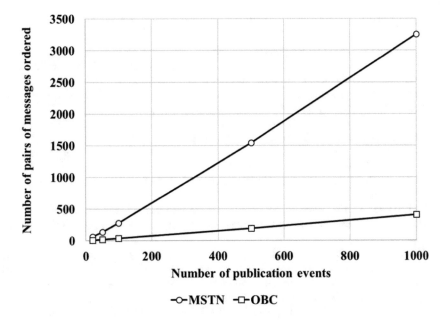

Fig. 3. Number of ordered messages.

about 410 pairs of messages are ordered by the OBC precedent relation and about 3300 pairs of messages are ordered by the TBC precedent relation for 1,000 publication events. This means, about 87% of messages ordered in the MTBC protocol are meaningless.

7 Concluding Remarks

In this paper, we discussed how to causally deliver related messages in the P2PPS model. A pair of messages which carry a same object are defined to be related in this paper. A message m_1 OBC-precedes a message $m_2(m_1 \Rightarrow m_2)$ iff the message m_1 causally precedes the message m_2 and the messages m_1 and m_2 carry a common object. Then, we defined the TBC-precedent relation $\Rightarrow$ among messages in terms of topics. If $m_1 \Rightarrow m_2, m_1 \Rightarrow m_2$. We proposed the MTBC protocol where messages are ordered by using the topic vector. A message m_1 precedes a message $m_2(m_1 \prec m_2)$ in the MTBC protocol if m_1 TBC-precedes $m_2(m_1 \Rightarrow m_2)$. However, even if $m_1 \prec m_2, m_1 \Rightarrow m_2$ may not hold. Here, $m_1 \Rightarrow m_2$ may not hold either. In the evaluation, we showed the number of pairs of messages unnecessarily ordered in the MTBC protocol in terms of objects.

References

1. Google alert. http://www.google.com/alerts
2. Google cloud pub/sub. https://cloud.google.com/pubsub/docs/overview?hl=ja
3. Arridha, R., Sukaridoho, S., Pramadihanto, D., Funabiki, N.: Classfiation IoT-big data analytic for smart environment monitoring and analytic in real-time system. Int. J. Space-Based Situated Comput. (IJSSC) **7**(2), 82–93 (2017)
4. Eugster, P., Felber, P., Guerraoui, R., Kermarrec, A.: The many faces of publish/subscribe. ACM Comput. Surv. **35**(2), 114–131 (2003)
5. Lamport, L.: Time, clocks, and the ordering of event in a distributed systems. Commun. ACM **21**(7), 558–565 (1978)
6. Mattern, F.: Virtual time and global states of distributed systems. In: Parallel and Distributed Algorithms, pp. 215–226 (1988)
7. Nakamura, S., Enokido, T., Takizawa, M., Ogiela, L.: An information flowcontrol models in a topic-based publish/subscribe systems. J. High Speed Netw. **24**(3), 243–257 (2018)
8. Nakayama, H., Duolikun, D., Enokido, T., Takizawa, M.: Selective delivery of event messages in peer-to-peer topic-based publish/subscribe systems. In: Proceedings of the 18th International Conference on Network-Based Information Systems (NBiS-2015), pp. 379–386 (2015)
9. Nakayama, H., Duolikun, D., Enokido, T., Takizawa, M.: Reduction of unnecessarily ordered event messages in peer-to-peer model of topic-based publish/subscribe systems. In: Proceedings of IEEE the 30th International Conference on Advanced Information Networking and Applications (AINA-2016), pp. 1160–1167 (2016)
10. Nakayama, H., Ogawa, E., Nakamura, S., Enokido, T., Takizawa, M.: Topic-based selective delivery of event messages in peer-to-peer model of publish/subscribe systems in heterogeneous networks. In: Proceedings of the 18th International Conference on Network-Based Information Systems (WAINA-2017), pp. 1162–1168 (2017)
11. Saito, T., Nakamura, S., Enokido, D.D.T., Takizawa, M.: Object-based selective delivery of event messages in topic-based publish/subscribe systems. In: Proceedings of the 13th International Conference on Broadband and Wireless Computing, Communication and Applications (BWCCA-2018), pp. 444–455 (2018)
12. Saito, T., Nakamura, S., Enokido, D.D.T., Takizawa, M.: Evaluation of TBC and OBC precedent relations among messages in P2P type of topic-based publish/subscribe system. In: Proceedings of the Workshops of the 33rd International Conference on Advanced Information Networking and Applications (WAINA-2019), pp. 570–581 (2019)
13. Saito, T., Nakamura, S., Enokido, T., Takizawa, M.: A causally precedent relation among messages in topic-based publish/subscribe systems. In: Proceedings of the 21st International Conference on Network-Based Information Systems (NBiS-2018), pp. 543–553 (2018)
14. Setty, V., van Steen, M., Vintenberg, R., Voulgais, S.: Poldercast: fast, robust, and scalable architecture for P2P topic-based pub/sub. In: Proceedings of ACM/IFIP/USENIX 13th International Conference on Middleware (Middleware-2012), pp. 271–291 (2012)
15. Tarkoma, S.: Publish/Subscribe System: Design and Principles, 1st edn. Wiley, Hoboken (2012)
16. Tarkoma, S., Rin, M., Visala, K.: The publish/subscribe internet routing paradigm (PSIRP): designing the future internet architecture. In: Future Internet Assembly, pp. 102–111 (2009)

A Comparative Analysis of Neural Networks and Enhancement of ELM for Short Term Load Forecasting

Rahim Ullah[1,2], Nadeem Javaid[1(✉)], Ghulam Hafeez[1,3], Salim Ullah[1],
Fahad Ahmad[4], and Ashraf Ullah[1]

[1] COMSATS University Islamabad, Islamabad 44000, Pakistan
nadeemjavaidqau@gmail.com
[2] Higher Education Department, KP, Peshawar 25000, Pakistan
[3] Wah Engineering College, University of Wah, Wah Cantt 47040, Pakistan
[4] International Islamic University, Islamabad 44000, Pakistan
http://www.njavaid.com

Abstract. Smart grid is the evolved form of traditional power grid with the integration of sensing, communication, computing, monitoring and controlling technologies. These technologies make the power grid reliable, efficient and economical. Smart grid enable users to make bidding on the basis of demand side management models. Demand side management can be made responsive and efficient by effective and accurate load forecasting. Accurate load forecasting is an important but challenging task because of irregular and non-linear consumption of individual users and industrial consumers. Different approaches have been proposed for load forecasting, but artificial intelligent models, specifically ANNs perform well for short, medium and long term load forecasting. The main focus of this paper is to present a comparative analysis of NNs for short term load forecasting. NYISO dataset is used for experiments. XGboost and decision tree are used for features importance calculation and RFE is used for features extraction on the basis of score assigned. Three basic techniques (CNN, MLP and ELM) are used for forecasting. Furthermore, ELM is enhanced and E_ELM is proposed for STLF. Moreover, results are evaluated on four statistical measures (MAPE, MAE, MSE and RMS).

Keywords: Smart grid · Load forecasting · Neural networks ·
Features · Features engineering · Classifiers ·
Machine learning techniques · XGboost · DT · ELM ·
E_ELM · MLP · CNN

1 Introduction

Short term load forecasting is very important for the trust-able, secure and efficient execution of electricity system. Forecasted load can help in improving important decisions for economic activities. Such decisions include flow of

© Springer Nature Switzerland AG 2020
L. Barolli et al. (Eds.): CISIS 2019, AISC 993, pp. 73–86, 2020.
https://doi.org/10.1007/978-3-030-22354-0_7

load, transaction evaluation, generator unit commitment, coordination of thermal units, fuel allocation, analysis of network, short and long term maintenance, protection of the system, load balancing and contingency planning. As load forecasting possesses a certain amount of error, which can cause greater loss. Moreover, as industries have started relying on forecasting, the importance has accuracy increased. An accurate forecasting technique can help in developing an efficient action plan. Optimal action plan can help in the reduction of risk and improvement of economical benefits of management [1].

Table 1. List of abbreviations used in this paper

Abbreviation	Explanation
STLF	Short Term Load Forecasting
LTLF	Long Term Load Forecasting
ARIMA	Autoregressive Integrated Moving Average
FWPT	Flexible Wavelet Packet Transform
CMI	Conditional Mutual Information
NLSSVM	Non-linear Least Square Support Vector Machine
ABC	Artificial Bee Colony
CFS	Correlation-based Feature Selector
AI	Artificial Intelligence
ANN	Artificial Neural Network
RFE	Recursive Features Elimination
NYISO	New York Independent System Operator
RNN	Recurrent Neural Network
ML	Machine Learning
MLP	Multi Layer Perceptron
CNN	Convolutional Neural Network
ELM	Extreme Learning Machine
E_ELM	Enhance Extreme Learning Machine
MTR	Model Tree Rule
RMS	Root Mean Square
MSE	Mean Squared Error
MAE	Mean Absolute Error
MAPE	Mean Absolute Percent Error

Accurate STLF is a challenging job. Electricity load has complex and non-linear cycles on daily, weekly and annual basis. There can be some random components as well because of irregular, unplanned and unpredicted use of electricity by individual users, building consumers and large industries. Irregularities may also be caused by long term operation of industries, sudden weather change, special events and holidays which make load forecasting challenging [2]. List of abbreviations is given in Table 1.

There are four classes of load forecasting on the basis of horizon: short term is for days to weeks ahead forecasting, very short term is for minutes to hours ahead forecasting, medium term is for weeks to months ahead and long term is for years ahead load forecasting. This paper focus on short term load forecasting for New York on the basis of NYISO dataset. It can be used by market operators for setting demand requirements and by the consumer to make bids. Smart grid further enhances the importance of STLF by demand response mechanism and time varying price, as they require prediction at short intervals [3].

Load forecasting can be performed by three methods: traditional statistical method, advanced machine learning models and hybrid models. As illustrated earlier, short term load data has many non-linear and non-stationary components which cannot be captured well by statistical models [4]. On the other hand, a number of ML techniques have shown better performances in different scenarios [5]. In ML, there are a number of techniques which can be used for load forecasting; however, ANNs possess many advantages over the others such as: capability to capture non-linear relationship between input features and predictor variables, making of patterns instead of assumptions and tolerance to noise.

Because of such advantages, ANNs are used in too many studies, but there are a number of flavors of them. It is not evident which one is suitable for short term load forecasting. The main focus of this papers is to give a comparative analysis of three NNs (CNN, MLP and ELM) for STLF. Furthermore, we enhance the performance of ELM by tuning the parameters iteratively and proposed E_ELM for STLF.

In this paper, a comparative study of NNs for STLF is given using NYISO dataset. Dataset has sixteen features; however, all features are not contributive to output. First we extract the important features. For features extraction, the importance of features are calculated by XGboost and DT. Afterward, features are extracted by RFE on the basis of score assigned. The dataset was then split into training and evaluation sets and models were trained by training set while evaluated on testing set. The results are evaluated on four statistical evaluators (MAE, MAPE, RMS and MSE).

The remaining paper is organized as, Related Work is discussed in Sect. 2, Problem Statement and Contribution is briefly discussed in Sect. 3, Sect. 4 presents Proposed System while Sect. 5 explains Simulation Setup. Section 6 is dedicated to Critical Analysis and Results and Discussion are elaborated in Sect. 7. Paper is concluded in Sect. 8 while References are presented at the end of paper.

2   Related Work

With the introduction of competitive electricity market, the importance of STLF is increased. Smart grid has boosted further the importance of STLF. It can be performed with the help of three main approaches, traditional based on statistical tools, advanced machine learning based and hybrid. ARIMA and exponential growth are the two very prominent techniques in traditional approach [6]. There are a bunch of techniques in second approach. Because of their advantages and efficiency, these are used in many studies from different perspectives [5].

In [7], a hybrid model for load and price forecasting with demand side management has been given. Simultaneous price and cost forecasting is performed by using different datasets in which the linear components are fed to its suitable model and non-linear are fed to the other. The proposed model has three main components, the first one consists of FWPT which decomposes the signal into many components having distinct frequencies. CMI, a new model for features selection is used for the selection of useful and valuable features. Afterward, ARIMA is used to forecast the linear component while NLSSVM is used for non-linear components. Moreover, to optimize the parameters of SVM, an improved version of ABC based on time varying coefficient and stumble generator operator is used. The proposed system performed very well for various real market datasets in forecasting price and load.

A load forecasting system based on data pre-analysis and weight co-efficient optimization is presented in [8]. Experiments are performed with three datasets having half-hourly updated from state of New South Wales, state of Victoria and State of Queensland for almost three consecutive years. A number of algorithms are used for data pre-processing, feature selection and forecasting such as, cuckoo search algorithm is used for optimized weight assignment. A number of ANNs are used for load forecasting. Obtained results are compared with benchmark, ARIMA. The proposed system outperforms ARIMA for all the datasets and also overcomes the problem of instability and unitary problem.

Koprinska et al. in [2], presented a study for load forecasting. The main focus of this paper is on the selection of important features. Four methods are used for the selection of appropriate features in order to give a comparative analysis of these features selectors. Among the features selection methods, one is the traditional statistical method (autocorrelation), while, the other three are advanced ML approaches (MI, RReliefF and CFS) for features selection integrated with forecasting algorithms. Load is forecasted on three ML algorithms (NN, Linear Regression and MTR) for two years Australian market data. Results show that all features selectors select a subset from the features set; however, AC and ANN performed very well. The proposed system is compared with existing work and some industry benchmarks.

Electricity load forecasting using RNN is presented by Zheng et al. in [9]. According to author's the load data of power system has non-linear, non-seasonal and non-stationary patterns which cannot be predicted accurately by statistical techniques and simple AI techniques. Authors proposed the use of RNN for long short term load forecasting. Long STLF deals with the short term load forecasting having long term dependencies. The system is evaluated for different horizons and found comparably efficient.

In [10], the authors proposed complex load decomposition for forecasting the aggregated load of a campus. The main aim is that different components of campus have different load profile and forecasting load as a whole is illogical. The focus of this article is to decompose the overall load into its representative components. Afterward, cluster the loads with the similar profiles and perform the day ahead forecast for each cluster. It is proved that the system can performs better for LTLF if data of long term is available. The system is evaluated on different statistical measures and found well as compared to other state of the art techniques.

Nowotarski et al. in [11] aimed to improve the accuracy of short term load forecasting by combining sister forecasts. According to the authors, all forecasting algorithms are sisters and combining them will improve load forecasting. Eleven algorithms are combined with two features selection techniques. Sister forecasts are demonstrated with two case studies. Sisters forecast outperforms the benchmark algorithms in term of accuracy and MAPE. Authors suggest that sisters load forecast has high accuracy for academic and practical values.

Beside the above mentioned studies, too many other studies can be found using NNs for load forecasting; however, no study can be found for short term load forecasting and comparing the mentioned techniques.

3   Problem Statement and Contribution

Electricity load forecasting is an important activity for supply markets as well as consumers. Three approaches are in practice for load forecasting, elaborated in next section. ML has significant benefits over traditional models [12]. ML has a number of techniques each with some unique characteristics, but NNs have shown better performance in too many studies. A number of flavors of NNs are available and used in too many studies; however, it is not evident which technique is suitable for STLF. Furthermore, a comparison of NNs for STLF is not available.

The main contribution of this paper can be summarized as follows.

- The use of an integrated model for features importance calculation and features selection.
- Presentation of a comparative analysis of NNs for STLF.
- Enhancement of ELM for short term load forecasting by iterative parameters tuning. In other words, proposed E_ELM for short term load forecasting which beats ELM.

4   Proposed System Model

Short term load forecasting with high accuracy has prominent importance in electricity market. It is fruitful for decisions regarding reliable, secure and optimal use of electricity [13]. On the other hand, smart grid planning, transactions and investment are also using load forecasting for optimal decisions. However, because of diverse factors such as change in weather and change in social behavior of consumers, load forecasting is a challenging task. A number of models have been proposed for load forecasting in recent years. These models are categorized as: AI models, time series models and hybrid models which has the good features of both.

Time series model is used by a number of studies [14–16]; however, time series models do not capture the diverse factors of power system, that is why cannot produce good accuracy [4]. In contrast, artificial intelligence models can capture the diverse features and the non-linear behavior in a very good manner and produce good result. Therefore, it is used by a number of studies for load forecasting in so many studies [17–19]. That is why, this paper focus on second approach for STLF.

The proposed system mainly consists of four main components, as shown in Fig. 1. The first component at left most is the dataset which is NYISO dataset of three years (2015, 2016 and 2017), publicly available. The details of using dataset is given in section-IV, simulations setup. Second part is feature engineering in which we try to extract the important features, which can help in good forecasting. The main goal of features engineering is to increase classification accuracy, reduce the dimensions of data so that complexity can be avoided and time of processing can be reduced [20]. Three techniques are used for features engineering in the proposed system. Firstly, the importance of features are calculated by XGboost and decision tree. Afterward, RFE is used to discard the low important features from the dataset. The features with high importance are left, which can help in achieving good accuracy.

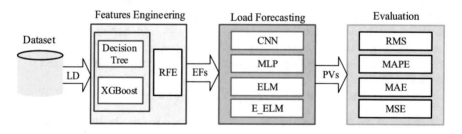

Fig. 1. System model of the Proposed System where LD stands for Loaded Data, RFE for Recursive Features Elimination, EFs for Extracted Features and PVs for Predicted Values

The third part of the system model deals with load forecasting. Before feeding data to classifiers, it is divided into training and testing sets with ratio 3:1. The data of first nine months of each year is used as training set and the rest of the three months is used as testing set. Four classifiers such as CNN, MLP, ELM and E_ELM are used. We have enhanced ELM and proposed E_ELM for load forecasting. A number of studies have used different ANN techniques, but it is not evident which one is suitable for load forecasting. The core purpose of this paper is to present a comparative analysis of ANN techniques for short term load forecasting and enhance ELM.

Fourth part is about the evaluation of the proposed system. Four statistical measure are used for evaluation such as MAPE, MAE, RMS and MSE. The details of theses errors are discussed in detail in Sect. 6.

5 Experimental Setup

Experiments are performed in Python. The three years data of NYISO is taken for experiments [21]. Data has sixteen features and 1095 instances, consists of system load recorded for each day and many other attributes described in Table 2 below. As all the features are not supportive to load forecasting, therefore, in first step we extract the important features. For features extraction, the importance of features are calculated by XGboost and decision tree which are state of the art and most efficient technique. A score is assigned to each feature separately by each technique. The score assigned by XGboost is shown in Fig. 2, while that of DT is shown in Fig. 3 as follow. Afterward, features are selected by RFE on the basis of score assigned to it. The combined score of DT and XGboost is used for features extraction. Among sixteen features, System_Load is used as label, eight important features (DA_Demand, RT_Demand, DA_LMP, DA_EC, RT_EC, Dry_Bulb, Dew_Point, Reg_Cap_Price) are extracted and seven (DA_CC, DA_MLC, RT_LMP, RT_CC, RT_MLC, Reg_Ser_Price) are discarded.

Table 2. Features of dataset and its description

Attribute	Description
Date	Date of Data Recording
DA_DEMAND	The Day Ahead Demands
RT_DEMAND	The Day Locational Demands
DA_LMP	The Locational Marginal Price One Day Ahead
DA_CC	The Congestion Part of Day Ahead Price
DA_EC	The Energy Part of Day Ahead Price
DA_MLC	The Marginal Loss Part of the Day Ahead Price
RT_LMP	The Dynamic Locational Marginal Price
RT_EC	The Energy Part of the Dynamic Price
RT_CC	The Congestion Part of Dynamic Price
RT_MLC	The Marginal Loss Part of Dynamic Price
Dry Bulb	The Dry Bulb Temperature (F°)
Dew Point	The Dew Point Temperature (F°)
Reg_Cap_Price	Regional Capacity Price
Reg_Ser_Price	Regional Service Price
System Load	The Per Day Load of the System

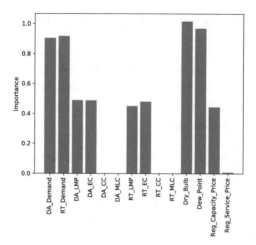

Fig. 2. Importance of features calculated by XGboost

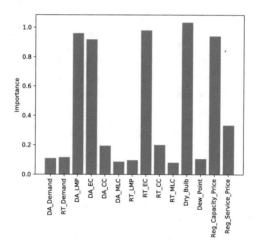

Fig. 3. Importance of features calculated by decision tree

After features extraction, the next step is to feed the extracted features to classifiers, but before giving data to classifiers, the data was split into training and evaluation sets. Data of nine months of each year is kept in training set while the rest of three months of each year is kept as evaluation set. Afterward, the models are trained with training set. The primary justification of this paper is to present a comparison of basic NNs and one enhanced ELM for STLF. These three techniques (MLP, CNN and ELM) are being used frequently for load forecasting, but it is not clear which one is the most suitable for short term load forecasting. After training, the performance of classifiers were evaluated by testing set.

For evaluating the accuracy of classifiers, four statistical measures such as RMS, MSE, MAPE and MAE are used. Built in functions of MSE and MAE are available while functions are defined for RMS and MAPE.

6   Simulation Results and Discussion

In this section, we give a discussion about the results obtained. In the first phase, features engineering is performed to extract best features which contributes a lot in good accuracy. Two features scoring techniques such as XGboost and DT are used which assign score to each feature of the dataset. Afterward, RFE eliminates the unimportant features and extracts the important features. Eight features are kept and the rest are eliminated. This makes the system simple, accurate and less time consuming. A number of techniques are available for features selection and extraction; however, in our case, the above mentioned techniques performed very well.

Electricity load can be forecast by a number of techniques which are categorized as traditional statistical methods, AI models and hybrid models. As power data consists of non-linear behavior, which cannot be well captured by statistical model alone. On the other hand ML algorithms and hybrid model have shown good results presented in many studies. Moreover, a number of AI techniques are available, but ANNs has many advantages over other techniques.

Neural networks are the most popular load and price forecasting techniques. These techniques are widely used by both industries and research forecasters [2]. NNs possess many advantages over traditional statistical methods (LR and ARIMA) such as (1) capable to model non-linear relationship between features and predictor variables. (2) Capable to make patterns from given examples instead of making assumptions. (3) NNs are tolerant to noise. There are a number of techniques available in NNs; however, it is not evident which technique is suitable for short term load forecasting.

In this study we focus to use three basic NNs such as CNN, MLP, ELM and one enhanced version of ELM(E_ELM) for short term load forecasting and find which one can perform well. Here we present the comparative analysis of the afore mentioned techniques for NYISO dataset. The trend of prediction is shown in Fig. 4 as under. The trend of forecasting is compared with actual trend and it can be seen clearly that CNN and E_ELM follow the actual trend very well. It means these two algorithms are performing very well. Furthermore, E_ELM beats the ELM in following trend.

Moreover, the comparison is shown on the basis of different statistical measures as well. Mean absolute error is shown in Fig. 5. As shown in figure, MAE of CNN is very low, it means it produces very low error as compared to other techniques. Furthermore, the processing time of CNN is also very short in our scenario. MAE for E_ELM is ranked on second and is better than ELM. Mean absolute percent error is also calculated for the forecasting of short term load as shown in Fig. 6. The error rate of CNN in MAPE is a bit unpredictable, it is higher than E_ELM. Here E_ELM performs well and outperforms all the techniques. The performance of CNN may increases with increase in the number of

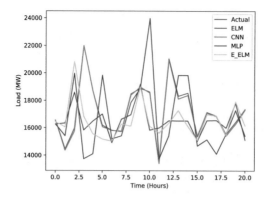

Fig. 4. Trend of forecasting by different techniques

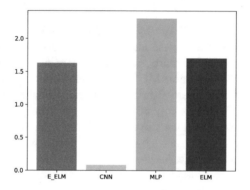

Fig. 5. Mean absolute error score

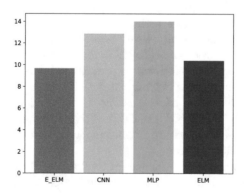

Fig. 6. Mean absolute percent error score

instances. E_ELM works well and has a very short emergence time. The performance of RMS is shown in Fig. 7. Root meas square has a bit same pattern as MAPE. The performance of E_ELM is better than others. ELM stands second

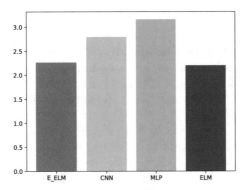

Fig. 7. Root mean square score

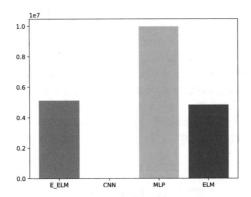

Fig. 8. Mean square error score

here. For calculating RMS, a function is defined using the standard formula of RMS. In last, the score given by mean squared error is shown in Fig. 8. The built-in function of sklearn are used for MAE and MSE. The MSE for CNN is about zero, which indicates that it has very low mean squared error. Again E_ELM stands second. The score of ELM is also a bit identical to E_ELM, but E_ELM performs outstanding here as well. From the above detailed discussion about results, it is clear that CNN and E_ELM are very good for short term load forecasting because of easy use, fast emergence and very good accuracy. Moreover, CNN and E_ELM can capture the non-linear and noisy data of load forecasting very well. Furthermore, with the course of time, the data of load will increase, but the performance of CNN will not decrease, rather will increase. So, on the basis of our experiments, we suggest the use of CNN and E_ELM for short term load forecasting among before mentioned NNs.

7   Critical Analysis

Beside performance, NNs have certain advantages over other AI and traditional model for load forecasting such as: tolerance to noise, making of pattern instead of assumption, handling non-linearity and ease of use [22]. Specifically, we used three techniques (CNN, ELM and MLP). Reasoning for the use and performance of each individual technique is briefly elaborated here.

CNN has a multi layered architecture which allows the learning structure to learn complex relationships between input and predicting variables and extracts complex patterns from data. Thus, CNN has revolutionized too many fields including load forecasting [23]. Although CNN is basically for images but have shown significant performance for load forecasting in too many studies.

MLP is suitable for mapping N-dimensional input to M-dimensional output with high accuracy. It has multi layers and as the learning process proceeds, it learns implicit and salient features from data, identifies error signal and improves accuracy. It is appropriate algorithm for data having high dimensions and load data has very high dimensions [24].

ELM is a single layer feed forward neural network which chose the input weights randomly and analytically determines the output weights which extremely enhance the accuracy and is very fast [25]. In our scenario, it gives good results in a very short time. Furthermore, we have enhanced the basic ELM and have proposed E_ELM which further improves the performance and beats ELM.

8   Conclusion and Future Work

This paper simulated short term load forecasting on NYISO dataset with four machine learning techniques. XGboost and decision tree are used for features importance calculation and RFE is used to extract important features. Three ML techniques (CNN, MLP and ELM) are used for short term load forecasting. Moreover, ELM is enhanced (E_ELM) for short term load forecasting which has beaten ELM in all cases. Results are evaluated on four statistical measures. It is found that CNN and ELM performed well in all the cases and we suggest the use of these two techniques for STLF.

In future, we plan to use the above mentioned techniques for combined load and price forecasting for short term. Furthermore, it is planned to evaluate the system on some other datasets as well as compare the system with some industrial bench marks.

References

1. Önkal, D., Sayim, K.Z., Lawrence, M.: Wisdom of group forecasts: does role-playing play a role? Omega **40**(6), 693–702 (2012)
2. Koprinska, I., Rana, M., Agelidis, V.G.: Correlation and instance based feature selection for electricity load forecasting. Knowl.-Based Syst. **82**, 29–40 (2015)

3. Chan, S.-C., Tsui, K.M., Wu, H.C., Hou, Y., Wu, Y.-C., Wu, F.F.: Load/price forecasting and managing demand response for smart grids: methodologies and challenges. IEEE Sig. Process. Mag. **29**(5), 68–85 (2012)
4. Che, J., Wang, J., Wang, G.: An adaptive fuzzy combination model based on self-organizing map and support vector regression for electric load forecasting. Energy **37**(1), 657–664 (2012)
5. Alfares, H.K., Nazeeruddin, M.: Electric load forecasting: literature survey and classification of methods. Int. J. Syst. Sci. **33**(1), 23–34 (2002)
6. Taylor, J.W.: An evaluation of methods for very short-term load forecasting using minute-by-minute British data. Int. J. Forecast. **24**(4), 645–658 (2008)
7. Ghasemi, A., Shayeghi, H., Moradzadeh, M., Nooshyar, M.: A novel hybrid algorithm for electricity price and load forecasting in smart grids with demand-side management. Appl. Energy **177**, 40–59 (2016)
8. Xiao, L., Wang, J., Hou, R., Wu, J.: A combined model based on data pre-analysis and weight coefficients optimization for electrical load forecasting. Energy **82**, 524–549 (2015)
9. Zheng, J., Xu, C., Zhang, Z., Li, X.: Electric load forecasting in smart grids using long-short-term-memory based recurrent neural network. In: 2017 51st Annual Conference on Information Sciences and Systems (CISS), pp. 1–6. IEEE (2017)
10. Park, K., Yoon, S., Hwang, E.: Hybrid load forecasting for mixed-use complex based on the characteristic load decomposition by pilot signals. IEEE Access **7**, 12297–12306 (2019)
11. Nowotarski, J., Liu, B., Weron, R., Hong, T.: Improving short term load forecast accuracy via combining sister forecasts. Energy **98**, 40–49 (2016)
12. Zhang, J., Wei, Y.-M., Li, D., Tan, Z., Zhou, J.: Short term electricity load forecasting using a hybrid model. Energy **158**, 774–781 (2018)
13. Hu, R., Wen, S., Zeng, Z., Huang, T.: A short-term power load forecasting model based on the generalized regression neural network with decreasing step fruit fly optimization algorithm. Neurocomputing **221**, 24–31 (2017)
14. Papalexopoulos, A.D., Hesterberg, T.C.: A regression-based approach to short-term system load forecasting. IEEE Trans. Power Syst. **5**(4), 1535–1547 (1990)
15. Mbamalu, G.A.N., El-Hawary, M.E.: Load forecasting via suboptimal seasonal autoregressive models and iteratively reweighted least squares estimation. IEEE Trans. Power Syst. **8**(1), 343–348 (1993)
16. Chen, J.-F., Wang, W.-M., Huang, C.-M.: Analysis of an adaptive time-series autoregressive moving-average (ARMA) model for short-term load forecasting. Electric Power Syst. Res. **34**(3), 187–196 (1995)
17. Rahman, S., Bhatnagar, R.: An expert system based algorithm for short term load forecast. IEEE Trans. Power Syst. **3**(2), 392–399 (1988)
18. Pai, P.-F., Hong, W.-C.: Forecasting regional electricity load based on recurrent support vector machines with genetic algorithms. Electric Power Syst. Res. **74**(3), 417–425 (2005)
19. Pandian, S.C., Duraiswamy, K., Rajan, C.C.A., Kanagaraj, N.: Fuzzy approach for short term load forecasting. Electric Power Syst. Res. **76**(6–7), 541–548 (2006)
20. Chitsaz, H., Zamani-Dehkordi, P., Zareipour, H., Parikh, P.P.: Electricity price forecasting for operational scheduling of behind-the-meter storage systems. IEEE Trans. Smart Grid **9**(6), 6612–6622 (2018)
21. NYISO: NYISO Electricity Market Data. http://www.nyiso.com/. Accessed 8 May 2019
22. Hayati, M., Shirvany, Y.: Artificial neural network approach for short term load forecasting for Illam region. World Acad. Sci. Eng. Technol. **28**, 280–284 (2007)

23. Dong, X., Qian, L., Huang, L.: Short-term load forecasting in smart grid: a combined CNN and K-means clustering approach. In: 2017 IEEE International Conference on Big Data and Smart Computing (BigComp), pp. 119–125. IEEE (2017)
24. Mori, H., Yuihara, A.: Deterministic annealing clustering for ANN-based short-term load forecasting. IEEE Trans. Power Syst. **16**(3), 545–551 (2001)
25. Huang, G.-B., Zhu, Q.-Y., Siew, C.-K.: Extreme learning machine: a new learning scheme of feedforward neural networks. Neural Netw. **2**, 985–990 (2004)

Cognitive Personal Security Systems

Marek R. Ogiela[1(✉)] and Lidia Ogiela[2]

[1] Cryptography and Cognitive Informatics Research Group,
AGH University of Science and Technology, 30 Mickiewicza Ave,
30-059 Kraków, Poland
mogiela@agh.edu.pl
[2] Department of Cryptography and Cognitive Informatics,
Pedagogical University of Krakow, Podchorążych 2 St., 30-084 Kraków, Poland
lidia.ogiela@gmail.com

Abstract. In this paper will be described possible applications of cognitive systems for security purposes and cryptographic protocols. In particular will be presented the ways of using advanced cognitive approaches and personal, cognitive features for creation of secure protocols, oriented on user authentication or secure data management in Cloud and Big data resources. The new computing area called cognitive cryptography will be also introduced.

1 Introduction

In advanced modern cryptography we can find many interesting solutions and protocols, which are based on using cognitive features and approaches. One of the most important example can be connected with application of behavioral features for creation of security lock, or application of gesture for creation of cryptographic encryption keys. Such protocols create a new branch of modern cryptography called personalized or cognitive cryptography [1, 2]. Among many existing personally oriented security solutions, we can also define a new classes of authentication procedures, which allow to perform user authentication, and resembles of CAPTCHA codes [3, 4]. CAPTCHA are very popular in many authentication protocol, but it guarantee the secure authentication, based on assumption that changes or distortions, which can be introduced into original visual patterns, makes the recognition or understanding procedure much more difficult than analysis of original images. In this paper we try to define new classes of cognitive CAPTCHA, which will be connected with special cognitive abilities or skills, necessary during secure user authentication [5–7]. Such solution is not a typical cryptographic authentication protocol, but very special procedure, oriented on selected group of users, who poses particular cognitive skills or knowledge, necessary to understand or guess the meaning of visual pattern used during authentication. The main purpose of definition of such innovative protocols is to create a new user oriented solutions, which allow to determine users according theirs cognitive skills. It seems that such authentication procedures will have a broad range of possible applications in Future Internet, IoT, smart technologies, and services management [8].

© Springer Nature Switzerland AG 2020
L. Barolli et al. (Eds.): CISIS 2019, AISC 993, pp. 87–90, 2020.
https://doi.org/10.1007/978-3-030-22354-0_8

2 Fuzzy Perception CAPTCHA

Cognitive skills are very often connected with possibilities of proper interpretation of visual patterns, especially fuzzy ones. Below will be described such type of cognitive CAPTCHAs, which require the proper understanding or interpretation of fuzzy visual codes.

Examples of such codes should be connected with analysis of CAPTCHA having several different perception thresholds. In such approach it is possible to present for recognition different visual patterns softened in such way, that the main objects are difficult to recognize for not familiar users. Authenticated person has to understand the content of fuzzy patterns, what is possible only when he has some special skills or experiences in recognition of similar patterns. Such special abilities or knowledge, allow him to guess in proper manner the final answer, required for successful authentication. Such authentication can be performed in several stages presenting patterns with different perception thresholds, and changed during verification according cognitive and perceptual skills. In Fig. 1 presents an example of image content recognition with different perception thresholds.

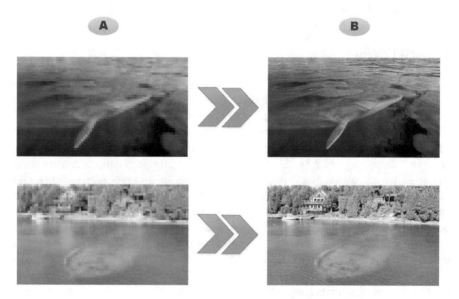

Fig. 1. Examples of images with fuzzy perception thresholds. In column A user should recognize a fuzzy patterns. In column B more detailed recognition and interpretation is available.

Described authentication procedures should use any visual pattern, which presents not a typical object or situation, and is additionally softened or blurred. In subsequent stages such patterns can be presented to authenticated users with different softened or fuzziness levels, evaluated for specific groups of persons, who have particular recognition skills. The only problem may arise in fast selection of a great number of unusual visual patterns, which presents not typical objects, unexpected by authenticated users.

Such authenticating approach can be applied for authorization users or instances in distributed networks, or prevent access to large databases and repositories, as well as Cloud Computing infrastructures.

The most important feature of fuzzy perception CATCHA is possibilities to gain access for services or information at different levels, depending on perception thresholds, and required accessing grants [9, 10]. Generation of softened visual patterns with sequences of possible answers for authenticating protocols, can be conducted with application of cognitive information systems, especially those ones oriented for visual pattern understanding [11]. Such cognitive systems allow to create a semantic records describing the meaning of presented visual patterns, and such description can be used for final authentication procedure.

3   Security Analysis

During verification stages are presented to users softened or blurred visual patterns with barely visible objects. Such patterns are difficult for fast recognition and understanding by humans, who should guess the proper content, based on the analysis of the whole image. This causes that also computer systems cannot manage such task in fast and correct manner. CAPTCHTA verification codes are secure due to the difficulties of recognition items on distorted images, so the same problem we observe with relation to cognitive CATCHA. Such codes require having very specific cognitive skills and knowledge to successfully pass verification protocol. Even if computer try to use thematic databases and advance image processing approaches, it will not be able in short time to quickly perform an image analysis with semantic classification towards successfully passing CAPTCHA authentication. This is the main feature, which guarantee the high level of security of such codes, and allow to pass verification protocols only for human users.

4   Conclusions

In this paper has been described new approach for creation of cognitive authentication protocols, which apply CAPTCHA codes. Presented solution perform authentication procedure in several iterations, presenting softened or blurred patterns, which can be recognized at different perception thresholds, and cognitive skills of users. Such codes works in secure manner because require advanced semantic interpretation of the visual content, and fast meaning evaluation by human users [12]. Presented solutions are oriented for groups of users having particular cognitive or perceptual skills. Additionally such procedures can use cognitive vision systems while generating authenticating patterns. This solutions extend classical cryptographic protocols towards new classes of security procedures connected with cognitive cryptography, which join traditional encryption approaches [13, 14] with cognitive functions and application of personal features.

Acknowledgments. This work has been supported by the National Science Centre, Poland, under project number DEC-2016/23/B/HS4/00616.

This work has been supported by the Pedagogical University of Krakow research Grant No BS-376/M/2019.

References

1. Ogiela, M.R., Ogiela, L.: On using cognitive models in cryptography. In: IEEE AINA 2016 —The IEEE 30th International Conference on Advanced Information Networking and Applications, pp. 1055–1058. Crans-Montana, Switzerland, 23–25 March, 2016
2. Ogiela, M.R., Ogiela, L.: Cognitive keys in personalized cryptography. In: IEEE AINA 2017 The 31st IEEE International Conference on Advanced Information Networking and Applications, pp. 1050–1054. Taipei, Taiwan, 27–29 March, 2017
3. Meiappane, A., Premanand, V.: CAPTCHA as Graphical Passwords - A New Security Primitive: Based on Hard AI Problems. Scholars' Press (2015)
4. Osadchy, M., Hernandez-Castro, J., Gibson, S., Dunkelman, O., Perez-Cabo, D.: No bot expects the DeepCAPTCHA! Introducing immutable adversarial examples, with applications to CAPTCHA generation. IEEE Trans. Inf. Forensics Secur. **12**(11), 2640–2653 (2017)
5. Ogiela, L.: Cognitive computational intelligence in medical pattern semantic understanding. In: Guo, M.Z., Zhao, L., Wang, L.P. (eds.) In: ICNC 2008: Fourth International Conference on Natural Computation, vol. 6, Proceedings, 18–20 October, Jian, Peoples R China, pp. 245–247 (2008)
6. Ogiela, L.: Cognitive Information Systems in Management Sciences. Elsevier, Academic Press, London, UK (2017)
7. Ogiela, L., Ogiela, M.R.: Bio-Inspired cryptographic techniques in information management applications. In: IEEE AINA 2016—the IEEE 30th International Conference on Advanced Information Networking and Applications, pp. 1059–1063. Crans-Montana, Switzerland, 23–25 March, 2016
8. Ogiela, U., Ogiela, L.: Linguistic techniques for cryptographic data sharing algorithms. Concurr. Comp. Pract. E. **30**(3), e4275 (2018). https://doi.org/10.1002/cpe.4275
9. Ogiela, L., Ogiela, M.R.: Insider threats and cryptographic techniques in secure information management. IEEE Syst. J. **11**, 405–414 (2017)
10. Ogiela, M.R., Ogiela, U.: Secure information management in hierarchical structures. In: Kim, T.-h., et al. (eds.) AST 2011. CCIS, vol. 195, pp. 31–35 (2011)
11. Ogiela, L., Ogiela, M.R., Ogiela, U.: Efficiency of strategic data sharing and management protocols. In: The 10th International Conference on Innovative Mobile and Internet Services in Ubiquitous Computing (IMIS-2016), 6–8 July, Fukuoka, Japan, pp. 198–201 (2016), https://doi.org/10.1109/imis.2016.119
12. Ogiela, L.: Advanced techniques for knowledge management and access to strategic information. Int. J. Inf. Manag. **35**(2), 154–159 (2015)
13. Easttom, Ch.: Modern Cryptography: Applied Mathematics for Encryption and Information Security. McGraw-Hill Education, New York (2015)
14. Schneier, B.: Applied Cryptography. Wiley, Indianapolis (2015)

Incremental Patent Semantic Annotation Based on Keyword Extraction and List Extraction

Xu Chen[1], Weixian Zong[1], Na Deng[2], Shudong Liu[1], and Yipeng Li[1(✉)]

[1] The School of Information and Safety Engineering,
Zhongnan University of Economics and Law, Wuhan, China
chenxu@whu.edu.cn, zongweixian1313@qq.com,
liumu1321@zuel.edu.cn, lyp2357@163.com
[2] Computer School, Hubei University of Technology, Wuhan, China
iamdengna@163.com

Abstract. At present, there is a lack of in-depth processing and indexing of Chinese patents in China, which makes the patent data retrieval inaccurate and incomplete, leading to duplication of applications and waste of resources. Aiming at the problem of lacking annotated patent data in Chinese patent indexing, this paper studies an incremental patent annotation method. By using co-training method, keyword extraction and list extraction can cooperate with each other and iteratively annotate the functional clauses, which achieves the effect of obtaining much more annotated data through a small quantity of training data. Experiment results indicate this method can gradually improve the recall without sacrificing much precision.

Keywords: Patent annotation · Keyword extraction · List extraction · Co-training

1 Introduction

Patent is one of the most important information resource for technical analysis [1]. If enterprises can make full use of patent documents to guide technological innovation, they can effectively save 40% of research funds and 60% of research time, so its importance is continuously recognized by enterprises.

According to the World Intellectual Property Organization (WIPO) survey, patent literature contains 90% to 95% of the latest scientific research achievements in the world every year. Now, patent data worldwide has reached 15 million, and is growing at a rate of more than 6% [2].

Faced with the problem of massive patent data storage, in the 1990s, the first patent database system began to be established. The establishment of patent citation data provided effective basic data for the discovery of core patents and the analysis of patent technology development routes.

Patent annotation is of great significance to the construction of patent database. Patent annotation refers to extracting and identifying the characteristic information of a

© Springer Nature Switzerland AG 2020
L. Barolli et al. (Eds.): CISIS 2019, AISC 993, pp. 91–101, 2020.
https://doi.org/10.1007/978-3-030-22354-0_9

patent, which is of great significance to patent retrieval, analysis and mining. It can facilitate more accurate and intelligent patent retrieval, transform unstructured patent text into structured feature items, and help establish semantic association between patents. Patent semantic annotation is similar to named entity recognition in information extraction and natural semantic processing. There are three main methods: the manual template-based approach, the machine learning approach and the combination of the two methods [3].

The manual template-based approach shows its advantage of high accuracy for patent annotation of small data sets. In 2009, Peter Parapatics [4], defined the manufacturing method, usage method, detail description and composition structure template of patent declaration. The accuracy of this method has reached more than 90% after testing. Wang Peiyan [5] used the patent title and the statement that contains the information of subject to build templates, then used the template to extract the patent subject. This kind of method can be achieved easily, the accuracy of the result is higher, especially for the words of low-frequency. However, this method is closely related to the domain and linguistics of patents, and is not suitable for the annotation of a large number of patents.

Based on machine learning method, the labeled patents are used as training set to automatically learn the patent patterns, which is more suitable for the situation where there are a large number of labeled patents. Hidetsugu Nanba [6] used support vector machine to annotate the function of the patented technology. Although this machine learning method does not require manual participation, its accuracy is not high. The accuracy of the above SVM-based method is only about 50%. Machine learning can be automated, but it requires 100,000-level labeled data, especially for Chinese patents, it is very time-consuming and labor-consuming to establish a huge set of patent data training.

The hybrid method combines the advantages of both methods, including the automation process of machine learning and the precision process of manual annotation. Gui Jie [7] first learned some rules by using conditional random fields, and then screens these rules manually. The experimental comparison showed that the accuracy of this method is 10% higher than that of using conditional random fields alone.

In China, the attention to patent annotation is relatively lagging, until recent years, it has attracted the attention of a small number of domestic enterprises and scholars. The State Intellectual Property Office and Dongfang Lingdun [8] have carried out in-depth processing of the Chinese medicine patent database and the world traditional medicine patent database, which greatly improved the efficiency of patent retrieval and lay a foundation for patent analysis. Zhu [9] applied the technology of information mining to the patent literature, and proposed a new method for the automatic extraction of patent subject words. Zhang Bopei [10] used hidden Markov model to mining efficacy information of patent text, which greatly improved the accuracy of efficacy word recognition.

This paper will study an incremental method of patent semantic annotation, in which "incremental" refers to using the data of the previous annotation to conduct the next annotation. Due to the abstracts of patent often highly summarize the main information [11], the main object of this article's analysis is the patent abstract. The co-training method based on keyword extraction and list extraction is used to realize the annotation of functional fragments. and it is compared with the machine learning method of SVM.

2 Method

2.1 Features of Patent Semantic Annotation

Patent functional clause refers to the clause that can embody the complete functional semantics of patent in the Chinese patent abstract text. The clause is a sentence that divides symbols by commas or semicolons.

Patent annotation is able to extract and identify patent functional clauses in Chinese patent abstract text, which is expressed by XML tags. For the functional clause in the patent, <Effect> </Effect> is used to express the effect of patent; <Attribute> </Attribute> is used to express the object that the patent produces effects, usually a noun; <Value> </Value> is used to express the effect of patent, usually a verb or adjective; <Attribute> and <Value> are sublabels of <Effect>.

Through the analysis of Chinese patent literature, we find that Chinese patent functional clauses have three distinct characteristics:

- Functional clauses are often a series of sub-bureaus, and they are concentrated in the abstract.
- Functional clauses usually contain fixed collocation patterns of several keywords.
- Owing to the habit of writing by the same author, patent sentences have certain regularity and similarity in writing, and the same author tends to use the same or similar keywords to describe the characteristics of efficacy.

According to these characteristics of patent sentences, two methods of extracting functional clauses are proposed: keyword extraction and list extraction.

2.2 Keyword Extraction

Keyword extraction plays an increasingly crucial role in information retrieval, natural language processing and other several text related researches [12]. It is based on the patented functional clauses often have fixed collocation patterns of keywords, which can be extracted into functional clauses by identifying the fixed collocation of keywords. In this paper, we use a binary representation of keyword collocation patterns, which can be mapped to <attribute> and <value> tags in essence. The content of <attribute> tags corresponds to the object of effect produced by patents, which is also called functional attribute words; and <value> tags correspond to the effect words produced by patents, which are called functional value words.

According to the flexibility and complexity of Chinese words, each word has synonyms that can be freely replaced, so the collocation of keywords for patent efficacy is flexible and changeable. Therefore, a functional semantic dictionary can be established to store the corresponding collocations. The dictionary contains three aspects:

the functional attribute words and their synonyms, the functional value words and their synonyms, and the corresponding relations between the functional attribute words and the functional value words. They correspond to the ellipse from left to right in Fig. 1 below.

The set of functional attributes words: $A = \{a_1, a_2, \ldots\ldots, a_n\}, a_i(1 < i < n)$, denotes a function attributes word.

The set of synonyms of functional attributes words: $B_i = \{b_{i1}, b_{i2}, \ldots\ldots, b_{im}\}$, there are m synonyms in the function attribute word a_i.

The set of functional value words that collocate with a functional attribute word a_i: $V_i = \{v_{i1}, v_{i2}, \ldots\ldots, v_{ip}\}$, indicating that there are a total of p functional value words collocated with a functional attribute word a_i. $v_{ij}(1 < j < p)$ denotes a functional value word that is collocated with the functional attribute word a_i.

The set of synonyms of the function value word v_{ij}: $C_{ij} = \{c_{ij1}, c_{ij2}, \ldots\ldots, c_{ijq}\}$ denotes that there are q synonyms in the function value word v_{ij}.

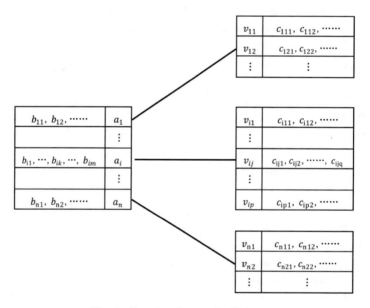

Fig. 1. Functional semantic dictionary

2.3   List Extraction

Since most of the functional clauses are clustered together, they can be regarded as a linked list of serial links, so several sequential functional clauses can be extracted at the same time. There are two main types of list extraction: edge extraction and coordination extraction.

Edge Extraction.

The principle of edge extraction is based on the fact that functional clauses are usually connected together. For example:

<Effect>**A**</Effect>, **B, C, D,** <Effect>**E**</Effect>。

A and E at the beginning and end of this paragraph are marked as functional clauses respectively, while the middle clauses are not marked, so all the middle sub-bureaus can be marked as functional clauses. The main steps of boundary extraction are:

(1) Get the serial number of each clause in the set S containing a patent' clauses which have been annotated as functional clauses;

(2) For each pair of neighboring functional clauses in S, if they are in the same sentence and the clauses between them are not in S, then annotate these clauses are functional clauses (Fig. 2).

```
def border(patent):
    pat=patent.replace('；',',')
    sen=pat.replace('。',',')
    str=sen.split('，')
    fun=[]
    for index,eff in enumerate(str):
        if eff.startswith("<effect>") and eff.endswith("</effect>"):
            fun.append(index)
    add=[]
    for x in range(0,len(fun)-1):
        if fun[x]+1==fun[x+1]:
            continue
        pattern = re.search(str[fun[x]]+'(.*?)'+str[fun[x+1]],patent,re.S).group(1)
        if pattern.find('。')==-1:
            add.append(str[fun[x]+1:fun[x+1]])
    return add
```

Fig. 2. Edge extraction

Coordination Extraction.

The working mechanism of coordination extraction is that all coordination sentences of a functional clause are also functional clauses. The main steps in the coordination extraction are:

(1) For each clause C in the set S containing a patent' clauses which have been annotated as functional clauses, find out its all neighboring clauses in the patent. The adjacent clauses here refer to the two clauses that are physically adjacent in the patent text. They can be in the same sentence or in different sentences.

(2) If our neighboring clause is the coordination sentence of C and it is not in S, then annotate this clause as functional clause. The two clauses are coordinate sentence if they satisfy the following conditions at the same time:

(a) The two clauses are in the same sentence;

(b) The two clauses are adjacent;

(c) The two clauses are connected by conjunction "并且", "而且", "还", "不仅" or "并" (Fig. 3).

```
def compound(patent):
    if patent[-1]=='。':
        patent=patent[:-1]
    pat=patent.replace('；',',')
    sen=pat.replace('。',',')
    str=sen.split(',')
    fun=[]
    add=[]
    for index in range(0, len(str)-1):
        eff=str[index]
        if eff.startswith("<effect>") and eff.endswith("</effect>"):
            if str[index+1].startswith("<effect>") and str[index+1].endswith("</effect>"):
                continue
            else:
                pattern = re.search(str[index]+'(.*?)'+str[index+1], patent, re.S).group(1)
                if pattern.find('。')==-1:
                    if str[index+1].startswith(("并且",'而且','还','不仅','且','或者')):
                        add.append(str[index+1])
    return add
```

Fig. 3. Coordination extraction.

2.4   Representation of Functional Fragments

In the same sentence in the Chinese Patent Digest, several sequential connected patent functional clauses constitute a patent functional fragment.

In this paper, triple (sentencePosition, clausePosition, indexLength) is used to represent functional fragments. sentencePosition is used to indicate the indexing position of the sentence in which the functional fragment is located in the whole patent abstract text. clausePosition is the index position of the functional fragment in its sentence, and indexLength represents the index length of the functional fragment. The specific definitions are as follows:

$$sentencePosition = \begin{cases} s, & s < n/2 \\ s - n - 1, & s \geq n/2 \end{cases}$$

$$clausePosition = \begin{cases} -e - 1, & f \geq e \\ f, & f < e \end{cases}$$

$$indexLength = \begin{cases} length, & clausePosition \geq 0 \\ -length, & clausePosition < 0 \end{cases}$$

Figure 4 of the line segment diagram indicates the positive and negative according to the position of the 'effect', in which 'n' represents the number of sentences in the entire Patent digest text; 's' represents the sentence number of the sentence in which the functional fragment is located in the entire patent summary text; 'f' indicates the number of clauses in front of the functional fragment in the sentence where the functional fragment is located; 'e' indicates the number of clauses behind the functional fragment in the sentence in which the functional fragment is located; 'length' indicates the length of the functional fragment.

Fig. 4. Triple representation of functional fragment

Since the triple specifically represents the position of the functional sentence, the maximum common triple can be obtained from the triple of several functional fragments, which represents a common feature of the position of the functional fragment in the patent. Placing it in other patents to match the corresponding location of the fragment, it is likely that this fragment is a functional fragment. The main step in the maximum common triple:

(1) N functional segments' triples are denoted as $(a_i, b_i, c_i), 0 < i < N$. From the set $\{a_i\}, 0 < i < N$, find the most frequent element, if there are multiple elements with the same occurrence number, choose the one with the least absolute value. If there are multiple elements with the same occurrence number and the same absolute value, choose the negative one. We denote the chosen element a_r;

(2) From the set $\{(a_i, b_i, c_i)\}, 0 < i < N$, we get its subset with the first element is a_r. Consider the set of the second element of this subset, we find the most frequent element. If confronting the problem of multiple elements, we handle it using the method in Step 2. The ultimate element is denoted as b_p;

(3) From the set $\{(a_i, b_i, c_i)\}, 0 < i < N$, we get its subset with the first element is a_r and the second element is b_p Consider the set of the third element of this subset, we find the element with the least absolute value. If there are multiple elements with the same absolute value, choose the negative one. The ultimate element is denoted as c_q;

(4) The maximum common triple of these N triples is (a_r, b_p, c_q).

In these three steps, this chapter determines that the index of the three elements of the maximum common triple is the maximum number of occurrences, the minimum absolute value and the negative number. This is because the occurrence of the first and second elements indicates that this position is the common feature of the functional fragments represented by these triples. The preference to choose the first and second elements of the minimum absolute value is based on the following considerations: the functional clauses usually appears at the beginning and end of the abstract and the text of the patent; The first, second, and third elements tend to be negative because the utility statement appears more at the end of the patent than at the beginning of the patent.

2.5   Function Annotation Algorithm Based on Co-Training

Co-training method is based on the premise that high-yield and high-quality patent inventors have fixed writing habits and styles within a specific topic. When writing patents in a certain field, they like to use the same or similar collocations to describe the efficacy of the patents invented, and according to their personal habits, they like to use the merits of the patents. The specific position of validity in the patent abstract, such as the beginning or the end of the article, etc. In addition, patents in the same field usually have the same or similar efficacy.

Based on the above premise, this paper starts with list extraction, and its flow chart is shown in Fig. 5. The specific steps corresponding to the label in the figure are explained as follows:

(1) Firstly, the scope of function annotation is a patent for a specific topic, which can be a specific IPC classification number or all patents containing a keyword.

(2) The inventors in this topic are arranged in descending order according to the number of their invention patents, and the first K inventors with more invention patents are selected. The reason for this is that the patents of high-yielding inventors are generally of high quality. These inventors' patents on this topic usually contain most of the power combinations in this field. Moreover, because inventors have fixed writing habits and styles, we can find the implicit writing habits and styles by analyzing the patents of high-yield inventors.

(3) For each inventor, starting with list extraction, his N patents are randomly selected, and all the functional clauses in the N patents are marked manually. Using triple to represent these functional clauses.

(4) Compute the maximum common triples of these triples in the previous step.

(5) Based on the fact that patent functional clauses are usually linked together, the maximum common triple is used to annotate the functional clauses in the inventor's other patents.

(6) Perform a judgment on the efficacy statement found in the previous step, and manually filter out the statement that does not indicate function, otherwise, it will be marked as functional clauses.

(7) Manual method is used to extract the functional collocations. The clauses containing these functional collocations are labeled as functional sentences. The extracted new functional collocations are stored in the functional semantic dictionary.

(8) For each new functional collocation in the functional semantic dictionary, conduct functional clauses retrieval in all patents of the inventor to search for the patent fragments containing the collocation.

(9) When several function statements have been marked in the patent, list extraction can be started. List extraction can be 9.1 boundary extraction or 9.2 coordination extraction. For 9.1 boundary extraction, clauses between two already marked functional clauses are regarded as new functional clauses. The 9.2 coordination extraction is based on the clause juxtaposed with a functional sentence, which is also regarded as a functional clause.

(10) Make a judgment on the functional clauses extracted through the chain, and manually filter out those sentences which are not expressing function. Otherwise, they are annotated as functional clauses. For the newly added functional clauses extracted by chain, the function matches are manually extracted and added to the functional semantics dictionary, and the steps (7) to (10) are executed circularly until no new functional clauses are generated.

(11)   For all patents of other inventors, use the existing functional semantic dictionary containing a lot of functional collocation to mark functional collocation and functional clauses.

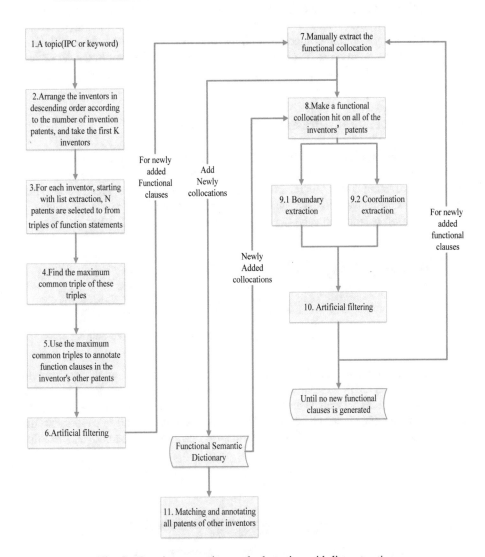

Fig. 5.   Function extraction method starting with list extraction

In the definition and description of the functional semantic dictionary, we know that through the functional semantic dictionary, we can find those functional collocation words which are synonymous with a certain functional collocation. In step (8) above, for each newly added functional collocation, the functional clause hits in all patents of the inventor, where "hitting" refers to hitting at the semantic level. The implementation of step (8) is as follows:

(a) To get the functional attribute words and functional value words of the functional collocation to be matched.
(b) According to the dictionary of functional semantics, the semantic classes of functional attribute words and functional value words are obtained respectively.
(c) Calculate the Cartesian product of the two semantic classes in the previous step, that is, the synonymous functional collocation that is intended to be matched.
(d) For each clause in a patent, check whether it contains an element in the set of synonymous functional collocations. If it does, annotate the clause as a functional clause.

3   The Experimental Results and Conclusions

In this experiment, accuracy, recall and F-measure are used as evaluation criteria. As shown in Fig. 6, the results of each iteration of co-training using keyword extraction and chain extraction without manual filtering are shown in the figure. It can be seen that recall rate is constantly increasing and accuracy has not decreased much. Of course, if manual filtering is added to each iteration, the accuracy may reach 100%.

PRECISION	90%	88%	83%	81%	80%	75%
RECALL	38%	41%	48%	61%	79%	81%
F-MEASURE	60%	71%	78%	81%	82%	82%

Fig. 6. Experimental results of functional extraction based on co-training

We also use the machine learning method to mark the patent, and use the support vector machine (SVM). The training features include the verb part and the noun part of the functional phrase, as well as the four words around the functional phrase. As shown in the following Fig. 7, it can be seen that the co-training method is superior to the existing machine learning method in terms of performance. Moreover, machine

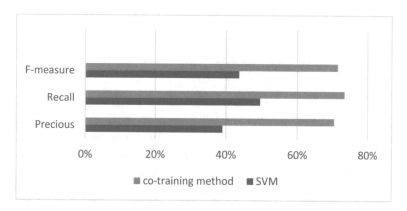

Fig. 7. Comparison of co-training method and SVM method in extracting functional clauses

learning requires a large amount of data as the training set in the early stage, which is far more than the initial amount of data for co-training, so co-training can greatly save human participation.

Acknowledgments. This work was supported by the Zhongnan University of Economics and Law (2722019JCT035, 2722019JCG074), the National Natural Science Foundation of China (61602518), and the Fundamental Research Funds for the Central Universities National Social Science Fund of China (NO:16CXW019).

References

1. Zhang, J., Liu, M.-J., Zhai, D.-S.: Technology topic in RFID based on patent co-word analysis. Scientific Management Research, Oct. 2013
2. Gurry, F.: World intellectual property indicators—2011 Edition [EB/OL], 7 Apr 2014. http://www.wipo.int/export/sites/www/freepublications/en/intproperty/941/wipo_pub_941_2011.pdf
3. Fa-guo, Z.H.O.U., Ying-long, W.A.N.G., Bing-ru, Y.A.N.G., et al.: Research on key technologies of unstructured information extraction. Comput. Eng. Appl. **45**(14), 1–6 (2009)
4. Parapatics, P., Dittenbach, M.: Patent claim decomposition for improved information extraction[M]. In: Parapatics, P., Dittenbach, M. (eds.) Current Challenges in Patent Information Retrieval, pp. 197–216. Springer, Berlin Heidelberg (2011)
5. Wang, P.-Y., Zhang, G.-P., Cai, D.-F.: An automatic generation method for patent keyword extraction template. Journal of ShenYang Institute of Aeronautical Engineering **27**, 46–49 (2010)
6. Nanba, H., Kondo, T., Takezawa, T.: Hiroshima city university at NTCIR-7 patent mining task[C]. In: Proceedings of the 7th NTCIR Workshop Meeting, pp. 369–372, 2008
7. Gui J., Li P., Zhang C., et al. Integrating crf and rule method for knowledge extraction in patent mining task at NTCIR-8[C]. In: Proceedings of the 8th NTCIR Workshop Meeting, pp. 341–344, 2009
8. Sun Yan-ling, Liu Hua-bing, Wang Hai-hong, et al. Deep indexed chinese pharmaceutical patent database[j]. Chin. J. Med. Guide **10**(1), 22–24 + 26 (2008)
9. Zhu, L., Lv, X., Xu, L.: Patent subject words extraction based on integrated strategy method [C]. In: International Symposium on Parallel & Distributed Computing (2017)
10. Bopei, Z., Yongping, D., Wenjian, M.: Efficacy word recognition based on hidden markov model [J]. Inf. Eng. **1**(03), 81–89 (2015)
11. Guangpu, F., Xu, C., Zhiyong, P.: A rules and statistical learning based method for Chinese patent information extraction[C]. In: Eighth Web Information Systems & Applications Conference. IEEE Computer Society, 2011
12. Chen, Y., Zhou, R., Zhu, W., et al.: Mining patent knowledge for automatic keyword extraction. J. Comput. Res. Dev. **53**(8), 1740–1752 (2016)

Blueprint of Driving Without Emission: EV with Intelligent Charging Stations Network

Xu Chen[1], Deliang Zhong[1], Shuhong You[1], Shuqi Yang[1], Na Deng[2(✉)], Shudong Liu[1], and Yipeng Li[1]

[1] Zhongnan University of Economics and Law, 182 Nanhu Avenue, Hongshan District, Wuhan, Hubei, China
chenxu@whu.edu.cn, 18770770508@163.com, 627623528@qq.com, 781020094@qq.com, liumu1321@zuel.edu.cn, lyp2357@163.com
[2] Computer School, Hubei University of Technology, Wuhan, China
iamdengna@163.com

Abstract. The improvement of infrastructure is conducive to promoting the development and popularization of electric vehicles, and different countries and regions have different situations. We have developed the Development Strategy Selection Model (DSM) for countries to judge their own situation and choose the best development strategy, and use the Strategic Effectiveness Evaluation Model (SEM) to measure the effectiveness of the strategy. We think that people's travel demand can be divided into short-distance travel demand and long-distance travel demand, so the demand of charging station can also be divided into fast charging demand and normal charging demand. We have developed point-based planning model (POM) and path-based programming model (PAM) to calculate these two needs. We select the cases of China and the United States for analysis, and the results show that our model has a good estimation effect.

1 Introduction

1.1 Background

With the ever increasing problem of environmental pollution and the oil crisis, the electric vehicle (EV) with saving energy and decreasing emissions has become more and more popular. Because of the EV technology's improvement and the government's support, the EV market (like Tesla) grows constantly. However, the imperfect EV charging infrastructure network inhabits the further promotion of the EV. So, building a good infrastructure network is very essential for promoting the use of electric vehicles.

Compared with traditional fuel vehicles, electric vehicles have two obvious disadvantages which are also the key factors for the network building: short travel range and long charging time, so how to decide the sufficient number and location of charging stations is really critical. At the moment, electric vehicles are no doubt a hot spot: Governments formulate a series of relevant policies to promote its development, consumers are eager to migrate to it, also at the same time, all big manufacturers are

© Springer Nature Switzerland AG 2020
L. Barolli et al. (Eds.): CISIS 2019, AISC 993, pp. 102–113, 2020.
https://doi.org/10.1007/978-3-030-22354-0_10

making great efforts to ensure their company is at the forefront of the development of electric vehicles.

However, the migration to the driving without emission is by no means of something that can happen overnight, considering that electric vehicles have two obvious disadvantages when compared with traditional fuel vehicles: short travel range and long charging time, how to decide the sufficient number and location of charging stations to build an intelligent network is really crucial. Our work is to achieve these goals, building up the related models and verify it.

1.2 Literature Review

An extensive literature exists on the refueling infrastructure problem. With the development of EVs, many scholars have done research on the location problem of electric vehicle charging stations from many angles. However, it may be due to the fact that some factors of charging station location are more prominent, so from the overall point of view, these researches of the location mainly focus on the following aspects:

Ren et al. (2011) considered the charging station layout optimization model based on hard time window constraints under the condition of the lowest operating cost. Liu et al. (2012) did a research which was to maximize the benefits of operators, using quantum particle swarm optimization algorithm to obtain the optimal planning scheme for charging stations; In addition, there are still many factors to consider in the location of electric vehicle charging stations. Power factors such as grid capacity, load, and available capacity (Feng et al. 2012) were also taken into account. Xi et al. (2013) develops a simulation-optimization model that determines where to locate electric vehicle chargers to maximize their use by privately owned electric vehicles. Chu et al. (2015) proposed gradual coverage model based on time satisfaction to simulate location selection. Zhu et al. (2018) did a research on competitive location problem of charging station considering gravitational factors and time satisfaction.

Most of these studies have started to explore the location of charging stations from several key factors while most of them do not take gdp, population density and amount of cars into consideration.

1.3 Objectives and Contributions

In fact, gdp, population density, and vehicle ownership will affect people's demand for electric vehicles. Widely distributed charging requirements and charging stations are obviously attractive to each other, and the degree of attraction has a significant relationship with the distance between the two.

The objective of this paper is to propose a mathematical model to predict the location of charging stations. The contribution of this paper can be summarized as follows. We propose point based planning model, path based planning model, and double gravity model to measure the degree of attraction between electric vehicle users and charging stations. And then predict the need and location of charging stations.

2 Assumption and Model

People's travel plans are divided into long-distance travel and short-distance travel, which correspond to two different charging modes. Long-distance travel uses super charging stations to supplement energy sources, while short-distance travel is not sensitive to charging time. Conventional charging stations are used to control costs. Our model is based on that the current road network will not change much. All the energy sources of new energy vehicles are electricity. All long-distance trips have been planned and prepared in advance, so electric vehicles have 100% electricity before they leave.

Based on this, we propose two sub-models: Point Based Planning Model (POM) and Path Based Planning Model (PAM).

Because of limited resources, we should give priority to meeting the needs of regions that can bring more significant results. Different regions have different situations, so we designed Development Strategy Selection Model (DSM). In order to measure the validity of our model, we designed Strategic Effectiveness Evaluation Model (SEM).

2.1 Point Based Planning Model (POM)

The POM model is mainly used to solve the short distance charging demand, in which case users usually spend hours or overnight charging. So we can ignore the charging time and consider only the demand point, the total demand, and the candidate charging station location. As for the determination of demand point and aggregate demand, we use the GDP per capital, population density and vehicle ownership to stratify according to the thermodynamic chart of the class.

We can assume that electric vehicles that have a daily charge demand account is 40% of the total. The total number of cars in the planning area is N so the number of charging stations is:

$$N_{changer} = \frac{N \bullet 40\% \bullet 50}{n \bullet \beta \bullet 24} \tag{1}$$

In the above formula, n is the number of cars that can be charged at the Tesla charging station at the same time. We take n = 6 and β as the power of the charging pile for Tesla destination.

Now, the demand point, the total demand and the candidate locations have been determined. In order to enable users to detour at least and get the charging service in time, we use the P- median model to optimize the charging station planning. The objective function and constraint conditions of the model are as follows:

$$Min(Z) = \sum_{i \in I} \sum_{j \in J} U d_{ij} x_{ij} \tag{2}$$

S.t:

$$\sum_{i \in I} x = 1 \qquad \forall j \in J$$

$$\sum_{j \in J} y_i = q \qquad \forall j \in J$$

$$y_i \geq x_{ij} \qquad \forall i \in I, \forall j \in J$$

$$x_{ij}, y_{ij} \in (0, 1) \qquad \forall i \in I, \forall j \in J$$

Where x_{ij} is the demand point I accepts service at the service point J. $x_j(y_q)$ is the value of $x_j(y_q)$ is 1 if there is charging station set up on the node, else 0.

2.2   Path Based Planning Model (PAM)

There are three main factors affecting the number of long-distance travel charging stations: distance, traffic flow and electric vehicle mileage. In order to simplify the problem, our mileage is calculated according to the mileage of the local highway. The mileage of the electric vehicle is represented by the mileage of Tesla. As for the estimation of traffic flow, we introduce a dual force model (Zhu et al. 2018): economic attraction (EF) and tourism cultural attraction (TF). Economic attraction brings human resources and goods transportation demand, and tourism cultural attraction brings private tourism transportation demand.

$$G = \beta_1 \bullet EF + \beta_2 \bullet \tag{3}$$

$$EF = \frac{K_1 \bullet M_i \bullet M_j}{\alpha^2} \tag{4}$$

$$TF = \frac{K_2 \bullet (T_i \bullet C_i) \bullet (T_j \bullet C_j)}{\alpha^2} \tag{5}$$

$$NC = G \bullet \alpha / P_{max} \tag{6}$$

Where G is total attraction between two cities, β is weight value of two attractions. K is adjustment coefficients in both places. M_i is the GDP of i city without Tourism Culture Income. T_i is the car ownership, C_i is the Tourism Culture Income. P_{max} is maximum service volume per unit charging station. NC is the number of charging station.

2.3   Development Strategy Selection Model (DSM)

There are many factors that affect government and enterprises' judgement the main influencing factors are population density distributions and wealth distributions. Many scholars believe that geographic topography is also an important factor, but we believe that the impact of topography is more on the livability of land, which has been indirectly reflected by population density.

We use correlation coefficient r to measure the impact of economic and population density on the demand for electric vehicles. Then use the coefficient β to judge the current significant influencing factors and choose the appropriate development strategy. We think that when the coefficient $\beta \leq 20\%$, we can temporarily ignore the smaller influencing factors.

$$\beta = \frac{r_1 - r_2}{r_1 + r_2} *100\% \tag{7}$$

$$r = \frac{\frac{1}{n}\sum (x - \bar{x}) \bullet (y - \bar{y})}{\sqrt{\frac{\sum (x-\bar{x})^2}{n}} \bullet \sqrt{\frac{\sum (y-\bar{y})^2}{n}}} \tag{8}$$

2.4   Strategic Effectiveness Evaluation Model (SEM)

The choice of any strategy is to better promote the popularization and development of electric vehicles, so the criterion to measure the effectiveness of the strategy should be whether the popularization speed of electric vehicles is accelerated after the implementation of the strategy. Firstly, we use the first step of AHP to establish the general framework of the influencing factors (Table 1).

Table 1. Factors affecting the development of electric vehicles

Primary factor	Secondary factor	Notation
Speed of electric vehicle	Extension mileage	x_1
	Charging speed	x_2
	Driving speed	
	Car price	x_3
Infrastructure construction	Number of charging station	x_4
	Power grid construction	
	Public transportation construction	
Government support policy	Preferential policies for manufacturers	
	Preferential policies for consumers	x_4
	Policies on non-electric travel tools	
Others	Per capital GDP	x_5
	Population density	x_6

We numbered these factors and built Multiple Linear Regression Model. The prediction model is: $\hat{y} = a_0 + a_1x_1 + a_2x_2 + a_3x_3 + a_4x_4 + a_5x_5 + a_6x_6$

The regression coefficients of the model are: $a_0, a_1, a_2, a_3, a_4, a_5, a_6$

Let: $Y = (y_1,\ y_2,\ y_3,\ y_4,\ y_5,\ y_6)^T$

$$A = (a_0 \ \cdots \ a_6)^T$$

$$M = \begin{bmatrix} 1 & x_{11} & \cdots & x_{16} \\ 1 & x_{21} & \cdots & x_{26} \\ \vdots & \vdots & \ddots & \vdots \\ 1 & x_{61} & \cdots & x_{66} \end{bmatrix}$$

Then $Y = MA$

According to the least square method, we can get: $A = (M^T M)^{-1} M^T Y$.

3   Case Study

The development of electric vehicles in the United States is relatively mature, so we choose the United States for case analysis.

3.1   The Result of POM Model

First of all, we deal with the point based planning of the charging stations and analyze the distribution of electric cars in the United States.

As we can see from Fig. 1, the greatest influence factor in the United States is the population density. With the increasing density of the population, the number of vehicle ownership is decreasing, especially in areas with high population density and high developed economy.

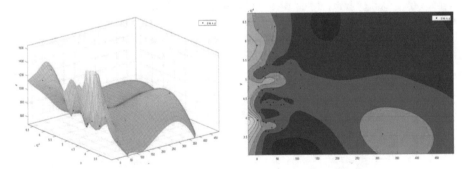

Fig. 1. The density map of the total number of private cars in the United States (x-Population Density, y-Per Capital GDP, z-Car Total)

In fact, the public transportation of these areas is more developed and private travel is inconvenient. In combination with the population and car ownership in the United States, we predict a total need $N_{changer} = 2777778$ to fill the power station.

3.2   The Result of PAM Model

Then we analyze the number of charging stations based on the path. By the end of 2015, the total mileage of the United States was 4115462 miles (Figs. 2 and 3), so we need

24208 supercharging stations. To sum up the number of two charging stations, we can get a total of 2801986 charging stations in the United States, of which 64% are in cities, 23% in suburbs and 13% in rural areas.

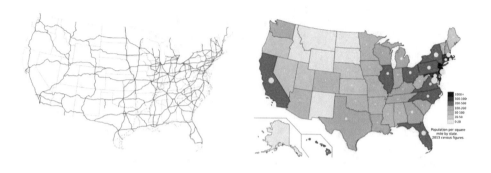

Fig. 2. Road map of United States

Fig. 3. Gravitational maps of American states

3.3   The Result of DSM Model

We analyze the correlation between per capital GDP, population density and car ownership in the United States, and draw the conclusion that $\beta = 0.2199 > 20\%$, we can ignore the impact of economic factors on car distribution. In fact, although there is a gap between rich and poor in the United States, this difference will not have a great impact on whether individuals own cars. As can be seen from the chart below, car ownership increases with population growth in the United States when population density is less than 300 people per square kilometer. When this threshold is exceeded, people tend to reduce private car ownership by public travel (Fig. 4).

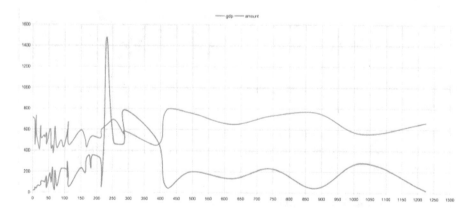

Fig. 4. The change of American car ownership with economy and population density

3.4   The Result of SEM Model

We use the grey Markov chain to predict the development of electric vehicles in the current state and get the initial contrast curve, as shown in Fig. 5. After applying our intervention, we get the development curve after intervention. As can be seen from the figure, the market share will increase significantly after the intervention, and then the production enterprises will recover part of the cost and further invest in technological research and development. Therefore, the emergence of sudden technological progress of electric vehicles ahead of time, further accelerating the market share of electric vehicles. So our intervention is effective.

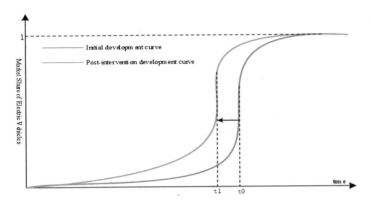

Fig. 5. Development curve of electric vehicle market occupancy

4   Sensitivity Analysis

In the process of development of electric vehicles, there are many factors that will affect them. Typically, there are two kinds: government intervention and technological breakthroughs. Under the influence of these factors, is our model still applicable? In order to verify the validity, we have made sensitivity analysis on government subsidies to Manufacturers and Charging Speed of Charging Station.

4.1   Government Subsidies to Manufacturers

Direct economic subsidy to electric vehicle manufacturers is a common policy in many countries. This policy directly affects the sales price of electric vehicles, but also has a certain impact on enterprise research and development. In order to study the effect of this impact, we use the SEM model to evaluate the penetration rate of EV after the implementation of the strategy. We can see from Fig. 6 that when the government subsidies are moderate, they can have a more obvious positive impact on the market share of electric vehicles in the short term, but in the long run, they have little impact on the time of technological breakthroughs. It is shocking that when the government subsidies are excessive, the short-term impact on market share is very significant, but in the long run, this behavior has delayed the arrival of technological breakthroughs, and the time for electric vehicles to fully occupy the market has also been delayed.

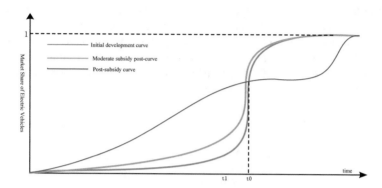

Fig. 6. Development curve after government subsidies to manufacturers

Therefore, we do not recommend that governments directly subsidize producers, which will have a negative impact on market competition.

4.2   Charging Speed of Charging Station

The charging speed mainly meets the needs of people for long-distance travel, that is to say, it improves the endurance of electric vehicles. We use the SEM model for analysis.

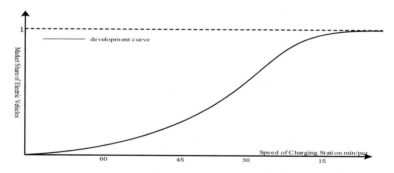

Fig. 7. Effect of charging speed on market share of electric vehicles

As shown in Fig. 7, when the charging speed is controlled within 15 min, it can basically meet all users'long-distance travel needs, that is to say, the charging speed will no longer have a great impact on the popularity of electric vehicles. The revelation of this result is that the charging speed of Tesla Super Charging Station has approached the customer's demand within 30 min, and the large-scale popularization of electric vehicles will occur when the cost is controlled and the production capacity is adequate.

When the charging speed of electric vehicles reaches 15 min, investing a large amount of R&D power in the field of fast charging technology can not bring significant benefits.

5 Prediction of Expressway Traffic Flow by Various Prediction Methods

In this section, we evaluate the prediction accuracy of the dual force model in PAM model. We compare it with BP neural network, RBF neural network and ELNAM neural network.

Two cities in China were selected for model prediction and verification. A city 19391 yuan per person, per capita tourism income 6451 yuan, B city 14663 yuan per person, per capita tourism income 4057 yuan, the relative distance between A and B 6320.

Table 2. Error comparison of different prediction methods

Time	1	2	3	4	5	6	7	8	9	10
Actual value	5019	5206	5018	4539	4887	4952	4936	4959	5234	4869
PA forecast	5122	5517	5500	4665	5109	5082	5227	5092	5448	5282
Error rate (%)	2.05	5.97	9.60	2.79	4.54	2.62	5.90	2.68	4.09	8.48
GM(1, 1) forecast	4193	4176	4160	4144	4128	4112	4096	4080	4069	4033
Error rate (%)	16.5	19.8	17.1	8.69	15.5	16.6	17.0	17.7	22.3	16.9
BP forecast	4617	5061	4625	5046	5174	5538	5539	5590	5363	5378
Error rate (%)	8.1	2.8	7.85	11.2	5.9	11.8	12.3	12.7	2.5	10.4
RBF forecast	4680	5195	5219	4689	4765	4859	4358	4289	5112	5100
Frror rate (%)	6.75	0.21	4.01	3.30	2.50	1.88	11.6	13.5	2.34	4.75
ELMAN forecast	4539	4824	4835	4891	4891	5012	4751	4789	4956	5030
Error rate (%)	9.5	7.34	3.65	7.75	0.08	1.22	3.74	3.43	5.3	3.31

Our formula for calculating the error rate r is:

$$r = \frac{|\,\text{predictedValue - realValue}\,|}{\text{realValue}} * 100\%$$

As can be seen in the Table 2, our dual force model has played a better prediction effect. Unlike other methods, our model predictions are generally larger than the actual values, which may be due to the fact that there are other uncounted roads or uncounted traffic flows between the two cities. In the early stage of development, this excessive forecast will lead to more investment in resources, which can further stimulate the consumer market. But when the market tends to be saturated, it will result in waste of resources.

6 Strategy Suggestion

In our DSM model, we mentioned that the bottleneck of developing new energy vehicles in different countries is different because of different national conditions. In the case study, we analyze the situation of the United States, Ireland, Australia and China, and draw three typical bottlenecks: population density, wealth distribution and altitude. In view of these bottlenecks, we hypothesize many strategies, and through the evaluation of SEM model, we give the best development strategy.

6.1 Population Density Based Planning Approach

When the population density is the main factor that affects the consumption of electric vehicles, we should give priority to the development of charging facilities in more populous areas, so we first build the charging stations in the urban area. In addition, if charging facilities are widely built at this stage, a lot of charging stations will be idle, so it is suitable to encourage consumption and then improve infrastructure.

6.2 Wealth Distributions Based Approach

When economic factors become the main factors that affect the promotion of electric vehicles, it means that the promotion of electric vehicles depends on the purchasing power of the consumers. So at this time, we should give priority to the construction of the charging pile in the city and promote and encourage the consumption first, and then promote the construction of the charging facilities.

6.3 Altitude Based Planning Approach

When we say that time geography as the main factor, it means the desire of buying electric car will not change too much for the reason of economic. This time both city residents and rural residents or the suburbs have Car Buying demand, so that we should built Station at everywhere has enough people living. At this time, the residents have the ability to consume. It should be perfected to stimulate the desire of consumption by improving the related facilities. So the charging station should be repaired first and encourage consumption.

7 Directions for Future Research

Like all studies, our findings have limitations. First of all, our strategy selection model uses relative influence to judge the current main bottleneck. In fact, these factors are based on our subjective assumptions and then compared. This may cause some unknown but influential factors to be omitted by us, thus affecting our accurate decision-making. Secondly, in the process of our case study, there are many inaccurate data, and it is very difficult to obtain, which has caused great obstacles to our research. On the first issue, we hope that future researchers can use some improved methods to measure the impact of some potential factors effectively; on the second issue, we suggest that future researchers can conduct a large number of effective experiments, through the experimental data for subsequent analysis and demonstration.

Acknowledgement. This work was supported by the Zhongnan University of Economics and Law (2722019JCT035,2722019JCG074), the National Natural Science Foundation of China (61602518), and the Fundamental Research Funds for the Central Universities National Social Science Fund of China (NO:16CXW019).

References

Ren, Y., Shi, L., Zhang, Q., et al.: Study on optimal distribution and scale of electric vehicle charging station. Autom. Electr. Power Syst. **35**(14), 53–57 (2011)

Liu, Z., Zhang, W., Wang, Z.: Optimal layout of urban electric vehicle charging station based on quantum particle swarm optimization algorithm. China Electr. Eng. **32**(22), 39–45 (2012)

Feng, C., Zhou, B., et al.: Application of integrated comprehensive evaluation method in optimal decision of electric vehicle charging station location. Power Autom. Equip. **3**(9), 25–29 (2012)

Xi, X., Sioshansi, R., Marano, V.: Simulation-optimization model for location of a public electric vehicle charging in- frastructure. Transp. Res. Part D: Transp. Environ. **22**(4), 60–69 (2013)

Chu, Y.J., Ma, L., Zhang, H.Z.: Location-allocation and its algorithm for gradual covering electric vehicle charging stations. Math. Pract. Theory **10**, 101–106 (2015)

Zhu, J., Wang, H., Li, Q.: Research on competitive location problem of charging station considering gravitational factors and time satisfaction. Math. Pract. Theory **48**(24), 59–65 (2018)

https://www.statista.com/

http://www.stats.gov.cn/

Realtime Road State Decision System Based on Multiple Sensors and AI Technologies

Yoshitaka Shibata[1(✉)], Akira Sakuraba[1(✉)], Goshi Sato[2(✉)], and Noriki Uchida[3(✉)]

[1] Iwate Prefectural University, Takizawa, Japan
{shibata,a_saku}@iwate-pu.ac.jp
[2] Resilient ICT Research Center, Sendai, Japan
sato_g@nict.go.jp
[3] Fukuoka Institute Technology, Fukuoka, Japan
n-uchida@fit.ac.jp

Abstract. This paper introduces a realtime road state decision system in both urban and country roads based on various typed environmental sensors. Those sensors are installed on the vehicle and collects the sensor data with various road surfaces conditions while running along the street. From the collected sensor data, the road surface states are decided such as dry, wet, snowy, icy by sensor server. The decided road states can be share with the many vehicles through V2 V and V2I communication protocols. In this paper, the system configuration and decision method of road state conditions in realtime are introduced. Performance evaluation of the prototype with the proposed system is carried out to verify the effect of our suggest method.

1 Introduction

As progress of aging and declining birthrate in Japan, the number of population is rapidly decreasing particularly in local areas [1, 2]. Although the scale of social and economic activities is shrinking and public transportation services getting worse, many old people in local areas still continue to stay in their original places. Therefore, transportation or mobility is the most important means to maintain safe and reliable daily life and economic activity. For this reason, even the people are getting old and driving ability are getting worse, they have to drive their cars to continue their daily lives. In fact, the number of the traffic accident by aging population more than 75 years old increases on bad road condition year and year and serious problem.

Secondly, in the cold or snow countries, such as northern countries in Japan, most of the road surfaces are occupied with heavy snow and iced surface in winter and many slip accidents occurred even though the vehicles attach snow specific tires. In fact almost more 90% of traffic accidents in northern part of Japan is caused from slipping car on snowy or iced road. In those cases, traffic accidents are rapidly increased. Safer and more reliable road monitoring and warning system which can transmit the road condition information to drivers is indispensable before passing through the dangerous road area.

L. Barolli et al. (Eds.): CISIS 2019, AISC 993, pp. 114–122, 2020.
https://doi.org/10.1007/978-3-030-22354-0_11

On the other hand, recently self-driving cars have been emerged and tested running on both public ordinal roads and highways in well developed countries. On the road condition of those self-driving is ideal with flat, clear condition, clear lane with good weather condition without no obstacles on road. However, on the other hand, the road infrastructures in many developing countries are not well maintained compared with well developed countries due to luck of regular road maintenance for such cases where falling objects from other vehicles, overloaded trucks or buses are frequently occurred. Therefore, in order to maintain safe and reliable self-driving, the vehicles have to detect and avoid those obstacles in advance when they pass through (Fig. 1).

Fig. 1. Road condition in snow country

In order to resolve those traffic problems, we introduce a new generation wide area road surface state information platform based on crowd sensing and V2X technologies [3, 4]. In crowd sensing, many data from many vehicles with various environmental sensors including accelerator, gyro sensor, infrared temperature sensor, quasi electrical static sensor, camera and GPS data are precisely detected the various road surface states and identify the dangerous locations on GIS in reatime. The road information are transmitted to the neighbor vehicles and road side server V2X communication network [5–7].

In V2X communication on the actual road, both the communication distance and the total size of data transmission must be maximized at the same time when vehicles are running on the road. The conventional single wireless communication such as Wi-Fi, IEEE802.11p, LPWA, cannot satisfy those conditions at the same time. In order to resolve such problems, N-wavelength cognitive wireless communication method is newly introduced in our research [8, 9]. Multiple next generation wireless LANs including IEEE802.11ac/ad/ah/ai/ay in addition to the current popular LANs with different wavelengths are integrated to organize a cognitive wireless communication. The best link of the cognitive wireless is determined by Software Defined Network (SDN).

Thus, the sensor data and the road state information can be exchanged to the neighbor vehicles and road side server and indicated on the graphical interface as GIS data. By checking the road state on the graphical user interface, the driver can safely and reliably run on the road by paying attention or avoiding the dangerous point. The detected data are collected to the global cloud computing system as bigdata. By combining those road state data and open data with weather and geological data, wide area road state can be predicated and indicated as GIS system.

In the following, general system and architecture of Road Surface State Information Platform are explained in Sect. 2. The V2X communication system and its function are explained in Sect. 3.

The sensing system and its functions with various sensors are precisely shown in Sect. 4. The prototype system to evaluate function and performance of the proposed sensing system is explained in Sect. 5. Finally conclusion and future works are summarized in Sect. 6.

2   Road Surface State Information Platform

In order to resolve those problems in previous session, we introduce a new generation wide area road surface state information platform based on crowd sensing and V2X technologies as shown in Fig. 2. The wide area road surface state information platform mainly consists of multiple road side wireless nodes, namely Smart Relay Shelters (SRS), Gateways, and mobile nodes, namely Smart Mobile Box (SMB). Each SRS or SMB is furthermore organized by a sensor information part and communication network part. The sensor information part includes various sensor devices such as semi-electrostatic field sensor, an acceleration sensor, gyro sensor, temperature sensor, humidity sensor, infrared sensor and sensor server. Using those sensor devices, various road surface states such as dry, rough, wet, snowy and icy roads can be quantitatively decided.

On the other hand, the communication network part integrates multiple wireless network devices with different N-wavelength (different frequency bands) wireless networks such as IEEE802.11n (2.4 GHz), 11ac (5.6 GHz), 11ad (28 GHz), ah (920 MHz) and organizes a cognitive wireless node. The network node selects the best link of cognitive wireless network depending on the observed network quality by Software Defined Network (SDN). If none of link connection is existed, those sensing data are locally and temporally stored until approaches to another mobile node or road side node, and starts to transmit sensor data by DTN Protocol. Thus, data communication can be attained even though the network infrastructure is not existed in challenged network environment such as mountain areas or just after large scale disaster areas.

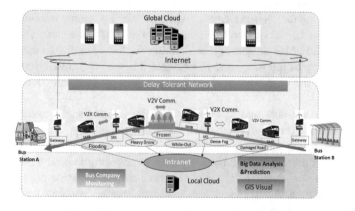

Fig. 2. Road surface state information platform

3 V2X Communication Between SMBs and SRS

In Fig. 3 shows V2X communication method between the SMB of a vehicle and the SRS of road side server. First, one of the wireless networks with the longest communication distance can first make connection link between SMB and SRS using SDN function. Through the this connection link, the communication control data of other wireless networks such as UUID, security key, password, authentication, IP address, TCP port number, socket No. are exchanged. As approaching each other, the second wireless network among the cognitive network can be connected in a short time and actual data transmission can be immediately started. This transmission process can be repeated during crossing each other as long as the longest communication link is connected. This communication process between SMB and SRS is the same as the communication between SMB to other SMB except for using adhoc mode.

In our system, SRS and SMB organize a large scale information infrastructure without conventional wired network such as Internet. The SMB on the car collects various sensor data including acceleration, temperature, humidity and frozen sensor data as well as GPS data and carries and exchanges to other smart node as message ferry while moving from one end to another along the roads.

On the other hand, SRS not only collects and stores sensor data from its own sensors in its database server but exchanges the sensor data from SMB in vehicle nodes when it passes through the SRS in road side wireless node by V2X communication protocol. Therefore, both sensor data at SRS and SMB are periodically uploaded to cloud system through the Gateway and synchronized. Thus, SMB performs as mobile communication means even through the communication infrastructure is challenged environment or not prepared.

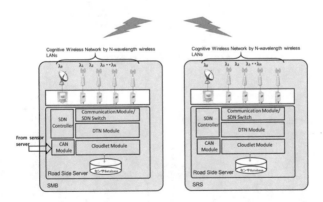

Fig. 3. V2X communication method between the SMBs and SRS

4 Sensing System with Various Sensors

In order to detect the precise road surface conditions, such as dry, wet, dumpy, showy, frozen roads, various sensing devices including accelerator, gyro sensor, infrared temperature sensor, humidity sensor, quasi electrical static sensor, camera and GPS are integrated to precisely and quantitatively detect the various road surface states and determine the dangerous locations on GIS in sensor server as shown in Fig. 4. The sensor server periodically samples those sensor signals and performs AD conversion and signal filtering in Receiver module, analyzes the sensor data in Analyzer module to quantitatively determine the road surface state and learning from the sensor data in AI module to classify the road surface state as shown in Fig. 5. As result, the correct road surface state can be quantitatively and qualitatively decided. The decision data with road surface condition in SMB are temporally stored in Regional Road Condition Data module and mutually exchanged when the SMB on one vehicle approaches to other SMB. Thus the both SMBs can mutually obtain the most recent road surface state data with just forward road. By the same way, the SMB can also mutually exchange and obtain the forward road surface data from road side SRS.

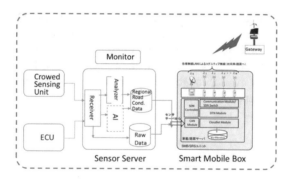

Fig. 4. Sensor server system

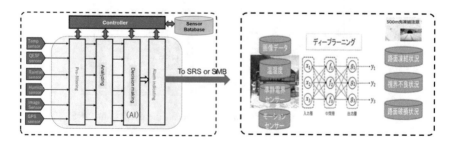

Fig. 5. Analyzer module by AI

5 Prototype System and Evaluation

In order to verify the effects and usefulness of the proposed system, a prototype system is constructed and those functional and performance are evaluated. The prototype system with sensor server system and communication server system of both SRS and SMB for two-wavelength communication is shown in Fig. 6. We currently use OiNET-923 of Oi Electric Co., Ltd. for 920 MHz with, WI-U2-300D of Buffalo Corporation for of 2.4 GHz and T300 of Ruckus for 5.6 GHz as the prototype of two-wavelength communication. OiNET-923 is used as control data communication channel including UUID, security key, password, authentication, IP address, TCP port number, socket No. are exchanged. On the other hand, both WI-U2-300D and T300 are used for sensor data transmission channel. Raspberry Pi3 Model B+ is used for N-wavelength cognitive communication to perform cognitive controller and SDN function. Intel NUC Core i7 is used for sensor data storage and data analysis by AI based road state decision.

On the other hand, in sensor server system, several sensor including BL-02 of Biglobe as 9 axis dynamic sensor and GPS, CS-TAC-40 of Optex as far-infrared temperature sensor, HTY7843 of azbil as humidity and temperature sensor and RoadEye of RIS system and quasi electrical static field sensor for road surface state are used. Those sensor data are synchronously sampled with every 10 ms. and averaged every 1 s to reduce sensor noise by another Raspberry Pi3 Model B+ as sensor server. Then those data are sent to Intel NUC Core i7 which is used for sensor data storage and data analysis by AI based road state decision. Both sensor and communication servers are connected to Ethernet switch.

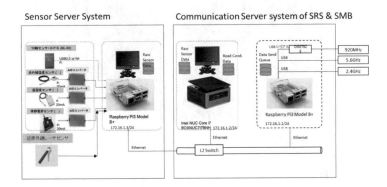

Fig. 6. Realtime road surface data exchange system

In order to evaluate the sensing function and accuracy of sensing road surface condition, both sensor and communication servers and various sensors are set to the vehicle as shown in Fig. 7. We ran this vehicle about 4 h to evaluate decision accuracy in realtime on the winter road with various road conditions such as dry, wet, snowy, damp and icy condition around our campus in winter. In order to evaluate the road surface decision function, the video camera is also used to compere the decision state and the actual road surface state. The Fig. 8 and 9 show a typical image shot of the graphical user interface on test running and their sensor values. Through the test running video and other sensor data, our sensing system could identify the real winter road states more than about 80% as accuracy. We also output the friction of the road surface. This is very important for drivers to easily understand weather the running the winter road is slippery or not.

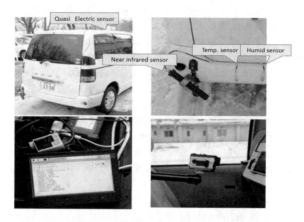

Fig. 7. Sensing vehicle in winter road

Fig. 8. Screen shot of graphical user interface

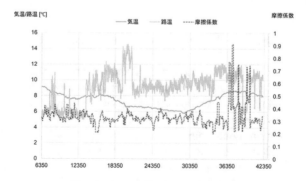

Fig. 9. Screen shot of sensor data

6   Conclusions and Future Works

In this paper, we introduce a realtime road state decision system for both urban and local roads based on various typed environmental sensors. Those sensors can be easily attached to any types vehicles to collect the sensor data with various road surfaces conditions while running along the street for all seasons. From the collected sensor data, the road states can be real-timely and easily decided such as dry, wet, damp, snowy, icy or even friction rate on road surface by AI based decision making. The decided road states information can be shared with the many vehicles through V2 V and V2I communication protocols. On order to verify the suggested system, prototype system is constructed and ran the vehicle on the actual road in winter. Through the performance evaluation, the effects and usefulness of the proposed system could be verified.

As the future works of our research, more sensing data for wider areas, in local and urban areas, on different weather conditions should be corrected as training data for AI decision model to improve the decision accuracy. Social experiment for the actual driving operations such as bus, taxi, renter car enterprises is also required. Furthermore, wide area road state GIS by combining those road state data and open data with weather and geological data should be be predicated and indicated as GIS system near future.

Acknowledgement. The research was supported by Strategic Information and Communications R&D Promotion Program Grant Number 181502003 by Ministry of Affairs and Communication.

References

1. A 2016 Aging Society White Paper. http://www8.cao.go.jp/kourei/whitepaper/w-014/zenbun/s1_1_1.html
2. A 2016 Decling Birthrate White Paper. http://www8.cao.go.jp/shoushi/shoushika/whitepaper/measures/english/w-2016/index.html
3. Shibata, Y., Sato, G., Uchida, N.: A new generation wide area road surface state information platform based on crowd sensing and v2x technologies. In: The 21th International Conference on Network-Based Information Systems (NBiS2018), Lecture Notes on Data Engineering and Communications Technologies book series (LNDECT, volume 22), pp. 300–308, September 2018
4. Shibata, Y., Sakuraba, A., Sato, G., Uchida, N.: IoT based wide area road surface state sensing and communication system for future safety driving. In: The 33rd International Conference on Advanced Information Networking and Applications (AINA2019), pp. 1123–1132, March 2019
5. Ito, K., Hirakawa, G., Shibata, Y.: Experimentation of V2X communication in real environment for road alart information sharing system. In: IEEE AINA2015, pp. 711–716, March 2015
6. Otomo, M., Sato, G., Shibata, Y.: In-Vehicle cloudlet computing based delay tolerant network protocol for disaster information system. In: Advances on Broad-Band Wireless Computing, Communication and Application Applications, Vol. 2 of the series Lecture Notes on Data Engineering and Communications Technologies, pp. 255–266, October 2016

7. Shibata, Y., Ito, K., Noriki, G.: Mobility information infrastructure by a new n-wavelength wireless communication method and IoT road condition technology. In: The 16th ITS Asia-Pacific Forum Fukuoka 2018, ITS24, 133, May 10, 2018
8. Shibata, Y., Ito, K., Uchida, N.: A new V2X communication system to realize long distance and large data transmission by n-wavelength wireless cognitive network. In: The 32nd IEEE AINA 2018, pp. 587–592, Krakow, Poland, May 2018
9. Shibata, Y., Sato, G., Uchida, N.: A prototype system for V2X communication platform for safety driving by IoT technology. In: The 13th International Conference on Advances on Broadband and Wireless Computing, Communication and Applications, Lecture Notes on Data Engineering and Communications Technologies book series (LNDECT, volume 25), pp. 381–390, October 2018

From BPMN Models to SoaML Models

Abderrahmane Leshob[1,2]([✉]), Redouane Blal[1,2], Hafedh Mili[1], Pierre Hadaya[2],
and Omar Khadeer Hussain[3]

[1] LATECE Laboratory, University of Quebec at Montreal, Montreal, Canada
{leshob.abderrahmane,mili.hafedh}@uqam.ca, blal.redouane@courrier.uqam.ca
[2] UQAM School of Management (ESG UQAM), Montreal, Canada
hadaya.pierre@uqam.ca
[3] School of Business, University of New South Wales, Canberra, Australia
o.hussain@adfa.edu.au

Abstract. Organizations build information systems to support their business processes. Today's business processes often cross the organizations' boundaries and become increasingly complex. Therefore, information systems that automate these business processes must take into account collaborative and complex scenarios involving distributed partners. Designing such systems is not trivial considering: (i) the complexity of the cross-organizational business processes, and (ii) the large gap between business processes and information systems. To address this gap, this paper relies on the service-oriented architecture (SOA) paradigm to propose an end-to-end method to design SOA-based information systems from business process models. More precisely, this paper proposes to generate SOA design models expressed in SoaML from the specifications of a collection of organizations' private processes expressed in BPMN.

1 Introduction

Today's global markets environment tends to increase the complexity of business collaborations. Business processes have become more complex and often cross the boundaries of organizations [1]. The processes that cross the organizations' boundaries are called *Cross-organizational business processes* (COBP). Within the context of global supply chains, an end-to-end sales process will typically involve several business partners–besides the buyer and the seller–including carriers, brokers, financial institutions, and the like. To handle the complexities of such processes, organizations need to build information systems (IS) that should at least, manage the information flows between the business partners, and ideally, automate many of the tasks of these interorganizational processes. Despite many advances in ERP systems and IS engineering, developing such systems remains costly and time-consuming. Researchers and practitioners alike, have recognized that similarities between business processes –be they private or cross-organizational– typically translate into similarities between the information systems that support them. A number of approaches have capitalized on this relationship to build *abstract high-level* IS models from the *business/process* models that they need to support [2–8]. However, to the best of our knowledge,

© Springer Nature Switzerland AG 2020
L. Barolli et al. (Eds.): CISIS 2019, AISC 993, pp. 123–135, 2020.
https://doi.org/10.1007/978-3-030-22354-0_12

there is no end-to-end method that specifies *concrete detailed* IS models from the specifications of organizations' private business process[1] models.

In [9,10], we proposed a semi-automatic method to build architectural models using the SOA paradigm from ready-to-use business process models expressed in BPMN. The present work aims to go a step further. Its contribution is twofold: (1) from a usability point of view, this research improves the work presented in [9,10] by eliminating the complex step that decomposes the collaborative process models and simplifying SOA services identification; (2) from a structural perspective, this research proposes an end-to-end method that takes as input a collection of organizations' private process models instead of ready-to-use collaborative process models.

The remaining of this paper is organized as follows. Section 2 describes the approach to identify and specify SOA-based services out of a collection of private process models. Section 3 illustrates the proposed approach in the context of a B2B collaboration. Section 4 surveys related work. Section 5 draws our conclusions and summarizes our envisioned further work.

2   Proposed Approach

As shown in Fig. 1, the proposed method is composed of two phases. The first phase builds the COBP model out of a collection of private process models and prepares the resulting process model for the next phase. The second phase identifies the SOA-based services and specifies them using the SoaML language. The following subsections detail each of these two phases.

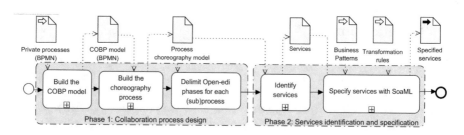

Fig. 1. Overall process for SOA services specification from BPMN models.

2.1   Phase 1: Collaboration Process Design

The goal of the first phase is to design the collaborative process (COBP) model and transform it to facilitate the identification and the specification of SOA services. This phase starts by connecting together the input private process models to build a COBP model that allows the organizations involved to collaborate.

---

[1] A private business process is a process that occurs within the boundaries of the organization.

Then, it uses the obtained COBP model to build a choreography model in order to allow the identification of services. Finally, it delimits the *Open-edi phases* [11] to enable the specification of the identified services.

COBP Modeling: According to Hammer and Champy [12], a business process is a set of activities that, together, produce a result of value to the customer. Process-oriented organizations rely on business processes to support their business functions. Today's organizations use their business processes to collaborate through what we call COBP. Designing a COBP model is complex and requires that designers have extensive experience, especially when the private processes of the organizations involved are incompatible [13]. To build a COBP model from a collection of organizations' private process models, we use the approach presented in [13]. This approach consists of four steps that connect the private processes together to build a new COBP allowing partners to collaborate in the context of a shared business goal. The first step analyzes the private processes in order to connect them using messages. The second step identifies adaptation patterns[2] to resolve any interoperability issues found during the first step. The third step applies these patterns to resolve identified issues, if any. The last step builds the COBP model that will allow the organizations to collaborate.

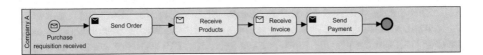

Fig. 2. Company A private process model.

Fig. 3. Company B private process model.

Let us consider the procurement process of Fig. 2 from company A and the sales process of Fig. 3 from company B. Figure 4 shows the model of the cross-organizational business process that supports the collaboration between Company A and B after connecting the two private processes of Figs. 2 and 3.

---

[2] In [13], we identified six process adaptation patterns that resolve incompatibilities when connecting private processes to build COBP models.

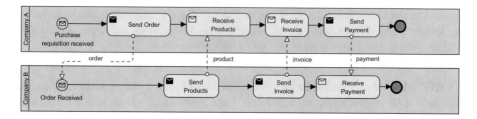

Fig. 4. Cross-organizational business process model.

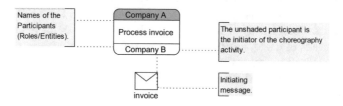

Fig. 5. A choreography task.

Choreography Process Model Design: A choreography model is a type of process that formalizes the way business partners coordinate their interactions [14]. This model focuses on the exchange of information (Messages) between the partners rather than the orchestration of the activities performed by each partner [14]. Choreography models have activities (Tasks) that consist of one or more interactions between partners. A *Choreography Task* represents one or more message exchanges. A single choreography task may use more than one message. When that is the case, it is important to know the first (initiating) message of the interaction. For return messages, the symbol of the message is shaded with a light fill [14]. Figure 5 shows an example of a choreography task representing the exchange of the entity INVOICE between Company A and Company B of Fig. 4.

To build the choreography model from a BPMN COBP model, we simply capture the messages exchanged between the BPMN participants (pools and swimlanes). The orchestration of the choreography tasks corresponds to the sequence flow defined in the BPMN COBP model.

Open-Edi Phases Demarcation: According to the International Organization for Standardization (ISO) *Open-edi model* [11], a business transaction moves through the following five phases: *planning, identification, negotiation, actualization, and post-actualization.* The planning and identification phases involve the activities that identify the partners and the resources that will be offered/consumed during the collaboration. The purpose of the negotiation phase is to achieve an explicit contract for the business transaction. The actualization phase encompasses the activities needed to exchange the resources. The post-actualization phase involves the activities after the exchange.

Our approach uses the Open-edi reference model to map SOA services to business patterns in order to enable their specification. For that, we have identified and adapted patterns from [15–17] for each Open-Edi phase. To delimit the Open-edi phases for each (sub)process/business transaction, our approach uses a semi-automatic approach that asks the user (e.g., a business analyst) to:

1. Annotate the BPMN choreography model to distinguish the resources exchanged (products, services, etc.) and the exchange contracts (e.g., purchase order, distribution order) of each business transaction.
2. Answer a question about the existence of an agreement that governs the exchange and another about the last activity that establishes the one-to-one link between the partners for each business transaction.

Then, we proceed as follows for each process. The actualization phase boundaries are set using the start and end of resource transfer activities (resources are annotated). The post-actualization phase starts after all resource transfers are finished. If activities are detected before the actualization phase, unless an agreement governs the actual process, the activity that marks the establishment of a *one-to-one link* between the partners determines the start of the negotiation phase. The end of the negotiation phase matches the start of the actualization phase. The planning/identification[3] phase includes activities performed before the negotiation phase.

2.2 Phase 2: Identification and Specification of Services

Identification of Services: A service is a *value* delivered to a business partner in exchange for some value [18]. Services allow organizations to get something done without doing it or even without knowing how to do it, enabling organizations to be more efficient and agile [18]. A service results in work provided to one party by another. Its value is delivered and available to partners through an interface.

We had two options to identify services. The first option consists of extracting a service per message from the BPMN model. This design option however identifies too many nano-services (fine grained services) with granular responsibilities. According to [19], this is a design anti-pattern called FINE GRAINED SERVICES. The second option identifies one service per Open-edi phase (e.g., a service for the actualization phase). This option identifies services that may contain a large number of methods. According to [19], this is a design anti-pattern called GOD OBJECT SERVICES. Thus, we discarded these two ineffective and bad design solutions.

To identify services with adequate coverage within the boundaries of each collaboration phase while avoiding both anti-patterns (i.e., FINE GRAINED SERVICES and GOD OBJECT SERVICES), our solution identifies a *single service per*

---

[3] In this work, we merged the planning and identification phases into a single open-edi phase.

choreography task. This design option is motivated by the fact that each choreography task represents a set of one or more exchanged messages ("One Way" or "Request-Response").

To allow the specification of identified services in the next step, each service has to be linked with one or more business patterns based on the *Open-edi reference model*. Therefore, services supporting choreography tasks that exchange resources during the actualization phase (i.e., products, services, etc.) are associated with the EXCHANGE PATTERN (see [17]). Services that support the claim[4] management (e.g., invoicing) during the actualization phase use the CLAIM/CLAIM MATERIALIZATION PATTERN (see [17]). Services supporting the exchange contract activities (e.g., Purchase order, Sales order) between the partners during the negotiation phase use the COMMITMENT and CONTRACT PATTERNS (see [17]). These mapping between services and business patterns are not fully automated yet. For the time being, the mapping between actualization phase services and their corresponding patterns is automated. For the remaining services, we rely on user annotations.

Specification of Services: Our approach uses the Service oriented architecture Modeling Language (SoaML) to specify the identified services. The SoaML specification provides a metamodel and a UML profile for the specification and design of services within SOA [18]. There are three approaches to specify a service with SoaML [18]:

1. UML SIMPLE INTERFACE: It focuses on a one-way interaction provided by a participant on a port represented as a UML interface (e.g., RPC style web services).
2. SERVICE INTERFACE: Allows to specify bi-directional services i.e., there are "callbacks" from the provider to the consumer as part of a conversation between the parties. A service interface is defined in terms of *the provider of the service*. It specifies: (i) the interface that the provider offers, and (ii) the interface, if any, it expects from the consumer. The latter agrees to use the service as defined by its service interface, and the provider agrees to provide the service according to its service interface.
3. SERVICE CONTRACT: Specifies how partners (providers, consumers and other roles) work together to exchange values. The service contract defines the roles that each partner (participant) plays in the service and the interfaces they implement. The service contract represents an agreement between the participants. It defines how the service is provided and consumed. This agreement includes the choreography, interfaces and any conditions. The fundamental difference between the contract and interface based approaches is that the interactions between participants are defined separately from the participants

---

[4] When the exchange of the resources between the partners do not happen simultaneously, the exchange is out of balance for a certain period of time. This temporary imbalance results in a claim between partners. An Invoice is an example of claim materialization.

in a SERVICE CONTRACT while these interactions are defined individually in each participants' service and request using the SERVICE INTERFACE approach.

To specify the identified services, we rely on the structure and behavior of a set of business patterns from [15–17]. Each service is associated with one or more business patterns. Thus, to specify each service, we use transformation rules that rely on the specification of its associated business patterns. Each service specification results in transformations that model its interfaces through SoaML SERVICE INTERFACES approach.

3   Example

This section illustrates the proposed method in the context of a B2B collaboration between the Woods company (the supplier) that provides the Construction company (the purchaser) with raw materials (woods). Figures 6 and 7 show the private business process of the Woods company and that of the Construction company respectively.

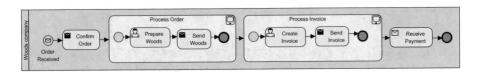

Fig. 6. Woods company sales process.

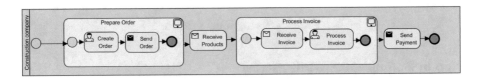

Fig. 7. Construction company purchase process.

Step 1: Build the Collaboration Model

To build the COBP model, we need to connect the private processes of Figs. 6 and 7 together using the four-steps approach presented in [13]. In this case, we simply need to connect the sending activities to the corresponding receiving activities[5].

Figure 8 shows the BPMN process model that handles the collaboration between the Woods company and the Construction company.

---

[5] The confirmation is a reply message. We had the option of either ignoring it by applying the SEZE pattern (see [13]) or connecting it to the (sub)process containing the activity that sends the original message. We chose the second option.

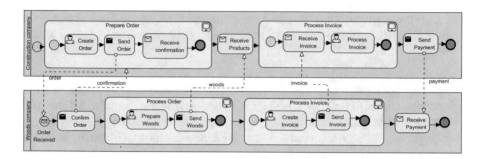

Fig. 8. Collaboration process model.

Step 2: Choreography Process Model Design

The approach to build the choreography model from a COBP model consists of analyzing the messages exchanged between the BPMN participants. Thus, we create a choreography task for each of the exchanged messages between the collaborating partners. Figure 9 shows the choreography model that corresponds to the collaboration process of Fig. 8.

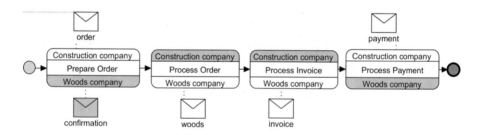

Fig. 9. Choreography process.

Step 3: Delimit OPEN-Edi Phases

To delimit the Open-edi phases, we use the semi-automatic approach presented in Sect. 2.1. First, the user annotates the choreography model to distinguish: (i) the exchanged resources (i.e., woods and payment), and (ii) the exchange contract (i.e., order). Second, the user answers two questions: the first is about the existence of an agreement that governs the exchange between the partners while the second identifies the activity that establishes the one-to-one link between the partners. In this process, the Construction company sends a purchase order directly to the Woods company without looking for potential partners through, for example, a bidding process. Thus, the activity that establishes the one-to-one link between the partners is not needed. The identification phase is therefore empty. The actualization phase boundaries are set using the start and end of the resources (i.e., woods and payment) transfer activities. The negotiation phase encompasses all activities before that. Figure 10 shows the boundaries of the Open-edi phases.

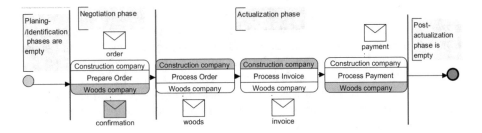

Fig. 10. Open-edi phases.

Step 4: Identify Services

Our approach to identify the services uses the choreography diagram. Each choreography task is supported by a single service. Thus, four services are identified. One service to manage the negotiation phase and three services for the actualization phase: two of these services support the activities that exchange economic resources (i.e., woods and payment) and one service to manage the invoicing process. The services that manages the resources exchange activities during the actualization phase will be linked to the exchange pattern (see [17]). The service that manages the order during the negotiation phase is linked to the contract pattern (see [17]). The service that manages the invoicing activity during the actualization phase will be linked to the patterns Claim and Claim Materialization (see [17]). Note that, the process of linking the business patterns to the identified services is not fully automated yet. Table 1 shows the four services that support the collaboration process and their corresponding patterns.

Table 1. Identified services.

Service	Open-edi Phase	Patterns
Order Service	Negotiation	Contract/Commitment
Woods Transfer Service	Actualization	Exchange
Payment Transfer Service	Actualization	Exchange
Invoice Service	Actualization	Claim/Claim Materialization

Step 5: Specify Services

The previous step identified the services and their associated patterns. In this step, we will show how to specify the Order Service using the SoaML SERVICE INTERFACE approach. The service interface is defined from the perspective of the service provider (i.e., Woods company). It shows: (i) the provided and required Interfaces, (ii) the ServiceInterface class, and (iii) the Behavior.

To specify the Order Service, we use the Contract and Commitment patterns [17]. A contract is a collection of commitments and terms. A commitment in a business transaction is a promise of a participant to provide or receive a resource in the future. Terms are potential commitments that are created if certain conditions are met, such as when a commitment is not being fulfilled. Figure 11 shows the Service Interface definition of the ORDER SERVICE from the perspective of the service provider. The classifiers ORDER RECEIVER and ORDER CONSUMER, represent respectively the interfaces for the provider (Woods Company) and the consumer (Construction company). Figure 12 shows the service behavior using a UML communication diagram.

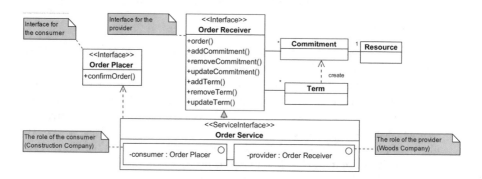

Fig. 11. Service Interface.

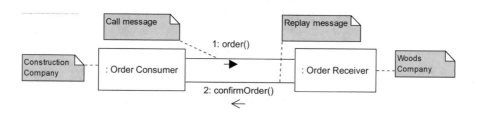

Fig. 12. Service Behavior.

4   Related Work

These last few years, many research efforts were conducted to develop methods that build software models from the specification of business processes. In [8], Rodríguez *et al.* proposed a model-driven development (MDD) approach based on transformations from (*Computation Independent Model*, CIM) models defined in BPMN to derive analysis-level classes and use cases. In [20], Coad *et al.* proposed a question-based development method to build UML models from generic enterprise-component models (ECM). ECM models define a fundamental model shape for supporting some aspects of business processes. In [3], Cruz *et al.* proposed an approach to identify system functional requirements by extracting

UML use case models from BPMN models. Later, in [4], Cruz *et al.* proposed a new approach to generate aggregated data models expressed in UML from a set of interrelated BPMN models.

Authors in [2], proposed a semi-automatic method to support software designers in the process of analyzing collaborative business process models in order to identify functionalities that can be exported as services. The proposed method starts with an analysis of value and task dependencies of the process to identify candidate services. Candidate services are refined afterward before being conciliated. However, this method is limited to service identification. In [21], Gonzalez-Huerta *et al.* presented an interesting three-steps transformation approach to derive software models from refined BPMN models. The method elaborates the to-be process model by applying process re-engineering patterns to the original BPMN models and then generates software models. However, this approach offers limited software models derivation options.

Authors in [5] presented an MDD approach including a tool to support automated transformations based on a set of one-to-one mapping between BPMN models and SoaML. In [6], authors proposed a similar but semi-automated method where CIM-level BPMN models are analyzed through a set of mapping rules in order to obtain PIM-level models presented as SoaML service constructs and architectures models. Nevertheless, service models obtained using model-to-model transformations based on mapping rules are still lacking business semantics. In [7], authors proposed an approach for translating BPMN models into *Business Process Execution Language* (BPEL). The approach consists of a more comprehensive an improved method of existing BPMN-to-BPEL translations. Their method relies on mapping between identified block-structured patterns from the BPMN models and the block-structured BPEL constructs.

5 Conclusion and Future Work

Today's organizations collaborate through the use of complex business processes. To support these processes, organizations build ISs that must take into account complex collaborative scenarios. Designing such IS is a difficult and time consuming task considering the large gap between business processes and IS [10]. An emerging approach to design such IS is to use the SOA paradigm given its many benefits.

The purpose of this work is to assist organizations in the process of designing SOA-based IS taking as input a collection of private business process models. To this end, we propose a model-driven development method to design SOA services using the SoaML language from the specification of business processes expressed in BPMN. This work improves the approach presented in [9,10] by proposing: (i) a method that is easy to understand and implement and (ii) an end-to-end method that takes a collection of private process models as inputs instead of ready-to-use collaborative process models.

This work is part of a long-term research program that consists of providing organizations with a method and tools that characterize the transformations from process models to SOA models. The next challenges we face are: (i) evaluating the method, (ii) automating the mapping between identified services and business patterns that allow their specification, (iii) developing a web tool that supports the method, and (iv) extend the method to support flexible processes.

Acknowledgments. This research was supported by the Natural Sciences and Engineering Research Council of Canada (NSERC).

References

1. Zeng, Q., Sun, S.X., Duan, H., Liu, C., Wang, H.: Cross-organizational collaborative workflow mining from a multi-source log. Decis. Support Syst. **54**(3), 1280–1301 (2013)
2. Bianchini, D., Cappiello, C., De Antonellis, V., Pernici, B.: Service identification in interorganizational process design. IEEE Trans. Serv. Comput. **7**(2), 265–278 (2014)
3. Cruz, E.F., Machado, R.J., Santos, M.Y.: From business process models to use case models: a systematic approach. In: 4th Enterprise Engineering Working Conference, EEWC 2014. LNBIP, vol. 174, pp. 167–181 (2014)
4. Cruz, E.F., Machado, R.J., Santos, M.Y.: Deriving a data model from a set of interrelated business process models. In: Proceedings of 17th International Conference on Enterprise Information Systems, ICEIS 2015, vol. 2, pp. 49–59 (2015)
5. Delgado, A., De Guzman, I.G.R., Ruiz, F., Piattini, M.: From BPMN business process models to SoaML service models: a transformation-driven approach. In: Proceedings of 2010 2nd International Conference on Software Technology and Engineering, ICSTE 2010, vol. 1 (2010)
6. Elvesaeter, B., Panfilenko, D., Jacobi, S., Hahn, C.: Aligning business and IT models in service-oriented architectures using BPMN and SoaML. In: 1st International Workshop on Model-Driven Interoperability, MDI 2010, pp. 61–68 (2010)
7. Ouyang, C., Dumas, M., Van Der Aalst, W.M.P., Ter Hofstede, A.H.M., Mendling, J.: From business process models to process-oriented software systems. ACM Trans. Softw. Eng. Methodol. **19**(1), 2:1–2:37 (2009)
8. Rodríguez, A., Fernández-Medina, E., Piattini, M.: Towards obtaining analysis-level class and use case diagrams from business process models. In: Lecture Notes in Computer Science (Including Subseries Lecture Notes in Artificial Intelligence and Lecture Notes in Bioinformatics) (2008)
9. Blal, R., Leshob, A.: A model-driven service specification approach from BPMN models. In: Proceedings - 14th IEEE International Conference on E-Business Engineering, ICEBE (2017)
10. Blal, R., Leshob, A., Gonzalez-Huerta, J., Mili, H., Boubaker, A.: From interorganizational business process models to service-oriented architecture models. Serv. Oriented Comput. Appl. **12**(3–4), 227–245 (2018)
11. ISO. Information technology: Business operational view- Part 1: Operational aspects of open-edi for implementationl. ISO ISO/IEC 15944-1:2011, International Organization for Standardization, Geneva, Switzerland (2011)
12. Hammer, M., Champy, J.: Reengineering the Corporation: A Manifesto for Business Revolution. Harper Business (1993)

13. Aouachria, M., Leshob, A., Gonzalez-Huerta, J., Ghomari, A.R., Hadaya, P.: Business process integration: how to achieve interoperability through process patterns. In: Proceedings - 14th IEEE International Conference on E-Business Engineering, ICEBE 2017 - Including 13th Workshop on Service-Oriented Applications, Integration and Collaboration, SOAIC 2017 (2017)
14. OMG. Business Process Model and Notation (BPMN) (2011)
15. Kartseva, V., Gordijn, J., Tan, Y.H.: Designing value-based inter-organizational controls using patterns. In: Lecture Notes in Business Information Processing (2009)
16. Geerts, G.L., McCarthy, W.E.: The ontological foundation of REA enterprise information systems. In: Annual Meeting of the American Accounting Association, pp. 1–34, March 2000
17. Hruby, P.: Model-Driven Design Using Business Patterns. Springer, Heidelberg (2006)
18. OMG. Service Oriented Architecture Modeling Language (SoaML) (2012)
19. Palma, F., Moha, N., Tremblay, G., Guéhéneuc, Y.G.: Specification and detection of SOA antipatterns in web services, vol. 8627. LNCS, pp. 58–73 (2014)
20. Coad, P., De Luca, J., Lefebvre, E.: Java Modeling in Color with UML Enterprise Components and Process. Prentice Hall PTR (1999)
21. Gonzalez-Huerta, J., Boubaker, A., Mili, H.: A business process re-engineering approach to transform BPMN models to software artifacts. In: Aïmeur, E., Ruhi, U., Weiss, M. (eds.) E-Technologies: Embracing the Internet of Things, pp. 170–184. Springer, Cham (2017)

Message Dissemination Using Nomadic Lévy Walk on Unit Disk Graphs

Koichiro Sugihara[1] and Naohiro Hayashibara[2(✉)]

[1] Graduate School of Frontier Informatics, Kyoto Sangyo University, Kyoto, Japan
i1788106@cc.kyoto-su.ac.jp
[2] Faculty of Computer Science and Engineering, Kyoto Sangyo University,
Kyoto, Japan
naohaya@cc.kyoto-su.ac.jp

Abstract. Lévy walk is a family of random walks whose distance of a walk is chosen from the power law distribution. There are lots of works of Lévy walk in the context of target detection in swarm robotics, analyzing human walk patterns, and modeling the behavior of animal foraging in recent years. According to these results, it is known as an efficient method to search in a two-dimensional plane. On the other hand, mobile agents which move based on Lévy walk diffuse messages efficiently by opportunistic communication on unit disk graphs. In this paper, we focus on Nomadic Lévy Walk which is a variant of Lévy walk, and propose a base movement strategy for Nomadic Lévy Walk and analyze the impact of the base repositioning strategy of Nomadic Lévy Walk on the efficiency on message dissemination by measuring the number of steps for diffusing messages by opportunistic communication on unit disk graphs. Our simulation results indicate that the proposed strategy is significantly efficient to message dissemination.

1 Introduction

Recently, Lévy Walk has attracted in various areas. Lévy Walk is used as a model of the foraging behavior of albatrosses, honeybee and so on. It is a kind of Random Walks that the move length obeys the power law distribution. It is more efficient regarding message dissemination by opportunistic communication compared to Random Walks [6].

Fujiwara and Miwa proposed Homesick Lévy Walk (HLW) which is a variant of Lévy Walk. It is based on a human mobility model obtained empirically and has a behavior of returning to home (or a base) position in addition to the behavior of Lévy Walk. It is more suitable as human behavior compared to Lévy Walk in the point of view of the encounter probability. Thus, it is worth to use a behavior model for mobile wireless networks.

On the other hand, this type of movement pattern is also useful for multi-hop mobile wireless sensor networks with mobile sinks. There are several works based on this type of system model [1,4,10]. In this network, a mobile sink moves with mobile sensors. Mobile sensors are devices with moving, sensing,

© Springer Nature Switzerland AG 2020
L. Barolli et al. (Eds.): CISIS 2019, AISC 993, pp. 136–147, 2020.
https://doi.org/10.1007/978-3-030-22354-0_13

processing and transmitting functions. They are low-cost devices in general. Mobile sinks have storage and a recharging capability in addition to moving, processing and transmission functions. Mobile sensors periodically go back to the sink to accumulate sensor data and recharge their batteries. Therefore, the homing behavior of HLW is suitable for the routing of mobile sensors in this type of system model.

In this paper, we propose Nomadic Lévy Walk (NLW) that is an extension of HLW. It assumes the sink (base) repositioning in addition to the homing behavior and ballistic trajectory. Furthermore, NLW includes the base repositioning strategy. It depends on the application and improves its performance such as the search efficiency.

We also analyze the impact of the base repositioning strategy on the efficiency of message dissemination. We assume *agents* as mobile entities (e.g., mobile sensors) and they diffuse messages with opportunistic communication. It means that messages are stored, carried and forwarded among agents. We assume a broadcast-based communication. It might be useful for updating firmware and a configuration change in mobile sensor networks.

We conduct the simulations regarding message dissemination on unit disk graphs. Most of the work in Lévy Walk and its variants have been done on a continuous plane. However, the movement is limited by obstacles. Unit disk graph that we assume as an underlying environment is an intersection graph with a constant size of circles in the plane. It is used to model for a road network.

2   Related Work

We now introduce several research works on Lévy walk in a continuous plane and on random walks in a graph.

Valler et al. analyzed the impact of mobility models including Lévy walk on epidemic spreading in MANET [9]. They adopted the scaling parameter $\lambda = 2.0$ in the Lévy walk mobility model. From the simulation result, they found that the impact of velocity of mobile nodes does not affect the spread of virus infection.

Thejaswini et al. proposed the sampling algorithm for mobile phone sensing based on Lévy walk mobility model [8]. Authors showed that proposed algorithm gives significantly better performance compared to the existing method in terms of energy consumption and spatial coverage.

Fujihara et al. proposed a variant of Lévy walk which is called Homesick Lévy Walk (HLW) [3]. In this mobility model, agents return to the starting point with a homesick probability after arriving at the destination determined by the power-law step length. As their result, the frequency of agent encounter obeys the power-law distribution.

There are several works related to Lévy walk on graphs. Most of the works related to Lévy walk assume a continuous plane and hardly any results on graphs are available. Shinki et al. defined the algorithm of Lévy walk on unit disk graphs [6]. They also found that the search capability of Lévy walk emphasizes according to increasing the distance between the target and the initial position of the searcher. It is also efficient if the average degree of a graph is small [5].

Sugihara et al. measured the cover ratio of Homesick Lévy walk on a square grid graphs [7]. They also measured the area covered by an agent moving with the behavior pattern. In fact, it covers a wide area in contrast with the cover ratio. On the other hand, it repeatedly goes through the same nodes on the way to the sink. We conclude that one of the reasons why the cover ratio is low is a fixed sink position. According to the result, we improve the area that covered by an agent with this type of behavior models.

3   System Model and Problem Description

3.1   System Model

In this paper, we assume a unit disk graph $UDG = (V, E)$ with a constant radius r. Each node $v \in V$ is located in the Euclidean plane and an undirected edge (link) $\{v_i, v_j\} \in E$ between two nodes v_i and v_j exists if and only if the distance of the link is less than $2r$. Note that r is the Euclidean distance in the plane. We assume that any pair of nodes in the graph has a path (i.e., connected graph).

We also assume computational entities that are called *agents*. Each agent can move between nodes in the graph. They can communicate with other agents only in the same node. Practically speaking, they have the short-range communication capability such as Bluetooth, ad hoc mode of IEEE 802.11, Near Field Communication (NFC), infrared transmission, etc.

Also, we assume that each node knows own position (e.g., obtained by GPS) which is accessible from agents, and each agent has a compass to get the direction of a walk. Each node has a set of neighbor nodes and those information (i.e., positions of neighbors) is also accessible. Moreover, every agent has no prior knowledge of the environment.

3.2   Problem Description

First, we assume to have agents initially located at random positions on a unit disk graph. Each of them can communicate with other agents when they are on the same node. It means that agents have some short-range communication capability such as Bluetooth, ad hoc mode of IEEE802.11, NFC, and infrared communication.

The goal of the problem is to spread an identical message to all agents. In the beginning, only one agent has the message and then it is gradually diffused by encountering one another.

There are many simulation results on the encounter of mobile entities using Lévy walk such as [2,3,8,9]. However they are on continuous fields, and hardly any results on graphs are available, so far.

4   Nomadic Lévy Walk

Lévy walk is known as an efficient way to explore sparse targets and propagate messages on unit disk graphs [5,6]. Sugihara et al. empirically confirmed the ability to explore in a unit disk graph of Homesick Lévy walk [7]. As a result, it had been lost the property of the wide area search which is a particular feature of Lévy walk. However, it is still attractive because of its homing behavior. The behavior is useful as a model of the animal and human behavior in bio-logging and MANETs. It is also useful for exploring an afflicted area stricken by an earthquake, where cars and people carefully move around while securing safety.

4.1   Definition

We now define a random walk called Nomadic Lévy walk, which is a variant of Homesick Lévy walk [3], to improve the ability of the broad area search while preserving the homing behavior. It holds the following properties in addition to Homesick Lévy walk.

Nomadicity: Each agent moves the position of its base (e.g., sink) with the given probability γ.

Base movement strategy: The trajectory of a sink is like Lévy walk. The next position of the sink is decided by a particular strategy, for example, to cover the untracked area based on the information exchanged with other agents.

Figures 1 and 2 present the trajectory of Homesick Lévy walk and Nomadic Lévy walk in a two-dimensional plane. The difference is that the base is fixed or not. The agent obeying Homesick Lévy walk moves radially from the base. On the other hand, the agent obeying Nomadic Lévy walk explores a broader range of the field comparing to the one with Homesick Lévy walk.

All agents move along Homesick Lévy walk. Thus, their trajectory is radi-ally from their base. Moreover, they move their base at the probability γ in Nomadic Lévy walk. In fact, the fixed base restricts the area that each agent explores. The property of Nomadicity is expected to improve coverage by each agent.

4.2   The Base Movement Strategy

NLW has two different trajectories on agents and their base. First, each agent set the base position as an initial position and then it changes the base position with the given probability γ. Thus, the path of an agent includes that of its base.

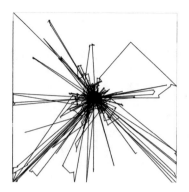

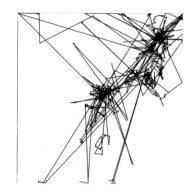

Fig. 1. The trajectory of Homesick Lévy walk.

Fig. 2. The trajectory of Nomadic Lévy walk.

The coverage of a graph by agents depends on the repositioning strategy of the base position. We focus on the following two base repositioning strategy of NLW in this paper and clarify the impact of those strategies on the efficiency of message dissemination.

4.2.1   Lévy Walk (LW) Strategy

The next position of the base is changed obeying the Lévy Walk movement pattern. The orientation o of the next base is selected at random from $[0, 2\pi)$ and the distance d of it from the current position is decided by the power-law distribution.

4.2.2   Reverse Prevention (RP) Strategy

In this strategy, each agent assumes a set of the history of the base positions B_{hist} with the size h to store the coordinates of the nodes' position at which its base has been located in the past. It calculates the reverse orientation of each position in the set. Figure 3 shows the example of the calculation of the next base position based on the past base positions b_1 and b_2. For each position $x \in B_{hist}$, the line L_x that is orthogonal to the line between x and the current base position b is computed. It also indicates the range of the orientation of the next base. The intersection of the ranges $\bigcap_{x \in B_{hist}} L_x$ is defined as θ_{cand}. The orientation of the next base is determined randomly from θ_{cand}. The procedure baseReposition() for calculating the orientation of the next base based on the RP and LW strategies is shown in Algorithm 1.

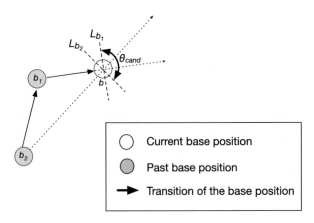

Fig. 3. The RP strategy of NLW.

Algorithm 1. baseReposition($strat$, B_{hist})

> **Initialize:**
>> $b \leftarrow$ the position of the base
>> $B'_{hist} \leftarrow B_{hist} \setminus \{b\}$
>> $\theta_{cand} \leftarrow [0, 2\pi)$

2: **if** $strat = \text{'}LW\text{'}$ **then**   ▷ For the LW strategy.
>> *return* the orientation selected randomly from $[0, 2\pi)$

4: **else if** $strat = \text{'}RP\text{'}$ **then**   ▷ For the RP strategy.
>> **for** $x \in B'_{hist}$ **do**
6:            **if** $\theta_{cand} \cap L_x \neq \emptyset$ **then**
>>>> $\theta_{cand} \leftarrow \theta_{cand} \cap L_x$ ▷ L_x is the angle π formed by b and x.
8:            **end if**
>> **end for**
10:      *return* the orientation selected randomly from θ_{cand}
> **end if**

4.3   Algorithm on Unit Disk Graphs

We now explain the detail of the algorithm of Nomadic Lévy walk on unit disk graphs in Algorithm 2.

Algorithm 2. A walk using Nomadic Lévy Walk on Unit Disk Graphs

---

1: **Initialize:**
    $b \leftarrow$ the position of the base
    $c \leftarrow$ the current position
    $o \leftarrow 0$ ▷ orientation for a walk.
    $brFlag \leftarrow False$ ▷ Flag for the base repositioning procedure.
    B_{hist}.initialize(h)   ▷ The sliding window on base positions is
    initialized with the size h.
    $strat \in \{LW, RP\}$ ▷ The given strategy of NLW.
    $PN(c) \leftarrow$ the possible neighbors to move.
2: **if** Probability: α **then**
3:     $d \leftarrow$ the distance to b
4:     $o \leftarrow$ the orientation of b
5: **else**
6:     d is determined by the power-law distribution
7:     o is randomly chosen from $[0, 2\pi)$
8: **end if**
9: **if** Probability: γ **then**   ▷ The base repositioning procedure.
10:     $brFlag \leftarrow True$
11:     move to b ▷ going back to the base position.
12:     B_{hist}.enqueue(b)   ▷ add the base position into the sliding window.
13:     d is determined by the power-law distribution
14:     $o \leftarrow$ baseReposition($strat, B_{hist}$)   ▷ change the base position.
15: **end if**
16: **while** $d > 0$ **do**
17:     $PN(c) \leftarrow \{x|abs(\theta_{ox}) < \delta, x \in N(c)\}$
18:     **if** $PN(c) \neq \emptyset$ **then**
19:         $d \leftarrow d - 1$
20:         move to $v \in PN(c)$ where v has the minimum abs(θ_{ov})
21:         $c \leftarrow v$
22:     **else**
23:         $brFlag \leftarrow False$
24:         **break**   ▷ no possible node to move.
25:     **end if**
26: **end while**
27: **if** $brFlag = True$ **then**
28:     $b \leftarrow c$ ▷ The base position is updated.
29:     $brFlag \leftarrow False$
30: **end if**

At the begging of the algorithm, each agent holds the initial position as the base position b. In every walk, each agent determines the step length d by the power-law distribution and selects the orientation o of a walk randomly from $[0, 2\pi)$. It can obtain a set of neighbors $N(c)$ and a set of possible neighbors $PN(c) \subseteq N(c)$, to which agents can move, from the current node c. In other words, a node $x \in PN(c)$ has a link with c that the angle θ_{ox} between o and the link is smaller than $\frac{\pi}{2}$.

In unit disk graphs, it is not always true that there exist links to the designated orientation. We introduce δ that is a permissible error against the orientation. In this paper, we set $\delta = 90$. It means that agents can select links to move in the range ± 90 with the orientation o as a center.

In a given probability α, the agent goes back to the base node (line 2 in Algorithm 2). In this case, it sets the orientation to the base node as o and the distance to the base as d.

Each agent changes its base position with the given probability γ (line 10 in Algorithm 2). The next position of the base will be determined by the given strategy stated in Sect. 4.2.

5   Performance Evaluation

The aim of the performance evaluation is to clarify the impact of the base movement strategy of NLW on the efficiency of message dissemination on unit disk graphs. We assume HLW and the two strategies of NLW, the Lévy Walk (LW) strategy, and the Reverse Prevention (RP) strategy. We measure the time (simulation time) to distribute a message to every agent in the simulations. We implemented the discrete event-based simulator in C++ for the simulations.

5.1   Environment

We set the radius $r \in \{35, 50\}$ to form unit disk graphs. r is correlated to the average degree of a graph. The average degree is 14.9 with $r = 35$ and 28.9 with $r = 50$. It means that r is proportional to the average degree. As a result, the flexibility of the movement of agents decreases according to a decrease of r.

5.2   Parameters

We configure the following parameters for our simulations.

Number of agents k: It is the number of agents located in a graph. Each agent is located at a unique node and the position of it is determined at random. We set $k = [2, 50]$.

Scaling parameter λ: It is a parameter for the Lévy Walk family to determine the trajectory of agents. The ballistic trajectory is emphasized if λ increases. We set $\lambda = 1.2$ because it is efficient for a search on unit disk graphs [5] and it can cover a wider area on a graph.

The probability α for returning the base: With the given probability α, agents go back to their base position. We set $\alpha \in \{0.4, 0.6, 0.8\}$. The frequency that agents go back to their base increases according to an increase of α.

Base repositioning probability γ: Each agent changes its base position with the probability γ. The frequency that agents change their base increases if γ increases. We set $\gamma \in \{0.2, 0.5, 0.8\}$. The next position of their base will be determined by the base repositioning strategy.

5.3   Result on $r = 50$

Figures 4, 6, and 8 show the average simulation time for message dissemination using HLW and NLW with the LW strategy by the number of agents. As observed in these figures, NLW is efficient than HLW with few agents (e.g., $k \leq 20$). It is obvious that agents that move obeying NLW cover a wider area because they change their base position with the given probability γ. On the other hand, the difference between them is getting smaller by an increase of k and they are almost the same in the case of $k \geq 20$. It means that the number of agents is enough to forward the message to all agents. There is almost no difference in the probability α in the simulation.

We also analyze the difference between the strategies of NLW. Figures 5, 7, and 9 show the average simulation time for message dissemination using RP strategy of NLW by the number of agents. Compared to Figs. 4, 6 and 8, RP strategy is significantly efficient than LW strategy. In particular, the RP strategy efficiently diffuses the message even with few agents (e.g., $k \leq 10$). It means that this strategy improves the probability of encounter with other agents.

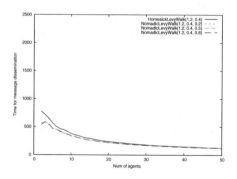

Fig. 4. HLW and NLW with LWS and $\alpha = 0.4$.

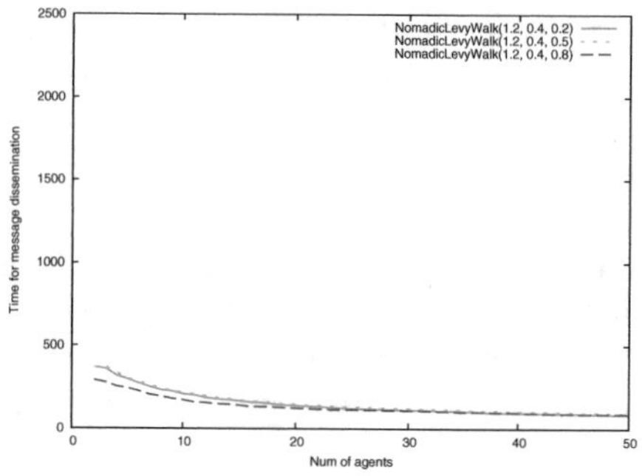

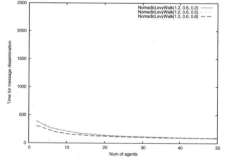

Fig. 5. The trajectory of Nomadic Lévy walk.

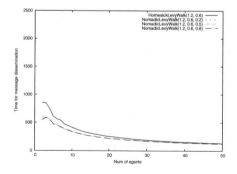

Fig. 6. The trajectory of Homesick Lévy walk.

Fig. 7. The trajectory of Nomadic Lévy walk.

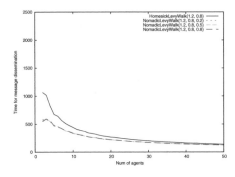

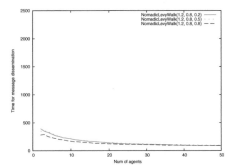

Fig. 8. The trajectory of Homesick Lévy walk.

Fig. 9. The trajectory of Nomadic Lévy walk.

5.4   Result on $r = 35$

Figures 10, 12, and 14 show the average simulation time for message dissemination using HLW and NLW with the LW strategy, and Figs. 11, 13, and 15 show that using NLW with RP strategy in the same manner as the results with $r = 50$. The former ones indicate that the tendency is the same as the result of $r = 50$. However, the time for distributing the message using NLW using the LW strategy increases 80% compared to the result of $r = 50$. This is because the movement of agents in the graphs with $r = 35$ is more strict than that of $r = 50$.

On the other hand, Figs. 11, 13, and 15 show that the RP strategy of NLW improves the efficiency of message dissemination significantly. Moreover, the base repositioning probability γ has a substantial impact on the efficiency in the case of $r = 35$ compared to that of $r = 50$. Interestingly, the efficiency of the RP strategy with $r = 35$ is almost the same as that with $r = 50$ in terms of $\gamma \in \{0.5, 0.8\}$ even though the restriction of the movement is limited. The message transmission efficiency gets worse according to a decrease of γ.

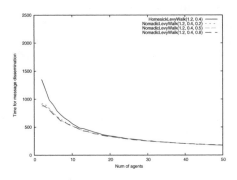

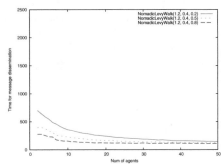

Fig. 10. The trajectory of Homesick Lévy walk.

Fig. 11. The trajectory of Nomadic Lévy walk.

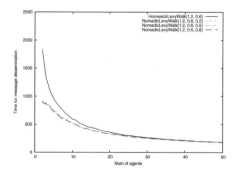

Fig. 12. The trajectory of Homesick Lévy walk.

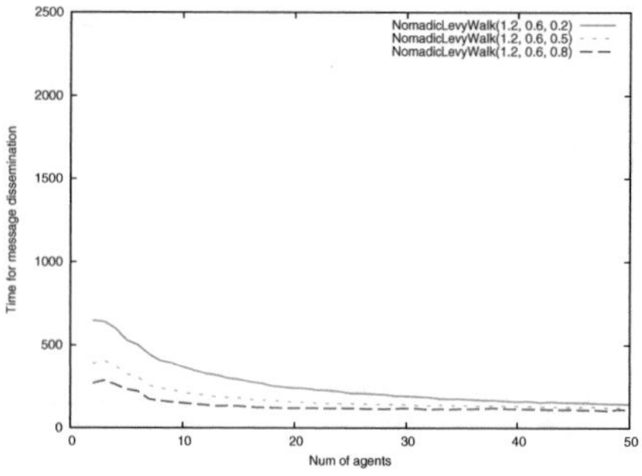

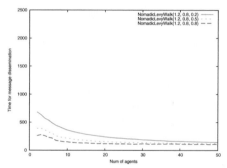

Fig. 13. The trajectory of Nomadic Lévy walk.

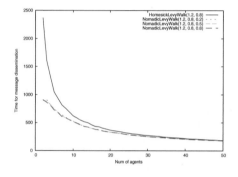

Fig. 14. The trajectory of Homesick Lévy walk.

Fig. 15. The trajectory of Nomadic Lévy walk.

6   Conclusion

In this paper, we proposed the two strategies of NLW and measured the efficiency of message dissemination using HLW and NLW to clarify the impact of the base repositioning and the NLW's strategies. According to the simulation results, the base repositioning is efficient regarding message dissemination with few agents. On the strategies of NLW, the RP strategy is significantly efficient than the LW strategy and the efficiency emphasizes on the graph with $r = 35$ rather than that with $r = 50$.

Acknowledgement. This work was supported by JSPS KAKENHI Grant Number JP16K00449.

References

1. Akkaya, K., Younis, M., Bangad, M.: Sink repositioning for enhanced performance in wireless sensor networks. Comput. Netw. **49**(4), 512–534 (2005). https://doi.org/10.1016/j.comnet.2005.01.014
2. Birand, B., Zafer, M., Zussman, G., Lee, K.W.: Dynamic graph properties of mobile networks under levy walk mobility. In: Proceedings of the 2011 IEEE Eighth International Conference on Mobile Ad-Hoc and Sensor Systems, MASS 2011, pp. 292–301. IEEE Computer Society, Washington (2011). https://doi.org/10.1109/MASS.2011.36
3. Fujihara, A., Miwa, H.: Homesick lévy walk and optimal forwarding criterion of utility-based routing under sequential encounters. In: Proceedings of the Internet of Things and Inter-Cooperative Computational Technologies for Collective Intelligence 2013, pp. 207–231 (2013)
4. Marta, M., Cardei, M.: Improved sensor network lifetime with multiple mobile sinks. Pervasive Mob. Comput. **5**(5), 542–555 (2009)
5. Shinki, K., Hayashibara, N.: Resource exploration using lévy walk on unit disk graphs. In: the 32nd IEEE International Conference on Advanced Information Networking and Applications (AINA-2018), Krakow, Poland (2018)
6. Shinki, K., Nishida, M., Hayashibara, N.: Message dissemination using lévy flight on unit disk graphs. In: the 31st IEEE International Conference on Advanced Information Networking and Applications (AINA 2017), Taipei, Taiwan ROC (2017)
7. Sugihara, K., Hayashibara, N.: Collecting data in sensor networks using homesick lévy walk. In: the 11th International Workshop on Advanced Distributed and Parallel Network Applications (ADPNA-2017), pp. 779–786, Toronto, Canada (2017)
8. Thejaswini, M., Rajalakshmi, P., Desai, U.B.: Novel sampling algorithm for human mobility-based mobile phone sensing. IEEE Internet Things J. **2**(3), 210–220 (2015)
9. Valler, N.C., Prakash, B.A., Tong, H., Faloutsos, M., Faloutsos, C.: Epidemic spread in mobile ad hoc networks: determining the tipping point. In: Proceedings of the 10th International IFIP TC 6 Conference on Networking - Volume Part I, NETWORKING 2011, pp. 266–280. Springer, Berlin (2011). http://dl.acm.org/citation.cfm?id=2008780.2008807
10. Vincze, Z., Vass, D., Vida, R., Vidács, A., Telcs, A.: Adaptive sink mobility in event-driven multi-hop wireless sensor networks. In: Proceedings of the First International Conference on Integrated Internet Ad Hoc and Sensor Networks, InterSense 2006. ACM, New York (2006). https://doi.org/10.1145/1142680.1142697

Modelling a Smart Motorway

Edward Richardson[1]([✉]), Philip Davies[2], and David Newell[2]

[1] Capgemini, London, UK
ed.richardson@capgemini.com
[2] Bournemouth University, Bournemouth, UK
{daviesp, dnewell}@bournemouth.ac.uk

Abstract. The increasing number of vehicles in the UK is putting strain on the motorway infrastructure. Smart Motorways have been implemented to reduce congestion. To test systems before they are physically built, they are simulated using objects models and GIS data. This paper has modelled a new smart traffic light system for a slip road joining a smart motorway. The smart traffic light models have been simulated alongside driver behavior models which together show that a smart traffic light system can adapt to continuously fluctuating traffic flow. The smart traffic light system reduced the number of congestion alerts by 80% on the smart motorway.

1 Introduction

In the UK, traffic congestion puts increasing strain on the motorway infrastructure. The increasing number of cars has resulted in heavy congestion and motorways reaching capacity. Smart technology is considered the best option to alleviate congestion with the smallest impact on the public and the environment [7]. One application is Smart Traffic Lights (STL) on slip roads that communicate with motorway sensors to aid traffic flow for joining traffic.

Smart Motorways (SM) are modelled by simulators to provide an insight into how smart technology such as STL impacts traffic flow. Traffic simulators use Geography Information System (GIS) data to create accurate base maps for the simulator and GIS shape files are used to model the dynamics of an object such as a vehicle or a pedestrian. Miller et al. [13] have shown that two main types of model are required, a traffic model and a driver behaviour model.

1.1 Vehicle Models

Zhao et al. [19] created a simulator to predict marine incidents predicting ship collisions. The simulation of ship traffic flow has the same issues as motorway traffic flow due to the variety in vehicle sizes and speeds. Their approach was to create shape files in GIS that could be used in a simulator with the exact shape of different ships. This would allow the simulator to be able to predict more accurately the interaction between vessels of different sizes rather than having different vessels as generic object shapes.

© Springer Nature Switzerland AG 2020
L. Barolli et al. (Eds.): CISIS 2019, AISC 993, pp. 148–158, 2020.
https://doi.org/10.1007/978-3-030-22354-0_14

1.2   A Review of Macroscopic and Microscopic Viewpoints

The modelling of a simulator is controlled by the level of the simulator view which can be macroscopic, mesoscopic or microscopic. The level of the simulator view is commonly macroscopic [1, 8, 11, 18]. However the lack of detail is a problem in complex heterogeneous systems such as simulating built-up urban areas. A macro-traffic flow approach has been taken by Zhao et al. [19] who specified the shape of the vessels but did not include behavioural models and could not show the individual behaviours of ships, only simulating the action of ships leaving/entering ports. Sasso and Biles [14] simulated a large geographical area, the Panama Canal, where only a macroscopic view would be able to simulate it. Ships and vehicles were simulated as 2D objects, with ships as small rectangles and their individual behaviour not being simulated. Their approaches focused on the overall flow of traffic, however, they assumed all objects act in the same way which is not the case for vehicles with different drivers

Huang and Pan [8] built a simulator for the prediction of the best route for emergency responders to an incident. They used a macroscopic view point which provided a good overall summary but did not allow for the behaviour of drivers at junctions and the impact upon the emergency responders route.

Liao and Yen [11] simulated a comparable emergency event for an earthquake evacuation for sudden heavy traffic flow. They used a micro-traffic flow simulator which allowed a focus on specific road junctions but did not show the overall traffic flow which may be heavily impacted by a road junction

Khalesian and Delavar [9] used a microscopic view point and showed that the driver's behaviour on a motorway is an important factor to consider, especially when considering multiple lanes and behavioural models need to be taken into account. Their approach is one that can be applied to any simulator and provides each vehicle with an individual behaviour model which allows for the vehicles to interact. In a similar way Dallmeyer et al. [4] modelled the cars using an object orientated simulator where objects can act independently. They additionally modelled a prediction method for the direction of the traffic in an urban setting but did not include in their behaviour models the different vehicles in an urban environment and how they will behave differently.

The models required to determine how the objects interact is a crucial part of any simulator and what can be seen from this brief review is a gap in approach with a lack of traffic and behaviour models for simulators using GIS data. We therefore propose a new traffic and behaviour model incorporating GIS data and will test the approach using the following hypotheses to see if this approach to modelling SM improves traffic flow.

2   Method

We propose two types of model; traffic and behavioural. The models will be built in Simulink and random number generators (RGN) will be used to create realistic vehicle parameters with a set mean and variance. The number of vehicles joining from the slip road and on the motorway will be based upon peak traffic flow numbers on the M3 [7]. The simulation will be at a microscopic level as the models are object orientated and the simulator is designed to see how the STL adapts. To simulate vehicles with different properties an RNG that has a fixed mean and variance will be used for determining each model parameter. The mean and variance are based upon real data. The STL can be refined to allow more vehicles through depending on the impact on the motorway traffic flow. We now set out some clear definitions of the terms used.

2.1 Smart Motorway (SM)

A SM is defined as a motorway that uses smart technology to monitor and control traffic flow. SM's vary in design and are adapted to the setting of the motorway especially in locations where several motorways merge. The SM is monitored by the Motorway Incident Detection and Automatic Signalling (MIDAS) system that can differentiate between congestion and queued traffic [10]. The MIDAS system controls the variable speed limits (VSL) shown on the gantries above a motorway and are designed to help with the traffic flow by reducing the stop/start traffic conditions from occurring.

2.2 Traffic Models

The term 'Traffic Model' is used in a variety of different contexts and therefore has multiple definitions. Song et al. [15] define it as a model that influences how objects behave and interact with each other and their environment. Boxill and Lu [2] define a traffic model as a model that defines the environment that an object interacts with. Taplin [16] define a traffic model as modelling the flow of traffic based on dynamic models. We use the term 'Traffic Model' to refer to a model that defines the environment of the simulator. This includes models such as the STL and SM models. In developing our Traffic Model we explicitly combine a STL and a driver behaviour model to simulate how it impacts congestion on a SM by controlling the slip road traffic.

2.3 Ramp Metering

Ramp metering is the term used for traffic lights controlling the traffic flow joining a motorway where a STL calculates when to release the traffic [10]. To simulate a STL controlling the flow of traffic, the models require live information provided from both the slip road and the SM. Cai et al. [3] carried out the modelling of STL for an intersection that has traffic approaching from four different directions. The Queue Tracker Model (QTM) shown in Eq. 1 calculates the queue length at each traffic light. The model decides which traffic queue to release next based on the queue length to reduce congestion at the junction.

$$O = \frac{1}{T} \sum_{i=1}^{C} \frac{l_i + l_{loop}}{v_i} \tag{1}$$

$$\mu = \frac{\alpha\sqrt{ql - qt}}{vl} \qquad (2)$$

The model has been adapted to provide the STL with information regarding the number of vehicles in the queue and calculates how many vehicles are queued on the slip road by measuring individual vehicle size. Equation 2 calculates the number of vehicles queued on the slip road. This model will be continuously calculating the length of the vehicles in the traffic queue on the slip road (Tables 1 and 2). The STL requires an additional model to calculate if there are any gaps in the motorway traffic.

Table 1. QTM model

Model element	Element definition
O	Time Occupancy (Seconds)
T	Time (Seconds)
C	Vehicle Count
L_i	Vehicle Length (Metres)
L_{loop}	Length of Induction Loop in Road (Metres)

Table 2. QTM model

Model element	Element definition
μ	Number of vehicles in the queue
A	Mean Velocity (M/S)
q_l	Queue Length starting from traffic lights (M)
q_t	Queue Time (S)
Vl	Vehicle Length (M)

2.4 The Gap Analysis Model-GAP

The role of the STL is to calculate the motorway's current capacity and calculate the most efficient time to release vehicles. The STL is designed to calculate the size of the gap in the motorway traffic and consider the size of the vehicle that is on the slip road. The STL can detect if a large vehicle such as a lorry is present and wait for a gap suitable for the lorry to join however the driver still has the final decision.

The QTM is used alongside the GAP to determine how quickly the queue is moving and if there are vehicles in the queue. The STL requires information from the motorway regarding the traffic density to calculate gaps in the traffic on the SM The output from the model is the size of the gaps in the traffic based upon the traffic density (Eq. 3, Table 3).

$$Ga = \frac{Li + Lloop}{C} \times (O \times Vi) \tag{3}$$

Table 3. GAP model

Model element	Element definition
Ga	Size of gap in traffic (Meters)
Li	Vehicle Length (Meters)
L_{loop}	Length of Induction Loop in Road (Meters)
V_i	Vehicle Velocity (M/S)
C	Vehicle Count

3 Behaviour Models

The driver behaviour model calculates the possible action a driver can take based on the traffic conditions and their ability. A vehicle's performance is part of a driver's behaviour model. We require multiple models that can predict the behaviour of an object to simulate the complex decision processes of a driver. We follow Toledo's [17] identification of driver behaviour models for a microscopic simulator.

3.1 Car Following Model

The car following model is a non-linear model based on the idea that the lead vehicle controls the velocity and acceleration of the following vehicle [5]. This sets the likelihood of the driver to change lanes dependent upon their sensitivity and desire to change their vehicle's velocity. This model does not include the driver's ability to keep following the vehicle in front or the driver's reaction time (Eq. 4) (Table 4).

$$a_n(t) = \alpha \frac{V_n(t)^\beta}{\Delta X_n(t - \tau_n)^\gamma} \Delta V_n(t - \tau_n) \tag{4}$$

Table 4. Car following model

Model element	Element definition
Gf	Lag Distance (metres)
g_{0f}	Lag-space gap at start of lane changing (metres)
v_f	Speed of lag vehicle (M/S)
D_t	dv/b_f
b_f	Deceleration of target lag vehicle (M/S)
Dv	Level of aggressiveness of lag vehicle
Vs	Speed of subject vehicle (M/S)
Dl	Desire to change lane

3.2 The Lane Changing Gap Model (LCM)

The LCM is based on the model proposed by Hidas [6]. The model has been adapted to include the driver's level of desire to change lane. The output is the size of the gap for the driver to change lane (Eq. 5). The greater the desire to change lane results in the size of gap perceived by the driver to increase in size. Another factor included in the model is the level of aggression a driver has. This presumes that a more aggressive driver is more likely to change lane than a non-aggressive driver (Table 5).

$$gf = g0f - \left(VfDt - \frac{Bf}{2Dt^2} \right) + VsDt \times Dl \tag{5}$$

Table 5. LCM

Model element	Element definition
$\Delta Vn(t - \tau n)$	Leader relative speed (M/S)
$\Delta Xn(t - \tau n)$	Spacing between subject and lead vehicle (Metres)
an(t)	Acceleration of the subject vehicle (M/S)
Vn(t)	Speed of the Subject Vehicle (M/S)
α, β and γ	Parameters

In the middle lane a driver can decide to change into either the left or the right lane so we require an additional model to determine which lane they move to. It is presumed that a driver in the middle lane is in this lane to overtake other vehicles and that drivers intend on being in the left lane (Eq. 6). The decision to change lane is combined with the lane changing model as the driver may want to change lane but be unable to do so (Table 6).

$$Ld = \frac{Dl \times \frac{da}{va}}{Ls} \tag{6}$$

Table 6. Lane desire model

Model element	Element definition
Ld	Desire for left or right lane (Positive for left, negative for right)
Dl	Desire to change lane
Da	Driver ability to change lane
Va	Vehicle ability to change lane
Ls	Probability of driver selecting the left lane

3.3   Slip Road Velocity Model (RVM)

The RVM is designed to model whether a driver decides join the motorway. Even if STL calculates there is a gap for the vehicle to join the motorway the driver may decide otherwise. This is based on a model developed by Michaels and Fazio [12] which calculates the angular velocity for a vehicle joining a motorway. This model does not include the length of the vehicle and presumes that all vehicles have the same rate of acceleration (Eq. 7). It simulates a driver's decision to join the motorway from the slip road, which may not be determined by their vehicle ability (Table 7).

$$W = k(V_f - V_r)/L^2 \tag{7}$$

Table 7. RVM

Model element	Element definition
W	Angular velocity
V_f	Motorway vehicle speed (M/S)
V_r	Ramp vehicle speed (M/S)
L^2	Distance separation (metres)
K	Lateral offset (metres)

3.4   Hypotheses

Using these set of models, we will test the following hypotheses:

1. The STL will reduce the number of congestion alerts in comparison to a free-flowing slip road.
2. The STL will reduce the number of congestion alerts in the left lane in comparison to a free-flowing slip road.
3. The STL will reduce the number of congestion alerts in the middle lane in comparison to a free-flowing slip road

4   Simulation-MatLab Simulink

The slip road will be controlled by the STL which is receiving live information. The sensors at the end of the motorway measure the output of all three lanes of traffic and the motorway will be monitored based on the traffic flow which Simulink displays. In the simulation, the information received by the STL comes before the driver behaviour model which allows for the free flow vehicles after the traffic readings have taken place. Figure 1 shows a conceptual design of the SM. The simulation is set up so that one unit of time is the equivalent to 10 s. This allows the models to adapt in a realistic manner and they can be validated over a longer time. The models for driver behaviour will be applied to a group of vehicles resulting in groups of vehicles changing lane. The total simulation time is 1.5 h.

4.1   Data Analysis

The hypotheses were tested by comparing a SM and a normal motorway in the simulator. The normal motorway does not include the STL and has been replaced in the simulator by a RNG. The parameters for the RNG are based upon ensuring a comparable output from the STL. The parameters for all the driver behaviour models have been kept the same for both simulations to be comparable. The SM simulator includes a congestion alert system when the motorway capacity reaches 95% within a 250 m part of the SM. 250 m was selected because this is the distance that the current SM in England measure traffic flow [7]. Four parts of the simulator were tested to see if they trigger congestion alerts, one for each of the three lanes at the end of the simulator and one at the point where the traffic joins from the slip road. Table 8 is an overall summary of results from the congestion alert system and shows that overall the STL reduced the number of congestion alerts.

Table 8. Congestion alert results

	Number of congestion alerts		
	Normal slip road	SM-STL	% Decrease
Slip road-Motorway	15	0	100%
Left Lane	119	5	95.8%
Middle Lane	77	32	58%
Right Lane	1	1	0%
Total	212	38	82%

An Anova analysis was used. In order for hypothesis to be rejected the P value needs to be greater than the Alpha value in order to be statistically significant. This was set at 0.05. The variance around the mean is relatively high for both groups however this is expected due to the different lanes having varying numbers of vehicles as a result of the driver behaviour models. The STL is shown to have reduced the number of congestion alerts on the SM and therefore improved the overall traffic flow.

Table 9 shows the STL has reduced the number of times congestion occurs by 82%. The STL has zero impact on the right lane but does reduce congestion by 58% in the middle lane. It was expected that the STL would impact the slip road and the left lane, but it was not expected to impact the middle lane. This validates that STL and the driver behaviour models both work as expected.

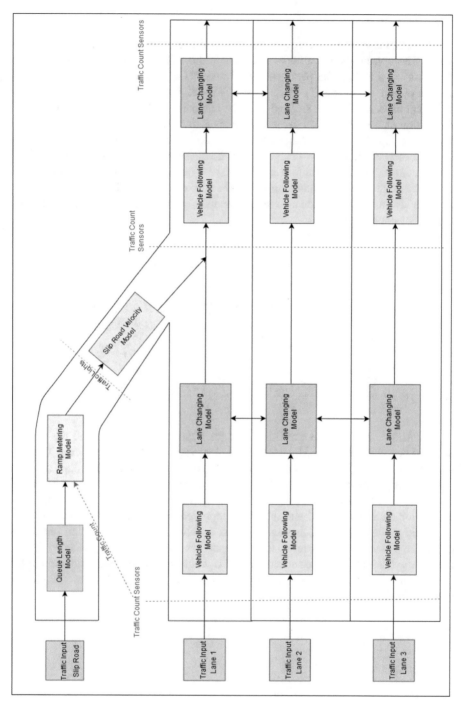

Fig. 1. SM conceptual design

Table 9. Statistical analysis

Source of variation	Sum of Squares	Mean of Squares	P-Value
Between Groups	3784.5	3785	0.1782
Within Groups	9769	1628	
Total	13553.5		

4.2 Discussion

Hypothesis H1 can now be accepted as there was an overall reduction in congestion alerts of 82% on average across the SM. This result is as expected and provides validation that the STL models proposed in this paper can improve the overall traffic flow; showing a difference between the number of congestion alerts with and without a STL being on the slip road. Hypothesis H2 is also accepted as the congestion alerts reduced by 95.8%;which is higher than expected. Hypothesis H3 can also be accepted due to a 58% reduction in the number of congestion alerts. This result is higher than expected and can be explained by fewer vehicles changing to the middle lane as the STL controls the flow of traffic from the slip road.

5 Conclusion and Future Work

One of the main issues for previous simulators has been the lack of driver behaviour models. This paper has adapted and optimised four driver behaviour models to work alongside the STL models, allowing the system to improve traffic flow. The STL was successfully developed from originally being an intersection model to being validated as a STL on a SM. The driver behaviour models have been validated in the simulator. The RVM has been optimised to work with the STL models. The RVM ramp vehicle speed parameter was altered to be more realistic and to allow vehicles to join the motorway when the STL identifies the gap in traffic.

The models that have been demonstrated in this paper can be applied to the next stage which would include GIS data, specifically the shape files representing different vehicles and a base map simulating a motorway. The driver behaviour models could be applied to the specific shapes and have different parameters dependent upon the vehicle and driver. The models can now run continuously and change as the vehicle changes lane. In the future, the simulator could be extended by adding SM features such as variable speed limits. This could reduce the number of congestion alerts by increasing the capacity of the motorway.

References

1. Bhaskar, A., Chung, E., Kuwahara, M.: A Multi-agent based traffic network micro-simulation using spatio-temporal GIS. In: Development and Implementation of the Area Wide Dynamic Road Traffic Noise (DRONE) Simulator, vol. 5, no. 12, pp. 371–378 (2007)
2. Boxill, S., Yu, L.: An Evaluation of Traffic Simulation Models for Supporting ITS Development. Texas Southern University, Houston (2000)

3. Cai, C., Hengst, B., Ye, G., Huang, E., Wang, Y., Aydos, C., Geers, G: On the performance of adaptive traffic signal control. In: 2nd International Workshop on Computational Transportation Science, Seattle (2009)

4. Dallmeyer, J., Lattner, A., Timm, I: From GIS to mixed traffic simulation in urban scenarios. In: 4th International ICST Conference on Simultion Tools and TEchniques, pp. 134–143 (2011)

5. Gazis, D., Herman, R., Rothery, R.: Nonlinear follow-the-leader models of traffic flow. Oper. Res. **9**:545-567 (1961)

6. Hidas, P.: Modelling vehicle interactions in microscopic simulation of merging and weaving. Transp. Res. Part C: Emerg. Technol. **1**(13), 37–62 (2005)

7. Highways England: Smart Motorways Programme [Online] (2018). http://www.highways.gov.uk/smart-motorways-programme/

8. Huang, B., Pan, X.: GIS coupled with traffic simulation and optimization for incident response. Comput., Environ. Urban Syst. **2**(31), 116–132 (2007)

9. Khalesian, M., Delavar, M.: A Multi-agent based traffic network micro-simulation using spatio-temporal GIS. In: The International Archives of the Photogrametry, Remote Sensing and Spatial Information Science (2008)

10. Li, Y.: Modelling and optimisation of dynamic motorway traffic. University College London, London (2015)

11. Liao, D., Yen, P.: A linkage tool for analysing earthquake traffic impact in micro level based on seismic risk assessment and traffic simulation. In: 1st International Conference and Exhibition on Computing for Geosptail Research & Application, Washington (2010)

12. Michaels, R., Fazio, J.: Driver behaviour model of merging. University of Chicago (1989)

13. Miller, J., Peng, H., Bowman, C.: Advanced tutorial on microscopic discrete-event traffic simulation. In: Winter Simulation Conference (2017)

14. Sasso, D., Biles, W.: An object-orientated programming approach for a GIS data-driven simluation model: of traffic on an inland waterway. In: 2008 Winter Simulation Conference (2008)

15. Song, D., Tharmarasa, R., Zhou, G., Florea, M., Duclos-Hindie, N., Kirubarajan, T.: Multi-vehicle tracking using microscopic traffic models. In: IEEE Transactions on Intelligent Transportation Systems, pp. 1–13 (2018)

16. Taplin, J.: Simulation models of traffic flow. In: The 34th Annual Conference of the Operational Research Society (1999).

17. Toledo, T.: Driving behaviour: models and challenges. In: Massachusetts Institute of Technology (2003)

18. Zhang, H., De Farias, O.: City traffic simulator using geographical information systems and agent-based simulation. In: 3rd IET International Conference on Intelligent Environments (2007)

19. Zhao, M., Yao, X., Sun, J., Zhang, S., Bai, J: GIS-based simulation methodology for evaluating ship encounters probability to improve maritime traffic safety. In: IEEE Transactions on Intelligent Transportation Systems, pp. 1–15 (2018)

Personalized Protocols for Data Division and Knowledge Management

Lidia Ogiela[1]([✉]), Makoto Takizawa[2], and Urszula Ogiela[1]

[1] Pedagogical University of Krakow, Podchorążych 2 Street,
30-084 Kraków, Poland
lidia.ogiela@gmail.com, uogiela@gmail.com
[2] Department of Advanced Sciences, Hosei University, 3-7-2,
Kajino-Cho, Koganei-Shi, Tokyo 184-8584, Japan
makoto.takizawa@computer.org

Abstract. In this paper will be proposed new personalized cryptographic protocols dedicated for secure information sharing and data management. Such new procedures will be based on mathematical formalisms and application of selected personal or behavioral patterns, which will be involved in the division and management processes. Such new management procedures can be created with application of different formal data representation classes. Possible application of such method will be also presented.

Keywords: Data division protocols · Cloud computing and secret management · Secure information management

1 Introduction

The data division processes can be realised with application of cryptographic protocols such as splitting and sharing algorithms [2, 7, 10]. These protocols allows to split data (secret information) between trusted parts and distribute them between holders (protocol participants). The main idea of such solution is to divide – separation of the secret – and not to pass it to one owner. This allows to maintain neutrality and independence from one-man actions on secret management processes [3]. The process of secret management, dependents on a certain group of secret trustees (at the stage of creating an algorithm).

In the class of cryptographic data division protocols there are classical splitting and sharing techniques [2, 4, 5, 8, 13]. These classes of secret division were extended by the linguistic and biometric threshold schemes [1, 6, 12]. Both of the new classes of threshold schemes can introduce additional security stages against unauthorized transfer (distribution) of secret part and unauthorized access to it.

In linguistic protocols, the use of semantic secret division has been proposed, implemented on the basis of the knowledge protocol owned by the participants. In this class of protocols it is possible to use such a way of dividing the secret, which will take into account different levels of knowledge – from the simple general and public knowledge also specialist, till to expert knowledge. In this class of protocols there are

© Springer Nature Switzerland AG 2020
L. Barolli et al. (Eds.): CISIS 2019, AISC 993, pp. 159–164, 2020.
https://doi.org/10.1007/978-3-030-22354-0_15

used linguistic methods for secret description as well as formalisms of mathematical linguistics, based on various structures, i.e. sequence, tree or graph [1, 6, 10].

The second type of new classes of cryptographic division schemes are biometric threshold schemes [9]. In the biometric threshold schemes, the personal-biometric marking stage of individual secret shadows, was introduced. Personal (individual) marking shadows using a selected type of biometrics allows to clearly assign both a shadow to its holder, but also a participant of the protocol to the right part of the secret. Shadow marking can be done using various types of biometric features such as DNA code, hand, facial, eyes, and voice biometrics etc. The choice of biometrics in the process of secret's marking and sharing depends on the arbitrator or the supervisor the implementation of the protocol.

In order to strengthen the proposed solutions, a class of protocols combining the presented linguistic and biometric solutions, has also been introduced. In this class of protocols is possible the both of the linguistic division of the secret and biometrics marking of individual secret parts. It is therefore a class of protocols that:

- allows reached secrecy of a secret by its division and distribution of secret parts between participants of the protocol,
- allows to apply a linguistic description of a shared secret that takes into account various levels of access depending on participant's knowledge,
- allows to apply various biometric features in the processes of division, distribution and reconstruction of the secret.

This paper presents the possibilities of using behavioral features used in hybrid threshold schemes.

2   Behavioral Features in Hybrid Threshold Schemes

In the hybrid threshold schemes, different techniques of personalised marking the entire secret (or secret parts) by personal features, can be used. The selection of biometric features is realise from a set of all possible to define biometrics. This process can be carried out:

- arbitrarily – by choosing one type of biometric that will be used in the marking process of secret shadows, or
- randomly – by randomly indicating the biometrics used in the currently implemented marking process.

In hybrid threshold schemes, is also possible in the data labelling process the use of behavioural features. In this approach, it is possible to use the characteristics of:

- gestures,
- movements,
- the rhythm of behavior,
- repetitive behaviors,
- gait,
- words,
- way of writing, etc.

In this class of personal characteristics, particular importance will be attributed to the personal characteristics associated with the various (typical for a given secret holder) behaviour. Analysis of behavioural features will be used to determine typical and unusual behaviours characteristic of individual secret holders.

In proposed hybrid threshold schemes we can apply behavioural features for marking secret parts. The difference between biometric threshold schemes, where classic biometrics are used for the process of marking individual parts of the secret, consists in the fact that an extraction of behavioural features will unambiguously assign the secret holder to the characteristic feature. Behavioural features allow for an unambiguous "identification" of their holder. They are unique in a complete degree, and their repetitions cycles (multiples) characteristic for a given holder allow to confirm their proper attribution to right holder. Behavioural features as a group of personal characteristics (broadly understood biometrics) allow for proper identification of their owners.

Figure 1 presents an idea of used behavioural features for labelling process in hybrid threshold schemes.

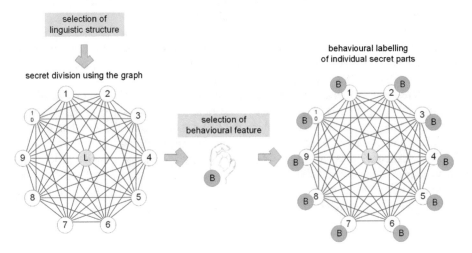

Fig. 1. An idea of behavioral labelling in hybrid threshold schemes.

The selected type of behavioral features (features of the gesture O.K.) is therefore used to designate a part of the shared secret. This secret is shared using a graph structure between 10 participants of the protocol with an additional linguistic shadow. In this example all normal secret parts are marking by biometric.

3   Possible Application of Proposed Methods

The proposed data sharing solutions have a wide range of practical applications. The most important of them are data encryption and secret management [3, 11]. Data encryption processes are implemented through the use of secret sharing protocols, including linguistic and biometric procedures in hybrid threshold schemes. Secret management processes are implemented by applying the discussed procedures in different structures and at different levels of management.

In the procedure supporting knowledge (secret) management processes, different levels of data management are considered, which include Cloud, Edge, Fog and the structure of a given entity. Secret management can take place at different levels and the shadow distribution process can be carried out at different levels. The shadow distribution process can occur within a given level (Cloud, Edge, Fog) but it can also get off to the higher level. The first case is characteristic of layered structures, while the second is an example of hierarchical structures. While the shadow distribution process performed both within a higher level and the basic level (in the case of a privileged distribution) is characteristic of mixed structures.

Figure 2 presents examples of distribution secret parts between the Cloud and Edge levels. Figure 2 presents two different situation. Generally secret was split by the 10 participants with one additional part – linguistic shadow. Secret parts can be distributed by different schemes at different management levels.

At the Cloud level it is necessary to put the smallest number of shadows (in this case it is 5 shadows). Combination of all five secret parts can reveal the secret. In this case we can see situation in which the shadows number 1, 3, 5, 7, 9 reproduced the secret (without the use of a linguistic shadow).

Different situation presents the second example – used of hybrid threshold scheme at Edge level. At this level is necessary to combine six secret parts to reveal the secret. In this case we can see situation in which the six selected parts – shadows number 2, 4, 5, 7, L, 10 reproduced the secret (included the linguistic shadow).

The presented data sharing techniques support knowledge management (secret management) processes within any structure and at any management level. They are universal and can be used independently of the entity in which the data security processes are implemented.

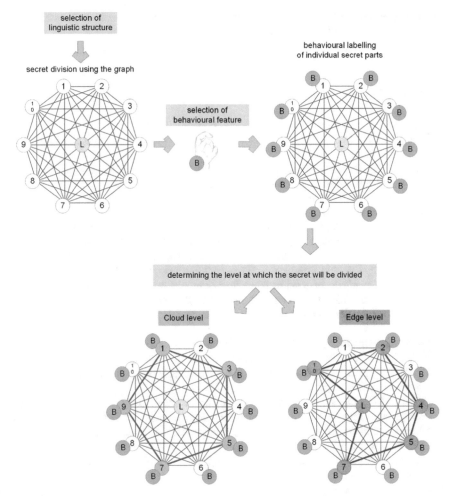

Fig. 2. Examples of possible used hybrid threshold schemes at Cloud and Edge levels.

4   Conclusions

In this paper new methods for data security have been proposed. These methods based on the application of secret sharing techniques, have been enriched with the possibility of using behavioral features in biometric coding modules. Biometric labelling of information is a required secret securing process carried out in biometric cryptographic protocols. At the same time, in the hybrid threshold schemes, the stages of semantic interpretation of the secret are implemented on protocol participants' knowledge.

These data security schemes can be used in secret management processes, which are carried out at various levels inside and outside the structures. The ability to protect data beyond the basic structure applies to the Cloud, Edge and Fog areas.

Acknowledgments. This work has been supported by the National Science Centre, Poland, under project number DEC-2016/23/B/HS4/00616.

This work was supported by JSPS KAKENHI grant number 15H0295.

References

1. Chomsky, N.: Syntactic Structures. Mouton, London (1957)
2. Gregg, M., Schneier, B.: Security Practitioner and Cryptography Handbook and Study Guide Set. Wiley, Hoboken (2014)
3. Laudon, K.C., Laudon, J.P.: Management Information Systems – Managing the Digital Firm (7th ed.), Prentice-Hall International Inc. (2002)
4. Menezes, A., van Oorschot, P., Vanstone, S.: Handbook of Applied Cryptography. CRC Press, Waterloo (2001)
5. Nakamura, S., Ogiela, L., Enokido, T., Takizawa, M.: Flexible synchronization protocol to prevent illegal information flow in peer-to-peer publish/subscribe systems. In: Barolli, L., Terzo, O. (Eds.), Complex, Intelligent, and Software Intensive Systems, Advances in Intelligent Systems and Computing, 611, 82–93. Springer, Cham (2018)
6. Ogiela, L.: Cognitive computational intelligence in medical pattern semantic understanding. In: Guo, M.Z., Zhao, L., Wang, L.P.: ICNC 2008: Fourth International Conference on Natural Computation, vol. 6 Proceedings. Jian, Peoples R China, pp. 245–247, October 18–20, 2008
7. Ogiela, L.: Cryptographic techniques of strategic data splitting and secure information management. Pervasive and Mobile Computing. **29**, 130–141 (2016)
8. Ogiela, L., Ogiela, M.R.: Insider threats and cryptographic techniques in secure information management. IEEE Syst. J. **11**(2), 405–414 (2017)
9. Ogiela, M.R., Ogiela, L., Ogiela, U.: Intelligent Bio-inspired Approach for Secrecy Management in the Cloud, Ninth International Conference on P2P, Parallel, Grid, Cloud and Internet Computing 3PGCIC, Guangzhou, Peoples R China, pp. 286–289, 8–10 Nov, 2014
10. Ogiela, M.R., Ogiela, U.: Secure information splitting using grammar schemes, new challenges in computational collective intelligence. Stud. Comput. Intell. **244**, 327–336 (2009)
11. Ogiela, M.R., Ogiela, U.: Secure information management in hierarchical structures, advanced computer science and information technology. Commun. Comput. Info. Sci. **195**, 31–35 (2011)
12. Ogiela, U., Takizawa, M., Ogiela, L.: Classification of cognitive service management systems in cloud computing. In: Barolli, L., Xhafa, F., Conesa, J. (eds.) Advances on Broad-Band Wireless Computing, Communication and Applications BWCCA 2017, Lecture Notes on Data Engineering and Communications Technologies, vol. 12, pp. 309–313, Springer International Publishing AG 2018. https://doi.org/10.1007/978-3-319-69811-3_28
13. Yan, S.Y.: Computational Number Theory and Modern Cryptography. Wiley, Hoboken (2013)

The Spatiotemporal Prediction Model of Opioids Spread Trend Based on Grey Correlation

Tingting Rao, Caiquan Xiong[(⊠)], Yi Liang, and Shishuang Deng

School of Computer Science, Hubei University of Technology,
Wuhan 430068, China
rao.tting@gmail.com, x_cquan@163.com

Abstract. With the rapid increasing of opioids abuse, it is critical to determine the influence factors and to predict the opioids spread trend. In this paper, a prediction method combining spatiotemporal characteristics and grey relational analysis model is proposed, and the spatiotemporal prediction model of opioids spread trend based on the grey correlation of multifactor is established by integrating various panel data. The time series model is used to identify and fit the multifactor panel data. The gray relational prediction model is established combining the spatial influence factors by Principal Component Analysis (PCA). Results of the simulated experiment show that the method is accurate and the model is feasible and reasonable.

1 Introduction

The rapid increasing in the use of opioid drugs causes crisis in the United States. How to predict and control the crisis has become a problem. The grey relational degree analysis is widely used for predicting infectious, i.e. the Grey Incidence Model proposed by YU [1]. The traditional grey correlation model only analyses influence factors in single dimension time series. In practice, we need to consider more complex cases. In order to overcome this shortcoming, we propose a grey prediction model based on the principle of linear regression with spatiotemporal multivariate features. Wang [2] improved grey correlation analysis combined with the Analysis Hierarchy Process (AHP). We are inspired from this method and innovate grey correlation model with the Principal Component Analysis (PCA) to obtain weight values of indicators. These indicators are time variation, location of spread, and personal characteristics.

Therefore, we devise the spatiotemporal prediction model based on grey correlation to predict the overdose trend of opioid drugs such as Oxycodone and Heroin. We also identify that the origins are Cabell and Kanawha in West Virginia from which opioids spread to adjacent counties within several years. According to the specific characteristics of the dispersion pattern and the spread rule [3], our model combines time variation with space distribution. The threshold value of opioids proportion in use is 0.75, concluded from the model prediction test. By measuring and calculating the proportion of the spatial and temporal evolution similarity with the panel data in NFLIS

© Springer Nature Switzerland AG 2020
L. Barolli et al. (Eds.): CISIS 2019, AISC 993, pp. 165–175, 2020.
https://doi.org/10.1007/978-3-030-22354-0_16

of America, the characteristics of special opioid spread trend can be accurately described and predicted with the absolute correlation degree of 99.17%.

Our work consists of the following four steps. First, we deeply analyze the socio-economic factors of the U.S. Census Bureau and remove the default indicators or objects and retained 115 relevant indexes. In order to conform our model, the original data need to be standardized [4]. Secondly, we assume the 115 standardized indexes as potential factors and applied the Principal Component Analysis (PCA) [5] to acquire 20 maximum weight of index. According to the similarity of these indexes, we obtain 6 criteria indexes by R-means Cluster Algorithm. Thirdly, we validate how to combine the original indexes with these important socioeconomic factors through analogue simulation. We select two parameters: time variation and the spatial variation of spread trend for sensitivity analysis. The prediction is more precise with the identification coefficient of 0.05 and the development coefficient of −0.6. Finally, we reach our conclusion and predict the next ten years of opioids overdose trend. The significant parameter boundaries of the result dependence are determined.

2   Spread Rules of Opioids

In order to describe the spread trend of the opioids, we need to combine time dimension and the spatial dimension simultaneously. Firstly, we calculate the proportion of drug report for all opioids medicine in each state. As shown in the Table 1, we discover the overdose proportion of opioid is generally increasing. The highest proportion of states is West Virginia (WV).

Table 1. The proportion of opioid reports in five states from 2010 to 2017

Year	KY	OH	PA	VA	WV
2010	0.3533	0.2776	0.2202	0.2095	0.3334
2011	0.3638	0.2852	0.2303	0.2330	0.3513
2012	0.3899	0.2710	0.2540	0.2428	0.3580
2013	0.4157	0.2864	0.2831	0.2448	0.4465
2014	0.4092	0.3043	0.3221	0.2801	0.4736
2015	0.3822	0.3401	0.3404	0.3167	0.4810
2016	0.3427	0.3684	0.3615	0.3040	0.4714
2017	0.3254	0.3863	0.4057	0.2824	0.4395

Due to the complexity of geographic location information and numerous statistics, we use the software ARCGIS to perform cluster analysis of locations on the corresponding map. We observed the distribution and trend of opioid overdose data and use different colors to distinguish the proportion of incident among all check reports in every country. Experiments results show that the origin is the countries of West Virginia (WV). The nearer to the origin, the higher proportion of opioid overdose is. Combining the data and geographic information in the map, we conclude the origin are Cabell and Kanawha in West Virginia, US. The spread trend pattern and range of opioids inspired us to consider time and spatial variation these two main dimensions in the Grey Correlation Model.

Then, we calculate the proportion of every opioid in the drug report. As shown in the Fig. 1, the top two opioids are Heroin and Oxycodone. We select these two opioids which were used most in the later research.

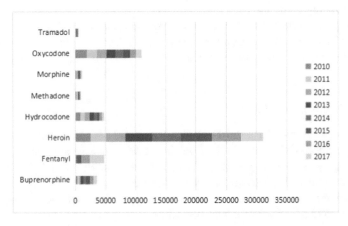

Fig. 1. Eight most reported cases of opioids

Based on the analyses above, we found the origin where specific opioid overdose might have started in the five states and summarized the rule of the spread trend of special opioids overdose in time and spatial dimensions. In this paper, we will propose a prediction model based on spatiotemporal factors and grey correlation to describe and predict the spread trend characteristics of the opioid overdose.

3 Spatiotemporal Grey Correlation Opioids Prediction Model

The model proposed in this paper is called Spatiotemporal Grey Correlation Opioids Prediction Model (SGCOPM), which can describe and predict the spread trend of the opioid overdose. According to the data preprocessing and analysis, the transmission of opioids mainly spread in time and space dimensions. The indexes observation values of each sample at different time points can be set as a data sequence. The panel data is a set composed of multiple samples in different dimensions [6]. We selected counties of states, the specific opioids, and the proportion of overused opioids in drug reports within ten years as indexes of the model [7].

Assume that $x_i(t)$ is the amount (observation value) of opioids in the i-th country in every t year. At the same location, the initial data sets in different time is represented as:

$$A_i^{(0)}(t) = \left(x_i^{(0)}(1), \ldots, x_i^{(0)}(n) \right) \tag{1}$$

At the same time, the initial data sets across different countries is represented as:

$$B_i^{(0)}(t) = \left(x_1^{(0)}(t), \ldots, x_N^{(0)}(t) \right) \tag{2}$$

The n is the maximum length of a time series and the N is the total number of samples in the panel data [8–10]. We set $A_i^{(0)}(t)$ as the observation sequence in the same countries at different time t, and $B_i^{(0)}(t)$ is the observation sequence at the same time t in different countries [11].

We analyze the incidents of illegal opioid overdose in five states based on the data, and established the Spatiotemporal Grey Correlation Opioids Prediction Model (SGCOPM), which take the time dimension and the spatial dimension into consideration.

3.1   The Time Dimension

In order to analyze the characteristics of opioid spread over time in a county. We can compare the trend of the proportion of opioids at different time in a certain county and establish a gray correlation prediction model about the time series [12]. We accumulated and generated the data of each year. According to the above formula, the predicted value of a county at different times can be obtained by subtracting consecutively the initial formula. The result is represented as:

$$A_i^{(1)} = \left(A_i^{(1)}(1), A_i^{(1)}(2), \ldots, A_i^{(1)}(n) \right), (i = 1, \cdots, N) \tag{3}$$

$$A_i^{(1)}(p) = \sum_{t=1}^{p} x_i^{(0)}(t), p = 1, 2, \ldots, n \tag{4}$$

Establish the linear differential equation of $A_p^{(0)}$:

$$\frac{dA_i^{(1)}}{dt} + aA_i^{(1)} = b \left(A_i^{(1)} \right)^2 \tag{5}$$

The undetermined coefficients a and b are called development coefficient and grey action respectively, set vectors $u = (a, b)^T$ as parameter. If $(R^T \cdot R)^{-1}$ exist $\widehat{u} = (\widehat{a}, \widehat{b})^T = (R^T R)^{-1} R^T Y$ then:

$$R = \left\{ \begin{matrix} -z_2^{(1)}(t) & \left(z_2^{(1)}(t) \right)^2 \\ -z_3^{(1)}(t) & \left(z_3^{(1)}(t) \right)^2 \\ \vdots & \vdots \\ -z_N^{(1)}(t) & \left(z_N^{(1)}(t) \right)^2 \end{matrix} \right\}, \quad Y = \left\{ \begin{matrix} x_P^{(1)}(2) \\ x_P^{(1)}(3) \\ \vdots \\ x_P^{(1)}(n) \end{matrix} \right\} \tag{6}$$

The R is the data matrix and the Y is the data vector. According to the above formula, we subtracted consecutively the initial formula. Then, the predicted value of a county in different time can be obtained as follow:

$$\hat{x}_i^{(1)}(t+1) = \frac{\hat{a}x_i^{(0)}(1)}{\hat{b}x_i^{(0)}(1) + \left(\hat{a} - \hat{b}x_i^{(0)}(1)\right)e^{\hat{a}t}} \tag{7}$$

3.2   The Spatial Dimension

In order to analyze the characteristics of opioid spread over locations at the same time. We compare the proportion of opioids overdose by "horizontal distance [13]", "delta distance", "variation distance" to measure similarity of panel data in time and space evolution. It can accurately depict spatial characteristics of panel data. According to formula above, the correlation degree between counties at a certain time node can be obtained:

$$V\left(B_0^{(1)}, B_i^{(1)}\right) = \frac{1}{n}\sum_{i=0}^{n} V\left(B_0^{(1)}(t), B_i^{(1)}(t)\right) \tag{8}$$

3.3   Experiment and Analysis

Step1: Data processing. In the remaining indicators, default data or data significantly higher than the remaining measurement data are considered as invalid data. The average value is used to reflect the level of the indicator. Considering that the primary indicator is affected by the secondary indicator, the factor analysis method only use the primary indicator without the secondary indicator. This step avoids the repetitive correlation between the primary indicator and the secondary indicator. Then, there are 115 indicators left.

Step2: Data standardization. Because different data may have different properties and different dimensions in practical problems, in order to make the original data meet the requirements of our model, it is necessary to standardize the original data. We use the Z-score standardization [14]. The formula is obtained:

$$x'_{ij} = \frac{x_{ij} - \bar{x}_j}{S_j} \tag{9}$$

$\bar{x}_j$ is the average value of the j-th indicator of each county, S is the standard deviation. The standardized values appearing below are always replaced by x_{ij}.

Step3: We apply the Principal Component Analysis (PCA) algorithm [15] to set 115 indicators $(x_1, x_2, \cdots, x_{115})$ for the opioid spread trend as potential factors. In the 115-dimensional space of the original variables, 115 new indexes were composed of eigenvalues.

$$\begin{cases} y_1 = w_{11}x_1 + w_{21}x_2 + \cdots + w_{m1}x_m \\ y_2 = w_{12}x_1 + w_{22}x_2 + \cdots + w_{m2}x_m \\ \quad\vdots \\ y_m = w_{1m}x_1 + w_{2m}x_2 + \cdots + w_{mm}x_m \end{cases} \tag{10}$$

$w_j = (w_{1j}, w_{2j}, \cdots, w_{mj})$ is the eigenvector of m indicators, and the corresponding eigenvalue is $\lambda_1, \lambda_2, \cdots \lambda_m$. Factor analysis is used to explain the information contained in most of the variances in the original data with less than 115 new variables. The influence of the other variances is small.

The feature vector and eigenvalue of the feature variance are the 115 mutually orthogonal directions that maximize the variance value and the variance along these directions. We consider the contribution rate of the principal component in the total trend.

$$C_j = \frac{\lambda_j}{\sum\limits_{k=1}^{m} \lambda_k}, j = 1, 2, \cdots, m \tag{11}$$

Then the cumulative contribution rate of the principal component is $y_1, y_2, \cdots y_p$. The number of principal components is extracted according to the contribution degree in combination with a component having an eigenvalue greater than one as a principal component. A total of 20 principal components were extracted from the indicators by using SPSS software. The results were shown in the Table 2.

Step4: Cluster analysis. We observe the principal components extracted from the principal components as a class, and merge them gradually according to the distance or similarity between the two classes until they are merged into one class. Then, we reduce the dimensions of the indicators, and use R-means cluster analysis [16]. The 20 factors extracted from the principal component analysis of the 115 indicators constitute a 464-dimensional vector. Standardize them, and set the vector of the influencing factors. The result is:

$$U_j = (x_{1j}, x_{2j}, \cdots, x_{464j}), \quad j = 1, 2, \cdots 20 \tag{12}$$

We define the distance between squares of Euclidean distance as:

$$D(U_i, U_j) = \sum_{k=1}^{464} (x_{ki} - x_{kj}) \tag{13}$$

Table 2. Indicators from principal component analysis

Element	Eigenvalue	Contribution	Cumulative contribution (%)
1	22.034	19.160	19.160
2	12.802	11.132	30.292
3	9.596	8.344	38.636
4	5.502	4.784	43.420
5	4.599	3.999	47.419
6	4.289	3.729	51.148
7	3.244	2.821	53.969
8	2.630	2.287	56.256
9	2.600	2.261	58.517
10	2.368	2.059	60.576
11	2.034	1.769	62.344
12	1.915	1.666	64.010
13	1.771	1.540	65.550
14	1.633	1.420	66.970
15	1.584	1.377	68.347
16	1.554	1.351	69.698
17	1.421	1.236	70.933
18	1.356	1.179	72.112
19	1.308	1.137	73.250
20	1.215	1.056	74.306

4 Model Verification and Precision Analysis

According to data preprocessing and our model, we choose the proportion of drug report of West Virginia, US form 2010 to 2017 as reference data. We designed three models to test our Spatiotemporal Grey Correlation Opioids Prediction Model (SGCOPM) below.

(1) *Residue conformity model*

Residue is the difference value between estimated value and the actual value. The difference of degree provided by the residue is usually used to examine the rationality of the model hypothesis and the reliability of the data. We set residue as:

$$\varepsilon_i(t) = \frac{x_i^{(0)}(t) - \hat{x}_i^{(0)}(t)}{x_i^{(0)}(t)} \tag{14}$$

The relative error is: $\Delta_t = \left| \frac{\varepsilon_t}{x_i^{(0)}(t)} \right|$

The average relative error is: $\bar{\Delta} = \frac{1}{n} \sum_{t=1}^{n} \Delta_t$

We define the simulation precision of point t is: $1 - \Delta_t$

If the $\overline{\Delta} < \sigma$ and that prove our model is eligible.

(2) *Correlation degree conformity model*

We define grey absolute correlation degree as:

$$g = \frac{1 + |s_0| + |s_i|}{1 + |s_0| + |s_i| + |s_i - s_0|} \tag{15}$$

$$|s_i| = \left| \sum_{t=2}^{n-1} x_i^0(t) + \frac{1}{2} x_i^0(n) \right| \tag{16}$$

$x_i^0(t)$ is the starting point of x_i, if $g_0 > 0$ and $g > g_0$, that prove our model is eligible.

(3) *Mean variance ratio conformity model*

Mean variance ratio is the sum of squares of distances from the true value of each data, which is also the sum of squares of errors. The mean square error can be used to determine the effectiveness of the model and reflect the relationship between the predicted data series and the true value.

$$S_1^2 = \frac{1}{n} \sum_{t=1}^{n} \left(x_i^{(0)}(t) - \bar{x} \right)^2, \quad S_1^2 = \frac{1}{n} \sum_{t=1}^{n} (\varepsilon_k - \bar{\varepsilon})^2 \tag{17}$$

Mean square variance ratio is $C = S_1/S_2$, if $C_0 > 0$ and $C < C_0$, that prove our model is eligible.

(4) *Model Verification Result*

By simulation with MATLAB, we obtain the value as shown in the Fig. 2 below.

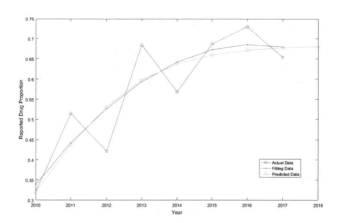

Fig. 2. Fitting of actual and predicted value of opioid proportions in WV.

Before the three test models and computer simulation, we observed that the threshold was positively correlated with the spatiotemporal factors and the value is 0.70, which not accurate. As shown in the Fig. 2, the blue discrete points are the actual values before model verification, the red line is the fitting model result. The results of our model modification are as follows.

Table 3. Forecast value and error of WV state from 2010 to 2017

Year	Initial data	Predict data	Residue	Relative error
2010	0.3334	0.3334	0.0000	0.0000
2011	0.3513	0.3651	−0.0138	0.0392
2012	0.3580	0.3920	−0.0339	0.0948
2013	0.4465	0.4140	0.0325	0.0728
2014	0.4736	0.4314	0.0421	0.0890
2015	0.4810	0.4450	0.0360	0.0749
2016	0.4714	0.4553	0.0161	0.0342
2017	0.4395	0.4630	−0.0235	0.0535

As shown in the Table 3, we calculate the relative error is 5.543% in average and correlation degree g = 0.9917. According to the accuracy test level of gray correlation model, our model is in the level second, which indicate the high conformity and proves the accuracy and rationality of our Spatiotemporal Grey Correlation Opioids Prediction Model (SGCOPM). Finally, we predict the proportion of opioid Spread trend in next ten years in the West Virginia.

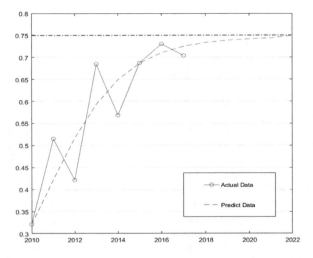

Fig. 3. Prediction result of opioid overdose proportions in WV.

After our verification and precision, the dotted curve in the Fig. 3 is the precise spatiotemporal opioid spread trend prediction characteristics values, which is our model prediction results. In our verification model, the identification threshold levels occur in Cabell country of West Virginia, US in 2022, which match perfectly with the data from NFLIS of America. The proportion of opioids overdose is the highest in the origin. In the time series, the proportion is getting higher, and the maximum threshold is 0.75. The opioids will spread from the origin Cabell and Kanawha in West Virginia to surrounding country and states in the space dimension.

5   Conclusion

In this paper, we improved the traditional gray correlation method and established the spatiotemporal prediction model of opioids spread trend based on grey correlation. Based on the high correlation among the indicators, the indicators affecting opioids were downgraded, and six representative benchmark indicators were extracted to describe the influencing factors of opioid overdose spread trend. Results show that the spatiotemporal prediction model of opioids spread trend is able to describe and predict accurately the spread trend and characteristics of the opioid overdose between five states and counties over time and location. The proposed method contributes to put forward the possible strategy for countering the opioid crisis.

Acknowledgments. This research is supported by Chinese National College Students' innovation and entrepreneurship training programs under grant number 201810500032.

References

1. Yu, Y., Nie, S.: Research progress on prediction and model selection of infectious diseases. Public Health Prev. Med. **29**(05), 89–92 (2008)
2. Wang, Y., Li, K.: Evaluation method for Green jack-up drilling platform design scheme based on improved grey correlation analysis. Appl. Ocean Res. **85**(08), 119–127 (2019)
3. Mohamadi, A., Chan, J.: Risk factors and pooled rate of prolonged opioid use following trauma or surgery. J. Bone Joint Surg. **100**(15), 1332–1340 (2018)
4. Qian, W., Wang, Y.: Grey matrix correlation model based on multi-index panel data and its application. Syst. Eng. **31**(10), 70–74 (2013)
5. Yu, T., Zhou, Y.: Time series prediction based on gray GM(1,1) model. Microcomput. Appli. **31**(13), 65–67 (2012)
6. Su, B., Cao, Y.: Grey prediction model of multi-factor time series. J. xi 'an Univ. Archit. Techno. (natural science edition). **39**(02), 289–292 (2007)
7. Shen, M., Xue, X., Zhang, X.: Selection of resolution coefficient in grey relational analysis. J. Air Force Eng. Univ. (natural science edition). **4**(01), 68–70 (2003)
8. Dang, Y., Shang, Z.: A novel grey incidence model for the relationship between indicators of panel data and its application. Control and Decision **18**(10), 15–38 (2018)
9. Jiang, K., Cai, Z., Lu, Z., Anterograde joint prediction model of chaotic time series based on RBF neural network. J. Wuhan Univ. Technol. (traffic science and engineering edition) **2**(02), 259–261 + 340 (2007)

10. Ding, S., Sang, Y.: Multivariable grey prediction model based on interaction and its application. Syst. Eng. Electron. Technol. **40**(03), 595–602 (2008)
11. Li, Y., Zhu, S.: Multi-attribute grey target decision method with three-parameter interval grey number. Grey Syst. **6**(2), 270–280 (2016)
12. Dan, R., Wang, S.: Research on combination forecasting model based on time series model and grey model. J. Yanshan Univ. **36**(01), 79–83 (2012)
13. Liu, R., Gao, X.: Uncertain multiple attribute decision making method with interval index and weight based on grey entropy model and its application. Control and Decision **10**(13), 13–19 (2018)
14. Higham, S., Bernstein, L.: Opioid distributors sued by West Virginia counties hit by drug crisis. Washington Post **6**(02), 80–88 (2017)
15. Fang, Y., Li, G.: Decision-making evaluation method for regional rail transit system based on grey entropy. Syst. Eng. **33**(02), 152–158 (2015)
16. Quan, J., Zeng, B.: Maximum entropy methods for weighted grey incidence analysis and applications. Grey Syst. **8**(2), 144–155 (2018)

Transferring Informal Text in Arabic as Low Resource Languages: State-of-the-Art and Future Research Directions

Ebtesam H. Almansor[1,2(✉)], Ahmed Al-Ani[1], and Farookh Khadeer Hussain[1]

[1] School of Computer Science, Faculty of Engineering and Information Technology,
University of Technology Sydney, Sydney, Australia
ebtesamhussain.almansor@student.uts.edu.au,
{ahmed.al-ani,farookh.hussain}@uts.edu.au
[2] Community College, Najran University, Najran, Saudi Arabia

Abstract. Rapid growth in internet technology lead to increase the usage of social media platforms which make communication between users easier. Through the communication users used their daily languages which considered as non-standard language. The non-slandered text contains lots of noise, such as abbreviations, slang which used more in English languages and dialect words which are widely used in Arabic language. These texts face challenging using any natural language processing tools. Therefore, these texts need to be treated and transferred to be similar to their standard form. According to that the normalization and translation approach have been used to transfer the informal text. However, using these approach need large label or parallel datasets. While high resource languages such as English have enough parallel datasets, low resource languages such as Arabic is lack of enough parallel dataset. Therefore, in this paper we focus on the Arabic and Arabic dialects as a low resource language in the era of transferring non-stander text using normalization and translation approach.

Keywords: Transferring informal text ·
Machine translation and normalization · Low resource languages

1 Introduction

In the era of advanced technology, the internet considered as a key source for various information. Valuable information can be found on different types of media that are available in the internet, such as images, texts, and videos [1,2]. Users are sharing their experiences and showing their opinions about products, news, and events when they used these sites [3]. Importantly, users can use their own languages which make communication between users has become easier. The text written in these sites is usually informal and is different from the formal

© Springer Nature Switzerland AG 2020
L. Barolli et al. (Eds.): CISIS 2019, AISC 993, pp. 176–187, 2020.
https://doi.org/10.1007/978-3-030-22354-0_17

text written in newspapers and books [2]. These text considered as non-standard text which has huge of noise, such as abbreviations slang and dialect words [1,4].

The Natural Language Processing (NLP) tools are designed for standard text which make no-standard text face challenging using these tools [5]. Based on that, the importance of treading and transferring theses text are increase. There are two approach that have been used for transferring these texts namely normalization and translation. Using these approaches required large parallel datasets [6].

Parallel corpus is the main important component of machine translation and normalization; however, collecting these data are time consuming and costly. The availability of large dataset are easy defined in high resource languages such as English. Whereas, the low resource languages that suffer from sufficient dataset such as Arabic has no much available datasets which make it difficult to used translation or normalisation approaches [6,7]. Low resource languages lack large parallel datasets, and encounter difficulties in using machine translation. More recently, these languages received attention from the machine translation community [8,9].

The best of our knowledge, no article review the literature from the respective of the transforming approach between low resource languages (Arabic as use case) with high resource language (English). Therefore, we address this issue by comparing both languages based on the transformation approach what used for non-standard text. This paper, is the first paper that highlighted the importance of transferring dialectal Arabic as low resource language. In addition, we proposed a taxonomy for the existing approach that used to transfer the non-standard text and more details has been explained in the next sections.

This paper is organized as follows: the first section presents the proposed taxonomy for transferring approaches; the second section presents a detailed of normalization approach; the third section explain translation approach; the fourth section highlights the importance of the existing model in low resource language, finally, fifth section presents the conclusion and future work.

2   Normalization

Normalization can play an important role in processing non-standard (informal) text that is used in social media platforms. Non-standard words add noise to the data as they increase the amount of unknown words [11]. There are numerous issues that cause non-standard words, for example, use of capitalization to emphasise some words (e.g. "LOOK AT this" and "WOOW"). Pennell and Liu [11] found that capitalization can affect some neural language processing (NLP) tasks such as name recognition, as this issue makes the identification process for the names more difficult. Capitalization helps with recognizing the names in some languages such as English [11,12]. Additionally, abbreviations and slang that are widely used by the young generation in social media increase non-standard words [11,13]. These texts are translated using different methods (e.g. character level "cu" which means "see you", word level "l8r" which means "later") [11]. There are a number of techniques that have been proposed for the implementation of normalization in NLP that are shown in Fig. 1. Inaddtion, The use of social media in the Arab World has increased recently. There are two types of the Arabic language. The first one is Modern Standard Arabic (MSA) which is the formal Arabic language used in newspapers and books; and the second type is colloquia or dialect (DA), which is a spoken language that is used in daily life and in social media [14].

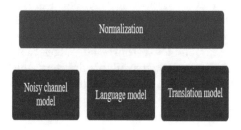

Fig. 1. Approaches that are used for normalization

Normalization is an important aspect for the Arabic language as it is used as a pre-processing step that helps to increase readability and machine processing capability. Simple normalization steps are used in Arabic; for example, an algorithm that can normalize the texts in social media has been developed to normalize the Alef (أ ، آ ، إ) to (ا), the Alef maksoura (ى) to (ي), Ta Marbota (ة) to (ه) and remove the diacritics as a basic normalization process [15]. In addition, it eliminates any repeated unnecessary letters in a word, and then tries to choose the most frequent words in corps (e.g., كثييييير it will be normalized to كثير).

3   Inference Approach for Noisy Text

3.1   Noisy Channel Model

Noisy channel model aims to find the most applicable candidate for handling noisy text. Choudhury et al. [16] proposed a supervised noisy channel model using Hidden Markov Model (HMM) and this is used to normalize the abbreviation in the Short Message Service (SMS). Similarly, Deepak and Subramaniam [17], developed a supervised model that can learn from a pair of data such as (childe, chld), and this model was applied to clean SMS messages. Cook and Stevenson [18] attempted to improve the previous methods with the aim of developing unsupervised noisy channel model, where the authors categorized the abbreviations into eleven types. In addition, another noisy channel model that depends on orthographic edit distance has been used on the web to collect noisy text automatically and use it for spelling suggestion [11,19]. Beaufort et al. [20] developed a combination of rule-based finite–state with noisy channel model that was applied to French SMS.

The noisy channel model is based on an equation that is stated below, where e is the correct word and f is the noisy word [21].

$$\hat{e} = argmax P(e|f) \tag{1}$$

$$= argmax \frac{P(f|e)P(e)}{P(f)} \tag{2}$$

$$\hat{e} = argmax \quad \underbrace{P(f|e)}_{translation model} \quad . \quad \underbrace{P(e)}_{language model} \tag{3}$$

3.2 Language Model

Language models (LM) are based on conditional probabilities that are assigned to sequence of words or tokens over a fixed vocabulary. LM presents the probability of the current words given the previous words [22]. Han and Baldwin [23] used a combination of language models, lexical and phonetic edit distance strategy to generate the candidate (correct form) for the out-of-vocabulary (OOV) or ill- formed words used in Twitter and SMS messages. Pennell and Liu [24] developed a deletion based system that is based on the LM and statistic model. This model aims to identify deleted characters from words.

3.3 Translation Model

Translation model is a technique that transfers the source sentence into the target sentence and it can be trained at word, phrase and character level. Bangalore et al. [25] used a consensus translation to bootstrap, which helps to translate the abbreviations that are commonly written in messages or chat rooms. Aw et al. [26] proposed a phrase statistic model in order to normalize SMS messages written in English. The research of Hernández [27] aimed to normalize Twitter messages by training the translation system on three different dictionaries. Contractor et al. [28] used statistic machine translation (SMT) but they faced difficulties in collecting and annotating a large parallel corpus. Character level MT and statistical classifiers were used to determine if characters were removed from words.

4 Machine Translation

Machine translation has three main categories: rule based, statistical and neural machine translation. Figure 2 shows overview of the classification of the Machine Translation approaches along three dimensions.

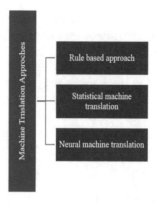

Fig. 2. Machine translation approaches

4.1   Rule-Based Approach

The rule-based approach utilizes a set of linguistic rules to translate text from the source to target languages. The rules are usually constructed by a language expert [29]. This approach also relies on bilingual dictionaries. Rule-based approach is used to translate between Arabic and other languages. Salem et al. [30] presented a system called UniArab that used Role and Reference Grammar (RRG) to support the development of a rule-based lexical framework for the Arabic language [10,30]. As a morphological language, Habash and Sadat and, Lee [31,32] showed the importance of tokenization for Arabic translation. Furthermore, Attia [33] implemented a morphological analyser that was based on rules and dealt with the stem as the main form of the Arabic words. Similarly, in 2007 Attia et al. [34], proposed a rule-based tokenization that can be used as a pre-processing step in machine translation. Phillips and Cavalli-Sforza [35] used Example Based Machine Translation (EBMT) to translate Arabic to English. The authors also used Buckwalter Arabic Morphological Analyser (BAMA) due to the rich morphological features of the Arabic language. Alansary et al. [36] developed a system that contains four dimensions of analysis, including morphological, lexical, syntactic and semantic, and this system has been tested on an Arabic corpus. In the same year, Köprü and Miller [37] implemented a morphological analyser and generator for Arabic; this system is based on a combination of rule-based and statistic machine translation. Habash [38] discussed the issue which is also related to Arabic script and machine translation. He focused on the Arabic orthography, phonology, morphology, semantic and syntax. Bisazza and Federico [39] presented a chunk-based reordering technique that can identify and move a clause-initial verb in the Arabic side. This approach only focused on reordering the verbs in verb-subject-object (VSO) sentences between Arabic-English translation. Later on, Carpuat et al. [40] used a syntactic parser to reorder the VS constrictions into the SV for word alignment between Arabic-English, and this approach had improved the overall SMT. In addition, a system called "Napae-Rbmt" translation was developed to translate Arabic noun phrases into English [41]. Another transfer-based approach was developed to build a translation model from Arabic-English with a focus on noun and verb phrases [10].

Rule-based approach is currently the most popular method for of the translation the dialectal Arabic. Habash et al. [42] implemented a parser for Levantine (LA) Arabic using MSA Treebank, then LA-MSA lexicon; syntactic and morphological rules were used to map from LA-MSA. Another study also focussed on transferring the Egyptian (EGY) texts to MSA based on the rule- and corpus-based approaches [5]. Bakr et al. [43] proposed hybrid methods using the EGY-MSA lexicon and morphological analysis to translate EGY-MSA. Sawaf [44] used a rule-based approach at character level and morphological analysis, and this approach was applied to translate the EGY to MSA. Salloum and Habash [45] reduced the out of vocabulary (OOV) of the dialect word by mapping the affixed morphemes to their MSA. On the other hand, Mohamed at el. [46] built a translation model that translates MSA to EGY using the rule-based approach. Another lexicon for the EGY words and their MSA synonym was built based on the semantic and part-of-speech (POS) tag features [47]. Boujelbane at el. [48] developed a translation tool for building Tunisian lexical. More recently, El-taher at el. [49] built a translation model from the Egyptian to the Modern Standard Arabic using different rules and DA-MSA dictionary.

While rule based approaches are widely used, however, it requires a large dictionary and a good set of linguistic rules. It is extremely difficult to build a comprehensive dictionary that covers the vast majority of words. Building less comprehensive dictionaries that are relatively large is still expensive and time consuming. In addition, a comprehensive set of rules requires a linguistic expert and is also expensive to construct [50]. Due to the these issues, the practical application of rule-based approaches is challenging.

4.2 Statistical Approach

The main difference between rule-based and statistical approaches is that the statistical approach is based on probability without any linguistic knowledge. Statistical machine translation has been widely used in Arabic and other languages. There are various methods that have been developed. For example, Stalls and Knight [51] adapted the statistical approach to translate between Arabic and English. A morphological static model was also implemented to translate between Arabic and English [52]. Other than English, Hasan et al. [53] proposed a statistic approach that can translate from Arabic to French in the medical field. Habash and Hu [54] enhanced the Arabic-Chinese statistic machine translation by using English as pivotal language. Also, segmentation of Arabic text showed improvement in the statistical translation system between Arabic and English [55].

Also, dialects have been translated to English and Arabic using the statistic machine translation approach. Riesa and Yarowsky [56] developed a statistical morphological segmenter for Levantine and Iraqi dialects. Similar to the previous work, Mansour [57] developed a new segmentation method and compared it with different segmenters. The aim for this study was evaluating Arabic segementers for Arabic statistical machine translation. Sajjad at el. [58] built static MT that translate between Egyptian (EGY) and English. As a pre-processing step, the author used character level translation model to translate the EGY to MSA. Sajjad et al. [59] implemented the statistical translation model that translates between Egyptian to English; this model used additional features, such as those produced by the sequence model.

4.3   Neural Machine Translation

In recent years, there has been an increasing interest in neural machine translation (NMT). This approach trains a neural network to compute the probability of the target sentence [60]. Sutskever et al. [61] has used seq2seq model to map sequence of the input to vectors of fixed dimension and it used to translate English to French languages. Bahdanau et al. [60] introduced the idea of the attention mechanism which gives the decoder a way to pay attention to input. In addition, the decoder can also select any part of the input sentence for translation. Attention is calculated using each encoder output and the current hidden state, resulting in a vector the same size as the input sequences This vector is termed as attention weights.

In neural machine translation, there are many proposed alternative methods which support low resource languages. For example, instead of using parallel datasets to train the NMT model, alternative methods that attempt to use monolingual datasets to train the seq2seq model have been developed [70]. Also, an unsupervised pre-trained language model that is trained on monolingual corpus has been proposed to translate German to English [71]. The adaptive attention-based neural network has been used to translate Mongolian-Chinese as low resource language [9].

Most of the neural machine translation research was applied to English and some other languages. There are few studies that used the neural approach to translate Arabic text. Almahairi et al. [73], developed a neural translation model between Arabic and English. There are some studies that used neural features in SMT [74,75]. However, Arabic lacks comprehensive parallel corpus especially for its dialects, which is a main challenge for the utilization of NMT.

From this review we observed that the translation of Arabic and dialectal Arabic has not been thoroughly investigated, particularly using neural network and deep learning approaches. This research issue was addressed by Almansour et al. [76]. As a low resource language, Arabic and dialectal Arabic lack sufficient parallel datasets. Word embedding approach was used as translation model for Arabic and dialectal Arabic [6]. This model overcome the problem of parallel data as it trained on monolingual datasets.

Recent advances in neural machine translation have facilitated investigation dealing with low resource languages. However, most of the models still rely on pre-trained model or large datasets. Moreover, it has been advised that dealing with morphologically rich languages is very challenging, particularly for using neural machine translation [77]. Also, they used RNN as basic encoder and decoder which has limitation to deal with long sentences and using the word as the basic unit in the translation lead to increase the issue of the out of vocabulary.

5   Conclusion and Future Work

This paper has given a comprehensive review of the existing research that has been conducted about non-standard text. We classify these approach into two main categorise namely; normalization and translation. In addition, details have been given about neural machine translation, and its utilization in the translation of low resource languages. This review has focused on Arabic and dialectal Arabic as low-resource languages.

Further research may focuses on these point as future directions:

- The lack of comprehensive parallel datasets which can help the research to investigate translation task using advanced approach.
- Translating dialectal Arabic still in early stage they used basic approaches such as rule based or statist approach, however, there is an initial work that used neural network. Further, research that used neural net need more investigate.

References

1. Han, B.: Improving the utility of social media with natural language processing. Ph.D. thesis, University of Melbourne, Department of Computing and Information Systems (2014)
2. Almansor, E.H.: Translating Arabic as low resource language using distribution representation and neural machine translation models. Ph.D. thesis (2018)
3. Hidayatullah, A.F.: Language tweet characteristics of Indonesian citizens. In: 2015 International Conference on Science and Technology (TICST), pp. 397–401. IEEE (2015)
4. Ghareb, A.S., Hamdan, A.R., Bakar, A.A., Yaakub, M.R.: Hybrid statistical rule-based classifier for Arabic text mining. J. Theoret. Appl. Inf. Technol. **71**(2) (2015)
5. Shaalan, K., Bakr, H., Ziedan, I.: Transferring Egyptian colloquial dialect into modern standard Arabic. In: International Conference on Recent Advances in Natural Language Processing (RANLP–2007), Borovets, Bulgaria, pp. 525–529 (2007)
6. Almansor, E.H., Al-Ani, A., Al, A.: Translating dialectal Arabic as low resource language using word embedding. In: RANLP, pp. 52–57 (2017)
7. Zoph, B., Yuret, D., May, J., Knight, K.: Transfer learning for low-resource neural machine translation. arXiv preprint arXiv:1604.02201 (2016)
8. Firat, O., Sankaran, B., Al-Onaizan, Y., Vural, F.T.Y., Cho, K.: Zero-resource translation with multi-lingual neural machine translation. arXiv preprint arXiv:1606.04164 (2016)
9. Wu, J., Hou, H., Shen, Z., Du, J., Li, J.: Adapting attention-based neural network to low-resource Mongolian-Chinese machine translation. In: International Conference on Computer Processing of Oriental Languages, pp. 470–480. Springer (2016)
10. Alqudsi, A., Omar, N., Shaker, K.: Arabic machine translation: a survey. Artif. Intell. Rev. 1–24 (2014)

11. Pennell, D.L., Liu, Y.: Normalization of informal text. Comput. Speech Lang. **28**(1), 256–277 (2014)
12. Sproat, R., Black, A.W., Chen, S., Kumar, S., Ostendorf, M., Richards, C.: Normalization of non-standard words. Comput. Speech Lang. **15**(3), 287–333 (2001)
13. Han, B., Cook, P., Baldwin, T.: Automatically constructing a normalisation dictionary for microblogs. In: Proceedings of the 2012 Joint Conference on Empirical Methods in Natural Language Processing and Computational Natural Language Learning, pp. 421–432. Association for Computational Linguistics (2012)
14. ElSahar, H., El-Beltagy, S.R.: A fully automated approach for Arabic slang lexicon extraction from microblogs. In: CICLing (1), pp. 79–91 (2014)
15. Darwish, K., Magdy, W., Mourad, A.: Language processing for Arabic microblog retrieval. In: Proceedings of the 21st ACM International Conference on Information and Knowledge Management, pp. 2427–2430. ACM (2012)
16. Choudhury, M., Saraf, R., Jain, V., Mukherjee, A., Sarkar, S., Basu, A.: Investigation and modeling of the structure of texting language. Int. J. Doc. Anal. Recogn. **10**(3), 157–174 (2007)
17. Deepak, P., Subramaniam, V.: Correcting SMS text automatically (2012)
18. Cook, P., Stevenson, S.: An unsupervised model for text message normalization. In: Proceedings of the Workshop on Computational Approaches to Linguistic Creativity, pp. 71–78. Association for Computational Linguistics (2009)
19. Whitelaw, C., Hutchinson, B., Chung, G.Y., Ellis, G.: Using the web for language independent spellchecking and autocorrection. In: Proceedings of the 2009 Conference on Empirical Methods in Natural Language Processing, vol. 2, pp. 890–899. Association for Computational Linguistics (2009)
20. Beaufort, R., Roekhaut, S., Cougnon, L.A., Fairon, C.: A hybrid rule/model-based finite-state framework for normalizing SMS messages. In: Proceedings of the 48th Annual Meeting of the Association for Computational Linguistics, pp. 770–779. Association for Computational Linguistics (2010)
21. Aransa, W.: Statistical machine translation of the Arabic language. Ph.D. thesis, Université du Maine, Le Mans, France (2015)
22. Bengio, Y., Ducharme, R., Vincent, P., Jauvin, C.: A neural probabilistic language model. J. Mach. Learn. Res. **3**(Feb), 1137–1155 (2003)
23. Han, B., Baldwin, T.: Lexical normalisation of short text messages: Makn sens a# Twitter. In: Proceedings of the 49th Annual Meeting of the Association for Computational Linguistics: Human Language Technologies, vol. 1, pp. 368–378. Association for Computational Linguistics (2011)
24. Pennell, D., Liu, Y.: Toward text message normalization: modeling abbreviation generation. In: 2011 IEEE International Conference on Acoustics, Speech and Signal Processing (ICASSP), pp. 5364–5367. IEEE (2011)
25. Bangalore, S., Murdock, V., Riccardi, G.: Bootstrapping bilingual data using consensus translation for a multilingual instant messaging system. In: Proceedings of the 19th International Conference on Computational Linguistics, vol. 1, pp. 1–7. Association for Computational Linguistics (2002)
26. Aw, A., Zhang, M., Xiao, J., Su, J.: A phrase-based statistical model for SMS text normalization. In: Proceedings of the COLING/ACL on Main Conference Poster Sessions, pp. 33–40. Association for Computational Linguistics (2006)
27. Hernández, A.: A ngram-based statistical machine translation approach for text normalization on chat-speak style communications (2009)
28. Contractor, D., Faruquie, T.A., Subramaniam, L.V.: Unsupervised cleansing of noisy text. In: Proceedings of the 23rd International Conference on Computational Linguistics: Posters, pp. 189–196. Association for Computational Linguistics (2010)

29. Okpor, M.: Machine translation approaches: issues and challenges. Int. J. Comput. Sci. Issues (IJCSI) **11**(5), 159 (2014)
30. Salem, Y., Hensman, A., Nolan, B.: Implementing Arabic-to-English machine translation using the role and reference grammar linguistic model (2008)
31. Habash, N., Sadat, F.: Arabic preprocessing schemes for statistical machine translation. In: Proceedings of the Human Language Technology Conference of the NAACL, Companion Volume: Short Papers, pp. 49–52. Association for Computational Linguistics (2006)
32. Lee, Y.S., Papineni, K., Roukos, S., Emam, O., Hassan, H.: Language model based Arabic word segmentation. In: Proceedings of the 41st Annual Meeting on Association for Computational Linguistics, vol. 1, pp. 399–406. Association for Computational Linguistics (2003)
33. Attia, M.: Developing a robust Arabic morphological transducer using finite state technology. In: 8th Annual CLUK Research Colloquium, pp. 9–18 (2005)
34. Attia, M.A.: Arabic tokenization system. In: Proceedings of the 2007 Workshop on Computational Approaches to Semitic Languages: Common Issues and Resources, pp. 65–72. Association for Computational Linguistics (2007)
35. Phillips, A.B., Cavalli-Sforza, V.: Arabic-to-English example based machine translation using context-insensitive morphological analysis. Journées d'Etudes sur le Traitement Automatique de la Langue Arabe (2006)
36. Alansary, S., Nagi, M., Adly, N.: Towards analyzing the international corpus of Arabic (ICA): progress of morphological stage. In: 8th International Conference on Language Engineering, Egypt, pp. 1–23 (2008)
37. Köprü, S., Miller, J.: A unification based approach to the morphological analysis and generation of Arabic. In: Farghaly, A., Megerdoomian, K., Sawaf, H. (eds.) 3rd Workshop on Computational Approaches to Arabic Script-based Languages at MT Summit XII. IAMT, Ottowa (2009)
38. Habash, N.: Introduction to Arabic natural language processing. In: Tutoriel in the ACL 43th Annual Meeting (2005)
39. Bisazza, A., Federico, M.: Chunk-based verb reordering in VSO sentences for Arabic-English statistical machine translation. In: Proceedings of the Joint Fifth Workshop on Statistical Machine Translation and MetricsMATR, pp. 235–243. Association for Computational Linguistics (2010)
40. Carpuat, M., Marton, Y., Habash, N.: Improving Arabic-to-English statistical machine translation by reordering post-verbal subjects for alignment. In: Proceedings of the ACL 2010 Conference Short Papers, pp. 178–183. Association for Computational Linguistics (2010)
41. Shirko, O., Omar, N., Arshad, H., Albared, M.: Machine translation of noun phrases from Arabic to English using transfer-based approach. J. Comput. Sci. **6**(3), 350 (2010)
42. Habash, N.Y., Rambow, O.C., Chiang, D., Diab, M., Hwa, R., Sima'an, K., Lacey, V., Levy, R., Nichols, C., Shareef, S.: Parsing Arabic dialects (2006)
43. Bakr, H.A., Shaalan, K., Ziedan, I.: A hybrid approach for converting written Egyptian colloquial dialect into diacritized Arabic. In: The 6th International Conference on Informatics and Systems, infos2008, Cairo University (2008)
44. Sawaf, H.: Arabic dialect handling in hybrid machine translation. In: Proceedings of the Conference of the Association for Machine Translation in the Americas (AMTA), Denver, Colorado (2010)

45. Salloum, W., Habash, N.: Dialectal to standard Arabic paraphrasing to improve Arabic-English statistical machine translation. In: Proceedings of the First Workshop on Algorithms and Resources for Modelling of Dialects and Language Varieties, pp. 10–21. Association for Computational Linguistics (2011)
46. Mohamed, E., Mohit, B., Oflazer, K.: Transforming standard Arabic to colloquial Arabic. In: Proceedings of the 50th Annual Meeting of the Association for Computational Linguistics: Short Papers, vol. 2, pp. 176–180. Association for Computational Linguistics (2012)
47. Al-Sabbagh, R., Girju, R.: Mining the web for the induction of a dialectical Arabic lexicon. In: LREC (2010)
48. Boujelbane, R., Khemakhem, M.E., Belguith, L.H.: Mapping rules for building a Tunisian dialect lexicon and generating corpora. In: IJCNLP, pp. 419–428 (2013)
49. El-taher, F.E.Z., Hammouda, A.A., Abdel-Mageid, S.: Automation of understanding textual contents in social networks. In: 2016 International Conference on Selected Topics in Mobile & Wireless Networking (MoWNeT), pp. 1–7. IEEE (2016)
50. Charoenpornsawat, P., Sornlertlamvanich, V., Charoenporn, T.: Improving translation quality of rule-based machine translation. In: Proceedings of the 2002 COLING Workshop on Machine Translation in Asia, vol. 16, pp. 1–6. Association for Computational Linguistics (2002)
51. Stalls, B.G., Knight, K.: Translating names and technical terms in Arabic text. In: Proceedings of the Workshop on Computational Approaches to Semitic Languages, pp. 34–41. Association for Computational Linguistics (1998)
52. Lee, Y.S.: Morphological analysis for statistical machine translation. In: Proceedings of HLT-NAACL 2004: Short Papers, pp. 57–60. Association for Computational Linguistics (2004)
53. Hasan, S., El Isbihani, A., Ney, H.: Creating a large-scale Arabic to French statistical machine translation system. In: Proceedings of the Fifth International Conference on Language Resources and Evaluation (LREC), pp. 855–858 (2006)
54. Habash, N., Hu, J.: Improving Arabic-Chinese statistical machine translation using English as pivot language. In: Proceedings of the Fourth Workshop on Statistical Machine Translation, pp. 173–181. Association for Computational Linguistics (2009)
55. Badr, I., Zbib, R., Glass, J.: Segmentation for English-to-Arabic statistical machine translation. In: Proceedings of the 46th Annual Meeting of the Association for Computational Linguistics on Human Language Technologies: Short Papers, pp. 153–156. Association for Computational Linguistics (2008)
56. Riesa, J., Yarowsky, D.: Minimally supervised morphological segmentation with applications to machine translation. In: Proceedings of the 7th Conference of the Association for Machine Translation in the Americas (AMTA06), pp. 185–192 (2006)
57. Mansour, S.: MorphTagger: HMM-based Arabic segmentation for statistical machine translation. In: International Workshop on Spoken Language Translation (IWSLT) 2010 (2010)
58. Sajjad, H., Darwish, K., Belinkov, Y.: Translating dialectal Arabic to English. In: ACL, vol. 2, pp. 1–6 (2013)
59. Sajjad, H., Durrani, N., Guzman, F., Nakov, P., Abdelali, A., Vogel, S., Salloum, W., Kholy, A.E., Habash, N.: Egyptian Arabic to English statistical machine translation system for NIST OpenMT 2015. arXiv preprint arXiv:1606.05759 (2016)
60. Bahdanau, D., Cho, K., Bengio, Y.: Neural machine translation by jointly learning to align and translate. arXiv preprint arXiv:1409.0473 (2014)

61. Sutskever, I., Vinyals, O., Le, Q.V.: Sequence to sequence learning with neural networks. In: Advances in Neural Information Processing Systems, pp. 3104–3112 (2014)
62. Chung, J., Cho, K., Bengio, Y.: A character-level decoder without explicit segmentation for neural machine translation. arXiv preprint arXiv:1603.06147 (2016)
63. Ling, W., Luís, T., Marujo, L., Astudillo, R.F., Amir, S., Dyer, C., Black, A.W., Trancoso, I.: Finding function in form: compositional character models for open vocabulary word representation. arXiv preprint arXiv:1508.02096 (2015)
64. Ballesteros, M., Dyer, C., Smith, N.A.: Improved transition-based parsing by modeling characters instead of words with LSTMs. arXiv preprint arXiv:1508.00657 (2015)
65. Zhang, X., Zhao, J., LeCun, Y.: Character-level convolutional networks for text classification. In: Advances in Neural Information Processing Systems, pp. 649–657 (2015)
66. Santos, C.D., Zadrozny, B.: Learning character-level representations for part-of-speech tagging. In: Proceedings of the 31st International Conference on Machine Learning (ICML 2014), pp. 1818–1826 (2014)
67. Kim, Y., Jernite, Y., Sontag, D., Rush, A.M.: Character-aware neural language models. In: AAAI, pp. 2741–2749 (2016)
68. Bahdanau, D., Chorowski, J., Serdyuk, D., Brakel, P., Bengio, Y.: End-to-end attention-based large vocabulary speech recognition. In: 2016 IEEE International Conference on Acoustics, Speech and Signal Processing (ICASSP), pp. 4945–4949. IEEE (2016)
69. Chan, W., Jaitly, N., Le, Q., Vinyals, O.: Listen, attend and spell: a neural network for large vocabulary conversational speech recognition. In: 2016 IEEE International Conference on Acoustics, Speech and Signal Processing (ICASSP), pp. 4960–4964. IEEE (2016)
70. Gulcehre, C., Firat, O., Xu, K., Cho, K., Barrault, L., Lin, H.C., Bougares, F., Schwenk, H., Bengio, Y.: On using monolingual corpora in neural machine translation. arXiv preprint arXiv:1503.03535 (2015)
71. Ramachandran, P., Liu, P.J., Le, Q.V.: Unsupervised pretraining for sequence to sequence learning. arXiv preprint arXiv:1611.02683 (2016)
72. Zhao, S., Zhang, Z.: An efficient character-level neural machine translation. arXiv preprint arXiv:1608.04738 (2016)
73. Almahairi, A., Cho, K., Habash, N., Courville, A.: First result on Arabic neural machine translation. arXiv preprint arXiv:1606.02680 (2016)
74. Devlin, J., Zbib, R., Huang, Z., Lamar, T., Schwartz, R.M., Makhoul, J.: Fast and robust neural network joint models for statistical machine translation. In: ACL, vol. 1, pp. 1370–1380 (2014)
75. Setiawan, H., Huang, Z., Devlin, J., Lamar, T., Zbib, R., Schwartz, R., Makhoul, J.: Statistical machine translation features with multitask tensor networks. arXiv preprint arXiv:1506.00698 (2015)
76. Almansor, E.H., Al-Ani, A.: A hybrid neural machine translation technique for translating low resource languages. In: International Conference on Machine Learning and Data Mining in Pattern Recognition, pp. 347–356. Springer (2018)
77. Costa-Jussà, M.R., Fonollosa, J.A.R.: Character-based neural machine translation. CoRR **abs/1603.00810** (2016)

Consumers' Attitude Toward Cloud Services: Sentiment Mining of Online Consumer Reviews

Asma Musabah Alkalbani[(⊠)]

College of Applied Sciences, Ibri, Sultanate of Oman
asmam.ibr@cas.edu.om

Abstract. Automatically generated cloud services users' experiences summaries could aid potential consumers in selecting cloud services. This study proposes a novel methodology for analysing consumer's attitude toward cloud services by applying sentiment mining on online consumer reviews. The cloud services were collected across different web platforms, then analysed using sentiment analysis to identify the attituded of each cloud services review. The analysis conducted using a data mining tool namely RapidMiner and the proposed model is based on fours supervised machine learning algorithms: Nave Bayes, K-Nearest Neighbour (K-NN), Decision Tree and support vector machine. The results show that the prediction accuracy of the SVM-based TF-IDF approach (10-fold cross validation testing) and Naive Bayes TF-IDF approach (10- fold cross validation testing) is 88.29%. This indicates that Naive Bayes and SVM perform better in determining sentiment than in determining other classifiers.

Keywords: Cloud services reviews · Sentiment mining ·
Cloud polarity dataset · Supervised machine learning · Opinion mining

1 Introduction

Predictive Analytics is growing tremendously in most of the computer science and information technology disciplines as it answers, "What is likely to happen?" [19]. One such discipline is Sentiment mining, which is an active research domain that is concerned with identifying the opinion from the text written by the human in natural language using Artificial Intelligence (AI) [18]. Recently, Sentiment mining (which is also known as Opinion mining) has attracted the attention of the researchers in the different sectors, such as stock market prediction and marketing. Sentiment mining includes the use of machine learning algorithms, which is a type of AI, to classify opinions related to a service/product, such as consumer review [13]. The Sentiment mining method is essential for potential consumers to get sufficient information about the quality of services provided by the service provider; and ensuring the competitiveness of their products. Potential consumers can benefit from Sentiment mining method by having access to the experiences of the previous consumers that allows them to make a better choice when buying a service or a product. Online consumer review

© Springer Nature Switzerland AG 2020
L. Barolli et al. (Eds.): CISIS 2019, AISC 993, pp. 188–199, 2020.
https://doi.org/10.1007/978-3-030-22354-0_18

(which is also known as business review) is an essential element of a marketing strategy as it impacts on the buying decision stated that the second most trusted source of product information is the consumer review after the recommendations from friends and family [15]. According to [6], the consumer review is more user-oriented provides the evaluation of product/service from consumer's perspective. However, it is becoming an increasingly difficult task to read online comments and understand the consumers' opinions due to the large number of online reviews posted across different web platforms. Finally, it is challenging for businesses to derive business insights based on these reviews. In the field of cloud computing, there is an increasing number of cloud consumer reviews for different cloud services. These reviews are posted online across various reviews web portals, such as getApp.com and serchen.com [2, 3]. Such web portals can be an excellent source to understand the cloud consumers' satisfaction. Therefore, there is a need for automating the analysis of consumers' online reviews of cloud services. Furthermore, none of the existing studies considers proposing any means by which cloud consumers can choose a cloud service with the best QoS. According to [1, 4], the Quality of Service information provided by cloud providers are insufficient and cannot guarantee the Quality of Services. With absence of QoS information at the time of making the buying decision by potential consumers, it is important to develop an intelligent yet reliable method to analyse consumers' reviews and determine the intention of the cloud reviewers. Such an approach can give the potential consumer access to the previous buyers' experiences when buying the cloud service. To address these shortcomings, in this paper, we present 'Cloud Trust Derived Cloud Intelligence' methodology to analyse cloud consumers' reviews that reflect the user's experience with cloud services. Such analysis can assist potential consumers in buying cloud services. The objectives of this methodology include:

1. Automatically harvesting cloud services reviews from several web portals.
2. Constructing the world's first cloud reviews dataset.
3. Automatically analysing the sentiment of cloud reviews and generating the cloud polarity dataset. The cloud polarity dataset has a collection cloud reviews labeled with positive, negative, neutral, which is a training dataset for supervised machine learning.
4. Building machine learning classifiers for automatic prediction of consumer's review intention: positive, neutral or negative in the future. Knowing the review intention can indicate the real quality of service based on previous users' experiences.

The rest of this paper is organized as follows: Sect. 2 presents the methodology used in this work. Section 3 demonstrates the workflow of data analysis tools. Section 4 explains the data analysis phases. Section 5 explains the experiments conducted to evaluate our methodology. Section 6 presents the results and the evaluations and Sect. 7 concludes the paper.

2 Cloud Services Trust Derived Intelligence Framework

In this section, we introduce an intelligent method 'Cloud services trust derived intelligence' framework for collecting online consumers' reviews related to cloud services as shown in Fig. 1. We then analyse and classify the sentiment of the reviews.

This framework is the first work done in this area and it tackle a fundamental issue of automating the sentiment analysis for cloud services, which is called sentiment polarity classification. In our framework, we follow the data analysis phases, which are described in detail in Sect. 2. The data analysis consists of five main phases: identifying the problem; designing data requirements; pre-processing data; performing the data analysis; and visualizing data. As shown in Fig. 1, Task one is to identify the data analysis problem that this research considers which is related to automating sentiment analysis for cloud consumers' reviews. Task two involves creating the cloud services reviews dataset. Task three involves cleaning the data in order to avoid misleading results. Task four involves using sentiment analysis to generate the polarity dataset, in which each review is labelled as 'positive', 'negative' or 'neutral'. This polarity dataset is used as training dataset for the data classification process. Finally, task five is the classification process for building machine learning classifiers to automatically classify the consumer reviews into (positive, neutral or negative) in the future and drawing the conclusion about the overall consumers' experiences with cloud services. Sentiment analysis is a well-known natural language processing method which classifies the intention of the reviews as either "positive" (admiration), "negative" (criticize), or "neutral". To determine the sentiment of online consumers' reviews, Machine Learning (ML) is essential. ML techniques can be categorized as either supervised or unsupervised. This research focuses on applying supervised ML, which is based on the fact that the dataset acts as a guide to teach the ML algorithms what conclusions it should come up with. The dataset usually consists of known input data and known expected outputs for guiding and training the algorithm. Because this research aims to predict the sentiment of online consumers reviews, in the first phase, we need a training dataset. This is a polarity dataset of reviews that has the reviews' text along with the sentiment label (positive, neutral or negative). All existing data sentiment features are combined in a prediction model that can predict whether the attitude of new reviews is positive, neutral or negative.

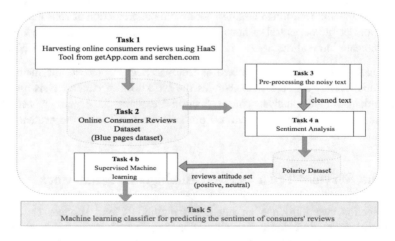

Fig. 1. Cloud services trust derived intelligence framework

3 Data Analysis Tools

Nowadays, there are several tools available for data analytics, such as R, Weka, RapidMiner and KNIME. A comparative study was undertaken by [9] to compare RapidMiner, Weka, R and Knime from different aspects, including volume of data used, response time, ease of use, price tag, and analysis algorithm and handling. The result of this study shows that KNIME is recommended for the beginner users, and that Weka is very similar to KNIME [5]. The study shows that RapidMiner is more for experts who need programming skills and less visualization; while the R Tool is the leading tool for visualization. Also, RapidMiner is the only tool which has statistical and predictive analytics capabilities, so it can be easily used and implemented on any system. Moreover, RapidMiner integrates the maximum algorithms of the other mentioned tools. A study undertaken by [12] stated that compared with the other data analysis tools such as R and KNIME the RapidMiner is more famous to use in business and marketing analysis. Therefore, in our research, we used RapidMiner for conducting the experiments as our focus is to analyse business reviews related to cloud services. RapidMiner provides an interactive user environment for machine learning and data mining processes and it is open source. Also, it has more than hundreds learning schemes for clustering, classification, and regression tasks.

4 Data Analysis Phases

There are five sequential phases in data analysis process which are performed to get the analysis result [7]. The phases are as follows:

a. Identifying the problem
b. Designing data requirements
c. Pre-processing data
d. Performing the data analysis
e. Visualizing data.

In our research, we follow the above-mentioned phases to carried out the analysis of online consumers reviews:

a. **Identifying the problem**: We perform the data analytic techniques in order to automate the sentiment analysis of online consumers reviews related to cloud services/products. This analysis can provide the potential consumers with consumers' experiences as well as assisting them in making the most appropriate purchase decision.
b. **Designing data requirements**: To perform the data analysis process, we need a dataset of online consumer reviews related to cloud services. Therefore, we collected 9270 reviews over a period of 3 months (January–March 2019) from getApp.com and serchen.com, using harvesting tool [10]. Each posted review reflects the cloud consumers experience with cloud services written in English. Each post contains the text of the review and other attributes, for example, the names of reviewers and the dates that reviews were posted online.

c. **Pre-processing data**: Before using the cloud reviews dataset, we need to pre-process the data to translate it into a specific format before applying the data analysis process. This pre-processing task include data cleaning and data sorting. This step will help avoid generating misleading results.

d. **Performing data analysis**: In this step, we are applying the sentiment analysis to the cloud reviews data, to generate the polarity dataset, which has each review labeled to positive, negative or neutral. We then use the polarity dataset as training dataset for data classification process. In the data classification process, we use four data classification algorithms in our experiment: Decision Tree, Naive Bayes, K-Nearest Neighbour and Support Vector Machines.

e. **Visualizing data**: Deriving these steps we visualize the data.

5   Experiments

5.1   Sentiment Analysis Using RapidMiner

In this study, RapidMiner Studio 8.1.001 is used to apply document-level sentiment analysis on consumer text reviews related to cloud services/products. RapidMiner supports various data analytic API, one of which is AYLIEN API. The AYLIEN Text Analysis API easily extracts and analyses insights from text. The API is capable of performing document-level sentiment analysis, as well as feature-based or aspect-based sentiment analysis. The API supports four different domains for aspect-based analysis such as cars, hotels, airlines and restaurants. For this research, AYLIEN Text Analysis by AYLIEN 0.2.0 extension is installed inside RapidMiner Studio. The primary operator (which is named Analyse Sentiment) is used to perform document- level sentiment analysis on consumer reviews. The operator extracts sentiments as being positive, neutral or negative from the consumer reviews that we supply as an input to this operator. The operator requires the following mandatory input parameters:

1. Connection: An application ID and application key needs to be set up to authenticate the user;
2. Input attribute: The customer review needs to be analysed is set in this parameter.

The pre-processing data phase is a very important phase for improving the model's accuracy. This phase primarily involves formatting and cleaning data. For the data preparation and processing phase, we used two RapidMiner extensions: "Text processing" and "wordnet". The sentiment analysis workflow in RapidMiner which is explained as follows:

1. **Step 1 Tokenize**: This operator splits the review text into a sequence of tokens. There are various ways to split the sentence into words like non-letters, specify character, and regular expression. For this research, the non-letter mode is used to split review text.
2. **Step 2 Transform Cases**: This operator transforms all the characters of the tokens into lowercase or uppercase. For example, if the review text contains words like amazing or "Amazing" or "AMAZING", then all these words are converted into the same case and are all treated the same.

3. **Step 3 Filter Tokens (by length):** This operator filters tokens based on a specified minimum and maximum character limit. Using this operator, we can filter out unnecessary tokens such as "at", "for", "an" and "of". In this study, we have filtered all small words with length less than four characters to improve the model execution time.

4. **Step 4 Extract Sentiment (English):** This operator extracts sentiments based on the SentiWordNet dictionary. The output value of the sentiment is between -1 and 1, where -1 means the text is very negative, 1 means the text is very positive, and 0 means the text is neutral. Document-level sentiment is calculated based on the average sentiment value of all tokens.

5. **Step 5 Open WordNet Dictionary:** This operator seeks the path of the dictionary so that the Extract Sentiment operator can use this dictionary to calculate sentiment level.

For this experiment, a free plan of AYLIEN API is used. This plan allows 1000 hits per day, which is not enough for our research purpose as our review dataset has around 10000 records. Moreover, the plan performs only 60 hits per minute, which is very slow, and there is a missing domain related to software services or IT-related products for performing aspect-based sentiment analysis. We have used 10000 reviews and a system of AMD6 processor with 4 GB of RAM to make this experiment.

5.2 Classification Using RapidMiner

In this section, we present the experiments which have been undertaken by RapidMiner software. Our basic mechanism for predicting the sentiment of online consumers' reviews involves using the learning supervised techniques. Our model consists of three main steps: Step 1: preparing the dataset, which has been described in detail in data collection section; Step 2: the training process; and Step 3: the prediction process. Firstly, we downloaded the polarity dataset. We then examined the dataset manually to ensure the quality of the data. This dataset has three sentiment labels (positive, neutral and negative). In our experiments, we used the negative, neutral and positive reviews. Secondly, we trained different algorithms using the training dataset with the known data (reviews) and known responses (labels) as input, and the prediction model as the output. Finally, we tested the prediction models to evaluate their ability to predict the sentiment of any new reviews.

5.3 Data Classification Algorithms Used in the Experiments

In our experiments, we applied five different supervised machine learning algorithms: k-Nearest Neighbour, Naive Bayes, Naive Bayes (kernel), Support Vector Machines, Decision Tree, and Random Forest. A brief description of each algorithm is below:

- **k-Nearest Neighbour Model/Classifier**: k-Nearest Neighbour algorithm (which is also known as lazy learning algorithm) is the simplest algorithm of all the machine learning models. This algorithm compares a given example "x" with training examples which are similar to it [20].

- **Naive Bayes Model/Classifier**: Naive Bayes classifier is a popular algorithm that is used for text classification in different domains [11]. It is a simple probabilistic classifier applying Bayes theorem that is used to predict the class of a new document. In contrast to the other classifiers, a Naive model is efficient since it only requires a small training dataset to estimate the variance and means that need for document classification [14].
- **Support Vector Machines Model Classifier**: This algorithm is a traditional text classification model that performs well in different domains. It was developed by [17] for binary classification, however, it has been applied successfully to many applications (e.g. [8]). In this study, the Linear Support Vector Machine algorithm is used for sentiment classification. It is a hyperplane represented by the vector that separates the negative and positive training vectors with a maximum margin.
- **Decisions Tree Model Classifier**: The Decision Tree model is similar to an inverted tree as it has a root at the top and it grows downwards. This algorithm identifies different ways to split a dataset into segments, like branches, and these segments form the decision tree. Decision trees have been widely used for data mining and text classification, such as in [16].

6   Results and Evaluation

Table 1 shows distribution of reviews into positive, neutral and negative, with level of threshold. The results show that there is a specific range threshold for each polarity category as follows: positive between (0 and 1), negative between (−1 and 0) and neutral is zero. The result of this analysis demonstrates that cloud services reviews are more likely to share positive experience in relation to cloud products (80.9%). Also, a less that 20% of reviewers gave negative feedback about their experience with cloud services (13.7%), whereas a small number of reviewers posted neutral comments (5.3%). In summary, the result of this stage is a polarity data set, which used in the next stage as training dataset to build machine learning classifiers.

We use the RapidMiner application to build supervised machine learning classifiers k- Nearest Neighbour Algorithm, Naive Bayes Algorithm, Support Vector Machines, Decisions Tree Algorithm and Random Forest Algorithm were performed in the data set. For the validation purpose of each classifier, we conducted a 3-fold cross validation. This means that the data is divided into three, one being the testing set and the other two being the training sets. Then, we conducted 5-fold cross validation, which means the data is divided into five, one being the testing set and the other four being the training sets. Also, we conducted 10-fold cross validation. This means that the data is divided into five, one being the testing set and the other nine being the training sets. We used TF-IDF approach to generate the word vectors. To evaluate the sentiment classification of our classifier, we used the common index for text classification including accuracy, classification error, precision, and recall for each classifier. The results are presented in the form of tables shown below. Best accuracy was achieved by two classifiers: SVM and Naive Bayes.

Table 1. Summary of sentiment analysis using rapid miner

	True negative	True neutral	True positive	Class precision
Pred. negative	730	44	197	75.18%
Pred. neutral	30	216	164	52.68%
Pred. positive	510	232	7118	90.56%
Class recall	57.48%	43.90%	95.17%	

Table 2. Performance of linear SVM classifier

	True negative	True neutral	True positive	Class precision
Pred. negative	776	32	180	78.54%
Pred. neutral	31	241	157	56.18%
Pred. positive	463	219	7142	91.28%
Class recall	61.10%	48.98%	95.49%	

Table 3. Confusion matrix of linear SVM classifier (5 fold)

	True negative	True neutral	True positive	Class precision
Pred. negative	730	44	197	75.18%
Pred. neutral	30	216	164	52.68%
Pred. positive	510	232	7118	90.56%
Class recall	57.48%	43.90%	95.17%	

Table 4. Confusion matrix of linear SVM classifier (10 fold)

	True negative	True neutral	True positive	Class precision
Pred. negative	776	32	180	78.54%
Pred. neutral	31	241	157	56.18%
Pred. positive	463	219	7142	91.28%
Class recall	61.10%	48.98%	95.49%	

Table 5. Performance of K-NN classifier

k folds cross validation	Recall	Precision	Accuracy	Classification error
3 folds	46.58% ± 1.07%	70.52% ± 1.30%	24.29% ± 0.25%	75.71% ± 0.25%
5 folds	49.73% ± 1.96%	70.33% ± 0.57%	26.69% ± 0.66%	73.31% ± 0.66%
10 folds	52.21% ± 2.36%	70.44% ± 1.12%	28.45% ± 1.40%	71.55% ± 1.40%

Table 6. Confusion matrix of K-NN classifier (k = 8) (5 folds)

	True negative	True neutral	True positive	Class precision
Pred. negative	1261	308	6447	15.73%
Pred. neutral	1	178	5	96.74%
Pred. positive	8	6	1027	98.66%
Class recall	99.29%	36.18%	13.73%	

Table 7. Confusion matrix of K-NN classifier (k = 8) (10 folds)

	True negative	True neutral	True positive	Class precision
Pred. negative	1260	279	6310	16.05%
Pred. neutral	1	206	6	96.71%
Pred. positive	9	7	1163	98.64%
Class recall	99.21%	41.87%	15.55%	

Table 8. Performance of decision tree classifier

k folds cross validation	Recall	Precision	Accuracy	Classification error
3 folds	46.32% ± 1.25%	70.49% ± 1.25%	24.20% ± 0.20%	74.90% ± 0.20%
5 folds	49.43% ± 1.66%	70.13% ± 0.47%	26.62% ± 0.63%	72.98% ± 0.66%
10 folds	51.98% ± 2.16%	70.28% ± 1.10%	28.35% ± 1.40%	71.50% ± 1.38%

Table 9. Confusion matrix of decision tree classifier (5 folds)

	True negative	True neutral	True positive	Class precision
Pred. negative	729	43	195	74.85%
Pred. neutral	28	214	162	52.02%
Pred. positive	508	230	7116	90.15%
Class recall	57.07%	43.25%	94.78%	

Table 10. Confusion matrix of decision tree classifier (10 folds)

	True negative	True neutral	True positive	Class precision
Pred. negative	774	30	179	78.07%
Pred. neutral	30	240	155	55.76%
Pred. positive	460	217	7140	90.55%
Class recall	60.46%	48.38%	95.08%	

Table 11. Performance of Naive Bayes classifier

k folds cross validation	Recall	Precision	Accuracy	Classification error
3 folds	61.53% ± 1.17%	69.39% ± 1.49%	85.92% ± 0.41%	14.08% ± 0.41%
5 folds	65.52% ± 1.92%	72.90% ± 1.61%	87.26% ± 0.54%	12.74% ± 0.54%
10 folds	68.52% ± 2.66%	75.50% ± 1.85%	88.29% ± 1.02%	11.71% ± 1.02%

Table 12. Confusion matrix of Naive Bayes classifier (5 folds)

	True negative	True neutral	True positive	Class precision
Pred. negative	730	44	197	75.18%
Pred. neutral	30	216	164	52.68%
Pred. positive	510	232	7118	90.56%
Class recall	57.48%	43.90%	95.17%	

Table 13. Confusion matrix of Naive Bayes classifier (10 folds)

	True negative	True neutral	True positive	Class precision
Pred. negative	776	32	180	78.54%
Pred. neutral	31	241	157	56.18%
Pred. positive	463	219	7142	91.28%
Class recall	61.10%	48.98%	95.49%	

Table 14. Classifiers performance evaluation and comparison

	Accuracy	Classification error	Recall	Precision
SVM	88.29%	12.74%	75.50%	68.52%
K-NN	28.45%	71.55%	52.21%	70.44%
Decision tree	28.35%	71.50%	51.98%	70.28%
Naive Bayes	88.29%	11.71%	68.52%	75.50%

7 Conclusion

In this paper, we introduce the "Cloud services trust derived intelligence" framework. This framework will be used to collect online consumers' reviews related to cloud services, and then analyse and classify the sentiment of the reviews. This framework is the first work done in this area and it tackles a fundamental issue of automating the sentiment analysis, called sentiment polarity classification. In our framework, we follow the data analysis phases. These phases are identifying the problem, designing data requirements, pre-processing data, performing the data analysis, and visualizing data. Task 1: involves identifying the data analysis problem that this research considers which automating sentiment analysis for cloud consumers' reviews. Task3 involves cleaning the data in order to avoid misleading results. Tasks 4 and 5 involve applying sentiment analysis to generate polarity dataset, which has each review labelled to positive, negative or neutral. We then use the polarity dataset as a training dataset for data classification process. The classification process is for building machine learning classifiers to automatically classify the consumer reviews into "positive", "neutral" or "negative" in the future.

References

1. Abdelmaboud, A., Jawawi, D.N., Ghani, I., et al.: Quality of service approaches in cloud computing: a systematic mapping study. J. Syst. Softw. **101**, 159–179 (2015)
2. Alkalbani, A.M., Ghamry, A.M., Hussain, F.K., et al.: Blue pages: software as a service data set. In: 2015 10th International Conference on Broadband and Wireless Computing, Communication and Applications (BWCCA). IEEE, pp. 269–274 (2015)
3. Alkalbani, A.M., Ghamry, A.M., Hussain, F.K., et al.: Predicting the sentiment of SaaS online reviews using supervised machine learning techniques. In: 2016 International Joint Conference on Neural Networks (IJCNN). IEEE, pp. 1547–1553 (2016)
4. Alkalbani, A.M., Hussain, F.K.: Quality CloudCrowd: a Crowdsourcing platform for QoS assessment of SaaS services. In: International Conference on P2P, Parallel, Grid, Cloud and Internet Computing. Springer, pp. 235–240 (2017)
5. Berthold, M.R., Cebron, N., Dill, F., et al.: KNIME-the Konstanz information miner: version 2.0 and beyond. ACM SIGKDD Explor. Newsl **11**, 26–31 (2009)
6. Casas, P., Schatz, R.: Quality of experience in cloud services: survey and measurements. Comput. Netw. **68**, 149–165 (2014)
7. Cios, K.J., Pedrycz, W., Swiniarski, R.W. Data mining and knowledge discovery. In: Data mining methods for knowledge discovery. Springer, pp. 1–26 (1998)
8. Cristianini, N., Shawe-Taylor, J. An introduction to support vector machines and other kernel-based learning methods. Cambridge University Press (2000)
9. Dwivedi, S., Kasliwal, P., Soni, S. Comprehensive study of data analytics tools (RapidMiner, Weka, R tool, Knime). In: 2016 symposium on colossal data analysis and networking (CDAN). IEEE, pp. 1–8 (2016)
10. Ghamry, A.M., Alkalbani, A.M., Tran V et al. Towards a public cloud services registry. In: International Conference on Web Information Systems Engineering. Springer, pp 290–295 (2017)
11. Joachims, T. Text categorization with support vector machines: Learning with many relevant features. In: European conference on machine learning. Springer, pp. 137–142 (1998)
12. Jovic, A., Brkic, K., Bogunovic, N. An overview of free software tools for general data mining. In: 2014 37th International Convention on Information and Communication Technology, Electronics and Microelectronics (MIPRO). IEEE, pp. 1112–1117 (2014)
13. Liu, B.: Sentiment analysis and opinion mining. Syn. Lect. Human Lang. Tech. **5**, 1–167 (2012)
14. Mccallum, A., Nigam, K. A comparison of event models for naive bayes text classification. In: AAAI-98 workshop on learning for text categorization. Citeseer, pp. 41–48 (1998)
15. Park, D.-H., Lee, J., Han, I.: The effect of on-line consumer reviews on consumer purchasing intention: the moderating role of involvement. Int. J. Elec. Com. **11**, 125–148 (2007)
16. Quinlan, J.R.: Induction of decision trees. Mach. Learn. **1**, 81–106 (1986)

17. Rajendran, V., Swamynathan, S. A novel approach for semantic service discovery in cloud using broker agents. In: International Conference on Advances in Computing, Communication and Information Science (ACCIS-'14), pp 242–250 (2014)
18. Sun, L., Dong, H., Hussain, F.K., et al.: Cloud service selection: state-of- the-art and future research directions. J. Netw. Comp. Appl. **45**, 134–150 (2014)
19. Thiyagarajan, K., Kodagoda, S., Van Nguyen, L., et al.: Sensor failure detection and faulty data accommodation approach for instrumented wastewater infrastructures. IEEE Access **6**, 56562–56574 (2018)
20. Weinberger, K.Q., Saul, L.K.: Distance metric learning for large margin nearest neighbor classification. J. Machine Learn. Res. **10**, 207–244 (2009)

HPC, Cloud and Big-Data Convergent Architectures: The LEXIS Approach

Alberto Scionti[1(✉)], Jan Martinovic[2], Olivier Terzo[1], Etienne Walter[3],
Marc Levrier[3], Stephan Hachinger[4], Donato Magarielli[7], Thierry Goubier[6],
Stephane Louise[6], Antonio Parodi[9], Sean Murphy[5], Carmine D'Amico[1],
Simone Ciccia[1], Emanuele Danovaro[9], Martina Lagasio[9], Frederic Donnat[10],
Martin Golasowski[2], Tiago Quintino[8], James Hawkes[8], Tomas Martinovic[2],
Lubomir Riha[2], Katerina Slaninova[2], Stefano Serra[11], and Roberto Peveri[11]

[1] Advanced Computing and Applications, LINKS Foundation, Torino, Italy
alberto.scionti@linksfoundation.com
[2] IT4Innovations, VSB - Technical University of Ostrava, 17. listopadu 15/2172,
708 00 Ostrava-Poruba, Czech Republic
{jan.martinovic,martin.golasowski,tomas.martinovic,lubomir.riha,
katerina.slaninova}@vsb.cz
[3] ATOS, Paris, France
[4] Leibniz Supercomputing Centre (LRZ), Garching, Germany
[5] Cyclops Labs, Zurich, Switzerland
[6] CEA List, Paris, France
[7] AvioAero, Torino, Italy
[8] European Centre for Medium-Range Weather Forecasts (ECMWF), Reading, UK
[9] CIMA Research Foundation, Savona, Italy
[10] Outpost24, Nice, France
[11] TESEO, Torino, Italy

Abstract. High Performance Computing (HPC) infrastructures (also referred to as supercomputing infrastructures) are at the basis of modern scientific discoveries, and allow engineers to greatly optimize their designs. The large amount of data (Big-Data) to be treated during simulations is pushing HPC managers to introduce more heterogeneity in their architectures, ranging from different processor families to specialized hardware devices (e.g., GPU computing, many-cores, FPGAs). Furthermore, there is also a growing demand for providing access to supercomputing resources as in common public Clouds. All these three elements (i.e., HPC resources, Big-Data, Cloud) make "converged" approaches mandatory to address challenges emerging in scientific and technical domains.

The LEXIS project aims to design and set up an innovative computing architecture, where HPC, Cloud and Big-Data solutions are closely integrated to respond to the demands of performance, flexibility and scalability. To this end, the LEXIS architecture leverages on three main distinctive elements: (i) resources of supercomputing centers (geographically located in Europe) which are seamlessly managed in a federated fashion; (ii) an integrated data storage subsystem, which supports Big-Data ingestion and processing; and (iii) a web portal to enable users to

© Springer Nature Switzerland AG 2020
L. Barolli et al. (Eds.): CISIS 2019, AISC 993, pp. 200–212, 2020.
https://doi.org/10.1007/978-3-030-22354-0_19

easily get access to computing resources and manage their workloads. In addition, the LEXIS architecture will make use of innovative hardware solutions, such as burst buffers and FPGA accelerators, as well as a flexible orchestration software. To demonstrate the capabilities of the devised converged architecture, LEXIS will assess its performance, scalability and flexibility in different contexts. To this end, three computational highly demanding pilot test-beds have been selected as representative of application domains that will take advantage of the advanced LEXIS architecture: (*i*) Aeronautics – Computational Fluid Dynamics simulations of complex turbo-machinery and gearbox systems; (*ii*) Earthquake and Tsunami – acceleration of tsunami simulations to enable highly-accurate real-time analysis; and (*iii*) Weather and Climate – enabling complex workflows which combine various numerical forecasting models, from global & regional weather forecasts to specific socio-economic impact models affecting emergency management (fire & flood), sustainable agriculture and energy production.

1 Introduction

High Performance Computing (HPC) and Big-Data technologies are the two key elements enabling the so called "fourth paradigm" [1] of scientific discoveries. Unlike resorting to physical experiments or mathematical models to study phenomenon, the fourth paradigm leverages on the availability of massive data-sets and the capability of analyzing them using powerful computing infrastructures (aka supercomputers). In this context, the demand for ever more computing resources is growing, so that new technologies are rapidly adopted to speed up simulations and data analysis.

Among the others, architectural specialization is used to offer the best in class performance for various types of workflows, ranging from simulations requiring very high precision arithmetic (e.g., the massive use of GPU computing and many-core devices such as Intel Xeon-Phi [10]) to application exploring deep learning (DL) techniques [13]. Hardware heterogeneity offers also a solution for coping with the problem of keeping overall power consumption under control. Indeed, hardware accelerators sport higher FLOPS-Watt ratios compared to general purpose CPUs. Pure computing capabilities are not the only factor that matter in achieving high performance; also, interconnection and storage subsystems play a key role, as well as the software that is used to govern all the infrastructure [14]. Very high-speed interconnections (e.g., low-latency multi-gigabit per second interconnection, such as Infiniband [15]) are common in HPC infrastructures. Conversely, storage capability is generally offered through a dedicated secondary storage system which is accessed by HPC nodes through an internal high-speed network.

Recently, to cope with the growing demand of supporting applications crunching massive data-sets, HPC architectures started being enriched with dedicated nodes equipped with local high-speed storage. These so called *Burst Buffers* (BB) [16] enable applications to temporally cache data, that will be later on transferred to the secondary storage. To this end, BB employ local storage that uses solid-state drives (SSDs) attached to PCI-Express bus to provide higher read/write throughput. Even more, memory technologies replacing conventional NAND-flash (NVM – Non Volatile Memories) [2] will push performance of BB higher than those allowed by current technology. For instance, among the others, Intel 3D-Xpoint technology [3,4] combines the advantage of NVM (persistency) and the advantage of traditional DRAMs (high speed access, endurance, single word addressing), which make convenient to populate nodes with NVM-based DIMMs. Also, such technology allows to keep large amount of persistent data closer to the processors.

Moving further in the direction of specialized chip fabrics, Field Programmable Gate Arrays (FPGAs) [17] offer the largest trade-off among performance, power consumption and flexibility. Thanks to modern high level synthesis (HLS) tool-set (application kernels can be written using high-level languages such as C/C++), programming such devices has become easier than in the past. FPGAs are becoming attractive also for HPC and Big-Data/ML applications [5–7], since their flexibility, combined with ever higher performance, provide massive computing power without impacting on overall power consumption. Also, FPGAs allow designer to create customized circuit for supporting acceleration of low level functions such as on-the-fly data encryption (at line rate) and data compression. Similarly, customized network protocols can be easily implemented on top of FPGAs. All such functions become valuable when considering integration of FPGA technology into BB.

Nowadays, large HPC infrastructures generally operate in an isolated fashion, so that access to computational resources is granted to users accounted to the infrastructures, although they are not easily reachable from the outside. Conversely, Cloud Computing (CC) model goes in the opposite direction. In CC, computing and storage resources are made easily accessible from the users, which can acquire (and release) them on-demand. Such management model offers great flexibility in the way resources are allocated to diverse requests of users. Following the road of evolution, the convergence between HPC and Cloud domains is the enabling element for offering HPC resources in a more flexible and convenient way to a much broader userbase (i.e., HPC-as-a-Service – HaaS). Among the various technologies, *orchestration software* represents a key element in any Cloud architecture. Its primary objective is to allocate resources according to certain policies, as well as satisfying input user requests. Interestingly, emerging HPC infrastructures tightly coupled with in-situ Cloud infrastructures pose new challenges from the workflow orchestration viewpoint, especially when multiple of such hybrid infrastructures are federated (e.g., as in the case of LEXIS project). The second technological pillar of CC model is "virtualization", i.e., a software mechanism to abstract a certain hardware component. The result is a set of virtual resources that can be easily allocated, deallocated, moved dynamically. Through a careful design of the infrastructure, it will be possible to seamlessly integrate pure HPC resources and Cloud resources in a single platform. Furthermore, resources geographically distributed may be made accessible as they were located within the same physical infrastructure. However, one of the critical point of such platforms is the management of user authentication and authorization process. In addition, designing such kind of platform without considering any potential security issue, risks to compromise the functioning of the platform. Finally, an accurate billing system is mandatory to any provider of any ⋆-as-a-service.

The LEXIS (*Large-scale EXecution for Industry & Society*) project received funds from European Commission to devise, design, implement and set up a platform for supporting the execution of workloads coming from diverse application domains, and providing resources through a flexible HPC-as-a-Service model. LEXIS project will build an advanced engineering platform at the confluence of HPC, Cloud [12] and Big-Data, which will leverage large-scale geographically-distributed resources from existing HPC infrastructures (provided by three national supercomputing centers, IT4Innovations, LRZ, and ECMWF), employ Big-Data analytics solutions and augment them with Cloud services. Driven by the requirements of its pilot test beds, the LEXIS platform will build on best of breed data management solutions and advanced, distributed orchestration solutions (TOSCA), augmenting them with new, efficient hardware capabilities in the form of Burst Buffers and federation, and monitoring and accounting/billing supports to realize an innovative solutions.

2   Project Objectives

The LEXIS project, with a duration of 30 months (the project started in January 2019) brings together a consortium with the skills and experience to deliver a complex multi-faceted project, spanning a range of complex technologies across seven European countries, including large industry, flagship HPC centers, industrial and scientific compute pilot users, technology providers and SMEs. Federated authentication and authorization infrastructure across these HPC service and data providers will also be a key. To this end, the LEXIS project is organized around three main objectives to be achieved during the project lifetime. Such objectives are as follows.

- **Foundation:** design and build a distributed HPC infrastructure for Big-Data analytics to provide applications and services useful for the industrial sectors of the developed test-beds;
- **Innovation:** validate the platform within three different test-beds and generate valuable outcomes for test-beds stakeholders, combining data assets and innovative technological solutions. This objective aims to exploit big-data assets available for the execution of test-beds and to improve the performance of analyses, both in terms of computational time, results accuracy and relevance for stakeholders.
- **Extension**: extending the use of the LEXIS services to external stakeholders, developing suitable solutions for interoperability, security, resources orchestration, interaction and visualization.

Besides these three main objectives, LEXIS selected a set of additional objectives which are strictly connected to the three pilot test-beds, as well as it defined a set of technological objectives to achieve during the project lifetime. Concerning the former, LEXIS considers the following objectives.

- **Aeronautics:** Reduce the running time of computer-aided engineering simulations to improve engineering productivity and design process quality. This objective will address CPU-demanding, data-intensive, and time-consuming computational fluid dynamics simulations referred to aircraft engines turbomachinery and rotating parts. It will be achieved by investigating the industrial applicability of newly designed HPC and Big Data platforms.
- **Earthquake and Tsunami:** Provide near real-time earthquake and tsunami damage/loss assessments and estimate of the tsunami inundation through simulations based on earthquake parameters, ensuring the delivery of expected consequences in time for fast response planning by emergency dispatchers. Optimal mapping of different stages of processing workflow to various resources in LEXIS will allow us to improve accuracy and computation speeds of simulations and the damage/loss assessment codes.
- **Weather and Climate:** Increase the timeliness and quality of prediction and analyses. Simplify the access to such services from the Cloud, in order to expand the downstream markets: emergency management, sustainable food and energy production, air quality.

Concerning the technological objectives, the LEXIS consortium selected the following ones as the most challenging:

- Providing a ready-to-be-used HPC infrastructure that offers HPC-as-a-Service capabilities without incurring in performance/efficiency slowdowns.
- Implement a heterogeneous data storage management system providing simplified access to huge amounts of data. LEXIS will build a scalable Big-Data storage platform, based on modern data-management technologies.
- Providing an improvement in executing CPU intensive and data/memory intensive algorithms aiming at supporting real-time decision making. The exploitation of innovative programming models for HPC heterogeneous infrastructures will be the key challenge for the achievement of this objective. High Performance Data Analytics (HPDA) techniques will be applied to the processing of extremely huge amounts of data, taking advantage of an innovative infrastructure that empowers different workloads.
- Optimize data management operations and analytic algorithms that exploit the underlying infrastructures at their best to eventually extract outputs from data that help stakeholders improve their businesses.
- Provide simple and secure HPDA service provisioning, through Cloud technologies, for the pilot test-beds, accessible also for other users. This objective will be achieved through the implementation of different user services that will guarantee interoperability, security, resources orchestration, interaction and visualization.
- Guarantee interoperability with external data sources and seamless integration with external systems.

3   The LEXIS Architecture

This section describes the LEXIS platform architecture, which considers the integration of computing and storage resources across different supercomputing infrastructures, as well as the use of innovative hardware systems (e.g., Burst Buffers, FPGAs, large GPU-based nodes, novel memory hierarchy which includes NVMe SSDs and NVDIMMs) and orchestration software. The LEXIS platform refers to the integration of two main infrastructural elements, namely the *LEXIS Computing Infrastructure* (LEXIS-CI) and the *LEXIS Data Infrastructure* (LEXIS-DI), which are described in Sects. 3.1 and 3.2 respectively. The overall architectural design of the LEXIS platform is depicted in Fig. 1. Actually, only two out of three supercomputing centers (IT4Innovations and LRZ) will be directly integrated in the LEXIS federated infrastructure, since the ECMWF data center is used in production services.

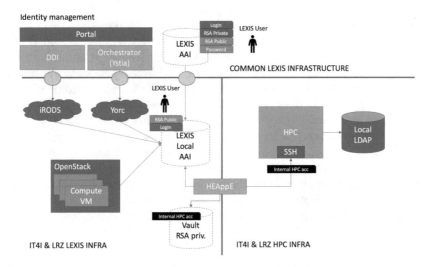

Fig. 1. LEXIS overall architecture: portal and authentication mechanism highlighted.

3.1   LEXIS Computing Infrastructure

Supercomputing infrastructures provide computing, storage and networking resources for supporting the execution of pilot test-beds applications. More specifically, two out of three of such supercomputing infrastructures (i.e., IT4Innovations and Leibniz Supercomputing Center – LRZ) provide resources according to both HPC and Cloud models, except ECMWF, which provides only resources for managing data ingestion of new in-situ observations (indeed, ECMWF computing resources are used in production to create weather forecasts four times per day). HPC resources are based on available supercomputers in the above mentioned centers, augmented with specialized high-memory

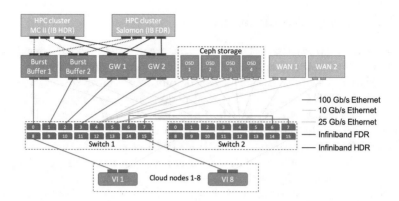

Fig. 2. LEXIS architecture: IT4Innovations Cloud resource integration proposal.

GPU-based nodes (e.g., Nvidia DGX-2) to better support highly parallel workloads, as well as deep learning applications. Cloud resources are based on a set of separated nodes, and the allocation of jobs on the HPC or Cloud parts is done through an orchestrator called YSTIA [8]. This latter is an orchestration tool designed to manage workflows in complex and heterogeneous environments, such as that envisioned by the LEXIS platform. YSTIA orchestrator is workflow driven, meaning that it does not contain any hard-coded life-cycle logic. This allows to fully customize applications behaviour and to execute custom workflows at run-time. It is designed for large-scale and is built with a tasks/stateless workers model in mind allowing to scale it horizontally easily. To interact with diverse infrastructural elements, YSTIA orchestrator interacts with other infrastructure management software, such as batch schedulers (e.g., PBS, SLURM, etc.) used by HPC partitions and other resource orchestration tools available on the Cloud partitions (e.g., OpenStack, Kubernetes, etc.), as well as it can drive the resource allocation in a bare-metal fashion. Furthermore, YSTIA code is made modular and easily extensible, so that in LEXIS it will be possible to integrate dedicated solutions for accounting and authentication of the platform users (a specific solution will be designed and integrated in the LEXIS platform portal to enable mapping LEXIS users to internal HPC users which are granted to actually access HPC resource both on IT4Innovations and LRZ sites). Within the scope of the LEXIS project, IT4Innovations will offer an additional layer of security in the form of a HEAppE Middleware application framework [11] which is responsible for the actual job submission, monitoring and overall management instead of a direct user access to the HPC infrastructure. The YSTIA orchestrator also integrates components to correctly access LEXIS-DI API. YSTIA orchestrator is also TOSCA [9] native to allow handling complex applications in a standard way. Finally, its comprehensive REST API and a modern CLI allow using, deploying and interacting with application at run-time. LEXIS foresees the potential use of FPGAs to improve capabilities of nodes used as burst buffers (i.e., Accelerated Burst Buffer –ABBs) providing acceleration for low level functions, such as on-the-fly data compression and encryption, as well as to accelerate specific application parts (depending on the specific use case, such type of acceleration will be evaluated if feasible). As typical for HPC infrastructures, secondary storage is managed through dedicated nodes, while storage abstraction is provided by parallel file-systems such as Lustre. Burst buffer approach provides a state-of-the-art approach to benefit from ultra-fast I/O capabilities (faster than the HPC parallel file systems themselves). To this end, burst buffers can be seen as a cache in front of the parallel file systems involving NVM devices to allow unprecedented I/O performance. LEXIS project aims also to extend functionalities of burst buffers to maximise offered performance and flexibility. At the time of writing this document, Cloud partitions are planned as extensions of the currently available HPC resources. Figure 2 shows the architectural design of the Cloud partition and its integration with the HPC resources as planned at IT4Innovations. As a minimal configuration, the planned Cloud partition will integrate from 6 to 8 nodes, 2 dedicated ABB nodes, fully redundant 100 Gbps Ethernet network including connection towards HPC resources, fully redundant 10 Gbps Ethernet connections towards the storage subsystem (cloud storage based on CEPH), and a WAN connection using 4×10 Gbps, although it is planned to extend it to 100 Gbps in near future.

Figure 3 shows the architectural design of the Cloud partition and its integration with the HPC resources as planned at Leibniz Supercomputing Centre (LRZ). The LRZ will provide access to its "Compute-Cloud" infrastructure for LEXIS (125 nodes with OpenNebula management, 64 nodes with OpenStack), to be used in shared mode with other projects, while setting up two dedicated ABB nodes for LEXIS. For the ABB nodes, a redundant 100 Gbps connectivity is envisaged, while the common Compute-Cloud nodes are equipped with at least 10 Gbps cards (with a 100 Gbps connection from each rack e.g. to the storage systems). The interconnectivity between cloud and HPC partition is over LRZ's network infrastructure with a redundant bandwidth of >100 Gbps (shared with other projects).

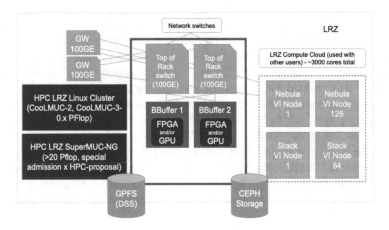

Fig. 3. LEXIS architecture: Leibniz Supercomputing Center Cloud resource integration proposal.

3.2 LEXIS Data Infrastructure

The LEXIS data layer runs on top of the LEXIS-CI and includes the architectural components aimed at the collection, management, storage, retrieval and provisioning of data. Cloud services use APIs provided by the data layer (LEXIS Data System APIs) and by the LEXIS-CI to provide users with suitable interfaces for accessing the platform and executing analytic jobs. To this end, LEXIS platform includes a mechanism (user interface) through which the platform users can request the scheduling of an analytic applications, visualize results, and explore available data-sets.

The LEXIS data layer consists of two major components: (i) the (distributed) LEXIS Data Infrastructure (LEXIS-DI) and Weather & Climate Data API (WCDA); (ii) the Data System API (DSA), offered as HTTP REST APIs. In addition, the data layer includes monitoring functionality to support Cloud services (LEXIS Monitoring System) and interoperability (LEXIS Data Interoperability).

The LEXIS-DI leverages the low-level/site-specific storage systems, connecting them with a middleware (e.g., iRODS, GridFTP). The middleware makes the different storage systems into a unified "virtual" file-system accessible from all sites. This file system can contain extended meta-data and fine-grained rules also for storage distribution and data management in general. In parallel to this, the WCDA will provide a highly-specialized weather & climate data service which will facilitate the structured and efficient transfer of voluminous meteorological data, specifically for the weather and climate pilot.

The distributed data system (DSA) is augmented by APIs for data exploration, search, staging, retrieval, transfer and pre-processing (aggregation, filtering, format conversion). These APIs simplify everyday tasks in the LEXIS platform and allow the data system to connect to the YSTIA orchestrator through the LEXIS portal. The LEXIS Monitoring System is responsible for tracking users access and usage of both data and computing resources and store the information into the monitoring DB. The billing system will then connect to the LEXIS Monitoring System via a suitable monitoring API, enabling a fine-grained accounting and billing for usage or according to any other model.

Data-Transfer API (DTA) is interfaced with external data sources to prepare data for the subsequent analyses. It provides data staging and retrieval functionality, i.e., it will extract data from sources, supporting protocols for both batch and real-time streaming. The data layer adapts to specific sources, for instance supporting protocols for IoT devices like MQTT or AMQP or being implemented through a data stream-process platform like Apache Kafka, which could be used also for the following steps. Data Interoperability API (DIA) is implemented through: (i) data filtering to remove erroneous data; (ii) data fusion to synchronize data from multiple sources; (iii) data harmonization to apply a standard common format to data, in order to guarantee interoperability also with external system.

4   LEXIS Pilots

The LEXIS consortium intends to demonstrate the potentiality and capabilities offered by the LEXIS platform through three pilot test-beds. Applications belonging to each pilot's domain will be used to assess the performance, scalability and flexibility offered by LEXIS platform, highlighting its capability in managing different type of workflows. For each of the use cases, LEXIS will pay great attention to all the security aspects to ensure correctly granting access to the infrastructure resources. In the following the three LEXIS pilots are described.

4.1   Aeronautics Test-Bed

In the context of the Aeronautics pilot, two use cases (UCs) have been selected as a mean of demonstration of the capability of the LEXIS platform: the *Avio Aero Turbomachinery* UC and the *Avio Aero Rotating parts* UC. **Avio Aero Turbomachinery UC**. The challenge of Avio Aero is to develop new technologies enabling to design and produce aeronautical components and modules for next generation greener and quieter engines. Identified tasks will focus on multistage Low Pressure turbine modules simulation, where accurate prediction of flow physics has been and is nowadays the subject of intensive research to achieve better engines' efficiency, stability and operability. By leveraging Lexis Technology, we intend to obtain a marked step-change: less time consuming HW/SW coupling, opening the doors to the "real time" design approach. Furthermore, the big data produced as a result will require proper solutions to be put in place for quick data access, management and post-pro. **Avio Aero Rotating parts UC**. This project aims to develop enablers supporting the digital (r)evolution being applied to design and development of Gearboxes that are key products for Avio Aero business. The aim is to investigate advanced numerical modeling capabilities of complex Computational Fluid Dynamics (CFD) analyses applied to simulate mechanical parts rotating in presence of air and oil. Nowadays, this kind of simulation is at the leading edge of numerical technology and needs to leverage GPU-based computing resources. Hence advanced HW solutions are envisaged: new GPU-accelerated infrastructure and big data extensive treatment will open new scenery to mechanical parts design and optimization, vital for assuring next generation engines' performance and reliability. Avio Aero team will work to set up a new CFD methodology applied to Gearboxes engineering, to deploy and test the related numerical solver on LEXIS infrastructure, and test and validate the benefits of this new engineering analysis approach.

4.2   Earthquake and Tsunami Test-Bed

The *Earthquake and Tsunami* UC is about building a time-constrained HPC-based emergency response simulation and decision support flow. It combines many elements: (*i*) fast, rough on-line simulations, giving immediate, but only approximate, results; (*ii*) accurate compute intensive simulations, giving iteratively more precise results about the extent and damage; (*iii*) offline scenario data, pre-computed scenarios to be used by experts as support for decision making; (*iv*) time-constrained flow control, ensuring that at pre-set deadlines, the flow provides the best information it has computed so far; (*v*) live updates, so that both background information and new, ground or sensor information is immediately propagated into the flow. External data sources (remote sensing) will also be called upon in the process, both on flow outcomes and as flow inputs; and (*vi*) emergency support data products processing: maps, damage and loss estimates.

This pilot brings together among the best components in the field, covering: (*i*) on-line, on-time precise simulations with TsunAWI (tsunami simulations); (*ii*) damage exposure and estimate (OpenBuildingMap); (*iii*) the ability to express time constraints at which point preliminary data must be made available; (*iv*) the ability to acquire data, and to deliver data products on time: warnings, estimates, maps; (*v*) optimisation analysis and deployment of optimisation techniques to improve the ability of the system to provide accurate decision support on time.

4.3   Weather and Climate Test-Bed

The *Weather and Climate* UC represents a complex system designed to provide a diverse set of forecasts for weather, flood, fire, energy and air pollution. Several complex workflows will be piloted, each consisting of various hydrometeorological components. These components include conventional observation data sources (satellite, radar, etc.), unconventional observation data sources (mobile phones, personal weather stations and other IoT devices), global and regional weather models running on various HPCs and several cloud-based applications.

Each step of these workflows will produce large data sets of forecast data, ranging from 100+ TiB global weather models, to MiB at the decision maker level. Both the assimilated data (inputs) and data produced by each layer (outputs) will be made available through the Weather and Climate Data API (WCDA), which will facilitate the interchange of data between the aforementioned components and complement the LEXIS data infrastructure.

The system will be applied to several test cases, such as application models for flash-flood and forest fire risk, and socio-economic impact models for renewable energy forecast and urban air quality in cities forecast, spanning all the available forecasts, to fine tune the system and demonstrate the innovative value.

5   Background

The LEXIS project has as a notable ambition to contribute to accelerate the pace of digitization and the innovation potential in Europe's key industrial and societal sectors like aeronautics manufacturing, weather- & climate-related information services and civil protection. Up to now, the Big-Data, HPDA, Cloud and

HPC landscape with its big societal impact is fragmented. In particular, the community of traditional, solver/simulation-based HPC is split from the European Data Management community, which itself again is split at least in two parts – one with focus on large amounts of data (Big-Data, as it appears, for example in the ClimEx project 2 generating one of the largest Climate data sets of current time), the other on large amounts of (sometimes small) datasets which must be carefully procured and equipped with metadata (typically library websites or community-data projects such as TIPTOPbase 3 or PANGAEA 4). Bringing these two aspects of data management together with HPC and HPDA systems is essential for a European leap forward in the big data field. The BDVA SRIA 5 identifies the following challenges, in the fields of HPC for Big Data, real-time analytics, and HPDA services provisioning, that will be addressed in LEXIS.

Looking at Aeronautics pilot, CFD simulations currently are hosted by internal HPC resources. The main issue is correlated to the impressive need to scale to bigger data-sets, strongly limiting, in cascade, the numbers of optimization loops. Concerning the Weather and Climate pilot, forecast and impact models have generally run on dedicated compute clusters or HPC facilities. However, cloud systems are used for a large proportion of access to, and further processing of, large data sets. The lack of effective interoperability between HPC and cloud systems poses an obstacle to effective use of model data. Current operational systems are well suited to assimilate traditional and conventional observations. However, they are not ready to handle large volumes of unstructured observations which are becoming available from IoT and Edge-Computing technologies. Finally, concerning the Earthquake and Tsunami pilot, tsunami simulations (TsunAWI) have until now only be used offline on HPC clusters, to calculate scenarios that are later integrated into a database; however, there is a strong pressure for being able to run such simulations online with well specified deadlines.

With regards current status for the three mentioned pilots, LEXIS will strongly ameliorate this situation, by leveraging the skills and hi-tech facilities of partners involved in the development of next generation HPC hardware.

6   Conclusions

There is a growing interest for HPC, Big-Data and Cloud architectures that originates from the ever large availability of massive data-sets, as well the demands for accessing high performance systems through a flexible on-demand mechanism. To address such challenges, the LEXIS consortium will design and implement a converged platform relying on the computing capabilities of three Supercomputing centers. By leveraging on advanced technologies (Smart Burst Buffers, FPGAs, smart orchestration tool), the LEXIS platform will be able to provide unprecedented level of performance and flexibility, as well as the capability of ingesting and analyzing local in-situ data, in the context of three different pilots.

Acknowledgments.

This project receives funding from the EU's Horizon 2020 research and innovation programme (2014–2020) under grant agreement no. **825532**.

References

1. Hey, T., Tansley, S., Tolle, K.: The Fourth Paradigm: Data-Intensive Scientific Discovery. Microsoft Press (2009)
2. Micheloni, R.: 3D Flash Memories. Springer, Netherlands (2016)
3. Bourzac, K.: Has Intel created a universal memory technology?[News]. IEEE Spectr. **54**(5), 9–10 (2017)
4. Hady, Frank T., et al.: Platform storage performance with 3D XPoint technology. In: Proceedings of the IEEE 105.9, pp. 1822–1833 (2017)
5. Moss, D.J.M., et al.: High performance binary neural networks on the Xeon + FPGA^{TM} platform. In: 2017 27th International Conference on Field Programmable Logic and Applications (FPL). IEEE (2017)
6. Nurvitadhi, E., et al.: Can FPGAs beat GPUs in accelerating next-generation deep neural networks? In: Proceedings of the 2017 ACM/SIGDA International Symposium on Field-Programmable Gate Arrays. ACM (2017)
7. Li, Y., et al.: A 7.663-TOPS 8.2-W energy-efficient FPGA accelerator for binary convolutional neural networks. In: FPGA (2017)
8. ATOS-Bull: Ystia Orchestrator. https://ystia.github.io/
9. Breitenbücher, U., et al.: Combining declarative and imperative cloud application provisioning based on TOSCA. In: 2014 IEEE International Conference on Cloud Engineering. IEEE (2014)
10. Xia, Y., et al.: A GPU-accelerated package for simulation of flow in nanoporous source rocks with many-body dissipative particle dynamics. arXiv preprint arXiv:1903.10134 (2019)
11. HEAppE. http://heappe.eu
12. Mell, P., Grance, T.: The NIST Definition of Cloud Computing. https://csrc.nist.gov/publications/detail/sp/800-145/final
13. Li, P., et al.: Heterospark: a heterogeneous CPU/GPU spark platform for machine learning algorithms. In: IEEE International Conference on Networking, Architecture and Storage (NAS). IEEE (2015)
14. Duro, J., et al.: Workload characterization for exascale computing networks. In: 2018 International Conference on High Performance Computing & Simulation (HPCS). IEEE (2018)
15. Li, A., et al.: Tartan: evaluating modern GPU interconnect via a multi-GPU benchmark suite. In: IEEE International Symposium on Workload Characterization (IISWC). IEEE (2018)
16. Liu, N., et al.: On the role of burst buffers in leadership-class storage systems. In: 28th Symposium on Mass Storage Systems and Technologies (MSST). IEEE (2012)
17. Gaide, B., et al.: Xilinx adaptive compute acceleration platform: versal (TM) architecture. In: Proceedings of the 2019 ACM/SIGDA International Symposium on Field-Programmable Gate Arrays. ACM (2019)

Knowledge Sharing System Database Architecture for Global Knowledge Sharing

Alsaleh Saad[1](✉) and Haryani Haron[2]

[1] Hail Health Affairs, Ministry of Health, Riyadh, Kingdom of Saudi Arabia
saabalsaleh@moh.gov.sa
[2] Faculty of Computer and Mathematical Sciences,
University Technology MARA (UiTM), Shah Alam, Malaysia
haryani@tmsk.uitm.edu.my

Abstract. This paper offers a new approach to ascertain the characteristics of the knowledge sharing network (KSN) existing among academicians and to develop a knowledge sharing system database (KSSDB) architecture to suit the discovered network. A case study qualitative method was adopted to gather meaningful insights into the academic institutions with which Malaysian academics share their knowledge. The study data were collected from fifteen academic participants in the largest university in Malaysia through direct face-to-face interviews over a period of six months. The study found that the Malaysian academicians shared their knowledge with other academicians in two networks consisting of, respectively, fifteen local academic institutions (LAI), and twelve international academic institutions (IAI) located in six countries including Japan, India, Germany, South Korea, Brunei Darussalam, and Indonesia. Based on the qualitative findings, a KSSDB architecture was developed to accommodate these two KS networks. This research contributes to the field of knowledge sharing systems (KSS) by providing a better understanding of the academic contacts with whom academicians share their knowledge globally and how to develop an appropriate KSSDB. The results would help administrators from similar academic institutions to make decisions that could yield benefits from such sharing and enhance the advantages of KS through support of KS activities at both national and international levels.

1 Introduction

Knowledge is the most valuable resource for any organization, especially, academic institutions. If the knowledge is shared among the institution's members, the institution will gain a competitive edge [1, 2]. Further, information Technology (IT) is now playing an important role in creating and distributing knowledge among individuals regardless of where they are located because it crosses the boundaries of both space and time [4, 5]. IT is a key factor in manipulating the knowledge commodity [3]. However, according to a study [6], a large amount of knowledge within organizations is not codified and is, therefore, not easily transferable. It is especially important for individuals within academic institutions to engage and share knowledge in their knowledge

© Springer Nature Switzerland AG 2020
L. Barolli et al. (Eds.): CISIS 2019, AISC 993, pp. 213–223, 2020.
https://doi.org/10.1007/978-3-030-22354-0_20

sharing network (KSN). According to [7], creating a KSN is the most valuable activity which knowledge management should focus on.

Hence, this study examined the KSN existing among academicians in academic institutions, both nationally and internationally and developed a knowledge sharing system database (KSSDB) architecture which took both types of KSN into consideration.

2   Literature Review

2.1   Knowledge Sharing: Concept and Importance

Knowledge sharing (KS) is the process of exchanging knowledge among individuals. KS is viewed as a procedure where knowledge gets created and shared through interactions [8] in which "individuals share organizationally relevant information, ideas, suggestions, expertise with one another" [9, p. 65]. Many factors affect KS activities including institutional climate, interpersonal trust, KS behavior [10], perceived organizational support [11], social norms and customs [12] and organizational incentive systems [13]. Intensive KS activities are a sign of a viable institution, particularly academic institutions. The success of any institution is usually linked with the KS behavior among its employees. According to [14] KS among workers allows organizations to capitalize on knowledge-based resources. Additionally, if the knowledge is used and shared properly the organization will acquire a competitive edge [15].

2.2   Enhance of MHealth Services Quality

Any success health organization must continue to improve and enhance its services quality. The mHealth quality importance has been demonstrated in many studies [13, 15, 29]. According to [11] "the role of service quality in fostering the growth of mHealth services has gained much attention in the academic and practitioner communities" (182). Researchers [40] emphasized that the health service "is considered as the major element of fourth-generation health systems and the key to their success and future evolution" (547). Poor quality leads to complications and the need for additional care, which raises costs substantially [16]. There are three parameters of service quality in mHealth settings were suggested by [12] which are: Knowledge and competence of the provider (training and orientation for healthcare professionals), capacity of access and monitoring devices (aware of the limitation of mobile devices), operational compatibility among multiple platforms and interoperability of information systems. A study [17] found that the service quality that provided via a mobile platform is influenced by either IS, mobile network or information itself. This study focused only on one element, namely; the information. A prior study found that there is an association between service quality and the quality of health life perception [18].

2.3   Knowledge Sharing Network

Recently, KSNs have received increased attention in academia. The word "networks" can be understood as relationships among individuals, groups, or organizations [16].

A knowledge network is "a set of actors connected by a set of repeated interactions of formal and/or informal ties" [17, p. 4]. The actors, according to [17] encompass both institutions and the people within them. The relationships among those actors are described as ties. The relationships among actors can be classified according to contents (including products, services, emotions), form of relationship, closeness, and communication frequency [16].

For the purpose of the present research, a knowledge network refers to individuals who cooperate and connect together with the aim of sharing knowledge. As explained by [18], the network leads to intensification and dissemination of knowledge to a greater extent than could otherwise be reached by organizations or individuals. One main benefit of creating an effective network is that it enhances organizational efficiency [7].

2.4   Knowledge Sharing Database

A database can be defined as a collection of data that can be accessed, organized and stored in an electronic form. There are several types of database including relational databases, hierarchical databases, graph databases, object-oriented databases and network databases. This study focused on the network database. Related studies have classified three types of network databases: spatial network databases, graph databases, and spatio-temporal databases [19]. In general, a network database is used on a large number of computers. A network database is more efficient in terms of the number of connections that can be made between different data types [20].

Many large organizations develop a complex official site over the Internet for their workers in order to facilitate their KS activities. For example, one big international company set up a sophisticated website to enable employees from its overseas offices to share knowledge [21, p. 76]. Many databases are merely collections of files assembled for specific purposes and as they grow in size, users of such databases face difficulties to find the data they are seeking [22]. Additionally, there are institutions that cannot adapt their systems to changing needs because of limits on the database [23]. This study proposes a unique database architecture which is in line with current real types of users who share their knowledge locally and internationally.

3   Research Methodology

The aim of this qualitative study was to propose and develop a KSSDB architecture for academic institutions. As a descriptive qualitative research with a case study approach, the study procedure began with a content analysis of the data. In the coding process, interviewees' taped responses were transcribed into a textual form and then reviewed many times to determine the themes and categories relevant to an academic institution. The analysis and coding of responses was carried out line-by-line to identify the categories. The main categories were identified through free coding into subcategories. The categorization process was carried out based on groups of similar participants answering together. According to [24], the researcher in qualitative research "reviews all of the data and makes sense of it, organizing it into categories or themes" (p. 45). In addition, the qualitative researcher engages in interpreting the data [24]. In the present study, the analysis resulted in an accurate identification of local and international universities with which academicians share their knowledge.

This study falls within the interpretive paradigm, which supports the view that there are numerous truths and multiple realities. This type of paradigm focuses on the holistic perspective of the personalities involved. According to [25], the interpretive paradigm is commonly associated with methodological approaches that provide an opportunity for the voices, concerns, and practices of research participants to be heard [25]. Table 1 summarizes the research paradigm.

Table 1. Research paradigm

N	Characteristic	Interpretive view
1	Purpose	The researcher will interview the participants to understand and describe their knowledge sharing network
2	Beliefs	Participants share knowledge in different academic institutions
3	Research method	Qualitative
4	What study data is based upon	Descriptive and contextual words of interview data
5	Study sample	Representatives who are have a high level of knowledge and are able to provide answers to the research questions. Participants should hold at least either positions as either associate professor or professor

The research was conducted in four main stages as shown in Fig. 1.

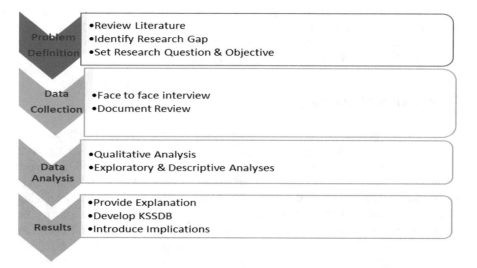

Fig. 1. Main stages of research

A sequence of interviews was conducted with target participants who were academicians in a large public university to identify the knowledge network in which they share knowledge. Data were obtained through semi-structured interviews with open-ended questions. The data were collected from a purposive sample of fifteen academic staff. The main criterion for participants to be considered as a primary source for the study was that the respondent should be well-informed and hold a senior position in the institution.

All the participants in this research were holding positions as either associate professor or professor. Invitations to participate were made through a phone call with those who had been selected as potential participants. A face-to-face interview schedule was prepared to conduct interviews with those who accepted the invitation. The face-face semi-structured interviews were conducted and administered over a period of approximately six months.

4 Data Analysis and Findings

4.1 Data Analysis

The study procedure established a coding protocol to analyze responses of the participants. The research question was "With which universities do you share your knowledge as an academician within an academic environment?" In her answer, Participant 4 mentioned that she shares her knowledge with lecturers in some other universities. She stated "I share my knowledge with lecturers from other universities... lecturers in Terengganu and Johor... I have networking already, like University Islam Sultan Sharif Ali (UNISSA), and University Science Malaysia (USM)". Participant 8 revealed that he shares his knowledge with a quite number of academic institutions: "At this moment, for example, several universities such as University Putra Malaysia (UPM), University Technology Malaysia (UTM) and University of Perlis". He also revealed that he was currently collaborating with another university, Nottingham University, Malaysia. Participant 11 had a wide network of academicians with whom he shared at many universities: "I have connections and network with people at the University of Malaya (UM), University Kebangsaan Malaysia (UKM), UPM, UTM, and USM... Usually we exchange research expertise, research experience and consult, mostly about research activities".

Participant 9 explained that he is involved in research projects with academicians in other universities "like USM, UKM, so they are also involved in our research project. That is how we get them to be involved with us, and they sometimes invite us to their universities". Participant 10, a computer science professor, gave one example about her sharing of knowledge and expertise, in which she gave a keynote lecture as presenter to introduce her research findings in one university: "For example, just last month, we went to UKM to a seminar on information retrieval and management where I gave some keynote lectures to tell what is our progress and what have I found out".

It was found that the academicians shared their expertise not solely to exchange knowledge, but also to evaluate other academic institutions' program standards, using their knowledge and expertise. For instance, Participant 13: "Because I am the external examiner of one of the programmes, so usually when I go there, I assess the standard of their programme, their syllabi and the quality of their students. I have to write reports about the standards of their programmes". Participant 12 demonstrated in-depth his sharing of knowledge with several higher learning institutions that need his expertise in media literacy, which is his area of expertise: "I attended and presented papers at other conferences organized by other universities. For example, I am presenting at University UTARA, at the International Islamic University (IIUM) and some other universities as well. Also, I was invited as guest lecturer and visiting professor in other universities like UPM, UKM, even some other private universities… I have been sharing my knowledge in communication media with, for example, Tunku Abdul Rahman University (UTAR). They invited me to give lectures on film studies. I was invited to give lectures on Media Literacy at University of Selangor (UNISEL)… with other universities that lack certain expertise; they do not have experts. For example, one university there is a lack of expertise in the area of what we call Media Literacy, so I share with them."

Participant 15 stated that she was invited to give talks at UNISEL. Participant 14, who is a professor of mathematics, pointed out that: "Being an academician, we have friends in other institutions locally… From another university like UTM, USM, UM, UMT (University of Management and Technology) … I do have colleagues working in the same area with me from other universities, especially in fuzzy mathematics and complex analysis". Participant 4 explained her sharing of knowledge internationally: "I share internationally, like in Japan with Tokyo University, like Riau Indonesia University and University Islam Indonesia. We have a project with our expertise. I went to the United Kingdom to visit the professor. That is why I would say, you see real experts in an area then, you can learn another quality of knowledge, the way different things are being managed". Participant 5 explained his KS with an international University: "With engineering society, the international engineering society, I gave a talk in Brunei in UNISSA University". Participant 6 added: "I have got an invitation from Noorul Islam University in India; they did a seminar on research advances in civil engineering this last September, quite recently… I also went to Thailand for a presentation and before that I went to Singapore, also for a presentation; so we also go overseas for that". Some academicians share their knowledge as a visiting professor or through attending conferences. Participant 7 highlighted that: "I am also at the Manipal University, as visiting professor to India. I have been in other countries to share knowledge through conferences such as Indonesia, Singapore, Thailand, and the Philippines".

In his response, Participant 9, the dean of a faculty, mentioned that: "We do organize professional talks. We invite speakers, international speakers to give a special talk to our faculty… from overseas professors, well-known professors, to present their expertise. We have an MOU (Memorandum of Understanding) with Indian universities like Nour Al Islam University, B. S. Abdur Rahman University. In Japan also, we have a conference with them; we have joint conferences with Germany's Hannover University; Daegu University in South Korea, we also went there; so I made a connection with Daegu University and we have an MOU with them; we do collaborate with them". Participant 9's response was unique since no other respondents mentioned the MOU term. According to him, his faculty has an MOU with a few international universities.

Participant 11 mentioned her overseas sharing of knowledge activities. As she put it: "I share my knowledge in terms of presenting at conferences on an international level... the most recent one was in Pakistan. I have been to several countries, Thailand, Indonesia, United Kingdom (UK)". There is also collaboration between academicians with other academicians from European countries in writing books together. For example, Participant 13 explained that: "I have some friends in the UK, because I got my degree from the UK. So I still have friends there. One thing is we are writing a book together". Participant 14 described his external effort to share his knowledge and expertise in mathematics. He stated that: "I have friends from Germany and Turkey... I attended some international conferences in Germany, for example. Then, I went to Indonesia recently to share about how fuzzy can be applied in various areas. In Germany, I was talking about quite a theoretical aspect of fuzzy, and in Jakarta recently I talked on how fuzzy can be applied to tourism problems in order to enhance the tourism industry".

Participant 8 mentioned several universities; especially in Japan, that he has collaboration and sharing activities with: "In Japan I have many, specially Tohoku University, Waseda University, Tsukuba University, Tokyo University, and Kyushu University. I have collaborations with them, and just this year we held an international conference, and I was the main person for that conference from this faculty. I invited twelve professors as keynote speakers. They came here and they helped us to make our conference successful".

4.2 Findings

It can be deduced from the participants' responses that they share their knowledge with academicians in many local academic institutions (LAI) and also international academic institutions (IAI). Tables 2 and 3 summarizes all LAIs and IAIs.

Table 2. Malaysian local academic institutions

N	University name
1	University Technology MARA (UiTM)
2	University Malaysia Terengganu
3	University Technology Malaysia (UTM)
4	University of Malaya (UM)
5	Melaka Manipal University
6	University Putra Malaysia (UPM)
7	University Malaysia Perlis
8	The University of Nottingham
9	University Kebangsaan Malaysia (UKM)
10	International Islamic University of Malaysia (IIUM)
11	University UTARA Malaysia (UUM)
12	Tunku Abdul Rahman University (UTAR)
13	University of Selangor (UNISEL)
14	University Science Malaysia (USM)
15	University of Management and Technology (UMT)

Table 3. International academic institutions

N	Country	University
1	Japan	1. Tokyo University 2. Kyushu University 3. Nagoya University 4. Tsukuba University 5. Waseda University
2	Indonesia	1. Indonesia Islamic University 2. Riau Indonesia University
3	India	1. B.S. Abdur Rahman University 2. Manipal University 3. Noorul Islam University
4	Germany	1. Hannover University
5	South Korea	1. Daegu University
6	Brunei Darussalam	1. University Islam Sultan Sharif Ali

The KSSDB architecture is divided into four layers. The first layer is for the registration of end users (academicians). The second layer encompasses the system functions, including knowledge creation, distribution (sharing), storing, and searching. The third layer is the application layer where the user can insert, delete and update his contribution to the system. The fourth layer is the database layer. It includes two databases: one allocated for the academicians in the LAIs, and the other for the academicians in the IAIs. Figure 2 illustrates these layers and shows how the user can use the knowledge processes within different databases, each of which are allocated for the different types of academic location.

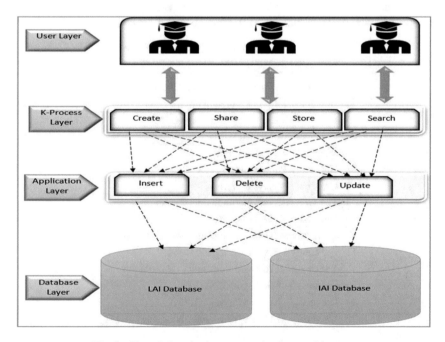

Fig. 2. Knowledge sharing system database architecture

In the proposed KSSDB architecture, there is a database for each type of academic institution; namely, LAI or IAI. Academicians can manipulate knowledge through four processes or functions that include: create knowledge, store knowledge, share knowledge, and search knowledge.

In the *create* process, the academician can add any types of knowledge they are interested to post. By using the *store* process, the academician can store the knowledge that he has posted in the database. Using the *share* process, the academician can send the knowledge he added to a specific other user, and/or to a group of people and/or to all system members. Through the *search* process, the academician can search for any type of knowledge that has been already stored by others by typing a related term or phrase in the search engine box.

Academicians who want to register and use the system function, must provide their personal and registration information. This information includes the registrant's name (both first and last names), birth date, gender, email address, contact number, postcode, and address. He must input his workplace information including ID, current position, field of expertise, faculty name, and university name. Finally, the academician must select a user name and password. Figure 3 shows the registration interface snapshot for academicians.

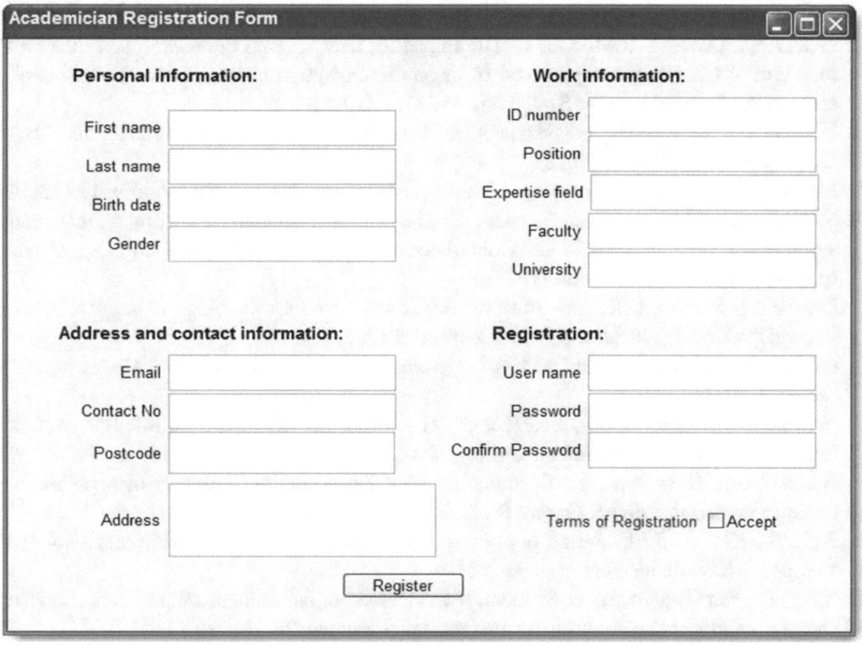

Fig. 3. Academicians' registration interface

5 Conclusion

This case study provides an example of the knowledge sharing networks at both national and international level among a sample of academicians. The research is a multi-disciplinary study as it comprises a study of KS behavior among academicians and the development of a system architecture for KS in academic institutions. Findings from this qualitative study provide insights into the KS networks of persons and institutions, at both local and international levels, with whom Malaysian academicians share their knowledge. It is recommended that university managements should highlight the advantages of such local and global KS collaboration among their highly educated academicians. This study focuses on academic staff; non-academic staff are not included. Future research could investigate knowledge sharing among non-academic staff members with a view to organizational enhancement.

Acknowledgment. The authors of this study extend their gratitude to the Saudi Ministry of Health for its moral encouragement and support.

References

1. Mohammad, M.T.F., Alajmi, S.A., Ahmed, E.A.R.D.: Motivation factors toward knowledge sharing intentions and attitudes. Int. J. Bus. Admin. **9**(4), 110–126 (2018)
2. Huang, Q., Davison, R.M., Gu, J.: The impact of trust, guanxi orientation and face on the intention of Chinese employees and managers to engage in peer-to- peer tacit and explicit knowledge sharing. Inform. Syst. J **21**, 557–577 (2011)
3. Hislop, D.: Knowledge management in organizations: a critical introduction. Oxford University Press, Oxford (2013)
4. Singh, S.: What are we managing–knowledge or information? VINE **37**, 169–179 (2007)
5. Salloum, S.A., Al-Emran, M., Shaalan, K.: The impact of knowledge sharing on information systems: a review. In International Conference on Knowledge Management in Organizations (pp. 94–106). Springer, Cham (2018)
6. Davison, R.M., Ou, C.X.J., Martinsons, M.G.: Information technology to support informal knowledge sharing. Inform. Syst. J **23**, 89–109 (2013)
7. Büchel, B., Raub, S.: Building knowledge-creating value networks. Eur. Manag. J. **20**(6), 587–596 (2002)
8. Newell, S., Robertson, M., Scarbrough, H., Swan, J.: Managing knowledge work and innovation. Palgrave Macmillan, Basingstoke (2009)
9. Bartol, K.M., Srivastava, A.: Encouraging knowledge sharing: The role of organizational reward systems. J. Leader. Organ. Stud. **9**(1), 64–76 (2002)
10. Park, S., Kim, E. J.: Fostering organizational learning through leadership and knowledge sharing. J. Knowledgement Manag. (2018)
11. Yang, H., van Rijn, M. B., & Sanders, K.: Perceived organizational support and knowledge sharing: employees' self-construal matters. Int. J Human Res. Manag, 1–21 (2018)
12. Witherspoon, C., Bergner, J., Cockrell, C., Stone, D.: Antecedents of organizational knowledge sharing: a metaanalysis and critique. J. Knowledge Manag. **17**(2), 250–277 (2013)

13. Kang, Y., Kim, S., Chang, G.: The impact of knowledge sharing on work performance: an empirical analysis of the public employees' perceptions in South Korea. Int. J. Public Admin. **31**(14), 1548–1568 (2008)
14. Cavaliere, V., Lombardi, S., Giustiniano, L.: Knowledge sharing in knowledge-intensive manufacturing firms: an empirical study of its enablers. J Knowledge Manag **19**(6), 1124–1145 (2015)
15. Mohammad, M.T.F., Alajmi, S.A., Ahmed, E.A.R.D.: Motivation factors toward knowledge sharing intentions and attitudes. Int J. Bus. Admin. **9**(4), 110–126 (2018)
16. Seufert, A., Von Krogh, G., Bach, A.: Towards knowledge networking. J. Know. Manag. **3**(3), 180–190 (1999)
17. Kühne, B., Lambrecht, E., Gellynck, X.: Network types and their importance for knowledge exchange and innovation in the agri-and horticultural sector. In Proceedings. Presented at the 21st Annual world symposium of the International Food and Agribusiness Management Association, International Food and Agribusiness Management Association (IFAMA) (2011)
18. Stone, D.: Knowledge networks and global policy. In Global knowledge networks and international development (pp. 109–125). Routledge (2004)
19. George, B., Kim, S., Shekhar, S.: Spatio-temporal network databases and routing algorithms: A summary of results. In International Symposium on Spatial and Temporal Databases (pp. 460–477). Springer, Berlin, Heidelberg (2007)
20. Arjun Panwar.: Types of database management systems. Retrieved from https://www.c-sharpcorner.com/UploadFile/65fc13/types-of-database-management-systems/ on 26 Jan 2019
21. McDermott, R., O'dell, C.: Overcoming cultural barriers to sharing knowledge. J. Knowledge Manag. **5**(1), 76–85 (2001)
22. George, B., Kim, S., Shekhar, S.: Spatio-temporal network databases and routing algorithms: A summary of results. In International Symposium on Spatial and Temporal Databases (pp. 460–477). Springer, Berlin, Heidelberg (2007)
23. Atkinson, M.P., Buneman, O.P.: Types and persistence in database programming languages. ACM Computing Surveys (CSUR) **19**(2), 105–170 (1987)
24. Creswell, J.W.: Qualitative inquiry and research design: choosing among five approaches. SAGE Publications (2007)
25. Cole, M.: Qualitative research: a challenging paradigm for infection control. British Journal of Infection Control **7**(6), 25–29 (2006)

An Intelligent Predictive Analytics System for Transportation Analytics on Open Data Towards the Development of a Smart City

Abdul-Rasheed A. Audu[1], Alfredo Cuzzocrea[2],
Carson K. Leung[1(✉)] (iD), Keaton A. MacLeod[1], Nibrasul I. Ohin[1],
and Nadège C. Pulgar-Vidal[1]

[1] University of Manitoba, Winnipeg, MB, Canada
kleung@cs.umanitoba.ca
[2] University of Trieste, Trieste, TS, Italy

Abstract. As time is a precious asset, bus riders would desire to get accurate information about bus arrival time. Although different research approaches have been developed to correctly predict bus arrival time, very few of them produce highly precise and accurate results based on open data. In this paper, we present an intelligent system designed for transportation analytics on open data such as bus delay data. Specifically, the system accesses open data to analyze public transport data—such as historical bus arrival time—for urban analytics; it then conducts data analytics and mining to discover frequent patterns. Based on the discovered patterns, the system makes predictions on whether the bus arrives on time or is being late. Evaluation on real-life open data provided by a Canadian city show the effectiveness and prediction accuracy of our intelligent system in transportation analytics on open data. The results are encouraging towards the goal of developing smart cities.

Keywords: Intelligent system · Transportation analytics · Open data · Public transportation · Bus · Bus delay · Data analytics · Frequent pattern mining · Predictive analytics · Smart city

1 Introduction

As we are living in the era of big data [1–5], big data are everywhere. With advances in technologies, huge volumes of a wide variety of valuable data—which may be of different levels of veracity (e.g., precise data, imprecise and uncertain data)—can be easily generated or collected at a high velocity. They can be originated from a wide variety of data sources in various real-life applications. These include Internet of Things (IoT) data [6], music [7], stock prices [8], meteorological data [9], web data [10], and urban data (e.g., public transit data).

Public transit has become the daily mode of commuting for a large portion of the population due to its convenience in many scenarios [11, 12]. This includes the low cost of commuting (cf. owning and operating a private vehicle), as well as the environmental effect perceived by individuals who are concerned about the amount of carbon being emitted into the atmosphere daily [12, 13]. Due to the ease of data storage techniques, data relating to transit usage and transit delays have been collected and

© Springer Nature Switzerland AG 2020
L. Barolli et al. (Eds.): CISIS 2019, AISC 993, pp. 224–236, 2020.
https://doi.org/10.1007/978-3-030-22354-0_21

made accessible in the form of open data in multiple cities across the world. For instance, in Canada, from (Pacific) coast to (Atlantic) coast, many Canadian provinces and territories, as well as municipalities, have joined the conversation on *open government*, in which residents can find and easily access information about open government activities in their jurisdiction and/or across the country. Specifically, so far, nine provincial governments and 66 municipal governments have set up open data portals. Examples include the following:

- City of Toronto Open Data Portal[1], and
- City of Winnipeg Open Data Portal[2].

Among the available data in these open data portals, we focus on transit data in this paper as transit data are closely related to most residents. In particular, we examine bus delay data[3] from Toronto Transit Commission (TTC).

In general, *transit delay* is a regular occurrence and its patterns take an interesting form [14]. In this paper, we analyze bus delay patterns over time, study the patterns that exist, and observe delay correlations which can be used to predict the possibility of a future delay given previous scenarios. These results can give an insight into existing problems in the infrastructure, bus routes and bus schedules; they can also give an insight into future planning and development of a safer, more rider-friendly, more rider-convenient transit systems.

The transit system is an integral part to the lives of many people in modern society. It facilitates a safe way for people to move across their cities for a variety of reasons (e.g., commuting to work, school, or shopping mall; visiting friends; going to recreation centers, sports games, or concerts). It also provides services for both frequent bus riders (who usually hold weekday passes, weekly passes, or monthly passes) and occasional bus riders.

As seasons come and go, weather changes, and events of different scales happening in the cities, there may be problems of maintaining the schedule of some buses due of one factor or a combination of factors [14]. To prevent and reduce the number of delays that occur, knowing (a) the causes of delays and (b) the delay trends that appear frequently can serve as a starting point for policymakers in the affected cities. Transit riders are limited to knowing only when and where their buses are supposed to arrive, without any other reliable information. In this paper, our *key contribution* is our intelligent system for analyzing public transport data on bus arrival time and predicting on whether or not bus arrives on time or late. Specifically, our system conducts frequent pattern mining for predictive analytics on open data for transportation analysis. To evaluate our intelligent system, we apply it to real-life bus delay data provided by the TTC. It predicts future bus delay data based on the frequently mined patterns from previous years. Evaluation results show that the system accurately predicts bus delay data for test data (for year 2017) based on the frequent pattern mining of the historical bus delay data (from years 2014–2016).

---

[1] https://www.toronto.ca/city-government/data-research-maps/open-data/open-data-portal/.

[2] https://data.winnipeg.ca/.

[3] https://portal0.cf.opendata.inter.sandbox-toronto.ca/dataset/ttc-bus-delay-data/.

The remainder of this paper is organized as follows. Next section discusses related works. Section 3 describes our intelligent system for transportation analytics of bus data and predictive analytics for bus delay. Evaluation on Sect. 4 shows the efficiency of our intelligent system in predicting bus delay for real-life situations for buses operated by TTC. Finally, Sect. 5 draws conclusions.

2 Related Works

There has been works on predictive analytics [15–19] on various real-life applications. In particular, there has been research works done to accurately predict bus arrival times. For instance, Sun et al. [20] focused on predicting the correct bus arrival time based on the geographic information system (GIS)-based map-matching algorithm, which is used to project each received location onto the underlying transit network, thus pinpointing bus location to predict its arrival. Lin et al. [21] also focused on predicting the correct bus arrival times, but based on global positioning system (GPS) data and automatic fare collection (AFC) system data with the help of *artificial neural network* (*ANN*). Rajput et al. [12] analyzed New York open data, and applied *clustering* to identify (a) highly congested areas and (b) areas with less bus stops to provide suggestion for bus stops. Many other researchers [22–26] used *Kalman filter* or *time series models* to predict future travel times, under the assumption of a direct relationship with previous travel times. In other words, most of the aforementioned approaches

- use auxiliary information such as GIS, GPS, and/or AFC system data (which may not be easily accessible by general public); and/or
- apply data mining tasks like classification and/or clustering—via techniques like ANN, Kalman filter, and/or time series models.

In contrast, in our current paper, we use *open data* (which are freely accessible by general public) and apply a different data mining task of *frequent pattern mining* (e.g., using the FP-growth algorithm) on the bus delay data of Toronto to generate a predictive model to predict possible bus delays based on day and time of the delay, delay duration, delay severity, and delay type.

3 Construction of Our Intelligent System

To conduct transportation analytics on open data (specifically, predictive analytics on bus delay data), our intelligent system first cleans the input data. It then analyzes the cleaned data to discover interesting patterns and to make accurate predictions.

3.1 Data Understanding

The TTC provides delay data for the following modes of ground transportation served in the City of Toronto:

- buses,
- streetcars (i.e., trams, trolleys), and
- subways (i.e., underground rapid transit rails).

In the current paper, we focus on the bus data. However, it is important to note that the knowledge learned from the current paper on TTC bus delay data can be transferred to delay data for other transportation modes such as streetcar and/or subway—via *transfer learning*. Along the same direction, such knowledge can also be transferred to other jurisdictions and/or other transportation data via transfer learning.

For bus delay data (such as the TTC bus delay data available on the City of Toronto Open Data Portal), they usually contain the following information:

- report date, which captures the date when the delay-causing incident occurred;
- route, which captures the number of the bus route;
- time, which captures when the delay-causing incident occurred;
- day, which captures day of the week;
- location, which captures the location of the delay-causing incident occurred;
- incident, which captures the description of the delay-causing incident;
- minimum delay, which captures the delay (in minutes) to the schedule for the following bus;
- minimum gap, which captures the total scheduled time (in minutes) from the bus ahead of the following bus (i.e., time gap between successive buses running the same route);
- direction, which captures the direction of the bus route; and
- vehicle, which captures the vehicle number.

The data are usually provided on a yearly basis—in an easily accessible format (e.g., Excel files)—and updated on a regular basis (e.g., multiple times per month). An example view of the data provided is in Table 1. Here, for the "direction" attribute, five possible values were expected:

- N/B, which indicates the northbound route;
- S/B, which indicates the southbound route;
- E/B, which indicates the eastbound route;
- W/B, which indicates the westbound route; and
- B/W, which indicates both ways.

Table 1. Sample bus delay data.

Report date	Route	Time	Day	Location	Incident	Min delay	Min gap	Direction	Vehicle
01-Jan-19	39	12:13:00 AM	Tuesday	NECR	Mechanical	9	18	W/B	1794
01-Jan-19	111	12:15:00 AM	Tuesday	Eglington	Mechanical	15	30	S/B	8065
01-Jan-19	35	12:18:00 AM	Tuesday	Finch	Mechanical	9	18	S/B	3275
01-Jan-19	25	12:30:00 AM	Tuesday	Don Mills Rd/ Eglinton Ave E	Mechanical	9	18	N/B	8840
01-Jan-19	36	12:40:00 AM	Tuesday	Humberwood	Investigation	9	18	E/B	9119
...	...	...	...	...	...	...	...	...	...

3.2   Data Cleaning

In the first step, our intelligent system removes NULL and incomprehensible values for each attribute/parameter name. Specific details are described below.

For the attribute/parameter "time" (i.e., time when the delay-causing incident occurred), our system converts the 12-hour time representation to 24-hour time representation. It then discretizes the data and bins the values into the following four equal-size intervals of 6 h each:

- Night, for 23:00–5:00 (i.e., 11:00PM–5:00 AM);
- Morning, for 5:00–11:00 (i.e., 5:00 AM–11:00 AM);
- Afternoon, for 11:00–17:00 (i.e., 11:00 AM–5:00 PM); and
- Evening, for 17:00–23:00 (i.e., 5:00 PM–11:00 PM).

Such a step is beneficial for the ease of frequent pattern mining and association rule mining as it helps us to get insights about (a) the frequency of delay intervals and (b) relationships among delay-causing incidents. Figure 1 shows the frequency of time and its corresponding discretized interval. The figure also reveals that most delays occur at 15:00, 8:00, and 14:00.

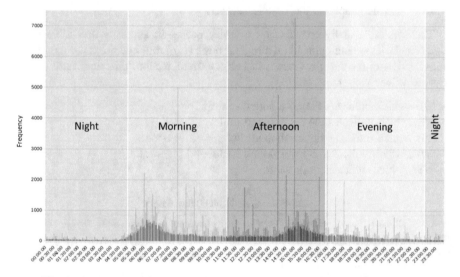

Fig. 1. Frequency of time of the day when the delay-causing incident occurred

For the attribute "location" (i.e., the location of the delay-causing incident occurred), it seems to be manually entered into the dataset. Consequently, this leads to inconsistencies in syntax—especially, in terms of spelling, special characters, and capitalization. Our system fixes this problem by imposing a consistent naming standard between all the location names in the database. It does so by utilizing related auxiliary datasets (say, data about routes and schedules[4]) that contain information for routes and stop locations—represented by bus stop names, as well as other identifiers (e.g., IDs, latitude and longitude coordinates, etc.) of bus stops. With standard stop names, our system conducts pairwise comparisons between any problematic bus stop name entry (i.e., any bus stop name that is unavailable from the dataset containing standard stop names) with standard ones. Specifically, the system applies fuzzy string matching to compare differences between the pairs of bus stop names and find the most similar match (of the stop name). As there may be more than a single similar bus stop name for a given problematic bus stop name entry, our system associates each potential match or similar bus stop name with a similarity value. The one with the highest similarity value is considered as the closest match for that problematic bus stop entry.

For the attribute "minimum gap" (i.e., the delay in minutes to the schedule for the following bus), our system discards insignificant delays (e.g., delays of less than 5 min). These very short delays are usually unavoidable but tolerable by bus riders. The remaining delays (i.e., those delayed for at least 5 min) are then discretized and binned into the following five categories, which indicate the delay severity:

- short delay of at least 5 min but less than 10 min,
- medium delay of at least 10 min but less than 20 min,
- long delay of at least 20 min but less than 30 min,
- severe delay of at least 30 min but less than 60 min, and
- crippling delay of at least 60 min.

We select these bins based data distribution (see Fig. 2) and general perception of bus delay times by transit riders.

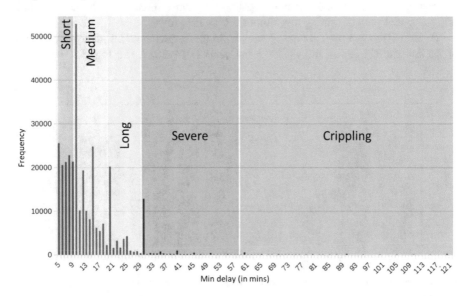

Fig. 2. Frequency of the minimum delay time

---

[4] https://portal0.cf.opendata.inter.sandbox-toronto.ca/dataset/ttc-routes-and-schedules/.

For the attribute "direction" (i.e., direction of the bus route), we know there supposes to be only five possible values (namely, N/B, S/B, E/B, W/B, and B/W). However, similar to the attribute "location", the direction also seems to be manually entered into the dataset. Consequently, this also leads to inconsistencies in syntax—especially, in terms of spelling and capitalization (e.g., "NB", "nb", "N"). Our system fixes these spelling and capitalization problems by finding the closet matches among the five possible values for direction. However, it can be challenging when the problematic direction entry shows "W". Does "W" mean westbound or both ways? Our system partially fixes this problem by again utilizing related auxiliary datasets (say, data about routes and schedules) that contain information for routes. Based on the routes, the system can determine that "W" means both ways if the route is a north-south one. However, if the route is an east-west one, the system may require additional information to precisely determine whether "W" means both ways or just westbound.

3.3 Data Analytics via Frequent Pattern Mining for Checking the Feasibility of Predictions

Once the data are cleaned, our intelligent system conducts data analytics to find evidence for supporting the conjecture that "patterns discovered from bus delay data in previous years are sufficient to predict bus delays in the future". To do so, we examine a few frequent pattern types, where the type of a frequent pattern is defined by the rows it includes, and count the occurrences of each type across years to get some insights about the data. For instance, we examine correlation between different attributes and severity of delay incurred. Moreover, we also examine the intersection of patterns between years and counted number of patterns by types. Frequent patterns are mined using frequent pattern mining algorithm such as FP-growth.

By using frequent pattern mining, we obtain the frequency of bus delays by (a) *time* of the day and (b) *day* of the week for the years 2014–2017. See Fig. 3.

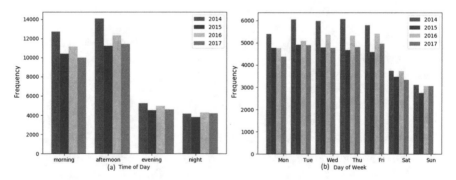

Fig. 3. Frequency of the (a) time of the day and (b) day of the week when the delay-causing incident occurred.

Observed from Fig. 3(a), bus delays throughout time of the day were consistent throughout all the years from 2014 to 2017. Moreover, bus delays were more frequent in the morning and afternoon than in the evening and at night. These patterns reveal that, as a majority of people commute to school/work in the morning, more congestion on the road, which leads to more bus delays. Similar scenarios for the afternoon when people come back from school/work in the afternoon.

Observed from Fig. 3(b), bus delays were most frequent in year 2014. Moreover, bus delays were more frequent during weekdays than weekends. These patterns reveal that people commute to work on weekdays, which results in higher probability of congestion on the roads, which leads to delayed buses.

As the patterns are consistent throughout all the years for bus delays by both (a) day of week and (b) time of the day, *patterns discovered from bus delay data in 2014–2016 are reasonably sufficient to predict bus delays in 2017.*

3.4   Feature Selection via Frequent Pattern Mining

Our intelligent system conducts frequent pattern mining on historical bus delay data to select feature for predicting future bus delays. Specifically, the system iterates through each year of bus delay data to mine frequent patterns in each of the historical years using FP-growth. Any attributes within our dataset can be taken into account when mining frequent patterns. Among the mined results, interesting results are obtained by using the following parameters:

- report date (or report month),
- route,
- time (or time interval),
- day,
- location (or standard bus stop name),
- incident, and
- minimum delay (or delay severity).

These attributes provide insight into a plethora of interesting delay factors, which we could further analyze to determine interesting results as to when, why, and where delays were frequently occurred.

Once these frequent patterns consisting of sets of parameters were computed, the frequent patterns were partitioned into two categories: (a) *previous* years and (b) a single *future* year. Next, we identify all of the frequently occurring patterns, which exist in every previous year. These were the patterns that occur most consistently between years, and they are good candidates—as *selected features*—for making accurate predictions for future year. See Table 2 for frequency of frequent patterns mined from (a–c) three previous years 2014–2016, (d) future year 2017, and (e–g) their three combinations of previous & future years within 2014–2017.

Table 2. Sample frequent patterns (i.e., selected features for predictive analytics).

Frequent patterns	Frequency						
	Previous years			Future year	Intersections of previous & future years		
	2014	2015	2016	2017	$2016 \cap 2017$	$2015 \cap 2016 \cap 2017$	$2014 \cap 2015 \cap 2016 \cap 2017$
1. {Route, time, delay severity}	200	160	170	152	128	101	97
2. {Route, delay severity}	174	148	161	149	138	125	123
3. {Route, direction, delay severity}	172	148	159	136	125	106	98
4. {Route, incident, delay severity}	160	134	147	120	104	83	77
5. {Month, incident, delay severity}	138	130	132	134	127	125	124
...	...	...	...	...	...	...	...
15. {month, delay severity}	52	52	54	53	51	48	45
16. {day, delay severity}	35	35	35	35	35	35	35
17. {incident, delay severity}	25	24	25	25	24	24	24
18. {Time, delay severity}	20	20	20	20	20	20	20

Observed from Table 2, simpler patterns consisting of two features (e.g., frequent patterns 15–18) are constantly frequent with (almost) the same frequency across individual years from 2014 to 2017. They are reliable features to be selected for predictive analytics. Moreover, more complex patterns consisting of more features (e.g., frequent patterns 1–5, each consist of three features) are also constantly frequent within a reasonable range of frequency across individual years from 2014 to 2017. They are also reliable features to be selected for predictive analytics.

As expected, frequency of the intersections does not increase when more years are added to the intersections. When frequency of the intersections of three previous years 2014–2016 is high, the patterns are very likely to occur in future year 2017 and thus are good predictive features. Along this direction, they would also be good predictive features in the true future (say, 2018, 2019, etc.) as well.

3.5 Predictive Analytics via Decision Tree Induction

Once the features are selected by frequent pattern mining, our intelligent system conducts predictive analytics. Specifically, it builds a decision tree, which stores the probability that a bus delay would occur given a user query with the following input parameters:

- a specific date including (a) month and (b) day of the week;
- time of the day; and
- bus stop location.

Our system stores the decision tree on disk, which gives us a viable solution that does not require computation/classification each time a query is made. For more accurate/time-relevant predictions, the system incorporates a time-fading model so that more recent bus delay data carry heavier weights than those older bus delay data. Also, whenever a new data entry comes in, the tree is updated (or a new tree is built) to reflect the new value. By doing so, the system would be able to provide real-time predictive functionality.

4 Validation of Our Intelligent System with Real-Life Open Data

To evaluate our intelligent system and validate its prediction results, we used the aforementioned TTC bus delay data available on the City of Toronto Open Data Portal.

Specifically, to evaluate the prediction accuracy of our system, once all of patterns are identified (i.e., features are selected), we examine the frequently occurring patterns of the future year, and create two lists:

- Frequent patterns occurring in all years, both previous and future (i.e., successful predictions); and
- Frequent patterns occurring in all previous years, but not the future year (i.e., unsuccessful predictions).

Table 3 shows some examples of successful and unsuccessful predictions. For instance, based on bus delay data in previous years 2014-2016, our intelligent system predicts that "mechanical problems occur on a Tuesday afternoon in January are likely to cause a short delay" in future year 2017, and appeared frequently in 2017. This is one of many successful predictions returned by our system.

Table 3. Sample predictions.

Frequent patterns (with feature values)	Delay severity	Previous years			Future year
		2014	2015	2016	2017
S1. {Jan, afternoon, Tue, mechanical}	Short	73	41	26	53
S2. {May, afternoon, Wed, mechanical}	Short	43	44	31	51
S3. {Mar, afternoon, Wed, mechanical}	Short	73	38	34	43
U1. {Mar, morning, Tue, mechanical}	Short	49	53	31	Infrequent
U2. {Mar, afternoon, Tue, mechanical}	Medium	49	36	32	Infrequent

In terms of prediction accuracy, our intelligent system is accurate. For instance, among those 1553 frequent patterns of length 4 (i.e., each frequent pattern consists of values for selected features/parameters like those shown in Table 3) returned by our system, 1,378 were successful predictions (i.e., 89% of the generated predictions are *true positives*) and only 175 were unsuccessful predictions (i.e., 11% of the generated predictions are *false positives*).

5 Conclusions

As we are living in the era of big data, big data are everywhere. With advances in technologies, huge volumes of a wide variety of valuable data—which may be of different levels of veracity—can be easily generated or collected at a high velocity. They can be originated from a wide variety of data sources in various real-life applications. Public transportation data or urban data are examples of big data.

Timely public transportation is important because it encourages more people to utilize transit and reduce the congestion in the roads. Delayed public transportation (e.g., delayed bus) has significant negative impacts on the quality of life for the general inhabitants in the city as people end up losing their precious time waiting for buses due to bus delay. In this paper, we present an intelligent predictive analytics system for transportation analytics on open data towards the development of a smart city. Specifically, our system analyzes *open data* (which are freely accessible by general public)—namely, the TTC bus delay data from the City of Toronto Open Portal—and apply a data mining task of *frequent pattern mining* (e.g., using the FP-growth algorithm) to show that historical bus delay data are relevant to making predictions for future bus delays. Our decision tree based model accurately predicts the severity of possible bus delays based on input parameters like the month, time, day, time, incident and/or location of the delay. In other words, our key contribution is our intelligent system for analyzing public transport data on bus arrival time and predicting on whether or not bus arrives on time or late.

Please note that, although most of the city transits have GPS (which makes real-time scheduling or prediction easier than using other historical data), these GPS data may be out of reach by general public. This explains why we focus on open data, which are freely accessible by general public, in this paper. However, as ongoing work, we are interested in examining how additional information (e.g., GPS data if available) would affect the prediction accuracy and the ability of real-time prediction.

Moreover, as real traffic delays often determine by many other external factors outside the normal traffic congestion (e.g., road accidents, events, road maintenance, temporary closure of routes, etc.). As a second direction for ongoing work, we are interested in incorporating other data sources (e.g., Google Maps, Waze)—for additional navigation information, real-life travel times, and route details—into our intelligent systems.

Furthermore, as a third direction for ongoing work, we are exploring future enhancements to our intelligent systems. For instance, we are exploring the use of random forests (rather than the decision tree) for predictive analytics.

In addition, as a fourth direction for ongoing work, we are also exploring *transfer learning* from predictive analytics on bus delay data to other public transit modes such as predictive analytics on streetcars and/or subways. Specifically, we are exploring the transfer of knowledge learned from the current paper on TTC bus delay data to delay data for other transportation modes such as streetcar and/or subway via transfer learning. Along the same direction, we are also exploring the transfer of knowledge to other jurisdictions and/or other transportation data via transfer learning.

Acknowledgments. The project is partially supported by (i) NSERC (Canada) and (ii) University of Manitoba.

References

1. Bellatreche, L., Leung, C., Xia, Y., Elbaz, D.: Advances in cloud and big data computing. Concurrency Comput. Prac. Exp. **31**(2), e5053:1–e5053:3 (2019)
2. Hoi, C.S.H., Khowaja, D., Leung, C.K.: Constrained frequent pattern mining from big data via crowdsourcing. In: BigDAS 2017. AISC, vol. 770, pp. 69–79 (2017)
3. Leung, C.K.: Big data analysis and mining. In: Advanced Methodologies and Technologies in Network Architecture, Mobile Computing, and Data Analytics, pp. 15–27 (2019)
4. Leung, C.K., Deng, D., Hoi, C.S.H., Lee, W.: Constrained big data mining in an edge computing environment. In: BigDAS 2017. AISC, vol. 770, pp. 61–68 (2017)
5. Leung, C.K., Hoi, C.S.H., Pazdor, A.G.M., Wodi, B.H., Cuzzocrea, A.: Privacy-preserving frequent pattern mining from big uncertain data. In: IEEE BigData 2018, pp. 5101–5110 (2018)
6. Abdalaal, R.M., Ho, C.N.M., Leung, C.K., Ohin, N.I., Ur-Rehman, S.H.: A remotely control dimming system for LED lamps with power factor correction. In: IEEE ECCE 2018, pp. 4721–4727 (2018)
7. Barkwell, K.E., Cuzzocrea, A., Leung, C.K., Ocran, A.A., Sanderson, J.M., Stewart, J.A., Wodi, B.H.: Big data visualisation and visual analytics for music data mining. In: IV 2018, pp. 235–240 (2018)
8. Camara, R.C., Cuzzocrea, A., Grasso, G.M., Leung, C.K., Powell, S.B., Souza, J., Tang, B.: Fuzzy logic-based data analytics on predicting the effect of hurricanes on the stock market. In: FUZZ-IEEE 2018, pp. 576–583 (2018)
9. Cox, T.S., Hoi, C.S.H., Leung, C.K., Marofke, C.R.: An accurate model for hurricane trajectory prediction. In: IEEE COMPSAC 2018, vol. 2, pp. 534–539 (2018)
10. Leung, C.K., Jiang, F., Souza, J.: Web page recommendation from sparse big web data. In: IEEE/WIC/ACM WI 2018, pp. 592–597 (2018)
11. Aoki, E., Otsuka, S., Ikenaga, T., Yatsuzuka, M., Tokiwa, K.: Study on regional transportation linkage system that enables efficient and safe movement utilizing LPWA. In: CISIS 2018. AISC, vol. 772, pp. 968–977 (2018)
12. Rajput, P., Toshniwal, D., Aggarwal, A.: Improving infrastructure for transportation systems using clustering. In: BDA 2017. LNCS, vol. 10721, pp. 129–143 (2017)
13. Kennedy, C., Steinberger, J., Gasson, B., Hansen, Y., Hillman, T., Havránek, M., Pataki, D., Phdungsilp, A., Ramaswami, A., Mendez, G.V.: Greenhouse gas emissions from global cities. Environ. Sci. Technol. **43**(19), 7297–7302 (2009)
14. Morency, C., Trépanier, M., Agard, B.: Measuring transit use variability with smart-card data. Transp. Policy **14**(3), 193–203 (2007)

15. Bañeres, D., Serra, M.: On the design of a system to predict student's success. In: CISIS 2018. AISC, vol. 772, pp. 274–286 (2018)
16. Chen, R., Tong, Y., Yang, J., Wu, M.: Residual reconstruction algorithm based on sub-pixel multi-hypothesis prediction for distributed compressive video sensing. In: CISIS 2018. AISC, vol. 772, pp. 599–605 (2018)
17. Khan, M., Javaid, N., Iqbal, M.N., Bilal, M., Zaidi, S.F.A., Raza, R.A.: Load prediction based on multivariate time series forecasting for energy consumption and behavioral analytics. In: CISIS 2018. AISC, vol. 772, pp. 305–316 (2018)
18. Kulla, E., Morita, S., Katayama, K., Barolli, L.: Route lifetime prediction method in VANET by using AODV Routing Protocol (AODV-LP). In: CISIS 2018. AISC, vol. 772, pp. 3–11 (2018)
19. Phankokkruad, M., Wacharawichanant, S.: Prediction of mechanical properties of polymer materials using extreme gradient boosting on high molecular weight polymers. In: CISIS 2018. AISC, vol. 772, pp. 375–385 (2018)
20. Sun, D., Luo, H., Fu, L., Liu, W., Liao, X., Zhao, M.: Predicting bus arrival time on the basis of global positioning system data. Transp. Res. Rec. **2034**(1), 62–72 (2007)
21. Lin, Y., Yang, X., Zou, N., Jia, L.: Real-time bus arrival time prediction: case study for Jinan. China J. Transport. Eng. **139**(11), 1133–1140 (2013)
22. Chien, S.I., Kuchipudi, C.M.: Dynamic travel time prediction with real-time and historic data. J. Transport. Eng. **129**(6), 608–616 (2003)
23. Kormáksson, M., Barbosa, L., Vieira, M.R., Zadrozny, B.: Bus travel time predictions using additive models. In: IEEE ICDM 2014, pp. 875–880 (2014)
24. Shalaby, A., Farhan, A.: Prediction model of bus arrival and departure times using AVL and APC data. J. Public Transport. **7**(1), 41–61 (2004)
25. Vanajakshi, L., Subramanian, S.C., Sivanandan, R.: Travel time prediction under heterogeneous traffic conditions using global positioning system data from buses. IET Intel. Transport. Syst. **3**(1), 1–9 (2009)
26. Williams, B.M., Hoel, L.A.: Modeling and forecasting vehicle traffic flow as a seasonal ARIMA process: theoretical basis and empirical results. J. Transport. Eng. **129**(6), 664–672 (2003)

Multi-criteria Group Decision Making and Group Agreement Quotient Analysis Based on the Delphi Method

Song Lin[1], Li Shen[2], Caiquan Xiong[2(✉)], and Xuan Li[3]

[1] Strategic Teaching and Research Section, Naval Command College,
Nanjing 210016, China
[2] School of Computer Science, Hubei University of Technology,
Wuhan 430068, China
x_cquan@163.com
[3] Beijing North-Star Digital Remote Sensing Technology Co., Ltd,
Wuhan 430074, China

Abstract. In order to overcome the limitations of individual decision makers' knowledge, a multi-criteria decision-making method based on Delphi method is proposed. This method uses the anonymity mechanism and feedback mechanism of the Delphi method. In the first round of decision-making process, the host collects the opinions given by the experts on the scheme set. Then through the calculation, obtain the consensus of the group. When the consensus of the group does not reach a certain threshold, the system feeds back the consensus degree between the experts based on the criteria and the detailed information of each scheme evaluation, so that the experts can make the next round of decision until the consensus is reached. This method makes each expert revise their opinion based on anonymous feedback information and the process of consensus-building is accelerated, avoiding endless discussion.

1 Introduction

The Delphi method is a well-known group decision-making method, in which a group of experts or stakeholders engage in an argumentation about a certain topic and finally reaches a consensus [1]. Since its introduction in the 1960s, it has been used in various fields, such as predicting mobile broadband traffic [2], identifying relevant features of a chronic disease type [3], and understanding cloud computing application issues [4], etc. Delphi method can be used for both prediction and evaluation. Numerous application examples show that Delphi method makes full use of human knowledge, experience and wisdom, and becomes an effective means to solve unstructured problems [5].

With the development of society, people are facing more and more complex decision-making problems, which promotes the development of multi-criteria decision-making technology. Multi-criteria decision-making technology has developed rapidly, many fruitful research results on multi-criteria group decision-making have been

L. Barolli et al. (Eds.): CISIS 2019, AISC 993, pp. 237–246, 2020.
https://doi.org/10.1007/978-3-030-22354-0_22

proposed: (1) Analytic Hierarchy Process (AHP) method, as in the literature [6], it aims at solving the defects of traditional analytic hierarchy process method, proposed an improved AHP method based on the exponential interval number scaling of group decision-making. (2) Bayesian probabilistic method [7, 8], as in the literature [7], for solving the risky multi-criteria group decision-making problem with unknown criterion weight and random criterion value, it proposes two kinds of Bayes models by aggregating decision makers and experts' subjective probability. It used Monte Carlo simulation method to obtain the ranking of schemes. (3) Fuzzy evaluation method, as in the literature [9], it uses the hesitant fuzzy set to extend the classical VIKOR, and proposes a hesitant fuzzy VIKOR method and solved the multi-criteria decision problem.

These methods have done a lot of work on the aggregation of individual information, but they have not defined the degree of consistency of expert opinions. They are only interested in the fusion of expert preference information to sort the schema set, and finally get the result set, but it cannot determine if it was the best result set. Based on this, this paper introduces Delphi's feedback mechanism and anonymity mechanism, proposes a multi-criteria decision-making model based on Delphi method, the group consistency analysis is introduced to the decision results [10], and add expert individual criterion consistency calculation in the feedback link. With the detailed scoring information based on the criteria, experts can make results better based on the other experts' opinion, after several rounds of constant self-correction, through an example obtain that the final decision result have a good convergence.

2   Multi-criteria Decision Making and Group Consistency Analysis

2.1   Multi-resolution Solution

A multi-criteria decision problem can be described as: limited group set $E = \{e_1, e_2, \cdots, e_s\}(s \geq 2)$, and the limited scheme set $X = \{X_1, X_2, \cdots, X_m\}(m \geq 2)$, among the scheme set, judge the limited criteria set $A = \{a_1, a_2, \cdots, a_n\}(n \geq 2)$. Hypothesis the e_k's judgement from group E value to scheme i's criteria is v_{ij}^k, ($v_{ij}^k \geq 0$, $1 \leq k \leq s$, $1 \leq i \leq m, 1 \leq j \leq n$), then we can get the e_k's preference vector matrix:

$$V^k = \begin{bmatrix} v_{11}^k & \cdots & v_{1n}^1 \\ \vdots & \ddots & \vdots \\ v_{m1}^k & \cdots & v_{mn}^k \end{bmatrix}$$

Definition 1: The scheme x_k's calculation formula is:

$$S_k = \sum_{j=1}^{n} \sum_{q=1}^{s} W_j V_{kj}^q. \tag{1}$$

w_j is the weight of criterion a_j

In order to ensure the group preference vector is coincident, the group consistency needs to be analyzed before assembling the preference vectors. All preferences are actually a vector matrix of m-dimensional space (suppose there are m alternatives). In fact, group consistency analysis is to analyze the consistency of s vector matrices (suppose its number is s).

2.2   The Consistency of Preference Vector

The analysis of group agreement quotient is generally based on the calculation of the consistency of the two pairs of preference vectors. There are many calculations for vector consistency. For example, Cook et al. [11] proposed the Euclid distance method, Cook and Kress [12] proposed the L1 space norm number method, Hamer et al. [13] proposed the cosine method of the angle between two vectors, and Basilersky et al. [14] et al. proposed the sine method of the angle between two vectors. Herrera-Viedma et al. [15] also proposed a method for determining preference consistency based on the ordering relationship of alternatives in each given preference vector is proposed, which overcomes the consistency of preference dimensions. However, these methods need to be aggregated to obtain the temporary evaluation value. It is not as accurate as the initial evaluation information. This paper first calculates the consistency of the expert in pairs on each criterion according to the initial evaluation information, and then collects all the criteria. The degree of consistency calculates the degree of agreement between the expert in pairs.

Definition 2: The consistency of criterion a_j about expert e_t, e_k is $d_j(e^t, e^k)$ the calculation formula as follow: The preference vector of a_j about the expert in pairs e_t, e_k is $v_j^t = (v_{1j}^t, v_{2j}^t, \cdots, v_{mj}^t)$. m is the number of the plans.

$$d_j\left(e^t, e^k\right) = \left(v_j^t, v_j^k\right) / \left(\left\|v_j^t\right\| \circ \left\|v_j^k\right\|\right). \tag{2}$$

(v_j^t, v_j^k) represents the vector inner product of vector v_j^t and vector v_j^k. $\|v_j^t\|$ and $\|v_j^k\|$ represent the norm of vector v_j^t and vector v_j^k (the length). The greater value of $d_j(e^t, e^k)$, the greater consistency about the plans of criterion a_j, and $d_j(e^t, e^k)$ content the following conditions:
  (1) $0 \le d_j(e^t, e^k) \le 1$;
  (2) $d_j(e^t, e^k) = d_j(e^k, e^j)$;
  (3) $d_j(e^t, e^t) = 1$;

Definition 3: The consistency of all expert e_t, e_k is $d(e^t, e^k)$, The calculation formula is:

$$d(e^t, e^k) = \sum_{i=1}^{n} w_j d_j(e^t, e^k). \tag{3}$$

Where n is the number of criteria, w_j is the weight of a_j's criterion.

2.3   The Analysis of Group Agreement Quotient

Definition 4: Group Agreement Quotient (GAQ) means the similarity of group preference vector matrix:

$$GAQ = \frac{\sum_{i=1}^{s-1} \sum_{j=i+1}^{s} d(e^i, e^j)}{C_S^2}. \tag{4}$$

The smaller of GAQ means the smaller of the group consistency.

In the actual research process, we set the end condition of the discussion as threshold of group strong consistency and the maximum number of decision rounds. When the group consistency does not reach the specified threshold and the current number of evaluation rounds is less than the maximum number of evaluation rounds, the host proposes to invite experts to rethink and give new decisions. At the same time, all the scoring details (hidden name, identity information), the consistency of the experts' criteria, the group consistency, the criteria score details and other information are fed back to each expert as a reference for the next round of decision-making.

3   Research on Decision-Making Process Based on Delphi Method

3.1   Delphi Method

Delphi method is a feedback anonymous inquiry method. It makes all the members of the expert group do not meet directly, but communicate through correspondence, in this way, the influence of authority can be eliminated. This is an important feature of this method. Experts who engaged in solving complex problems do not know each other. They exchange their ideas in a completely anonymous situation. If they have no identical ideas, make experts rethink in next round, until they reach an agreement. Another important feature is feedback. This method needs several rounds of information feedback. In each feedback, the expert group can conduct in-depth research, so that the final results can basically reflect the basic ideas, and the results are more objective and credible. Group members communicate by answering the organizer's questions, usually through several rounds of feedback to complete the prediction.

The process is shown in the diagram (Fig. 1):

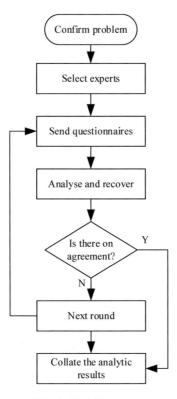

Fig. 1. Delphi process

3.2 Decision-Making Process Based on Delphi Method

The decision consensus is different from the general group decision-making. The ultimate goal is to make the decision-making opinions consistent. Therefore, this article combines multi-criteria group decision-making and group agreement quotient analysis with Delphi method, when the group consistency is poor, a feedback mechanism is needed, in order to present the individual opinions and the group consistency status in real time. Promote experts rethinking and making decisions. The discussion process as follows:

Step 1: The host give the decision-making theme, then choose the expert set $E = \{e^k | k \in s\}$, scheme collection set $X = \{x_i | i \in m\}$, clear scheme limited criteria set $A = \{a_i | i \in n\}$, set the threshold η of group consistency and the Maximum number of decision cycle *MAXCYCLE*. The cycle initial value *CYCLE* = 0;

Step 2: Make every expert think independently based on the Delphi method's anonymity, make their own decision. At last Submitted as a utility value.

Step 3: Take multi-rounds based on Delphi method's feedback mechanism, make *CYCLE* = *CYCLE* + 1, and then calculating *GAQ*.

Step 4: If the group consistency index is less than η, and *CYCLE < MAXCYCLE*, Feedback group consistency and criteria-based consistency indicators and criteria-based scoring details between the experts themselves and other experts, then turn to step 2;

Step 5: Aggregate the preference matrix vectors of each expert, calculate the final score of each scheme, and obtain the decision result.

This process is shown in the diagram (Fig. 2):

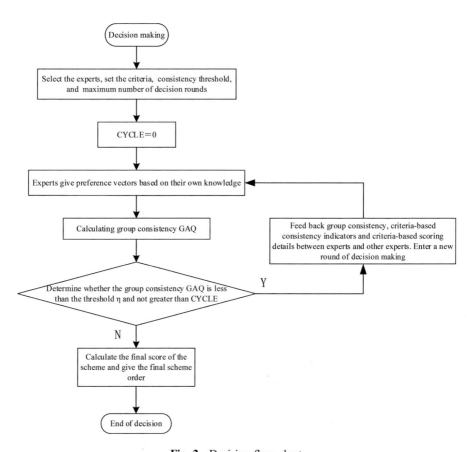

Fig. 2. Decision flow chart

When the group consistency does not meet the threshold requirement, and the number of decision rounds is not greater than the maximum decision theory, the feedback information generated by the previous round of decision-making will serve as an important basis for the next round of decision-making by experts. It can be seen that the content of the feedback information directly affects the next round of decision-making results.

The feedback information includes expert consistency indicators and scheme score details. In the case of expert e, the expert consensus indicator describes the degree of agreement between experts e and other experts on the criteria, expressed in decimal numbers between 0 and 1, and the closer to 0, the closer of opinions of the two experts on a certain criterion are to each scheme; the expert e can find the evaluation criteria a_1 that is the least consistent with other experts. Then, according to the plan evaluation details, the expert e can specifically check the scoring situation of the criterion a_1, and through further analysis, fine-tune the original score in a round of decision.

4 Case Analysis

Suppose there are 6 decision makers $E = \{e_1, e_2, e_3, e_4, e_5, e_6\}$, to the summer internship options for students $T = \{t_1, t_2, t_3\}$ to make decision, the scheme decision criteria is $A = \{a_1, a_2, a_3, a_4\}$, the decision criteria weight is $W = \{w_1, w_2, w_3, w_4\}$.
Step1: The host set decision theme, the Maximum number of decision rounds cycle *MAXCYCLE* = 3, the group consistency threshold $\eta = 0.95$, Scheme set X, Set of criteria A, Selection of decision makers E.

The first round of decision making begins, the process is as follows:
Step2: Expert give the decision matrix $V^k = (v_{ij}^k)(k = 1, 2, 3 \cdots 6)$:

$$V^1 = \begin{bmatrix} 6 & 5 & 5 & 6 \\ 4 & 7 & 3 & 8 \\ 8 & 7 & 8 & 7 \end{bmatrix} V^2 = \begin{bmatrix} 7 & 3 & 8 & 5 \\ 3 & 6 & 2 & 6 \\ 9 & 7 & 8 & 6 \end{bmatrix} V^3 = \begin{bmatrix} 3 & 5 & 5 & 6 \\ 8 & 3 & 3 & 6 \\ 4 & 5 & 8 & 4 \end{bmatrix}$$

$$V^4 = \begin{bmatrix} 10 & 6 & 8 & 4 \\ 8 & 2 & 5 & 3 \\ 8 & 7 & 5 & 8 \end{bmatrix} V^5 = \begin{bmatrix} 4 & 3 & 4 & 4 \\ 4 & 5 & 4 & 5 \\ 4 & 4 & 6 & 6 \end{bmatrix} V^6 = \begin{bmatrix} 7 & 8 & 10 & 10 \\ 8 & 7 & 5 & 9 \\ 9 & 9 & 9 & 9 \end{bmatrix}$$

Step3: The host collects the opinions of all experts, then *CYCLE* + 1, use formula 1, 2, and 3, calculate the degree of consistency between experts and other experts, all criteria consistency and group opinion consistency *GAQ* = 0.53. We can see the *GAQ* less than the group consistency threshold and *CYCLE* less than *MAXCYCLE*. The host returns feedback to each expert. Take expert e_1 as an example, the feedback as follows:

1. About the consistency of the four scoring criteria from expert e_1 and other 5 experts as the picture shows (Fig. 3):

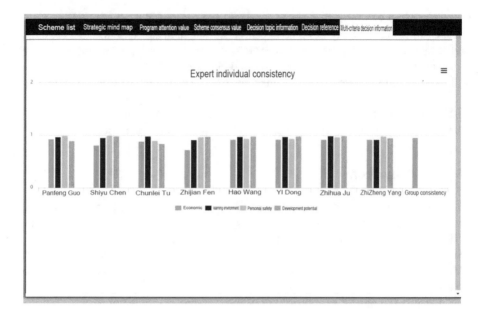

Fig. 3. e_l individual consistency

2. Based on the details scoring of the 6 experts and the scheme score details table, this model gives the average score and variance of criterion for each scheme, reflects the expected value of the score and the degree of dispersion of the score.

Step 4: Based on the feedback, the experts fine-tune their initial opinions and make a second round of decision-making.

$$V^1 = \begin{bmatrix} 4 & 6 & 3 & 6 \\ 3 & 7 & 3 & 8 \\ 6 & 7 & 8 & 7 \end{bmatrix} \quad V^2 = \begin{bmatrix} 6 & 5 & 8 & 5 \\ 3 & 5 & 3 & 3 \\ 9 & 8 & 10 & 7 \end{bmatrix} \quad V^3 = \begin{bmatrix} 3 & 5 & 5 & 6 \\ 7 & 3 & 3 & 6 \\ 6 & 5 & 8 & 4 \end{bmatrix}$$

$$V^4 = \begin{bmatrix} 9 & 8 & 8 & 6 \\ 5 & 3 & 6 & 6 \\ 7 & 7 & 7 & 9 \end{bmatrix} \quad V^5 = \begin{bmatrix} 3 & 4 & 4 & 5 \\ 6 & 4 & 4 & 5 \\ 3 & 4 & 5 & 5 \end{bmatrix} \quad V^6 = \begin{bmatrix} 7 & 7 & 8 & 7 \\ 7 & 6 & 6 & 8 \\ 8 & 9 & 8 & 8 \end{bmatrix}$$

Step 5: Host collects expert's evaluation opinions and then make current number of decision rounds $CYCLE + 1$. Through consistency analysis we can get $GAQ = 0.65$ equal the η, reach the group consistency threshold. It explains that experts have reached consensus on the evaluation of each scheme. At this time, according to formula 1, the final score of each scheme can be obtained, and the scheme is sorted according to the score. The result is $x_5, x_6, x_1, x_4, x_3, x_2$.

Table 1. Comparison with traditional decision support system

	DSS	Delphi method
Individual preference	Strong	Strong
Group preference	Weak	Strong
Anonymity mechanism	Weak	Strong
feedback mechanism	Medium	Strong

5 Result Analysis

From the case of this paper, we can conclusion that the greatest contribution of this method is to combine Delphi method with traditional multi-criteria decision-making, so that experts will not be affected by the authoritative experts in the decision-making process. After many rounds of feedback and decision-making, the final results will be obtained. Compared with traditional decision support system (DSS), as shown in Table 1, this method adopts anonymity mechanism and feedback mechanism in the feedback process. The advantage of this method is that each expert can think independently and make judgments. After several rounds of feedback, new conclusions based on the feedback information can be drawn. This method allows each expert to fully refer to the opinions of others, and then combine their own ideas, achieve group consistency, so that the decision-making results are more representative after combining the opinions of all experts.

6 Conclusion

Multi-criteria group decision-making and group agreement quotient analysis is important not only in decision support system, but also in most decision-making occasions. This article combines the traditional decision support system with the Delphi method, through learning its anonymity mechanism and feedback mechanism, making the decision process more efficient and accurate and avoiding having a significant impact on the results from the opinions of individual experts. Through the example, the result indicated that the improved model can generate more decision attribute, thus verified the effectiveness of the model.

Acknowledgments. This research is supported by National Key Research and Development Scheme of China under grant number 2017YFC1405403, and National Natural Science Foundation of China under grant number 61075059, and Green Industry Technology Leading Project (product development category) of Hubei University of Technology under grant number CPYF2017008.

References

1. Jones, C.: The Delphi Method: Techniques and Applications, pp. 155–161. Addison-Wesley, Reading (1975)
2. Lee, S., et al.: Forecasting mobile broadband traffic: application of scenario analysis and Delphi method. Expert Syst. Appl. **44**, 126–137 (2016)
3. Agell, N., et al.: A consensus model for Delphi processes with linguistic terms and its application to chronic pain in neonates definition. Appl. Soft Comput. **35**, 942–948 (2015)
4. El-Gazzar, R., Hustad, E., Olsen, D.H.: Understanding cloud computing adoption issues: a Delphi study approach. J. Syst. Softw. **118**, 64–84 (2016)
5. Jun, T., et al.: The integrating model of expert's opinion based on Delphi method. Syst. Eng.-Theory Pract. **1**(1), 57–62+69 (2004)
6. Qu, B., Xiao, R., Lin, J.: Application of improved AHP and group decision theory in bridge assessment. J. Cent. S. Univ. (Sci. Technol.) **46**, 4204–4210 (2015)
7. Bi, W., Chen, X.: Risky multicriteria group decision approach based on Bayesian theory and monte carlo simulation. Syst. Eng. Electron **32**(5), 971–975 (2010)
8. Emmerich, M.T., Deutz, A.H., Yevseyeva, I.: A Bayesian approach to portfolio selection in multicriteria group decision making. Procedia Comput Sci. **64**, 993–1000 (2015)
9. Liao, H., Xu, Z.: A VIKOR-based method for hesitant fuzzy multi-criteria decision making. Fuzzy Optim. Decis. Making **12**(4), 373–392 (2013)
10. Xiong, C.Q., Li, D.H., Zhang, Y.: Clustering analysis of expert's opinion and its visualization in hall for workshop of meta-synthetic engineering. Pattern Recognit. Artif. Intell. **22**(2), 282–287 (2009)
11. Cook, W.D., Seiford, L.M.: Priority ranking and consensus formation. Manage. Sci. **24**(16), 1721–1732 (1978)
12. Cook, W.D., Kress, M.: Ordinal ranking with intensity of preference. Manage. Sci. **31**(1), 26–32 (1985)
13. Hamers, L.: Similarity measures in scientometric research: the Jaccard index versus Salton's cosine formula. Inf. Process. Manage. **25**(3), 315–318 (1989)
14. Basilevsky, A.: Applied Matrix Algebra in the Statistical Sciences. Courier Corporation, Mineola, NY (2013)
15. Herrera-Viedma, E., et al.: A consensus model for group decision making with incomplete fuzzy preference relations. IEEE Trans. Fuzzy Syst. **15**(5), 863–877 (2007)

Graph-Based Semantic Query Optimization for Intensional XML Data

Abdullah Alrefae[(⊠)], Jinli Cao, and Eric Pardede

Department of Computer Science and Information Technology,
La Trobe University, Melbourne, VIC 3086, Australia
afalrefae@students.latrobe.edu.au,
{j.cao,e.pardede}@latrobe.edu.au

Abstract. The evolving of graph database's techniques and tools has granted this type of connected databases more popularity in data storage and query answering techniques. These techniques include structural indexing, knowledge representation, and logic or ontology reasoning. In this paper, we propose a logical data model in graphical structure, which logically structure XML nodes and semantic relationships between them. In addition, we propose a semantically annotated XML database model, to reflect the logics and semantics inferred from the graph database. The reason is that we apply a third dimension to the tree pattern query (TPQ) matching process to traverse XML entities based on the logical data structure between database entities. The other two dimensions are the traditional predicates of TPQ matching of structure and values between the query and document. The semantic annotations, accordingly, represent these logical data structure in the XML database.

1 Introduction

The evolving of graph database's techniques and tools has granted this type of connected databases more popularity in data storage and query answering techniques [1]. The connected graph data model, and its query languages such as SPARQL and Cypher, provide a powerful, agile, and flexible NoSQL data model [2, 3]. Therefore, various types of data management systems implement this labelled property graph data models at different stages of data processes for different purposes. These implementations includes structural indexing [4], knowledge representation [5], and logic or ontology reasoning [6].

In general, Graph database provides a promising technique for online transaction processing (OLTP) because of the fast query response and flexible schema. However, it is not the most effective data model for data mining and online analytical processing (OLAP).

In our previous work [7], we proposed a graph database layer to optimize the query processing over XML database. In this paper, we will show the semantic connections between the graph database and XML database layer. What types of the XML data will be graphically connected; and how this graph connection can improve the traversing process of the XML document by indexing the XML nodes in graph structure.

L. Barolli et al. (Eds.): CISIS 2019, AISC 993, pp. 247–256, 2020.
https://doi.org/10.1007/978-3-030-22354-0_23

The following of the paper will be as follows. In Sect. 2, we provide a brief background explanation of intensional XML document, the nominated framework of XML data models. Following that, in Sect. 3 we provide definitions of preliminaries concepts and their implementation in data integration and exchange including semantic data models, ontology data management, and RDF. In Sect. 4, we represent the contribution of semantic data model, which include semantic graph database, semantic annotations of XML data, and the conceptual framework of the contribution, before we conclude the paper of experiment result to show the query precession optimization we have achieved.

2   Research Background and Context

Intensional XML data model [8] relies on intensional parts of the document that represent service calls to the relevant resource of the respective data. These service calls will be materialized passively when needed by the hosting system or shared with other peers in the domain, to extract the real-time values on-demand. The Active XML model [9] provides a framework for such intensional documents to exchange, materialize, and query the intensional data. In our model [10], we examined the technical possibility of implementing these intensional XML documents for real-time decision making. We proposed the intensional XML metadata structural framework for Web-based Real-time Decision Support System (WB-RT-DSS). Afterwards, we defined the query optimization model over the intensional XML database [7]. The proposed technique included the definition of graph database layer and mapping algorithm between the XQuery and Cypher graph query.

In this paper, we continue the approach of graph-based query optimization over intensional XML data. We propose the semantically annotated XML database model, to reflect the reasonings of the logic-based graph database. The reason is that we apply a third dimension to the tree pattern query (TPQ) matching process, which traverse document's content based on semantics inferred from the logical data structure. The other two dimensions are the traditional predicates of TPQ matching of structure and values between the query and document. Therefore, the semantic annotations represent the logical data structure between database entities. In addition, the semantics also annotat the intensional parts of the document to automate the service call activation process.

A sample of an annotated XML document tree is shown in Fig. 1. The intensional XML document includes data of metropolitan hospitals in Melbourne, Australia. It has static nodes such as name and address details, and other active nodes for operational data like availability status and crowdedness measurements. The "*isAccessedVia*" annotations is showing the intensional parts of the document, that are responsible for the corresponding active nodes.

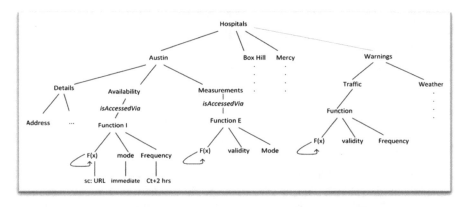

Fig. 1. Tree representation of intensional XML data with semantic annotations

3 Preliminaries

The web-based connection between the intensional XML database and the various resources of real-time data feeds requires a systematic automation of the process. The idea of semantic annotation has been applied in semantic web model for different applications and provided various advantages in data integration, as described below.

3.1 Traditional XML Model vs. Semantic Model

Enriching domain's data with semantics enhances the automation of sharing and exchanging the data between the heterogeneous and independently designed resources.

To show the advantages of semantic annotations in the sharing and exchanging of the domain knowledge, the following differences, in Table 1, can be highlighted between the traditional XML model and the semantic model [11].

Table 1. Traditional XML model vs. Semantic web model.

Traditional XML model	Semantic web model
The collaboration between the domain resources requires the agreement of the data format	To collaborate, the domain resources can apply a common vocabulary or base ontology to describe the domain knowledge
The adding of new data sources will require a similar human involvement to interpret the meaning and formats in the new resources	The defining of the standard domain knowledge is not always required, as there are various standard ontologies already available to be adopted

There are two related types of semantic heterogeneity in the context of data integration and exchange. First, the semantic heterogeneity in the definition of terms in the system domain and the usage of these terms. Second, the semantics can also refer to the interpretation of information and knowledge that can be inferred from rules and logics in the data and its connections [12].

In this paper, our semantic heterogeneity refers to the second type. In the context of decision support and recommendations making, the interpretation of information is needed to find the most relevant and accurate alternatives to the decision maker. Ontology management languages and techniques are implemented to deal with this semantic heterogeneity. Resource description framework (RDF) and web ontology language (OWL) are examples of languages to represent knowledge and map relations between objects, as described in the following subsections.

3.2   Ontology-Based Data Management

Ontology heterogeneity between distributed databases is an important issue when there is an attempt to automate the process of data retrieval and integration. One of the techniques for managing ontology heterogeneity is to apply "*mapping*" between ontologies to provide interoperability among applications in the system domain [13]. There are three main ontology approaches for data integration to overcome semantic heterogeneity [14] as shown in Fig. 2.

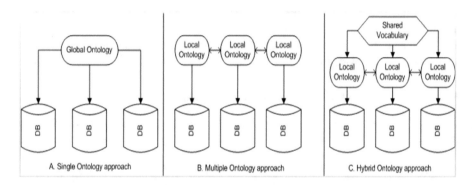

Fig. 2.   Ontology heterogenity mapping approaches

The single ontology approach (Fig. 2A) provides a single view of the domain and combines different ontologies in one global ontology to represent the conceptualization of the domain. However, this approach lacks the flexibility to add and remove data sources. This drawback motivates the development of the multiple ontology approach (Fig. 2B). In this approach, adding and removing of sources or applying changes in any of the ontologies is applicable without any ontology restrictions because there is no common ontology that all sources must follow. However, in the multiple ontology approach, it is difficult to apply comparisons between different ontologies because of the absence of a shared vocabulary in the domain. Therefore, the need for a shared vocabulary over the different ontologies has led to the development of the hybrid ontology approach (Fig. 2C). In this approach, it is possible to compare ontologies using the shared vocabulary. However, adapting this approach requires the redevelopment of existing ontologies to adapt the shared vocabularies in the domain [14].

The development of Semantic Web, as a framework for combining data web data from heterogeneous data sources, influenced the publishing of various standard vocabularies metadata to provide shared vocabularies for almost every domain of knowledge. These vocabularies represent metadata for different subjects and are free to use and adapt in any domain by defining them in the system's ontology and without the need for rebuilding the ontologies.

3.3 Resource Description Framework (RDF)

RDF is a metadata framework for Web Services. It is recommended by W3C for Semantic Web activity to describe the content of Web pages and other information, such as title, author, and time if updates [15]. RDF is written in the XML language syntax, which gives is the flexibility to be exchanged online between Web applications. One of the purposes of RDF and RDF schema is to represent the semantics between the web service entities, based on its ontology. Most likely in triple format (subject, object, predicate). Accessing the distributed databases via web services will need ontology mapping between the domain ontology and the data source ontology that is described in their RDF, as shown in Fig. 3.

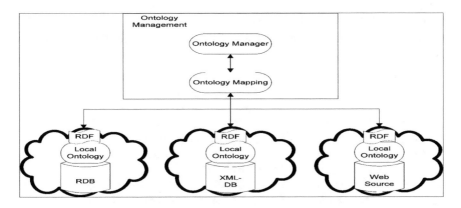

Fig. 3. Ontology management and mapping via RDF

Practically, RDF represents the sources ontology and can show some ruled semantics in data entities. However, when it comes to infer knowledge and query these semantics, RDF data does not deliver the best results. The syntax of RDF and its query language SPARQL requires explicit definition of semantics and rules but will not deduce related knowledge by reasoning and logics. Next, we introduce our technique of labelled property graph layer to query the semantics in XML data.

4 Semantics Representation and Query Over Labelled Property Graph

In order to represent logic semantics, we implement a graph database layer for indexing principal element nodes and connect them via relevant logical relationships. In other words, the graph database will be the first layer in query processing plan to store and query the domain's master data. For example, we have an XML database shows suburbs and hospitals details in Melbourne, Australia. In this case, the main elements are suburb and hospital. For simplicity, we will show only three relations that defined the graph schema, as show in Fig. 4.

Fig. 4. Graph schema

4.1   Semantic Annotations

Annotating RDF document with semantic attributes provides interoperability in web services. It shows a linked data model to the resource information. These semantic annotations are becoming more applicable with the availability of online ontology, which can be simply adapted and shared in domain sources. A simple of XML\RDF data to reflect the reasoning of semantics defined in the graph database can be as follow.

```
<root prefix="dc: http://purl.org/dc/elements/1.1/ foaf:
http://xmlns.com/foaf/0.1/ cld:http://purl.org/cld/terms/">
  <Austin about="http://ww.example.org/austins_nearby_Hospital"
rel="foaf:based_near"> Austin's nearby Hospital.
    <Hospital property="dc:name" content="Mercy"> Mercy
      <Availability property="dc:isAccessedVia">
      <sc>performance.health.vic.gov.au/Home/Emergency-
department-status("Mercy")</sc></Availability>
    </Hospital>
  </Austin>
  <Austin about="http://ww.example.org/austins_nearby_Hospital"
rel="foaf:based_near"> Austin's nearby Hospital.>
    <Hospital property="dc:name" content="BoxHIll"> Mercy
      <Availability property="dc:isAccessedVia">
      <sc>performance.health.vic.gov.au/Home/Emergency-
department-status("BoxHill")</sc></Availability>
    </Hospital>
  </Austin>
</root>
```

In Fig. 5, we show the representation of this semantically annotated RDF\XML data.

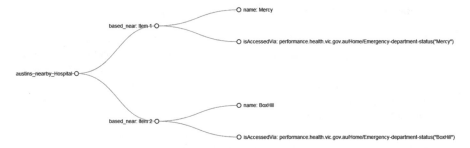

Fig. 5. Representation of semantic annotations

The main purpose of annotating XML documents, in our case, is to define the active parts of the document, as an access point to the data values for the relevant parts of the document.

4.2 Framework Components and Description

The graph database represents a data model to query semantics and logics, that need also to be reflected in the XML database. The XML annotations represent the ontology reasoning in XML database. In Fig. 6, we show the conceptual framework of the proposed system. After receiving the XPath\Query, the query processor produces the corresponding graph query using the mapping algorithm proposed in [7]. The graph database provides the ability to query the semantic relations and structure of the data.

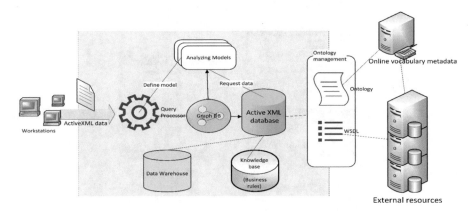

Fig. 6. Conceptual framework of system components

In addition, the graph database provides an inverted index to the labelled XML data. The query of the graph data shows the connected nodes and their degree of connections based on query properties, as shown in the implementation below. After we define the needed nodes tags, we continue to use the semantic annotations in the document to traverse to these nodes only.

4.3   Labelled Property Graph to Query Semantic Data

The possibility to query semantics of the data can improve the query processing evaluation. It provides index structure based on the semantic connection between the database nodes. In the implementation below (see Fig. 7), we show two different processing steps for the same query to find hospitals that provide support for same suburbs as Austin Hospital. On the left-hand side, we show a processing of the query on the predefined data of each hospitals and suburbs it supports. On the right-hand side, we defined an arbitrary relation based named *nearby* to define just that if suburbs support same areas then they are nearby hospitals. As we can see the processing time and database hits has improved from 50 ms with total of 1305 database hits to become 1 ms with only 21 database hits.

In addition, the result shows only the related nodes to be matched and traversed in the XML document. It avoids traversing through large XML documents and materialize included intensional elements with service calls to external resources.

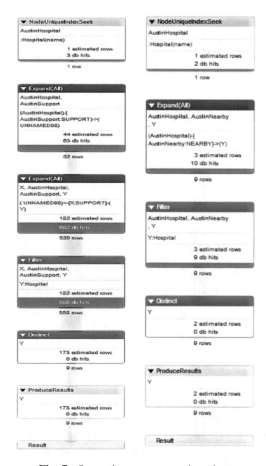

Fig. 7. Semantic query processing plans

4.4 Graph-Based Index for XML Nodes

The semantic rule of finding related node can be defining through the formula:

let H = {set of all nodes with label Hospital}
and S = {set of all nodes with label Suburb}
for each x ∈ F and y ∈ F, x is related to y if and only if

$$|N(x) \cap N(y)| > 0 \tag{1}$$

where N(x) is the set of S_i, where x is connected to i
and N(y) is the set of S_i, where y is connected to i

The higher the value of (1) the more semantic relation between nodes x and y, as shown in the index structure in Fig. 8.

"Austin Health"	"Western Health"	0.0
"Austin Health"	"Western Hospital"	0.0
"Austin Health"	"Williamstown Hospital"	0.0
"Austin Health"	"Austin and Repatriation Hospitals"	62.0
"Austin Health"	"Austin Health - Austin Hospital"	62.0
"Austin Health"	"Austin Health - Heidelberg Repatriation Hospital"	62.0
"Austin Health"	"Royal Talbot Rehabilitation Centre"	62.0
"Austin Health"	"Mercy Health - O'Connell Family Centre"	62.0
"Austin Health"	"Mercy Hospital for Women"	62.0
"Austin Health"	"Caritas Christi Hospice Ltd"	62.0
"Austin Health"	"St George's Health Service"	62.0
"Austin Health"	"Sunshine Hospital"	62.0

Fig. 8. Inverted index structure using graph database

5 Conclusion

In this paper, we defined a semantic model to represent and query master data of intensional XML-based real-time system. The logics and semantics knowledge are represented in a graph database layer, which provide semantically linked data platform for fast and structure free query interface. The resulted nodes will be tracked in the XML database with the service of ontology and semantic annotations in the XML/RDF data.

The proposed technique will optimize the tree pattern query process in XML with intensional parts define the service calls to external data source for the relevant data.

In the future work, we will define the query taxonomy based on when to use attributes and when to use relations for query matching process.

Acknowledgements. This paper is part of corresponding author's PhD research. The research degree is financially sponsored by the Ministry of Higher Education in Saudi Arabia.

References

1. Miller, A., Justin, J.: Graph Database Applications and Concepts with Neo4j (2013)
2. Webber, J., Robinson, I.: How graph databases relate to other NoSQL data models. In: Graph Databases. O'Reilly, Sebastopol (2013)
3. NoSQL Database Explained (cited 2017). https://www.mongodb.com/nosql-explained
4. Yan, X., Yu, P.S., Han, J.: Graph indexing: a frequent structure-based approach. In Proceedings of the 2004 ACM SIGMOD International Conference on management of data, pp. 335–346. ACM, Paris, France (2004)
5. Chein, M., Mugnier, M.-L., Croitoru, M.: Visual reasoning with graph-based mechanisms: the good, the better and the best. Knowl. Eng. Rev. **28**(3), 249–271 (2013)
6. Zeng, Q., Jiang, X., Zhuge, H.: Adding logical operators to tree pattern queries on graph-structured data. Proc. VLDB Endow. **5**(8), 728–739 (2012)
7. Alrefae, A., Cao, J., Pardede, E.: Graph Database Indexing Layer for Logic-Based Tree Pattern Matching Over Intensional XML Document Databases. Springer International Publishing, Cham (2018)
8. Abiteboul, S., Benjelloun, O., Milo, T.: The active XML project: an overview. VLDB J. **17**, 1019–1040 (2007)
9. Viet Phan, B., Pardede, E.: Active XML (AXML) research: survey on the representation, system architecture, data exchange mechanism and query evaluation. J. Netw. Comput. Appl. **37**, 348–364 (2014)
10. Alrefae, A., Cao, J.: Intensional XML-enabled web-based real-time decision support system. In: 2017 International Conference on Computing Networking and Informatics (ICCNI) (2017)
11. Antoniou, G., et al.: The semantic web vision. In: A Semantic Web Primer, p. 296. The MIT Press, Cambridge (2012)
12. Blomqvist, E.: The use of semantic web technologies for decision support - a survey. Semant Web J **5**(3), 177–201 (2014)
13. Haase, P., Motik, B.: A mapping system for the integration of OWL-DL ontologies. In: Proceedings of the first international workshop on interoperability of heterogeneous information systems. ACM, Bremen, Germany (2005)
14. Wache, H., et al.: Ontology-based integration of information - a survey of existing approaches. Proceedings of the IJCAI-01 workshop on ontologies and information sharing, pp. 108–117 (2001)
15. w3 schools the world's largest web development site. http://www.w3schools.com/

Design and Implementation of Security System for IPv6 Sensor Networks

Di Luo[1], Hao Wang[1(✉)], Shengwei Yi[2], and Qiao Wang[1]

[1] Chongqing University of Posts and Telecommunications,
Chongqing 400065, China
1422554376@qq.com, wanghao@cqupt.edu.cn,
704454617@qq.com
[2] China Information Technology Security Evaluation Center,
Beijing 100085, China
yishengwei@foxmail.com

Abstract. This paper analyzes the security requirements for IPv6 Sensor Networks (IPv6 SNs). Aiming at the security threats existed in IPv6 SNs, a security system for IPv6 SNs is designed. In this system, security association, key management, multicast security transmission, attribute-based access control and network security detection are used to ensure the security of IPv6 SNs. The test results show that the system can authenticate the identity of the nodes joining the network, complete the issuance of security materials, and achieve secure multicast. With the security association mechanism, the system also can resist replay attacks and man-in-the-middle attacks, which improves the security of the networks.

1 Introduction

Wireless Sensor Networks (WSNs) is a new network which integrates sensor technologies, micro-electromechanical systems technologies, wireless communication technologies and distributed information processing technologies. With the widespread application of WSNs, the IP of wireless sensor nodes and network architecture is an important trend in the development of wireless sensor networks technology. In view of the exhaustion of IPv4 addresses in the Internet and the huge demand for objects identification in WSNs, it is an inevitable choice to adopt IPv6 technology with rich address resources and fine mobility when introducing IP mechanism into WSNs.

With the introduction of IPv6 protocol, sensor nodes can take advantage of various IP-based security methods of the existing Internet. This brings some convenience. However, considering that the sensor nodes are resource-constrained device with constrained storage capacity and computing capacity, some original security mechanisms and methods are not completely suitable for the use of sensor nodes. Therefore, it is necessary to develop security solutions suitable for IPv6 SNs according to the characteristics of sensor nodes, which meets the security requirements of actual network deployment. Only in this way can IPv6 technology be widely used and promoted in WSNs [1–4].

In view of the resource constraints of wireless sensor networks nodes, a lightweight uncertified signature algorithm and key agreement mechanism for sensor networks are

© Springer Nature Switzerland AG 2020
L. Barolli et al. (Eds.): CISIS 2019, AISC 993, pp. 257–268, 2020.
https://doi.org/10.1007/978-3-030-22354-0_24

proposed in [5]. The online/offline mechanism is adopted in the signature process to avoid bilinear pairings calculation, which makes the signature algorithm more efficient in WSNs. The key negotiation mechanism based on the discriminability of signature algorithm can avoid man-in-the-middle attacks. Zhang et al. [6] proposed a key update mechanism based on secret information. The node broadcasts the session key material encrypted by the global key related to the secret information to its neighbor nodes, and establishes the session key between the two nodes through the Diffie-Hellman key exchange method. In [7], a low-overhead asymmetric broadcast authentication algorithm is proposed, which uses the Chinese Remainder Theorem (CRT) to complete the node broadcast authentication. The above security mechanisms and methods have a positive effect on the improvement of the security of the sensor networks. However, in practical application scenarios, especially in IPv6 SNs, it is necessary to further solve the adaptation of protocols, security threats, availability and stability in large-scale environments.

Based on the security technology of sensor networks and the lightweight of IPv6 network security technology, a variety of IPv6-based security mechanisms for WSNs is proposed in this paper. Combined with specific hardware facilities, an IPv6-based security system for WSNs is designed and implemented. The network security is effectively guaranteed by the cooperation of multiple security mechanisms.

2   Overview of IPv6 Protocol

IPv6 is an implementation of the next generation Internet Protocol, which designed by the IETF to replace the IPv4 protocol. IPv6 supports a 128-bit network addressing architecture. The rich address space solves the problem of address shortage in the large-scale node deployment of the sensor networks, and the simplified header improves the routing switching performance. The IPSec support is mandatory to ensure the security of the network layer. In mobile scenarios, the original IPv6 address can still be maintained to enable end-to-end seamless communication independent of the access mode. The comparison of the IPv4 and IPv6 header structures is shown in Fig. 1 [8].

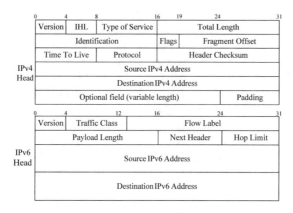

Fig. 1. Header structure of IPv4 and IPv6

3   Design of System

IPv6 SNs combine 6LoWPAN adaptation technology and ROLL (Routing over Lossy and Low-power Networks) protocol, which is essentially capable of interconnecting with IPv6 Ethernet. That is, remote IPv6 network users can directly interact with IPv6 nodes. When the sensor networks accesses the IPv6 networks, the original secure data transmission will not meet the adaptation requirements of the IPv6 networks, and the overhead caused by the existing security mechanism of the IPv6 networks is not suitable for the direct use of the sensor networks [9]. The security system described in this paper considers the characteristics of two networks and the adaptation requirements. A variety of security mechanisms for IPv6 SNs are designed to ensure the security of network establishment, data collection, data transmission and data processing.

The IPv6 SNs security system consists of three parts: IPv6 SNs, security manager and user. It completes the process of data collection, data security transmission and processing. The legitimate users can access the node information of IPv6 SNs through the gateway. The system structure is as shown in Fig. 2.

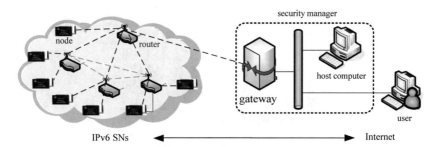

Fig. 2. Security system structure

(1)   There are nodes and routers in the IPv6 SNs. The routers are responsible for completing the routing establishment process, realizing data forwarding and node security management. The nodes are responsible for completing data collection and transmission, realizing its own key management.

(2)   The security manager composed of gateway and host computer is responsible for completing the security management and real-time status monitoring for the entire network. The gateway is responsible for ensuring the security of the data transmission from the IPv6 SNs to the Internet. At the same time, it satisfies the Internet communication protocol to complete the network protocol parsing, data forwarding and processing. The host computer realizes real-time monitoring for the entire security system, and can issue related command to acquire the current device information and change the network running behavior and status.

(3)   After the user complete registration and authorization at the security manager, the information in the IPv6 SNs can be accessed through the gateway.

4   Design of Hardware and Software

In order to improve the network security intensity, based on the characteristics of the sensor networks, combined with the requirements of IPv6 protocol adaptation and the performances of hardware device, appropriate hardware facilities are selected and reasonable security mechanisms are designed to ensure network security operations.

4.1   Design of Hardware

In order to satisfy a variety of application scenarios and different environmental requirements, the IPv6 SNs security system consists of four sub-networks with different frequency bands. Based on CC2530 and STM32 chips, the sensor nodes are responsible for gathering the information of four frequency bands, and sending these information to the specified location according to the network requirements. With ARM9 platform based on S3C2440 and Linux system with 2.6.24 kernel, the gateway completes the processing and transmission of the information, and undertakes part of network management and security management functions (Table 1).

Table 1. Selection of hardware

Device name	Microcontroller	RF Chip
433 MHz nodes	STM32F103	CC1101
470 MHz nodes	STM32F103	CC1101E
780 MHz nodes	STM32F103	AT86RF212
2.4 GHz nodes	CC2530	
Multi-band gateway	ARM9 platform based on S3C2440 and Linux system with 2.6.24 kernel	

4.2   Design of Software

The software design is based on the principle of low computational overhead and low communication overhead. In the light of the security threats inherent in the sensor networks itself and the introduction of the IPv6 protocol, the reasonable security mechanisms and strategies are designed to ensure the safe operation of the IPv6 SNs.

4.2.1   Security Association

The security association is the basis of the IPSec. The two parties complete the negotiation of the public information through this security association. The security association uses the negotiation strategy based on the pre-shared key authentication mechanism to complete the identity authentication of the nodes. The trusted security association is achieved by the use of the low-cost HMAC key negotiation.

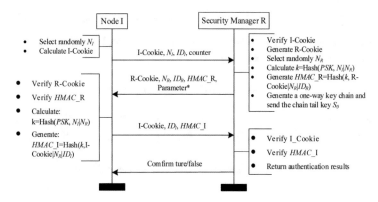

Fig. 3. Process of security association

The process of security association is shown in Fig. 3. The communication parties complete the security association process through four interactive packets. The information negotiated during the interaction includes the session key and security parameters used to perform multicast security transmission protection.

4.2.2 Key Management

Data encryption is the most common method to protect sensor data, and key is the necessary material in the encryption process. Therefore, the key becomes the core of protection. The management and maintenance of the key is decisive for ensuring the safe operation of the sensor networks. For IPv6 SNs with numerous nodes, the number of key becomes huge and difficult to maintain, so key management and maintenance is particularly important. The key management of this system is divided into three parts, mainly: key pre-configuration, key establishment and key update.

In the phase of key pre-configuration, the deployment server configures a master key and a join key for each node. The key establishment process includes key negotiation and key transmission, and the key update process adopts a low overhead update scheme based on hash and XOR, as is shown in Fig. 4.

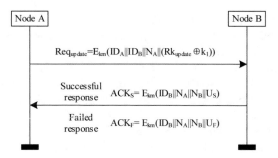

Fig. 4. Process of key update

Node A initiates a session key update to Node B. The process is as follows:

(1) Node A inverts the current session key between A and B to Rk_1 ($k_1 \rightarrow Rk_1$) by bit, and calculates the new key k_{update} by formula (1) and stores it.

$$k_{update} = HMAC(k_1, ID_A||ID_B||Rk_1||N_A) \tag{1}$$

ID_A is the identifier of node A, ID_B is the identifier of node B, N_A is the random number generated by node A, and k_1 is the current session key between nodes A and B.

(2) Node A constructs a key update request Req_{update} sent to B, where k_m is the primary key.

$$Req_{update} = E_{k_m}(ID_A||ID_B||N_A||Rk_{update} \oplus k_1) \tag{2}$$

(3) After receiving the request, B uses k_m to decrypt and obtain N_A and $Rk_{update} \oplus k_1$, and calculates Rk_{update}, then inverts the bit to get k_{update} and finally calculates k'_{update}.

$$k'_{update} = HMAC(k_1, ID_A||ID_B||Rk_1||N_A) \tag{3}$$

Comparing k'_{update} with k_{update}, if they are consistent, the update is successful and the successful response ACK_S is returned, otherwise the failed response ACK_F is returned.

$$ACK_S = E_{k_m}(ID_B||N_A||N_B||U_S) \tag{4}$$

$$ACK_F = E_{k_m}(ID_B||N_A||N_B||U_F) \tag{5}$$

(4) After decryption, it is checked whether it is consistent with the random number sent in the previous step (2). If they are consistent, the successful response is returned, otherwise the failed response is returned. N_B is the serial number randomly generated by B, which can be used to prevent replay attacks.

4.2.3 Multicast Transmission Security

Routing information, neighbor discovery, and secure communication in IPv6 SNs require multicast communication support. Therefore, ensuring the transmission security of multicast information is of great significance to ensure the security for the entire network. Referring to the idea of hidden communication, using the theory that Chinese Remainder Theorem congruence equations have unique solutions, the multicast message is embedded into the ordinary message authentication code for broadcast transmission. In this case, only the corresponding nodes can extract the hidden message to protect the confidentiality and integrity of the multicast data. The specific process is shown in Fig. 5.

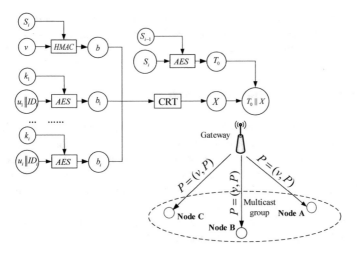

Fig. 5. Process of security multicast

Where S_n ($n = 0, 1, \cdots, i$) is the master key, k_i is the key shared between the gateway and node i, v is the ordinary message, and u_i is the multicast hidden message that needs to be sent to node i.

4.2.4 Attribute-Based Access Control

IPv6 protocol is introduced into sensor networks, which brings the IP-based security method to the sensor networks and exposes the sensor networks to the traditional network security threat. The interconnection for heterogeneous network also provides higher requirements for the security of the sensor networks.

A third-party authentication and attribute-based access control scheme are designed and implemented in this system. By adopting third-party authentication method, the security manager can authenticate the identity of users, which satisfies the security intensity of the IP network authentication, and effectively avoids the overhead brought by the process to sensor networks nodes. The attribute-based authority management scheme sets the access permissions and time limits according to the type of the users and the resources, so as to realize the reasonable protection for the sensitive resources and information in the sensor networks.

4.2.5 Network Security Detection

In order to cope with the common attacks on the sensor networks, the system uses the attack feature library matching method to detect network intrusion behavior, and achieves automatic processing and interface alarm. By making full use of the characteristics of the sufficient resources of gateway, a common attack feature library is established. The gateway can detect the abnormal behavior in the network by using the matching algorithm and the configured network parameters, and uses the pre-configuration scheme to process and report the abnormal behavior, which is given by the interface of the host computer.

5   Test and Verification of System

Based on the security system structure of the IPv6 SNs, combined with the specific hardware equipment and software environment, the IPv6 SNs security system shown in Fig. 6 is constructed. The system includes a multi-band gateway, 400 sensor nodes (100 in each of 4 bands) and system monitoring and management interface, which accomplish collaboratively 5 functions: device security association, key management, multicast security transmission, access control and network security detection. The interface includes 5 main modules: network topology, security association, key management, multicast security test, and access control.

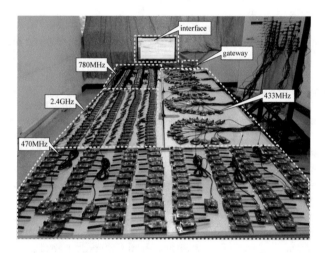

Fig. 6. IPv6 SNs security system

5.1   Network Topology

The monitoring and management interface can monitor the operation status of the system in real time, display the current network structure and collect the data information acquired by the sensor nodes. The 2.4 GHz network topology is shown in Fig. 7. By clicking on the buttons of different frequency bands in the left band selection area, the current topology and the number of online nodes of each band can be observed.

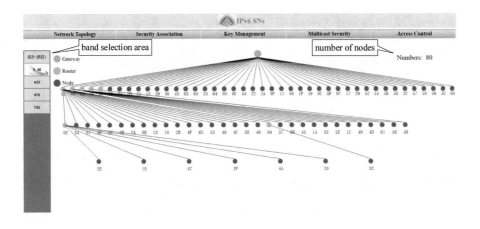

Fig. 7. Network topology of 2.4 GHz

5.2 Security Association

Normal nodes perform the node authentication and network access process through the four-time security association interactive packets, and receives the relevant security parameters delivered by the gateway, such as the prime number used for multicast security transmission. At the same time, this process can realize the recognition and processing of the replay attacks and the man-in-the-middle attacks, and alarm these to the host computer. The process of security association is shown in Fig. 8.

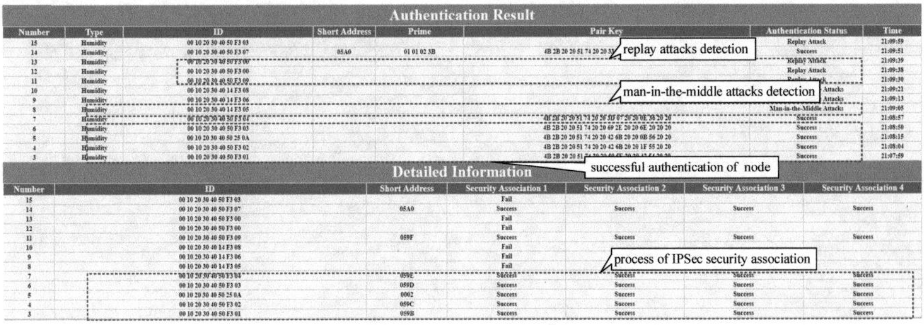

Fig. 8. Security association

5.3 Key Management

After the nodes join the network, the gateway reports the node key information to the host computer, so that the key information can be observed in real time through the key management module of the monitoring management interface, which is shown in Fig. 9. The key of the specified node can be updated by the key update submodule.

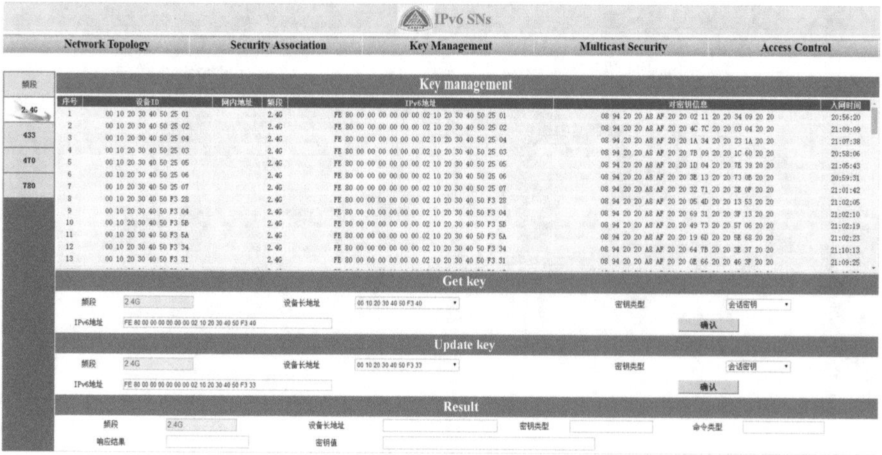

Fig. 9. Key management

5.4 Multicast Security Test

After the nodes complete the security association, the large prime number allocated by the gateway for multicast security transmission is obtained. When the security multicast is required, the gateway uses the Chinese Remainder Theorem and the large prime number to calculate the signature information and delivers the information to the nodes for response. The test interface is shown in Fig. 10.

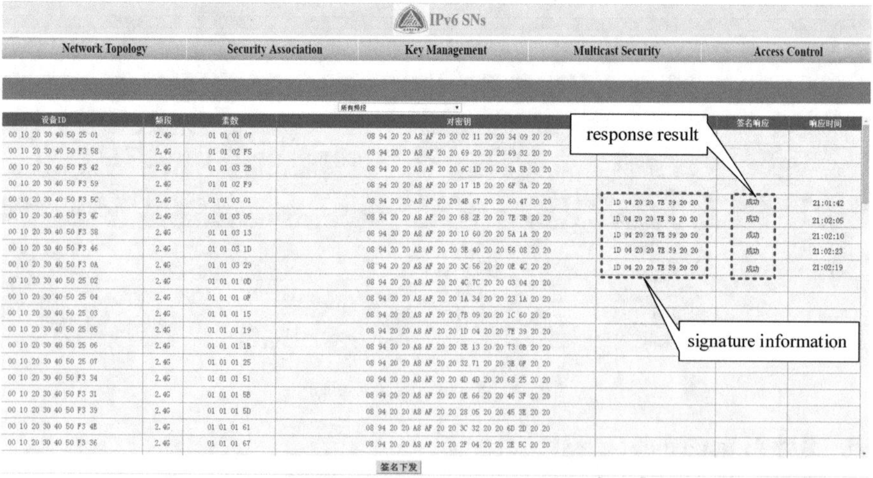

Fig. 10. Multicast security testing

5.5   Access Control

Internet users need to register and be authorized by the security manager before accessing the IPv6 SNs. The users can access the sensor networks within the access authority. The monitoring management interface can monitor and display the access status of the users in real time, which is convenient for managers to view. The monitoring interface is shown in Fig. 11.

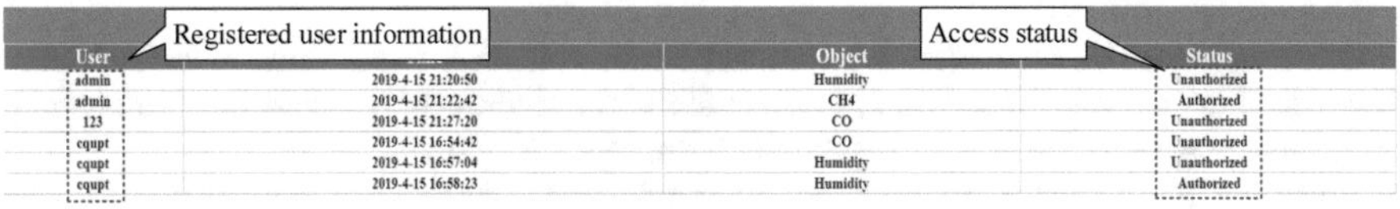

User	Registered user information	Object	Access status Status
admin	2019-4-15 21:20:50	Humidity	Unauthorized
admin	2019-4-15 21:22:42	CH4	Authorized
123	2019-4-15 21:27:20	CO	Unauthorized
cqupt	2019-4-15 16:54:42	CO	Unauthorized
cqupt	2019-4-15 16:57:04	Humidity	Unauthorized
cqupt	2019-4-15 16:58:23	Humidity	Authorized

Fig. 11. Access control monitoring

6   Conclusion

With the wide application of wireless sensor networks, the management about network and node in large-scale deployment environments have become the hotspots of current research. The management of sensor networks nodes based on the management mode of the Internet has been gradually researched and applied. IP-based sensor networks nodes and network architecture are an important direction of current research and application. Therefore, it is of great significance to solve the security problems brought by the introduction of IP technology and ensure the safe operation of the sensor networks.

On the basis of fully considering the security requirements and resource characteristics of sensor networks, combined with the security requirements and adaptation requirements introduced by the IPv6 protocol, a variety of security mechanisms suitable for IPv6 SNs are designed and an IPv6 SNs security system that can meet a variety of application scenarios is built from the perspective of practical application. The test results show that the system satisfies the security requirements of the sensor networks, and adapts the characteristics of the IPv6 protocol to achieve end-to-end secure communication between the Internet network hosts and the sensor networks nodes. The operation of the IPv6 SNs is ensured effectively by cooperation among security mechanisms.

Acknowledgments. This work is financially supported by the Research and Experimental Verification of Key Technical Standards for Time-Sensitive Networking ([2018] 281), Research, Development and Application Demonstration of Key Technologies of Internet of Things Security for Smart Cities (cstc2018jszx-cyztzxX0012).

References

1. Zhu, Y.H., et al.: Latency aware IPv6 packet delivery scheme over IEEE 802.15.4 based battery-free wireless sensor networks. IEEE Trans. Mob. Comput. **16**, 1691–1704 (2017)
2. La, V.H., Fuentes, R., Cavalli, A.R.: A novel monitoring solution for 6LoWPAN-based wireless sensor networks. In: Asia-Pacific Conference on Communications, pp. 230–237 (2016)
3. Peng, W., Yuankun, L., Wenpeng, L., et al.: Research on 6LoWPAN wireless sensor network and IPv6 network interconnection in power distribution system. In: China International Conference on Electricity Distribution, pp. 1–4 (2016)
4. Jun, L.I., Ding-Hai, G.: A survey on security issues in IPv6. J. Hechi Univ. **9**, 98–102 (2008)
5. Wang, Hao, Zhang, Xiao, Wang, Ping, et al.: Lightweight certificateless signature and key agreement scheme for WSNs. J. Jilin Univ. **44**, 465–470 (2014)
6. Min-Qing, Z., Wen-Hua, F., De-Long, L.: A new key management scheme based on secret information for WSN. In: 2011 IEEE 3rd International Conference on Communication Software and Networks, pp. 518–521 (2011)
7. Wang, H., Fang, W., Zhang, X., et al.: Real-time broadcast authentication algorithm for wireless sensor networks. J. Comput. Inform. Syst. **9**, 1051–1059 (2013)
8. Hinden, R., Deering, S.: RFC 3513-Internet protocol version 6 (IPv6) addressing architecture [j]. Network Working Group (2003). http://www.faqs.org/rfcs/rfc3513.html
9. Haoxing Wang.: The IPSec for WSNs based IPv6. Chongqing University of Posts and Telecommunications (2014)

A UML Profile for the Service Discovery in the Enterprise Cloud Bus (ECB) Framework

Misbah Zahoor[1($\boxtimes$)], Farooque Azam[1], Muhammad Waseem Anwar[1],
Nazish Yousaf[1], and Muhammad Kashif[2]

[1] Department of Computer and Software Engineering,
College of Electrical and Mechanical Engineering,
National University of Sciences and Technology (NUST), Islamabad, Pakistan
{mzahoor17, nazish.yousaf15}@ce.ceme.edu.pk,
{farooq, waseemanwar}@ceme.nust.edu.pk
[2] Department of Electronics and Computer Engineering,
Istanbul Sehir University, Istanbul, Turkey
muhammadkashif@std.sehir.edu.tr

Abstract. Enterprise Cloud Bus (ECB) framework is the cloud computing based service model that references to model of web services, which provide services on request over the network. Agent based cloud computing has increased significantly today due to rapid growth and quick increase in number of clouds and their services. With the increasing number of cloud services to specify the performance issues, ECB framework is introduced. There is semi-formal representation of service discovery in ECB for agent based technique. Consequently, the design and implementation complexity of service discovery has significantly increased. Therefore, there is dire need to simplify the service discovery process in ECB framework for large and complex system by exploiting the concepts of Model Driven Architecture (MDA). In this article, a UML profile is proposed to model service discovery requirements in ECB. Particularly, provider agent, client, meta service registry, cloud agent and scheduler agent stereotypes are proposed to simplify the specification of publishing services to meta registry, providing services to clients and receiving requests from cloud agent. This delivers the bases or foundation to convert source models into required target models. The applicability of proposed profile is validated through flight reservation system case study. The results prove that the proposed profile significantly reduces the design and implementation complexity of service discovery in ECB.

1 Introduction

Cloud computing infrastructure has developed now a days that helps quick delivery of computing resources. Cloud computing provides services on demand access along with offering benefits like speed, cost and effectiveness. Cloud computing also provides services to distributed environment. So, the requirement for the organizations to gain cloud computing as their integration policies is rising exponentially.

The response for cloud computing services is increasing day by day. Enterprise Service Bus (ESB) technology delivers the generalization on execution of an enterprise

© Springer Nature Switzerland AG 2020
L. Barolli et al. (Eds.): CISIS 2019, AISC 993, pp. 269–279, 2020.
https://doi.org/10.1007/978-3-030-22354-0_25

messaging system, but there are still some limitations like it rises upstairs and slows down the communication speed due to increasing number of cloud and cloud services recently. There is semi-formal representation of service discovery using ECB framework for agent based technique. The purpose is to show the communication process between different components and using the MDA concepts we can modeled the requirements of system and transform to gain the model with different low level details. MDA represents a new framework to comprehend, examine, design and maintain all phases using different type of models. Modeling is helpful and needed when to interconnect the preferred organization and behavior of process or system and to seize the dynamicity of software architecture. So, there is need to model the ECB framework in profile in order to show proper process for the service discovery process.

In this paper, UML profile for service discovery in Enterprise Cloud Bus framework is proposed and presented which covers the overall requirement for the process in order to model the system accordingly to show the dynamic behavior of system. The proposed solution intends to make the system easily understandable and usable by developers, designers and users. The aim of the paper is to design UML profile to standardize the service discovery process which offers operative procedure for agent based web services application and to provide proper UML profile that can be applicable to any service discovery process.

The proposed UML profile shows the complete picture of ECB framework for service discovery. This also shows how the ECB services is available to clients or customers who made request for services. At the end, the case study is presented to signify that UML profile is implemented on it and it meets all the critical requirements.

The paper is organized in a way that in Sect. 2 related work or literature is presented. Section 3 provides details on the proposed UML profile. Section 4 presents validation using flight reservation system case study. In Sect. 5, discussion section is presented. Finally, Sect. 6 concludes the paper.

2   Related Work

The research by Bramantyo Adrin and Lukman [1] is about the improvement of cloud infrastructure service providers that works on virtual machine supply mechanism. Cloud services are increasing day by day and challenges are more for service discovery. The authors in [2] have proposed ECB framework as abstraction layer that is based on agent based cloud computing and in [3] they discussed and presented how services can be revealed and identified dynamically using relational model. The parameters related to quality are presented, Quality of Service (QoS) shows vibrant character because the delivered services of cloud computing integrate QoS and measured accordingly to ensure the availability of services [4].

Many companies have moved to deploying their services in cloud computing environment hence, improving the consistency and trustworthiness of cloud services has become a thought-provoking and inspiring research problem. In Cloud computing, reliability is significant and reliability improvement is needed. Although the techniques like reliability and fault tolerance are extensively considered in different distributed systems. The reliability and fault tolerance techniques also premeditated in high routine computing, cloud computing and exceptional architecture [5]. In [6] authors proposed Configurable Cloud Service Discovery and Selection System that helps in retrieving the IaaS Cloud resources to cloud users. The author proposed reliability based cloud approach that depends on different layer of cloud architecture. Cloud providers deliver cloud web services SaaS to their customers. The quality of services defined by the Service Level Agreements (SLAs) [7].

Services can conjoin and collaborate to provide broader functionality and complete task rapidly in cloud based Service Oriented Architecture (SOA) environment [8]. In [9] authors proposed a dynamic cloud service model based on service level agreement and privacy awareness. Around the performance and availability models of an IaaS Cloud optimization framework is developed [10]. Different things are managed by service broker which include performance and distribution between different providers and consumers and also discuss relationships. Brokers rely on different type of principals like architectural principals etc. Two platforms and applications are discussed by author as a result of classification [11]. The quality of cloud services are defined by service level agreement and different services are offered to different consumers. The services that are delivered by cloud providers towards different consumers satisfy the Service Level agreement SLA [12].

In [13] the author proposes the automatic framework which aimed at monitoring and benchmarking various cloud services that are offered by different cloud service providers. The authors also proposed some QoS compliance of different web services. Dynamically choosing the services that are related to their preferences is complex for cloud users. The author proposed CBRSM model that is specifically for the service selection [14]. In [15] the authors proposed an optimal service selection scheme that work both for local and public cloud for mobile devices. The functionality for working both in local and public is considered in hybrid cloud. So, it is based not only on service selection but also on cloud resource scheduling algorithm.

In [16] authors proposed abstraction layer and ECB has been modelled using UML diagram based on multi agent technology. The authors show the dynamicity of cloud services discovery that helps in identification of cloud services in dynamic environment with the help of relational model. In [17] authors used model driven approach for arbitrary cloud services and defined meta-model Cloud service (COPS) that is using sequential models for the description of service composition of different types of elements.

From the related work, it is analyzed that there is dire need to show the whole service discovery process in some model driven approach based on profile and there is no proper profile or meta-model specifically for service discovery that is helpful for designers. It is also significant to use some standardize language and define profile to show the process for cloud services requested by different clients and to model and test the whole system. Although in [18], authors show the working of cloud services using UML diagram, but it is semi-formal representation and still there is need to model the process with the help of profile according to MDA specification.

2.1   Description of ECB Framework

In cloud computing environment, Enterprise Cloud Bus and Enterprise Service Bus are abstraction layers of software as a service. Cloud Enterprise Service bus is the Enterprise Cloud bus extension and boosts the ESB's to catalogue the services for cloud platform [18, 19]. The provider agent publishes the request received from client into corresponding Meta Service registry (CUDDI). Service providers of cloud computing deploys cloud agent that collect all services from CESB and provide it to the extended meta service registry (HUDDI). The scheduler agent performs its duty by receiving services from the meta service registry and move forward towards the service mapper which is a software tool to map the services as per QoS parameters. The mapped services then move towards service scheduling log. The architecture of ECB framework is presented in Fig. 1.

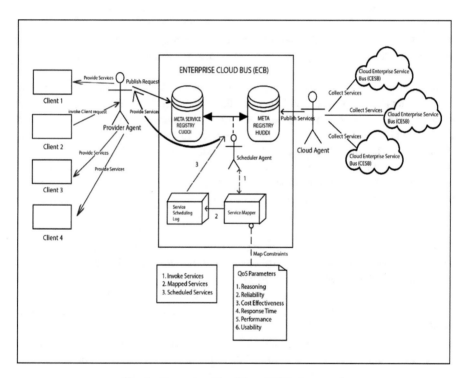

Fig. 1.   Enterprise Cloud Bus (ECB) framework

3   Proposed Profile

Model Driven Architecture (MDA) is used frequently in the development of composite application and it is used to simplify complex design and system verification [20]. In MDA the concept of UML profiling is one of the chief concepts and we have used model based approach based on profile. UML is part of Unified Modeling Language extension mechanism. UML profile consist of some main components:

1. Meta-classes represent the classes.
2. Interface represents all the interfaces in UML profile.
3. Stereotype and tagged values are main part of profile.
4. Operation meta-class represents the functionality of profile.
5. Association meta-class represents the relationship.
6. Instance Specification represents the different instances.

In this paper, we have used structure based modeling for UML profile which can be represented through class diagram that concerns with attributes and relationships that helps to solve the complex problems into approximately low level. We have used model driven approach based on profile for the process. Therefore, for structural representation we used class diagram, stereotypes and tagged values. The proposed UML profile is shown in Fig. 2.

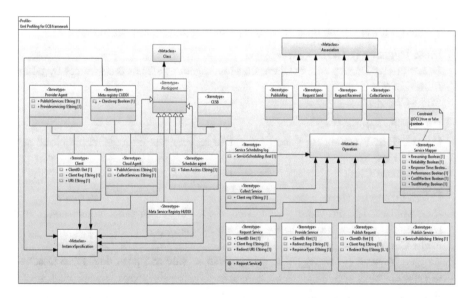

Fig. 2. UML Profile for service discovery using ECB framework

3.1 Description of Stereotype

Client Stereotype
Description: It is the participant that request for services needed which belong to cloud services.
Meta-Class: Class
Tagged Values: Client Meta class has following three tag values:
(1) Client ID: EInt[1] which is known by browser
(2) Client Req: EString[1] which is sent by client to web browser
(3) Uri: EString[1] which can be sent by client to web browser

Cloud Agent Stereotype

Description: The Cloud Agent collect services from CESB and publish into Meta Service Registry (HUDDI).

Meta-Class: Class

Tagged Values: Cloud Meta class has following two tag values:

(1) Collect Services: EString[1]

(2) Publish Services: EString[1]

Scheduler Agent Stereotype

Description: The Scheduler agent performs its duty by receiving services from the Meta service registry and move forward towards the service mapper.

Meta-Class: Class

Tagged Values: Token Access: EString[1]

Meta Service Registry (HUDDI) Stereotype

Description: The repository where the provider agent put the service that is to be requested.

Meta-Class: Class

Tagged Value: Client Req: EString[1] which is the request from client.

Meta Registry (CUDDI) Stereotype

Description: The repository where the cloud agent publishes the services by taking from CESB.

Meta-Class: Class

CESB Stereotype

Description: It stands for Cloud enterprise service bus through which cloud services can be taken.

Meta-Class: Class

Provider Agent Stereotype

Description: The Provider Agent lying under the meta class which sends and receives the request service to the respective client.

Meta-Class: Class

Tagged Values: It has tagged values publish service and provide service which is of type EString.

Request Service Stereotype

Description: It represent the response for the service to be requested.

Meta-Class: Operation

Tagged Values: It has tagged values Client ID, Client Req and redirect request which is of type EString.

Provide Service Stereotype

Description: It represents the response for the service to be provided.

Meta-Class: Operation

Tagged Values: It has tagged values for Client Req, Response type and redirect request which is of type EString.

Publish Request Stereotype
Description: It represent the response for the request to be published.
Meta-Class: Operation
Tagged Values: It has tagged value Client Req which is of type EString.

Publish Service Stereotype
Description: It represent the response for the service to be published.
Meta-Class: Operation
Tagged Value: It has tagged value ClientID, Client req and redirect request.

Collect Services Stereotype
Description: It represents the collected service from CESB.
Meta-Class: Operation
Tagged Value: It has tagged value service publishing.

Service Mapper Stereotype
Description: The Service Mapper maps the services as per Quality of Services (QoS) parameters that are represented as constraints including Time performance, Response time etc. It deactivates when the results move towards the service scheduling log.
Meta-Class: Operation
Tagged Values: It has tagged values that have type Boolean to check the certain parameters are included when the request is sent towards the client these are Reliability, CostEffective, Performance, Reasoning, Response Time etc.

Service Scheduling Log Stereotype
Description: Service Scheduling log receive the services from service mapper and after sending the scheduled services to service mapper, it goes in deactivation mode.
Meta-Class: Operation
Tagged Value: It has tagged value service publishing.

Publish Req Stereotype
Description: The request is publishing through this association.
Meta-Class: Association

Req Send Stereotype
Description: The request is send through this association.
Meta-Class: Association

Req Received Stereotype
Description: The request is receive through this association.
Meta-Class: Association

Collect Services Stereotype
Description: The request is collected through this association.
Meta-Class: Association

Constraint Stereotype
Description: It returns true or false for the QoS parameters.
Meta-Class: Service Mapper

Instance Specification Stereotype
Description: It helps in creating the instances.
Meta-Class: Instance Specification

4   Validation

To validate the approach, we demonstrate the applicability of ECB framework through flight reservation system case study. A profile is applied to case study in order to validate the profile. The case study is used to uphold flight detail and booking process. Flight details contain the name and seat detail. Flight reservation maintains ID, ticket number, passenger name & passport number, seat detail, travel charge, destination and date of travel are held in reserve. Finally, at the end flight detail, flight status and reservation tickets a report is generated.

Airline Reservation system provided following:

- Flight details Inquiry: The user can make inquiry about the flight detail like fair details.
- User Registration: The user can do registration as a new user to book flight etc.
- Booking of Flight: The user can book the flights.
- Flight Cancellation: The user can cancel the booked flight.
- Administration: The system allows admin to handle the flight details.

4.1   Case Study Description

The flight reservation system working is as follows. Firstly, the client will send request like flight reservation detail or booking detail to web browsers. The <<client>> stereotype is applied to the client class. The web browser that act as provider agent and the stereotype <<provider agent>> is applied on it, will receive the request from the client and it will move forward to the Meta registry (CUDDI) with applied stereotype. <<Meta registry CUDDI>> If the service is not available it will move forward the request towards HUDDI that is another service registry with applied stereotype <<Meta service registry HUDDI≫. It collects services from all CESB with applied stereotype <<CESB>> and give back to HUDDI via cloud agent. Meta registry then provide services to flight reservation server via scheduler agent with applied stereotype <<scheduler agent>> that perform different functions including: Service mapping via service mapper, service scheduling via service scheduling log and after doing these tasks it will provide services back to web browser. Web browser then provide service to respective client based on their client ID shown in Fig. 3.

4.2   Case Study Model

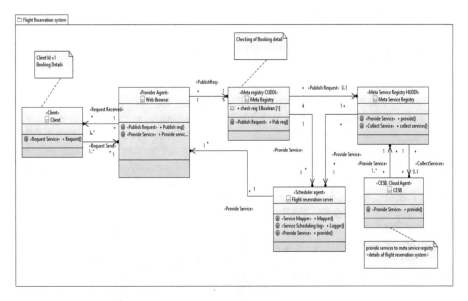

Fig. 3. Flight reservation system case study with applying profile

5   Discussion

This paper presents UML profile for service discovery in ECB framework that provide on demand services to cloud users. The whole process shows how the client request for cloud services. The inspiration to the research paper is to systematize the framework of ECB in relations of services which offers operative procedure for agent based web services application and provide proper UML profile that can be applicable to any service discovery process. This signifies with the help of UML profile, it is probable to model the infrastructures and instances of cloud.

The UML profile shows some confines like it is impracticable to make realistic environment for testing consequently simulation based testing is conceded out in order to check the whole process for verification. UML profile contains a set of UML extensions which complements different models into some particular domain. It is significant to use some standardize language and define profile to show the process for cloud services requested by different clients and to model and test the whole system. The standard unified modeling language like UML condense the cost requisite for understanding and modeling the cloud based services that is major benefit of this profile.

The proposed UML profile shows the complete picture of ECB framework for service discovery. This also shows how the ECB services is available to clients or customers who made request for services. The basic purpose is to show that according to MDA specification, by following a well-defined method, we modeled the system requirements and transform to get the same model but with low level details. At the end the case study is presented to signify that UML profile is implemented on it and it meet all requirements that are critical for the large systems. The case study for flight reservation system is practical in this context and profile is applied on it.

6 Conclusions

This paper familiarizes and categorizes the requirements necessary to represent the ECB framework process and presents UML profile for service discovery in ECB framework. The aim of the paper is to design UML profile to standardize the service discovery process which offers operative procedure for agent based web services application and to provide proper MDA concepts based on profile that can be applicable to any service discovery process. In this paper standard modeling language is used to model system requirements of the cloud services which is likely to test the entire process with unified language and transform to get the model with low level information. Consequently, UML based approach significantly reduces the implementation complexity of cloud computing ECB framework. The profile then validated through flight reservation system case study.

We intend to improve the search results with the help of some data engineering technques like K-means and K-medioid algorithms to achieve the relevant result for the service discovery process among the collections of web services and designed them with the help of MDA concepts.

References

1. Adrian, B., Heryawan, L.: Analysis of K-means algorithm for VM allocation in cloud computing. In: International Conference on Data and Software Engineering (2015)
2. Khan, G., Sengupta, S., Sarkar, A., Debnath, N.C.: Modeling of inter-cloud architecture using UML 2.0: multi-agent abstraction based approach. In: Proceedings of 23rd International Conference on Software Engineering and Data Engineering (SEDE), New Orleans, Louisiana, USA, pp. 149–154, 13–15 October 2014
3. Khan, G., Sengupta, S., Sarkar, S.: WSRM: a relational model for web service discovery in enterprise cloud bus (ECB). In: 3rd International Conference on Ecofriendly Computing and Communication Systems (ICECCS), NITK Surathkal, Mangalore, 18–21 December 2014
4. Mollah, M.B., Azad, M.A., Vasilakos, A.: Security and privacy challenges in mobile cloud computing: survey and way ahead. J. Netw. Comput. Appl. **84**, 38–54 (2017)
5. Zhou, A., Wang, S., et al.: On cloud service reliability enhancement with optimal resource usage. IEEE Trans. Serv. Comput. **4**(4), 452–466 (2016)
6. Hajlaoui, J.E., Omri, M.N., et.al.: QoS based framework for configurable IaaS cloud services discovery. In: IEEE 24th International Conference on Web Services, pp. 460–467 (2017)
7. Ibrahim, A.A.Z.A., Varrette, S., Bouvry, P.: On verifying and assuring the cloud SLA by evaluating the performance of SaaS web services across multi-cloud providers. In: 48th Annual IEEE/IFIP International Conference on Dependable Systems and Networks Workshops (2018)
8. Fernando, R., Ranchal, R., Bhargava, B., Angin, P.: A monitoring approach for policy enforcement in cloud services. In: IEEE 10th International Conference on Cloud Computing (2017)

9. Wang, Y., Wen, J., Zhou, W., Luo, F.: A novel dynamic cloud service trust evaluation model in cloud computing. In: 17th IEEE International Conference on Trust, Security and Privacy in Computing and Communications/12th IEEE International Conference on Big Data Science and Engineering (2018)

10. Ghosh, R., Longo, F., Xia, R., Naik, V.K., Trivedi, K.S.: Stochastic model driven capacity planning for an infrastructure-as-a-service cloud. IEEE Trans. Serv. Comput. **7**(4), 667–680 (2014)

11. Fowley, F., Pahl, C., Jamshidi, P., Fang, D., Liu, X.: A classification and comparison framework for cloud service brokerage architectures. IEEE Trans. Serv. Comput. **6**(2), 358–371 (2018)

12. Ibrahim, A.A., Kliazovich, D., Bouvry, P.: On service level agreement assurance in cloud computing data centers. In: Proceedings - 2016 IEEE 9th International Conference on Cloud Computing, pp. 921–926 (2016)

13. Ibrahim, A.A.Z.A., Varrette, S., Bouvry, P.: PRESENCE: toward a novel approach for performance evaluation of mobile cloud SaaS web services. In: IEEE International Conference of Information Networks (ICOIN), Chiang Mai, Thailand (2018)

14. Wang, X., Cao, J., Xiang, Y.: Dynamic cloud service selection using an adaptive learning mechanism in multi-cloud computing. J. Syst. Softw. **100**, 195–210 (2015)

15. Li, C., Yanpei, L., Youlong, L.: Efficient service selection approach for mobile devices in mobile cloud. J. Supercomput. **72**(6), 2197–2220 (2016)

16. Ahmad, F., Sarkar, A., Debnath, NC.: Analysis of dynamic web services. In: IEEE International Conference on Computing, Management and Telecommunications (ComMan-Tel 2014), Vietnam, 27–29 April 2014

17. Masteli, T., García, A., Brandi, I.: Towards uniform management of multi-layered cloud services by applying model-driven development. J. Syst. Softw. **121**, 358–371 (2016)

18. Khan, G., Sengupta, S., Sarkar, A.: Modeling of services and their collaboration in enterprise cloud bus (ECB) using UML 2.0. In: International Conference on Advances in Computer Engineering and Applications (ICACEA) IMS Engineering College, Ghaziabad, India, pp. 207–213 (2015)

19. Kamali, A., Mohammadi, S., Barforoush, A.A.: UCC: UML profile to cloud computing modeling. In: 7th International Symposium on Telecommunications (2014)

20. Anwar, M.W., Rashid, M., Azam, F., Kashif, M.: Model-based design verification for embedded systems through SVOCL: an OCL extension for SystemVerilog. J. Des. Autom. Embed. Syst. **21**(1), 1–36 (2017). http://link.springer.com/article/10.1007/s10617-017-9182-z

HPC-as-a-Service via HEAppE Platform

Vaclav Svaton[1], Jan Martinovic[1], Jan Krenek[1(✉)], Thomas Esch[2],
and Pavel Tomancak[3]

[1] IT4Innovations, VSB – Technical University of Ostrava, Ostrava, Czech Republic
`{vaclav.svaton,jan.martinovic,jan.krenek}@vsb.cz`
[2] German Aerospace Center, Earth Observation Center, Oberpfaffenhofen, Germany
`thomas.esch@dlr.de`
[3] Max Planck Institute of Molecular Cell Biology and Genetics, Dresden, Germany
`tomancak@mpi-cbg.de`

Abstract. The HPC-as-a-Service concept is to provide users with simple and intuitive access to a supercomputing infrastructure without the need to buy and manage their own physical servers or data centers. This article presents the commonly used services and implementations of this concept and introduces our own in-house application framework called High-End Application Execution Middleware (HEAppE Middleware). HEAppE's universally designed software architecture enables unified access to different HPC systems through simple object-oriented web-based APIs, thus providing HPC capabilities to users without the necessity to manage the running jobs forms the command-line interface of the HPC scheduler directly on the cluster. This article also contains the list of several pilot use-cases from a number of thematic domains where the HEAppE Platform was successfully used. Two of those pilots, focusing on satellite image analysis and bioimage informatics, are presented in more detail.

1 Introduction

HPC-as-a-Service is a well-known term in the area of high performance computing. It enables users to access a High Performance Computing (HPC) infrastructure without the need to buy and manage their own infrastructure. Through this service, academia and industry can take advantage of the technology without upfront investment in the hardware. This approach further lowers the entry barrier for users who are interested in utilizing massive parallel computers but often do not have the necessary level of expertise in the area of parallel computing.

Nowadays, very few companies or scientific research groups utilize the capabilities provided by Cloud [2,9] or HPC [16] infrastructures. For this reason, there is a number of service providers who offer their application frameworks, middlewares, or APIs [14] to their users to improve their knowledge about this technology and mainly to provide them with an easy-to-use access to HPC computational resources or Cloud [13,17].

The major representatives of the commercial service providers are Google LLC with *Google Cloud* [7] and Amazon Inc. with the *Amazon Web Services*

© Springer Nature Switzerland AG 2020
L. Barolli et al. (Eds.): CISIS 2019, AISC 993, pp. 280–293, 2020.
https://doi.org/10.1007/978-3-030-22354-0_26

(AWS) [6,8]. For academic use there is *EMU-Cluster* [4] hosted by eResearch South Australia.

Google Cloud service provides complex brand of services ranging from HPC computations and big data analysis to machine learning use cases. *Google High Performance Computing* service provides a number of computation options to meet the standards of the users (multi-instance or single-instance virtual machines). *Google Compute Engine* is a product which provides customizable Virtual Machines (VMs) with a number of features and the option to deploy a user code or application directly to a VM or via a container management system.

Amazon Web Service are web-based cloud services that provide computing services for all types of data. This service enables the users to utilize the computational environment based on the virtual infrastructure or HPC infrastructure. AWS also provides intuitive workflow manager *AWS Step* for the ability to create computational pipelines consisting of a number of interconnected computational tasks. The actual resource allocation is selected automatically based on the chosen block of operations and the type of the input data. AWS Step also provides the information about the state of every running computation from each executed pipeline.

For academic purposes and eRSA researchers, *eRSA EMU-Cluster* provides service to access and use HPC computational resources. This service uses *Torque job management system* to submit and monitor the cluster's jobs. This service is able to allocate a maximum of 16 compute jobs (136 cpu cores) for a single user.

The common problem with the above mentioned services and approaches is that they are either commercial services which do not disclose their internal workflows or they are tailored for a specific type of infrastructure. As *IT4Innovations national supercomputing center (IT4Innovations)* is in the role of an HPC provider, there is also a problem with the internal control mechanisms based on *ISO standards* and other *certifications* and *security mechanisms*. It is almost impossible to adopt third-party software as a middleware to access HPC infrastructure directly instead of the users while retaining the same quality of service provided by a standard supercomputer support staff. Due to this reason we have developed our own open-source implementation of an HPC-as-a-Service concept called High-End Application Execution Middleware (HEAppE Middleware) [3]. IT4Innovations currently operates two HPC clusters, and thus also wanted to provide them as a service for users.

This paper is organized as follows. Section 2 describes the overall general architecture of the developed HEAppE Middleware. It also describes the main internal processes and mechanisms regarding the user and cluster accounts, data management, job submission and application interfaces. Section 3 illustrates several use-cases where the HEAppE platform was successfully used, including domains of earth-observation image analysis in the *ESA's Urban Thematic Exploitation Platform* and biological image data analysis in the *SPIM image processing plugin for FIJI*. Finally, Sect. 4 presents the conclusion, lesson learned, and future work in the context of this project.

2   HEAppE Middleware

IT4Innovations supercomputing center as a HPC provider and DHI Group[1], a nation-wide company developing hydrologic software MIKE[2] powered by DHI, were a collaborators during the design and implementation phase of the HEAppE Middleware (formerly known as an HPC as a Service Middleware). At the beginning of the project, the HEAppE was used in the area of hydrological modelling in a decision support system for crisis management [15]. Since then the HEAppE was successfully used in a number of projects from a different thematic domains like satellite image analysis [10] and biological image analysis [11]. A more detailed description of these use-cases is located in Sect. 3.

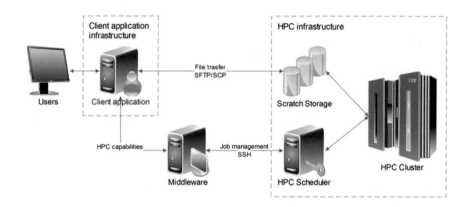

Fig. 1. HEAppE Middleware general architecture

2.1   Middleware General Architecture

There was the need for an architecture, which would allow users to run complex and computationally demanding calculations on a supercomputer directly from the user interface of a client application without the necessity to connect directly to the HPC cluster and manage the jobs from the command line interface of the HPC scheduler. Therefore, IT4Innovations[3] has developed an application framework called HEAppE Middleware. HEAppE manages and provides information about submitted and running jobs and their data between a client application and an HPC infrastructure. This middleware is able to submit the required computation or simulation to the HPC infrastructure, monitor the progress, and notify the user if needed. The Scheme of a general architecture is displayed in Fig. 1.

---

[1] https://www.dhigroup.com/.

[2] https://www.mikepoweredbydhi.com/.

[3] https://www.it4i.cz/.

Platform features:

- Providing HPC capabilities as a service to client applications and their users
- Unified middleware interface for different operating systems and schedulers
- Authentication and authorization for the provided functions
- Monitoring and reporting of executed jobs and their progress
- Current information about the state of the clusters
- Job accounting and job reporting for a single user or a user group
- Secure data migration between different jobs
- Prepared job templates for different computational tasks

The internal middleware architecture is separated into a number of interconnected layers. To simplify this architecture, there are three main layers called *ServiceTier*, *BussinesLogicTier* and *HpcConnectionFramework*.

ServiceTier represents high-level view of the middleware provided services. This layer is used for simple integration into different user interfaces or client applications. BussinesLogicTier contains concrete implementation of the provided functions and is therefore invoked from the ServiceTier layer. HpcConnectionFramework contains implementation of direct access to an HPC infrastructure. This could be considered as the lowest layer, which is responsible for the actual job submission and monitoring. The current implementation of the HpcConnectionFramework uses *PBS Manager* [1] as a main HPC job manager. If there is the need to support a different kind of HPC job manager, e.g. *Slurm* [12] which could be available at a specific HPC center the simple extension of this layer is all that is needed to use the HEAppE Middleware in a different HPC center.

Independently of these layers, there are five main categories into which the functions provided by the middleware are divided. These categories are as follows:

- *User and Limitation Management* provides several authorization and authentication services to the users and management of user limitation parameters.
- *Cluster Information* provides information about the state of the connected clusters, available types of processing queues, and cluster node usage.
- *Job Management* ensures comprehensive job management functionality to create, submit, cancel, delete, and monitor the state of computational jobs.
- *File Transfer* provides the users with the necessary methods for data management between the client application and an HPC infrastructure.
- *Job Reporting* is used for reporting statistics. It provides a comprehensive reports for a specific user or a group of users in terms of submitted job and used resources.

2.1.1   User Accounts

HEAppE Middleware introduces an additional layer of abstraction in remote access to an HPC infrastructure from the security point of view. Middleware shields the users from direct access to HPC computing resources and all internal processes that are associated with direct access to a supercomputer and its storage. Therefore, the middleware does the mapping between *external user accounts* and *internal cluster accounts* (see Fig. 2). Passwords stored within the middleware's internal database are encrypted using *PBKDF2 salted password hashing* [5].

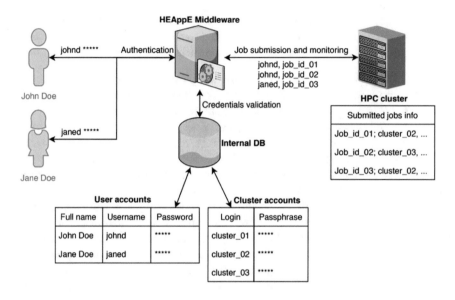

Fig. 2. Middleware's mapping mechanism

External user accounts are provided by HEAppE Middleware to users and is used by them to authenticate themselves via middleware to access its functionality. Therefore, the user is only able to access the functions offered by the middleware and not the HPC infrastructure itself.

Internal cluster accounts are used by the middleware to access an HPC infrastructure and perform the job submission instead of the user. These accounts are basically the same as the one acquired by the user at the end of the successful standard authentication process within the HPC center.

Middleware keeps track of all submitted and running jobs, and via its monitoring and reporting services it is able to provide the users with an up-to-date information about running jobs and used resources.

2.1.2 Template Preparation

For security purposes, HEAppE Middleware enables the users to run only a prepared set of so-called *Command Templates*. Each template defines an arbitrary script or executable file that will be executed on the cluster, any dependencies or third-party software it might require, and the type queue that should be used for processing (type of computing nodes to be used on the cluster).

The template also contains the set of input parameters that will be passed to the executable script during run-time. This is achieved via *wildcard parameters* "%%{}". Thus, the users are only able to execute pre-prepared command templates with the pre-defined set of input parameters. The actual value of each parameter (input from the user) can be changed by the user for each job submission.

Id	Name	Description	Code	Executable File	Command Parameters	Preparation Script	Cluster Node Type
1	TestTemplate	Desc	Code	/scratch/temp/ HaasTestScript/test.sh	"%%{inputParam}"	module load Python/2.7.9-intel-2015b:	7

Fig. 3. HEAppE command template

Example of command template with one custom user attribute *inputParam* is illustrated in the table in Fig. 3. This Command Template allows the users to run only the prepared script *test.sh* which is already prepared on a cluster storage. The *Preparation Script* column defines any third party software or dependency that should be loaded before the actual script execution. The last parameter called *Cluster Node Type* denotes the type of the processing queue to be used on the cluster.

2.1.3 Data Management

The software architecture of the HEAppE Middleware requires the user to copy the input data to a cluster storage before processing and to download or remove the data from the cluster storage when the processing is finished. Each computation job is considered as an independent unit by the middleware.

However, this poses a problem if one computation job is just a part of the processing pipeline where the output of one job is used as an input of the following job. Therefore, there was the need for functionality that would provide the users with a secure way to migrate the data between different computation jobs. Based on this requirement, the middleware enables the users to copy the selected data from a finished computation job to a temporary user's storage and to copy this data from the storage to an input of a newly created job thus mitigating the data transfer bottleneck. This methods utilize the secure token called *sessionCode*, which is extensively used by the middleware to authenticate the user as a key for *CopyToStorage* and *CopyFromStorage* functionality. Using the valid token, the user can store the data to a temporary storage and later use the same token to retrieve the data from the storage in a different computational job.

In the case of a standard computational job the user usually needs to upload the input data to a cluster storage for processing. For data management purposes, middleware offers the users the *File Transfer* methods as described in

Sect. 2.1. The purpose of this functionality is to provide a secure way for the user to upload the data. Internally, the middleware creates a *temporary SSH key* for the user to access the job folder in the cluster storage to upload the input data. This data transfer is performed through a standard *Secure Copy Protocol (SCP)* without involvement of the middleware as a bottleneck. After successful data upload, this key is removed from a cluster and the job is prepared for submission to a job management system.

2.1.4 APIs

There is a number of ways to access the functionalities provided by the HEAppE Middleware. The most commonly used type of access is through standard *web services*. This approach includes an easy accessible web interface and programming language interoperability in case that these web services are invoked from a custom client interface, e.g. web page, desktop application, or a mobile application. Web services are simple to use and give the middleware user the freedom to develop his/her own application interface, i.e. wrapper build upon these services. The list of web services available to users via the middleware relates closely to the six main categories of the middleware's functions listed in Sect. 2.1.

One of the examples of a custom API wrapper is *Jupyter Notebook*[4]. The Jupyter Notebooks are live editable documents that allow mixing of *live* programming codes, equations, formulas, visualizations, and texts. Because of their flexibility and available kernels in many different programming languages they have become useful solution for demonstrations and education. Mixing the explanatory and instructional text with the programming code in Jupyter Notebook documents makes it generally easier for users without comprehensive knowledge about the topic. They have become popular in various scientific domains like Artificial Intelligence and Machine Learning in science area.

Two Jupyter Notebooks were created as an example for the users to explain the basic abilities of the HEAppE Middleware and to help these users learn how to use it. These notebooks (Java and Python based kernels) were implemented as wrappers on top of the standard web service API. An example of the HEAppE Java Client is shown in Fig. 4.

2.1.5 Middleware Deployment

The deployment procedure can be divided into two separate tasks; deployment of the HEAppE Middleware application framework itself and preparation of the HPC infrastructure. Middleware deployment is pretty straightforward. As it is a standard web service application it just has to be set up in a web server environment. Upon a successful deployment to a web server, the middleware's services should be visible on a specific web address. Middleware also automatically creates its own internal database to store the user account information, information about submitted jobs, and much more.

---

[4] http://jupyter.org/.

Fig. 4. HEAppE java client in jupyter notebook

For the middleware to be able to submit jobs to a cluster queue, there is the need for a set of cluster accounts that need to be bound to a specific *computation project*. Therefore, a set of cluster service accounts should be created for a specific computation project ID and project name. These accounts will be used by the middleware (have to be also included into a middleware's database) for the job submission on the cluster's side of the architecture.

The second task consists of cluster environment setup. This mainly includes preparation of so-called *key scripts* that contain various internal mechanisms e.g. job directory creation in a cluster storage, generation of a temporary ssh key for the file transfer methods, etc. These scripts will be invoked by the middleware during the execution of a specific computational job to simply prepare the cluster environment for the job submission.

It is a good practice to deploy independent instance of HEAppE Middleware for each computational project. Every instance contains its own physically separated database with a new set of user credentials used to authenticate the external users via the HEAppE Middleware, thus allowing them to remotely submit or manage their computation jobs and also a new set of internal cluster accounts to separate the computational project's jobs at an HPC infrastructure level.

2.2   HEAppE at IT4Innovations

IT4Innovations national supercomputing center operates supercomputers Salomon (2 PFLOP/s) and Anselm (94 TFLOP/s). The supercomputers are available to academic community within the Czech Republic and Europe and industrial community worldwide. Both supercomputers are available to users via HEAppE Middleware. The scheme of the HPC architecture with the Salomon and Anselm clusters is displayed in Fig. 5. The detailed description of the IT4Innovations' HPC infrastructure and the hardware specification of both clusters can be found in the official documentation[5].

HEAppE's universally designed software architecture enables unified access to different HPC systems through a simple object-oriented client-server interface using standard web services, REST API, and Jupyter Notebooks. Thus providing HPC capabilities to the users yet without the necessity to manage the running jobs from the command-line interface of the HPC scheduler directly on the cluster.

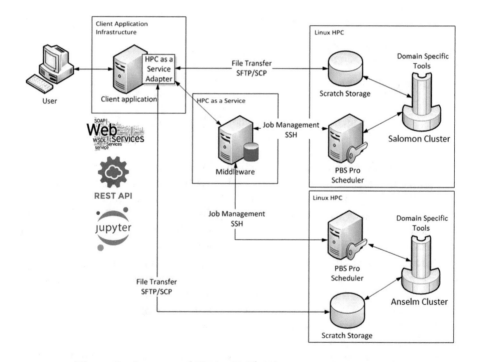

Fig. 5. Deployment of HEAppE Middleware at IT4Innovations

---

[5] https://docs.it4i.cz.

3   Integration Use Cases

HEAppE Middleware has already been successfully used in a number of public and commercial projects where there is a need to remotely manage computational jobs without directly accessing the HPC infrastructure itself.

HEAppE was used in the crisis decision support system *Floreon+* [15] for What-If analysis workflow utilizing HPC clusters; in the *Urban Thematic Exploitation Platform* (Urban-TEP) [10] financed by the ESA as a middleware enabling sandbox execution of user-defined docker images on the cluster; in the *H2020 project ExCaPE*[6] as a part of a Drug Discovery Platform enabling the execution of drug discovery scientific pipelines on a supercomputer; in the area of molecular diagnostics and personalized medicine in the scope of the *Moldimed project*[7] as part of the Massive Parallel Sequencing Platform for analysis of NGS data; and in the area of bioimage informatics as a integral part of the *FIJI plugin* [11] providing unified access to HPC clusters for image data processing.

Two of these use-cases are introduced in the following subsections in more detail to illustrate that the user usually sees only the dedicated graphical interface and is completely shielded from the functionality provided by the HEAppE Middleware. *Submitting a job to a supercomputer is as easy as clicking the button on a website.*

3.1   Urban Thematic Exploitation Platform

The main goal of the *Urban Thematic Exploitation Platform (Urban TEP)* project is implementation of an instrument that helps address key research questions and societal challenges arising from the phenomenon of global urbanization. The created *Urban TEP* platform[8] is a workplace for a scientific community and commercial subjects alike to access a number of thematically oriented processing services and data products. These products and services are created with a focus on the domain of global urbanization and provide the users with an easy and intuitive way of how to access and analyze satellite image data from a number of space missions.

The main role of the *IT4Innovations* center in the project is to provide state-of-the-art technology and expertise in high performance computing. Thus the IT4Innovations center provides the platform with the processing services and data storage services needed to access, analyse, and visualize geospatial data and derived products (see Fig. 6). Via the platform's processing services the users are able to submit their analysis into the IT4Innovations processing center.

This access to the IT4Innovations' HPC infrastructure is achieved via a simple integration of the HEAppE Middleware into the job management processes of the platform. The results of these remotely invoked computations are automatically visible directly in the geobrowser of the platform immediately after the

---

[6] http://excape-h2020.eu.

[7] https://www.imtm.cz/moldimed.

[8] https://urban-tep.eu.

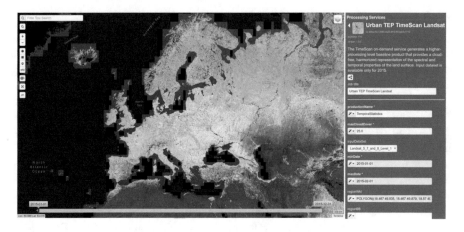

Fig. 6. Urban TEP Platform: Landsat TimeScan product and on-demand processing service

computation is done, thus completely shielding the users of the web platform from a somehow complicated direct HPC infrastructure access.

3.2 SPIM Image Processing Pipeline

State-of-the-art imaging devices, such as light sheet microscopes, produce datasets so large that they can only be effectively analyzed by employing methods of image processing on high-performance computing clusters. To address this issue, an HPC plugin for Fiji[9], one of the most popular open-source software tools for image processing, has been developed. The plugin enables end users to make use of HPC clusters to analyze large scale image data remotely and via the standard Fiji user interface (see Fig. 7).

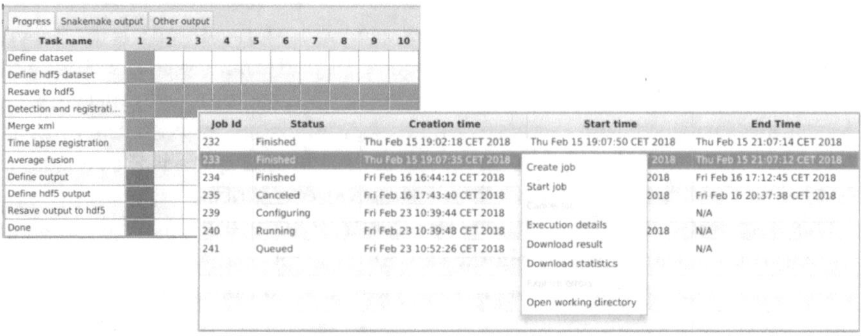

Fig. 7. Fiji plugin: SPIM image processing pipeline interface

---

[9] http://fiji.sc.

This Fiji plugin utilizes the HEAppE Middleware for remote execution of *SPIM image processing pipeline* on selected HPC infrastructures. The created framework will form the foundation for parallel deployment of any *Fiji/ImageJ2* command on a remote HPC resource, greatly facilitating big data analysis.

4   Conclusion and Future Work

The HEAppE Middleware is a powerful tool for the users who wish to utilize the *HPC-as-a-Service* concept. It provides a way how to easily access an HPC infrastructure without upfront investment in the hardware itself, necessary services associated with the maintenance of a data processing and data storage infrastructure, or training of specialized personnel.

The first version of the middleware has been extensively used within the IT4-Innovations center in a number of internal or public projects for the scientific community or private sector. Whenever there is a need for an easy-to-use access or integration of HPC capabilities, the HEAppE Middleware is a go-to alternative for the users who wish to use the supercomputer infrastructure yet unwilling to spend too much time learning how to both access it and work with it.

Since 2018, HEAppE Middleware has been released as an *open source under the GNU General Public License v3.0.* The source codes are available for everyone via the project's git repository at http://heappe.eu.

In terms of future work this project is under constant development. As the current implementation of this application framework is based on .NET Framework and uses Microsoft-SQL database it can only be deployed on the windows operated machines. We are currently preparing the .NET Core multi-platform version of the middleware with a simplified middleware deployment and versioning functionality by utilizing the Docker packages.

The second main task is to provide the users with a simple and easy-to-use graphical interface in the form of a workflow manager with the ability to create custom analytical pipelines (a sequence of interconnected computational tasks) to run on the cluster as a single computational job while retaining the ability to monitor the running tasks in near real time.

Aknowledgments. This work was supported by The Ministry of Education, Youth and Sports from the National Programme of Sustainability (NPS II) project 'IT4Innovations excellence in science - LQ1602', by the European Regional Development Fund in the IT4Innovations national supercomputing center - path to exascale project, project number CZ.02.1.01/0.0/0.0/16_013/0001791 within the Operational Programme Research, Development and Education, partially supported by The Ministry of Education, Youth and Sports from the Large Infrastructures for Research, Experimental Development and Innovations project 'IT4Innovations National Supercomputing Center –LM2015070' and by the SGC grant No. SP2019/108 'Extension of HPC platforms for executing scientific pipelines', VSB - Technical University of Ostrava, Czech Republic. The project 'Urban Thematic Exploitation Platform' was funded by European Space Agency (ESA) of ESA Contract No. 4000113707/15/I-NB.

References

1. Altair Engineering, Inc.: PBS Professional. https://www.pbsworks.com/PBSProduct.aspx?n=Altair-PBS-Professional&c=Overview-and-Capabilities. Accessed 28 Feb 2019
2. Baun, C., Kunze, M., Nimis, J., Tai, S.: Cloud Computing Web-Based Dynamic IT Services. Springer, Heidelberg (2011). ISBN 978-3-642-20917-8
3. IT4Innovations National Supercomputing Center. High-End Application Execution Middleware. http://www.heappe.eu. Accessed 28 Feb 2019
4. eResearch South Australia. eResearch SA EMU-Cluster. https://www.ersa.edu.au/service/cloud/clusters-in-the-cloud. Accessed 9 Feb 2019
5. Ertaul, L., Kaur, M., Gudise, V.A.K.R.: Implementation and performance analysis of PBKDF2, Bcrypt, Scrypt algorithms. In: Proceedings of the International Conference on Wireless Networks (ICWN), Athens, pp. 66–72 (2016)
6. Fusaro, V.A., Patil, P., Gafni, E., Wall, D.P., Tonellato, P.J.: Biomedical cloud computing with Amazon Web Services. PLOS Computat. Biol. (2011). https://doi.org/10.1371/journal.pcbi.1002147
7. Google LLC. Google Cloud HPC Computing. https://cloud.google.com/solutions/hpc. Accessed 28 Feb 2019
8. Halligan, B.D., Geiger, J.F., Vallejos, A.K., Greene, A.S., Twigger, S.N.: Low cost, scalable proteomics data analysis using Amazon's cloud computing services and open source search algorithms. J. Proteome Res., 3148–3153 (2009). https://doi.org/10.1021/pr800970z
9. Armbrust, M., Fox, A., Griffith, R., Joseph, D.A., Katz, R., Konwinski, A., Lee, G., Patterson, D., Rabkin, A., Stoica, I., Zaharia, M.: A view of cloud computing. Commun. ACM **53**(4), 50–58 (2010)
10. Esch, T., Asamer, H., Bachofer, F., Balhar, J., Böttcher, M., Boissier, E., Hirner, A., Mathot, E., Marconcini, M., Metz-Marconcini, A., Permana, H., Soukup, T., Svaton, V., Üreyen S., Zeidler, J.: New prospects in analysing big data from space - the urban thematic exploitation platform. In: 2018 IEEE International Geoscience and Remote Sensing Symposium, IGARSS 2018, pp. 8193–8196 (2018)
11. Krumnikl, M., Moravec, P., Kozusznik, J., Klimova, J., Bainar, P., Svaton, V., Tomancak, P.: SPIM workflow manager for HPC. Bioinformatics (2019). https://doi.org/10.1093/bioinformatics/btz140
12. Yoo, A.B., Jette, M.A., Grondona, M.: SLURM: simple linux utility for resource management. In: Job Scheduling Strategies for Parallel Processing, pp. 44–60 (2003). https://doi.org/10.1007/10968987_3
13. Schadt, E.E., Linderman, M.D., Sorenson, J., Lee, L., Nolan, G.P.: Computational solutions to large-scale data management and analysis. Nat. Rev. Genet. 647–657 (2010). https://doi.org/10.1038/nrg2857
14. Richardson, L., Ruby, S.: RESTful Web Services, 1st edn. O'Reilly Media, Newton (2007). ISBN 978-0596529260
15. Svaton, V., Podhoranyi, M., Vavrik, R., Veteska, P., Szturcova, D., Vojtek, D., Martinovic, J., Vondrak, V.: Floreon+: a web-based platform for flood prediction, hydrologic modelling and dynamic data analysis. In: GIS OSTRAVA 2017, pp. 409–422 (2018). https://doi.org/10.1007/978-3-319-61297-3_30

16. Vecchiola, Ch., Pandey, S., Buyya, R.: High-performance cloud computing: a view of scientific applications. In: 10th International Symposium on Pervasive Systems, Algorithms, and Networks, pp. 4–16 (2009). https://doi.org/10.1109/I-SPAN.2009.150
17. Youseff, L., Butrico, M., Silva, D.D.: Toward a unified ontology of cloud computing. In: Grid Computing Environments Workshop, pp. 1–10, November 2008

A Distributed Environment for Traffic Navigation Systems

Jan Martinovič[1(✉)], Martin Golasowski[1], Kateřina Slaninová[1],
Jakub Beránek[1], Martin Šurkovský[1], Lukáš Rapant[1], Daniela Szturcová[1],
and Radim Cmar[2]

[1] IT4Innovations, VŠB - Technical University of Ostrava,
17. listopadu 15/2172, 708 00 Ostrava-Poruba, Czech Republic
{jan.martinovic,martin.golasowski,katerina.slaninova,jakub.beranek,
martin.surkovsky,lukas.rapant,daniela.szturcova}@vsb.cz
[2] Sygic a.s., Twin City C, Mlynské Nivy 16, 821 09 Bratislava, Slovak Republic
rcmar@sygic.com

Abstract. Effective navigation and distribution of traffic flow in large cities has become a hot topic in recent years. The authors have developed an advanced server side routing system which, together with client side navigation systems, is able not only to navigate cars according to their routing requests, but also to distribute traffic flow within a city. The main goal of the paper is to propose a distributed environment used for an advanced server side navigation system with a focus on effective usage of computational resources for different tasks that need to be solved within the system. A combination of cloud and high performance computing resources in one environment is proposed. The authors also developed a simulator for testing these distributed computational resources. The system and the simulator were tested on the infrastructure at IT4Innovations National Supercomputing Center in the Czech Republic.

1 Introduction

Today's cities are experiencing an unprecedented increase in the volume of traffic. This increase raises a number of problems both from the socio-economic and environmental perspective. There are roughly two ways to handle this increase. The first one is the restriction of traffic in cities. While it can prove to be very effective, it will never be widely popular. The other possibility is to try to optimise the traffic in cities as a whole. From our point of view this approach is much more interesting because it offers a number of challenges that can be solved to improve the quality of life without imposing any harsh restrictions. It also fits snugly into the concept of smart city [2]. A smart city is an urban area (i.e. a district, a city, or a conurbation) utilizing various types of automatic and electronic data collection sensors to provide information which can be utilised to improve the management of the assets and resources of a given area more efficiently (in our case, the transportation network).

© Springer Nature Switzerland AG 2020
L. Barolli et al. (Eds.): CISIS 2019, AISC 993, pp. 294–304, 2020.
https://doi.org/10.1007/978-3-030-22354-0_27

Several works have already dealt with the problem of global traffic optimisation. For example, Jeong et al. [7] present a solution based on the Dijkstra algorithm extended with a metric for delay calculation of individual edges and vertices. However, their approach is simplified by the fact that once the car receives its route, it does not update it unless it leaves it. This means that there can potentially be a better road available later on, but the individual cars will not be aware of it. Also interaction between routed cars is minimal. One of the other approaches is represented by the paper written by Ma et. al. [8]. Their proposed solution is based on a multi-agent model combined with numerical optimisation of their flow. Their model has some advantages and shortcomings. It takes into account other cars, and changes their routes according to the actual traffic conditions. It also models their interaction. However, numerical optimisation quickly becomes very problematic once we want to solve the problem with thousands of cars. Based on these two examples, we can state qualities we would expect from our model:

- Easy application on a large number of cars
- Some form of interaction among the optimised cars
- The possibility of reacting to actual developments on the road network

Therefore, we should be inspired by the better parts from both presented approaches while countering their disadvantages to propose our own solution.

2   Global Optimisation and Self-adaptive Routing

The routing of a single car is at the heart of our approach. This routing, if properly implemented, can improve a travel time of a given car by taking into account both the length of the path and dynamically changing traffic conditions. These traffic conditions can be derived from the data fusion of available data sources. For example, this data fusion could be done by combining the information about current speed from traffic sensors with information about traffic incidents and with weather forecast for the given area. The first part of this routing has already been thoroughly researched, and a number of algorithms such as Dijkstra or A* have already been proposed. To capture the influence of dynamically changing traffic conditions, we utilised the Probabilistic approach presented in our previous papers [13] and [5]. This Probabilistic time-dependent routing algorithm (PTDR) in combination with a k alternative variant of former ones can produce a number of the shortest paths with a probability distribution of their duration. Based on user preferences, the best path is then chosen. This last part of the proposed approach is called reordering.

One of the greatest benefits of this approach is that it can be extended to optimise all cars in the traffic network. It can be done by introducing one more element into the reordering calculation. This element is based on the number of cars that are traversing, or are going to traverse, parts of a given route at the same time as the optimised car. This addition into the reordering step allows the individual cars to take into consideration the paths of other cars and chose the preferred road accordingly (possibly not the optimal one based on both usual traffic conditions and its length). A scheme of this approach can be seen in Fig. 1.

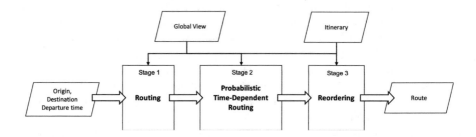

Fig. 1. Routing pipeline stages and Global view integration

This approach to traffic optimisation, however, has some particular shortcomings and challenges. The first one is the sheer amount of data it generates.

As shown in Fig. 1, this approach requires a description of spatial and temporal properties of the traffic network called the Global View. The Global View comprises information coming from various data sources. Map sources are at its base. They provide information about topology and physical distances in the traffic network. Next is information coming from the traffic sensors. These sensors are broadly divided into two categories: stationary sensors and floating car data (FCD) [10]. Stationary sensors are usually represented by either cameras or induction loops, and provide very reliable information about the traffic characteristics at specific locations of the traffic network. On the other hand, FCD provide more approximate information about the traffic conditions. However, they are theoretically available along the entire traffic network as long as there are some cars on each road. These data sources can be further enriched by information about traffic incidents. This information can be utilised to force the algorithm to reroute the traffic away from these spots. Traffic can also be influenced by some weather effects, for example squalls or snowstorms. These effects can also be propagated into the Global View. It is self-evident that getting all the data from the real world for the purpose of testing the optimisation is incredibly difficult.

Another problem is testing of this system. It requires very high penetration of the traffic to function properly. However, it is impossible to obtain a number of cars and drivers that would behave according to recommendations of this algorithm and monitor their routes.

The last challenge is closely related to the previous one. The proposed system must be able to process a very high number of routing requests within a reasonable time to keep the optimisation running smoothly. If it is not able to provide the optimal routes on time, the drivers would not be driving along the optimal routes, potentially creating unnecessary traffic jams.

These numerous problems have forced us to look to a virtual controlled environment that can simulate the traffic network, its characteristics, and cars travelling in it. It should also serve as a test-bed for whether the navigation system is able to respond within a reasonable time to the number of requests necessary for the optimisation. It is therefore desirable to utilise some form of traffic simulator.

There is a number of existing traffic simulators. However, simulators like VISSIM [3], SUMO [1], TRANSIMS [12] or MATSim [6] generally require tremendous amounts of actual data for their proper functioning because they are mainly based on microscopic approaches. Also, their potential deployment on an HPC infrastructure is complicated because of their existing implementation limits. Due to these complications associated with existing traffic models and simulators, we decided that it would be preferable to develop our own traffic simulator that can handle all tasks demanded for the testing of our proposed approach to the traffic optimisation. A more thorough description of our motivation and the basic ideas behind our simulator can be found in an article by Ptosek et al. [9]. Generally, our simulator comprises three parts: a simulator of cars, a virtual world, and server-side routing. The main aim of this article is the analysis of the last part of our developed simulator from the performance perspective, i.e. whether it is able to provide the routes for the individual cars in the simulator within defined limits for response.

3   Self-adaptive Navigation System

The server-side navigation system service provides optimal routes at any given time based on a global view of the traffic network, which is periodically refreshed by current data. The system user expects the best possible results in terms of response time and quality of the route provided by the service, independent of the load generated by other users.

The response time metric is defined by a Service Level Agreement (SLA) which can be relaxed compared to traditional on-line services. The SLA is typically a commitment of the service operator to the user that a given metric of the service quality will be within the specified limits. In our case, it corresponds to a selected percentile of the service response time.

The system tries to minimise the travel time globally for all users in the specified region. The quality of the routes is ensured by periodic refreshing of the global view with current data from multiple sources (traffic monitoring, weather prediction, etc.). The service exposes a simple HTTP-based API and provides the routing results in JSON format.

3.1   Service-Level Agreement in the Smart City Context

In the context of a smart city, there are two possible use cases of the service. The first one is drivers who use the service manually using a smartphone app or an in-car navigation system. The second case is autonomous vehicles using the service. In both cases, the response time of the service can be relaxed, since the reaction to a sudden change in the new route has to be smooth. It can be safely assumed that the path will change over the course of travel as the Global View reacts to the changing situation in the entire network. The reordering phase of the enhanced pipeline already implements a mechanism to reduce sudden changes in the provided paths.

Even if the new route is provided in a few hundred milliseconds, the driver or autonomous vehicle needs some time to assess the current traffic situation around the car and finally decide whether to take the new route or not as it may involve hard turns or changes in a speed.

The system always provides full optimal path from the current position to the destination. The rerouting interval has to be selected such that it is long enough to avoid sudden changes in the route, and short enough to reflect the changes in the traffic network state.

Based on this fact, the response time defined for the example by a 99% SLA can be relaxed to 30 s instead of mere milliseconds, meaning that the driver is unable to benefit from more frequent route updates. This relaxation also means that the amount of required compute resources can be significantly lower.

3.2   ANTAREX Autotuning and DSL Integration

The efficient operation of the service is supported by a set of tools developed in the ANTAREX project [11]. The autotuning framework is used to estimate the amount of compute resources needed by the service in order to comply with a given SLA. It is also used to dynamically reduce the running time of the PTDR algorithm, which is the most demanding stage of the routing pipeline, by making a trade-off between precision and time spent there. It is also used to estimate the amount of compute resources and running routing workers necessary to satisfy the required SLA.

As the amount of routing requests fluctuates over time, the amount of running routing pipeline workers has to change accordingly in order to minimise resource wastage. The autotuning approach, integrated within the mArgot autotuner[1] developed by Politechnica di Milano, automatically estimates the necessary number of workers per component of the routing pipeline needed to satisfy the actual request intensity, while keeping the response time of the entire pipeline within the given SLA requirement.

The goal of PTDR autotuning is to dynamically reduce the number of Monte Carlo samples based on the unpredictability feature extracted from prior execution of the total number of samples. The relationship between the unpredictability feature and the number of samples is obtained during the Domain Space Exploration (DSE) process which is executed on a representative sample of routes enriched by the probabilistic speed profiles. A detailed description of the process and analysis has been submitted to [14].

The domain specific language LARA is used to ease the effort needed for integration of the tools into the existing codebase and to generate parts of the data access API [4].

---

[1] http://antarex-project.eu/dissemination#tools.

4   Distributed Environment

We have implemented a distributed system that serves routing requests in real time. It was designed to handle a large number of clients simultaneously and thus all of its components are built to scale out to many nodes. The architecture of the system is depicted in Fig. 2. It consists of several independent components that communicate over the network.

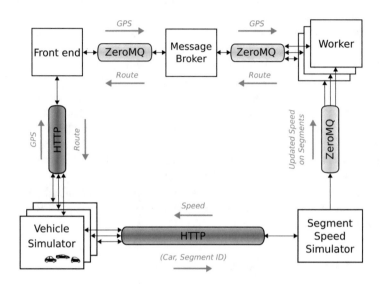

Fig. 2. Architecture of the routing system

The Front end component presents a public HTTP interface for the clients (vehicles). It is stateless and thus it may be scaled to multiple nodes and load balanced if necessary. The front end receives HTTP requests that consist of two GPS coordinates; origin and destination. It eventually returns a list of GPS waypoints that the vehicle should pass through. The requests from the clients are translated into Protobuf[2] messages that are then sent to the Message broker component. Communication between the front end and the message broker uses ZeroMQ[3] – a low latency distributed messaging framework. After the request is sent, the front end waits for a response for a given time (based on the SLA). A timeout error is returned to the client if no response arrives in time.

The message broker is a load balancing component that receives messages from the front end, distributes them amongst a set of workers using round robin load balancing, and then routes the responses from workers back to the corresponding front end component. The broker is a central piece of the routing

---

[2] https://developers.google.com/protocol-buffers.

[3] http://zeromq.org.

system, it has a fixed network address so that multiple front ends and workers may connect to it.

Workers are responsible for calculating the shortest route between two GPS coordinates. They are stateless and may be arbitrarily scaled, the Message broker handling the distribution of routing requests amongst all connected workers. Internally the workers translate the GPS coordinates into node identifiers in a routing graph, calculate the shortest route from the origin to the destination, and then translate the route waypoints into GPS coordinates and send them back to the message broker.

The segment speed simulator keeps track of how many vehicles are present on individual road segments and calculates a predicted speed of the vehicles based on this metric. The predicted speeds on the individual segments are periodically broadcast to all workers to refine their routing decisions.

To test the behavior of the system in scenarios with a large number of simultaneous client connections, we have developed a vehicle simulator that can simulate tens of thousands of vehicles on a single node. Each vehicle follows a specific route, while it periodically requests a route update from the routing system. It also requests the current predicted speed of the active road segment from the Segment speed simulator to adjust its driving speed.

The whole routing system can be deployed either on the cloud (using a containerised Docker build) or on an HPC cluster (we provide a PBS script for a simple deployment).

5   Validation Experiments

We have validated the approach implemented in the server–side navigation service by simulating a number of cars driving in a virtual world along routes provided by the navigation service. The virtual world uses a graph representation of a real-world traffic network while the speeds assigned to each segment are derived from an approximate model of the traffic load.

The performance tests were performed on a scenario located in the capital of the Czech Republic, Prague. We generated a set of origin and destination points in order to simulate a heavy traffic situation. The origin and destination locations are visible in Fig. 3. Destination points of the cars are selected such that each car passes through most of the city in the west-east direction. All cars depart within the first 30 s from the simulation's start. The rerouting interval, after which the car requests a new path, is also set to 30 s. Origin and destination locations of the cars simulated in the following experiments were sampled from this data set.

5.1   Scalability and SLA

The goal of this experiment was to verify the scalability of the distributed architecture. We expected that service response time would be reduced when number of compute resources was increased. The secondary goal of this experiment was

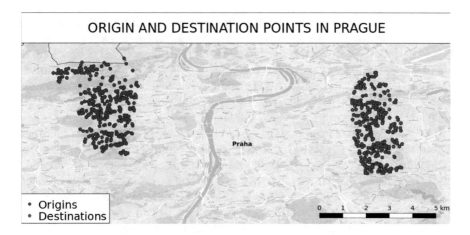

Fig. 3. Clouds of origin and destination points in Prague

to observe the SLA at various levels in order to estimate the operation cost of the service. Verification of the scalability of the routing service was demonstrated on three sets of experiments that differed in the number of simulated cars.

Table 1. Values for the three sets of experiments

Number of cars		Number of workers				
		1	2	3	4	5
10 thousand	Mean	48.00	30.41	24.10	22.06	20.98
	Std	56.71	71.34	31.95	29.20	32.35
	50.0%	24.00	18.00	16.00	16.00	15.00
	75.0%	64.00	28.00	23.00	22.00	21.00
	90.0%	113.00	66.00	45.00	37.00	35.00
	95.0%	154.00	94.00	74.00	64.00	60.00
	99.0%	281.00	171.00	134.00	125.00	102.11
	99.9%	502.00	420.92	346.00	381.06	401.00
20 thousand	Mean	67.76	35.58	27.38	21.39	21.62
	Std	87.20	54.20	38.01	33.32	30.87
	50.0%	42.00	18.00	16.00	14.00	15.00
	75.0%	82.00	35.00	26.00	20.00	20.00
	90.0%	135.00	80.00	63.00	35.00	35.00
	95.0%	201.00	120.00	89.00	62.00	62.00
	99.0%	448.00	245.00	186.00	149.00	144.00
	99.9%	918.98	673.00	384.00	417.00	395.00
30 thousand	Mean	108.77	43.13	31.45	27.10	19.44
	Std	216.26	129.10	52.83	42.73	28.77
	50.0%	62.00	18.00	15.00	15.00	13.00
	75.0%	113.00	40.00	26.00	23.00	19.00
	90.0%	180.00	93.00	75.00	65.00	31.00
	95.0%	324.00	132.00	118.00	97.00	59.00
	99.0%	1,087.00	345.00	240.00	208.00	136.00
	99.9%	2,720.76	2,436.40	647.00	468.00	381.00

The base scenario consisted of thousands of cars uniformly distributed across the Prague city road network (10, 20, and 30 thousand cars). The metric used to observe scalability was the response time of the routing service, i.e. how long did it take for the simulated car to obtain a new route from the system. The response time was monitored for an increasing number of workers for different amounts of cars. Selected statistics obtained from the individual runs are presented in Table 1 and visualised in graphs in Fig. 4.

The results confirm the scalability of the service. It is clear that the mean response time gets lower if the number of workers running on separate computing nodes is increased. The experiments also confirm that the service is capable of handling at least 30,000 cars driving in the traffic network of a major city. The worst value of the 99.9% SLA in this case is around 2.5 s, which is still significantly lower than the 30 s rerouting interval. The graphs in Fig. 4 also confirm that most of the requests were serviced in a much shorter time than the actual SLA value.

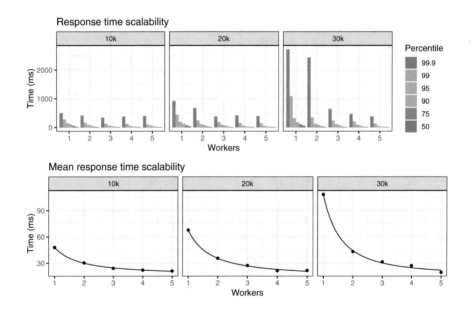

Fig. 4. Selected response time percentiles and mean for 10, 20, and 30 thousand cars, and 1 to 5 workers running on separate compute nodes

6   Conclusion

We have developed a distributed server-side navigation service which provides optimal routes based on a global view of the traffic network. The service can be used to optimise traffic in a smart city where it can easily be used by drivers via a smartphone app or an in-car navigation system. We have designed a traffic simulator with a virtual world which can be used to verify the optimisation capabilities of the service, and to test its performance. The focus of this paper was on the latter. We have deployed both the service and the simulator on an HPC cluster to test its performance scalability.

The service is designed to respond to the incoming routing requests in a matter of a few hundred milliseconds, operating on a major city road network (Prague or similar). We have stated that this requirement can be relaxed, since the cars do not need to obtain new routes more often than for example 30 s. We have verified this capability by running a set of traffic scenarios which mimic a real-world traffic situation in the city of Prague.

By observing the distribution of the response time we have confirmed that the service is scalable and is capable of providing short response times. Increasing the number of routing workers assigned to serve the requests caused the high percentile (99.9%) of the response time to decrease. Even in the worst scenario, with only 1 worker assigned, the system was able to serve 30,000 cars with response times that were significantly shorter than the rerouting interval. This result provides us with an opportunity to execute even larger and more complicated traffic scenarios in which we can artificially create obstacles in the network and observe the traffic behaviour.

In the future, we plan to focus on experimenting with the traffic optimisation itself. We plan to experiment with different route reordering strategies, and with the ratio of cars using the service that is necessary to attain successful traffic optimisation in a smart city.

Acknowledgements. This work was supported by The Ministry of Education, Youth and Sports from the National Programme of Sustainability (NPS II) project 'IT4Innovations excellence in science - LQ1602', by the IT4Innovations infrastructure which is supported from the Large Infrastructures for Research, Experimental Development and Innovations project 'IT4Innovations National Supercomputing Center – LM2015070', and partially by the SGC grant No. SP2019/108 'Extension of HPC platforms for executing scientific pipelines', VŠB - Technical University of Ostrava, Czech Republic.

References

1. Behrisch, M., Bieker, L., Erdmann, J., Krajzewicz, D.: Sumo – simulation of urban mobility an overview. In: SIMUL 2011, The Third International Conference on Advances in System Simulation, pp. 63–68 (2011)
2. Cowley, R., Joss, S., Dayot, Y.: The smart city and its publics: insights from across six UK cities. Urban Res. Pract. **11**(1), 53–77 (2018). https://doi.org/10.1080/17535069.2017.1293150
3. Fellendorf, M., Vortisch, P.: Microscopic traffic flow simulator VISSIM. In: Fundamentals of Traffic Simulation. International Series in Operations Research & Management Science, vol. 145 (2010)
4. Golasowski, M., Bispo, J., Martinovič, J., Slaninová, K., Cardoso, J.M.: Expressing and applying C++ code transformations for the HDF5 API through a DSL. In: IFIP International Conference on Computer Information Systems and Industrial Management, pp. 303–314. Springer, Cham (2017)

5. Golasowski, M., Tomis, R., Martinovič, J., Slaninová, K., Rapant, L.: Performance evaluation of probabilistic time-dependent travel time computation. In: IFIP International Conference on Computer Information Systems and Industrial Management, pp. 377–388. Springer, Cham (2016)

6. Horni, A., Nagel, K., Axhausen, K.W.: Introducing MATSim. In: The Multi-Agent Transport Simulation MATSim. Ubiquity Press (2016)

7. Jeong, J., Jeong, H., Lee, E., Oh, T., Du, D.H.C.: SAINT: self-adaptive interactive navigation tool for cloud-based vehicular traffic optimization. IEEE Trans. Veh. Technol. **65**(6), 4053–4067 (2016). https://doi.org/10.1109/TVT.2015.2476958

8. Ma, J., Smith, B.L., Zhou, X.: Personalized real-time traffic information provision: agent-based optimization model and solution framework. Transp. Res. Part C: Emerg. Technol. **64**, 164–182 (2016). https://doi.org/10.1016/j.trc.2015.03.004. http://www.sciencedirect.com/science/article/pii/S0968090X15000832

9. Ptošek, V., Ševčík, J., Martinovič, J., Slaninová, K., Rapant, L., Cmar, R.: Real time traffic simulator for self-adaptive navigation system validation (2018)

10. Seo, T., Bayen, A.M., Kusakabe, T., Asakura, Y.: Traffic state estimation on highway: a comprehensive survey. Annu. Rev. Control **43**, 128–151 (2017). https://doi.org/10.1016/j.arcontrol.2017.03.005. http://www.sciencedirect.com/science/article/pii/S1367578817300226

11. Silvano, C., Agosta, G., Bartolini, A., Beccari, A.R., Benini, L., Besnard, L., Bispo, J., Cmar, R., Cardoso, J.M., Cavazzoni, C., et al.: The ANTAREX domain specific language for high performance computing. arXiv preprint arXiv:1901.06175 (2019)

12. Smith, L., Beckman, R., Anson, D., Nagel, K., Williams, M.: TRANSIMS: transportation analysis and simulation system. In: National Transportation Planning Methods Applications Conference (1995)

13. Tomis, R., Rapant, L., Martinovič, J., Slaninová, K., Vondrák, I.: Probabilistic time-dependent travel time computation using monte carlo simulation. In: Kozubek, T., Blaheta, R., Šístek, J., Rozložník, M., Čermák, M. (eds.) High Performance Computing in Science and Engineering, pp. 161–170. Springer, Cham (2016)

14. Vitali, E., Gadioli, D., Palermo, G., Golasowski, M., Bispo, J., Pinto, P., Martinovic, J., Slaninova, K., Cardoso, J.M., Silvano, C.: An efficient Monte Carlo-based probabilistic time-dependent routing calculation targeting a server-side car navigation system. arXiv preprint arXiv:1901.06210 (2019)

Smart Scheduling Strategy for Lightweight Virtualized Resources Towards Green Computing

Alberto Scionti[(✉)], Carmine D'Amico, Simone Ciccia, Yuanyuan Li,
and Oliver Terzo

LINKS Foundation, Turin, Italy
{alberto.scionti,carmine.damico,simone.ciccia,yuanyuan.li,
olivier.terzo}@linksfoundation.com

Abstract. Modern cloud orchestrators are generally designed to make efficient use of resources in the data center, by consolidating the servers workload. Recently, energy efficiency has become critical factor to sustain the growth of cloud services; thus, more effective resource allocation and management strategies are required. The situation is exacerbated by introduction of HPC-oriented cloud services, where other aspects of the application execution are critical, such as the minimisation of the makespan. Although a short makespan allows for a rapid application execution, often the overall energy consumption of the whole cluster suffers, growing out of all proportion. Starting from the growing attention paid in recent years to the concept of "green computing" (or ICT sustainability), in this paper we propose a different type of resource scheduler, whose main objective is to maximise the (energy) power efficiency of the computational resources involved, while taking into account the overall application execution time. An artificial intelligence (AI) technique, in the form of population-based evolutionary algorithm, was used to develop the proposed scheduler, in order to find the best possible combination between tasks to be performed and usable nodes able to guarantee lower (energy) power consumption and, at the same time, the fulfilment of possible constraints related to tasks' execution. This paper focused on the implementation and evaluation of an evolutionary algorithm for efficient task scheduling. Experimental evaluation of such algorithm is discussed.

1 Introduction

Cloud computing (CC) paradigm constitutes the key technology element at the basis of any modern application and service. Cloud computing provides almost infinite computing and storage resources, which can be leveraged for processing and archiving massive amounts of data. Converged platforms, i.e., computing infrastructures which hardware and software stack is optimised for running both compute-centric (also referred to as HPC-oriented) and data-centric (also referred to as Big-Data-oriented) applications, are gaining momentum. Indeed, the large availability of low cost IoT devices, enables end users to continuously

© Springer Nature Switzerland AG 2020
L. Barolli et al. (Eds.): CISIS 2019, AISC 993, pp. 305–316, 2020.
https://doi.org/10.1007/978-3-030-22354-0_28

generate new data feeding cloud data centers. To cope with the ever growing demand for processing such massive data-sets, cloud providers are pushed to expand their infrastructures, as well as making them ever more heterogeneous. However, improving cloud infrastructures by means of additional (heterogeneous) hardware, comes at the cost of adding complexity to the management software stack. In this context, the role of "orchestration" software is changing over time: orchestrators, nowadays, are demanded to govern resources that are more diverse with regards to the past, and more distributed (also, geographically – Fog computing [1]).

Cloud computing brings the capability of exploiting massive computational resources to the masses, avoiding to incur high costs for acquiring and maintaining expensive machines. Aiming at supporting ever more diverse workflows and applications belonging to different domains, cloud providers and hyperscaler introduced dedicated offering based on dedicate hardware accelerators. Support for specialized hardware is growing, with hardware devices ranging from discrete GPUs to FPGAs and customised ASICs. For instance, Amazon AWS- EC2 [2] or Microsoft Azure [3] offer GPU-accelerated instances covering the most recently introduced architectures. Hyperscalers also started cooking their own custom chips to offer best in class performance for specific workloads (e.g., Google TPU [4] and Amazon Graviton processor [5]) and enabling energy saving, although their integration in the software stack pose several challenges [7]. Indeed, the added value of adopting such dedicate architectures at scale is twofold: (i) on one hand, they provide larger speedup in executing application code compared to general purpose processors (CPUs); (ii) they are far more energy efficient than their CPU counterparts.

Workload orchestration is a popular cloud approach to provide management features over: (i) reserving computing resources for the incoming jobs (i.e., virtual machines –VMs, Linux containers –LCs to launch); (ii) dynamically adapting the workload distribution to satisfy some specified criteria. Moving towards "green" management policies, cloud resource allocation algorithms need to be able to schedule VMs/LCs deployment in such way the overall (energy) power efficiency is maximised. Power (energy) management within allocation algorithms was firstly introduced to minimise the number of running servers [6]. A server in idle state consumes up to 65% of its peak power, thus, power consumption can be minimised by concentrating the workload on the minimum number of physical nodes and switching off unused ones. Following this approach, authors in [8] have proposed heuristics for dynamic adaptation of a VMs/LCs' allocation policy at run-time. According to the current resource utilisation, their heuristic applies live migration and idle node sleep mode switching. The aforementioned approach was tested on a data center composed of "heterogeneous" machines and took into account the characteristics of each physical node to reduce the overall energy consumption. Whenever dynamism is considered, also time required by a resource allocation algorithm to take decision becomes a critical factor. Due to the complex nature of the problem (resource allocation problem can be map on a NP-complete problem [9] – online bin-packing problem), heuristics can be used to quickly extract feasible solutions. The need of fast resource allocation algorithms is also emerging in relation to the ever large adoption of lightweight execution environments such as Kubernetes, where Linux containers (LCs) are preferred over traditional VMs to enable faster and more dynamic workflows execution. Despite several works tackled the problem of integrating (energy) power awareness into allocation algorithms, the most used ones still remain based on simple, fast approaches, such as First Fit (FF) and Best Fit (BF).

This paper presents a comprehensive study of an (energy) power-aware orchestration approach, which exploits the benefit and advantages of evolutionary computing techniques (inherently parallelization, historical behaviour embedding, etc.) to provide a fast and efficient scheduling for lightweight virtualized jobs (VMs/LCs). By mapping the allocation problem to the online bin-packing one, the proposed algorithm uses multiple concurrent agents to sample the optimisation search space. Proposed heuristic is also of interest as a potential candidate for being integrated as an (energy) power-aware scheduler into Kubernetes [10]. When compared to First Fit (FF) and Best Fit (BF) approaches, the proposed solution is better by far, allowing to save up to ~26.1% of (energy) power. Compared to the use of commercial solvers, the proposed algorithm provides near-optimal allocation in ~210 s.

2 Orchestrating Large-Scale Infrastructures

Aiming at supporting a new class of applications in modern data centers (i.e., deep learning and scientific computing applications), advanced architectural solutions comprising both hardware and software components at different levels must be leveraged. At the hardware level, this reflects in the massive adoption of specialized devices, as well as a large differentiation of the processor families. On the other hand, at the software level, heterogeneity reflects in emerging frameworks (e.g., TensorFlow, Caffe, etc.) and management components (e.g., OpenStack Cyborg). Despite these efforts, (energy) power efficiency management still remain a challenge. Techniques, both hardware (e.g., dynamic voltage and frequency scaling – DVFS) and software (e.g., dynamic migration of VMs/LCs) can be applied to reduce the power consumption without negatively impacting on the performance. Resource orchestrator (RO) is a key component in cloud data centers to efficiently manage the underlying resources, by means of ever more sophisticated algorithms. RO can be defined as the set of operations that cloud providers undertake (either manually or automatically via dedicated computer programs) for selecting, deploying, monitoring, and dynamically controlling the configuration of hardware and software resources, reflecting in a set of QoS-assured components seamlessly delivered to end users.

Figure 1 shows the main flow of operations performed by the RO. There are two main components that have a key role in managing the infrastructure: (i) the resource selection (RS) module; and the (ii) dynamic workload balancing (DWB). The former is responsible to find a match between available resources and the incoming VMs/LCs. To this end, the RS accesses to an internal database

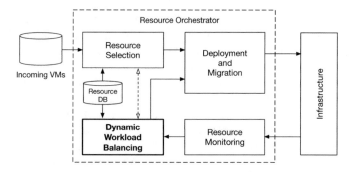

Fig. 1. Resource Orchestrator (RO): the main operation performed for dynamically controlling the Cloud infrastructure resources.

containing the updated description of the nodes in the infrastructure in terms of features (i.e., total number of cores, total amount of memory, storage, bandwidth, etc.) and the resources still available for allocating other VMs/LCs. Whenever the resource has been selected, a dedicated agent (deployment and migration – DM) performs all the necessary actions to boot-up the VM/LC on the node. The resource monitor (RM) is in charge of acquiring all the metrics needed to measure the resource usage, and updating the internal RO database. Information gathered by RM are used by the DWB component for periodically re-balancing the whole workload. Given the size of a large-scale infrastructure, dynamic reallocation of workloads requires fast efficient algorithms. The result of the DWB algorithms is the schedule of VMs/LCs that must be migrated. The periodicity of workload re-balance depends on several factors, including the number of nodes and VMs/LCs to take into account, the complexity of the DWB algorithms, and cloud provider specific constraints.

2.1   Data Center Infrastructure

A modern data center relies on several hardware and software components to ensure adequate capabilities for end user applications. The hardware level is composed by commodity systems (i.e., servers, storage nodes and network switches), with server nodes eventually augmented by accelerators. Interfacing with such accelerating devices is made through global memories and drivers integrated into the operating system. Global memory is implemented as a dedicated (on-board) block of fast memory. On top of the operating system, the virtualization layer is implemented, providing the functionalities that permit both the creation and migration of VMs/LCs, and to orchestrate resources. Resource orchestration is achieved by distributing the management logic, monitoring subsystem and deploying agents among various nodes.

3   Energy-Aware Cloud Resources Allocation Engine

To enable cloud orchestrators allocating data center resources according to a (energy) power saving policy (i.e., energy-aware resource allocation), we mapped the allocation problem to the online bin-packing problem [11] and implemented an evolutionary-based heuristic to solve it. This heuristic, namely the Energy-aware Cloud Resources Allocation Engine (ECRAE), considers that VMs/LCs have an associated cost that is expressed by the amount of resources to access (i.e., number of cores, amount of memory, storage, etc.) and the amount of instructions to be consumed (expressed in MIPS, i.e., millions of instructions per second). MIPS are also used to determine the VM/Linux container lifetime. The more MIPS the hosting node can process, the less time the VM/Linux container will last. VM/Linux container lifetime is used to measure the actual of amount of energy consumed by a given VM/Linux container when running on a specific node.

3.1   Evolving the Optimal Workload Schedule

The term *Swarm Intelligence* (SI) indicates the ability of a group of individuals to collaborate in order to solve a common problem. Each individual has local-level information about the solution space and it is guided by a set of rules to explore such environment. A series of interactions with other individuals (population), however, allows it to create a global knowledge about the problem addressed. This knowledge is what is regarded as "intelligent behavior". *Swarm-based optimization algorithms* (SOAs) are part of the broader domain of artificial intelligence algorithms (AI). At the basis of SOAs there is the idea of mimicking the concept of SI, using a stochastic approach and population-based heuristic. One important aspect of such stochastic-based heuristics is that they do not ensure to find the global optimum (i.e., maximum or minimum of the objective function) but are faster than other approaches in exploring the search space. Furthermore, since they are stochastic-based methods, different solutions can be discovered across multiple executions of the algorithm, even starting from the same initial conditions.

Bees Algorithms (BAs) belong to a particular class of SOAs that takes inspiration from the behaviour of bees, in particular on their foraging behaviour, when they need to find the best food sources. These are very similar to other nature-inspired classes of algorithms, the *Evolutionary Algorithms* (EAs) and the *Ant Colony* (AC) algorithms. One of the main advantages of using a BA is that it allows to perform both an exploration of the entire solution space and an exploitation of the most promising solutions found. In nature the foraging process involves the use of scout bees that look for promising flower patches (exploration phase). When these are found, the scout bees start to have a better look at these locations (exploitation phase). Then they come back to their hive to interact with each other, starting with another foraging phase based on the previous locations found (Fig. 2).

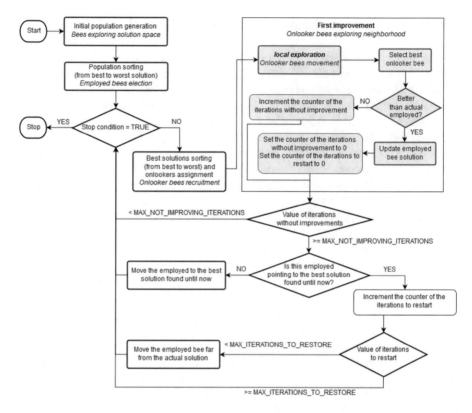

Fig. 2. Structure of the evolutionary algorithm (*bees algorithm*) used to dynamically optimise the resource allocation for a given input workload.

At the basis of the bees algorithm used in our work, there is a group of candidate solutions (*individuals* or *bees* in our case), each representing a possible workload schedule (i.e., a way of assigning tasks to available compute nodes, still fitting in the constraints of the optimisation problem formulation). In this work, the resource allocation problem has been formulated as a mixed integer linear programming (MILP) model [11]. The basic steps are as follows. Before, starting to explore the search space the population is initialised using a *greedy function*, that allows to have initial random solutions. These schedules are evaluated to calculate their corresponding *fitness* (i.e., in our case it corresponds to the overall (energy) power consumption of the data center). It follows the main loop, where through several iterations the best candidate solution is extracted. To this end, in each iteration all the solutions found in the previous iteration are ranked from the best to the worst, then each individuals are spread in the solution space, allocating the majority of them to the most promising locations, while keeping few ones still as *scouts*. The former exploit the locations previously found by performing one among the 5 possible mutations that are described below. The latter ones, instead, use the greedy function to create a new feasible solution, trying to further explore the whole solution space. Each new candidate is evaluated and if a new best ever solution is found, then current best solution is updated accordingly. The algorithm stops if one of the *stopping criteria* is met. As stopping criteria, we set the maximum number of iterations of the evolving loop (we also set a second stopping criteria concerning maximum elapsed time set to 300 s). Whenever the algorithm exits the main loop, the best ever found solution (*best candidate*) is compared with the initial allocation (*current schedule*) to determine the list of VMs/LCs to migrate. To ensure high quality of the solution, the algorithm also applies a final consolidation step: it tries to saturate server resources, thus to increase the number of idle nodes that can be switched off. If the consolidation resulted in a worse schedule, the best ever candidate solution is preserved and used. Also, the algorithm tracks the number of generations that are elapsed since the last best ever candidate found. Whenever such number of generations exceeds a predefined threshold, a steady state condition is detected. In that case, the current population is discarded, and it is reinitialised (population restart) using the greedy function to generate new candidates. Such new population is then mutated as described above.

Data structure. The proposed BA uses dedicated data structures to effectively represents a workload schedule and to keep track of allocated resources in the datacenter. To represent and efficiently manipulate candidate solutions, each individual is composed by 3 vectors:

- *Data center* **D**: is an array whose length equals the number of servers of the datacenter. Each element i of the array ($D_i = server$) contains a structure representing a server. A server is represented by the number of its CPU's cores, by its amount of RAM, by the maximum power that it can consumes and by the percentage of power consumed when it is in idle. Each server is identified by its index in the datacenter array.
- *Workload* **W**: is an array whose length equals the number of VMs/LCs to allocate. Each element i of the array contains a structure representing a task. A task is represented by the CPU and the RAM that it needs to be executed, and it is identified by its index in the workload array.

Greedy Function. One of the main requirements for having a good meta-heuristic is the possibility to generate random solutions, that allow to explore the whole solution space. In our work we implemented a greedy function for this purpose, which tries to create a new random feasible solution by selecting, for each task not yet allocated, the most promising server where it can be allocated. This last one is represented by the server that, executing the selected task, increases less the total power consumption of the whole datacenter, while preserving the feasibility of the solution.

Mutation Operators. Bin packing problem, such as the one concerning resource allocation, requires the implementation of dedicated mutation operators to allow the algorithm exploit a local solution. Unlike basic mutation operation used in evolution strategies and genetic algorithms, where each element of the solution array is changed by adding a small random quantity (i.e., generally, a random variable with Gaussian distribution is used), here the operators must preserve the correctness of the workload schedule. At each iteration only one of the implemented operators is used. The decision on which of these to choose is made using a *tabu list*. This last structure is updated from time to time by inserting those operators that have led to a worsening of the current solution, thus favouring the operators who have instead led to an improvement of the current scheduling. Given the data structure described above, we designed 5 different genetic operators that ensure to preserve correctness of the candidates:

- Task swapping (TSWP): this operator aims at discovering better VMs/LCs fitting configurations. To this end, the operator randomly selects one task from the W list and search among the others one task that can be exchanged with.
- Task first-fit consolidation (TFFC): this operator aims at finding allocation configurations in which tasks are more consolidated on less running servers. To this end, the operator randomly selects one task that can be moved on another server. The server is selected by scanning the list of all servers and selecting the first one that has enough resources to accommodate the moving task.
- Task best-fit consolidation (TBFC): the TBFC operator is very similar to TFFC operator, in which respect it uses the 'best node' among all the ones that have enough resources to accommodate the moving task. The best node is represented by the one with the largest unused amount of resources, both in terms of CPU cores and free memory.
- Server consolidation (SC): This operator aims at saturating the available resources of a given selected server. The idea behind its use is to make room on other servers for moving in a subsequent iteration larger tasks. To this end, the operator randomly selects one server on the server list D and iteratively check if a task (not yet allocated on the selected server) can be moved on it, still satisfying the constraints of the problem, both in terms of CPU and used memory. The tasks to move on the selected server are selected iteratively by scanning the **W** list.
- Server load reduction (SLR): This operator works in an opposite direction compared to SC operator. In fact, given a randomly selected server i, the operator tries to redistribute the whole load on other servers. To this end, for each task assigned to the selected server, its required resources R_c and R_m are compared to those of other servers (i.e., D_c, D_m for a server $k \neq i$). Thus, the server list D is scanned to search for a server k that can accommodate partially or the whole load of server i.

4   Simulations

Aiming at evaluating the effectiveness of the proposed allocation algorithm, we performed a set of experiments using an in-house simulation framework. The BA scheduling algorithm has been implemented using a modern high-level language, i.e., Go language [12] supporting concurrency and parallelization of the code. We opted for Go language since it allows to design code according to modern design patterns, and to generate compact and efficient binaries compiled natively for the host machine. Our simulation framework was composed of a software module used to generates initial workloads according to a given distribution of VMs/LCs and servers availability, a module to import such workloads within the BA scheduler, and the BA binary. The scheduling algorithm has been tested using different data center configurations, which are aimed at showing scaling capability of the proposed solution. During the workload generation process, different types of VMs/LCs were randomly picked up according to a given probability distribution and assigned to the nodes. The nodes were selected, among the ones that had enough resources to host the new VMs/LCs. According to Mazumdar et al. [11], the generation process iteratively selects nodes, one at a time, and tries to saturate their resources.

Various aspects influence the generation process: (*i*) *the distribution of VMs/Linux containers*: since small virtual instances (i.e., requiring less resources to run) are more frequent, we tuned the probability of selecting them higher than that of larger instances; (*ii*) *the resource saturation level*: in each simulated server, we reserved a fraction of its resources to run management level software (thus such resources were not available for scheduling VMs/Linux containers); (*iii*) *the type of VMs/Linux containers and type of servers*: we defined virtual instance (i.e., VMs/LCs) and server features in such way they were representative of real workloads and data centers; (*iv*) *optimization model constraints*: we restricted the constrained resources to only CPU and memory, since they are responsible for the majority of the (energy) power consumption in a real server machine.

4.1 Experimental Setup

We assessed the capability of the proposed BA-based scheduler by simulating the allocation of a variable number of VMs/LCs in a range from 150 to 1000+. Also, the size of the data center was varied in a range from 50 to 1000 nodes. For each run we launched the heuristic multiple times (trials) to cope with the stochastic nature of the BA algorithm; after the set of trials (we opted for running 10 trials for each configuration of the workload and data center) we kept the best overall solution as the final schedule. Worth to note, during the first trial we also generated the workload and saved it on disk. The subsequent trials reused the recorded initial workload schedule, since it was used as the initial starting condition. To generate the workloads we defined 15 types of hosting servers and 13 types of VMs/LCs. Servers had resources in the range of 2 CPU cores, 8 GiB of RAM to 20 CPU cores and 128 GiB of RAM. Power consumption of such nodes were in the range of 30 W to 500 W. Similarly, virtual instances have been defined in such way the required resources in a range from 1 CPU core and 1 GiB of RAM to 12 CPU cores and 48 GiB of RAM. We ran all the experiments on a server machine equipped with an Intel Xeon E5-2630 (supporting up to 20 parallel threads) and 128 GiB of RAM. The used server ran Linux CentOS 7 operating system.

4.2   Experimental Results

In order to assess the capability of the proposed scheduling algorithm to provide (energy) power improved solutions over traditional energy-unaware algorithms (e.g., First Fit (FF), Best Fit (BF), etc.), we performed a set of experiments aimed at verifying algorithm convergence and performance. Convergence
expresses the capability of the algorithm to provide solutions that are closer to
each others (and close to the optimum) among different runs (trials). Performance look at the capability of the algorithm to generate large (energy) power
saving with regards to the initial schedules. For all the experiments we set up
the maximum execution time to 300 s; interestingly most of the solutions (irrespective of the number of bee agents and threads used) have been reached in
less than 210 s, thus demonstrating the high efficiency of the proposed scheduler.
The average (energy) power saving ranged from 23.6% to 26.1% with regards
of the initial workload allocation (which is a good approximation of the feasible solutions provided by FF and BF algorithms), when a large workload was
considered (1000+ virtual instances). With small workflows the (energy) power
saving was even more pronounced. This demonstrate the effectiveness of the
proposed scheduling solution to improve energy efficiency and to tackle large
problem instances. The fast execution also makes the proposed scheduler as a
good candidate to replace standard Kubernetes scheduler in lightweight virtualization environments.

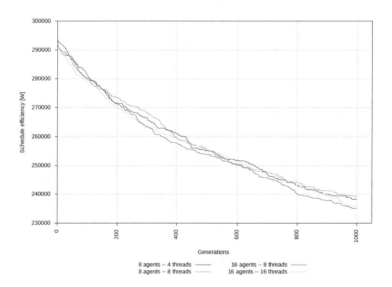

Fig. 3. Scheduler performance for different number of agents and execution threads.

Figure 3 shows the performance of the proposed BA scheduler when the number of bee agents and execution threads varied. As the 4 lines show, a more

pronounced difference in the generated solutions exists when the number of bee agents varies (yellow and blue lines vs. green and red lines). In that case, the higher the bee agent number was the larger power saving improvement was achieved. Indeed, more agents allowed for a better exploration of the search space, as well as reduced the likelihood of incurring in a local minimum of the optimisation function. Interestingly, when the number of agents was kept constant (irrespective of the number of running threads), the solutions generated by the scheduler were closer to each other in terms of power drawn of allocated virtual instances. This behaviour may be explained considering that the way Go language maps concurrent functions into system threads may hide performance gain of using more threads (Go maps n concurrent functions on m threads).

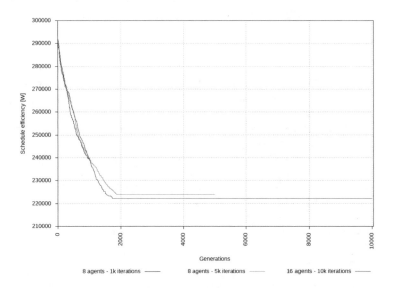

Fig. 4. Scheduler performance for different maximum number of iterations.

Figure 4 shows the performance of the proposed scheduling algorithm when the maximum number of iterations to perform varied. As expected, the larger the number of iteration was, the better the found solution was. However, it is worth to observe that the algorithm tended to reach the optimal solution very quickly (after $\sim$1900 iterations), spending the large fraction of the main iteration loop in a steady state (the algorithm also implements a steady state detection and restart mechanism). This analysis made over different workloads and data center configurations allowed us to tune the number of iterations (and thus the algorithm execution time) to its optimal value.

All these results clearly demonstrated the capability of the proposed scheduling algorithm to dynamically generate resource allocation schedules with a large improvement in terms of (energy) power saving and with a good stability over a series of runs.

5   Conclusion

The pace at which cloud services run by data centers and fog computing resources is growing, makes it critical finding effective approaches to keep (energy) power consumption of such large infrastructures under control. In order to correctly manage such vast computing and storage power, smart power (energy)-aware allocation policies should be implemented in the orchestration tools. This paper presents an energy-aware VMs/Linux containers allocation strategy. Being based on a population-based stochastic approach (bees algorithm – BA), the proposed approach allowed to achieve larger (energy) power saving when compared to broadly adopted policies, i.e., First Fit (FF) and Best Fit (BF). Experimental results confirm the effectiveness of the proposed algorithm when workloads of different sizes and types are executed on the data center.

References

1. Vilalta, R., et al.: TelcoFog: a unified flexible fog and cloud computing architecture for 5G networks. IEEE Commun. Mag. **55**, 36–43 (2017). https://doi.org/10.1109/MCOM.2017.1600838
2. Amazon Web Services (AWS) – Accelerated computing instances. https://aws.amazon.com/ec2/instance-types/
3. Microsoft Azure – GPU based instances. https://azure.microsoft.com/en-us/pricing/details/virtual-machines/series/
4. Jouppi, N.P., et al.: In-datacenter performance analysis of a tensor processing unit. In: 2017 ACM/IEEE 44th Annual International Symposium on Computer Architecture (ISCA). IEEE (2017)
5. Amazon AWS EC2 – Graviton Processor. https://aws.amazon.com/it/blogs/aws/new-ec2-instances-a1-powered-by-arm-based-aws-graviton-processors/
6. Pinheiro, E., Bianchini, R., et al.: Load balancing and unbalancing for power and performance in cluster-based systems. In: Proceedings of the Workshop on Compilers and Operating Systems for Low Power, pp. 182–195 (2001)
7. Beloglazov, A., Buyya, R.: Optimal online deterministic algorithms and adaptive heuristics for energy and performance efficient dynamic consolidation of virtual machines in cloud data centers. Concurr. Comput.: Pract. Exp. (CCPE) **24**(13), 1397–1420 (2012). John Wiley & Sons
8. Beloglazov, A., Abawajy, J., Ranjan, R.: Energy-aware resource allocation heuristics for efficient management of data centers for cloud computing. Future Gener. Comput. Syst. (FGCS) **28**(5), 755–768 (2012)
9. Casanova, H., Legrand, A., Yves, R.: Parallel Algorithms. CRC Press, Boca Raton (2011)
10. Burns, B., et al.: Borg, omega, and kubernetes (2016)
11. Mazumdar, S., Pranzo, M.: Power efficient server consolidation for cloud data center. Future Gener. Comput. Syst. **70**, 4–16 (2017)
12. Go Lang. https://golang.org

IoT Cloud Platform Based on Asynchronous Processing for Reliable Multi-user Health Monitoring

Nur Hakim Arif and Nico Surantha[(⊠)]

Computer Science Department,
BINUS Graduate Program – Master of Computer Science,
Bina Nusantara University, Jakarta 11480, Indonesia
{nur.arif,nico.surantha}@binus.ac.id

Abstract. Internet of Things (IoT) implementation has spread to various fields, including healthcare. Personal healthcare device such as ECG senses very sensitive voltage from heart and stream the data in high speed to the IoT platform. Data from ECG can be used for many disease diagnoses, including Arrhythmia and sleep disorder. In order to obtain the optimum diagnosis result, IoT platform for healthcare should be able to handle very rapid data stream from users or devices simultaneously and keep the successful connections at a high rate. In this paper, an IoT architecture for healthcare implementing asynchronous processing is proposed. Asynchronous process allows the system to process the data outside the main loop of process. The proposed system measures HTTP successful connections and analytics accuracy. By implementing asynchronous processing, the system can increase the successful HTTP connection up to 50% for 100 users testing scenario compared to the benchmark system. The increase of successful connections improves the analytics accuracy to 98.41%.

1 Introduction

Electrocardiography (ECG) signal is the representation of electrical activity in heart captured over time using electrode placed on the skin. There are several diagnoses or findings can be made based on ECG signal, e.g. arrhythmia and sleep quality detection. Arrhythmia is a symptom for abnormal rhythm of a heartbeat that can lead to heart failure [1]. Arrhythmia detection only can be performed by expert in this field, i.e. Cardiologist or Specialist Doctor by analyzing the printed signal from Electrocardiography (ECG) machine. ECG machine captures the electric signal produced by heart and will be drawn in PQRST graphic. It took a long time to examine the type of arrhythmia from a patient's signal. Cardiologist has to match the ECG graphic with the arrhythmia patterns. ECG machine can only be operated by a doctor or a well-trained professional so the patient has visit the hospital to be examined.

Technology used in healthcare field is more advanced. Information Technology for health can be implemented on various aspects, i.e.: remote monitoring, fitness program and elderly intensive care [2]. Medical devices like sensors, diagnosis equipment and

© Springer Nature Switzerland AG 2020
L. Barolli et al. (Eds.): CISIS 2019, AISC 993, pp. 317–330, 2020.
https://doi.org/10.1007/978-3-030-22354-0_29

image recording device can be regarded as smart devices or objects linked to a bigger system called Internet of Things (IoT). IoT implementation in health services will reduce cost, improve quality of life, and enrich the user interactivity. In medical professionals' perspective, IoT implementation in healthcare can help their daily operation and give better and accurate services. For patients, IoT implementation can deliver medical services in a faster way and serve as their personal medical data repository. By using IoT, it is possible to monitor human vital functions wherever they are and whatever they are doing [3]. The telemetry captured from patients can be sent to the remote with low cost and the expert can be aware of patients' physical status in near real-time.

An IoT-based system which can monitor the ECG signals remotely in a low-cost manner was proposed in [4]. Similar research was conducted by [3–6]. In [4], the system was tested with simultaneous connected user and the system can serve up to 20 connections simultaneously; as the simultaneous connections grow, the error increases up to 50–60%. This research has no analytical feature. The ECG signals are only presented in graphical format in a web browser. A survey was conducted by [7] to compare several classification methods. The most popular algorithm for classifying arrhythmia was Support Vector Machine (SVM) conducted in [8–10], Artificial Neural Network (ANN) [11, 12], and Linier Discriminant [13, 14]. Those algorithms can achieve classification accuracy of 98.8%–99.15%.

This research focuses on reducing HTTP request error on IoT platform. To reduce error and enhance the capabilities of the system, this system will implement asynchronous mechanism to process the tasks. This system also implements Convolutional Neural Network to classify the inputted ECG signal to measure the significance of data loss reduction by maintaining successful HTTP connection. This evaluation will indicate whether or not the successful connections are affecting the analytics result. The system was designed to have partial analytics platform. This kind of separation will ensure that computational power is reserved for analytics when the system is overwhelmed by data.

2 IoT Platform for Health Monitoring

Referring to International Telecommunication Union, Internet of Things is a global infrastructure of a group of information connected through physical or virtual devices, and develops that information by using information and communication technology [15]. Data from the real world can be collected by utilizing sensors in Internet of Things. The data were then transmitted to the other components to analyze and the insight of data can be concluded. Nowadays, IoT has been implemented in many fields, i.e.: smart cities, traffic management, waste management, security, health, emergency services, logistics, retail and industrial control [2]. This implementation is enabled by the supporting technologies such as: sensors technology, networking, RFID, cloud technology, Big Data, and Artificial Intelligent [16].

Internet of Things in healthcare has a big potential to improve many medical applications, such as remote health monitoring, fitness program, chronic disease and elderly care [2]. Another potential application is the compliance with treatment and medication at home. In IoT for healthcare, devices used such as medical devices, sensors, and diagnostic and imaging devices can be considered as smart devices. The implementation of IoT in healthcare is expected to reduce costs, increase the quality of life, and enrich the user's experience. To reduce device downtime, healthcare providers can use IoT technology to optimize the use of device through remote provision. By implementing IoT, the service providers can provide more efficient scheduling of limited resources by ensuring their best use and service to more patients.

In [4], the proposed system can be accessed by several users simultaneously via the internet network. The system consists of ECG hardware, transmission module based on Zigbee and web server for data storage and web application. The ECG signal taken from the patients' body is acquired by the ECG machine, and then raw data is sent serially to the computer server through Zigbee network. Furthermore, the data can be accessed by other authorized parties via the web pages for the purpose of treatment or consultation. The testing result shows that the system can handle up to 20 users without errors. Meanwhile, for 50–150 users, some errors occur due to insufficient bandwidth or high data traffic on the server. The system is limited to 50 concurrent user connections. The system scaling scenario has not been proposed in this research. Another drawback of this system is the processing of the streamed data which is still synchronous. So, there will be a bottleneck when the data streamed to the system is very large.

A similar system was proposed in [3], the system will assist practitioners to monitor patients' condition. Sampling frequency for ECG signals is set to 128 Hz which is typically used for Holter devices. This system has 4 modes of data transmission, i.e. Real-time continuous transmission, Continuous transmission in particular periods, Event-triggered transmissions, and Transmission on patients' demand. Those transmission methods will affect the amount of data sent to the server. This system is a closed system; the data should be monitored by practitioners. The analytics component does not exist in this system and the doctor or practitioner should be present while the patients send the data especially in patients' demand transmission mode.

3 Proposed System

This research aims to provide an IoT platform architecture that can serve users simultaneously with high HTTP successful connections rate. The architecture will utilize the asynchronous processing in its main system. This research uses Python Celery as its asynchronous framework. By using Python Celery, the system can distribute its tasks to smaller workers outside the main system. The processing of tasks would not effect the main system's resource usage [17]. The asynchronous scheme will also be used to provide communication between end-user to the system by using WebSocket. The system consists of: (1) ECG Sensor Network, (2) IoT Cloud Platform, and (3) User Interface. The proposed architecture can be seen in Fig. 1.

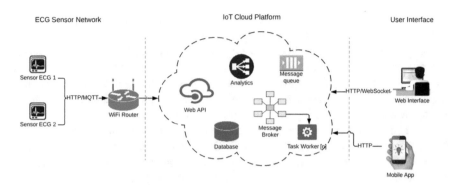

Fig. 1. Proposed system architecture

3.1 ECG Sensor Network

ECG Sensor used in this research is MySignals HW (eHealth Medical Development Shield for Arduino). This device deploys on top of Arduino Uno as a shield. This device is equipped with TFT LCD Display, an ESP8266 based WiFi module, Bluetooth Low Energy (BLE) module, and built in 17 sensors for measuring 20 biometrics parameters i.e.: SPO2 Pulse Oxygen in Blood Sensor, ECG (Electrocardiogram) Sensor, Airflow Sensor, Blood Pressure Sensor, Glucometer Sensor, Body Temperature Sensor, EMG (Electromyography) Sensor, Spirometer Air Capacity Sensor, GSR (Galvanic Skin Response) Sensor, Body Position Sensor, Snore Sensor, Body Scale BLE Sensor, SPO2 Pulse Oxygen in Blood BLE Sensor, Blood Pressure BLE Sensor, Glucometer BLE Sensor, MindWave EEG (Electroencephalogram) Sensor, Body Temperature BLE Sensor, Alarm/Emergency Button BLE Sensor, shown in Fig. 2. To communicate to the IoT Cloud Platform, this device uses WiFi transmission. RestAPI will be used for communication protocol to IoT Cloud Platform. For every signal captured by this device, a single ReSTAPI requested to the IoT Cloud Platform. A ReSTAPI request can be described as follow:

```
POST http://apiplatform:8080/api/measurement
Accept: */*
Cache-Control: no-cache
Content-Type: application/json

{
  "id":12,
  "value": -0.65,
  "time": 1542973287843,
  "user_id": 12,
  "session_id": 334
}
```

LCD
Display

BLE Module

WiFi Module

Sensor/Input

Fig. 2. Sensor module

3.2   IoT Cloud Platform

The IoT Cloud Platform in this research proposes asynchronous process to handle the inputted data. In order to enable the processing to become asynchronous, the system requires message broker to distribute the task among its Workers [17]. The Workers will process outside the main loop of the system. Message broker will communicate using Advanced Message Queueing Protocol (AMQP). The Message Broker will register the task that can be performed by Workers. When a task has been triggered, the Message Broker will broadcast the message to the workers, and that message will be consumed by idle Workers. The task will be stated as STARTING. When a task has been processed by a Worker, the same Worker will call back the Message Broker and return FINISHED state. In case of error, the returned state will be FAILED. Asynchronous communication between components in this system is shown in Fig. 3.

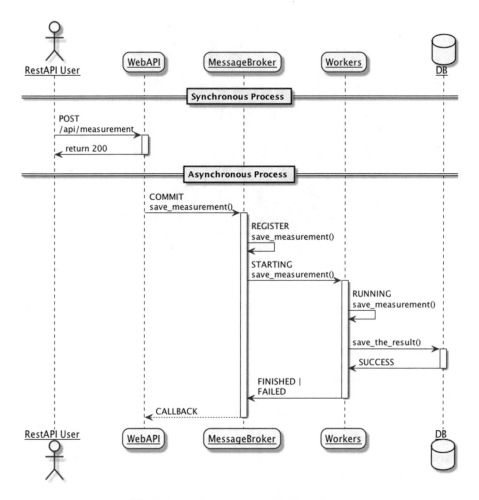

Fig. 3. Asynchronus communication scheme

To communicate with the end-user, system will provide HTTP and WebSocket protocol. HTTP protocol will be used for Request-Response communication scheme. Most web interface using this kind of communication. To get real-time communication, such as monitoring streamed ECG signal from ECG Sensors, this system uses Publish-Subscribe communication schema. For Publish-Subscribe communication schema, this system will use WebSocket. By utilizing WebSocket, data will be streamed to end-users without any interference from users or pushed by server without any user requests [18].

Analytics component in this architecture will diagnose the ECG signals. The streamed ECG signals will be segmented using Pan-Tomkin algorithm [19]. The segmentation process yields an array of signals of a QRS Segment. Analytics will classify the QRS segments into arrhythmia class, i.e.: Ischemic changes (Coronary Artery Disease), Old Anterior and Inferior Myocardial Infarction, Sinus Tachycardy,

Sinus bradycardy, Ventricular Premature Contraction (PVC), Supraventricular Premature Contraction, Left and Right bundle branch block, Left ventricular hypertrophy, Atrial Fibrillation or Flutter. For classification, this research implements Convolutional Neural Network as its classification algorithm. This algorithm can achieve 94.08% accuracy by training the model using Physiobank MIT-BIH data [20].

The network consists of 9 layers including 3 convolutional layers, 3 max-pooling layers and, 3 connected layers. On the convolutional layer (layer 1, 3, 5), these layers convolved with their respective kernels using the following formula

$$x_n = \sum_{k=0}^{n-1} y_k f_k - k \qquad (1)$$

After every convolutional layer, a max-pooling operation is applied to feature maps. The max-pooling process will reduce the size of feature map. The parameters for the kernel size are obtained through brute force technique while the stride for convolution and max-pooling operation is set at 1 and 2. The leaky rectifier linear unit (LeakyRelu) is used as an activation function for layers 1, 3, 5, 7, and 8. The fully-connected layers consist of 30, 20, and 5 output neurons in layer 7, 8, and 9, respectively.

3.3 User Interface Layer

User Interface layer in this system is an interface for the end-user to the system. User interface provides two main functions, i.e.: (1) Application Management and; (2) ECG Data monitoring. Application Management module consists of: (1) User/Patient Management, (2) Session Management, (3) Data Analysis, and (4) Reporting. HTTP protocol will be used in this module. So, end-user will request for a resource to get the response from system. User/Patient management manages User as operator and Patients as observation object. The differences are on their demographic data attached to the objects. User can login to CMS and assigned with proper privileges while the Patients will record for their age, gender, and any required additional data. Every registered patient will be recorded for his/her heartbeat signal and can be distinguished by requesting for a Session ID at the first time. This Session ID will be sent to the API Cloud Platform along with session_id value on the request body.

The ECG Data monitoring module utilizes WebSocket protocol to communicate to end-users. The protocol is used to provide end-user a near real-time communication. The system can push the data to end-user without end-user request. When the ECG Sensor stream data to the API Cloud Platform, the platform save the data to the database by invoking save_measurement() task on the worker. On the other hand, platform will push the latest data to the particular User by using WebSocket, and the received data will be populated into ECG diagram automatically. The web interface in this system is shown in Fig. 4.

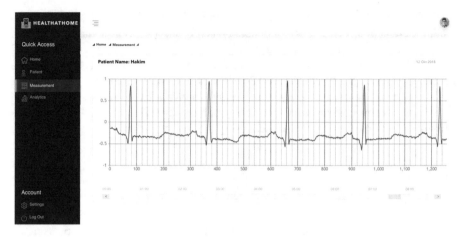

Fig. 4. ECG monitoring dashboard

4 Results and Discussions

4.1 Simulation Environment

In this research, benchmark system and proposed system will be deployed in the same environment and tested with the same dataset. Both proposed system and benchmark system are deployed into same machine. The specification of the machine is presented in Table 1. There are several deployment differences especially in the proposed system. The proposed system requires a Message Broker in order to run its asynchronous schema while the benchmark system does not. The benchmark system does not have Worker components; it is on its main program. The deployment schema for proposed system is shown in Fig. 5.

Table 1. Simulation machine specifications

Processor	Intel(R) Xeon(R) CPU @ 2.20 GHz
Memory	DIMM RAM 614 MB
Language	Python3.6 (Django, rabbitmq, WFDB Library)

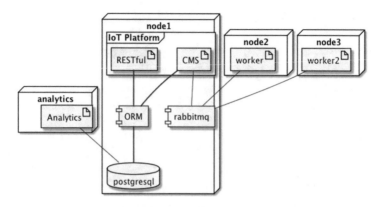

Fig. 5. System deployment program

4.2 Dataset

Dataset in this research consists of two kind of datasets: (1) Recorded signals from ECG sensor and; (2) MIT-BIH Dataset. Recorded signals from ECG are used to simulate the simultaneous connections from ECG Sensors to the IoT Cloud Platform. Recorded ECG Sensor dataset has no label or arrhythmia classes applied because this dataset is a raw data from the sensor. Recorded ECG Sensor has 5 minute-length data or equal to 108.000 signals. To reduce noises in this dataset, signal bandpass-filter is applied to the dataset. Figure 6. shows an example of raw and filtered Signal. To stream this dataset to the main system, the simple emulator was used by reading every signal in dataset and passing those data to the system using HTTP Requests.

MIT-BIH Dataset is a labeled dataset. Arrhythmia symptoms that occurred in every signal was annotated by cardiologists and stored in the record. In this research, this dataset is used for training the analytics model. This dataset comprised 48 half-hour excerpts of two-channel (two leads) ambulatory ECG recordings, obtained from 47 subjects [21]. The data are band-pass filtered at 0.1–100 Hz. The first-channel lead of 45 records is used, which is a Modified Lead II (MLII), leaving out records 102, 104 and 114, whose first-channel lead is not a MLII. The MIT-BIH database also provides annotations for each record.

In this research, Recorded ECG signal was used for evaluating the simultaneous connections test whilst the MIT-BIH dataset was used for testing the analytics components. The result of the simultaneous connections test is percentage of successful connection. For the analytic test, MIT-BIH dataset will be taken randomly according to the successful connection test result.

From Web API component, the streamed data will be pooled and segmented. The Analytics Component receives only filtered and segmented signal to be analyzed. For MIT-BIH data, every segment can contain one or more label/annotation, this annotation is used to mark the respective segment where an arrhythmia class occurred. The example of labeled/annotation ECG segment is shown in Fig. 7.

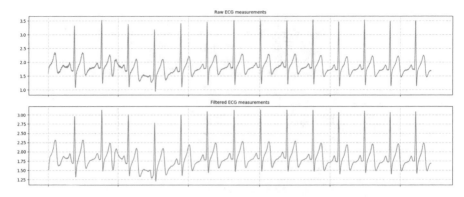

Fig. 6. Raw signals (top) and filtered signals (bottom)

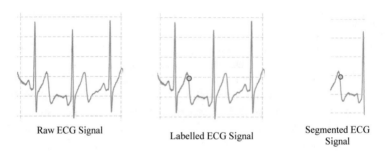

Raw ECG Signal   Labelled ECG Signal   Segmented ECG Signal

Fig. 7. Labelled/Annotated ECG segment

4.3   Simultaneous Connection Evaluation

To evaluate the system, a simulation from dataset was conducted. This simulation observes some parameters in the system, i.e. Successful Connection, and Response Time. These parameters were tested with 1–100 sensors that connect to system simultaneously. To perform this simulation, Apache JMeter set up in different machine in the same network. The streamed data is saved in internal database and preprocessed for analysis task. Each testing scenario becomes particular dataset and is trained separately, and the accuracy will be evaluated.

In this test, MIT-BIH data is streamed into IoT Cloud Platform. The number of simultaneous connections represents the number of streaming sensors to the system. Every HTTP Request to the server is considered successful when the server returns 200 response code; if returned 5xx response code, then it is counted as unsuccessful request. Based on the test result, Table 2 shows that the proposed system can handle 50 simultaneous users/sensors with only 2% error rate while the benchmark performance dropped into 20% successful connection only.

On 100 simultaneous connections, both systems falloff the performance, the proposed system reaches 60% successful connection and the benchmark system only reaches 11% successful connection. In the proposed system, the memory usage is raised when the queuing task increases. This is due to Message Broker saving the queue in memory by default. The message broker memory usage affects the system because it is consuming more than half of the system's memory.

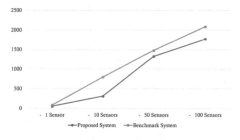

Fig. 8. Response time testing result

Table 2. Simultaneous connections test results

Num of Sensor	Proposed	Benchmark [4]
1	100%	100%
10	100%	94.65%
50	98.80%	21.69%
100	62%	11.41%

Another parameter measured in this test is Response Time. Response Time testing result is shown in Fig. 8. Both systems will become slower when the number of connected users increases. Benchmark System reaches 2 s in its response time when the simultaneous connection reaches 100 connections while the proposed one is faster 0.3 s. Response Time is measured from the beginning of HTTP Request to Client's acknowledge. Tester machine was in the same network with the main system machine so that the network latency could ignore in this test.

4.4 Accuracy Evaluation

In the previous test, streamed data from ECG sensor will be saved in internal database. Each scenario will become a particular dataset and will be trained using same analytics methods. For classification methods, this research implements Convolutional Neural Network with 50 filters initial convolutional layer, each of size (30, 1), followed by a maxpool layer of size (5, 1), a fully-connected layer with 100 units, and finally a sigmoid layer with 1 unit. Network learning rate set to 0.01 and decremented it by $(0.01–0.001)/1500 = 6 \times 10^{-6}$ at each of the 40 epochs. For training purpose, dataset was split into 80% for training and 20% for testing.

According to the simultaneous connection test result in Table 2, the MIT-BIH dataset was prepared for the analytics test. The prepared dataset was then named by the system and number of connections, i.e.: Proposed - 1, Proposed – 10, Proposed – 50, Proposed – 100, Benchmark – 1, Benchmark – 10, Benchmark – 50, Benchmark – 100. For 100% successful connections result (Proposed – 1, Proposed – 10, Benchmark – 1), the analytics test was only performed once and assumed to have the same result for the analytic test.

Table 3. Analytics accuracy test results

Scenario	Successful connection %	Max accuracy	Min accuracy	Epoch to max
Proposed − 1, Proposed − 10, Benchmark − 1	100%	98.56	88.91	29
Proposed − 50	98.8%	98.54	84.50	37
Proposed − 100	62%	98.41	74.26	25
Benchmark − 10	94.65%	98.52	84.51	40
Benchmark − 50	21.69%	96.41	0	29
Benchmark − 100	11.41%	96.68	0.2	36

The analytic accuracy test result is shown in Table 3 and the detail result in every epoch of each scenario is shown in Fig. 9. In scenario Propose − 1, Propose − 10 and Benchmark − 1, first epoch can reach 88.91% accuracy and it needs 29 epochs to reach its maximum accuracy of 98.56%. In a very low-quality data condition, like in Benchmark − 50 and Benchmark − 100, the data integrity is below 50%, the accuracy on the first epoch was 0 and 0.2, and it requires more than 29 epochs to reach its maximum accuracy. By considering those results, as the dataset shrinks due to the drop of successful connections, the accuracy will decrease respectively. The CNN algorithm in this research can reach its optimum accuracy within 15–20 epochs nevertheless the data integrity is still key to reach maximum accuracy. By maintaining the successful connection using asynchronous processing, the optimal accuracy of analytics can be achieved.

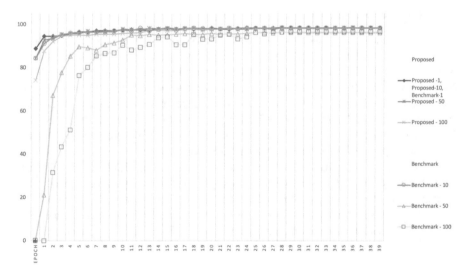

Fig. 9. Accuracy results per scenario

5   Conclusion

Asynchronous processing in this research is designed to separate time-consuming task to be done in other components of the system. By implementing asynchronous process, the proposed system can maintain successful HTTP connection from users/sensors. This research has proved that the design can improve the ability to handle HTTP request by 50% for 100 simultaneous users/sensors. In line with that, the analytics in this system can achieve 98.41% maximum accuracy in the same scenario. The number of epochs in CNN algorithm can improve the accuracy of the result but the quality of data is key to the end result.

Acknowledgments. This work is supported by the Directorate General of Strengthening for Research and Development, Ministry of Research, Technology, and Higher Education, Republic of Indonesia as a part of Penelitian Terapan Unggulan Perguruan Tinggi Research Grant to Binus University entitled "Prototipe dan Aplikasi Monitoring Kualitas Tidur Portabel berbasis Teknologi Cloud Computing dan Machine Learning" or "Portable Sleep Quality Monitoring Prototype and Application based on Cloud Computing Technology and Machine Learning" with contract number: 039/VR.RTT/IV/2019 and contract date: 29 April 2019.

References

1. Jatmiko, W., Setiawan, I.M.A., Ali Akbar, M., Eka Suryana, M., Wardhana, Y., Febrian Rachmadi, M.: Automatic Arrhythmia beat detection: algorithm, system, and implementation. Makara J. Technol. **20**(2), 82 (2016)
2. Islam, S.M.R., Kwak, D., Kabir, H.: The Internet of Things for health care: a comprehensive survey. IEEE Access **3**, 678–708 (2015)
3. Li, C., Hu, X., Zhang, L.: The IoT-based heart disease monitoring system for pervasive healthcare service. Procedia Comput. Sci. **112**, 2328–2334 (2017)
4. Nurdin, M.R.F., Hadiyoso, S., Rizal, A.: A low-cost Internet of Things (IoT) system for multi-patient ECG's monitoring. In: ICCEREC 2016 - International Conference on Control, Electronics, Renewable Energy, and Communications 2016, Conference Proceedings (2017)
5. Yang, Z., Zhou, Q., Lei, L., Zheng, K., Xiang, W.: An IoT-cloud based wearable ECG monitoring system for smart healthcare. J. Med. Syst. **40**, 286 (2016)
6. Yasin, M., Tekeste, T., Saleh, H., Mohammad, B., Sinanoglu, O., Ismail, M.: Ultra-low power, secure IoT platform for predicting cardiovascular diseases. IEEE Trans. Circuits Syst. I Regul. Pap. **64**(9), 2624–2637 (2017)
7. Luz, E.J.D.S., Schwartz, W.R., Cámara-Chávez, G., Menotti, D.: ECG-based heartbeat classification for arrhythmia detection: a survey. Comput. Methods Programs Biomed. **127**, 144–164 (2016)
8. Bazi, Y., Alajlan, N., AlHichri, H., Malek, S.: Domain adaptation methods for ECG classification. In: International Conference on Computer Medical Applications (ICCMA), March 2016 (2013)
9. Ye, C., Kumar, B.V.K.V., Coimbra, M.T.: Heartbeat classification using morphological and dynamic features of ECG signals. IEEE Trans. Biomed. Eng. **59**(10), 2930–2941 (2012)
10. Ye, C., Kumar, B.V.K.V., Coimbra, M.T.: Combining general multi-class and specific two-class classifiers for improved customized ECG heartbeat classification. In: Proceedings of the 21st International Conference on Pattern Recognition, no. Icpr, pp. 2428–2431 (2012)

11. Mar, T., Member, S., Zaunseder, S., Mart, J.P.: Optimization of ECG classification by means of feature selection. IEEE Trans. Biomed. Eng. **58**(8), 2168–2177 (2014)
12. Übeyli, E.D.: Combining recurrent neural networks with eigenvector methods for classification of ECG beats. Digit. Signal Process. A Rev. J. **19**(2), 320–329 (2009)
13. de Chazal, P., O'Dwyer, M., Reilly, R.B.: Automatic classification of heartbeats using ECG morphology and heartbeat interval features. IEEE Trans. Biomed. Eng. **51**(7), 1196–1206 (2004)
14. Llamedo, M., Martínez, J.P.: Heartbeat classification using feature selection driven by database generalization criteria. IEEE Trans. Biomed. Eng. **58**(3) PART 1, 616–625 (2011)
15. T. I. T. Union, Recommendation ITU-T Y.2060 : Overview of the Internet of Things (2012)
16. Li, S., Da Xu, L., Zhao, S.: The Internet of Things: a survey. Inf. Syst. Front. **17**(2), 243–259 (2015)
17. Lunacek, M., Braden, J., Hauser, T.: The scaling of many-task computing approaches in python on cluster supercomputers. In: Proceedings - IEEE International Conference Cluster Computing ICCC (2013)
18. Pardamean, B.: Asynchronous publish/subscribe architecture over WebSocket for building real-time web applications. Internetworking Indones. J. **7**(2), 15–19 (2016)
19. Pan, J., Willis, J.: A real-time QRS detection algorithm. IEEE Trans. Biomed. Eng. **32**(3), 230–236 (1985)
20. Acharya, U.R., et al.: A deep convolutional neural network model to classify heartbeats. Comput. Biol. Med. **89**, 389–396 (2017)
21. Azariadi, D., Tsoutsouras, V., Xydis, S., Soudris, D.: ECG signal analysis and arrhythmia detection on IoT wearable medical devices. In: 2016 5th International Conference on Modern Circuits and Systems Technologies, MOCAST 2016 (2016)

Towards Cloud-Based Personalised Student-Centric Context-Aware e-Learning Pedagogic Systems

Philip Moore[1], Zhili Zhao[1], and Hai Van Pham[2(✉)]

[1] School of Information Science and Engineering, Lanzhou University,
Lanzhou 730030, Gansu Sheng, People's Republic of China
`ptmbcu@gmail.com, zhaozhl@lzu.edu.cn`
[2] School of Information Technology and Communication,
Hanoi University of Science and Technology, Hanoi, Vietnam
`haipv@soict.hust.edu.vn`

Abstract. Pedagogic systems in higher education have often relied on the traditional approach using face-to-face tutorial sessions with an on-line presence to deliver information relating to the specific course of study. There is however a paradigm shift in pedagogic systems characterised by the availability of teaching materials *'anytime'* and *'anywhere'* using on-line e-learning applications. Moreover, there is a realisation that there is a need to introduce increasing levels of personalisation to address the diverse student population with differing learning styles while encouraging and maintaining student engagement. This paper considers the approach to personalisation and proposes an approach designed to accommodate the dynamic pedagogic requirements of the diverse student cohort. In this paper we consider the provision of higher education and vocational education and propose an approach to educational provision predicated on *personalisation* to enable monitoring of students' progress through testing and monitoring and, based on a student's progress, set new tasks until all available tasks are completed. We then consider extensions to the current proposed approach to introduce autonomous task grading and task assignment based on intelligent context-aware informatics implemented using fuzzy rule-based approach extended using hedge algebras with *Kansai* engineering to increase the granularity of fuzzy variables. We posit that our proposed system offers benefits for both the traditional student cohort and distance learning students.

Keywords: Personalisation · Intelligent informatics · e-learning · Student engagement · Context-aware learning · Student learning style · Pedagogic systems · Computer supported collaborative learning

1 Introduction

There has been a paradigm shift in the student population and their educational demands driven by financial and career demands which are reflected in the

© Springer Nature Switzerland AG 2020
L. Barolli et al. (Eds.): CISIS 2019, AISC 993, pp. 331–342, 2020.
https://doi.org/10.1007/978-3-030-22354-0_30

motivation for undertaking higher education (HE). Moreover, in HE there have been significant changes in university funding streams with degree and domain-specific short courses designed to meet specific academic and vocational demands.

Universities provide a broad and diverse range of courses which include: Bachelor of Science, Masters, and doctoral (Ph.D.) degrees. Additionally, maintaining a professional qualification may require continuing updating of knowledge relating to developments in a dynamically changing world, this is termed continuing professional development (CPD) and it is frequently the case that CPD is mandatory if the professional qualification is to be retained. Universities may provide highly focused domain specific courses to meet the needs of CPD.

There is a growing recognition that future employment will require high levels of training (including vocational training) including re-training. Moreover, there is a further realisation that there is a "skills gap" in many industries. Meeting the demands of vocational training requires both full-time and part-time courses. A further area where HE is delivered is where commercial organisations wish to up-skill employees; in such a case a university or further education college may provide short highly focused practical courses addressing specific skills.

The broad range of potential higher educational demands may employ a range of pedagogic systems. Pedagogic systems in HE (including vocational training) have historically been predicated on the traditional approach [to teaching] using face-to-face tutorial sessions with an on-line presence to deliver information relating to the specific course of study. While the use of on-line pedagogic systems is now well established there remains a place for the traditional *'face-to-face'* tutorial sessions supported by on-line pedagogic systems available *'anytime'* and *'anywhere'* [9] which have experienced rapid and far-reaching technological developments in e-learning systems including the affects of *disruptive technologies* [1] on pedagogic systems.

An issue in current HE provision is the broad range of abilities, learning styles, and the time students are availability for study; this applies to both face-to-face and on-line (distance) learning but the issues are more difficult to detect in the latter mode of delivery. The use of *'anytime'* and *'anywhere'* provision of teaching resources and materials generally addresses the time factor. Addressing the range of academic levels and learning styles represents a more difficult challenge in which students will progress at different rates; in such a scenario encouraging and maintaining student engagement including measuring progress and engagement is a significant issue.

In this paper we consider the provision of HE and vocational education and propose an approach to educational provision predicated on a *personalisation* where the goal is to use information systems to monitor students' progress through testing and monitoring and, based on a student's progress, set new tasks with each task graded automatically as tasks are completed. We then consider extensions to the current proposed approach to introduce autonomous task grading and task assignment based on intelligent context-aware informatics implemented using fuzzy rule-based informatics extended using hedge algebras to increase the granularity of fuzzy variables.

The remainder of this paper is structured as follows: Sect. 2 considers problem statement as it relates to the evolving HE landscape. Section 3 introduces computer supported cooperative mobile learning with information systems as they relate to pedagogic systems. Our proposed process model is set out in Sect. 4. Sections 5 and 6 consider practical 'real-world' pedagogic challenges experienced in Hanoi University of Science and Technology and The School of Information Science and Engineering at Lanzhou University) where teaching is provided to a diverse student cohort. There is a scenario-based evaluation set out in Sect. 7 and a discussion with consideration of implementation and potential future directions for future research to extend the current proposed mode is presented in Sect. 8; the paper closes with concluding observations in Sect. 9.

2 The Problem Statement

The traditional paradigm for HE landscape can be characterised by a pedagogic system predicated on *face-to-face* teaching (also termed *"talk and chalk"*) with an on-line presence to distribute information and resources. Distance learning pedagogies traditionally used postal provision for resource distribution and the return of tutor marked assignments; however, these methods have generally been replaced by on-line pedagogic systems [9].

To address the changing educational demands *'anytime'* and *'anywhere'* personalised e-learning utilising mobile technologies is becoming ubiquitous in the domain of HE, and increasingly e-learning is embracing Web 2.0 and Web 3.0 technologies to provide interactive networking functionality (on a social level) at both a pedagogic and personal level. Personalisation requires the creation of an individual's profile (termed a context), a context defining and describing a user's current state [7].

An issue that has affected education generally is student encouraging and measuring student engagement in the learning process. This is exacerbated by the use of Internet-based pedagogic systems where students have less (or even no) face-to-face contact with teachers. Moreover, there is a changing student cohort with widely diverging academic ability. In delivering a course curriculum there are potential issues where members of the student cohort has a widely differing academic levels and capabilities. In such a case there are students who find the course too easy while other students struggle to keep up with the course topics being taught. In both cases students will tend to "switch-off" and become disengaged [3, 10].

Personalisation in the provision of HE has gained traction driven by socio-economic, demographic, and employment changes in the student population. Concomitant with these changes is the evolving capability and ubiquity of mobile technologies. These developments have resulted in interest in e-learning to accommodate the diverse student population and leverage the power of mobile technologies. We consider HE with networking (in a collaborative and social networking sense). In this paper we consider the problem of maintenance and measurement of student engagement and propose the use of computer-supported collaborative learning and teaching (CSCLT) (intelligent groupware) to *stream* and *personalise* the teaching provision based on ongoing performance in course-work and the completion of example learning exercises.

3   Computer Supported Collaborative Pedagogic Systems

The concept of computer supported cooperative mobile learning [6] is not new. However, current approaches fail to address the need to personalised pedagogic provision to manage student engagement. The concept of personalisation [7] in its many forms has gained traction driven by the demands of computer-mediated interactions generally implemented in large-scale distributed systems and ad-hoc wireless networks.

Personalisation requires the identification and selection of entities based on a defined profile (a context); an entity has been defined as a person, place, or physical or computational object. Context employs contextual information that combines to describe an entities current state. Historically, the range of contextual information utilised (in context-aware systems) has been limited to identity, location, and proximate data; there has however been advances in the range of data and information addressed [7]. As such, context can be highly dynamic with inherent complexity. In addition, context-aware systems must accommodate constraint satisfaction and preference compliance.

The application of context-awareness in CSCLT provides an effective basis upon which personalisation in pedagogic systems can be realised to the mutual benefit of students who receive targeted personalised tuition and tutors who may eliminate (or at least mitigate) the issue of students becoming disengaged. Moreover, using the approach proposed in this paper my potentially improve the overall course results, an outcome that will benefit all stakeholders in the pedagogic system.

A central feature of pedagogic systems is information systems (IS) which provide the means by which performance and related metrics may be recorded for use by all stakeholders in an academic institution. Moreover, an IS provides essential support for managing student activities and for the proposed approach to manage student engagement as discussed in Sects. 4 and 7. Additionally, as discussed in Sect. 8, an IS forms an essential component in the proposed future research which addresses the investigation of an autonomic intelligent context-aware TMA grading system. In such a system a requirements specification will require a range of data which will be processed into useful information (or *CAPTA* [4]), Fig. 1 models the process and shows the data processing from raw data to *CAPTA*, information, knowledge, and ultimately wisdom as discussed in [8]. A detailed exposition of this process is beyond the scope of this paper, for a discussion on data processing in intelligent context-aware systems see [7] and [8].

In summary, as discussed in [4] the traditional process of data into information without proper selection results in too much information with much of it irrelevant information. The process set out in [4] introduces an intermediate stage

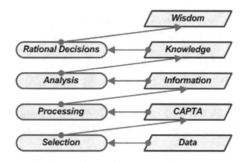

Fig. 1. A conceptual model illustrating the progression of data processing from the raw data to wisdom (rational decision-support). Shown is the intermediate stage as discussed in [4] where an initial selection of relevant and useful raw data (termed *CAPTA* [4]) is used.

in the processing of data, namely *CAPTA* in which relevant data is selected and processed into information useful in our proposed context-aware approach. The research published in [8] extends the process with additional processing as modelled in Fig. 1. It is envisaged that the process shown in this conceptual model will be used to create an intelligent context-aware decision-support system for the proposed approach presented in this paper.

4   The Process Model

The process model addresses a students performance in completing exercises and tasks assigned by tutors in the course of completing a course module. Consider a class of students $(S_{1...n})$ and a number of exercises or tasks $(T_{1...n})$ (termed: Tutor Marked Assignments (TMA)).

The proposed process may be modelled as follows for students successfully completing and submitting the TMA:

1. In the delivery of a module tutors will issue to students $(S_{1...n})$ TMA's $(T_{1...n})$ to assess the student understanding of the concepts and topics delivered in classes.
2. Upon receipt of a TMA $(T_{1...n})$ students $(S_{1...n})$ will work on the assignment.
3. Each TMA will have a submission date by which the completed TMA must be submitted.
4. *IFF* students submit the TMA on time the TMA will be graded and the students records updated with the correct grade.
5. The Module information system (IS) will check the TMA database for further TMA and *IFF* there are further TMA to be assigned the next TMA will be issued.
6. *IFF* there are no further TMA to be assigned the process will terminate.

The process may be modelled as follows for students who fail to complete and submit the TMA by the submission date:

1. In the delivery of a module tutors will issue to students $(S_{1...n})$ TMA's $(T_{1...n})$ to assess the student understanding of the concepts and topics delivered in classes.
2. Upon receipt of a TMA $(T_{1...n})$ students $(S_{1...n})$ will work on the assignment.
3. IFF Students fail to submit the TMA on time the student record will be updated.
4. Students will be offered additional tutorials to encourage the learning process with additional time to submit the TMA (possibly with capped grades due to late delivery).
5. On (late) submission of the TMA the student record will be updated with the (possibly capped) grade.
6. The Module information system (IS) will check the TMA database for further TMA and IFF there are further TMA to be assigned the next TMA will be issued.
7. IFF there are no further TMA to be assigned the process will terminate.

The proposed TMA process model has as its primary motivation the goal of encouraging student engagement in the pedagogic process. As discussed in Sect. 2 there is often a wide diversity in the academic abilities within a student cohort resulting in students becoming disengaged as the TMA may be either too easy or too difficult. This is an issue most tutors will have experienced in their teaching activities with student disengagement posing a major issue for all stakeholders in the HE system.

5   Hanoi University of Science and Technology

The e-learning system of Hanoi University of Science and Technology (HUST) is an on-line lecture resource between Korea and ASEAN countries; The HUST e-Learning system is based on common learning standards such as the Open educational resource (OER) and the *learning and management system* (LMS)/and the *learning content management system* (LCMS). The goal of the Korean-ASEAN network is the creation of partner schools in Korea and HUST is the real focal point of the ACU project for e-learning with regional schools implementing related activities.

From 2012–2018, the project has developed 70 e-learning courses with 89 classes and the participation of 9261 students. With that result, Hanoi University of Technology has actively participated in the ACU Project[1] with Yeungjin Cyber College partner school and CLMV schools (Cambodia, Laos, Myanmar and Vietnam). E-learning development and the application of information technology in providing pedagogic system support at the HUST with blended-learning; Building, researching and implementing LMS/LCMS system of blended-learning teaching in cooperation with HUST. With the current model, the form of e-learning contributes actively to training the traditional model of

---

[1] see: http://www.aseancu.org/pr/main.acu.

the school. Blended learning relates to an educational environment where students are able to spend their class time passing an engaging on-line course, collaborating on creative projects with their peers and getting on-time support and assistance from a teacher.

The ACU project has helped the HUST to build infrastructure for e-learning materials development such as: (a) facilities (recording studios, servers, workstations etc - i.e., CSCLT facilities ...), and (b) technical and teaching assistance staff (human resources) to support the required technical and pedagogic infrastructure for the delivery of on-line lectures using CSCTL systems.

As with the School of Information Science and Engineering at Lanzhou University empirical evidence supports the contention set out in the problem statement (see Sect. 2) that student engagement and disengagement represents an ongoing issue. We consider that the use of our proposed TMA approach (both as currently proposed and in its extended form) presents opportunities to address the issues identified around student engagement in pedagogic system.

The use of our proposed TMA approach, is considered from the perspective of HUST and the ACU project and the problem statement as set out in Sect. 2 (as it relates to Undergraduate students).

6   Lanzhou University

This section introduces the use of our TMA approach from the perspective of the School of Information Science and Engineering at Lanzhou University (LZU) and considers it from the perspective of the problem statement as set out in Sect. 2 as it relates to Undergraduate students. Potential undergraduate students face a selection process (an entrance examination) with a quantitative result identifying the students who will be selected to attend an undergraduate degree course.

A number of observations may be made regarding the undergraduate cohort. The level of freshmen (male and female) is uneven, and scores from the college entrance exam differ by dozens or more; this results in difficulties in course delivery and teaching. For example, in one class there are 116 students drawn from 6 provinces across China; the highest score achieved in the entrance exam is 629, but the lowest score is 535. As alluded to in the problem statement (see Sect. 2), in any one lecture the "good students cannot eat enough" and the less "able students also cannot eat enough".

Moreover, when freshmen enter the LZU, the students' computer foundation and application abilities are also very uneven. This is because some students have received a very good education in a high middle school where they may have learned basic computer knowledge and computer literacy before they entered LZU. Moreover, students family background may vary; for example some students come from wealthy families and have plentiful access to computers, computing devices, and applications such as video games etc, thereby having a high level of computer literacy. However, students from less affluent families often have received less exposure to computers and lack computer literacy. The disparity in the student cohort is exemplified in the teaching of computer science where there are students who are experienced in web site building and computer programming while others have very limited or no knowledge of the subjects.

Such problems pose difficulties in the basic computer teaching both in LZU and the pedagogical problem may be reflected in other Chinese universities and universities internationally. The proposed TMA approach may address (or at least mitigate) the issues around student engagement in pedagogic system. We appreciate there are potential ethical and regulatory challenges in our proposed TMA approach, we address these in the discussion in Sect. 8.

7   Scenario-Based Evaluation

To illustrate our proposed approach in a 'real-world' use-case we present a scenario-based evaluation to demonstrate how our approach may be implemented. Consider the two potential use-cases introduced in Sect. 4 which represent 'real-world' TMA conditions. In considering the two use-cases in the general approach the TMA's are often created to allow for the median student academic abilities. The are three potential use-cases in TMA's:

1. Given that in many courses an aim is to address students with median academic abilities. In such cases the level of challenge set by TMA's will be suitable for their abilities and such students will generally retain engagement in the pedagogic system as they may be adequately challenged.
2. In the case of the more academically able students within a class of or cohort the TMA's are often too easy and the more academically able students may complete the TMA quickly and easily but the lack of challenging TMA's will result in their disengagement.
3. Similarly, in the case of the less academically able students within a class or cohort the TMA's are often too difficult and too challenging, the result will again be disengagement.

By applying the proposed approach to TMA it is posited that by using the median approach to the creation of TMA's we may by incrementally increasing the TMA difficulty the more academically able students will retain engagement. Moreover, by introducing additional on-line tutorials the less able (and possibly for students with median academic ability) where more challenging TMA's are assigned the level of student engagement may be encouraged. While in a traditional university setting teachers may be able to mitigate the engagement issue to a limited degree, in on-line courses delivered outside the traditional university setting, the engagement issue is much more difficult to confront.

8   Discussion

In HE there is quantitative and empirical evidence from LZU and HUST to suggest that there is a wide diversity in the student cohort and as identified in Sects. 5 and 6 where there is a broad range of academic ability with students being represented in terms of an academic spectrum; moreover, there is a similar diversity in students experience. As discussed in Sect. 2 this can result in students lacking motivation and becoming disengaged in the pedagogic process. Empirical evidence suggests that teachers recognise the problem and struggle to design course curricula and assessment structures which satisfy the less able students, the more able students, and students with a median academic level and experience; the primary goal of teachers is to address (or at least mitigating) frustration by both students (manifested in terms of the level of engagement) and teachers (along with university administrators) where grades may not reach the anticipated level.

In this paper we have considered the issues around student diversity and the relates issue of engagement (or *disengagement* which is in fact the actual issue) and we have proposed an approach based around TMA's as discussed in Sect. 4 where our proposed process model is set out. In our proposed TMA model we are attempting to address the situation where the more advanced students become bored and disengaged and the less academically capable students are left behind (a situation teachers and HE providers are focused on avoiding). The students in the middle of the academic and experience spectrum tens to be satisfied and suitably challenged by the course curriculum.

Technological developments in terms of Internet technologies and cloud-based systems have enabled significant improvements in pedagogic systems using CSCLT with teaching resource provision being available and accessible *'anytime'* and *'anywhere'* as discussed in [9]. While this approach to teaching goes some way to keeping students engaged in the pedagogic process it generally fails to monitor and measure students progress. Our proposed approach is based on the concept of CSCLT with TMA's to measure students progress, challenge the more able students, and help less able students to understand the teaching materials delivered and improve their progress. Students with median abilities will generally manage the course but where they fail to do so the proposed TMA approach will identify them at an early stage (e.g., set coursework may not be returned of the grades are very low).

There are important ethical issues [14] and regulatory challenges in personalising educational provision based on academic ability or experience (often termed streaming). For example, if teaching and course delivery focuses on the less able students or alternatively the focus is on the students who find the taught topics easy this may result in regulatory constraints (students may complain that other students are receiving preferential treatment) and ethical issues may fall into the same bracket.

We posit that our proposed TMA method of curriculum delivery can address (or at least mitigate) the twin issues of student diversity and disengagement for students at either end of the academic and experience spectrum while maintaining an oversight on students in the median ability category and picking up any who begin the fall behind.

8.1 Future Directions for Research

Currently, the proposed TMA process model assumes that teachers will grade the completed TMA and manage the additional on-line tutorials; empirical evidence suggests that managing on-line CSCLT systems (such as Skype or Moodle) significantly increases the workload faced by teachers (e.g., moderation of user defined content and faster response times from teachers) and the increases imposed by our TMA approach may be considered to be unrealistic.

In traditional tertiary educational provision students may have a personal academic tutor who would provide mentoring to students to discuss issues and provide guidance. In on-line pedagogic systems, face-to face mentoring is not practical in all but very rare cases. However, we may consider on-line mentoring by a tutor as discussed in [2]. A potential approach may be to use CSCLT systems to provide some mentoring facilities; however, the time and workload issues are the same as for TMA. Such an approach will require evaluation and testing in a limited pilot program as the workload increases may not be realistic.

In considering future work there will be a need for a twin track approach based around a limited teacher led trial with the development of an autonomic implementation of the MA process model. The details are as follows:

1. To design and implement a limited trial of the TMA approach as currently envisaged with teachers implementing the process model as set out in Sect. 4.
2. A more realistic approach will involve the development of our proposed TMA process as an autonomic system (albeit with oversight by teaching staff. Such a trial would involve a small student cohort and would not impact their studies.
3. An autonomic system is envisaged as an intelligent context aware CSCLT (intelligent groupware) implemented using a fuzzy rule-based system as described in [7] extended to include hedge algebras [5,15] with semantic descriptors [11] and potentially *Kansai* engineering [12,13].

The potential difficulties in designing and implementing an autonomic TMA system are not underestimated. However, such an approach could radically change the traditional approach to teaching in the traditional university setting. Moreover, distance learning where tuition is provided on-line could be revolutionised with increased student engagement and related academic results. In both scenarios there are significant benefits for all stakeholders in the pedagogic system.

9 Concluding Observations

We have considered pedagogic systems and have observed that HE has often relied on the traditional approach using *face-to-face* tutorial sessions with an on-line presence to deliver information relating to a specific course of study. The paradigm shift in pedagogic systems is characterised by the availability of teaching materials *anytime* and *anywhere* using on-line e-learning applications.

Additionally, there is a need to introduce increasing levels of personalisation to address the diverse student population with differing learning styles while encouraging and maintaining student engagement.

Within LZU there are two campuses, one in the centre of Lanzhou City and one located out of the city some 50 km from Lanzhou. This scenario is repeated for many universities in China, for example the JInan Normal University has adopted the same strategy. In both cases there are significant logistical demands (measured in terms of both time and cost) as staff (generally teachers) and students travel between the City and out of town campuses for classes and other related activities. For the HUST such travel is not possible and the on-line approach is essential in enabling the ACU e-learning project to function. We may view the HUST ACU e-learning project in the same light. We consider that using our TMA approach would reduce the logistical load for Chinese universities and increase the pedagogic effectiveness to the mutual benefit of all stakeholders in the pedagogic systems.

In this paper we have presented a proposed approach to personalisation and proposes an approach designed to accommodate the dynamic pedagogic requirements of the diverse student cohort. The proposed approach targets the monitoring students' progress through TMA's assigned based on a student's progress with a new TMA being assigned as TMA's are completed. Our proposed approach is currently based on teachers implementing the TMA approach. However, this imposes significant increases in teacher workload. In this paper we have introduced our considered proposal for an autonomic TMA system based on CSCLT implemented using intelligent context-aware informatics in a fuzzy rule-based approach extended using hedge algebras, semantics, and *Kansai* engineering to increase the granularity of fuzzy variables.

We posit that our prosed system offers benefits for both the traditional student cohort and distance learning students. We posit that our prosed system offers benefits for both the traditional student cohort and distance learning students.

References

1. Adner, R.: When are technologies disruptive? A demand-based view of the emergence of competition. Strateg. Manag. J. **23**(8), 667–688 (2002). https://doi.org/10.1002/smj.246
2. Ana Rodriguez, M., Vicente Gabarda, M.: Improving students' competencies through perception of mentoring in an online master of teaching training. Soc. Sci. **5**(6–1), 7–13 (2016). https://doi.org/10.11648/j.ss.s.2016050601.12
3. Austin, R., Sharma, M., Moore, P., Newell, D.: Situated computing and virtual learning environments: e-learning and the benefits to the students learning. In: 2013 Seventh International Conference on Complex, Intelligent, and Software Intensive Systems (CISIS), pp. 523–528. IEEE (2013)
4. Checkland, P., Holwell, F.: Information, Systems and Information Systems: Making Sense of the Field. Wiley, London (1997)

5. Ho, N.C., Long, N.V.: Fuzziness measure on complete hedge algebras and quantifying semantics of terms in linear hedge algebras. Fuzzy Sets Syst. **158**(4), 452–471 (2007). https://doi.org/10.1016/j.fss.2006.10.023. http://www.sciencedirect.com/science/article/pii/S016501140600443X

6. Hu, B., Moore, P.: Context modelling to support location based cooperative mobile learning. In: IEEE Proceedings of the Tenth International Conference on Computer Supported Cooperative Work in Design (CSCWD 2006), Southeast University, Jiangsu, China, pp. 1372–1376. IEEE (2006). https://doi.org/10.1109/CSCWD.2006.253016

7. Moore, P., Pham, H.V.: Personalization and rule strategies in human-centric data intensive intelligent context-aware systems. Knowl. Eng. Rev. **30**(2), 140–156 (2015). https://doi.org/10.1017/S0269888914000265. Intelligent Computing in Large–Scale Systems

8. Moore, P., Pham, H.V.: On wisdom and rational decision-support in context-aware systems. In: The 2017 IEEE International Conference on Systems, Man, and Cybernetics (SMC2017), Banff, Alberta, Canada, pp. 1982–1987. IEEE (2017)

9. Moore, P.T.: Anytime-anywhere: personalised time management in networking for e-learning. eLearn Center Research Paper Series, pp. 48–59 (2011)

10. Newell, D., Davies, P., Austin, R., Sharma, M., Moore, P.T.: Models for an intelligent context-aware blended m-learning system. In: The 29th IEEE International Conference on Advanced Information Networking and Applications (AINA-2015), Gwangju, Korea. IEEE (2015). https://doi.org/10.1109/WAINA.2015.25

11. Nguyen, C.H., Tran, T.S., Pham, D.P.: Modeling of a semantics core of linguistic terms based on an extension of hedge algebra semantics and its application. Knowl.-Based Syst. **67**, 244–262 (2014). https://doi.org/10.1016/j.knosys.2014.04.047. http://www.sciencedirect.com/science/article/pii/S0950705114001774

12. Pham, H.V., Moore, P., Tran, K.D.: Context matching with reasoning and decision support using hedge algebra with Kansei evaluation. In: Proceedings of the Fifth Symposium on Information and Communication Technology (SoICT 2014), Hanoi, Vietnam, pp. 202–210. ACM, New York (2014). https://doi.org/10.1145/2676585.2676598

13. Salem, B., Nakatsu, R., Rauterberg, M.: Kansei experience: aesthetic, emotions and inner balance. Int. J. Cogn. Inform. Nat. Intell. **3**, 54–64 (2009)

14. Satterfield, D., Kelle, S.: Ethical issues in online education. In: Ahram, T.Z., Karwowski, W. (eds.) Advances in the Human Side of Service Engineering, pp. 257–266. Springer, Cham (2017). https://doi.org/10.1007/978-3-319-41947-3_24

15. Van Pham, H., Moore, P.: Robot coverage path planning under uncertainty using knowledge inference and hedge algebras. Machines **6**(4) (2018). https://doi.org/10.3390/machines6040046. http://www.mdpi.com/2075-1702/6/4/46

Learning Agile Scrum Methodology Using the Groupware Tool Trello® Through Collaborative Working

Nitin Naik[1(✉)], Paul Jenkins[1], and David Newell[2]

[1] Defence School of Communications and Information Systems, Ministry of Defence, Blandford Forum, UK
{nitin.naik100,paul.jenkins683}@mod.gov.uk
[2] Department of Computing and Informatics, Bournemouth University, Bournemouth, UK
dnewell@bournemouth.ac.uk

Abstract. Agile is a project management philosophy for collaborative working which consists of a set of values and principles that can be employed in any sector. It is adopted by the software development community a long time ago and now several methodologies based on agile principles have become established as a method of software development. These agile based software development methodologies have developed as an integral part of the software engineering and software development curricula for many computing degree courses. One such agile based methodology is Scrum which is widely used in the software industry and thus in teaching. Several agile Scrum tools are available for software development, however, for teaching and learning purposes, this would not be affordable for many institutions due to its cost or learning curve. This necessitates the requirement for a free or open-source Scrum tool without any learning curve. Trello which is a free project management and collaborative working tool but not particularly designed for Scrum. However, its functionality and features can be transformed to make it a basic Scrum tool for teaching and learning purposes at no cost. This paper presents a systematic development and application of Trello-based agile Scrum methodology not just for teaching and learning purposes but for real project development. It is employed in the delivery of the software engineering module in BSc courses and subsequently, in the development of the BSc final year project at Bournemouth University, UK. This implementation of Trello-based agile Scrum methodology in the project development is compared against the simple agile Scrum based project development practice to demonstrate the success and learning improvement of this proposed methodology.

Keywords: Agile Scrum methodology · Trello · Groupware · e-learning · Software engineering · Collaborative working

© Crown 2020
L. Barolli et al. (Eds.): CISIS 2019, AISC 993, pp. 343–355, 2020.
https://doi.org/10.1007/978-3-030-22354-0_31

1   Introduction

Software engineering methodologies and their application in project development is a key element of learning and accomplishing the computing degree. Today, agile is one of the most successful software development approaches employed by the majority of software development companies [1]. As a result, the teaching focus has moved towards the learning of various types of agile methodologies such as Scrum, eXtreme Programming (XP), Crystal and Dynamic Systems Development Method (DSDM), and away from the traditional methodologies [13,21]. Amongst all, Scrum is the most popular agile methodology employed in software development by a large number of software development companies [9,11,16]. While teaching software engineering concepts and these agile methodologies, it is equally important for students to understand the methodology in a way that they can employ in their actual project development at later stage. If the teaching and learning is based on practicals and utilises some tools or apps then it would be more effective than simply learning concepts without any learning aids, this is especially true for the kinaesthetic learner [5,12]. Teaching agile Scrum practicals is possible as there are several commercial tools available for use, however, they are mostly utilised by large software development companies due to their high price. This necessitates the requirement to resource a free or open source tool for teaching and learning the Scrum methodology. There are many free project management tools available; however, they are not particularly designed to implement agile Scrum methodology, nevertheless some can be utilised for it. One such tool is Trello, a project management and collaborative working tool which is not specifically designed for agile Scrum methodology. Nonetheless, it has potential to be utilised as an agile Scrum tool, which requires a careful transformation of the features of Trello into its corresponding components of the agile Scrum methodology. However, there is no proven research relating to Trello being used as an agile Scrum tool in the past including in the teaching and learning arena.

Therefore, this paper presents a systematic development and application of Trello-based agile Scrum methodology for both a teaching and learning and real-life project development. In this Trello-based agile Scrum methodology, three main components of Trello: Boards, Lists and Cards are transformed into agile Scrum components. The Trello board is considered as an agile Scrum project on which the entire Scrum team is working; Lists are utilised to design various backlogs: product backlog, release backlog and sprint backlog; and cards are used to represent the tasks or features (i.e. user stories). Subsequently, this Trello-based developed methodology is utilised in teaching of agile Scrum methodology in the Software Engineering module of the BSc course at Bournemouth University, UK. Finally, this Trello-based agile Scrum methodology is adopted for the BSc final year computing project which is sponsored by industry partners. The analysis of the project results revealed a significant improvement in the overall performance (i.e. each individual student) and this Trello-based practical and collaborative working assisted all students in understanding, clarifying and correcting their project activities in a timely manner.

The remaining paper consists of the following sections: Sect. 2 explains the agile Scrum methodology; Sect. 3 discusses Trello and its important features; Sect. 4 present the development and delivery of agile Scrum methodology using Trello; Sect. 5 presents the application of this Trello-based methodology in project development. Section 6 presents the results and analysis of this Trello-based agile Scrum methodology; Sect. 7 presents the summary of the paper and possible future work.

2   Agile Scrum Methodology

Scrum is one of the agile software development methodologies which was proposed by Sutherland and Schwaber [15]. It is a complete process from design through to the development of a product, all based upon a set of guidelines for managing that development. Combining the iterative and incremental process models to overcome the issues with the traditional models [14]. It focusses on the process adaptability and customer satisfaction by rapid delivery of working software products [15]. This is achieved by breaking the product into a number of small incremental builds [1]. These builds are provided in iterations. Each iteration has an elapsed time of typically 3–30 days (1–4 weeks), with iterations involving cross functional teams working simultaneously on various topics such as planning, requirements analysis, design, coding, unit testing, and acceptance testing. Each iteration ends in a Sprint where a working/shippable product is presented to the customer and important stakeholders (See Fig. 1). A significant issue in any software development project is changing requirements. Agile Scrum accepts the reality of change versus the desire for complete, rigid specifications [16], some of its main principles are:

1. User involvement is crucial
2. The development team must be empowered to make decisions
3. Requirements can change or evolve however the time scale is fixed
4. Requirements Capture is at a high-level, lightweight and visual process
5. Develop small, incremental releases and iterate
6. Frequent delivery of products is focussed on
7. Complete each feature before moving on to the next feature
8. Apply the 80/20 rule (i.e. Pareto Principal)
9. The project lifecycle integrates the testing throughout its development
10. A collaborative and cooperative approach between all stakeholders is essential

2.1   User Stories

Users describe the interaction of the user with the system through a mechanism of writing user stories, thus indirectly collecting the requirements [18]. It specifies the users and their roles both in the current and future system to be developed, describing the types of user, what they want and why. This process is not a highly

Fig. 1. Agile Scrum methodology

documented requirement but rather it should be very short narrative able to be mounted on a sticky note or card [18]. If the user stories are detailed they can be termed as an Epic or a Theme. An Epic is a short description containing a few stories, whereas a Theme is a detailed description that usually contains few epics. A good user story uses the INVEST model [4,18]:

☐ Independent: Reduced dependencies - Easier to plan
☐ Negotiable: Details added via collaboration
☐ Valuable: Provides value to the customer
☐ Estimable: Too big or too vague - Not estimable
☐ Small: Can be done in less than a week by the team
☐ Testable: Good acceptance criteria

The INVEST model is a commonly accepted set of criteria for assessing the quality of a user story. If the story fails to meet one of these criteria, then the Scrum team may want to reword it, or even consider a rewrite.

☐ Template/Syntax of User Stories:
- As a <role>, I want <feature> so that <reason> OR
- As a <types of user>, I want <goal> so that <receive benefit>

2.2 Product Backlog

The product backlog is an ordered list of everything (i.e. a wish list) which is identified to be necessary in the product. This is the single source of requirements for any changes to be made to the product [19]. Replacing the conventional software requirements specification (SRS) format. The tasks/features of a product backlog can have a technical nature or can be user-centric in the form of user stories [17]. Originally, a product backlog is developed with some initial requirements and it is never complete. This product backlog evolves as the product and the environment in which it will be used evolves. The product backlog is dynamic; it constantly changes to identify what the product requires in order to be appropriate, competitive, and useful [19]. If a product exists, its product backlog also exists. The responsibility for the product backlog lies with the product owner, including its content, availability, and ordering.

2.3 Release Backlog

A release backlog is a subset of the product backlog that is planned to be delivered in the next release, normally a three- to six-month horizon [2]. It would presumably contain the prioritised and selected features of the product backlog for the current release. In the agile Scrum development, in addition to the traditional product and sprint backlogs, the release backlog may or may not exist depending on the development requirement.

2.4 Sprint Backlog

The sprint backlog is a list of prioritised and selected features by the Scrum team to be completed during the current Scrum sprint. These features are selected from the product backlog (or release backlog if it exists) by the Scrum development team during the sprint planning meeting. The sprint backlog is a projection by the Scrum development team about what functionality will be in the next increment and the work required to deliver that functionality [20]. The Scrum development team makes changes to the sprint backlog throughout the sprint. During the development of a sprint, team members are expected to modify the sprint backlog as and when new information is available, but minimally once per day or at the time of daily Scrum meeting [20]. The sprint backlog is a extremely visible, real-time presentation of the work that the development team plans to accomplish during the sprint, and it belongs only to the development team [20].

2.5 Burndown Chart

The burndown chart is a graph that shows how quickly or slowly the Scrum development team are working (burning) through customer's user stories. Using a burndown chart is a means of visually depicting what work is left and the time left to complete it. The chart is a graphical illustration, presenting in an image what a thousand words might not be able to communicate as evidently. The burndown chart displays the total effort against the amount of work for each sprint. The quantity of work remaining is presented on a vertical axis, while the time that has passed since beginning the project is displayed horizontally on the burndown chart, illustrating the past and the future. The burndown chart is available to all stakeholders to update the status of the current sprint and the Scrum team updates it frequently to provide the most accurate information.

3   Trello - A Groupware Tool for Collaborative Working

Trello is one of the most popular project management and collaboration systems. The features available in Trello such as boards, lists, and cards enable the organization and prioritisation of projects in an engaging, flexible, and rewarding way [22]. Trello utilizes the concept of boards (which corresponds to projects) and within boards, there are lists (group of tasks) and cards (which represent tasks). The cards within lists can be used to track the progress of a project or to simply categorize events [3]. It allows integration of apps that the development team already uses directly into your workflow. This power of Trello converts its boards into living applications to meet the team's unique business needs [22]. Furthermore, it permits synchronisation across all of devices and sites. Offering real-time updates meaning everyone is able to share the same perspective. In terms of devices, it is available on a desktop browser and mobile devices whether online or offline. The GUI interface is very user-friendly and productive with a minimal interface so as not to impede the work. Some of the important features of Trello are given in Table 1 [3].

Table 1. Trello Features

No.	Features	No.	Features
1	Easy to learn and simple GUI	11	SSL encryption of data
2	Customisable boards for easy design	12	Easy uploading of files and attachments
3	Mobile functionality to access boards on the go	13	Archiving of card records (e.g. comments and changes)
4	Free or zero pricing for the basic service	14	Information retrieval and back-up
5	Native notifications	15	Voting feature and search function
6	In-line editing and data filtering	16	Deadline reminders and email notifications
7	Checklists, with progress meter	17	Task assignment and activity log
8	Visibility Scope	18	Power-ups features
9	Easy organization with features such as tags, labels and categories	19	Easy move, swap, drag and drop functionality
10	Quick overview on front and back of cards	20	Developer API

4   Development and Delivery of Agile Scrum Methodology Using Trello

Trello is a project management and collaboration system which can be used for various purposes where team working is required. Trello is not an agile Scrum tool but it has several similarities with the popular agile Scrum tools and can be a free alternative for expensive agile Scrum tools. However, there is no proven research of Trello being used as an agile Scrum tool in the past, even for teaching and learning purposes. Therefore, this section presents a systematic development of Trello-based agile Scrum methodology for teaching and learning purposes and for the real life project development. In this Trello-based methodology, three main components of Trello Boards, Lists and Cards are transformed into agile Scrum components. The Trello board is considered as an agile Scrum project on which entire team is working, Lists are utilised to design various backlogs: product backlog, release backlog and sprint backlog, and cards are used to represents the tasks or features (i.e. user stories).

Figure 2 shows a template of an agile Scrum project and its components developed on the Trello board. This example Scrum project consists of seven components in the form of seven lists and three Scrum team members. The first component is the product backlog which contains several example tasks from the user stories and prioritised as high, medium and low priority. Later, this complete product backlog is divided into two release backlogs and each release backlog is divided into two sprints. To aid learning about the Trello-based sprint design, an example sprint is shown in Fig. 3 explaining its associated components in the form of Trello features. It also contains the most critical sprint component a burndown chart to show the current progress of this sprint to everyone. Those who are not a member of the Scrum development team (e.g. product owner and other stakeholders) can subscribe to the complete sprint to get regular updates, therefore, they know the current status of the sprint. Finally, each task is designed on the Trello card utilising all the card features as shown in the example task in Fig. 4. It contains several features such as due date, checklist, attachment, description, watch and comment, which are very useful for managing and communicating about the task.

This developed Trello-based agile Scrum methodology is very useful in teaching software development methodologies through a practical approach, by students applying their learning of the agile Scrum methodology to the given scenario of the project case study in a collaborative manner. This Trello-based developed methodology is utilised in teaching of agile Scrum methodology in the Software Engineering module of the BSc course at Bournemouth University, UK. Students develop a complete Scrum project in Trello based on the given scenario as a part of their learning, therefore, they are able to apply this knowledge in their final year project at a later stage. This practical approach of learning the agile Scrum methodology provides the students with greater opportunities to work as team, communicate and discuss with each other solving the problem, and designing an acceptable solution in a limited time. The successful implementation of this practical approach based on Trello is demonstrated in the next section where its application for the BSc final year project is discussed.

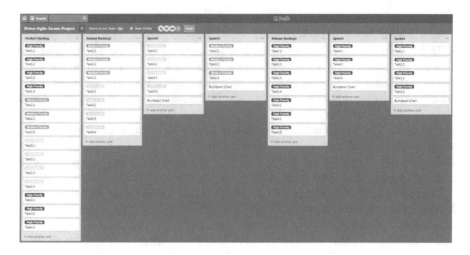

Fig. 2. Illustration of an example agile Scrum project and its components on the Trello board

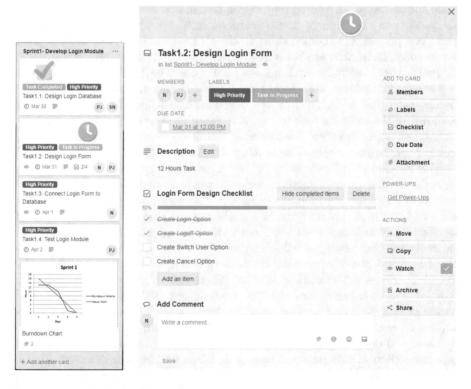

Fig. 3. Illustration of an agile Scrum sprint and its components on the Trello list

Fig. 4. Illustration of an agile Scrum task and its components on the Trello card

5   Practical Application of Trello Based Agile Scrum Methodology in the Development of Real Projects

This developed Trello-based agile Scrum methodology adopted for the BSc final year computing projects at Bournemouth University, and are sponsored by industry partners. Students complete these complex engineering or computing projects towards the partial fulfilment of the requirements to obtain their BSc degree alongside industry training. The project is a year-long assignment consisting of 2–3 students working as a group on a collaborative piece of academic work which produces a project prototype and final project report. The projects arise from suggestions proposed by industry partners, with the person proposing the project becoming the sponsor for the project team representing the industry partner. In addition to the industry partner, an in-house tutor is allocated to ensure that students gain practical industry skills by working on this extended project and that academic standards and a suitable learning environment is provided to allow the students to investigate an engaging, complex problem in a safe learning environment. This Project Based Learning (PBL) method assists students in developing teamwork and problem-solving skills in addition to the ability to communicate effectively with a variety of stakeholders at varying levels within the organisation [6, 7].

The agile Scrum development methodology is selected due to the time constraint for the development of the project. During this development, students are required to maintain a number of records in a collaborative working environment. Having been taught the developed Trello-based agile Scrum methodology during their Software Engineering module of their course, where students are taught the methodology. Trello was demonstrated as a groupware and web based tool for structuring and managing the stages and components of the project through the use of the Scrum methodology in a collaborative manner. This collaborative working requires regular interaction and communication amongst the project team which typically consists of the project sponsor (external- industry expert and can be considered as a product owner), the project tutor (internal- subject expert and can be considered as a Scrum master), and a number of students (can be considered as a Scrum development team). The first task is to prepare a proposal based on meetings with both external and internal supervisors and the team. They seek to establish the requirements and user stories, in order to begin the design process, producing the first component of the project, the Terms of Reference (ToR). The students utilise Trello to manage the communication amongst all the members of the project team and their internal and external sponsors. This means that when meetings are held with these stakeholders, everyone is informed of the progress to-date as Trello constantly sends updates of the progress made to the project team. This leads to more targeted meetings addressing meaningful issues thus reducing wasted time in the meetings.

As soon as they start their project development and build their user stories as a part of their product backlog in Trello, which is communicated to all the team members as they are uploaded to Trello. These user stories once verified by the team and the internal and external stakeholders forms the initial product backlog and once complete, means the project can progress to the next stage of the process. Subsequently, the product backlog is split into number of release backlogs and sprints depending on the requirement of the project with mutual negotiations. Every sprint is a shippable product and its progress is managed in Trello through the burndown chart. This burndown chart is created at the beginning of each sprint and updated regularly to inform of any change/ update in the sprint throughout the sprint life cycle. Upon completion of the sprint, it is reviewed by the external sponsor alongside the internal tutor during the sprint review meeting. If both are content and said DoD (Definition of Done) then the project team proceeds to the development of the next sprint. The same process is repeated for all the sprints until the project is complete. It cannot be underestimated the value of utilising Trello and the skills and experience gained from employing an agile Scrum software development methodology in a real-life project. The Trello software works well in this collaborative environment and due to its clean and user-friendly intuitive UI, it is effortless to learn and assimilate. Therefore, teaching time is minimal and students can produce meaningful information efficiently and effortlessly [8].

Furthermore, Students make use of Trello to maintain their Project Development Records (PDR), which are records of progress, thoughts and issues, held on Trello, these can easily be compiled into their PDR for presentation on project completion. In addition, internal and external tutor contact is maintained and where coaching intervention is required by academic staff, it can be provided promptly through Trello. Overall, the experience of using Trello for managing a complex, problem-solving challenge has been successful, with overall project marks improving from one year to the next. Moreover, it has brought together skills they have learned during their Software Engineering lectures and allowed the students to apply the agile Scrum methodology and free software tools (Trello) to the project work in a groupware and e-learning environment.

6   Results and Analysis

The previous section explained the successful development and delivery of the Trello-based agile Scrum methodology in the Software Engineering module, and later its application into the BSc final year projects. Having introduced this methodology, it was important to investigate the effect and success of this Trello-based methodology. Therefore, the project results when Trello-based methodology is employed were further analysed and compared against project results when agile Scrum methodology is also employed for the project but without Trello. This whole process was employed in practice for the two cohorts, therefore their projects results were compared against the previous cohort who developed their projects using agile Scrum methodology but without Trello based learning. There was a slight improvement in the overall project results ($<2\%$) for both cohorts but it was not a significant finding for Trello-based methodology because this improvement could be attributed to several other components of the project.

However, when the results for both cohorts were examined in more detail and a *grouped frequency distribution* analysis performed, there was an interesting pattern seen in both results when compared with and without Trello-based agile Scrum methodology. The use of Trello-based agile Scrum in project development showed a significant improvement in the overall performance (i.e. each individual student) of class with minimum standard deviation for the both cohorts as shown in Figs. 5 and 6. This was encouraging because it showed that the Trello-based practicals and collaborative working helped all students in understanding, clarifying and correcting their project activities in a timely manner. Trello-based agile Scrum methodology was one of the major components of the project throughout its life cycle, however, the completion of the actual project requires several other stages and is important for its success. For this reason, analysing the results and its improvement may not explicitly reflect the success of the Trello-based methodology but the improvement in the class performance (i.e. each individual student) is a strong indication of the success of this proposed methodology [10]. Nonetheless, in the future, it is important to explicitly evaluate the effect and success of this Trello-based methodology in the development of the project by creating a separate mechanism for its evaluation.

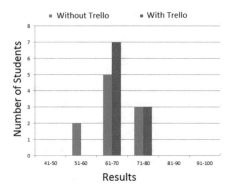

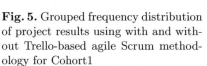

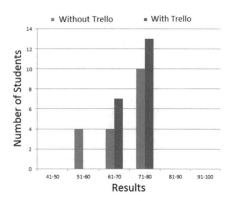

Fig. 5. Grouped frequency distribution of project results using with and without Trello-based agile Scrum methodology for Cohort1

Fig. 6. Grouped frequency distribution of project results using with and without Trello-based agile Scrum methodology for Cohort2

7   Conclusion

This paper presented the development of a Trello-based agile Scrum methodology, where Trello features were transformed to convert it to a basic Scrum tool for teaching and learning purposes at no cost. Later, it was employed in the delivery of the software engineering module in BSc courses and subsequently, in the development of the BSc final year project at Bournemouth University, UK. The effect and success of this Trello-based methodology was evaluated by comparing the project results of the two cohorts with and without Trello-based Scrum methodology. The comparative analysis of the project results revealed a significant improvement in the overall performance (i.e. each individual student) and this Trello-based practical and collaborative working helped all students in understanding, clarifying and correcting their project activities in a timely manner. Nonetheless, in the future, it is important to explicitly evaluate the effect and success of this Trello-based methodology in the development of the project by creating a separate mechanism for its evaluation. Additionally, it would be more useful to develop some case studies of this Trello-based agile Scrum methodology, which could be incorporated into this Trello-based method.

References

1. Beck, K., Beedle, M., Van Bennekum, A., Cockburn, A., Cunningham, W., Fowler, M., Grenning, J., Highsmith, J., Hunt, A., Jeffries, R., et al.: Manifesto for agile software development (2001)
2. Cohn, M.: Why there should not be a release backlog (2018). https://www.mountaingoatsoftware.com/blog/why-there-should-not-be-a-release-backlog
3. Financesonline.com: Trello review collaboration software (2019). https://reviews.financesonline.com/p/trello/
4. Hartman, B.: New to agile? INVEST in good user stories (2009). https://agileforall.com/new-to-agile-invest-in-good-user-stories/
5. Naik, N., Price, D.: Take it easy, it is games not mathematics. In: The 2nd HEA Future Directions for Higher Education in Wales Conference, Aberystwyth, UK (2014)
6. Naik, N.: A comparative evaluation of game-based learning: digital or non-digital games? In: European Conference on Games Based Learning, vol. 2, p. 437 (2014)
7. Naik, N.: Non-digital game-based learning in the teaching of mathematics in higher education. In: European Conference on Games Based Learning, vol. 2, p. 431 (2014)
8. Naik, N.: Non-digital game-based learning in higher education: a teacher's perspective. In: European Conference on Games Based Learning, p. 402 (2015)
9. Naik, N.: Crowdsourcing, open-sourcing, outsourcing and insourcing software development: a comparative analysis. In: 2016 IEEE Symposium on Service-Oriented System Engineering (SOSE), pp. 380–385. IEEE (2016)
10. Naik, N.: Dual powerpoint presentation approach for students with special educational needs and note-takers. Eur. J. Spec. Needs Educ. 32(1), 146–152 (2017)
11. Naik, N.: Software CROWD-sourcing. In: 2017 11th International Conference on Research Challenges in Information Science (RCIS), pp. 463–464. IEEE (2017)
12. Naik, N.: The use of GBL to teach mathematics in higher education. Innov. Educ. Teach. Int. 54(3), 238–246 (2017)
13. Pressman, R.S.: Software Engineering: A Practitioner's Approach. McGraw-Hill Education, New York (2014)
14. Schwaber, K.: Agile project management with Scrum. Microsoft Press (2004)
15. Schwaber, K., Beedle, M.: Agile software development with Scrum, vol. 1. Prentice Hall, Upper Saddle River (2002)
16. Schwaber, K., Sutherland, J.: The scrum guide. Scrum Alliance 21 (2011)
17. Scrum-institute.org: The scrum product backlog (2019). https://www.scrum-institute.org/The_Scrum_Product_Backlog.php

18. Scrumalliance.org: New to user stories (2019). https://www.scrumalliance.org/community/articles/2010/april/new-to-user-stories
19. Scrum.org: What is a product backlog? (2019). https://www.scrum.org/resources/what-is-a-product-backlog
20. Scrum.org: What is a sprint backlog? (2019). https://www.scrum.org/resources/what-is-a-sprint-backlog
21. Sommerville, I.: Software Engineering, 10th edn. (2015). ISBN-10, ISBN-13 **1292096136**
22. Trello.com: Trello lets you work more collaboratively and get more done (2019). https://trello.com/en

Self-Explanatory Capabilities in Intelligent Decision Support Systems in Resource Management

Maneerat Rumsamrong[(✉)], Andrew Chiou, and Lily Li

School of Engineering and Technology, Central Queensland University,
Rockhampton, QLD, Australia
{m.rumsarmong,a.chiou,l.li}@cqu.edu.au

Abstract. Self-explanations in decision support systems need to be presented in parallel with considering and understanding the outcomes of advice from the expert system. This advice can realise benefits such as increased user acceptance and confidence in the adoption of the improved system. There are numerous categories of explanation, including the following. In order for an expert system to reach a conclusion, there needs to be: (1) justification and a record of the inferential steps; (2) an explicit knowledge of the underlying argument, or (3) explanation of the rationale behind each inferential measure taken by the expert system. This recommendation will result in more persuasive justification and lead to satisfaction, and acceptance of advice. For this reason, it is proposed to announce a discourse semantics approach to an Intelligent Decision Support System framework by the inclusion of a discourse layer. In this paper, the discourse semantics layer approach will be demonstrated to show the mechanism of how a fuzzy logic based expert system utilises justification techniques for advice offered in the problem of control and resource management.

1 Introduction

An accessible intelligent system interface for serious games is regarded as being useful in advising participants in regard to deciding on particular strategies in order to accomplish specific goals. In a few circumstances, the end-user (e.g., human participant) may see an offered opinion or recommendation as not being the best available solution. The intelligent (serious games) system needs to be able to justify its outcome with options in assisting participants in making critical decisions during the action sequences. The discourse layer includes the discourse semantic functional element and the discourse output component. Due to the hybrid nature of the knowledge base data types, an explanation may not necessarily be provided in a primarily textual context. As a result, to handle all data types, the semantic discourse module in the proposed intelligent decision support system (IDSS) can be implemented as: (1) discourse using textual semantics, and (2) discourse using graphical visualisations. Limited research into explanatory capabilities in the area of fuzzy expert systems has been carried out. Hence, the need for such improvements in expert systems has long been recognised, and several frameworks provided in attempts to resolve shortcomings.

© Springer Nature Switzerland AG 2020
L. Barolli et al. (Eds.): CISIS 2019, AISC 993, pp. 356–367, 2020.
https://doi.org/10.1007/978-3-030-22354-0_32

This paper explores the design, technical aspects and outcomes of an IDSS shell in providing decision support by supplying expert advice over a range of recommendations and justification advice to support resolution in situations that involve conflict resolution and resource management. This paper presents the discourse semantics layer approach that has been undertaken in the development of CRISIS-Expert. This will be demonstrated to show the mechanism of how a fuzzy logic based expert system utilises justification techniques for advice offered in conflict resolution and resource management. In this case, it will present the problem area in the specific case of a realistic war game. Given that these types of problems are too complex and large to be approached realistically with existing resources and space, the alternative is to reduce the problem space into a manageable framework by scaling it down into a model with only the critical parameters representing the real world problem. The performance and results from the various core inference engines have been compared to the actual historical outcomes of a well-known conflict. The intention of providing game AI with similar ability to self-annotate and express the outcomes meaningfully to an intelligent human requires a mechanism that allows the system to be developed.

2 Background

Games with a purpose beyond providing engaging entertainment are termed as *serious games* and have become a focus area of recent research into game AI. AI has been commonly used in games, and AI algorithms have been innovatively developed as part of the effort to improve the quality of games. Design and development of games have, in turn, been significantly advanced through research into several AI methods that have largely occupied modern games. There are three main uses of AI in games [1]: (1) for playing games; (2) for generating content; and (3) for modelling players. Over the last decade the use of certain AI algorithms related to decision making and machine learning in serious games have been ongoing settled as a trend analysis games concerning the actual use of intelligent serious games [2]. However, it has not yet been indicated that decision-making processes based on conflict resolution and performance analysis of a serious and exciting game are best supported by computerised solutions (such AI systems are reasonably humble for individuals to develop but are difficult for us to describe how to do). Hence, there is a difference between what humans can do well and what machines are good at which was obvious in the early days of developing artificial human intelligence to and accomplish human-level problem solving or decision making skills. These problems were accessible to the machines as a set of formal mathematical notions within rather narrow and controlled spaces, which could be solved by some form of physical symbol manipulation or searches in symbolic space [1]. The most successful, formalised and symbolic representation allowed AI to succeed in many cases. Indeed, **board games** in particular were a well-known area for early AI challenges as they are formal and extremely constrained decision making environments while still being complex systems. A large part of the research community working in this field focuses on developing AI in several ways for **playing** *games* – as effectively as possible, or in the style of particular humans, or as autonomous entities. These styles are various explanatory frameworks for game AI which itself can be classified into three components, namely structured techniques, human-intuitive strategic skills, and computational intelligence (CI) [3].

Relatively recently, AI has correspondingly begun to use these components to analyse games and *model players* as they are participating in these games. The earliest use of incessant data gathering was an AI-supported analysis of the data and semi-automatic adaptation of the Facebook game *Farmville* by Zynga in 2009 and, more recently, Flying Mollusk developed the *Nevermind* game in 2016, which could track human-like behaviour and adapt the game accordingly [1]. AI research outcomes have been suggested as agents in games that have opened new perspectives in game AI. Further, game mechanics is one way of providing a training platform to measure ability and assist the design of human-like behaviour, and so are indebted to players and users of AI. This training platform is an optional of the Truing test in which judges must appropriately estimate whether an observed playing behaviour in a game is that of a human player and/or an artificial human intelligence. Therefore, in the context of this article, player modelling refers mainly to the use of AI methods for the production of CI models of players. In particular, the modelling of human-like behaviour and human experience is commonly termed *player modelling*, and this process has been underlined as an essential use of AI in games [1, 2].

AI in serious games is mostly helpful for modelling the behaviour of non-player characters in playing the game as a functional non-human entity, but that entity does not possess an inherent motivation for winning. Instead, serious games place importance on the actual process of playing that takes place. Due of the hybrid nature of the knowledge base data types in the game AI production shell, an explanation may not necessarily be implemented in a primarily textual context. Early research had indicated that computerised solutions provided by intelligent systems that offered explanations or justifications can make their recommendations more acceptable to end users [4–6]. This explanatory system was included in the main IDSS discussed in this paper.

3   Explanatory Capabilities

Previous research has established the requirement for expert system recommendations to be accompanied by explanations to assist understanding when considering offered advice [4, 5]). This explanatory capability can realise benefits, such as increased user acceptance and confidence in the adoption of the improved system. There are numerous categories of explanation, including the following. In order for an expert system to reach a conclusion, there needs to be: (1) justification and a record of the inferential steps; (2) an explicit knowledge of the underlying argument; or (3) explanation of the rationale behind each inferential measure taken by the expert system. Another way is to provide a better level goal structure that determines how the expert system uses its domain knowledge to perform a task.

Limited research into explanatory capabilities in the area of fuzzy expert systems has been carried out [5]. Hence, the need for improvements in this area of expert systems has long been recognised, and several frameworks have been provided in attempts to resolve shortcomings. Moreover, explanations may move beyond a simple tracing mechanism of what was involved in reaching particular decisions and should preferably conform to Toulmin's model of argumentation (Toulmin, 1985, cited in [5]), in providing adequate justification for the knowledge offered. This recommendation will result in more persuasive justification and lead to higher trust, agreement, satisfaction, and acceptance. With the development of discourse semantics over the last few years, an intelligent expert advisory system for control and management of parthenium weed infestation [8] has been developed. The discourse semantics approach has been demonstrated through empowering a fuzzy logic based expert system with justification techniques for advice offered on weed control and management strategies.

A user-friendly intelligent system interface for serious games is seen as convenient in advising participants in regard to choosing strategies/tactics in order to accomplish specific goals. In a few circumstances, the end-user (e.g., human participant) may see an offered opinion or recommendation as not being the best available solution. The intelligent (serious games) system needs to be able to justify its outcome with options which assist participants in making critical decisions during the action sequences. For this reason, it is proposed that a discourse semantics approach be introduced to an IDSS framework by the inclusion of a discourse layer. This layer includes the discourse semantic functional element and the discourse output component. Because of the hybrid nature of the knowledge base data types, an explanation may not necessarily be carried out in a primarily textual context. As a result, in order to handle all data types, the semantic discourse module in the proposed IDSS can be implemented as: (1) discourse using textual semantics; and (2) discourse using graphic visualisation and schematics. Implementation details of both capabilities are reported in the next two sections.

3.1 Discourse Using Textual Semantics

The conventional explanatory function often involves a format in the purely textual context of a conversation-interaction manner (e.g., printed English text). The theoretical base for the explanation function in conventional expert systems and its construction is explained by [9]. In the specific goals of the proposed IDSS framework, it is practical to extend textual explanations in conjunction with other forms of expression as part of its functionality [9].

3.2 Discourse Using Graphics and Semantics

In addition to textual interpretation, the proposed IDSS framework also needs to explain its inferential process using non-textual presentation. This need arises from the reality that the knowledge base of a few intelligent, serious games (e.g., military games) is composed of data types from board game scenarios and graphics. Therefore, a provision is made to include graphical and descriptive data types using symbolic schematics in its explanatory capabilities. Chiou and Wong [10] have proposed an auto-explanation method through symbolic reasoning. This method is presented in an implementation of game AI that is capable of describing its inferential process in order to help players understand the reasoning behind actions. The authors propose the representation of results on the forward prediction of possible movement situations. These semantics can show the results of the current position and assist in further explanation of how the system recommendations can, by judicious deployment of these currently positioned units, perform specific goals such as securing the area of conflict.

4 Methodology

The application design of the explanatory capabilities was discussed earlier, where it was shown how the individual system and subsystem components are consolidated into the IDSS framework. The core features of each component in the layered architecture were described. A detailed description of the methodology and implementation of the prototype intelligent system is provided in this section in order to establish the usefulness of the foundation of the IDSS framework as proposed in Sect. 3. The mechanics of the explanatory capabilities is the subject of this section. The overall methodology presented in this section covers four phases: pre-processing, thematic forecast, global forecast, and local forecast. All references made in the examples refer to the diagram in Fig. 1. The methodology requires each feature (e.g. terrain, fixed obstacles, removable obstacles, special forces) to be separately layered. The prediction of tactics used with assets to direct their actions are represented by using fuzzy logic applied to spatial imagery. Themes are used to simplify the assignment of appropriate fuzzy membership functions.

All sub-themes will be parallel processed via a multitasking sub-routine, where each sub-theme will attempt to predict the future direction of assets for the corresponding terrain found T_2 (Phase 2), the theme showing the forecasted direction. As shown in Fig. 2, each sub-theme will write its outcome onto the stack array for each hex in $\mathbf{B}_{T2}$. The time of writing to the stack is not linear, but takes place at random intervals as determined only by the processing time taken to infer each prediction. In order for the IDSS to provide explanatory capabilities, the *Enhanced Explanation using Visualisation Aids and Legend* (E^2VAL) technique has been designed and developed specifically for this task.

Basically, the mechanics of E^2VAL employ the use of *knowledge mapping tags* to keep a linear rule map of all relevant IF-THEN rules that have been instantiated. All rules in the knowledge base have a *tag* linked to each completed IF-THEN statement. These *tags* do not influence nor play any part in the inference process of the system. Their primary purpose is only to provide a pointer to the *entities* used in E^2VAL. These entities will hold all relevant discourse material that will subsequently be consolidated so as to provide the explanation and justification for the final recommendation.

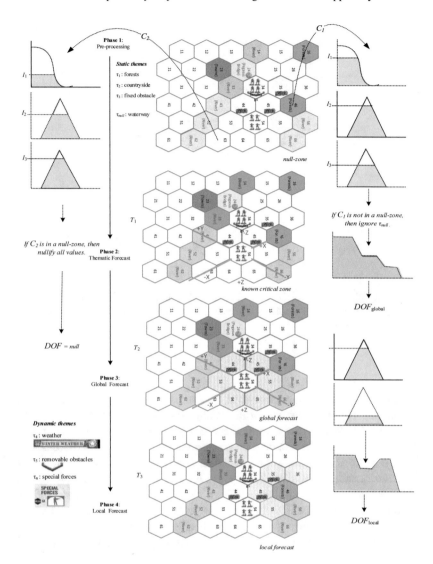

Fig. 1. Prediction of tactics used with assets to direct their actions are represented by using fuzzy logic applied to spatial imagery.

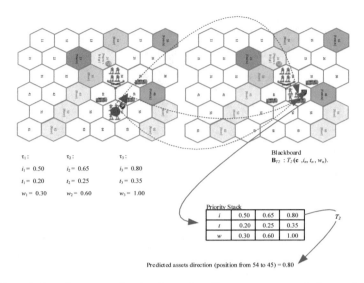

τ_1: $\quad\quad$ τ_2: $\quad\quad$ τ_3:

$i_1 = 0.50$ $\quad$ $i_2 = 0.65$ $\quad$ $i_3 = 0.80$

$t_1 = 0.20$ $\quad$ $t_2 = 0.25$ $\quad$ $t_3 = 0.35$

$w_1 = 0.30$ $\quad$ $w_2 = 0.60$ $\quad$ $w_3 = 1.00$

Blackboard

$\mathbf{B}_{T2} : T_2(\mathbf{c}, i_n, t_n, w_n)$.

Priority Stack

i	0.50	0.65	0.80
t	0.20	0.25	0.35
w	0.30	0.60	1.00

T_2

Predicted assets direction (position from 54 to 45) = 0.80

Fig. 2. To calculate the assets density in each cell, a concurrent priority stack operates simultaneously. This involves predicting the enemy's strength for each specific hex and inserting the current results in the stack.

5 Results

In the following test results, comparisons are made between the actual feedback given by the random human player, the artificial human experts, and the recommendations resulting from CRISIS-Expert's simulation. As regards the evaluation of the actual and simulated results, it should be noted that the outputs of CRISIS-Expert are descriptive and use symbolic schematics to help end-users make final *judgements*, in contrast to conventional quantitatively discrete outputs. As such, a level of qualitative interpretation is required in the evaluation of both sets of data. The primary goal of this simulation is not to test if CRISIS-Expert can provide a quantitatively accurate result, but rather to assess whether it can, in practice, provide highly useable advice and recommendations to end-users that are close to actual or best practice guidelines.

5.1 Zones of Fire

Zones of fire are areas delineated by the artificial human and CRISIS-Expert to show where the deployment strategy and subsequent directed actions are to take place. Zones of fire are not necessary for directed actions (e.g., opponents are usually unprotected or weak). The boundaries delineating these zones are indicated by a red border over the zone of interest. Sectoring of these zones is not carried out for random humans by simply identifying areas of directed action. Other factors, such as the existence of rivers, obstacles, and so on, need to be considered.

In Fig. 3(a) the artificial human has identified zones that are considered critical to the implementation of the suggested deployment strategy. Compare this to CRISIS-

Expert's output, which has identified four zones (Fig. 3(b)). Even though both sets of output have been similarly identified, the total area encompassed by the sector boundaries differs. Note that the random human prefers *close-fitting sectoring* over CRISIS-Expert's *shielded sectoring*. *Close-fitting* sectoring delineates as closely as possible the identified areas of interest. This type of sectoring normally results in fewer resources (e.g., infantry) being consumed. It is normally considered by the random human to be viable in terms of conservation of resources in the implementation of subsequent directed actions. *Shielded sectoring* ensures that there is a higher potential impact by preventing assets from being consumed within the same sector. Further outward shielded sectoring beyond this threshold is not as useful.

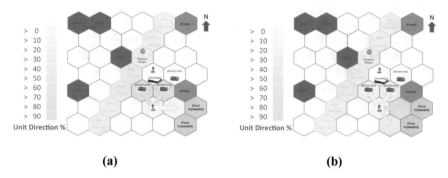

(a)   **(b)**

Fig. 3. (a). Critical sectors identified by an artificial human using *close-fitting sectoring* in order to indicate three directions of movement and focus in the battle. (b). Critical sectors identified by CRISIS-Expert.

Zones of fire sectoring plays an important role in the implementation of deployment strategies and how effective these strategies are. In this test case, CRISIS-Expert has a sectored area (Fig. 3(b)), which is the same area identified as a sector by the artificial human (Fig. 3(a)). Note that CRISIS-Expert's combination of two sectors overlaps. Depending on the subsequent recommendation of further actions to take, this overlap may be undesirable because of repeated treatment of the same area. From both attempts at critical sectoring, the artificial human and CRISIS-Expert's results were very similar and correctly identified.

5.2   Deployment Strategies and Potential Impact

The random player has suggested the *Defend type deployment strategy* (see Fig. 4(a)) for the corresponding critical sector (see Fig. 3(a)). As the areas of the *Zones of fire (critical sector)* were *close-fitting sectored*, the potential impact of the subsequent directed action is expected to be high, ranging from 50–80% effectiveness for the sector. In comparison, CRISIS-Expert recommends a similar *Defend* type *deployment* strategy for the sector (see Fig. 3(b)). There are four directions (as compared to Fig. 3 (a)), which have been identified using Shielded sectoring. Its potential impact for the corresponding sector in Fig. 3(b) is expected to be up to 80% effective. However,

CRISIS-Expert's recommended potential impact for sector 2 is a high 80–100% effectiveness. As mentioned, the sector in Fig. 3(b) is a *smaller fire zone* area that has been partitioned into a northern sector when compared to the sector in Fig. 3(a). In this way, a *deployment* strategy will have more impact because of its lesser fire zone area of coverage. This accounts for the high values assigned by CRISIS-Expert to the potential impact of the sector. Also, CRISIS-Expert's critical sector, which is the southern partition of the sector, now has a relatively more concentrated number of forces and battle range. This accounts for a lower potential impact of 50–75% effectiveness. For the sector as defined by the artificial human player (see Fig. 3(a)), the suggested *deployment* strategy is *Attack* the opponent's units. The opponent's protected units are found in this sector. Therefore, a general battle strategy cannot be applied which would result in *Defend* the protected opponent's units in the same sector. Instead, the suggested action is normal combat procedure (see the next section on directed actions). This would ensure that units are selectively reduced (e.g., when battling an enemy unit that is on a forest hex attacking with infantry units, roll one less die, and when attacking with tank units, roll two less dice). However, because of economic viability and the impracticality of such an action, its potential impact is only rated at 50%.

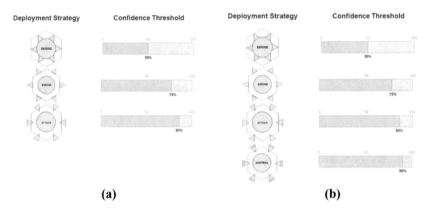

Fig. 4. (a). Control strategies suggested by an artificial human player with the expected level of confidence threshold corresponding to movement and the battle identified in Fig. 3(a). (b). CRISIS-Expert's recommended deployment strategies corresponding to critical sectors identified in Fig. 3(b).

With respect to this autonomous entity, CRISIS-Expert has recommended defended sectoring for sector 4, which is the identical critical sector suggested by the artificial human player (see Fig. 3(a)). In this case, CRISIS-Expert recommends maximum *Attack* because of the co-existing weakness in the opponent's defensive network. According to the historical record (available at www.memoir44.com under the Sword Beach scenario), the Allies are more likely to win. But, in fact, the Axis player most often secures the victory. As of the 1760th online battle report for Sword Beach (March 2013), the Axis powers had won 53% of the face-offs versus 47% for the Allies. There are traps on Sword Beach, and this result exposes the Axis powers with a view to giving the Allies the best chance of gaining a victory point in the game. CRISIS-Expert has anecdotal evidence that resource considerations make "The Death Zone," where the enemy artillery can fire on Allied units breaking into the central section, an area that the Allies must avoid at all costs unless they have a viable plan for destroying the artillery quickly (i.e., it is too costly for a default strategy). The potential impact of this control strategy is 50–100%. In *Defend*, because the Axis forces have the advantage of terrain (e.g., forests, villages, and bunkers: see Fig. 3(b)), the control strategy of "do nothing" may be difficult to achieve. The beach is very dangerous terrain because exposure may lead to an uncontrolled outcome. Thus, each player is going to have to make the most of the positives and compensate for the negatives if a potential impact at the lower end of 50% is to be achieved (see Fig. 3(b)).

In summary, the *Defend* control strategies recommended by both the human player experts and CRISIS-Expert were similar, with a potential impact of 50% and above. As for the advantage of terrain in the central section, it creates the potential for an Allied attack because of the advantage of being well-protected by bunkers, forests, and villages. The control strategy varied because of the application of different types of critical sectoring (i.e., *close-fitting* and *shielded sectoring*). However, both arrived at a similar conclusion to avoid this zone at all costs, unless there was a viable plan for destroying the artillery quickly. Here, it can be seen that CRISIS-Expert's results are comparable to the overall results provided by the expert human player.

5.3   Directed Actions and Explanations

After the human player expert and CRISIS-Expert identified the critical sections (the death zone) and control strategies, a directed action will be suggested with instructions on how to implement the follow-up actions. In comparing the symbolic schematics of the directed actions for outputs by the human player expert (see Fig. 4(a)) and CRISIS-Expert (see Fig. 4(b)), it is clear that the latter recommends the appropriate direction of Special Forces units when the control strategy employs maximum *Attack*. As attacking using Special Forces units is the most common type of attack, it is only logical that CRISIS-Expert should also come to that conclusion. This outcome is favourable when compared to the expert human player's suggestion (i.e., *Attack* using a Special Forces unit).

The directed actions recommended by CRISIS-Expert were also practical. Direction of the units focused on where the density of opponents was the strongest. As a result, there was minimal consumption and expenditure of resources. Note that in both the expert human player's and CRISIS-Expert's recommendations, the critical sector with protected skirmish indicated complete and maximum coverage over the entire sector. Complete coverage here is interpreted as "apply the recommended control strategy over the entire sector." In this case, the expert human player's strategy is to attempt to attack on the flanks, taking advantage of the relatively safe routes of passage on either side throughout the right section (see Fig. 3(a)), while CRISIS-Expert's strategy is to "take no action throughout the central section" (see Fig. 3(b)).

The textual instructions/explanation provided by the human expert for *Attack* control strategy (strategy 1 and 2 in Fig. 5(a)) is brief in comparison to the CRISIS-Expert (strategy 1, 2, and 3 in Fig. 5 (b)). In this regard, CRISIS-Expert has the benefit of referencing detailed databases which can be reproduced precisely. In this example, a list is simply reproduced as part of its *explanatory capabilities*. However, the expert human player's explanation, though brief, is more complex than that of CRISIS-Expert. CRISIS-Expert's predefined textual explanations were based solely on IF-THEN rules that have been instantiated and, hence, the explanation of the last output can be further improved. Note the conciseness of directions, and the explanation of the deployment strategy for the critical zone in Fig. 5(a) in particular. From the outcome of this test, it can be concluded that CRISIS-Expert is capable of providing explanations and comprehensive instructions by extracting detailed information from its database. Because of the unsophisticated implementation of its explanatory capabilities, it has limitations in providing explanations that are as complex as the explanations of human player experts. However, for the purpose of this application, CRISIS-Expert's explanations are, nonetheless, totally appropriate and applicable. Note the detailed guidelines and explanations are based on a predefined database. However, contrast this to (4) which has omitted any mention of other co-existing skirmishes (see Fig. 5(b)).

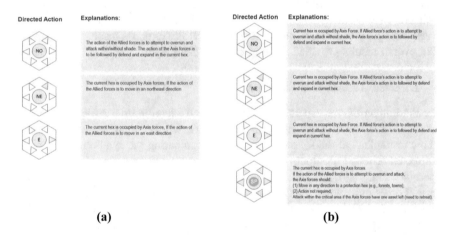

(a)   **(b)**

Fig. 5. (a). Directed actions and instructions as suggested by artificial human player. (b). CRISIS-Expert's version of directed actions and guidelines

6   Conclusion

This work set out to use CI in a prototype IDSS framework which was deployed as
   a serious game in order to demonstrate that the innovative framework was a workable proposition. The framework was implemented in an application which applied game AI to an historic war game, and provided autonomous entities with advice about operational actions and engagement strategies.

   Three primary intelligent mechanisms were adopted in this IDSS framework. These mechanisms were: fuzzy trans-consequent post adjustment operations; integration of player behaviour models into the fuzzy expert system rule base; and the ability to provide explanatory capabilities. The tests that were carried out have shown that the recommendations and advice generated by this IDSS framework compare favourably to the results provided by artificial humans about identical problem areas. As a result of the research undertaken, further work needs to be carried out on game AI with the aim of creating new intelligent systems which can provide the player with a comparatively real experience. Research may also be needed to improve the testing methodology currently used in developing serious games and to model game-playing AI with software capable of playing better than or similar to humans, even while bearing in mind that the end goal of such research will keep changing into the future.

References

1. Yannakakis, G.N., Togelius, J.: Artificial Intelligence and Games. Springer International Publishing, Cham, Switzerland (2018)
2. Frutos-Pascual, M., Zapirain, B.G.: Review of the use of AI techniques in serious games: decision making and machine learning. IEEE Trans. Comput. Intell. AI Games 9(2), 133–152 (2017). https://doi.org/10.1109/tciaig.2015.2512592
3. Rumsamrong, M., Chiou, A.: CRISIS-expert: conflict resolution and decision support in scaled-down models of serious resource management scenarios. In: Proceedings of the 2018 International Conference on Computational Intelligence and Intelligent Systems, Phuket, Thailand, pp. 14–19 (2018). https://doi.org/10.1145/3293475.3293478
4. Gregor, S.: Explanations from knowledge-based systems and cooperative problem solving: an empirical study. Int. J. Hum. Comput. Stud. 54(1), 81–105 (2001). https://doi.org/10.1006/ijhc.2000.0432
5. Gregor S., Yu, X. (2002). Exploring the explanatory capabilities of intelligent system technologies. In: Dimitrov, V., Korotkich, V. (eds.) Fuzzy Logic. Studies in Fuzziness and Soft Computing, vol 81. New York: Physica-Verl, Heidelberg
6. Ensmenger, N.: Is chess the drosophila of artificial intelligence? A social history of an algorithm. Soc. Stud. Sci. 42(1), 5–30 (2012). https://doi.org/10.1177/0306312711424596
7. Yannakakis, G.N.: Game AI revisited. In: Proceedings of the 9th Conference on Computing Frontiers, Cagliari, Italy, pp. 285–292 (2012). https://doi.org/10.1145/2212908.2212954
8. Chiou, A., Yu, X.: Industrial decision support system (IDSS) in weed control and management strategies: expert advice using descriptive schemata and explanatory capabilities. In: Proceedings of the 33rd Annual Conference of the IEEE Industrial Electronics Society, pp. 105–110 (2007). Retrieved from http://ieeexplore.ieee.org/stamp/stamp.jsp?tp=&arnumber=4460407&isnumber=4459874
9. Gregor, S., Benbasat, I.: Explanations from intelligent systems: theoretical foundations and implications for practice. MIS Q. 23(4), 497–530 (1999). https://doi.org/10.2307/249487
10. Chiou, A., Wong, K.W.: Auto-explanation system: player satisfaction in strategy-based board games. In: Nakatsu R., Tosa N., Naghdy F., Wong K.W., Codognet P. (eds.) Cultural Computing. ECS 2010. IFIP Advances in Information and Communication Technology, vol. 333, pp. 46–54. Springer, Berlin, Heidelberg (2010)

A Model-Driven Approach for Load-Balanced MQTT Protocol in Internet of Things (IoT)

Humaira Anwer[1,2(✉)], Farooque Azam[1],
Muhammad Waseem Anwar[1], and Muhammad Rashid[3]

[1] Department of Computer & Software Engineering, CEME,
National University of Sciences & Technology (NUST), Islamabad, Pakistan
humairaanwer13@ce.ceme.edu.pk,
{farooq,waseemanwar}@ceme.nust.edu.pk
[2] Department of Computer Science, Khwaja Fareed University of Engineering
& Information Technology, Rahim Yar Khan, Pakistan
[3] Computer Engineering Department, Umm al-Qura University,
Mecca, Saudi Arabia
mfelahi@uqu.edu.sa

Abstract. Internet of Things (IoT) has changed view of people about devices and has paved new ways of interaction. These devices form a powerful network that can be controlled from anywhere anytime. Though this is beauty of IoT but it comes with high price of vulnerability and high load. Broker based IoT protocols enable two way communication. One of these protocols is Message Querying Telemetry Transport (MQTT). Load balancing in IoT is huge problem when it comes to application of such publish subscribe models. This issue provides loophole for adversaries to exploit by attempting various attacks on network e.g. Denial of Service (DoS), Distributed Denial of Service (DDoS) etc. These attacks not only deprive network of its bandwidth but also are threat to confidentiality and privacy of any IoT based system. In this paper, we have tried to investigate load balancing problem by investigating different MQTT based threat models and proposing a UML profile for effective handling of load balancing in IoTs employing MQTT protocol. Intuition of proposed profile is to introduce lightweight extension that can provide a robust load-balanced version of MQTT protocol. Validation of profile is performed through case study of Proactive Technical Support System for HealthCare. From validation it is evident that proposed profile is robust in handling load balancing shortcomings in IoT ecosystems.

1 Introduction

Lately Internet of Things (IoT) has gained much popularity in field of information and communication because of its state of art services and ease of access. From households to industries, people are moving towards this technology for increasing quality of life by enjoying connections between virtually anything ranging from number of diverse devices like appliances, medical devices (e.g. MRI machines), automobiles, fire alarms, burglar alarms, sensors etc. With huge support provided for connecting diverse devices

© Springer Nature Switzerland AG 2020
L. Barolli et al. (Eds.): CISIS 2019, AISC 993, pp. 368–378, 2020.
https://doi.org/10.1007/978-3-030-22354-0_33

by IoT there are many associated challenges and issues attached; that are needed to be dealt with for smooth working of the aforesaid system.

Connections between devices are formed using various standard protocols that are specialized for use in IoT based infrastructure. By 2020 it is approximated that 20 billion devices are likely to be connected via IoT [1]. For several reasons IoT devices are potential target for attackers. Some possible reasons include; connections via internet [2], facility of being remotely controlled [3], lack of extensive security measures due to constraints on resources [2] and data content communicated and generated over IoT is usually sensitive and confidential [4]. Security is of last concern to vendors while developing IoT because of constraints like time to ship, cost, ease of access and user friendliness [5, 6].

Message brokers enable IoT to transfer and receive messages directly over internet [7]. Making IoT accessible over internet increases vulnerability of IoT devices and makes them any easy target for adversaries. Security risks are further fueled by the fact that security wise IoT devices are not been given importance due to resource constraints as discussed in [2]. To facilitate communication both in-between devices and in-between devices & servers, various protocols are available [8, 9]. One of the most adopted is Message Queue Telemetry Transport (MQTT) protocol [10]. MQTT has gained popularity in the field over the years due to its effective power consumption and comparatively low overhead [9, 10]. MQTT uses publish subscribe message broker based model to enable message exchange facility between clients [10]. Where clients are typical IoT devices.

Traditional clients/devices are rigid and have low dependability thus deploying a large scale IoT system becomes a challenge. Furthermore handling large number of requests from diverse clients simultaneously is a much studied problem in IoT [11]. Not only it hinders smooth working of IoT ecosystem but also gives rise to load balancing issues. This loophole of IoT is major fact that can be exploited by adversaries and attacks like Denial of Service (DoS) can be launched to halt the system and gain internal sensitive information. Thus it is very critical to make MQTT based IoT architecture robust against security threats and load balancing issues. Various studies are available which suggest using firewall plugins to make existent communication protocols robust against load balancing issues. Devising a generic firewall plugin that performs well in all IoT environments given the fact that IoT is very diverse is the problem statement considered in this study. We proceed by providing a UML profile based solution by enhancing existing MQTT model to make it robust against load balancing issues. The proposed profile incorporates lightweight plugins to ensure that model is resilient to high load requests.

This study is organized in following sections; in Sect. 2 we proceed by providing related literature review of the existing studies available in the area. Section 3 presents the proposed profile in detail and discusses validation of proposed study by providing profile application on a case study. Section 4 provides discussion on proposed model. Study is concluded in Sect. 5 which also discusses possibilities of future work in the area.

2   Literature Review

IoT ecosystem is designed in such a way that numerous (numbering to billions now-a-days) heterogeneous devices can communicate with each other. This could be considered as beauty of IoT as well as threat to security at the same time. Security and interoperability prove to be of crucial importance for smooth working of IoT operations. In this section we provide discussion on background related to architecture of IoT, its models of communication, MQTT protocol and security threats associated to it in IoT set-up.

2.1   IoT Architecture

According to [11, 12] vanilla architecture of IoT is comprised of three layers namely; Perception, Network & Application.

The first and foremost layer perception, as the name suggests is the physical layer that includes sensors and actuators which are responsible for collecting and pro-cessing data. Data collected by sensors is locally stored over cloud for future use. Network layer caters for communication that exists in between devices, in between devices & cloud or in between devices & gateways [5]. Layer responsible for security and reliability of IoT services is the application layer.

2.2   IoT Communication Model

There are different communication technologies that can be used as standalone or in combination with each other to facilitate connections to other devices and internet. These communication models are defined and tested by RFC 7452 standard [13] and are:

   i. Between Devices only (D&D)
  ii. Between Devices & Cloud (D&C)
 iii. Between Devices & Gateways (D&G)
  iv. Backend Data Sharing pattern

To facilitate communication over these models protocols are used. Many IoT protocols have been devised for the purpose [15]. For bidirectional communication environment publish subscribe architecture supporting protocols are considered suitable. Furthermore other environments that support large scale deployment of devices also call for the usage of publish subscribe protocols [16]. Publish subscribe supports way of messaging that facilitates message delivery from publishing module to the subscribing module following a predefined time constraint [17]. There are many publish subscribe protocols available for use in IoT but one of the mostly adopted protocol is MQTT. MQTT has many advantages over other protocols which makes it an ideal choice for IoT [18].

2.3 MQTT Protocol

Standardization of MQTT is credited to Organization for Advancement of Structured Information (OASIS) [9, 20]. Though originally MQTT was intended for remotely operated environments but it became basic IoT requirement for enabling communication because of its model simplicity, ease of use, and low overhead in terms of bandwidth [21]. MQTT architecture is designed on concept of publisher subscriber broker based environment. MQTT architecture uses fixed header of 2 bytes only. This is considered as one of the advantages of MQTT protocol as it results in low packet overhead. Furthermore MQTT has its basis on famous TCP protocol which makes it reliable in terms of delivery of messages. Control packets are of several types, some of them are SUBSCRIBE, SUBACK, PUBACK, PUBLISH, CONNACK, CONNECT and DISCONNECT [20]. Based on message delivery requirements there are three levels of Quality of Service (QoS) of MQTT protocol used by publishers [18]. Serious threats are posed by compromised messages brokers to the numerous IoT devices when they exchange messages. Cyber-attacks against infrastructure of IoT can be halted by developing some precautions and keeping an eye on the sensitivity of these applications.

DoS attacks in an IoT environments are usually caused by distorting message broker. Crashing of server or dribbling of messages is due to overtire broker resources. For large size payload messages received by IoT clients the requirement of resources increases. Denied services are the result of the overtire client and broker resources when they are being attacked. TCP based DoS attacks can be initiated against broker to overtire the broker bandwidth. Rival can utilize the QoS levels provided by MQTT protocol [18] to initiate DoS attacks. Level 2 published messages require more broker resources in contrast to QoS level 1 and 0 as discussed in [18].

3   Proposed Profile

Message brokers should implement measures to detect & block malicious client behavior. Our aim is to use MQTT protocol and devise a threat resilient load balancing model by using readily available capabilities of model. We are aiming to enhance the model by adding some check and balance parameters that have been effectively described in the proposed UML profile in Figs. 1 and 2.

One of the state of art techniques for development of applications is Model Driven Architecture (MDA). MDA is well known to significantly reduce complexity of design and verification of the system [21]. MDA provides developers with UML profile to create customized models for any specialized system. UML profile typically comprises of four main elements: (1) Metaclass to show classes, (2) Stereotypes to show extensions of metaclasses, (3) Interface to depict interfaces in profile, (4) Operation delivers functionality in profile. UML profile repository doesn't end here, it has a very good palate of almost all concepts available in original UML. In this study we have attempted to resolve load balancing problem in MQTT protocol by proposing a UML profile extension using model driven approach based on UML profile modeling. The proposed profile has its basis on MQTT model and attempts to enhance fundamental

MQTT concepts by introducing specialized modules to cater for load balancing. The proposed profile is based on class diagram to provide detailed structural view.

3.1 Description

Proposed profile is divided into two sub-profiles for simplicity and ease of understanding. These are:

  i. MQTT-Threat Model (please see Fig. 1)
 ii. MQTT-Message (please see Fig. 2)

MQTT Threat Model. Detailed information related to proposed profile stereotypes, base class, tagged values, operations and description pertaining to each class is provided in this sub-profile. Total 11 stereotypes have been used to devise load balancing specialized model for MQTT Threat Model shown in Fig. 1.

Broker Server. **Description:** It is part of MQTT protocol and is message broker responsible for routing messages. **Base Class:** Class. **Tagged Values:** It has one tagged value given as: *(1) Total_Subscriptions: EInt[1] keeps track of total devices subscribed to the broker.* **Operations:** It has one operation given as: *(1) ForwardMessage() it forwards messages to and from publishers and subscribers.*

Topic. **Description:** It is subject of interest to which subscribers can register. Topic is generalized stereotype of Broker Server. **Base Class:** Class. **Tagged Values:** It has one tagged value given as: *(1) TopicID: EInt[1] assigns unique ID to every Topic.*

IoT Device. **Description:** Device connected in the IOT network. **Base Class:** Class.

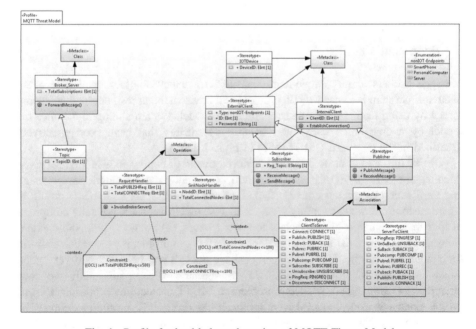

Fig. 1. Profile for load balanced version of MQTT Threat Model

Tagged Values: It has one tagged value given as: *(1) DeviceID: EInt[1] Identity of IoT device in the ecosystem.*

External Client. **Description:** Client outside the network trying to connect to the network. **Base Class**: Class. **Tagged Values:** It has three tagged values as: *(1) Type: nonIOT-Endpoints[1] this is enumeration which is defined to handle different types of endpoints e.g. smartphone, computer etc. (2) ID: EInt[1] each external client has unique ID (3) Password: EString[1] authentication code to connect to IoT network.*

nonIOT-Endpoints. **Description:** *This is an Enumeration to handle variations in types of Endpoint clients that can establish connection over IOT.*

Internal Client. **Description:** Client inside the system. **Base Class**: Class. **Tagged Values:** It has one tagged value given as: *(1) ClientID:EInt[1] each client has unique ID.* **Operations:** It has one operation given as*: (1) EstablishConnection() enables the client to establish connection with IoT system.*

Subscriber. **Description:** Entity receiving messages from broker. This stereotype is generalized from both internal & external client. **Base Class:** Class. **Tagged Values:** It has one tagged value given as: *(1) Reg_Topic:EString[1] topic that is subscribed by subscriber.* **Operations:** It has two operations given as: *(1) Re-ceiveMessage() enables subscriber to receive message routed by broker (2) SendMes-sage() enables subscriber to send message in response.*

Publisher. **Description:** Entity sending the messages to broker. This stereotype is generalized from both internal & external client. **Base Class:** Class. **Operations:** It has two operations given as: *(1) PublishMessage() enables publisher to publish message via broker (2) ReceiveMessage() enables publisher to receive message routed via broker.*

ClientToServer. **Description:** Connection points from client to broker. **Base Class:** Association. **Tagged Values:** It has 10 tagged values given as: *(1) Con-nect:CON-NECT[1] to send connect request from client to server. (2) Publish:PUBLISH[1] to send publish message (3) Puback:PUBACK[1] to send publish acknowledgement (4) Pubrec:PUBREC[1] to ensure receiving of publish re-quest (5) Pubrel:PUBREL [1] to release publish request (6) Pubcomp:PUBCOMP[1] message to ensure publish request is completed (7) Subscribe:SUBSCRIBE[1] request for subscription from client to server (8) Unsubscribe:UNSUBSCRIBE[1] request from client to server to unsub-scribe (9) PingReq:PINGREQ[1] request for PING (10) Disconnect:DISCONNECT[1] message for client disconnecting.*

ServerToClient. **Description:** Connection points from broker to client. **Base Class:** Association. **Tagged Values:** Most of the tagged values are similar to that of Cli-entToServer stereotype as shown in Fig. 1. Those are not being described here again and focus is on unique tagged values only. It has 4 unique tagged values given as: *(1) PingResp:PINGRESP[1] to send PING response back to client from server (2) Un-SuBack:UNSUBACK[1] connection point to send un-subscription acknowledgement back to client (3) SuBack:SUBACK[1] to send subscription acknowledgement back to client (4) Connack:CONNACK[1] to send successful connection establishment acknowledgement back to client.*

SinkNodeHandler. **Description:** Keeps records of total number of nodes in the net-work. **Base Class:** Interface. **Tagged Values:** It has 2 tagged values given as: *(1) NodeID: EInt[1] every node connected to the network has unique ID defined by this* *tagged value (2)TotalCONNECTEDNodes:EInt[1] keeps record of total nodes connected to the network.*

RequestHandler. **Description:** Request handler does the actual load balancing by keeping track of all the messages being sent to the broker. This is ensured by keeping record of total PUBLISH and CONNECT requests sent to the server by a single client or group of clients simultaneously. **Base Class:** Interface. **Tagged Values:** It has 2 tagged values given as: *(1) TotalCONNECTREQ:EInt[1] keeps record of total CONNECT requests to server (2) TotalPUBLISHREQ:EInt[1] keeps record of total PUBLISH requests to server.* **Operation:** It has one operation given as: *InvokeBrokerServer () that invokes broker server when CONNECT and PUBLISH requests are below than some selected threshold (say 100 and 500 respectively).*

MQTT Message. This sub-profile is specialized for explaining details pertaining to messages available in MQTT-protocol. This part of profile is devised for the purpose of handling receiving and dispatch of all types of messages over IoT using MQTT-protocol shown in Fig. 2. Additionally purpose of explaining messages in separate sub-profile is to reduce complexity of overall profile and to support reuse.

Some preliminary details of messages have already been discussed under ClientToServer and ServerToClient section of MQTT-Threat Model sub-profile. Here we will provide details that enable usage of these messages as connection points. This sub-profile supports 14 stereotypes and one enumeration.

CONNECT. **Description:** Connection point to send connect request from client to server. **Base Class:** Property. **Tagged Values:** It has 1 tagged value given as: *(1) encoding:EString[1] ensures that UTF encoding is enabled for optimization of communication of text messages.*

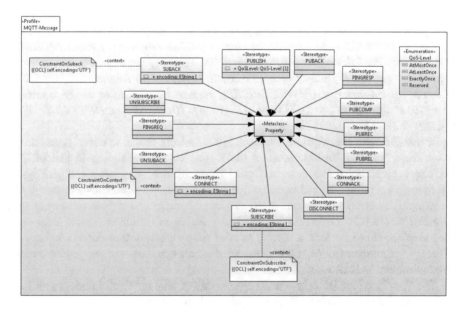

Fig. 2. Profile for MQTT Message

PUBLISH. **Description:** Connection point to send publish message. **Base Class:** Property. **Tagged Values:** It has 1 tagged value given as: *(1) QosLevel:Qos_Level[1] type of message to be published according to granted QoS Level.*

PUBACK. **Description:** Connection point to send publish acknowledgement from server to client. **Base Class:** Property.

PINGRESP. **Description:** Connection point to send PING response back to client from server. **Base Class:** Property.

PUBCOMP. **Description:** Connection point to ensure publish request is completed. **Base Class:** Property.

PUBREC. **Description:** Connection point to ensure receiving of publish request. **Base Class:** Property.

PUBREL. **Description:** Connection point to release publish request. **Base Class:** Property.

CONNACK. **Description:** Connection point to send successful connection establishment acknowledgement back to client. **Base Class:** Property.

DISCONNECT. **Description:** Connection point to send message for client disconnecting. **Base Class:** Property.

SUBSCRIBE. **Description:** Connection point to send request for subscription from client to server. **Base Class:** Property. **Tagged Values:** It has 1 tagged value given as: *(1) encoding:EString[1] ensures that UTF encoding is enabled for optimization of communication of text messages.*

UNSUBACK. **Description:** Connection point to send un-subscription acknowledgement back to client from server. **Base Class:** Property.

PINGREQ. **Description:** Connection point to send request for PING. **Base Class:** Property.

UNSUBSCRIBE. **Description:** Connection point to send request from client to server to unsubscribe. **Base Class:** Property.

SUBACK. **Description:** Connection point to send subscription acknowledgement back to client. **Base Class:** Property. **Tagged Values:** It has 1 tagged value given as: *(1) encoding:EString[1] ensures that UTF encoding is enabled for optimization of communication of text messages.*

QoS-Level. **Description:** This is an Enumeration to handle quality of service (QoS) levels of connection points [18]. This is to ensure that publish subscribe model supports original architecture level concepts of MQTT protocol as defined by standard.

3.2 Validation

In this section we provide validation of proposed profile through a case study of Proactive Technical Support System for HealthCare. This system is internet-based and provides communication between medical practitioner and internet enabled MRI machine. Please see Fig. 3 for case study used for validating proposed profile. This use case involves factors such as connection establishment, successful delivery and receiving of messages. We have taken case study of MRI Machine which is equipped with various sensors for information receiving and dispatch. Machine can also communicate with cloud services for saving and reporting information. Medical practitioner and MRI machine maintain constant network connectivity with cloud services during working hours of the clinic/hospital. For monitoring patient's status doctor/medical practitioner needs to access data from cloud even when office hours are over and in remote locations in case of emergencies. MRI machine must update data continuously to avoid any mishap due to technological shortcomings as medical field is considerably critical and needs to be accurate.

In discussed scenario as shown in Fig. 3, MRI System acts as IOTDevice and Inter nalClient of the system. MRI at one end is connected to SinkNodeHandler which keeps record of all active devices in the system at a single point in time. SinkNodeHandler is constraint enabled guard that allows for only specific number of devices to be in active state at a single point in time (say 100). At the other end MRI is connected to Internet server that acts as Subscriber of the system. MRI system sends CONNECT request to the Internet server that send back acknowledgement via CONNACK. At the other end Internet requests are directly received by Request-Handler which is also a constraint enabled guard that allows for a specific number of requests to be received (say 100). When constrained condition is satisfied, it invokes broker server and passes the request on to it via CONNECT message. If number of requests exceed the specific number (say 100), it restricts access to the broker server which in turn protects broker server from crashing down and consequently handles load balancing problem. Once condition is satisfied, PUBLISH request is sent from Server to Doctor. Doctor confirms by sending PUBACK request to the Server. Doctor acts as PUBLISHER of the scenario.

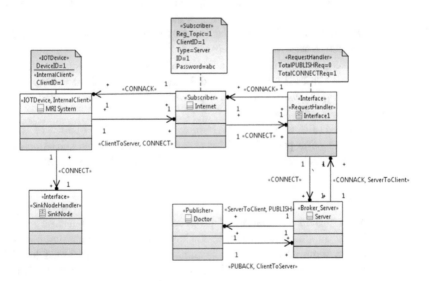

Fig. 3. Model of Proactive Technical Support System for Healthcare using Proposed Profile

4 Discussion

There are many underlying complexities that make IoT system unpredictable, complicated and complex [22]. Due to multidimensionality of complexities in IoT ecosystem it is not practically feasible to perform initial testing at early stages in real-time environment, thus validation can be performed using modeling techniques. There is dire need to cater for load-balancing issues in IoT which has proven to be one of the major pitfalls of the technology lately. Considering important role that MQTT plays in directing communication over IoT; a high abstraction layer for MQTT has been provided in the proposed profile. Underlying core concept for proposing profile is to introduce MQTT based model that facilitates control of load balancing in IoT with-out increasing complexity of existing model. Additionally, targeted and specialized MQTT implementations can also be generated automatically from proposed profile. Profile modeling enables rapid prototyping that results in effective decision making and improved quality.

5 Conclusion

Large number of enterprises are joining IoT due to its high efficiency. Hence there is dire need to protect IoT ecosystems from load balancing and misuse of resources. It is desirable to yield effective solutions that can strengthen existing IoT systems in terms of impeding and stopping attacks like DOS and handling load balancing issues. In this study we have attempted to present a UML profile based on typical IoT infrastructure that specifically emphasizes MQTT protocol based deployments in IoT. Proposed profile is easy to deploy and manages communication control based mechanisms particularly in load balancing scenarios. Total 25 stereotypes are supported by proposed profile model to effectively provide load balancing capabilities by introducing lightweight enhancements in existing MQTT protocol architecture. This capability enables application of profile to any target system for early verification. For validation purpose, we have applied profile to case-study of proactive technical support system for patient healthcare monitoring. The study is still under progress and as part of our future work we are aiming to investigate and model various attacks that can impact IoT environment particularly in cases where MQTT protocols are concerned.

References

1. Evans, D.: The internet of things: how the next evolution of the internet is changing everything (2011)
2. Heer, T., Garcia-Morchon, O., Hummen, R., Keoh, S., Kumar, S., Wehrle, K.: Security challenges in the ip-based internet of things. Wireless Pers. Commun. **61**(3), 527–542 (2011)
3. Li, S., Da Xu, L., Zhao, S.: The internet of things: a survey. Inf. Syst. Front. **17**(2), 243–259 (2015)

4. Sadeghi, A.-R., Wachsmann, C., Waidner, M.: Security and privacy challenges in industrial internet of things. In: 52nd ACM/EDAC/IEEE, Design Automation Conference (DAC) (2015)
5. Rose, K., Eldridge, S., Chapin, L.: The internet of things: an overview. The Internet Society (ISOC), pp. 1–50 (2015)
6. Penttinen, J.: Wireless Communications Security: Solutions for the Internet of Things. Wiley, Chichester (2016)
7. Waher, P.: Communication patterns for the internet of things (2016)
8. Cisco: Securing the internet of things: A proposed framework (2012)
9. Al-Fuqaha, A., Guizani, M., Mohammadi, M., Aledhari, M., Ayyash, M.: Internet of things: a survey on enabling technologies, protocols and applications. IEEE Commun. Surv. Tutor. 17(4), 2347–2376 (2015)
10. Waher, P.: Learning Internet of Things. Packt Publishing Ltd., Birmingham (2015)
11. Yang, Z., Yue, Y., Yang, Y., Peng, Y., Wang, X., Liu, W.: Study and application on the architecture and key technologies for IoT. In: International Conference on Multimedia Technology (2011)
12. Wu, M., Lu, T.-J., Ling, F.-Y., Sun, J., Du, H.-Y.: Research on the architecture of internet of things. In: 3rd International Conference on Advanced Computer Theory and Engineering (ICACTE) (2010)
13. Tschofenig, H., Arkko, J., Thaler, D., McPherson, D.: Architectural considerations in smart object networking, rfc 7452. Tech. Rep (2015)
14. Rose, K., Eldridge, S., Chapin, L.: The internet of things: an overview. The Internet Society (ISOC), pp. 1–50 (2015)
15. Zhang, Z.-K., Cho, M.C.Y., Shieh, S.: Emerging security threats and countermeasures in IoT. In: Proceedings of the 10th ACM Symposium on Information, Computer and Communications Security, New York (2015)
16. Olivieri, A.C., Rizzo, G., Morard, F.: A publish-subscribe approach to IoT integration: the smart office use case. In: IEEE 29th International Conference on Advanced Information Networking and Applications Workshops (2015)
17. Tarkoma, S.: Publish/Subscribe Systems: Design and Principles. Wiley (2012)
18. Syed, N.F., Zubair, B., Craig, V., Ahmed, I.: Modelling and evaluation of malicious attacks against the IoT MQTT protocol. In: IEEE International Conference on Internet of Things (iThings) (2017)
19. Banks, A., Gupta, R.: Mqtt version 3.1. 1. OASIS standard (2014)
20. Niruntasukrat, A., Issariyapat, C., Pongpaibool, P., Meesublak, K., Aiumsupucgul, P., Panya, A.: Authorization mechanism for mqtt-based internet of things. In: IEEE International Conference on Communications Workshops (ICC) (2016)
21. Anwar, M., Rashid, M., Azam, F., Kashif, M.: Model-based design verification for embedded systems through SVOCL: an OCL extension for System Verilog. J. Des. Autom. Embed. Syst. 21(1), 1–36 (2017)
22. Chernyshev, M., Baig, Z., Bello, O., Zeadally, S.: Internet of things (IoT): research, simulators, and testbeds. IEEE Internet of Things J. 5(3), 1637–1647 (2017)

From an Annotated BPMN Model to a Use Case Diagram: DESTINY Methodology

Nourchène Elleuch Ben Ayed[1]($\boxtimes$) and Hanêne Ben Abdallah[2]

[1] Higher Colleges of Technology, Abu Dhabi, UAE
nbenayed@hct.ac.ae
[2] Higher Colleges of Technology, Dubai, UAE
hbenabdallah@hct.ac.ae

Abstract. The necessity of aligning an enterprise's Information System (IS) model to its Business Process (BP) model is incontestable. However, the main difficulty of establishing/maintaining BP-IS models alignment stems from the dissimilarities in the knowledge of the information system developers and the business process experts. To face these predicaments, we propose a Model Driven Architecture (MDA) compliant methodology that aims to automate the generation of UML models from BPMN models. It allows mastering transformation from Computation Independent Model (CIM) to Platform Independent Model (PIM). The CIM level expresses the BP, which is modelled through the standard BPMN (Business Process Modelling Notation) and, at the PIM level represents the aligned, IS model is generated as use case diagram. Its originality resides in the CIM to PIM transformation which accounts for the BP structural and semantic perspectives in order to generate an aligned IS model that respects the best-practice granularity level of a use case.

1 Introduction

In the business world, the goals of any enterprise are accomplished through a set of coordinated activities called Business Processes (BP). A BP is described by a set of business models that encapsulate the core business logic in terms of strategies, tasks and policies. The activities of a business process manipulate and generate data that represent the daily transactions within the enterprise. To facilitate the management of this data, an enterprise relies on an information system.

The development of an Information System (IS) that supports the business activities and objectives of the enterprise has been an active niche of research in software engineering, in particular, under the scope of Process-Aware Information Systems (PAIS) [1]. The tight correlation between the IS and BP prompted researches [2] to consider deriving and extracting the PAIS requirements from business process models, rather than adopting classical requirements engineering techniques, such as brainstorming, interviews, etc.

On the other hand, the Model Driven Architecture (MDA) [4] is recognized as a promising approach for moving the complexity of system development from programming to modeling. It overcomes several development challenges through the separation of concerns, emphasizing modeling, and using model transformations.

© Springer Nature Switzerland AG 2020
L. Barolli et al. (Eds.): CISIS 2019, AISC 993, pp. 379–390, 2020.
https://doi.org/10.1007/978-3-030-22354-0_34

According to OMG, an MDA-based development life cycle can start by the creation of a Computation Independent Model (CIM) from which a Platform Independent Model (PIM) can be derived. The CIM depicts the business model of the system without presenting its construction details, and the PIM describes the functional requirements of the system independent of any platforms. The PIM is transformed to one or more Platform Specific Models (PSM) from which the code is generated.

In this paper, we propose a model-driven development approach called moDel-driven procESs-aware requiremenTs engineerINg methodology (DESTINY) that generates IS functional requirements from business models in order to build a PAIS Our approach accounts for the alignment between the BP and IS in order to provide a consistent way to generate a functional requirement model that fulfills the business needs and expectations. More specifically, we show in this paper how to generate a UML use case diagram, representing the PIM, from a BPMN model [3] representing the CIM. To do so, we tackle the differences in the semantics and granularity levels between UML and BPMN by defining a set of transformation rules that account for both the structural and informational perspectives of the BPMN model. Besides generating IS-BP aligned requirements, our approach has the merit of producing a use case diagram that satisfies best-practice quality criteria [9]. For instance, unlike the works of [10] and [11] which generates a use case from each BPMN task, our approach respects the best-practice granularity level of a use case, which recommends that a use case represents a non-atomic task covering one main flow and its variants.

The remainder of this paper is structured as follows: Sect. 2 analyses the main related work that deals with existing approaches for requirements engineering. Section 3 presents a detailed description of DESTINY methodology. Section 4 illustrates our methodology through a case study and evaluates the quality of the generated diagram by considering the recall and precision rates. Finally, in Sect. 5, we summarize the presented work and outline its extensions.

2   Related Work

To cover the three main phases of model-driven development in its MDA version, the functional requirements must first be modelled at the CIM level, from which the design (PIM) and ultimately the code (PSM) would be derived through a set of transformations. In the case of PAIS development, most works ignore the CIM level, and very few use the business requirements as a basis for functional requirements generation.

Table 1 summarizes the most relevant works related to CIM-to-PIM transformations, which we compared to answer the following research questions: What languages are used to define source and target models? Is the semantics of the modeling languages considered in the transformation? Is the transformation completeness considered? Is the transformation formalized?

The results of this comparison show that six out of seven methods use BPMN and UML for representing CIM elements, and one method [12] uses Data Flow Diagram (DFD). For the PIM level, all methods use UML to specify the software requirements from the structural and/or behavioral viewpoints. In addition, we note a

Table 1. Synthesis of related works.

	[11]	[8]	[14]	[13]	[12]	[10]
CIM model	BPMN	BPMN, UCD	BPMN	Enterprise model	DFD	AD
PIM model	UCD, CD, PD, & SD	CD & SD	UCD	UCD	UCD, AD, SSD, & CD	Comp D & CD
Mapping	1:1	1:1	n:m	n:m	1:1	1:1
Transformation completeness	No	No	Yes	No	No	No
Transformation formalization	ATL	QVT	ATL	QVT	No	No

predominance for the Use Case Diagram (UCD) to define the PIM model, except for [8] which uses it in the CIM level.

Most of the proposed transformations are syntactic, except the work of [14]. In addition, they propose neither requirements relationships derivation, nor requirements refinement. They rely on a 1-1 mapping between the elements of the source and target meta-models; for instance, the works of [10] and [11] transform any BPMN activity into a use case in spite of the different levels of granularity of the modeling languages. Consequently, the quality of the generated UCD does not respect best-practices guidelines [9].

The work of [13] defines eleven heuristics to semi-automate the generation of functional requirements from enterprise models. These heuristics rely on a many-to-many (n:m) syntactic mapping between the concepts of meta-models. However, none of the proposed heuristics considers the domain knowledge in the requirements refinement, even though the authors highlight the necessity to consider it.

Similar to our approach, the work of [14] also uses a (n:m) mapping that focuses on the derivation of relationships (include and extend) between use cases, from an annotated BPMN model. This works defines a set of BPMN patterns and their ATL transformations.

Finally, these related works show an increased interest in the application of MDA for requirements engineering in order to achieve the BP-IS alignment. However, most of them propose a syntactical CIM-to-PIM transformation without taking into account neither the domain knowledge nor the semantic differences of the modeling languages. In addition, none of the existing approaches clearly addresses the quality of the generated PIM. According to [15], the quality can be addressed from syntactic and semantic views. On the one hand, the concern at the syntactic level is the correctness of the obtained model, in terms of the model's conformity with its meta-model. On the other hand, the semantic level reveals two concerns, namely the validity of use cases' construction and composition as well as the completeness of use cases regarding the domain. We note that, in the existing works, the generated PIMs are validated from syntactic and semantics viewpoints.

The work presented in this paper attempts to overcome some of these limitations by: (1) defining a set of patterns for generating a UCD that considers the semantic of both source and target modeling languages; (2) implementing these patterns in terms of ATL rules; and (3) evaluating the quality of the generated model in terms of its syntactic conformity, the absence of anti-patterns [9], and through the calculation of recall and precision rates.

3 DESTINY Methodology

DESTINY is a model-driven and process-aware requirements engineering method that improves the IS effectiveness and reduces the risk of creating a requirements model that does not correspond to business needs and expectations. More specifically, it derives the functional requirements model (PIM) from a business process model (CIM) that is supposed to be representative of the real world of the enterprise. Towards this end, DESTINY accounts for both the structural and semantic perspectives of both models.

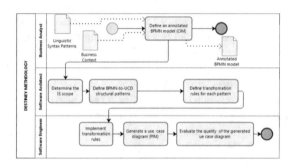

Fig. 1. Design methodology of DESTINY for CIM-to-PIM transformation

Figure 1 illustrates the DESTINY process for CIM-to-PIM transformation based on semantic and structural information. DESTINY involves three stakeholders who are the business analyst, software architect, and software engineer; and operates in three phases:

1. The The pre-processing phase during which the *Business Analyst* first defines the BPMN business model (CIM) according to BPMN syntactic meta-modeling rules. To guide the transformation and alleviate the complexity of the identification of use cases, we defined a set of linguistic syntax patterns to annotate the BPMN model as well as a business context to enhance it with semantic information related to the business logic and organizational aspect (see Sect. 3.1).
2. The transformation-definition phase during which the *Software Architect* defines the CIM-to-PIM transformations. In an MDA-compliant approach, the CIM-PIM transformation operates at the meta-model level. However, the 1:1 mapping between the CIM and PIM meta-model elements is not sufficient to preserve the

semantics of neither the business domain nor the modelling languages. To overcome this deficiency, the software architect should first identify and enumerate all patterns that respect the semantics of both the source and target languages (see Sects. 3.2, 3.3, and 3.4).

3. The transformation-implementation phase during which the *Software Engineer* formalizes/implements the transformation rules, which provides for the automated generation of the PIM model (a use case diagram).

3.1 Linguistic Patterns and Business Context

We defined a set of Linguistic Syntax Patterns (LSP) and a business context to guide the business expert in the annotation of the business processes. The LSPs are used to define the description field of a BPMN element (LSP1) and label the BPMN tasks (LSP2), while the business context is used to complement the BPMN elements with semantic information related to their functional and organizational perspectives.

Table 2. Linguistic syntax patterns

Linguistic pattern	Description
LSP1	BusinessObject + VerbalGroup + [Quantifier] + BusinessObject
LSP2	ActionVerb | CommunicationVerb + BusinessObject | NominalGroup + [[to ReceiverName] | [from SenderName]]

Table 2. summarizes the syntax of the linguistic patterns. We mean by *BusinessObject* any entity that describes the business logic. The *NominalGroup* is a set of pre/post-modifiers, which are centered around a *HeadWord* that constitutes the *BusinessObject*. The pre-modifiers (respectively post-modifiers) can be a noun, an adjective, or an ed/ing-participle (respectively, a noun, an adjective, or adverb). The *VerbalGroup* indicates the relationship type between *BusinessObjects*. The *Quantifier* gives an idea of the multiplicity. We note that the expression between brackets is optional.

Besides applying the linguistic patterns, the software analyst prepares the BPMN model by annotating it with its *business context,* which encapsulates the functional and organizational perspectives. The functional perspective represents the process elements being performed which are *Activities* (simple tasks or complex sub-processes). The organizational perspective represents *where* and *by whom* process elements are performed, which is mainly reflected by the *Pool* and *Lane* concepts.

Due to space constraints, we next present an excerpt of the business context of only BPMN activities. We note that the Pool/Lanes are also enhanced by a business context. We augmented the BPMN activities by an *actor identifier* who is responsible for performing it; an *actor description* indicating the relationships between the activity and the involved actors; *Extended attributes* describe the activity properties. Each attribute can be a pure value or a complex one representing a business entity. This distinction is extracted from their description; which indicates also the relationships between the business entities and/or the activity's extended complex attributes. The relationships' semantic follows the first linguistic pattern.

3.2 Transformation Definition Strategy

To provide for the automation of the first DESTINY phase, we defined at the meta-model level a set of patterns that may represent a use case. To do so, we used the structural and semantic perspectives of BPMN models to define BPMN model *fragments* representing user-system interactions. Recall that a use case represents a set of actions that the system(s) should or can perform in collaboration with one or more "external" users (*i.e.*, actors), and it should provide some observable result to the actors. In our case, we interpret the notion of "external" with respect to the BPMN Pool and Lane elements.

Structurally, we defined a *pattern* as a fragment F in a BPMN process model P, that is a connected, directed sub-graph of P starting at one activity and ending at another activity such that F contains the maximum number of activities between either two gateways, a start node and a gateway, or a gateway and an end node. We exclude the gateways from F because they do not represent a system activity.

In terms of the business domain semantics, a pattern may include special business event labels (*i.e.,* send, receive, acknowledge, etc.), activity's business objects (*i.e. request, document, invoice, etc.*), and lane/pool labels (*i.e. department, agent, unit, etc.*). These labels, along with their synonyms, guide the DESTINY transformations, as we will illustrate in the Sect. 3.4. To ensure a good quality of the BPMN fragmentation results, we suppose that the BPMN model is syntactical correct and semantically valid.

3.3 Source and Target Meta-Models

To simplify the definition of the transformation rules, we extended the source meta-model, *i.e.* BPMN [3], by adding six classes and some attributes in the original classes, as well as the target metamodel, i.e. UML use case meta-model [6] by adding two classes and four attributes. These additions are lightweight extensions; they do not modify the semantics nor the syntax of the standard BPMN and UML; they merely are used to facilitate the navigation of a BPMN model. The meta-classes marked in gray constitute the elements that we added (see Fig. 2).

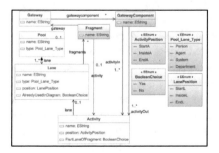 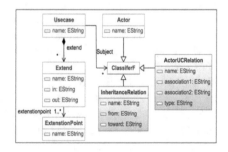

Fig. 2. Extract of the used BPMN and use case metamodel

The six added classes to the BPMN metamodel, are: *"fragment"* to classify the activities located between the gateways; *"GatewayComponent"* to identify the entry and the exit activities; *"Pool_Lane_Type"* to identify the type of the pool/lane and in case the type is *"department"*, a naming rule will be applied; *"LanePosition"* to identify the primary actor of the fragment; and *"ActivityPosition"* to identify the name of the first activity in the fragment; and *"BooleanChoice"* for the added attributes in the classes *"Activity"*.

The base meta-classes are *UseCase, Actor, Extend* and *Include*. Each Use Case might include or extend other Use Cases. We added the meta-class *"InheritanceRelation"* to facilitate the addition of an inheritance relationship between actors representing Lanes and actors representing Pools. We added also the meta-class *"ActorUCRelation"* to identify whether the relationship between the actor and the use case is a one-way or two-way relation.

3.4   Transformation Rules

We defined eight transformation rules that handle the structural aspect and the semantic information in a BPMN model. Each transformation rule operates on a canonical fragment *F* obtained from the decomposition of the BPMN model (see Sect. 3.2). The transformation rules are implemented using fourteen ATL matched rules and thirty-eight ATL helpers.

Rule 1: For each pool (respectively, lane), if its name is different from the to-be-system, then: *(1)* add a UML actor corresponding to the pool (respectively lane); *(2)* apply **Rule 6** to rename the actors; and *(3)* add an inheritance relation from the actor corresponding to the lane to the actor corresponding to the lane's pool (see Fig. 3).

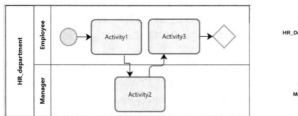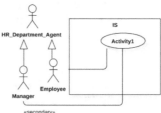

Fig. 3. Rule 1 and Rule 3.2 illustration

Rule 2: For each fragment *F* in the BPMN model *P*, which is composed by a set of activities that belong to the same lane, then: *(1)* create a use case *UC_F* with the name of the first activity *SA* of *F*; *(2)* add a two-way association between the actor whose Lane contains the activity *SA* and *UC_F*; and *(3)* add the remaining activities of *F* as the steps in the normal scenario describing *UC_F*. The succession between those activities determines the step order.

Rule 3: For each fragment *F* that has an activity *A* in a lane different from the start lane of *F*:

Rule 3.1: If the name of A *is* "receive x" (or any synonyms of receive), then add a one-way from *UC_F* to the Actor (as a secondary actor) whose Lane contains the activity A (see Fig. 3); else

Rule 3.2: add a two-way association between *UC_F* and the Actor (as a secondary actor) whose Lane contains the activity *A*.

Rule 4: For each gateway between two fragments *PF* (entry) and *NF* (exit) such that the activities of both fragments are in the same lane: *(1)* add an <<extend>> relationship from the use case *UC_NF* to the use case *UC_ PF*; and *(2)* add an extension named as the first activity's name of the second fragment (NF.SA) in the use case of the entry fragment PF (see Fig. 4).

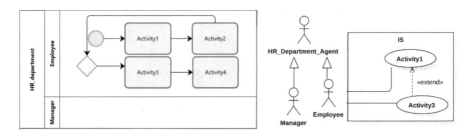

Fig. 4. Rule 4 illustration

Rule 5: For each gateway between two fragments *PF* (entry) and *NF* (exit) such that the activities of both fragments are in different lanes and:

Rule 5.1: if the name of the first activity in *NF* is *"send X"* (or any synonym of send), then: *(1)* rename the generated use case of *UC_NF* by applying the naming **Rule 7**; and *(2)* add a one-way association from the use case *UC_NF* to the primary actor of *UC_PF*.

Rule 5.2: if the name of the first activity of *NF* is *"send X to Y"* and *Y* is a pool or lane that is not transformed yet into an actor, then: *(1)* create a secondary actor *Y*; *(2)* apply **Rule 7** to rename the use case *UC_NF*; and *(3)* add one-way association from *UC_NF* to the secondary actor *Y*. If *Y* was already transformed into an actor, then apply only **Rule 5_2_3** and **Rule 8**.

Rule 5.3: if *NF* contains just one activity that is named *"receive X"*, then delete the use case *UC_NF* and all its associations and add a two-way association between *UC_PF* and the actor corresponding to *NF*. The single activity of *NF* will be a step in the normal scenario describing the use case *UC_PF*.

Rule 6: For each pool/lane whose label is synonym to "person", "agent" or "system", the corresponding actor name will be the pool/lane name. For each pool/lane whose label is metonymy of "department", "unit", "division" or "management", the corresponding actor name will be the concatenation of the pool/lane name and the word "Agent".

Rule 7: Each fragment *F* composed of only one activity labeled with "Send x" or "Send x to y", its corresponding use case *UC_F* will be named "Generate x".

Rule 8: If the first activity *SA* of a fragment *F* is labelled "Create x", then the corresponding use case *UC_F* will be named "Manage x".

3.5 Quality Evaluation of the Generated Diagram

As aforementioned, the construction of high quality use cases model has an impact on the development of the IS. We evaluated the syntactic and semantic quality of the generated use case model. The syntactic quality is evaluated through the study of the model conformity with its meta-model: Because our transformation process is meta-model-based, the generated model is conforming to the UML use case meta-model. The semantic quality deals with the validity of use cases' construction and composition as well as the completeness of the use cases regarding the domain. To this end, we used the anti-patterns defined by [9] to evaluate the validity of the use cases' construction and composition. These anti-patterns require the examination of the use case diagram and its textual description in order to look for any inconsistency or redundancy. In terms of the completeness of the set of transformations, our preliminary syntax-based investigation of the possible structures of the fragment patterns shows that the proposed set is complete. For more rigor, we are formalizing the completeness proof.

In addition, we examined the performance of the transformations experimentally through the calculation of recall and precision rates as well as the F-measure. These measures aim to compare the performance of our method to the human performance by analyzing the results given by our method to those supplied by the expert. We recall that the precision is the ratio of real elements generated by our transformation that were identified by the expert. It indicates how accurate the transformation rules are in the generation of UCD (usecase, actor, etc.) (see Formula 1). The recall is the ratio that indicates the capacity of our transformations to return all elements specified by the expert. High scores for both ratios show that the transformations return both an accurate UCD (high precision), and the majority of all relevant accurate UCD elements (high recall). It means that the generated UCD covers the whole domain precisely in accordance to the experts' perspective (see Formula 2). To have the harmonic mean of recall and precision, we have used the F-measure. F-measure has a parameter that sets the trade off between recall and precision. The standard F-measure is F1, which gives equal importance to recall and precision (see Formula 3). We calculate these rates according the following equations:

$$\text{Precision} = \text{TP}/(\text{TP} + \text{FP}) \tag{1}$$

$$\text{Recall} = \text{TP}/(\text{TP} + \text{FN}) \tag{2}$$

$$\text{F1} = (2 * \text{recall} * \text{precision})/(\text{recall} + \text{precision}) \tag{3}$$

Where:

- True positive (TP) is the number of existing real elements generated by our transformation;

- False Positive (FP) is the number of not existing real elements generated by our transformation;
- False Negative (FN) is the number of existing real elements not generated by our transformation.

4 Case Study

To illustrate the application of our transformation rules, we use the "Access Management" business process which is an official BPMN example from Bizagi.com. This process controls user privileges based on their roles and responsibilities. As illustrated in Fig. 5, the process starts when an *employee* requests some permissions. The *permission manager* decides: if this request requires approval, then he/she redirects it to the *employee's boss*; otherwise, he/she manages it by checking whether there are required documents to request from the *employee*, or there is a need for installing additional support requirements; in the latter case, the *permission manager* sends for support from the *support team*. Once all requirements are satisfied, the *permission manager* sends a notification and updates the employ's permissions. After approving all requested permissions, the *employ's boss* notifies the *permission manager*, and when some permissions are not approved, the *employ's boss* sends a rejection notification.

The first step in our method fragments the BPMN model according to patterns. Figure 5 shows in dotted red rectangles the seven fragments identified in the "Access Management" business process. Each of these fragments will be transformed into use cases by applying the transformation rules.

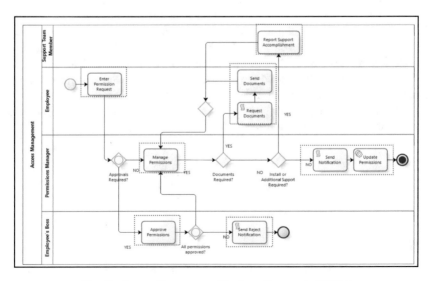

Fig. 5. *Access Management* Business Process in BPMN

Figure 6(a) shows the generated use case diagram model as follows: First, with Rule 1, the pool (*Access Management*) is the to-be system and it is transformed into Use case subject. All its lanes (*Employee, Employee's Boss, Permission Manager, Support Team Member*) are transformed to an actor. Second, by applying Rule 2, each of the seven identified fragments is transformed to a use case that is named based on the first activity in its corresponding fragment; the naming exception is the use case *"Generate Reject Notification"* for which Rule 7 is applied because its fragment has a unique "send X" activity (the last fragment in Fig. 11). Finally, by applying Rule 5, two unidirectional associations stereotyped <<*secondary*>> are added from *"Generate Notification"* to *"Employee"* and from *"Generate Reject Notification"* to *"Permission Manager"*.

We evaluated the quality of the generated model from different perspectives: (1) the syntactic quality was ensured by the conformity of the generated use case diagram to its meta-model; (2) the semantic quality was ensured by the absence of anti-patterns in the generated model; (3) the performance is experimentally evaluated through the calculation of the recall (71%), precision (78%), and F Measurement (74%). The high scores for both ratios mean that the generated use case diagram covers the whole domain precisely in accordance with the experts' perspective. We can deduce that the performance of our method approaches the human performance (see Fig. 6b).

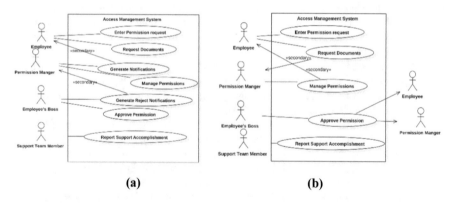

Fig. 6. (a) Generated use case diagram (b) Expert-defined use case diagram

5 Conclusion

This paper proposed a transformation-based approach to generate use case models from business process models. It provides for the generation of IS functional requirements that are aligned to the business logic. Compared to existing works, our approach has the merit of accounting for the granularity level of each modeling language, and the structural and semantic aspects of the business process model. This merit provides for the generation of refined use case diagrams that distinguish between primary and secondary actors as well as the different relationships among the use cases. In addition, it allows the generation of textual use case descriptions, artefacts that are important in the remaining IS development cycle. Furthermore, the proposed transformation approach is pattern-based, which provides for its automation and the formal assessment of the completeness of the transformation rules.

Ongoing work is oriented towards (1) enhancing the transformations in order to cover all target meta-model elements; and (2) completing the development of tool support.

References

1. Dumas, M., Van Der Aalst, W.M., Ter Hofstede, A.H.: Process-aware Information Systems: Bridging People and Software Through Process Technology. Wiley (2005)
2. Qazi, A., Rehman, M.S., Rehman, M.S., Maqbool Roa, N.: Enhanced model driven architecture software development life cycle with synchronized and consistent mapping. In: International Conference on Computer Communication and Management (ICCCM 2011), 2011
3. OMG: BPMN. Business Process Model and Notation (BPMN). formal/2011-01-03, 2011
4. OMG: The Fast Guide to Model Driven Architecture, The Basics of Model Driven Architecture (MDA), 2006
5. OMG: OMG Meta Object Facility (MOF) Core Specification. OMG Document Number: formal/2015-06-05, 2015
6. OMG: OMG Unified Modeling Language (OMG UML). formal/2015-03-01, 2015
7. Asadi, M., Ravakhah, M., Ramsin, R.: An MDA-based system development lifecycle. In: Second Asia International Conference on Modeling & Simulation. AICMS 08, pp. 836–842, 13–15 May 2008
8. Kriouile, A., Addamssiri, N., Gadi, T.: An MDA method for automatic transformation of models from CIM to PIM. Am. J. Soft. Eng. Appl. **4**(1), 1–14 (2015)
9. El-Attar, M., Miller, J.: Constructing high quality use case models: a systematic review of current practices. Requirements Eng. **17**(3), 187–201 (2012)
10. Kerraf, S., Lefebvre, E., Suryn, W.: Transformation from CIM to PIM using patterns and archetypes. In: 19th Australian Conference on Software Engineering. ASWEC 2008, 2008
11. Rhazali, Y., Hadi, Y., Mouloudi, A.: Model transformation with ATL into MDA from CIM to PIM structured through MVC. In: The 6th International Symposium on Frontiers in Ambient and Mobile SYstems (FAMS 2016), pp. 1096–1101 (2016)
12. Kardoš, M., Drozdová, M.: Analytical method of CIM to PIM transformation in model driven architecture (MDA). J. Inf. Org. Sci. **34**(1), 89–99 (2010)
13. Siqueira, F.L., Silva, P.S.M.: Transforming an enterprise model into a use case model in business process systems. J. Syst. Soft. **96**, 152–171 (2014)
14. Berrocal, J., García-Alonso, J., Vicente-Chicote, C., Murillo, J.M.: A Pattern-based and model-driven approach for deriving IT system functional models from annotated business models. In: José Escalona, M., Aragón, G., Linger, H., Lang, M., Barry, C., Schneider, C. (eds.) Information System Development. Springer, Cham (2014)
15. Kamthan, P.: A framework for understanding and addressing the semiotic quality of use case models. In: Model-driven Software Development: Integrating Quality Assurance. Christian Bunse (2009)

Intelligent Fall Detection with Wearable IoT

Farhad Ahamed$^{(\boxtimes)}$, Seyed Shahrestani, and Hon Cheung

Western Sydney University, Sydney, Australia
{f.ahamed, s.shahrestani, h.cheung}@westernsydney.edu.au

Abstract. Falls of older adults is a significant concern for themselves and caregivers as most of the times a fall leads to serious physical injuries. In the age of the Internet of things (IoT), connected smart homes and monitoring services have opened up opportunities for quality of life for the older adults. Detecting falls with wearable IoT devices can provide peace of mind for older adults and caregivers. Accelerometer based fall detection is investigated in this paper. Feed Forward Neural Network (FFNN) and Long Short Term Memory (LSTM) based Deep Learning network is applied to detect fall. LSTM network provides good accuracy based on the experiment. This experiment provides a promising indication that IoT-based fall monitoring can assure post-fall procedures to older adults and caregivers and this can increase the safety level and well-being of the older adults.

1 Introduction

Falls are detrimental events for the elderly. Falls among older people are a serious health issue and can result in hip fractures, traumatic brain injuries, and even fatalities. 25% of 65+ years older adults fall each year [1]. Out of this 25%, 20% of them get seriously hurt [1]. The conventional method for ensuring the safety of the elderly is through constant and vigilant observation of their daily activities under the eyes of hired nurses and caregivers. Falls related physical health hazard is one of the primary concerns for caregivers. The older adults living in residential care are also very concerned about the potential falls related danger. Based on another statistics of Noury et al., more than 33% of people aged 65 years and above experience one fall per year [1]. Falls cause physiological and psychological damage such as injury, restricted activity. On the other hand fall will also cause anxiety about living independently and anxiety of falling [2, 3].

In Australia, the cost of fall-related injuries in older people is more than $200 million per year and is increasing as the population ages [4]. Self-mobility is a vital issue for older adults to maintain self-dependency. Falls can reduce mobility and pose a significant health risk. Even the fear of falling can lead to reduced activity [6]. It places a substantial burden on the caregiver, to supervise and accompany the elderly at all times. As a result, one caregiver usually takes care of one or very few patients, which ultimately increases the personnel required in a nursing home. The cost of hiring such a staff privately at home or collectively in an older adults' shelter is very high.

© Springer Nature Switzerland AG 2020
L. Barolli et al. (Eds.): CISIS 2019, AISC 993, pp. 391–401, 2020.
https://doi.org/10.1007/978-3-030-22354-0_35

Furthermore, when the caregivers are not present, the elderly may fall and get seriously injured, because, it is near impossible to completely supervise anyone and at all times. Immediate medical attention would be required for them, but it may be delayed until the caregiver returns or receives knowledge of the accident. This type of delays raises a significant issue of complete reliability. Thus, more cost-effective and reliable options must be considered to provide safety and security to the elderly. Fall detection by wearable IoT can lead to overcoming this obstacle.

The sections are organized in the following ways. Section 2 narrates background of fall detection methods. Section 3 provides an overview of the dataset and the fall detection method used in this paper. Section 4 describes fall detection using FFNN and analyse the result. Section 5 describes fall detection using LSTM network and analyse the result. Finally, Sect. 6 provides concluding remarks with future work directions.

2   Background

Fall detection using IoT devices has been the focus of substantial recent research and systematic review [5]. Body-worn accelerometers detect impacts and changes in an orientation associated with falls. Accuracy may be improved by using multiple sensors. For example, video monitoring can identify fall and post movement after the fall. Current smartphones apps have also used gyroscopes, accelerometers and global position systems (GPS) to assist in detecting falls. These technologies aim to provide rapid detection of falls and, therefore, prevent frail older adults suffering "long lies" due to not being able to get up after a fall.

Hijaz et al. classified fall detection literature into three types [6]. (1) In computer vision (image processing) based method, real-time movement of the subject is monitored through video. Additionally, an algorithm is employed to determine the posture of the subject. (2) The acoustic-based method detects a fall in the frequency component of vibration produced by the fall. (3) In a worn sensor based method, kinematic sensors (accelerometers and gyroscopes) are used to distinguish a fall from the activities of daily life (ADL). All of these approaches reduce the number of personnel required to monitor the elderly and also minimise the caregiver's efforts.

There have been few others fall detection work based on PIR sensor. PIR sensors could be mounted on a wall or ceiling to detect human motion. In one study PIR sensors were mounted vertically to create an array. PIR sensors were providing motion related voltage reading from the movement of a human. The voltage graph represents the activities of the subject. Multiple filters and process are applied to isolate the graphs based on human activity, like, walking, sitting, standing and falling [7]. In acoustic and ambient sensors based technique to detect falls systems may include infrared sensors, microphones or vibration sensors. The results obtained from related studies show the proof of concept. Acoustic techniques also provide an unobtrusive way of monitoring persons of interest, just like the computer vision based techniques. The hardware and infrastructure required for such techniques are relatively inexpensive and straightforward, in most cases, when compared to computer vision based methods. Acoustic and ambience based systems consist of a set of acoustic or ambient sensors and a dedicated PC. The sensors gather data and send it to the PC for analysis. The PC, based on certain conditions and thresholds, decides on the detection of a fall event [8].

Fadel Adib et al. developed an enhanced wireless device to identify and trace human movement even behind a wall. They receive the feedback from the reflected wireless signal from the subject and by applying the machine learning technique, they can classify the movements of the body, posture, breathing, and heartbeat rate [9]. Wang et al. Developed a fall detection system called WiFall that claimed to achieve fall detection for a single person with high accuracy [10]. They have demonstrated by the experimental results, WiFall yields 90 per cent detection precision with a false alarm rate of 15 per cent on average using a one-class SVM classifier in all testing scenarios. It can also achieve an average of 94 per cent fall detection precisions with 13 per cent false alarm using the Random Forest algorithm. Wang et al. Developed RT-Fall, an indoor fall detection system using the commodity Wi-Fi devices [11]. The system used the phase and amplitude of the fine-grained Channel State Information (CSI) accessible in commodity Wi-Fi devices and attempted to segment and detect the falls automatically in real-time. They claimed their experimental results in four indoor scenarios outperforms the state-of-the-art approach WiFall with 14 per cent higher sensitivity and 10 per cent higher specificity on average. Our focus in this paper will be accelerator based fall detection with deep learning method. A comparison among most of the fall detection methods are presented in the Table 1.

Table 1. A comparison among most of the fall detection methods

	Fall detection type	Advantage	Shortcoming
1.	Computer vision, Camera-based, image processing, depth camera	Widely available, wide coverage, inexpensive	Night or darkness issue, occlusion
2.	Remote sensors (acoustic, Wifi-CSI based, PIR sensor, etc.)	Inexpensive	Coverage area is small, signal interference is high
3.	Body-worn sensors (mobile devices, smartwatch, implant devices etc.)	Wide coverage, reliable, mobile	Battery lifespan issue, always need to carry

In this work we focused on wearable sensor attached to body to detect falls as this can ensure fall detection inhouse as well as remote location. This method is free from the shortcoming of vision based method and very reliable compared to remote sensors.

3   Fall Detection Using Accelerometer

Recent development in wearable IoT caused accelerometer-based devices to gain immense popularity. It also opened the opportunity to detect fall more precisely. With the recent development in Deep learning method, neural networks can more precisely detect a fall. In this experiment, we used UR Fall detection dataset [12]. The accelerometer data of this dataset is not uniform. Hence, the dataset was not ready to feed into a training network. The dataset has 30 labelled fall events and 40 labelled ADL events. These events or data points have missing data, unsorted signals, wrong timestamps and uneven signal lengths. Therefore, at first it was required to prepare the data for machine learning purpose. In the dataset, there are video recordings of the fall as well as accelerator readings of the fall event from the sensor which was attached to body of the participants. The video recording and the accelerometer signals varied from 1.5 s to 10 s. The following sub-sections narrate the process to filter and prepare data to train the neural network.

3.1   Data Analysis and Preparation

The dataset consists of video recordings and accelerometer reading of the falls and ADL. We have compared the fall detection methods by time series based acceleration signal processing. The dataset consisted multiple fall scenario as

1) Forward fall from standing,
2) Forward fall from seating,
3) Forward fall from walking.

   The ADL data has the following actions

1) Walking,
2) Walking followed by lying down on the bed,
3) Picking up objects from the floor.
4) Walking in and seating in the room.

   The acceleration sensor provides the motion of the subject in X, Y and Z axis. In the case of the fall event, the most significant fluctuation of the acceleration happens for the Z axis. The geometric mean of the axis provides a mean signal value which is used as input for the machine learning. The mean value of the signal is retrieved based on the Eq. (1).

$$Signal, \quad s = \sqrt{X^2 + Y^2 + Z^2} \tag{1}$$

   From Fig. 1, it is evident that the mean value of the signals of 3-axes has close resemble Z-axis signal.

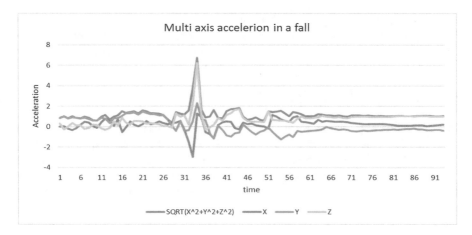

Fig. 1. The mean signal has a close relation with Z-axis signal.

The mean signal of the ADL and fall event differs almost all the cases. From the data visualization process, it stands out that the fall event has an inter-relation with rapid acceleration especially in Z-axis direction. On the other hand, almost all the ADL activities have relatively lower acceleration value compare with fall activities. Figure 2 compares one sample fall signal with an ADL signal. It can be quickly noted that the mean acceleration value is relatively higher in a fall event. From the dataset it is also discovered, there are a few data points in ADL activities which are very similar in values as a fall event signal. For example, when a person lay down on a bed or pick up something from the floor the accelerometer value gets very close to falling event. A signal interval (approximate 8 s) is chosen to feed the data into a machine learning algorithm to avoid these error scenarios.

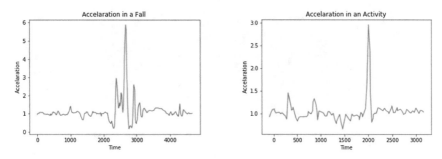

Fig. 2. Acceleration in a fall vs acceleration in an ADL

Furthermore, to overcome the challenge of uneven data in the dataset, all the samples were sorted based on timestamps. The signals contained few error timestamp values with negative timestamp randomly within data. These random timestamps were removed. Furthermore, for training purpose all the data points should be the same length of duration. The essential part of the dataset was the fall data points. Hence, all the fall data points were preserved with full length. However, if any ADL data point had longer signal duration, they were reduced in equal size of the longest fall data point. After these data filtering, the dataset was finally ready for training.

4   Applying FFNN to Detect Fall

One of the first deep learning nets was tested with the dataset was pattern recognition networks. They are two layered FFNN that can be trained to classify inputs according to target classes. In this method the fall detection was treated as a data classification problem using a two-layer feed-forward network. After the initial training of the network, the performance is evaluated using cross-entropy and per cent misclassification error.

4.1   Setup Training Options with FFNN

After initial data filtering the input file was prepared to feed an input size of 8715 data point of a signal and output classes were two:-

(a) fall class which was classified as '2'
(b) ADL class which was classified as '1'

The FFNN has two layers:- Hidden and Output. In this experiment, we have used 100 neurons in the hidden layer. It has an input layer to take a signal length of 8715 ms. The input layer is connected with the hidden layer that has 100 neurons. The hidden layer is connected to the Output layer. Finally, the output layer is connected with the binary layer as the final output. The Fig. 3 below shows the architecture of this pattern recognition network.

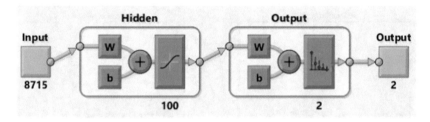

Fig. 3. The architecture of the pattern recognition network

In the training phase, "Scaled Conjugate Gradient Backpropagation" function was used. The backpropagation function updates weight and bias values according to the scaled conjugate gradient method. Conveniently, this training function uses less memory and suitable in low memory situations like wearable devices. To validate the performance of the network, "Cross-entropy" function was used. The whole dataset was divided into the training, validation and testing and they were split as 70%, 15% and 15% respectively.

4.2   Result Analysis: FFNN Based Fall Detection

After running over many iterations, the accuracy on training dataset was finally reached 97.9%, precision was 95.2% and the overall accuracy reached 94.3% and precision was 90%. The confusion matrix and error histogram is provided in Fig. 4.

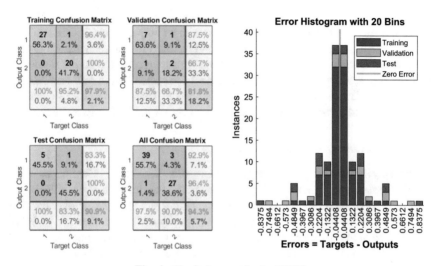

Fig. 4. Confusion matrix for FFNN

The histogram in Fig. 4 shows most of the instances aligned with correct predictions with few exceptions.

5   Applying LSTM to Detect Fall

LSTM deep learning network is widely used for prediction and identification of time series data for example stock market prediction, future events in a series, etc. LSTM networks can learn long-term dependencies between time steps of sequence data. In this experiment, initially we used bidirectional LSTM layer, to ensure the neural net looks at the sequence of the accelerometer signal in both forward and backward directions to identify a fall. As the input signals have one dimension each, we specify the input size to be sequences of size one. However, in the final architecture the input layer size was changed to two as two distinct features of the signals were used for better classification. Section 5.2 highlights that changes of the first architecture.

Fig. 5. The architecture of the LSTM based network to detect fall

We specified the bidirectional LSTM layer with an output size of 100 and outputted the last element of the sequence. This command instructs the bidirectional LSTM layer to map the input time series into 100 features and then prepares the output for the fully connected layer. As we have only two classes of expected output Fall and ADL, we specified two classes by including a fully connected layer of size 2, followed by a softmax layer and a classification layer. Figure 5 shows the final network architecture.

5.1 Setup Training Options with LSTM

Next we specify the training options for the classifier. We set the 'MaxEpochs' to 200 to allow the network to make 200 passes through the training data. A 'MiniBatchSize' of 1 directs the network to look at one training signals at a time. An 'InitialLearnRate' of 0.01 helps speed up the training process. We specify a 'SequenceLength' of 1000 to break the signal into smaller pieces so that the machine does not run out of memory by looking at too much data at one time. We also set 'GradientThreshold' to 1 to stabilise the training process by preventing gradients from getting too large. Then we specify 'Plots' as 'training-progress' to generate plots that show a graphic of the training progress as the number of iterations increases. We used the adaptive moment estimation (ADAM) solver. ADAM performs better with Recurrent Neural Networks (RNNs) like LSTMs than the default Stochastic Gradient Descent with Momentum (SGDM) solver.

5.2 Calculate Spikes and Spectral Entropy to Feed into LSTM

The first LSTM net with one input feature is not useful to generate better accuracy to detect falls. Hence, we redesigned the LSTM with two distinctive features of the signal. We used 100 data points to be considered for a feature map to check spikes in each signal. To check the spikes of the signal Instantaneous Frequency (IF) was calculated [13, 14]. The instantaneous frequency of a nonstationary signal is a time-varying parameter that relates to the average of the frequencies present in the signal as it evolves. The instantaneous frequency of a nonstationary signal is a time-varying parameter that relates to the average of the frequencies present in the signal as it evolves.

This method estimates the instantaneous frequency as the first conditional spectral moment of the time-frequency distribution of the input signal. It computes the spectrogram power spectrum $P(t, f)$, of the input using the *Pspectrum* function and uses the spectrum as a time-frequency distribution. *Pspectrum* function analyze signals in the frequency and time-frequency domains. Therefore, the instantaneous frequency function estimates the output using Eq. (2) [13, 14].

$$f_{inst}(t) = \frac{\int_0^\infty f\, P(t,f)df}{\int_0^\infty P(t,f)df}.$$ (2)

The second feature of the signal calculated was the spectral entropy (SE) [15]. SE of a signal is a measure of its spectral power distribution. The SE treats the signal's normalized power distribution in the frequency domain as a probability distribution, and calculates the Shannon entropy of it. This property can be useful for feature extraction in fault detection and diagnosis, speech recognition and biomedical signal processing. The equations for spectral entropy arise from the equations for the power spectrum and probability distribution for a signal. For a signal $x(n)$, the power spectrum is $S(m) = |X(m)|^2$, where $X(m)$ is the discrete Fourier transform of $x(n)$.

The probability distribution $P(m)$ is then [15]:

$$P(m) = \frac{S(m)}{\sum_i S(i)}$$ (3)

The spectral entropy H follows as [15]:

$$H = -\sum_{m=1}^{N} P(m) \log_2 P(m)$$ (4)

After calculating IF and SE, these features were used in the input layer of the LSTM architecture. As the ADL samples were 40 and Fall samples were 30, to avoid probability inclination towards ADL, ten fall samples were duplicated to make 40 + 40 samples of equal balance. Out of these 80 samples, 70 of them were used for training and ten was used for final testing.

5.3   Result Analysis LSTM Based Fall Detection

During the training phase, the accuracy and precision reached 97.1% and during testing the accuracy and precision was 100%. As mention in Sect. 5.2, the testing data was unique which was not used during the training period. In Fig. 6 it shows the prediction and accuracy were getting better starting from iteration number 8000 and finally finished with 97.1% accuracy after 14000 iterations.

LSTM network took a while finish calculation, however the accuracy and precision were very promising with 97.1%.

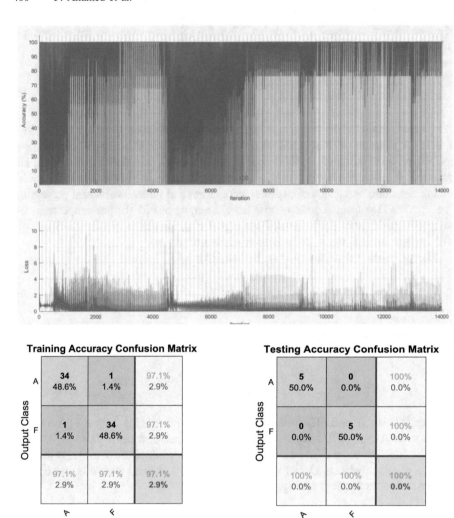

Fig. 6. Confusion Matrix of the LSTM based network to detect fall

6 Conclusion

The use of a single lightweight sensor to distinguish between different types falls and activities of daily living is a promising low-cost technology and clinical tool for long-term continuous monitoring of older people and clinical populations at risk of falls. However, currently the evidence is limited because these studies have primarily involved simulated laboratory events in young adults. Future studies should focus on validating fall detection in larger units and include data from (a) people at high risk of falling, (b) activities of daily living, (c) both near falls and actual falls, and (d) naturally occurring near falls. The proposed neural net can be used in wearable device like smartwatch or mobile phones to set emergency alarm in case of a fall detection aligning with other decisive factors such as age of the person, post fall movements, post fall response timer, etc. In this experiment, LSTM network provided auspicious result compare to FFNN for accelerator-based fall detection. Furthermore, the wearable device based fall detection has its application where the person is always wearing the device. However, to provide 100% coverage and assurance image based and sensor based multi-modal fall detection can play a vital role in aged care and residential care.

References

1. Noury, N.: A smart sensor for the remote follow up of activity and fall detection of the elderly. In: 2nd Annual International IEEE-EMB Special Topic Conference on Microtechnologies in Medicine & Biology, 2002, pp. 314–317. IEEE (2002)
2. Zhang, T., Wang, J., Liu, P., Hou, J.: Fall detection by embedding an accelerometer in cellphone and using KFD algorithm. Int. J. Comput. Sci. Netw. Secur. **6**(10), 277–284 (2006)
3. Mathie, M., Basilakis, J., Celler, B.: A system for monitoring posture and physical activity using accelerometers. In: Proceedings of the 23rd Annual International Conference of the IEEE on Engineering in Medicine and Biology Society, 2001, vol. 4, pp. 3654–3657. IEEE (2001)
4. Davis, J.C., Robertson, M.C., Ashe, M.C., Liu-Ambrose, T., Khan, K.M., Marra, C.A.: International comparison of cost of falls in older adults living in the community: a systematic review. Osteoporos. Int. **21**(8), 1295–1306 (2010)
5. Chaudhuri, S., Thompson, H., Demiris, G.: Fall detection devices and their use with older adults: a systematic review. J. Geriatr. Phys. Ther. **37**(4), 178–196 (2014)
6. Hijaz, F., Afzal, N., Ahmad, T., Hasan, O.: Survey of fall detection and daily activity monitoring techniques. In: 2010 International Conference on Information and Emerging Technologies (ICIET), pp. 1–6. IEEE (2010)
7. Sixsmith, A., Johnson, N.: SIMBAD: smart inactivity monitor using array-based detector. Gerontechnology **2**(1), 110 (2002)
8. Popescu, M., Li, Y., Skubic, M., Rantz, M.: An acoustic fall detector system that uses sound height information to reduce the false alarm rate. In: 30th Annual International Conference of the IEEE, EMBS 2008, Engineering in Medicine and Biology Society, 2008, pp. 4628–4631. IEEE (2008)
9. "New System Uses Low-Power Wi-Fi Signal to Track Moving Humans – Even Behind Walls," ed. Washington, DC (2013)
10. Wang, Y., Wu, K., Ni, L.M.: Wifall: device-free fall detection by wireless networks. IEEE Trans. Mob. Comput. **16**(2), 581–594 (2017)
11. Wang, H., Zhang, D., Wang, Y., Ma, J., Wang, Y., Li, S.: RT-Fall: a real-time and contactless fall detection system with commodity WiFi devices. IEEE Trans. Mob. Comput. **16**(2), 511–526 (2017)
12. Kwolek, B., Kepski, M.: Human fall detection on embedded platform using depth maps and wireless accelerometer. Comput. Methods Programs Biomed. **117**(3), 489–501 (2014)
13. Boashash, B.: Estimating and interpreting the instantaneous frequency of a signal. I. Fundamentals. Proc. IEEE **80**(4), 520–538 (1992)
14. Boashash, B.: Estimating and interpreting the instantaneous frequency of a signal—Part 2: algorithms and applications. Proc. IEEE **80**(4), 540–568 (1992)
15. Pan, Y., Chen, J., Li, X.: Spectral entropy: a complementary index for rolling element bearing performance degradation assessment. Proc. Inst. Mech. Eng. Part C: J. Mech. Eng. Sci. **223**(5), 1223–1231 (2009)

Peer-to-Peer Based Web of Things Resource Management

Yangqun Li[1,2(✉)]

[1] Jiangsu Key Laboratory of Broadband Wireless Communication
and Internet of Things, Nanjing University of Posts and Telecommunications,
Nanjing 210003, People's Republic of China
yqli@njupt.edu.cn
[2] School of Internet of Things, Nanjing University of Posts
and Telecommunications, Nanjing 210003, People's Republic of China

Abstract. More and more devices provide resources and services under the environment of Internet of Things (IoT). The location-based P2P distributed computing technology is adopted to implement the effective management of the Web of Things (WoT) resources, satisfying reliability, robustness and flexibility of the system. In order to effectively implement resource lookup in the context of massive resources, the probabilistic topic model is used to model IoT resource description and then the resulting topic model and geolocations are used as resource ID for resource storage and resource discovery, realizing distributed management of resources. And on this basis, the ROA style life cycle management for WoT resources under the P2P environment is provided. Finally, the feasibility of the system is proved by simulation experiments.

1 Background

The Internet of Things (IoT) is defined in ITU-T Y.2060 [1] as a global infrastructure for the information society that interconnects various devices to provide advanced services supported by existing or developing interoperable information and communication technologies. This concept reflects the main characteristics of the IoT which include globality, interoperability, connectivity, device diversity and service. Its essential attribute is to provide advanced service for the information society.

With the wide application of the IoT, more and more perceptual devices are deployed in real life. These numerous terminals with different functions are distributed around the world and belong to different enterprises, organizations and individuals. IoT technology connects these devices through various protocols to build a network with various resources and services. It faces many challenges that how the resources in the network effectively are used to develop service according to user requirements. These challenges include: how to organize and manage these resources effectively; How to efficiently discover and use resources; How to effectively manage the life cycle of resources and provide robust and reliable services.

For example, when a person plans to visit a scenic spot, she/he needs to know some basic information about the scenic spot, such as the current number of people in the scenic spot, the air quality of the scenic spot, the temperature and humidity of the

L. Barolli et al. (Eds.): CISIS 2019, AISC 993, pp. 402–416, 2020.
https://doi.org/10.1007/978-3-030-22354-0_36

scenic spot, and the noise level of the scenic spot. The information can be obtained by sensors in the scenic spot or the services provided by the sensors deployed at the scenic spot. For example, the number of people can be counted by the infrared sensors at the entrance of the scenic spot. In order to obtain the above information, users can inquire whether there are corresponding resources or services in the IoT according to their needs. The IoT requires effective management of existing resources to conduct effective resource search and service delivery. At the same time, it needs to find the most suitable resource/service to replace the failed node for ensuring the availability of the service.

Web of Things (WoT) [2] realizes the integration of various devices in heterogeneous network by using the World Wide Web (WWW) existing mature technology. For example, the data collection, data description, transmission, processing and application development of the IoT are implemented conveniently and efficiently by using the existing WWW technology. It virtualizes objects into WWW resources, and hence implements the integration between things and things and virtual resources [3] by adopting Web2.0 and Restful technologies. Although the Web technology provides a variety of Web means to enable various perceptual devices to interconnect in the Web tier, management problems of all kinds of resources in the IoT are not solved yet. Therefore, new methods and mechanisms are needed to meet WoT resources life cycle management, i.e. the connection of resources, the registration of resources, the storage of resources, the discovery of resources, effective security access to resources, the failure of resources and the availability of resources and so on.

Among these problems, a considerable part of current research work mainly focuses on resource search or discovery. At present, the number and types of resources in the large-scale IoT environment are huge. How to manage these resources and find the required resources accurately, quickly and effectively has become a concern of the academia and industry.

2   Resource/Service Discovery Related Work

Service discovery has been extensively studied in the Service Oriented Architecture (SOA), focusing on discovering services that meet requirements from services described in the Web Service Description Language (WSDL). These methods, based on service description information, combine semantic information, quality of service, input/output of the interface, and other factors to find services that match the requirements from the services stored in UDDI. The advantages of this method lie in: (1) reducing the complexity of service registration/storage and discovery by using the registry; (2) realizing service selection and matching by using complex algorithms. However, it also faces new problems, such as the unavailability of service description information caused by a single point of failure, more computing resources consumed by complex algorithms etc. These problems are even more prominent in the new IoT environment, where the computing and power capabilities of perceptive devices are limited.

2.1 Peer-to-Peer Technology

P2P (Peer-to-Peer) technology builds Overlay Network on the existing network by using DHT (Distributed Hash Table) technology. It is a new distributed computing technology that doesn't exist server compared with Client/Server model and has been widely used in resource sharing, virtual currency transactions and other fields. It has the following characteristics [4]:

(1) Excellent scalability. P2P is a decentralized structure and has low cost of system management and maintenance. The access and exit of nodes can be maintained automatically by the system without manual intervention.
(2) Excellent availability. P2P can overcome the problem of poor system availability caused by a single point of failure.
(3) The distribution and multi-point storage of resources improve the availability of resources and the efficiency of the system.

2.2 IoT and P2P

Under the environment of IoT, a variety of sensor resources and related services are distributed in different places. It is difficult to adopt the traditional centralized resource management method, which requires coordination and standardization among various agencies and organizations. Moreover, it lacks certain flexibility and has poor real-time performance in the life cycle management of resources. However, the advantage of this method is that it can realize the security audit of resources and ensure the accuracy and security of resources. In Table 1, we compare the characteristics of IoT applications and P2P technologies. It can be seen that the current situation of IoT applications is similar to the P2P network to some extent. It can be considered to combine the widely used P2P technology with the IoT technology to realize the effective management of IoT resources.

Table 1. P2P and IoT technology

	Resource type	Network scale	Management difficulty	Resource changes	Life cycle management	Resource publishing	Resource discovery
IoT	Variety of sensors or services	extensive and belonging to different owners	difficult	relatively frequent	No	By hand or hard coding, Not easy	Not easy
P2P	data/computing resource	extensive and belonging to different owners	distributed automated management	relatively frequent	P2P protocol	Easy	easy

At present, some research work has been done to combine P2P technology with IoT technology to solve the problem of resource storage and discovery. Hierarchical P2P architecture is adopted in literature [5] to realize EPCglobal resource discovery function. By comparing the hierarchical architecture with ordinary P2P architecture, the advantages of hierarchical architecture in terms of hops and query time can be analyzed. This hierarchical architecture takes advantage of geographical location to establish the top-level P2P architecture in different geographical locations, while the nodes in the same region establish the internal P2P network. However, in this paper, it does not indicate which node in the local P2P network participates in the top-level P2P network.

In reference [6], service discovery based on cache mechanism is proposed to improve the efficiency of service discovery in the context of IoT and the unstructured P2P network. This mechanism considers the problem of node mobility. Aiming to discovery of IoT resources, a storage and retrieval mechanism based on hierarchical Bloom Filter is proposed to reduce the storage space of routing table and search time. This method represents the resource model through the architecture of the semantic tree, analyzes the hierarchical relationship between the resource capabilities, and establishes the hierarchical relationship with the routing table structure to calculate the availability and stability of nodes. This service discovery algorithm changes the routing table structure in P2P nodes and stores capability entries in the semantic tree. The research focus on the efficiency of service discovery rather than the accuracy of service discovery. This method has not been verified in the real environment.

Aiming to service discovery, Q. He [7] proposed a P2P distributed service discovery mechanism Chord4S. DHT algorithm is improved for data availability, and a hierarchical service identifier mechanism is proposed. The service identifier is divided into two parts: the functional bit in front and the service provider bit behind. The function bit is divided into five parts and each part is a subclass of the previous part. The first part is the highest level, representing the class of the most abstract level. Each category is hashed to obtain the corresponding bit sequence. Therefore, the feature of this bit sequence is that the corresponding bit sequence in front of it is the same for the same type of service. Thus, the preceding parts of the identifier of services with the same functionality or services with a common ancestor are the same, which allows these services to be stored within same node range. The Chord4S can also support arbitrary query of services, even distribution of services among nodes, QoS based service search and multiple service queries.

2.3   WoT and P2P

InfraWoT [8] adopts hierarchical method to manage resources in small areas, such as houses, buildings or campuses. Tree structure is adopted to store and manage resources in spatial locations. This method cannot effectively integrate and manage the resources of different organizations. uBox [9] technology makes use of distributed resource management mechanism to realize cross-domain use of WoT resources. Gateway technology is used to manage local resources, but requests of mobile users can be routed to the WoT resources where the user used to be.

S. Cirani etc. [10] introduces the realization of resource registration and resource discovery in IoT by using P2P technology. The IoT gateway distributed in various regions will be used to maintain local sensor resources and is used to establish P2P network. In the implementation process, the physical location of the gateway is fully considered. The system makes full use of Web of things technology mechanism, CoAP/HTTP protocol to realize the connection between gateways and PUT/GET operation of RESTful style to realize resource registration and discovery. However, the resource discovery algorithm is not described in detail.

WoT.City [11] development framework aims to establish a WoT-based application development framework that supports P2P network connections and provides a Flow-based programming development environment. The framework is divided into three layers: the first layer is WoT layer which provides CoAP/HTTP/WebSocket transmission of IoT data. The layer also includes object description, request handler, URL Router and Event Emitter. The second layer is the Broker layer, which mainly includes events/data distribution, Callback/DHT/Chord p2p protocol and so on. The authors use the Broker in the MQTT protocol as a node in a P2P network to provide a service publication and subscription mechanism. The third layer is the business development layer, which provides the programming mode based on Flowchain and describes the input and output relationship between various components based on JavaScript to realize the development of IoT applications. The framework for this article is based on JavaScript engines, such as Node.js, V8 and JerryScript.

3   Web of Things Resource Management Based on P2P

The WoT can make use of the existing Web technology to construct the integration and development platform of various heterogeneous sensing devices, which has received wide attention. In this paper, P2P technology and the WoT are combined to realize the effective management of the WoT resources in large-scale network environment.

The main contributions of this paper include:

1. P2P technology is used to realize the distributed management of WoT resources. In the P2P network, WoT technology is used to realize the resource operation with resource-oriented architecture style.
2. The physical location of the resource in the WoT application is a very important factor. In the system, the nodes which are physically near to one another are made as adjacent nodes in P2P network.
3. Topic based resource description modeling. Top K most important topics are extracted from the resource description and are hashed to construct resource ID in combine with the information such as the geographical location, so as to realize the storage and search of resources.

The service management mechanism in reference [7] used Chord protocol to build a P2P network, hashed the service according to the service category for building the resource ID, and implemented the distributed storage and retrieval of resources.

For example, Hash operations are performed on service "Multimedia.Video.Encoder. AVI2RM" according to the hierarchy shown to get the whole resource ID. The precondition of this method is to establish a unified service hierarchy standard manually. In this paper, the service topics which are used to classify resources are automatically extracted by text analysis, and then are used combined with geolocation for service storage in a P2P network based on Kademlia.

4   WoT Resource Management System Architecture Based on P2P

In the practical application of the IoT, the heterogeneity of perceived resources, the universality of distribution and the huge number of resources require access management of these resources through the gateway. The functions of gateway include multi-protocol access, protocol/data format conversion, data cache and handling, security mechanism, and other functions. These complex and tedious functions are realized through the gateway, which simplifies the distributed management of WoT resources, data collection and application development. Therefore, WoT resources are connected to the IoT gateway, which is connected to the Overlay Networking established by P2P technology in a Web way, thus realizing the distributed resource management.

4.1   WoT Resource Management Architecture Based on P2P

Figure 1 shows the WoT resource management architecture based on P2P. On the basis of WoT resource management based on P2P, modules 5–8 in the architecture realize different functions for IoT application development. Module 5 realizes the resource discovery function. This module performs the resource matching algorithm according to the user's search request, and then returns the found resource rank list to the user according to resource matching score. Module 6 is a resource monitoring module, which realize the caching and monitoring of found resource lists. When the resources become invalid or the freshness of resources goes out of time, it will be removed from the resource list cache. When there is no resource list, module 5 will be called to find new resources from the P2P network. Module 7 is the service function scheduling module which provides services for module 8. When a certain resource used in the business process fails, the optimal resource that meets the requirements is selected from the resources for business process according to the scheduling algorithm to ensure the execution and the availability of the business process. Module 8 is a business process module, which is used by users and meets users' actual requirements by composing various resource services. The business process module includes the service compose module, service execution module and so on. The service compose module realizes the service compose process description, modeling and validation. The business execution module is responsible for the scheduling and monitoring of business processes.

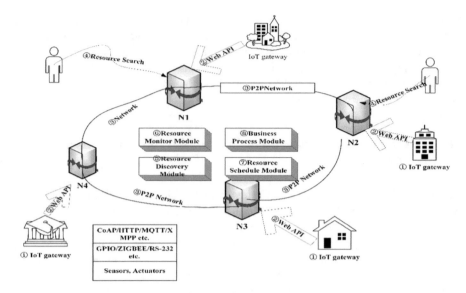

Fig. 1. P2P-based WoT resource management architecture

4.2 WoT Resource Management of P2P Based on Geolocation

In the P2P network, nodes with close physical locations are regarded as adjacent nodes. Therefore, when resources are stored, resources related to some geolocations should be stored in the corresponding location nodes as far as possible. Such advantages are as follows:

(1) Resources that are geographically close to each other can be stored in node spaces that are geographically close to each other.
(2) When a certain resource needs to be updated, it is more efficient and only a portion of the nodes in its geographical area are affected.

In P2P network, the node ID is calculated by formula 1, and then the distance between node IDs is calculated to establish the route between nodes.

$$\text{NodeID} = \text{Hash}(\text{IP address} + \text{UDP port}) \tag{1}$$

P2P network based on Kademlia protocol calculates the distance between nodes by XOR operations between IDs. In order to be adjacent in the network, the ID of nodes can be calculated by using formula 2. A node ID can contain multiple parts, where the geolocation can be placed in the first part. Therefore, if the geolocation of nodes is same, the XOR value of ID geolocation part is 0, which indicates that they are close.

$$\text{NodeID} = \text{Hash}(\text{geographical location}) \oplus \text{Hash}(\text{IP address} + \text{UDP port}) \tag{2}$$

Based on resource text description, the weight of top k topics and their weights are obtained by topic modeling algorithm. The most important topic is selected as a service ID and hashed with geolocation of the resource. The method is shown in formula 3.

$$\text{Resource Node ID} = \text{Hash(geolocation)} \oplus \text{Hash(service type)}... \qquad (3)$$

Figure 2 describes the process and method of building a P2P network based on geo location, which is implemented by Kademlia protocol. Formula (3) is used here for node ID generation. According to the Kademlia protocol and the identification generation method adopted here, a, b, c and d in Fig. 3 show the establishment process of P2P network based on geolocation. The first node in (a) generates its own ID firstly, and then inquires the nearest node is by contacting the bootstrap node. When started, there are very few nodes in the P2P network, the established network is shown in (a). In figure (b), the newly added node calculates its ID. It can be seen from the figure that when the nodes have same geolocation in the previous part of their IDs, their distance is smaller according to the XOR. Therefore, when routing is constructed based on such IDs, the nearest node in geolocation is selected as the neighbor node in P2P network. So the resulting topology is similar to figure (c). When more nodes join in, the final P2P network structure is shown in figure (d). Nodes close to each other geographically serve as neighbor nodes in P2P network.

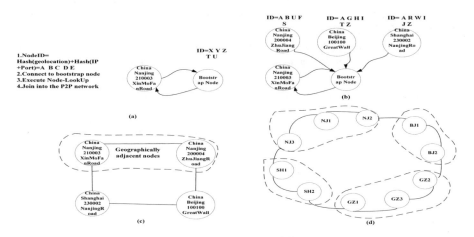

Fig. 2. P2P network build based on geolocation

4.3   WoT Resource Management Based on Probabilistic Topic Model

4.3.1   WoT Resource Publishing Based on LDA Model

Resources in WoT can be represented as tuples R→ (F,URI,Des,Opers,T,G,#). F is the function or data, URI is the resource location which is typically the Web URL of the resource. Des is a resource description which can be a text or semantically based description, OPER is a resource-oriented operation such as PUT/DELETE/UPDATE/ GET, of which GET is the most basic operation. G is the geolocation of the resource. T is the life-time model of the resource, and #represents the identity requirement for a resource consumer. The topic of the resource description is extracted and the hashed as shown in Fig. 5. Then the PUT operation of the Web API is called to store the resource in the P2P network.

Figure 3 depicts the generation of resource IDs in order to store resources in a P2P network. In step 1, the user or resource provider first gives the resource description. At the same time, the system selects different data sets and uses the topic model algorithm in step 2 to generate different topic models. These data sets can be generic, such as wikipedia-based document sets, or sensor-oriented document data sets. In step 1, the text in the data set needs to be preprocessed, which includes text segmentation, stop word removal and other operations. In step 2, Latent Dirichlet Allocation (LDA) algorithm is used to obtain LDA learning model M_{lda} by existing corpus. The number of topics K in the algorithm model needs to be adjusted and optimized according to the application scene, corpus and the number of texts. In steps 3–5, the segmentation information of the user-provided resource description is input into the learning model M_{lda} to obtain K topics of the resource and their probability in the document.

$$S_{topic} = \{topic_1 : w_1, \ topic_2 : w_2, \ topic_3 : w_3, \ topic_4 : w_4 \ldots \ldots topic_k : w_k\} \sum_{i=1}^{k} w_i$$
$$= 1$$

The topic with the highest probability is selected as the service type. In step 6, the geolocation identification and service type of the resource are hashed respectively as the most critical part of the resource ID, and the rest resource subtypes are hashed as the rest of the resource ID. Resource storage in this way enables geographically close and functionally similar resources to be stored on similar nodes as much as possible. When resource is stored, P2P node is searched according to resource ID. After finding the node, the tuple information of the resource is stored in the node.

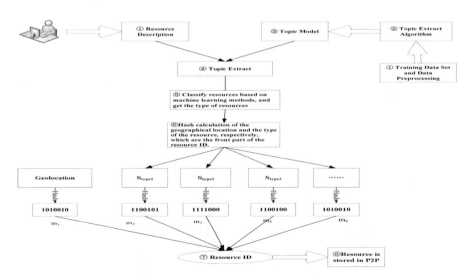

Fig. 3. WoT resource ID generation based on topic

WoT resource IDs based on geolocation and topic form the hierarchical ID space shown in Fig. 4. When a node in P2P network sends a request, the request is routed to the corresponding geospatial subspace, such as the geospatial region represented by the N2 node. In this space, multiple types of resources may be published, forming different subtype spaces, such as topic 1 space and topic 2 space. Under each topic space, it then may be divided into different subspaces according to resource subclasses.

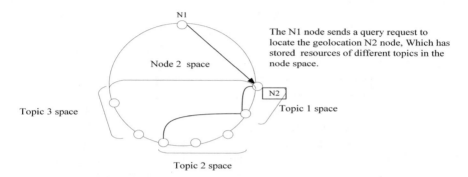

Fig. 4. WoT P2P resource space based on geography and topic

4.3.2 Resource Discovery

Users can enter information such as geolocation, topic or keyword to find the resource. Figure 5 shows the process of WoT resource discovery in P2P environment based on LDA model. This process is divided into two parts. The first part locates the node which is close to the requested resource according to the geolocation and service types. In the second part, the similarity between the returned resource and the resource requested by the user is compared and the result is ordered from high to low.

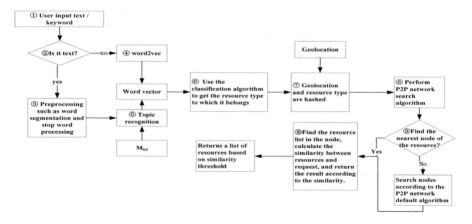

Fig. 5. P2P WoT resource discovery

5 Experiment and Result Analysis

5.1 Performance Analysis

In a P2P network with N nodes, the average time complexity of resource lookup is O (log N). It is assumed that the P2P network based on geolocation is constructed by K locations. According to the hierarchical relationship of geolocation, node in a sub-geolocation is built into P2P sub-network according to P2P protocol. We suppose that the average number of nodes in this subnetwork is N/K. The time of resource search includes two parts. Firstly, A certain geolocation in P2P network is located. Secondly, the specific nodes is searched in this subnetwork. so the average time complexity is O (log(N/K)) + O(log K) $\approx$ O(N).

That is, the average time complexity of the P2P network based on the geolocation is consistent with that of the ordinary P2P network. However, for IoT applications, it is necessary to make use of the sensing resources of different geographical locations. For example, when it looks for the resource at the geolocation W, it can cache the locations of several nodes related to the geographical area, namely {n1, n2, n3}. In the next resource discovery, a node can be quickly located to reduce the time complexity of resource discovery to O(log(N/k)). In addition to the above time, a resource matching time is also included, which is related to the number of resources stored on the node and the number of topics and attributes to be compared.

5.2 System Simulation

5.2.1 Simulation Based on Peersim

In this paper, a P2P simulation tool peersim [12] and Kademlia peersim implementation [13] were used for simulation. The operating environment is Thinkpad laptop T470, Windows10 operating system, Java virtual machine memory with 500 M memory space. We use the geolocation as the first part of the node ID according to the algorithm. The number of different geolocations is set to 100, the number of nodes is 10000, the number of token buckets is set to 20, and the alpha parameter is set to 40. The other parameters are set to default for performing network topology simulation. Figure 6 shows the comparison of the average hop count and delay of Kademlia based on geolocation with ordinary Kademlia. It can be seen that the performance between the two is close.

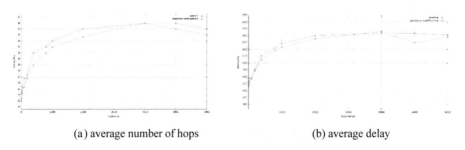

(a) average number of hops   (b) average delay

Fig. 6. Comparison of average hops and average delay

5.2.2   P2P WoT Resource Management Based on CoAP

This paper uses the CoAP protocol as transport layer here to establish a P2P network between IoT gateway nodes to implement the resource management mechanism of the WoT. The Peersim system architecture is shown in Fig. 7. The peersim system is launched by the PeerSim Simulator which calls four interfaces in peersim, namely Control, Initializer, Linkable, and Protocol. Their functions are as following [14]: Protocol interface defines the network protocol function interface, which can realize different P2P protocols and other network protocols; Control interfaces are used to manage and monitor the state of the network and perform various operations, such as resource discovery, resource lookup, and so on; Initializer interface is a special Control interface that is used for initialization of network startup and is executed only once; Linkable interface is used to store the network relationships between nodes. When the network was initialized, peersim installed the network protocol in each network node. When the network was running, peersim encapsulated operations in the network as events, which were stored in a queue. Then, dispatcher was used for scheduling, and corresponding processing was conducted by corresponding protocols in the network node to build P2P Overlay network. The Network in the figure stores the information of the entire Network node.

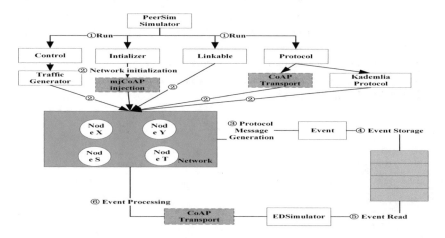

Fig. 7. PeerSim system architecture based on CoAP

In order to test the combination of P2P network and WoT and build the distributed IoT resource management, this paper partially extends the Peersim system architecture with CoAP protocol so that it can simulate the P2P network based on WoT. CoAP protocol is a lightweight UDP-based protocol for IoT applications in constrained environments. CoAP protocol was combined with P2P technology. On the one hand, CoAP protocol was used to transfer P2P network messages between network nodes to build P2POverlay network. On the other hand, CoAP itself can be used to realize resource storage and management of IoT nodes, which can make resource-limited devices, such as various terminal nodes, easy to access P2P network, and more convenient to integrate with Web applications to promote the development of IoT application.

Figure 7 implements a P2P protocol network simulation model based on CoAP protocol. The peersim architecture implements the P2P network construction process through the CoAP protocol, namely, the joining of nodes in the P2P network, the packet transmission between nodes, and the searching and storage of resources, for analyzing the P2P network overhead based on the CoAP protocol. Here, the mjCoAP open source is used to realize the CoAP function. mjCoAP [15] is a lightweight java-based CoAP library, which has performance advantages over Californium [16] that is also based on the Java language, and also supports mapping of CoAP to HTTP protocols. The Peersim environment implements the P2P algorithm. CoAP acts as the transport protocol. Each node contains a CoAP client and server to implement message transmission between nodes.

Figure 8 shows the performance comparison between the P2P transmission protocol using CoAP and peersim itself. First, the CoAP protocol overhead in ideal network condition is tested in non-blocking mode. Then, we tested the CoAP protocol overhead under ideal network condition in blocking mode. In this case, since the simulation environment peersim is running in event-driven single-thread mode and mjCoAP is multi-threaded model, the confirmation retransmission mechanism of CoAP is changed to non-retransmission mechanism for testing. In the figure, the blue curve is the delay of peersim protocol itself, while the yellow curve is the cost of CoAP protocol. As can be seen from the test results in these cases, in the network simulation environment, CoAP protocol overhead is relatively small ranging from a few tens to a hundred milliseconds, indicating the P2P network based on the CoAP protocol can not only satisfy performance requirements, but also provides the foundation for web application development.

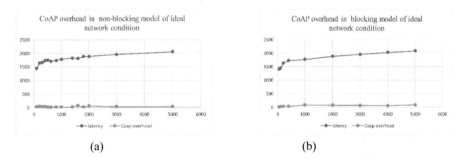

(a)   (b)

Fig. 8. Network latency of CoAP overhead in blocking and non-blocking model under ideal condition

6 Conclusion

WoT technology enables developers to take advantage of Web tools for rapid development of IoT applications. However, in a large-scale IoT environment, how to effectively manage resources has become a problem for WoT. By comparing the characteristics of IoT and P2P technology, we consider that P2P technology is very suitable for distributed management of WoT resources. WoT integrates with P2P technology to realize distributed management of resources by CoAP protocol. Considering the geographic location sensitivity of IoT applications, the construction of P2P networks takes into account the geolocation of the resources. At the same time, in order to manage the resources automatically, the LDA model is used to automatically extract the resource topics and classify them accordingly to realize the classified storage of resources. We use the peersim experimental environment to construct a p2p network based on WoT. The preliminary experimental results show that the CoAP-based WoT performance meets the requirements, and at the same time, it is convenient for users to develop Web-based P2P IoT applications.

Acknowledgments. This work was funded by the National Natural Science Foundation of China (No. 61502246), and the Research Project of Nanjing University of Posts and Telecommunications (No. XK0160915170).

References

1. Y.2060: Overview of the Internet of things. https://www.itu.int/rec/T-REC-Y.2060-201206-I . Accessed on August 2018
2. VladStirbu: Towards a restful plug and play experience in the web of things. In: IEEE International Conference on Semantic Computing, pp. 512–517 (2008)
3. Ostermaier, B., Kovatsch, M., Santini, S.: Connecting things to the web using programmable low-power WIFI modules. In: Proceedings of the Second International Workshop on Web of Things, pp. 1–6 (2011)
4. Balakrishnan, H., FransKaashoek, M., Karger, D., Morris, R., Stoica, I.: Looking up data in P2P systems. Commun. ACM **46**(2), 43–48 (2003)
5. Dahbi, A., Mouftah, H.T.: A hierarchical architecture for distributed EPCglobal discovery services. In: IEEE Global Communication Conference (GLOBECOM) 2015 (2015)
6. Moeini, H., Yen, I.-L., Bastani, F.: Efficient caching for peer-to-peer service discovery in internet of things. In: IEEE International Conference on Web Services, pp. 196–203 (2017)
7. He, Q., Yan, J., Yang, Y., Kowalczyk, R., Jin, H., Member, S.: A decentralized service discovery approach on peer-to-peer networks. Trans. Serv. Comput. **6**, 64–75 (2013)
8. Trifa, V., Guinard, D., Mayer, S.: Leveraging the web for a distributed location-aware infrastructure for the real world. In: Wilde, E., Pautasso, C. (eds.) REST: From Research to Practice. Springer (2011)
9. Namatame, N., Ding, Y., Riedel, T., Tokuda, H., Miyaki, T., Beigl, M.: A distributed resource management architecture for interconnecting web-of-things using uBox. In: Proceedings of the Second International Workshop on Web of Things, pp. 1–6 (2011)
10. Cirani, S., et al.: A scalable and self-configuring architecture for service discovery in the internet of things. IEEE Internet Things J. **1**, 508–521 (2014)

11. Chen, J.: Devify: decentralized internet of things software framework for a peer-to-peer and interoperable IoT device. SIGBED Rev. **15**, 31–36 (2018)
12. Montresor, A., Jelasity, M.: PeerSim: a scalable P2P simulator. In: 2009 IEEE Ninth International Conference on Peer-to-Peer Computing, pp. 99–100. Seattle, WA (2009)
13. Furlan, D., Bonani, M.: A Kademlia module for Peersim. http://peersim.sourceforge.net/code/kademlia.zip. Accessed on September 2018
14. Baresi, L., Derakhshan, N., Guinea, S.: WiDiSi: A Wi-Fi direct simulator study. In: Proceedings of the 14th IEEE Wireless Communications and Networking Conference (WCNC 2016). Doha, Qatar, April 2016
15. Cirani, S., Picone, M., Veltri, L.: mjCoAP: an open-source lightweight Java CoAP library for internet of Things applications. In: 22nd International Conference on Software, Telecommunications and Computer Networks, SoftCOM, Split, Croatia (2014)
16. Californium. https://www.eclipse.org/californium/. Accessed on October 2019

A Machine Learning Architecture Towards Detecting Denial of Service Attack in IoT

Yahya Al-Hadhrami$^{(\boxtimes)}$ and Farookh Khadeer Hussain

School of Computer Science, University of Technology Sydney,
Broadway, Ultimo, NSW 2007, Australia
yahya.s.al-hadhrami@student.uts.edu.au, farookh.hussain@uts.edu.au

Abstract. Internet of thing is part of our everyday life nowadays. Where millions of devices contented to the internet to collect and share data. Although IoT devices are evolving quickly to the consumer market where smart devices and sensors are becoming one of the main components of many households, IoT sensors and actuators have been also heavily used in the industry where thousands of devices are used to collect and share data for different purposes. With the rapid development of the Internet of Things in different areas, IoT is facing difficulty in securing overall availability of the network due to its heterogeneous nature. There are many types of vulnerability in IoT that can be mitigated with further research, however, in this paper, we have concentrated on distributed denial of Service attack (DDoS) on IoT. In this paper, we propose a machine learning architecture to detect DDoS attacks in IoT networks. The architecture collects IoT network traffic and analyzes the traffic through passing to machine learning model for attack detection. We propose the use of real-time data collection tool to dynamically monitor the network.

Keywords: IoT · Security · Machine learning

1 Introduction

Technology is becoming faster and more efficient day-by-day and moving towards an "always connected" model. This revolution makes each and every device capable of communicating and exchanging information between each other and fabricating the future internet [1]. This new concept of the future internet where millions of devices are connected to the internet to communicate and share information is known as the Internet of Things (IoT). IoT Networks can include many of our everyday life devices, which can range from simple motion sensors and coffee machines to smart cars that are connected to the Internet. The rapid increase in IoT devices can be explained by the decrease in memory, CPU, and battery costs in the last few years. Today, IoT has a wide variety of applications in different sectors such as in the energy, transportation, healthcare, retail, and military sectors. Such rapid implementation of IoT technologies can lead

© Springer Nature Switzerland AG 2020
L. Barolli et al. (Eds.): CISIS 2019, AISC 993, pp. 417–429, 2020.
https://doi.org/10.1007/978-3-030-22354-0_37

to security issues that can emerge due to a variety of factors. Some particular IoT security issues are privacy protection problems, heterogeneous network authentication, and access control problems [2]. These issues can be classified as communication, distributed denial-of-service (DDoS), node compromise, impersonation, and protocol-specific attacks [3]. Among these issues, DDoS attacks can hinder all of the activity in the network, which can lead to business loss due to distributive services. Therefore, it is important for IoT infrastructure to sense and analyze DDoS attacks properly and to take necessary security measures. Below, we explore some of the security goals, limitations, and challenges associated with IoT devices.

1.1 Security Goals

Different solutions have different security requirements. The most commonly used security and assurance model is the CIA triad model where it consists of three requirements detailed as follows:

1.1.1 Confidentiality

Ensuring sensitive data is protected from unauthorized entities, either when the data is in transit or in rest. IoT devices can have sensitive applications such as the health care system where personal information about the patient is critical and might be life-threatening. In such scenarios, confidentiality is crucial and must not be taken lightly.

1.1.2 Integrity

Data can be changed and altered when transmitted to the receiver, causing unreliable service in the IoT system. Ensuring integrity between IoT devices is essential in most scenarios and applications. Alteration and modification data, while in transit, can lead to serious negative implications in some IoT applications – for example, in the health sector where manipulating sensitive data (e. g., blood pressure, heart rate, etc.) means life or death for the patient [4].

1.1.3 Availability

Data can be changed and altered when transmitted to the receiver, causing unreliable service in the IoT system. Ensuring integrity between IoT devices is essential in most scenarios and applications. Alteration and modification data while in transit can lead to serious negative implications in some IoT applications – for example in the health sector where manipulating sensitive data (e.g., blood pressure, heart rate, etc.) means life or death for the patient [4].

1.2 Constraints and Security Challenges

Every kind of computer network has some kind of security requirements, and the IoT network is not different from these kinds of computer networks. However, security measurement for IoT must meet some criteria that might not be applicable to different networks due to its nature and resource constraints.

1.2.1 Resource Limitation

Every kind of computer network have some kind of security requirements and the IoT network is not different from these kinds of computer networks. However, security measurement for IoT must meet some criteria's that might not be applicable for different networks due to its nature and resources constraints.

1.2.2 Privacy and Data Confidentiality

In IoT, different applications have different privacy implications. The privacy level required for healthcare application, where patient health cannot be compromised, is different from the privacy level required for city temperature sensors. This is not to say that we should neglect the privacy concerns for some applications; on the contrary, what we are saying is that we should harness IoT devices more in situations where users' privacy is involved. IoT devices have already evolved to different aspects of our daily lives, such as smart vehicles and smart homes, which can provide sensitive user information, such as user location, health status, or maybe user home preferences, all of which can raise serious privacy concerns. The key idea about privacy and confidentiality is that the data should be kept private and accessible only by the authorized entity, whether human or machine. To achieve privacy and confidentiality a cryptography is a must and should be applied in a manner that doesn't affect the IoT devices' constraints and limitations.

1.2.3 Authentication

IoT Devices are generating a substantial amount of data every day. The data moving between entities must be securely transmitted, but this cannot be achieved without the use of an authentication mechanism. Unfortunately, there is no common standard used by all vendors for authentication. Different vendors use different authentication protocols, which raise security concerns among different platforms since there is no common way to authenticate. Therefore, the integration between these platforms is poor and can lead to security issues in the future [5].

1.2.4 Service Availability

Availability of services in IoT devices is prone to many denial of service attacks where nodes are compromised within the network itself or from outside intruders. This kind of attack can paralyze the whole IoT network and hinder all activities and services. Moreover, availability attacks usually try to consume all device resources, and since the IoT devices in many cases are battery powered this might cause the device to die. Ensuring the availability of a device or service is crucial since many applications are time and data sensitive such as in the healthcare system.

1.2.5   Data Management Challenge

IoT is all about the data, and with the increase in the amount of data generated by sensors and devices, data centers face an architectural challenge with how to cope with such data. Research has shown that the current data centers are not ready to handle such increases in data [6]. IoT at enterprise level generates a significant amount of big data that need to be processed, analyzed, and stored in real-time, which in this case will leave providers with security complications [7].

1.3   Contribution

In this study, we propose an intelligent architecture that utilizes machine learning in detecting service attacks in IoT networks. We believe that utilizing machine learning in such scenarios can help predict and mitigate such attacks without greatly compromising the difficulties and constraints of IoT devices explained above. In this paper our continuation can be summarized in the following key points:

- We introduce a state-of-the-art architecture that utilizes machine learning knowledge to detect denial of services attacks associated with IoT networks.
- Because of the limitation of current network security datasets, we have proposed building a collection tool to mine and gather IoT network datasets for security evaluation.
- We demonstrate the use of our model through an emulated environment implemented in Cooja emulator [8]

The rest of the paper is organized as follows. Section 2 goes through literature and related work. Intelligent DDoS Detection Framework is presented in Sect. 3. In Sect. 4 we implement an emulated real-world scenario to test the data collection model of our framework and show some results. In Sect. 5 we conclude the paper and explore future work and opportunities.

2   Related Work

Not many works have been written in the area of detecting attacks using machine learning in IoT. Although there are some works, they have focused on similar limited resources networks such as WSN and AD-HOC. In [9] the study focuses on developing a new IDS system that uses the Support Vector Machine SVM algorithm for a WSN network. The authors assume that the network is a clustered network that uses the base station to centrally contact with nodes. They propose the IDS be placed in the base station due to the limited resources in the WSN nodes and to avoid any resource exhaustion. The idea is to use one class SVM algorithm to classify the traffic based on two feature vectors only (the bandwidth and hop count). One class SVM has been chosen to avoid false alarms in unknown attacks to the SVM. Since the base station will be doing

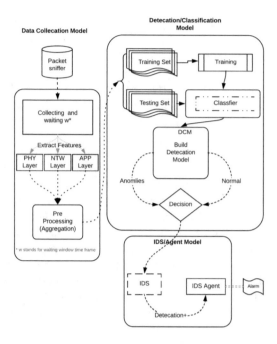

Fig. 1. Intelligent DDoS Detection Framework

all the computation, there will be no extra energy consumed in the end nodes. Another study utilized SVM for attack detection [10] they designed a hybrid distributed method for hierarchical WSN. The study focused on migrating the possibilities of anomaly detection IDS and signature-based IDS. The proposed method assumes the network is in a hierarchical mode where the cluster head has more power than the end nodes. Each node has the possibility of activating the integrated IDS. However, activating all nodes as an IDS node would have been a waste of energy, so the authors proposed a way to elect a few nodes as IDS in each cluster based on the density of the network and the range. The SVM was used in each IDS node to calculate the best possible support vector, which was sent back to the cluster head of each cluster where all support vectors were computed and sent back as a global support vector to the IDS node. Each IDS node used the global support vector to distinguish normal and abnormal traffic. Different methods have been proposed that utilize the intelligence of machine learning, such as the method proposed by the authors in [11] they proposed a new solution using a dynamic learning method. The basic idea is to use the difference between the destination sequence number of consecutive received PReps to calculate the mean vector that will be used in a multidimensional feature vector method. One advantage of this method is that the training data used to calculate the mean vector for each attack is always updated after each alarm is detected, which in this case helps in the detection accuracy. Moreover, it does not overwhelm the network with routing traffic [12]. The authors claim to implement

a solution that minimizes selective forwarding attacks by implementing a heartbeat protocol on top of the IPSEC function in the ipv6 protocol. The basic idea
is to send ICMPv6 messages from the 6BR router to all nodes in the network
and wait for the ICMPv6 echo reply from the nodes. This technique has been
implemented so that it sends ICMPv6 messages in an interval time, hence it is
called "the heartbeat." The authors claim that this technique helps to identify
which node has been filtered using the IPsec protocol, hence identifying any node
with possible attacks.

On the other hand, Kasinathan, and P., Pastrone propose an architecture
that integrates an IDS into a network frame that has been developed within
the EU FP7 project Ebbits [13]. The study proposes a solution that uses a
signature-based IDS that uses a predefined source of signatures and patterns
collected prior to implementing the solution. The author used probes in the
edge of the network to sniff packets that goes through the entire network and
analyses each packet to look for any suspected behavior, which is later on sent
to SenacIntrture IDS for further analysis. Likewise, the authors in [14] proposed
the same approach, which is dependent on the signature-based approach, but
the main focus of this study is to build an IDS system that addresses routing
attacks in an RPL network using a semi auto profiling technique. This technique
was used to gather and formulate a set of rules that will be integrated in the IDS
agent. Another method proposed by Cervantes and C., Poplade, they propose
INTI to identify sinkhole attacks on the routing services in IoT using reputation
and trust models to detect sinkhole attacks [15]. It reduced high false positives
and improved the mobility of SVELTE [16]. However, the limitation of both [15]
and [17] is that some critical QoS metrics were overlooked.

3   Framework

In developing any kind of solution for IoT scenarios, constraints and limitation
have to be considered. Developing any machine learning solutions requires the
acquisition of some kind of data related to the area researched. Unfortunately,
until writing this paper, and to our knowledge, there is no predefined dataset
that is designed for IoT networks specifically. The problem with the previous
dataset, the KDD dataset, is that it was designed and collected for a different
set of protocols that are not used for IoT networks. Therefore, in designing our
architecture we propose designing and collecting a new dataset for the purpose
of detecting attacks and encounters in the IoT networks. Figure 1 shows our
proposed architecture consisting of three models: (1) Data Collection Model
(DAM); (2) Classification Model (CM); and (3) Detection Model (DM). Each of
these are explained further below.

3.1   Data Collection Model

The main goal of this module is to collect IoT communication data, either in
a real network or in simulated 6lowpan and RPL network, but our architecture

Table 1. Data collection features

Feature	Description
Phy Layer Features	
Received Signal DBM	Mean Value of the Received signal at the mac layer
Transmission Signal DBM	Mean value of the Transmitted Signal at mac layer
RSSI Noise	Mean value of the noise recorded using RSSI
Beacon Interval	Mean value of the Beacon Interval
Network Layer Features	
LQI	Link quality indicator
EXT	Mean value of the Expected Transmission count
Number of Packets	Standard deviation of the number of packets transmitted
Number of lost Packets	Standard deviation Number of packet lost
Number of DIS messages	Number of DIS messages
Number of DIO messages	Number of the RPL DIO messages
route Metrics	Mean value of the Routing Matrices number of hops
RPL Rank	Number of Rank change over time
Hops	Number of Hops to the sink
PPR	Packet Reception Ratio
Number of Nabors	Number of Neighbours
Application Layer Features	
Temperature	Mean value of the temperature
Humidity	Mean value of the Humidity
Power Level	Mean Value of the Energy over time
Consumed Power	Mean value of consumed node power
Remaining Power	Mean Value of the reaming node power
Node ID	Node ID

is not limited to such protocols and can be generalized to any protocol. Prior to detecting any attack, we must acquire some data from the network. The acquisition process is divided into three categories: the physical layer features, the network layer features, and application layer features.

3.1.1 Physical Layer Features

Where the DCM extracts physical layer related features such as the received and transmitted signals at mac layer. This information is related to physical layer jamming attacks where they usually interfere with the transmitted signals.

3.1.2 Network Layer Features

Collection of network layer features are crucial for our IDS System since the features extracted are closely associated with many famous attacks, such as

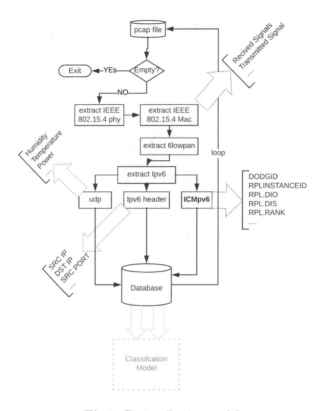

Fig. 2. Data collection model

deceased rank attack and blackhole attacks. In this category, features such as the number of DIS and DIO messages are extracted, and some other features related to the RPL protocol. Data transmitted through any network follow certain protocols. In IoT there are certain protocols available at every layer, and in this study we are focusing on RPL protocol on top of the 6lowpan protocol. This can be expanded to include extra features from other protocols.

3.1.3   Application Layer Features

At the application layer our module collects application-specific information such as the humidity, temperature, and node power level. The application-related features can be extracted programmable by programming the nodes to calculate the power consumption and other related features We chose to collect power consumption at the application layer because we could program the node to calculate the power level and send the result within the application-related information in the data frame payload.

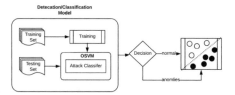

Fig. 3. Data detection model

3.1.4 Proposed Capture Method

To extract the network-related feature we have designed a specific tool that follows the 6Lowpan and RPL protocol structure. Figure 2 shows the process involved in extracting the specific features from the Pcap file captured from the IoT network. To acquire raw IoT network traffic we propose using the sensniff [18] sensor deployed to cover the entire network. The sniffed data are then fed to our DCM tool to extract the feature shown in Table 1. To extract 6lowpan and RPL feature, we have designed our tool to exploit both protocol structures. Figure 2 presents the detailed process of extracting packets transmitted in 6LOWPAN networks. At each layer encapsulation, there are a set of features extracted as we mentioned earlier. Furthermore, before even extracting the features, a time window has to be specified to aggregate data into records. This timeframe window will be used later to get the deviation or average for each node based on that time window. Depending on the IoT network application, the time window can vary accordingly. Designing a dataset for IoT electricity metering will have a different effect if the same dataset designed with the same time window for a parking sensors. Since each application has a different level of data generated during communication, determining the time window depends on the type of IoT application used and the type of protocols implemented. The main goal of this dataset is to train and test the machine learning algorithm and generate a detection model, which is explained in the next section.

3.2 Data Classification Module

The dataset generated from the DCM will be used for training and testing our machine learning algorithm. At this level we can experiment with different machine learning methods, and the one with the best result in terms of accuracy and performance will be selected. The quality of the result produced at this stage is heavily dependent on how the data was collected in the previous stage. Figure 3 shows an example of how to train and test a one class SVM for attack detection. The DCM is a two-stage scenario. The first stage is creating the machine learning model based on the dataset collected earlier. This model will be used for the detection process. The second stage is embedding the model generated from testing and training the machine learning algorithm into our IDS system. Figure 3 shows how real-time data is fed to the pre-processing unit, where statically, data will have generated and then been fed to the detection

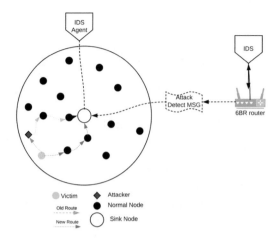

Fig. 4. Example IoT network

model. The machine leaning model then detects any abnormalities in the provided data and makes the decision. If any attack is detected by the IDS, an alarm message will be initiated and sent to the IDS agent. Otherwise, a message will be sent to the IDS agent indicating there was no attack detected. The data aggregation process is dependent on the time window defined earlier, where the DCM will take the online data and wait for a predefined time window and then extract the required features as shown in Fig. 1.

3.3 IDS Agent Model

This model will be built on the sink node since all nodes will be connected to the sink node either directly or through multi-hop. The idea of the model is to raise an alarm if an attack is detected. The main function of the module is to broadcast a warning message to all nodes in the network in case an attack is detected. The alarm message originates from the IDS system in the 6BR router after deep analysis of the traffic and sent to IDS agent. The agent then broadcasts the alarm message to the entire network. Then, the IDS agent initiates network topology reform in order to isolate the malicious node by establishing a new route to the sink from the victim node. All nodes will blacklist the malicious node, and all traffic from it is ignored and dropped. Figure 4 shows a simplified diagram of the whole process explained here.

4   Example Scenario for the IDS and Simulation

In this section, we provide a detailed explanation of our proposed system by gaining a detailed scenario and the possible distribution of the system. In this scenario, we assume we have an industrial IoT network that uses the 6lowpan protocol with the RPL-enabled protocol to collect temperature and humidity

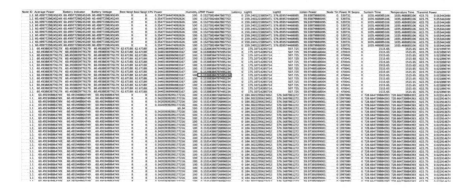

Fig. 5. Screenshot of the IoT-DDoS dataset

data across the site. The size of the network is 10 nodes which are connected to the sink node directly. Assuming the sniffers are placed at the edge of the network and cover the network half and half as shown in Fig. 4. The traffic data is transmitted to the IDS located in the border router, where we have two units—the data extraction unit and the detection model unit. The raw data then goes through the process of features extraction and aggregating them into a real-time data fed directly to the detection model to predict a decision about whether the data analyzed have an attacker on it or not. Once an attack is detected a flag message is sent to the IDS agent, and a broadcast MSG is sent to the entire network to isolate the malicious node. Simultaneously, the sink node will initiate network route reform and route the traffic of the affected node through the nearest neighbor in the network. If there is no neighbor located near the victim node, the node will be sent to sleep, and an alarm will be sent to the network administrator to manually locate a new route for the node. The network is exposed to the outside world using a 6BR. We have two main components of the IDS—the IDS and the IDS agent. The IDS is installed in the 6BR router, and all the heavy processing is usually done here due to the computation power associated with the edge node. From the above-mentioned scenario, we have built a simulated network in Cooja, [8] and we were able to extract the information explained in section A of the Intelligent IDS system by utilizing the use of the proposed DCM. Figure 4 presents the scenario network with a packet sniffer placed at the edge of the network. The network generated that form these nodes are exported to a pcap file to extract the related features from the proposed tool. Figure 5 shows a screenshot from our dataset, which will be used in the future for further processing and building our detection model.

5   Conclusion and Future Work

A comprehensive study of the literature has been performed in this study to explore all of the options available to detect DDoS attacks in IoT networks. From the literature we found that some solutions have been proposed, but unfortunately, little has been done by utilizing machine learning in IoT networks. Those in this paper have proposed a new framework to detect a DDoS attack in IoT networks using machine learning.

References

1. Mahmoud, R., Yousuf, T., Aloul, F., Zualkernan, I.: Internet of things (IoT) security: current status, challenges and prospective measures. In: 2015 10th International Conference for Internet Technology and Secured Transactions (ICITST), pp. 336–341, December 2015
2. Zhao, K., Ge, L.: A survey on the internet of things security. In: 2013 Ninth International Conference on Computational Intelligence and Security, pp. 663–667, December 2013
3. Gunduz, S., Arslan, B., Demirci, M.: A review of machine learning solutions to denial-of-services attacks in wireless sensor networks. In: 2015 IEEE 14th International Conference on Machine Learning and Applications (ICMLA), pp. 150–155, December 2015
4. Abomhara, M., Køien, G.M.: Cyber security and the internet of things: vulnerabilities, threats, intruders and attacks. J. Cyber Secur. 4(1), 65–88 (2015)
5. Conti, M., Dehghantanha, A., Franke, K., Watson, S.: Internet of things security and forensics: challenges and opportunities (2018)
6. Lee, I., Lee, K.: The internet of things (IoT): applications, investments, and challenges for enterprises. Bus. Horiz. 58(4), 431–440 (2015)
7. Rivera, J., van der Meulen, R.: Gartner says the internet of things will transform the data center (2014). Accessed 5 Aug 2014
8. Osterlind, F., Dunkels, A., Eriksson, J., Finne, N., Voigt, T.: Cross-level sensor network simulation with cooja. In: Proceedings of 2006 31st IEEE Conference on Local Computer Networks, pp. 641–648. IEEE (2006)
9. Kaplantzis, S., Shilton, A., Mani, N., Sekercioglu, Y.A.: Detecting selective forwarding attacks in wireless sensor networks using support vector machines. In: 2007 3rd International Conference on Intelligent Sensors, Sensor Networks and Information, ISSNIP 2007, pp. 335–340. IEEE (2007)
10. Sedjelmaci, H., Feham, M.: Novel hybrid intrusion detection system for clustered wireless sensor network. arXiv preprint arXiv:1108.2656 (2011)
11. Kurosawa, S., Nakayama, H., Kato, N., Jamalipour, A., Nemoto, Y.: Detecting blackhole attack on AODV-based mobile ad hoc networks by dynamic learning method. IJ Netw. Secur. 5(3), 338–346 (2007)
12. Wallgren, L., Raza, S., Voigt, T.: Routing attacks and countermeasures in the RPL-based internet of things. Int. J. Distrib. Sens. Netw. 9(8), 794326 (2013)
13. Kasinathan, P., Pastrone, C., Spirito, M.A., Vinkovits, M.: Denial-of-service detection in 6LoWPAN based Internet of Things. In: 2013 IEEE 9th International Conference on Wireless and Mobile Computing, Networking and Communications (WiMob), pp. 600–607. IEEE (2013)
14. Le, A., Loo, J., Chai, K.K., Aiash, M.: A specification-based ids for detecting attacks on RPL-based network topology. Information 7(2), 25 (2016)
15. Cervantes, C., Poplade, D., Nogueira, M., Santos, A.: Detection of sinkhole attacks for supporting secure routing on 6LoWPAN for Internet of Things. In: IM, pp. 606–611 (2015)

16. Raza, S., Wallgren, L., Voigt, T.: SVELTE: real-time intrusion detection in the Internet of Things. Ad Hoc Netw. **11**(8), 2661–2674 (2013)
17. Sonar, K., Upadhyay, H.: An approach to secure internet of things against DDoS. In: Proceedings of International Conference on ICT for Sustainable Development, pp. 367–376. Springer (2016)
18. Oikonomou, G.: Sensniff: live traffic capture and sniffer for iEEE 802.15.4 networks (2017). https://github.com/g-oikonomou/sensniff

Pipelined FPGA Implementation of a Wave-Front-Fetch Graph Cut System

Naofumi Yoshinaga$^{(\boxtimes)}$, Ryo Kamasaka, Yuichiro Shibata, and Kiyoshi Oguri

Graduate School of Engineering, Nagasaki University,
1-14 Bunkyo-machi, Nagasaki 852-8521, Japan
naofumi@pca.cis.nagasaki-u.ac.jp

Abstract. The current mainstream method for stereo vision is to find corresponding points of two-dimensional images obtained from multiple cameras and restore three-dimensional information using the principle of triangulation. However, the occlusion problem often makes it difficult to search for corresponding points. Therefore, a new approach has been proposed in which the three-dimensional space is directly considered as a three-dimensional graph instead of searching for corresponding points of two images. In this approach, a 3D grid graph is constructed based on luminance values obtained from the left and right cameras, and a highly likely object surface is obtained by cutting this graph. This paper proposes a pipelined architecture for 3D grid graph cut, aiming at a real-time stereo vision system. The system uses Wave-Front-Fetch algorithm, which is oriented for parallel processing. We achieved processing time of about 21 ms for a graph of $129 \times 129 \times 16$ nodes, resulting in a frame rate of about 49 fps. Our approach was about 19 times faster than a well-known graph cut software library.

1 Introduction

Stereo vision is one of the most active research fields in computer vision, which can be applied for environmental recognition of robots and automatic brake function of automobiles. The current mainstream method for stereo vision is to search for corresponding points on two images and restore three-dimensional information using triangulation. However, an occlusion problem, which is caused by a blind spot for one of the cameras, often makes it difficult to search for the corresponding points [7,8]. Therefore, instead of detecting corresponding points from 2D images, a new stereo vision approach called Gaze Line-Depth model Stereo (GLDS) has been proposed, in which points in a 3D space are directly considered [6]. In this method, a 3D graph is created based on information obtained from two cameras, and surface information of the target objects is estimated using graph cut.

Graph cut is one of the methods to solve energy minimization problems. It has been widely used in the field of image processing such as noise removal of binary image, image area segmentation, and stereo vision. However, since applying these algorithms in software takes a long processing time, real-time

© Springer Nature Switzerland AG 2020
L. Barolli et al. (Eds.): CISIS 2019, AISC 993, pp. 430–441, 2020.
https://doi.org/10.1007/978-3-030-22354-0_38

processing was difficult. To cope with this problem, FPGA acceleration of a Push-Relabel graph cut algorithm [2] was proposed and achieved better performance than CPU implementation [5]. However, this work only supports 2D grid graphs, therefore it cannot be used for stereo vision. Although an attempt to extend this FPGA acceleration approach to 3D grid graphs was also proposed [3], real-time processing performance was not achieved. On the other hand, a new graph-cut algorithm called Wave-Front-Fetch that takes advantage of the parallelism of hardware was proposed and implemented on an FPGA [4]. Although this implementation demonstrated real-time graph-cut processing for 3D grid graphs became feasible, hardware resource usage was enormous, so it can only process small size graphs. Therefore, in this paper, we propose a pipelined Wave-Front-Fetch graph cut FPGA system, which reduces required hardware resource usage while maintaining real-time performance.

2 Graph Cut

2.1 Flow Network and Residual Network

The graph used in the graph cut is called a flow network. Flow network is a kind of weighted directed graph. A flow capacity is set for each edge, and a flow exceeding this amount cannot exist on that edge. The node that supplies the flow is called the source node s, and the node from which the flow is discharged is called the sink node t.

In the graph, let V be a set of nodes and E be a set of edges. Suppose that a positive capacity $c(u, v)$ is set to an edge $(u, v) \in E$ with a finite directed graph $G(V, E)$. When the edge (u, v) is not connected, let $c(u, v) = 0$. When the flow flowing through the edge (u, v) is $f(u, v)$, the following holds.

$$0 \le f(u, v) \le c(u, v) \tag{1}$$

$$f(v, u) = -f(u, v) \tag{2}$$

$$\sum_{\omega \in V} f(u, \omega) = 0 (u \in V - \{s, t\}) \tag{3}$$

Expression 1 indicates that a flow flowing through an edge does not exceed its flow capacity. Expression 2 shows that the flow flowing from v to u is a anti-symmetric relationship with the flow flowing from u to v. Equation 3 shows that for one node u excluding the source node or the sink node, the total flow flowing in and the one flowing out are equal.

The residual capacity is a flow capacity that satisfies the following equations.

$$c_f(u, v) = c(u, v) - f(u, v) \tag{4}$$

$$c_f(v, u) = c(v, u) + f(u, v) \tag{5}$$

The flow network that uses the residual capacity as the flow capacity of each edge is called a residual network. Equation 4 shows that the residual capacity is the remaining capacity minus the flow flowing from the current capacity. Equation 5 holds from the anti-symmetry of the flow network, and the amount of flow that flowed is added to the residual capacity of the opposite edge. As a result, even in a general flow network that can flow only in one direction, there is a case that a flow in the opposite direction can be passed in the residual network. As described above, the residual network is a network having the residual amount of flows that can be passed to each edge, and is created from the flow network. Generally, graph cut algorithms handle a residual network as a graph.

2.2 Minimum-Cuts and Maximum-Flow Problem

In the graph $G(V, E)$, dividing the set of nodes V into two sets of nodes (S, T) is called cut. An edge (u, v) such that $u \in S$ and $v \in T$ is called the cut edge. Figure 1 shows the result of dividing the set of nodes $V = \{s, t, 1, 2, 3, 4\}$ into two $(S = \{s, 1, 2\}, T = \{t, 3, 4\})$. Here, $(1, 3), (2, 4) \in E$ are the cut edges.

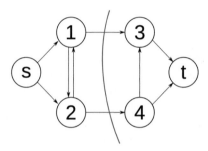

Fig. 1. Graph cut

In the flow network, the sum of the flow capacities of cut edges in cuts (S, T) is called cut capacity $c(S, T)$. Cutting that minimizes cutting capacity is called the minimum cut. The minimum cut problem is used to solve the energy minimization problem. Figure 2 shows the minimum cut at the edge from S to T in the flow network $G = (V = \{S, T\}, E)$. In this case, the cut capacity $c(S, T) = 20 + 5 + 3 = 28$.

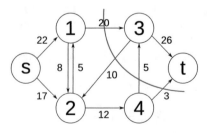

Fig. 2. Minimum-cut

The maximum flow problem is a problem of finding the maximum value of the flow that can flow from the source node to the sink node in the flow network. According to the maximum flow minimum cut theorem, the maximum flow value and the minimum cut capacity are equal. Therefore, the minimum cut problem can be solved by replacing it with the maximum flow problem. There are various algorithms for solving the maximum flow problem. In this work, the Wave-Front-Fetch method is used since it has a high degree of affinity with hardware implementation.

2.3   Wave-Front-Fetch Method

According to the flow conservation law in the ordinary flow networks, the inflow amount and the outflow amount are equal for one node. However, in the Wave-Front-Fetch (WFF) method, the preflow, where the inflow amount may exceed the outflow amount to the node, is used. When using preflow, there is a flow that remains unspilled at a certain node u. This flow is called excess flow $e(u)$.

In the Wave-Front-Fetch method, the wave number propagated from sink node is used. An image of propagation of wave numbers in WFF is shown in the Fig. 3. In this figure, nodes $\{1,2\}$ belong to wave number 1, and nodes $\{3,4\}$ belong to wave number 2.

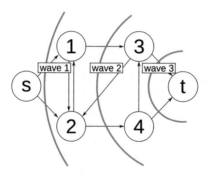

Fig. 3. Wave-Front-Fetch method

In WFF, the following operations are repeated for all nodes as long as any flows can move.

- Pull operation: If the wave number of node u is larger than the wave number of the connected node v, excess flow flows to the node u as long as the excess flow of the node v can flow.
- Update operation: If the wave number of node u is smaller than the largest wave numbers among connected nodes $w_{max}(u)$, the wave number of node u is updated to $w_{max}(u)$. However, if $u = t$, the wave number is incremented.

First, the following initialization is performed on the graph.

- Initialization of wave number: The wave number is initialized to 2 for the sink node t, 1 for the node adjacent to t, and 0 for otherwise.
- Initialization of excess flow: Preflow is made to flow from the source node to the node connected to the source as much as possible. Also, from the node connected to the sink to the sink node as well, preflow is flowed as long as the flow capacity allows.

A network initialized from Fig. 4 is shown in Fig. 5. The green numbers represent wave numbers and the red numbers represent excess flow.

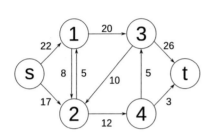

Fig. 4. Flow network before initialization

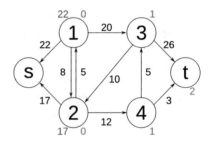

Fig. 5. Flow network after initialization

Next, in the network shown in Fig. 5, Pull and Update operations are performed for all nodes. Node 1 has 22 excess flows and the residual capacity $c_f(1,3) = 20$. Here, since node 3 has the wave number larger than that of node 1, node 3 can pull 20 excess flows from node 1. Similarly, node 4 can pull 12 excess flows from node 2.

An Update operation is performed on the network after the Pull operation. Since the edge $(3, t)$ is connected and node 3 has a smaller wave number than node t, the wave number of node 3 can be updated to the wave number of node t. Similarly, the wave number of node 4 can be updated. The wave number of the sink node t is incremented.

The network after the Pull operation is shown in Fig. 6, and the network after the Update operation is shown in Fig. 7. If the Pull operation and the Update operation are repeated until the flow stops flowing, it will finally become as shown in Fig. 8.

The node that has lost the connection with the sink side stops updating the wave number, and the flow stops flowing. All the flows that flowed into the sink side flow to the sink node, and finally, 28 flows are accumulated in the sink node. Thus, the maximum flow of this network is 28. In the network after application of the WFF algorithm, nodes traceable from the source node belong to a source side, and the other nodes are sink side.

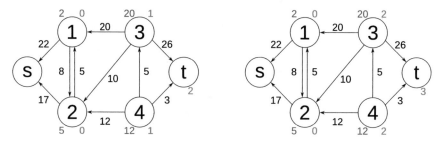

Fig. 6. Flow network after Pull operation   **Fig. 7.** Flow network after Update operation

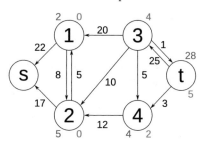

Fig. 8. After applying WFF method

3   Implementation

3.1   Graph and Data Structure

The graph used for this graph cut system is shown in Fig. 9. This graph has 10 neighbor nodes (8 neighbors in the xz plane and 2 neighbors in the y-axis direction) per node. In addition, each edge has any of the following attributes.

- Data: The reciprocal of the surface presence likelihood of the object determined from the difference between the left and right luminance values.
- Inhibit: A parameter to prevent the graph from being cut into the occlusion part.
- Penalty: A parameter to keep the object surface smooth.

In the system implemented in this research, the following initialization is performed to reduce graph data size.

- Move flows from the source node s to the node connected to s as much as the edge capacity allows.
- Move flows from the node connected to the sink node t to t as long as the edge capacity allows.
- Remove source and sink nodes from the graph.

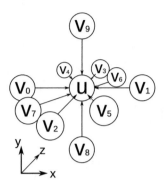

Fig. 9. Graph structure in this work

Initializing as shown in Fig. 10 reduces the graph size because source nodes and sink nodes are excluded from the graph.

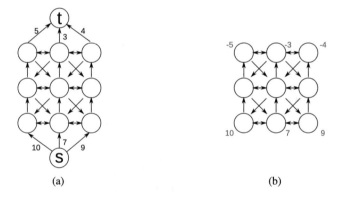

(a)   (b)

Fig. 10. Graph initialization (a) Graph before initialization, (b) Graph after initialization

In this architecture, the data structure possessed by node u is the following 72 bits in total.

- excess flow $e(u)$: 14 bits
- wave number $w(u)$: 10 bits
- residual capacity $c_f(v_i, u)|(i = 2, 3, 5, 7)$: 10 bits
- residual capacity $c_f(v_j, u)|(j = 1, 9)$: 4 bits

The Wave-Front-Fetch algorithm requires excess flow, wave number and residual capacity in data structure of each node. Residual capacity is required for 10 directions, but this architecture only has the above six to reduce resource usage, considering the characteristics of the graph as follows. The residual capacity in the x-axis and y-axis directions is set to 7 as the Penalty parameter at the initialization stage. Therefore, the sum of residual capacity in both directions between two nodes in the x-axis and y-axis directions is always 14, which is the double of Penalty. That is, by having the residual capacity in one direction as data between two nodes, the residual capacity in the opposite direction can also be determined, so resource usage can be reduced. The maximum value of the residual capacity in this directions is 14, so it can be expressed in 4 bits. In this implementation, the data structure has $c_f(v_1, u)$ as the residual capacity in the x-axis direction and $c_f(v_9, u)$ as the residual capacity in the y-axis direction.

In the xz plane diagonal direction, the Inhibit parameter is given to the edges from the sink side to the source side ($c_f(v_4, u)$ and $c_f(v_6, u)$ in Fig. 9), and this value is treated as infinity. Therefore, it is not necessary to have residual capacity as data.

3.2 Process Flow

In the proposed architecture, in order to reduce hardware resource usage, only a part of the graph is expanded in space on the processing system, and the part not being processed is stored in BRAM. In this implementation, the 3D graph is divided into xy planes, and each plane has BRAM. This BRAM is controlled as an FIFO, and pipeline processing is performed with node data deployed partially into the processing system. The processed node data is stored again in BRAM. The red rectangle area in Fig. 11 is the processing system. The xz and xy cross sections of this processing system are shown in Fig. 12.

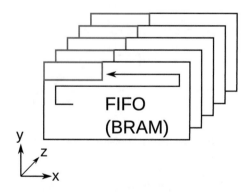

Fig. 11. Proposed architecture

This architecture operates according to the state transition diagram shown in Fig. 13. The data is initialized in the INIT state, and the data is deployed to the first processing system in the LOAD state. In the PROCESS state, a total of 11 stages of 10 direction pull operations and wave number update operations are assigned and processed.

The system performs operations in the y-axis direction from S0 to S9, operations in the x-axis direction in S1 and S2, operations in the z-axis direction in S3

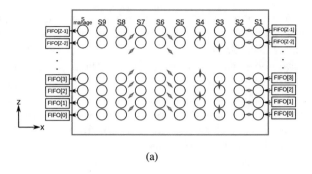

(a)

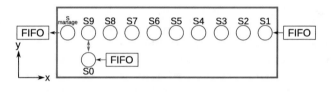

(b)

Fig. 12. Processing system: (a) X-Z cross section, (b) X-Y cross section

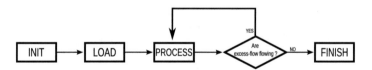

Fig. 13. State transition diagram

and S4, operations in the xz plane diagonally from S5 to S8 (Pull operation). It applies wave number update (Update operation) in the manage state and stores the result in BRAM. The pipelined process flow in the PROCESS state is shown in Fig. 14, which takes $XY + \alpha$ clock cycles to apply the pull-update operation once to all the nodes. The PROCESS state is repeated until the flow stops flowing, and then the state finally transits to the FINISH state. Since it is difficult to efficiently determine whether the flow is flowing in a parallel and distributed manner, in this implementation, the number of loops of the PROCESS state is used as the termination condition, which was obtained by software evaluation.

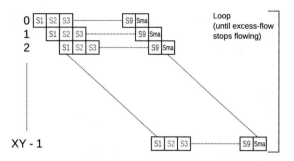

Fig. 14. Pipeline structure in PROCESS state

4   Evaluation and Discussion

The FPGA chip we used is Xilinx Virtex UltraScale xcvu095-ffva2104-2-e-es2, and the synthesis and mapping tool is Vivado 2017.2. The relationship between graph size and resource usage in the proposed architecture is shown in Fig. 15. Figure 15(a) shows that the resource usage is proportional to the size of Z in the graph for any of LUT, FF and BRAM resources. Since the size of the processing part of the proposed architecture depends on the size of Z in the 3D graph, it is considered that the LUT and FF used in the processing part are proportional to Z. Since the proposed architecture has BRAM FIFOs in each XY plane, the usage of BRAMs is also considered to be proportional to Z as shown in Fig. 15(a). Figure 15(b) reveals that the usage of LUT and FF stays almost the same, while the usage of BRAM goes up gradually for the increase in XY. Since the processing system part does not depend on the size in the X direction or Y direction, it is considered that the usage of LUT and FF is almost constant. On the other hand, since the capacity of each BRAM is fixed at 36 kB, it is considered that the number of BRAMs used in each XY plane increases according to the number of nodes in each plane. The rate of increase in BRAM usage is not

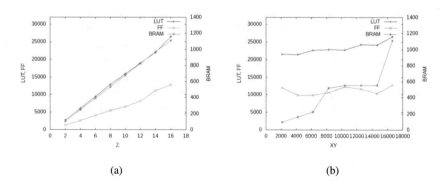

(a)   (b)

Fig. 15. Resource consumption of pipeline Wave-Front-Fetch graph cut system: (a) $129 \times 129 \times Z$, (b) $X \times Y \times 16$

constant. This is probably due to the fact that only a part of each BRAM is used because the BRAM capacity is fixed, and the usage does not increase until the capacity of each BRAM is fulfilled.

Table 1. Performance comparison at $129 \times 129 \times 16$ graph size

Graph				Frames per second		Speedup to CPU
No.	Max-flow	BK/CPU	P_WFF	BK/CPU	P_WFF/FPGA	P_WFF/FPGA
#1	545,133	400	16.84	2.50	59.38	23.75
#2	597,455	300	18.31	3.30	54.61	16.38
#3	1,621,425	480	26.69	2.55	37.47	17.98
Average		393	20.61	2.54	48.52	19.12

The maximum operating frequency of this architecture is approximately 200.96 MHz. The critical path exists in conditional logic of the BRAM write flag for inputting a value to the address bus. Table 1 summarizes the results of processing time by the CPU and FPGA of the following algorithm for 3 types of 3D grid graphs in $129 \times 29 \times 16$ graph size.

- Boykov-Kolmogorov (BK) method: C ++ software library 'maxflow-v3.04' [1]
- Pipelined Wave-Front-Fetch: FPGA implementation

The Wave-Front-Fetch method is an algorithm oriented for parallel processing, and applying this algorithm in software takes a long time to process. The software library of the BK method implemented by Vladimir Kolmogorov is known to be faster than the Push-Relabel method and is widely used in computer vision applications [1]. Therefore, we adopted this software for the comparison. The CPU is Intel Core i7-5960X (3.00 GHz) and the compiler is the GNU Compiler Collection (GCC) 4.4.7, with -O3 compile option.

As shown in Table 1, for the $129 \times 129 \times 16$ graph, P_WFF / FPGA is about 19 times faster than BK/CPU. By pipelining the process, we were able to achieve about 49 fps with a $129 \times 129 \times 16$ graph size while reducing resource usage. Considering the amount of BRAM resources, it is considered that graphs up to about 1.5 times the size of this graph can be loaded. However, with $640 \times 480 \times 80$, which is considered to be a practical graph size, it is likely to run out of the BRAM capacity. Therefore, it is necessary to devise an architecture so that it can handle larger graphs by reducing the amount of resources including BRAM.

5   Conclusion

This paper proposed a pipelined architecture for a 3D Wave-Front-Fetch graph cut algorithm. The implementation experiment on an FPGA demonstrated the proposed system can process $129 \times 129 \times 16$ graph in about 21 ms, achieving about 49 frames per second. This is about 19 times faster than the standard graph-cut software library. However, about 60% of BRAM is used for this graph size, and BRAM will be insufficient for practical stereo vision graphs such as $640 \times 480 \times 80$. One of our future work is to use not only the memory in the FPGA chip but also an off-chip memory such as a DRAM.

References

1. Boykov, Y., Kolmogorov, V.: An experimental comparison of min-cut/max-flow algorithms for energy minimization in vision. IEEE Trans. Pattern Anal. Mach. Intell. **26**(9), 1124–1137 (2004)
2. Goldberg, A.V., Tarjan, R.E.: A new approach to the maximum-flow problem. J. ACM (JACM) **35**(4), 921–940 (1988)
3. Kamasaka, R., Shibata, Y., Oguri, K.: FPGA implementation of a graph cut algorithm for stereo vision. In: Proceedings of International Symposium on Highly Efficient Accelerators and Reconfigurable Technologies (HEART), pp. 14:1–14:6 (2017)
4. Kamasaka, R., Shibata, Y., Oguri, K.: An FPGA-oriented graph cut algorithm for accelerating stereo vision. In: Proceedings of International Conference on Reconfigurable Computing and FPGAs (ReConFig) (2018)
5. Kobori, D., Maruyama, T.: An acceleration of a graph cut segmentation with FPGA. In: Proceedings of International Conference on Field Programmable Logic and Applications (FPL), pp. 407–413 (2012)
6. Oguri, K., Shibata, Y.: A new stereo formulation not using pixel and disparity models. arXiv:1803.01516 [cs.CV] (2018)
7. Smith, B.M., Zhang, L., Jin, H.: Stereo matching with nonparametric smoothness priors in feature space. In: Proceedings of IEEE Conference on Computer Vision and Pattern Recognition (CVPR), pp. 485–492 (2009)
8. Woodford, O., Torr, P., Reid, I., Fitzgibbon, A.: Global stereo reconstruction under second-order smoothness priors. IEEE Trans. Pattern Anal. Mach. Intell. **31**(12), 2115–2128 (2009)

A Self-partial Reconfiguration Framework with Configuration Data Compression for Intel FPGAs

Shota Fukui[✉], Yuichi Kawamata, and Yuichiro Shibata

Graduate School of Engineering, Nagasaki University,
1-14 Bunkyo-machi, Nagasaki 852-8521, Japan
fukui@pca.cis.nagasaki-u.ac.jp , shibata@cis.nagasaki-u.ac.jp

Abstract. In this paper, we implement and evaluate a self-partial reconfiguration (Self-PR) system, where configuration data for partial reconfiguration (PR) is stored in the hard macro memory on FPGA, and PR is performed from the inside of FPGA. As a result, in the case of the smallest PR region, the time required for Self-PR is about 2.8 ms, 97% less than using JTAG interface. The usage of hard macro memory blocks is about 21% of the total resources. We additionally implement and evaluate a mechanism that compresses the configuration data for PR, and the module that decompresses the data in the FPGA. As a result, the usage of hard macro memory blocks was reduced to 3% of the total resources. The increase in resource usage and the decrease in FMax due to addition of decompress circuit were quite limited, and there was no additional latency. Therefore, Self-PR with compressed configuration data can be performed in the same speed as when the compression mechanism was not utilized.

1 Introduction

Partial reconfiguration (PR) of FPGAs is a technology that reconfigures only a part of circuits while the entire circuit keeps operating. By using this technology, only the circuits necessary for processing can be configured on the FPGA, and the hardware scale and power consumption can be reduced. This approach is promising for FPGA accelerators in embedded systems, which are typically implemented into battery powered devices that have limited power available. In order to make the best use of the PR technology, self-partial reconfiguration (Self-PR) is also an important concept, where the FPGA itself initiates reconfiguration during the operation.

This paper proposes a framework mechanism of Self-PR for Intel FPGAs. Aiming at real-time reconfiguration for embedded applications such as image processing, configuration data for PR is stored into on-chip hard macro memory on an FPGA. In addition, to reduce the hardware resources required for the mechanism, a data compression mechanism for FPGA configuration is also proposed. As a preliminary evaluation, we will implement the Self-PR mechanism for small-scale circuits on an Intel FPGA and evaluate its performance.

© Springer Nature Switzerland AG 2020
L. Barolli et al. (Eds.): CISIS 2019, AISC 993, pp. 442–452, 2020.
https://doi.org/10.1007/978-3-030-22354-0_39

2   Partial Reconfiguration on Intel FPGAs

The Intel Arria 10 (10AX115N2F45E1SG) FPGA used in this work has the mechanism called Partial Reconfiguration (PR). In the following, a PR design flow for Arria 10 is described.

2.1   Overview

PR is a technology for reconfiguring only a part of the circuit on the FPGA while other circuits are operating. To create PR design, in addition to the normal circuit design, special circuit designs for PR are required. In this work, we use Quartus Prime Version 18.1 Pro Edition, an Intel development tool.

2.2   Procedure for PR Design

The design procedure of the PR circuit using Quartus consists of the following eight steps [7].

1. Create entire circuit design
2. Create Design Partition and LogicLock Region
3. Allocate Placement Region and Routing Region
4. Add PR Controller IP
5. Create Revision
6. Compile Base Revision
7. Output Database File
8. Compile each Persona Implementation Revision

The following describes each of the above steps.

2.2.1   Create Entire Circuit Design

Design circuits for the entire application.

2.2.2   Create Design Partition and LogicLock Region

Create a design partition by specifying an instance generated from the module which is partially reconfigured (PR module). Design partition is a partition in which a module is implemented as an independent circuit. Lock the created design partition at the specified position using Quartus's logiclock function. This region is called a logiclock region (PR region).

2.2.3   Allocate Placement Region and Routing Region

The logiclock region has a placement region and a routing region, and determines the position and size of each. The placement region is a region for arranging a PR circuit, and the routing region is a region for arranging a circuit connecting the outside of the logiclock region and the placement region. Since the minimum unit of a PR region is a logic block, it is not possible to specify a smaller logic element. Also, the size of the required routing region varies depending on the circuit, and can not be made smaller than the placement region. We can use the Quartus Chip Planner tool to determine the location and size of each region.

2.2.4   Add PR Controller IP

PR Controller IP, generated by Quartus IP Core, is the module which controls PR operations. When creating the PR design, only one PR Controller IP is required on the FPGA. Detailed specifications of the PR Controller IP is described in Sect. 3.

2.2.5   Create Revision

Quartus manages the design for each project. The project consists of design files and configuration files required to compile design. The revision is a component to manage multiple settings for a project. In creating PR design using Quartus, PR modules are designed separately for each revision. There are two revisions, base revision and persona implementation revision. The entire circuit is designed in the base revision, and only the PR modules are designed in the persona implementation revision. Also, a PR module is called a persona. Set the revision for the entire circuit to the base revision. Create a revision for the PR module, and set it to the persona implementation revision. A project is usually composed of one base revision and multiple persona implementation revisions.

2.2.6   Compile Base Revision

Compile base revision for synthesis, placement and routing, timing confirmation, and generating configuration file called SOF (SRAM Object File).

2.2.7   Output Database File

Output the database file which has synthesis data of base revision.

2.2.8   Compile Each Persona Implementation Revision

Compile each persona implementation revision for synthesis, placement and routing, timing confirmation, and generating configuration file for PR called RBF (Row Binary File) using database file.

3   Implementation

3.1   PR Controller IP

The interface of PR Controller IP is shown in Fig. 1 [4].

nreset is an asynchronous reset signal. *clk* is a clock signal, which supports up to 100 MHz. When *freeze* is high, it indicates that PR is in progress. PR starts only when *pr_start* is asserted while *freeze* is low. When starting PR, it is necessary to keep high for at least one clock. In addition, *pr_start* must be deasserted before PR finishes. *data_valid* indicates whether the input data is valid. When *data_ready* is high, it indicates that *data* is ready to receive configuration data. Configuration data is input to *data*. *data* width can be selected from 1, 8, 16, and 32 bits, but when using the JTAG interface, data width 16 bits is selected. In this

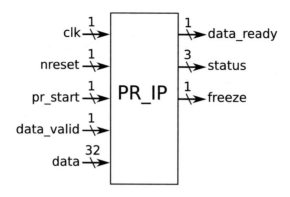

Fig. 1. PR Controller IP interface

research, 32 bits are used. *data* accepts input when *data_valid* and *data_ready* are high. *status* is a 3-bit error output signal that indicates the status of PR events. When using the JTAG interface, the PR Controller IP communicates with the JTAG interface, so input signals other than *nreset* are ignored.

3.2   On-Chip Memory

In this design, the configuration data for PR is stored in On-Chip Memory in order to perform PR from the inside the FPGA. The On-Chip Memory is the embedded memory on the FPGA, which can be configured with LUTs and hard macro memory blocks [3]. The hard macro memory blocks on Arria 10 used in this research is the M20K block, which is memory efficient and suitable for creating a large memory array [5]. In this research, to implement the On-Chip Memory, we used IP Core that generates 1-PORT ROM using M20K block. The interface is shown in Fig. 2.

The Quartus tool has a function called In-System Memory Content Editor (ISMCE), which can read and write On-Chip Memory during operation [2][6]. In this research, the configuration data for PR is stored using ISMCE in the ROM.

A Memory Initialization File (MIF) is required to initialize the On-Chip Memory. Therefore, we created a program for converting RBF to MIF using shell script. Since the maximum input data width of PR Controller IP is 32 bits, binary data in RBF is converted to MIF with 32-bit words.

3.3   Self-partial Reconfiguration System

The design of the Self-PR System is shown in Fig. 3. Furthermore, a timing chart of an input signal determined based on the specification of PR Controller IP is shown in Fig. 4.

PR is initiated from the inside of the FPGA by controlling *pr_start*, *data_valid* and *rom_addr*, using *start* and PR Controller IP output signals. *rom_addr* is an address of ROM IP and *start* is an input signal from outside. First, when *start*

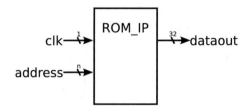

Fig. 2. 1-PORT ROM interface

becomes high, *pr_start* and *data_valid* become high. The value of the all registers
is updated at the next clock and input to the PR Controller IP. Then, when the
PR starts and *status* becomes "PR operation in progress," data input is kept
waiting until the PR Controller IP is ready to receive data. When *data_ready*
becomes high, the next address is input to *rom_addr*. Since ROM IP used in this
research has a latency of 2 clocks from the input of address to the output of
memory data, *data_valid* is made low during the period to delay data reception.
When input of configuration data for PR is completed and *data_ready* and *freeze*
become low, PR is completed and *status* becomes "PR operation passes." Then,
each register is made low to enable reconfiguration again.

3.4   Compression and Decompression of Configuration Data for PR

The size of RBF for a persona with the smallest PR region created in this work
was about 1.1 MByte. Also, the number of M20K blocks in Arria 10 is limited.
Therefore, it is difficult to create many ROM modules for storing RBF for multiple personas on the FPGA. In order to solve this problem, we implemented
the program that compresses the data during conversion to MIF by using the
fact that the same data often appears repeatedly in RBF, and the module which
decompresses the data in the FPGA when it is output from the ROM.

3.4.1   The Program to Compress Data

Data is compressed like the Run Length Encoding [1], using magic numbers.
First of all, it is confirmed that no magic number exists in RBF. If it exists,
another magic number is used. When there is data repeated in 32-bit units in
RBF, writing the magic number indicates the start of repetition. However, if
the number of repetitions is three or less, compression is not performed since
compression is not necessary. An example of compression is shown in Fig. 5.

3.4.2   The Module to Decompress Data

The design of a module for decompressing data output from an initialized ROM
with MIF that is compressed and converted, is shown in Fig. 6.

When the magic number is output from the ROM, *rep_en* is set to high,
and data output one clock before is stored in *rep_word*. The data output at the

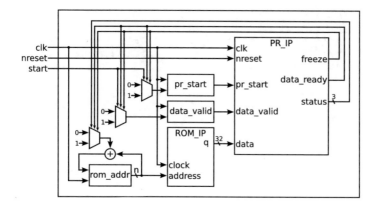

Fig. 3. Self-PR system

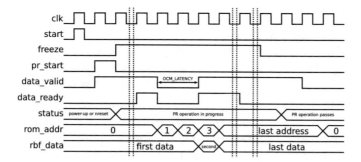

Fig. 4. Self-PR timing chart

next clock is stored in *rep_count*, and *rep_word* is output until *rep_count* and the counter value become equal. After repetition, the ROM address is updated.

4 Evaluation

4.1 Comparison Between JTAG Mode and Self-PR System

In the case where the PR region is configured with 1 logic block, JTAG mode is compared with Self-PR System using uncompressed MIF and compressed MIF.

4.1.1 Time Required for PR

The size of RBF and time required for PR (PR time) in each case are shown in Table 1. The PR time is the value obtained by dividing count of clock cycles while the PR Controller IP freeze signal is high by the input clock frequency of 100 MHz. Each value is an average time of 10 PRs.

There was not much difference in each size of RBF. In comparison with JTAG mode, Self-PR System reduced PR time by 97%. The difference between PR time

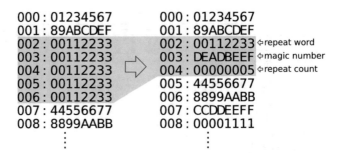

Fig. 5. Example of configuration data compression

when using uncompressed MIF and PR time when using compressed MIF was 0.01 µs.

4.1.2   Maximum Operating Frequency

The maximum operating frequency (FMax) in each case is shown in Table 2. In JTAG mode, FMax was significantly lower than the others. In the case of using compressed MIF, FMax was lowered by about 30 MHz as compared with using uncompressed MIF.

4.1.3   Resource Usage for Self-PR System

When using the JTAG interface, only PR Controller IP in the JTAG mode is implemented. When using uncompressed MIF, we implemented PR Controller IP, ROM IP, and circuits that control the input to PR Controller IP. In the case of using compressed MIF, circuits that decompresses MIF were added to the modules used for uncompressed MIF. Table 3 shows the resource available on the FPGA and the resource usage used for each PR configuration.

Overall, resource usage is less in JTAG mode than in the others. However, focusing on resource for PR Controller IP, the usage of LUTs, ALMs and registers was larger in JTAG mode than in the others. Also, when using compressed MIF, resource usage for PR Controller IP is not much different than when using uncompressed MIF. However, in resource usage for ROM IP, usage of LUTs, ALMs, M20Ks was significantly smaller, and usage of registers was smaller than when using uncompressed MIF. Focusing on usage of M20Ks, when using compressed MIF, 21% of the total resources was used, and when using uncompressed MIF, 3% of the total resources was used. In other words, compressing MIF reduced the usage of M20Ks by about 18 points.

4.2   Relation Between PR Region and Configuration Files

In order to investigate the change in compression rate as configuration data increases, the size of configuration data, the size of MIF and compression ratio were compared in the four cases where the PR regions consisting of six logic

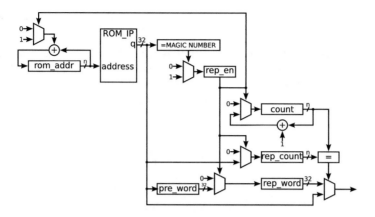

Fig. 6. Module for configuration data decompression

Table 1. Time required for PR

	RBF [Byte]	PR time [s]
JTAG mode	1131316	0.94258797
Uncompressed MIF	1136356	0.00284254
Compressed MIF	1136356	0.00284255

blocks are created, with width of 1 and height of 6, width of 2 and height of 3, width of 3 and height of 2, and width of 6 and height of 1. The comparison results are shown in Table 4.

The size of SOF, which is configuration data of the entire design, was almost the same regardless of the aspect ratio of the PR Region. The size of RBF increased as the PR Region height increased, and size of the compressed and uncompressed MIF also increased accordingly. When W1H6 and W6H1 are compared, there is a large difference in the size of the uncompressed MIF However, a difference in the compression ratio is only 1 point.

5 Discussions

5.1 Time Required for PR

According to Table 1, the Self-PR System reduced the PR time by about 97% as compared to the JTAG mode. The theoretical value of PR time in Self-PR is equal to the time required to input RBF. The PR Controller IP used in this research is generated with clock of 100 MHz and input data width of 32 bits, and the persona with the smallest PR region has RBF of 1136356 Bytes. Therefore, the theoretical value of the PR time is 2.84089 ms. The measured value is about 1.65 μs longer than the theoretical value. This is considered to be because it includes the waiting time of the start process and end process performed in PR Controller IP as shown in Fig. 4.

Also, there is a difference of only 0.01 µs between the PR time when using the compressed MIF and the PR time when using the uncompressed MIF. This is because the circuit added to decompress data has no additional latency, and PR completes in the same number of clock cycles as the module without decompression.

Table 2. Maximum operating frequency

	FMax [MHz]
JTAG mode	76.13
Uncompressed MIF	199.52
Compressed MIF	169.72

Table 3. Resource Usage for Self-PR System

Configuration	Module	LUT	ALM	Register	M20K
JTAG mode	PR Controller IP	284	200.8	311	0
Uncompressed MIF	Self-PR module	732	568.0	267	592
	PR Controller IP	131	95.2	138	0
	ROM IP	561	450.3	94	592
Compressed MIF	Self-PR module	447	303.5	336	98
	PR Controller IP	121	91.2	136	0
	ROM IP	129	91.3	77	98
Total Available			427200	1708800	2713

In theory, PR module can be reconfigured about 357 times per second, so it may be possible to realize real-time processing using Self-PR System, depending on the concurrently operating application.

5.2   Maximum Operating Frequency

According to Table 2, in the JTAG mode, the FMax is significantly low. This suggests the possibility that the circuit generated by PR Controller IP is a critical path when using a JTAG interface for PR.

Also, when using compressed MIF, the FMax is about 30 MHz lower than when using uncompressed MIF. However, this is not considered to be a major problem as it exceeds the clock frequency that PR Controller IP supports.

Table 4. Comparison among different PR region

	SOF [Byte]	RBF [Byte]	Uncompressed MIF [Byte]	Compressed MIF [Byte]	Compression ratio [%]
W1H6	36732064	3685588	21192211	3786888	17.87
W2H3	36732070	2246164	12915523	2231397	17.28
W3H2	36732072	1758292	10110259	1698717	16.80
W6H1	36732069	1244212	7154299	1134596	16.86

5.3 Resource Usage for Self-PR System

According to Table 3, the resource usage differs greatly between the JTAG mode and the other cases because the circuits generated by PR Controller IP are different.

When using the uncompressed MIF, the ROM IP uses more LUTs, ALMs, registers, and M20Ks than when using the compressed MIF. This is because a large amount of M20Ks is used to generate a ROM having a large memory size, and many LUTs, ALMs, and registers are used for memory access control.

In Self-PR System, resource usage is increased compared to the JTAG mode. Also, when using compressed MIF, the usage of the register is larger than when using uncompressed MIF. However, since the usage of LUTs, ALMs, and registers is less than 1% of the total resources on the FPGA, the implementation of the Self-PR System is considered to have little impact on other applications.

In the usage of M20Ks, 21% of the total resources is used when using uncompressed MIF, and 3% of the total resources is used when using compressed MIF. In other words, the usage of M20Ks was reduced by about 18 points. However, since it is impossible to implement a practical image processing circuit such as template matching with the smallest persona created in this research, it is necessary to consider a more efficient method of compressing configuration data.

5.4 Relation Between PR Region and Configuration Files

According to Table 4, while the size of RBF increases with the increase in height, the compression rate stays almost the same. From this, it can be understood that the data increased by the difference in height of the PR region can not be compressed very efficiently.

Also, comparing the persona with W1H6 and the persona with W1H1, the size of RBF and compression ratio are almost the same. Therefore, the size of RBF is deeply related to the height of the PR region. In addition to the compression method, we must be careful how to create the PR region.

6 Conclusion

In this paper, in order to realize real-time processing using PR, we implemented Self-PR system, where RBF is stored in ROM on the FPGA and PR is performed from the inside the FPGA. The time required for Self-PR is about 2.8 ms, 97% less than using JTAG interface. This is largely related to the specification of PR Controller IP used and the size of RBF, and a value almost equal to the theoretical value is realized. Therefore, it is proven that real-time processing using Self-PR System can be realized depending on the application processed simultaneously. Also, we implemented a function to compress configuration data for PR and store it in ROM and decompress data in FPGA. As a result, the usage of M20Ks was reduced to 3% of the total resources. Although the addition of the circuit to decompress the compressed configuration data lowered FMax by 30 MHz, the FMax still exceeded the maximum operating frequency supported by PR Controller IP. Our future work includes design and implementation of more efficient data compression methods for the Self-PR mechanism. Evaluation of the power consumption of the Self-PR system is also an interesting future work.

References

1. Golomb, S.: Run-length encodings (Corresp.). IEEE Trans. Inf. Theory **12**(3), 399–401 (1966)
2. Intel: Working with In-System Memory Content Editor Data. https://www.intel.com/content/www/us/en/programmable/quartushelp/15.0/mergedProjects/program/red/red_pro_import_export.htm
3. Intel: Embedded Memory Blocks in Intel Arria 10 Devices (2014). https://www.intel.co.jp/content/dam/altera-www/global/ja_JP/pdfs/literature/hb/arria-10/a10_memory_j.pdf
4. Intel: Partial Reconfiguration IP Core User Guide (2015). https://www.intel.com/content/dam/www/programmable/us/en/pdfs/literature/ug/ug_partrecon.pdf
5. Intel: Embedded Memory in Intel Stratix 10 User Guide (2017). https://www.intel.co.jp/content/www/jp/ja/programmable/documentation/vgo1439451000304.html#exy1480422175886
6. Intel: How To Use In-System Memory Content Editor (2017). https://www.youtube.com/watch?v=YI34AoA74_c
7. Intel: Partially Reconfiguring a Design on Intel Arria 10 GX FPGA Development Bord (2018). https://www.intel.com/content/www/us/en/programmable/documentation/ihj1482170009390.html

A Simple Heterogeneous Redundant Design Method for Finite State Machines on FPGAs

Takanori Itagawa, Ryo Kamasaka, and Yuichiro Shibata[✉]

Nagasaki University, 1-14 Bunkyo-machi, Nagasaki 852-8521, Japan
{itagawa,kamasaka}pca.cis.nagasaki-u.ac.jp, shibata@cis.nagasaki-u.ac.jp

Abstract. For FPGA applications in the application domains requiring a high degree of functional safety such as automotive and industrial infrastructure, simple homogeneous redundant logic design, where the same hardware modules are simply replicated, is not enough. In order to improve tolerance to common cause fault, heterogeneous redundant design, where different implementation approaches are taken for realizing the same logic functionality, is crucial. However, manual redundant design tends to place a burden on designers, reducing productivity of system development. To cope with this problem, this paper proposes a systematic heterogeneous redundant design approach for finite state machines on FPGAs, focusing on the diversity of state encoding methods. With this approach, designers can easily combine the state machines with different encoding to form heterogeneous redundancy, by inserting simple directives into RTL source code. In order to evaluate the effectiveness of the proposed approach, timing analysis of post-layout netlists is performed under an overclock situation as an example of common cause fault. The evaluation results demonstrate the proposed approach improves the error detection rate compared to conventional homogeneous redundant designs. It is also discussed how the choice of state encoding methods impacts the error detection rate.

1 Introduction

The importance of FPGA computing is expected to be actively increased in the application fields requiring a high degree of functional safety, such as automotive, industrial infrastructure, plant control and robotics. In such application domains, enough risk management in terms of the reliability is required to face with both systematic and random failures. A general approach is use of redundant design techniques such as triple module redundancy (TMR), where the same hardware modules are simply replicated. However, still all the redundant modules may be affected in the same way by a common cause fault. Therefore, it is recommended to introduce diversity in redundant design, that is, to take different approaches to implement the same functionality [3]. In other words, heterogeneousness is a crucial issue in redundant design for functional safety.

© Springer Nature Switzerland AG 2020
L. Barolli et al. (Eds.): CISIS 2019, AISC 993, pp. 453–461, 2020.
https://doi.org/10.1007/978-3-030-22354-0_40

Although many researches were carried out on redundant design for FPGAs [1,2,4], heterogeneous redundant design has not been addressed very much so far. For FPGAs, it is possible to manually design different circuits for the same logic functionality by introducing several different hardware structure or algorithms. However, this approach places a burden on designers, reducing productivity of logic design. We have also proposed a systematic heterogeneous redundant design method for FPGAs focusing on diversity of technology mapping [6,7]. However, this method is applicable only for combination circuits.

This paper proposes a novel approach to introduce diversity in redundant design of finite state machines, focusing on a variety of state encoding. In this approach, different encoding can be easily combined to form heterogeneous redundancy, by inserting a simple directive in SystemVerilog code.

The major contributions of this paper include:

- A novel heterogeneous redundant FPGA design approach for finite state machines, which introduces diversity of state encoding, is proposed.
- How the proposed design approach improves the error detection rates for timing fault is evaluated.
- How the choice of state encoding gives an impact on reliability of redundant designs is also discussed.

The rest of this paper is organized as follows. After presenting background of this work in Sect. 2, the proposed design approach is introduced in Sect. 3. In Sect. 4, evaluation results of the proposed methods are presented and discussed. Finally, the paper is concluded in Sect. 5.

2   Background

2.1   Redundant Design Techniques

Generally, in order to improve the reliability of circuits configure on FPGAs, redundant design, in which the same functional hardware modules are duplicated, is adopted. In such a redundantly designed system, if one of the module has a fault, it is possible to detect an abnormality by comparing the output values of the duplicated modules as shown in Fig. 1. Since the same hardware modules are duplicated, this is called homogeneous redundant design.

However, even with this approach, the redundant modules can output the same wrong values for the common cause fault such as power supply abnormality and clock signal abnormality. This is because the fault may affect the redundant modules in the same manner. This situations will result in overlooking of the error as illustrated in Fig. 2.

In order to cope with this situation, the functional safety standard IEC-61508 [3] requires to introduce diversity in redundant design, that is, to implement multiple modules with different structures or approaches to realize the same functionality. Having different structures makes the redundant modules easier to generate different wrong values for the common cause fault. Therefore, the error can be detected as shown in Fig. 3. This approach is called heterogeneous redundant design.

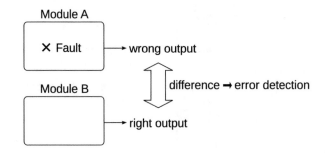

Fig. 1. Error detection with homogeneous redundant design

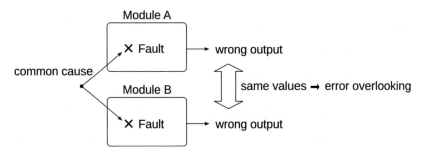

Fig. 2. Overlooking of common cause fault

2.2   Related Work

As a countermeasure of soft errors of FPGA circuits, homogeneous redundancy approaches have been widely addressed so far [1,2,4]. Although the effects of heterogeneous redundancy has been investigated for software design [5], heterogeneous redundant design approaches for FPGAs have not been discussed very much. We have proposed a heterogeneous redundant design method for FPGAs focusing on diversity of technology mapping [6,7]. In this method, by mapping the same logic functionality to different FPGA resources, error detection rate was improved compared to a conventional homogeneous redundancy approach. However, this method is applicable only for combination circuits. Therefore, in this paper, we proposed a novel heterogeneous redundant FPGA design approach for finite state machines which are a general framework of sequential circuits.

3   Proposed Design Method

For heterogeneous redundant design approaches, the following two points are desirable from a practical point of view. First, the negative impact on design productivity caused by introducing diversity in redundancy should be minimized. Second, the redundant design method should not require any special design tools. In terms of proven-in-use arguments in functional safety, only conventional FPGA design tools should be utilized.

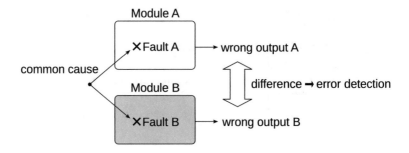

Fig. 3. Error detection with Heterogeneous redundant design

zwzwzw
```
typedef enum  {S0, S1, S2, S3, S4} state_t;
state_t state;
```

Fig. 4. State description example in SystemVerilog

zwzwzw
```
typedef enum  {S0, S1, S2, S3, S4} state_t;
(* fsm_encoding = "one_hot" *) state_t state;
```

Fig. 5. Directive to specify desired state encoding

Based on these requirements, we propose a novel heterogeneous redundant FPGA design approach for finite state machines, focusing on diversity in state encoding. As is well known, a finite state machine that has the same state transition diagram can be implemented with various different state encoding methods. In a general register transfer level (RTL) hardware design flow, the state encoding process is automatically performed by logic synthesis tool. However, by inserting a simple directive in RTL source code, designers can designate the synthesis tool to adopt a specific encoding method. By using this mechanism, designers can easily introduce design diversity without rewriting the RTL code. In addition, this redundant design method can be performed only using a conventional FPGA design tool chain.

Let us discuss RTL coding of finite state machines more concretely. In case of SystemVerilog, states for a finite state machine can be described as shown in Fig. 4. When a Xilinx Vivado tool compiles this code, the tool automatically selects a most preferable state encoding method among 'one-hot', 'sequential', 'johnson', and 'gray'. Although it is also possible to explicitly assign specific code for each sate in SystemVerilog, this method places a burden on designers, reducing productivity. Another option is to insert a directive into SystemVerilog code to designate a desired encoding as shown in Fig. 5. With the fsm_encoding directive, one of the aforementioned four encoding methods can be specified.

With this approach, designers can easily duplicate a finite state machine code with different state encoding by just adding a directive. In addition, there is no need to use any other special design tools.

4   Evaluation

4.1   Evaluation Method

In order to evaluate the effect of the proposed heterogeneous redundant design approach for finite state machines, we carried out simulation analysis of error detection. In this evaluation, a common cause fault was emulated with overclocking on the assumption that timing violation was made by clock signal distortion. As an evaluation target, the dk27 state machine was chosen from the MCNC benchmark [8]. As illustrated in Fig. 6, the dk27 state machine consists of seven states. The state machine was described in SystemVerilog and mapped on a Xilinx Kintex-7 xc7k325t-2 FPGA with a Vivado 2016.4 tool. By inserting the fsm_encoding directive, the four variants of the sate machine were implemented with different state encoding: 'one-hot', 'sequential', 'johnson', and 'gray'. The state code assigned by the Vivado tools is summarized in Table 1. A total of 10 combinations of redundant designs were evaluated, two of which were chosen from these four encoding methods.

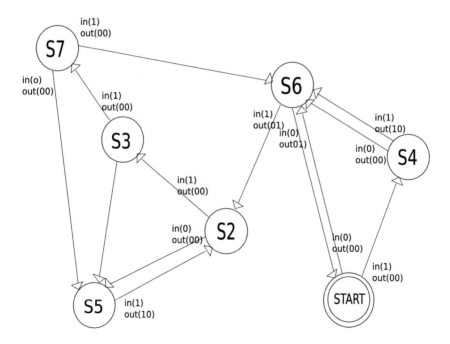

Fig. 6. State transition diagram for the dk27 state machine

Table 1. State encoding results for the `dk27` state machine

State	one-hot	sequential	gray	johnson
START	0000001	000	000	0000
S4	0000010	001	001	1000
S6	0000100	010	011	1100
S2	0001000	011	010	1110
S3	0010000	100	111	1111
S7	0100000	101	110	0111
S5	1000000	110	100	0011

After FPGA mapping, it was confirmed that the maximum clock frequency of the state machine was 100.51 MHz. Then, the post-map netlists with timing information was generated and simulated by changing the clock frequency from 205 MHz to 213 MHz by 1 MHz. For each frequency, a simulation for 100,000 clock cycles was preformed with random input values generated by the `$random` system task in SystemVerilog to evaluate the error detection rate.

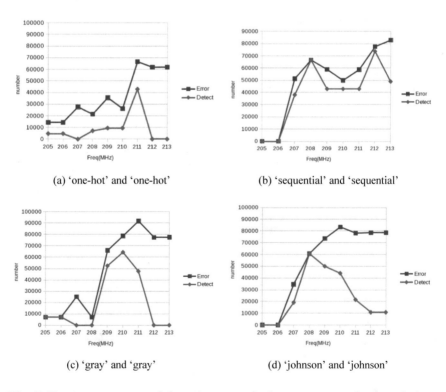

(a) 'one-hot' and 'one-hot'

(b) 'sequential' and 'sequential'

(c) 'gray' and 'gray'

(d) 'johnson' and 'johnson'

Fig. 7. Error occurrences and detection count for homogeneous redundant designs

4.2 Results and Discussion

Figure 7 shows how many errors occurred during the simulation and how many of them were detected for the homogeneous redundant designs. As the clock frequency increases, the number of occurrences of errors also increased. However, not all the errors were detected. For example, the combination of 'one-hot' and 'one-hot' overlooked all the errors at the frequency of 207, 212, and 213 MHz. The combination of 'gray' and 'gray' also overlooked all the errors at the frequency of 207, 208, 212, and 213 MHz. In Table 2, average error detection rates for the homogeneous redundant designs are summarized. Except for the 'sequential' and 'sequential' design, more than half errors could not be detected. The worst detection rate was shown for the 'one-hot' and 'one-hot' design, where only 23.72% of errors were detected. In total of the homogeneous redundant design, the average error detection rate was 48.75%.

Table 2. Error detection rate for homogeneous redundant designs

Design	Error detection rate (%)
'one-hot' and 'one-hot'	23.72
'sequential' and 'sequential'	79.80
'gray' and 'gray'	40.84
'johnson' and 'johnson'	44.46
Total average	48.78

Table 3. Error detection rate for heterogeneous redundant designs

Design	Error detection rate (%)
'one-hot' and 'sequential'	61.37
'one-hot' and 'gray'	58.16
'one-hot' and 'johnson'	58.93
'sequential' and 'gray'	71.00
'sequential' and 'johnson'	62.47
'gray' and 'johnson'	33.02
Total average	57.68

Figure 8 shows the evaluation results for heterogeneous redundant designs in which state machines with different state encoding were combined. Compared to the results for the homogeneous redundant designs, error detection rates were generally improved. As Table 3 shows, error detection rates of 50% or more were achieved except for the design with 'gray' and 'johnson'.

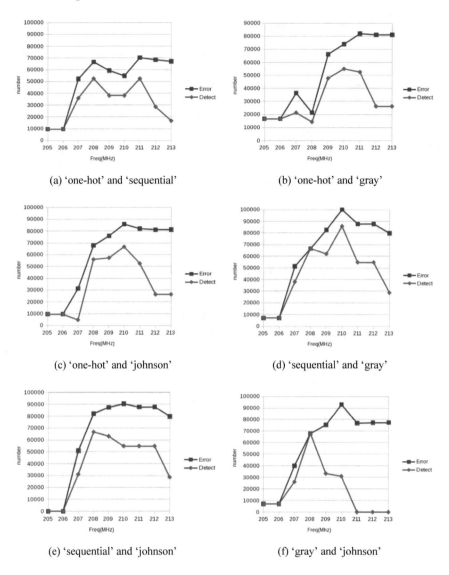

Fig. 8. Error occurrences and detection count for heterogeneous redundant designs

In total, the average error detection rate was improved from 48.78% to 57.68% by introducing heterogeneous redundancy. However, as aforementioned, the design with 'gray' and 'johnson' achieved only 33.03% of the detection rate. This result is considered to be due to the similarity of logic structure between a Gray code counter and Johnson counter. In addition, in spite of having homogeneous redundancy, the design with 'sequential' and 'sequential' achieved the best error detection rate. Also in the heterogeneous redundant designs, the state machines that include 'sequential' showed relatively high error detection rates.

This is considered to be due to the fact that multiple bits are different between adjacent states in binary counters. For example, when the state transits from '011' to '100', intermediate states such as '010' and '000' may be instantaneously passed, which is a phenomenon known as a logic hazard. Under the overclocking, these intermediate states may be sampled, but one state machine can get '010' while the other gets '000'. In other words, another diversity is introduced by logic hazard in error behaviors. On the other hand, in the other encoding methods, the Hamming distance between adjacent states is designed to be small. Although this is effective for enabling high speed operation, tolerance to common cause fault seems to be limited in terms of the diversity in error behaviors.

5   Conclusion

This paper proposed and discussed a novel heterogeneous redundant design approach for finite state machines focusing on the design diversity in state encoding. The evaluation experiments demonstrated the proposed approach improved average error detection rates for timing violation compared to conventional homogeneous redundant approaches. It was also suggested sequential state encoding has an advantage in terms of diversity in error behaviors. Our future work includes further evaluation with more practical circuits under various type of common cause fault and further analysis on preferable combinations of state encoding methods.

References

1. Hayek, A., Al-Bokhaiti, M., Borcsok, J.: Design and implementation of an FPGA-based 1oo4-architecture for safety-related system-on-chip. In: Proceedings IEEE 25th International Conference on Microelectronics (ICM), pp. 1–4 (2013)
2. Ichinomiya, Y., Tanoue, S., Ishida, T., Amagasaki, M., Kuga, M., Sueyoshi, T.: Memory sharing approach for TMR softcore processor. In: Reconfigurable Computing: Architectures, Tools and Applications, pp. 268–274. Springer (2009)
3. International Electrotechnical Commission: Functional safety of electrical/electronic/programmable electronic safety related systems. IEC 61508 (2000)
4. Konoura, H., Imagawa, T., Mitsuyama, Y., Hashimoto, M., Onoye, T.: Comparative evaluation of lifetime enhancement with fault avoidance on dynamically reconfigurable devices. IEICE Trans. Fundam. Electron. Commun. Comput. Sci. **97**(7), 1468–1482 (2014)
5. Lovric, T.: Systematic and design diversity — software techniques for hardware fault detection. In: Dependable Computing — EDCC-1, pp. 307–326. Springer (1994)
6. Morimoto, K., Shibata, Y., Shirakura, Y., Maruta, H., Tanaka, M., Kurokawa, F.: Diversity diagnostic for new FPGA based controller of renewable energy power plant. Int. J. Renew. Energy Res. **7**, 1403–1412 (2017)
7. Shirakura, Y., Segawa, T., Shibata, Y., Morimoto, K., Tanaka, M., Nobe, M., Maruta, H., Kurokawa, F.: A redundant design approach with diversity of FPGA resource mapping. In: Proceedings of International Symposium on Applied Reconfigurable Computing (ARC 2016). Lecture Notes in Computer Science, vol. 9625, pp. 119–131 (2016)
8. Yang, S.: Logic Synthesis and Optimization Benchmarks User Guide Version 3.0 (1991)

A Method of Collecting Four Character Medicine Effect Phrases in TCM Patents Based on Semi-supervised Learning

Deng Na[1], Chen Xu[2(✉)], Xiong Caiquan[1], Wang Chunzhi[1],
Zhang Mingwu[1], Ye Zhiwei[1], Li Desheng[3], and Yang Xuehong[4]

[1] School of Computer, Hubei University of Technology, Wuhan, China
iamdengna@163.com
[2] School of Information and Safety Engineering,
Zhongnan University of Economics and Law, Wuhan, China
chenxu@whu.edu.cn
[3] College of Mathematics, Physics and Information Engineering,
Anhui Science and Technology University, Fengyang, China
ldsyy2006@126.com
[4] China Unicom Research Institute, Beijing, China
yangxhl49@chinaunicom.cn

Abstract. As a result of historical reasons and writing habits, the effects of medicine in Traditional Chinese Medicine (TCM) patents are often described using four character phrases. These four character phrases are not easily identified by the Chinese word segmentation system, thus greatly affects the results of patent analysis and mining. This paper proposes a semi-supervised learning method to collect four character effect phrases from the abstracts texts of TCM patents, which can help enrich the lexicon of Chinese word segmentation system, and also provide support for semantic patent retrieval and analysis. The experimental results show the validity of the method.

1 Introduction

Traditional Chinese Medical science and Chinese herbal medicine have a long history, and they are the crystallization of the predecessors' wisdom for thousands of years. Today, western medicine becomes more and more developed; however, more and more people in the world choose treatment of Traditional Chinese Medicine (TCM). The reason is that compared with western medicine, traditional Chinese medicine is very suitable for the treatment of chronic diseases and difficult miscellaneous diseases, and moreover, Chinese herbal medicine is more safe and gentle.

With the rapid development of patent industry, the number of patent applications has risen year by year, and patents have become an important data source in the era of big data. As a special literal type that contains technology, law and economic information, patents contain a lot of valuable contents, which are of great significance for patent examiners, patent applicants and enterprises. For patent examiners, they need prior art searching to check the innovation of patents in audit; for patent applicants, on

© Springer Nature Switzerland AG 2020
L. Barolli et al. (Eds.): CISIS 2019, AISC 993, pp. 462–473, 2020.
https://doi.org/10.1007/978-3-030-22354-0_41

the one hand, they could get inspiration from published patents, on the other hand, it is very important, they need to make sure whether their own ideas have been applied by others, in order to avoid infringement; for enterprises, patent retrieval can help them make decision on the acquisition of patents. Therefore, finding accurate and comprehensive patents is very important for people of different roles, which is the job of patent retrieval. In addition to patent retrieval, patent analysis, patent classification, patent clustering, patent recommendation and patent early warning can further discover hidden information in patents and provide decision support. For Chinese patent analysis and patent mining, the first step is Chinese word segmentation. Its accuracy has a direct impact on the results of analysis and mining.

TCM patents often contain effect descriptions of medicines. Different from patents in other fields, the effect descriptions in TCM patents often appear in the form of four character phrases in classical Chinese. This is because the development of Traditional Chinese medical science and Chinese herbal medicine has lasted for thousands of years, in ancient China, people recorded the effects of prescription with classical style of writing, and this kind of writing is kept to the present.

This kind of phrases are short, only with 4 characters, but rich in meaning, and nearly all the professional TCM doctors can understand the meaning behind. However, the four character medicine effect phrases cannot be recognized by common Chinese word segmentation systems. In segmentation systems, these four character phrases are often split apart, seriously affecting the results of patent analysis and mining.

In order to solve this problem, this paper proposes a method to collect four character medicine effect phrases from the abstract texts of TCM patents. This method does not require too much training data, using co-training theory, starts from a small amount of annotated data, which is called as initial seeds, use the position features of four character medicine effect phrases, and gradually expand the scale of seeds, to learn and discover new target data from more data.

The collection of four character medicine effect phrases has the following advantages:

(1) It can enrich the lexicon of Chinese word segmentation system and greatly improve the accuracy of Chinese word segmentation of Chinese TCM patents.

(2) The identification of medicine effect phrases enables us to understand the effect of a patent in semantics and furthermore, to realize semantic patent retrieval.

(3) The identification of medicine effect phrases can help to build technology/effect matrix for TCM patents, and provide support for the innovation and application of patents.

2   Related Work

In recent years, as people pay more and more attention to patents, patent retrieval has become a hot research topic. Patent retrieval is a subdomain problem of information retrieval, but it is very different from common web search. The features of patent text, such as long text, semi-structured, metadata, vague expression, multi-figures and multilingualism, have brought great challenges to patent retrieval. Since 2002, NTCIR conferences of Japan's National Institute of Informatics has set up a special patent search symposium and released patent test data sets for English and Japanese [1]. CLEF (cross language evaluation Forum), an open evaluation platform of information retrieval special for European languages, has set up a special seminar CLEF-IP on patent retrieval since 2009 [2]. Recently, scholars started to look for methods of semantic patent retrieval. [3] used Word Net to implement semantic patent search. [4] utilized domain lexicon for query expansion of patent retrieval on CLEF-IP.

Patents are rich in technical information, and patent mining has also become a research direction in last several years [5]. [6] used Self Organizing Maps (SOM) and bibliographic coupling to cluster patents. [7] applied adaptive K-means into patent clustering. [8] presented an automatic patent classification system through a three-phase model, and used text processing, a document frequency matrix and boosted tree classifier to classify patents into two classes. [9] proposed an automatic patent classification method based on the functional basis and Naive Bayes theory. [10] predicted technology performance improvement rates by mining patent data. [11] integrated applicant citations and bibliographic information of patents into a heterogeneous citation bibliographic network and based on which, recommended patents to examiners.

Patent annotation, or patent information extraction, is an important step in patent analysis and mining. The accuracy of annotation directly affects the accuracy of analysis and mining. [12] proposed an ontology-based automatic semantic annotation approach which combines both structure and content characteristics of patents. [13] made use of machine learning to extract technologies, functions and their relations from patents.

The annotated data is a scarce resource in the era of big data. In order to provide high quality training sets and test sets for machine learning, people usually annotate through time consuming manual work. Co-training is a kind of semi-supervised machine learning method, and it is very suitable for the cases with large amounts of data, but no or less annotated data. In recent years, co-training has been applied to various kinds of data's annotation, such as images [14], web pages [15], emotion texts [16], etc. Its working principle is: firstly, two independent classifiers are built and a few labelled data are used to train one classifier; and then, labelled samples with high confidence are added into another classifier's training samples. These steps carry on alternately and gradually expand the scale of annotated data [17].

Because TCM patents cover two fields: traditional Chinese medicine and patent, it has the characteristics of both, thus increasing the difficulty of annotation. In addition, due to the complexity of Chinese language and the specificity of Chinese patent text, the annotation of TCM patents is still a blank area at home and abroad.

In the preliminary research work of the authors [18–20], we mainly focus on stop words' removal, the collection of effect clue words, the computation of similarities between TCM patents. We also made use of co-training to extract effect clauses from plain patent texts [21, 22]. However, since the particularity of the effect clauses in TCM patents, the previous methods cannot be directly applied into TCM patens.

3 Four Character Medicine Effect Phrases

In the abstract texts of TCM patents, there are usually statements describing medicine effect. As a result of historical reasons and cultural heritage, medicine effects often exist in the form of four character phrases. For example, "清热解毒(clearing heat and removing toxicity)", "活血化瘀(activating blood circulation to dissipate blood)". As shown in Table 1.

Table 1. Examples of four character medicine effect phrases.

Four character medicine effect phrases (in Chinese)	Meanings
驱风除湿	Dispelling wind and eliminating dampness
止咳化痰	Relieving cough and reducing sputum
健脾益肾	Invigorating the spleen and benefiting the kidney
活血益气	Invigorating blood and nourishing qi
祛腐生肌	Removing putrefaction and promoting tissue regeneration
平肝息风	Calm the liver and stop the wind
镇痛止血	Analgesic and hemostasis

4 Characteristics of Four Character Medicine Effect Phrases

4.1 Position Characteristics

Four character medicine effect phrases are often used to illustrate the function and effect of patents in TCM patent abstracts. In most cases, it is located in the latter part of the abstract text, and four character phrases are often relatively independent clauses and tend to appear continuously.

4.2 Part of Speech Characteristics

There are three types of common part of speech combinations for four character medicine effect phrases.

(1) Verb + Noun + Verb + Noun. The first character represents a verb, the second one represents a noun, the third one represents a verb, and the fourth one represents a noun. As shown in Table 1. This kind of part of speech combination conforms to the expression habit of classical Chinese language.

(2) Verb + Verb + Noun + Noun. The first character represents a verb, the second one represents a verb, the third one represents a noun, and the fourth one represents a noun. As shown in Table 2. This type is more in line with the expression habit of modern Chinese language.

Table 2. Examples of four character medicine effect phrases under Verb + Verb + Noun + Noun.

Four character medicine effect phrases (in Chinese)	Meanings
调理气血	Regulating qi and blood
促进消化	Promote digestion
增进食欲	Whet the appetite
缓解疲劳	Relieving fatigue
疏散风寒	Evacuate the wind and cold
消除痈疮	Eliminate carbuncle sore

(3) Noun + Noun + Verb + Noun. The first, second, and fourth characters represent nouns and the third one represents a verb. For example, "口舌生疮 (have ulcers in the mouth and on the tongue)".

5   Our Method

5.1   Definition

Definition 1: clause
When the abstract text of a TCM patent is divided with a comma, a period, a semicolon, a colon or any other punctuation marks, each part is called a clause.

5.2   The Idea of Our Method

The idea of collecting four character medicine effect phrases from the abstract texts of TCM patents is as follows. Since four character phrases have the location characteristics, we can consider extracting those four character clauses from patent abstract texts as the candidate dataset. Afterwards, we make use of the characteristics that they tend to appear continuously. If the surrounding clauses around some specific clause contain four character medicine effect phrases, this specific clause is more likely to be a four character medicine effect phrase.

In our previous research, we used co-training to collect clue words and effect clauses from plain Chinese patent abstract texts. Inspired by previous work, in this paper, we used co-training to collect four character medicine effect phrases. First of all, we extract all four character clauses from a small amount of patent abstract texts, after artificial selection, select high quality phrases as initial seeds; then, utilizing the characteristics that four character medicine effect phrases tend to appear continuously, in more patents, those clauses whose surrounding clauses contain seeds are also considered as four character medicine effect phrases; we check the quality of these newly obtained four character clauses, identify interference terms and develop a strategy to improve the accuracy of four character clause; the filtered high quality new medicine effect phrases are added to seeds thesaurus; repeat the above steps in more patents until seeds thesaurus reaches to a steady size.

5.3 Interference Items

When checking the quality of four character clauses, we find that many interference items influence the accuracy of four character clauses. Some interference items are domain related. Usually, there are several kinds of interference items as follows, but not limited to.

(1) In the abstract texts of TCM patents, sometimes, the proportions of Chinese herbal medicines may be listed. Therefore, there are often four character clauses like this: "当归5克 (Angelica 5 grams)", "桔梗3% (Platycodon grandiflorum 3%)", "3-5克 (3-5 grams)" and so on. The names of Chinese herbal medicine and the digits are both interference items.

(2) While dividing into clauses, there are often such four character clauses, "疗效显著(the curative effect is remarkable)", "配方如下(the proportions is as follows)" and so on.

In order to improve the quality of four character clauses, we will develop some strategies to remove these interference items.

5.4 Flow Chart

The flow chart of the method is shown in Fig. 1.

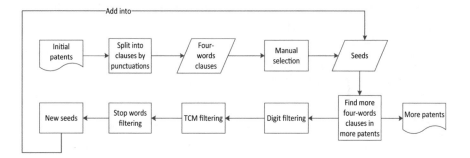

Fig. 1. Flow chart of our method.

As shown in Fig. 1, firstly, we select a small amount of patent abstract texts, split the texts into clauses with punctuations, extract all four character clauses, and select high quality ones as initial seeds by artificial screening; then, in more patent abstract texts, look for the clauses whose surrounding clauses contain seeds; and after digit filtering, Chinese herbal medicine name filtering and stop words filtering, high quality new seeds are obtained; add them into seeds thesaurus and repeat the above steps until seeds thesaurus reach to a steady size. The final seeds thesaurus is the collection of four character medicine effect phrases we have collected.

5.5   Pseudo Codes of Our Method

```
Algorithm 1: collection of four character medicine effect phrases
Input: patent abstract texts, denoted as PAT; window size, denoted as
WS
Output: the set of four character TCM effect phrases, denoted as seeds
1. seeds=∅
2. for each patent abstract text pt_i (0<i<=|PAT|)in subset of PAT:
3.     String[] clauses_i=pt_i.split(pt_i, "。|，|、|：|:|\\s+|;|: ");
4.         for each clause_ij in clauses_i:
5.             if(length(clause_ij)==4)
6.                 if clause_ij is a TCM effect phrases by manual judgment
7.             seeds=seeds U {clause_km}
8. for each patent abstract text pt_k (0<k<=|PAT|)in a greater subset of
   PAT:
9.     String[] clauses_k=pt_k.split(pt_k, "。|.|，|，|、|：|:|\\s+|;|: ");
10.        for each clause_km in clauses_k:
11.            if(length(clause_km)==4&&!containDigit(clause_km)&&
               !containTCM(clause_km)&&!containStopWord(clause_km))
12.                if(containSeedInWindow(clause_km, WS))
13.                    seeds=seeds U {clause_km}
14. repeat Step 8-13 until seeds reaches to a stable size.
```

Algorithm 1 is the main algorithm of our method. Step 1 to 7 is to collect initial seeds, and Step 8 to 13 iterate continuously to expand the seeds thesaurus gradually.

```
Algorithm 2: containDigit
Input: a String, denoted as str
Output: a Boolean result, that is, whether str contains any digit
1. Pattern p = Pattern.compile("[0-9]");
2. Matcher m = p.matcher(str);
3. if (m.find())
4.     return true;
5. else
6.     return false;
```

Algorithm 2 determines whether there is a digit in a string. Calling out Algorithm 2 in Algorithm 1 can remove the interference items containing any digit.

```
Algorithm 3: containTCM
Input: a String str; set of TCM names, denoted as {TCM}
Output: a Boolean result, that is, whether str contains any TCM name
1. for each TCM name TCM_i in {TCM}
2. if(str.indexOf(TCM_i)>=0)
3.     return true;
4. else
5.     return false;
```

Algorithm 3 determines whether there is a Chinese herbal medicine name in a string. Calling out Algorithm 3 in Algorithm 1 can remove the interference items containing any Chinese herbal medicine name.

```
Algorithm 4: containStopWord
Input: a String str; set of stop words, denoted as {StopWords}
Output: a Boolean result, that is, whether str contains any stop word
1. for each stop word sw_i in {StopWords}
2. if(str.indexOf(sw_i)>=0)
3.     return true;
4. else
5.     return false;
```

Algorithm 4 determines whether a stop word is contained in a string. Calling out Algorithm 4 in the Algorithm 1 can remove the interference items that contain stop words. The stop words here are not the same as the stop words mentioned before in text mining. In ordinary text mining, for example, web mining, news mining and so on, stop words refer to the frequently occurring words that are not meaningful to mining. For example, "is", "above", "however" and so on. In this paper, the word "stop words" refers to some frequently occurring, colloquial words that are not likely to appear in four character medicine effect phrases in TCM patents. For example, "have" "through" "following" "characteristics" and so on. In this paper, the stop words can be found and collected from those wrongly identified four character medicine effect phrases in several iterations.

```
Algorithm 5: containSeedInWindow
Input: a patent abstract text, denoted as pat; a clause in pat, denoted
as c; window size, denoted as WS
Output: a Boolean result, that is, whether the clauses around c contain
any seed word
1. split pat into clauses with punctuations, and store them into a
HashTable HT, the key is the sequence ID and the value is clause;
2. each clause on c's left in the window is denoted as LeftClause,
satisfying key(c)-key(LeftClause)<=WS;
3. each clause on c's right in the window is denoted as RightClause,
satisfying key(RightClause)- key(c)<=WS;
4. if any LeftClause or RightClause contain any seed word, return true,
else return false.
```

Making use of the characteristics that four character medicine effect phrases tend to appear continuously, we constitute a strategy: if the surrounding clauses of some specific four character clause contain any seed, then this clause is also considered to be a four character medicine effect phrase. We use the window to delineate the surrounding clauses. If we make the clause to be judged as the center, the clauses in both sides of the window are considered to be the surrounding clauses. When the size of the window is 0, it is an extreme case. In this case, only when the clause to be judged is the seed word, it will be considered to be a four character medicine effect phrase.

6 Experiment and Analysis

6.1 Data Source

This paper collected 10000 Chinese patents from patent searching system, using "清热解毒(clearing away heat and detoxifying)" as retrieval word. Each patent includes application number, inventor, title, abstract and IPC so on. As shown in Fig. 2.

Application Num	Application Date	Title	Application Type	Publication Date	First Applicant	Applicant	First Inventor	Inventors	Main IPC	IPC	Abstract
CN201510218	2015.04.27	清热解毒	注发明	2016.11.23	李林平	李林平	李林平	C12G3/04	C12G3/04	清热解毒的凉茶很多，但却没有清热解毒又能滋补身体的凉茶。本…
CN201610548	2016.07.13	山楂清热	注发明	2016.12.14	河北泰通	河北泰通	张学飞	张学飞	A23F3/34	A23F3/34	本发明公开了一种山楂清热解毒保健茶。其属于茶叶的技术领域。
CN201710115	2017.03.02	一种清热	注发明	2017.06.20	吴东涛	吴东涛	吴东涛	吴东涛	A23F3/34	A23F3/34	本发明提供一种清热解毒凉茶及其制备方法。由车前草30%?40%…
CN201610344	2016.05.16	清热解毒	注发明	2017.11.24	欧波	欧波	欧波	欧波	A23F3/34	A23F3/34	清热解毒的凉茶品种很多，但同时具有滋补作用的凉茶却很少。本…
CN201610936	2016.11.01	一种清热	注发明	2017.03.29	蔡少青	蔡少青	蔡少青	蔡少青	A23L23/0	A23L23/0	一种清热解毒菜汤，其特征在于，包括以下组分：麻菜20；桂枝10
CN201510095	2015.02.02	清热解毒	注发明	2015.06.17	李德伟	李德伟	李德伟	李德伟	A23F3/34	A23F3/34	传统的凉茶，只有清热解毒作用，不具备其它功能。本发明的清热
CN201710316	2017.05.08	一种清热	注发明	2017.08.11	楚雄医药	楚雄医药	陈琪珍	陈琪 姚荣	A23F3/34	A23F3/34	本发明公开了一种清热解毒茶及其制备方法它应用。所述的清热解
CN201610344	2016.05.16	具有清热	注发明	2017.11.24	欧波	欧波	欧波	欧波	C12G3/04	C12G3/04	清热解毒的凉茶品种很多，但却没有清热解毒又能滋补身体的药酒。本
CN201310260	2013.06.28	一种清热	注发明	2014.12.31	青岛联合	青岛联合	李阳	李阳	A23F3/14	A23F3/14	本发明提供了一种清热解毒茶。由以下组分按重量份数组成：野
CN201310356	2013.08.13	清热解毒	注发明	2013.12.04	李林平	李林平	李林平	李林平	A61K36/85	A61K36/85	清热解毒品系品种很多，如不上火凉茶、王老吉凉茶、但是清热解
CN201510952	2015.12.20	一种清热	注发明	2017.06.27	周跃贵	周跃贵	周跃贵	周跃贵	A23L7/10	A23L7/10	本发明提供一种均衡提供清热解毒的茶，原料包括水、大米、牛奶
CN200810027	2008.04.29	一种清热	注发明	2009.11.04	潘诚	潘诚	潘诚	潘诚	A23F3/34	A23F3/34	本发明公开了一种清热解毒茶，其特征是：组分包括半边莲、蛇王
CN201210486	2012.11.26	一种清热	注发明	2013.03.13	启东市大	启东市大	许洁	许洁	A61K36/85	A61K36/85	一种清热解毒丸，所述的清热解毒丸由下列组分按按比例配量份数组成
CN201010563	2010.11.29	一种清热	注发明	2012.05.30	潘亚琴	潘亚琴	不公告发	不公告发	A23F3/34	A23F3/34	本发明涉及一种清热解毒茶，它是将连翘12-14份、锦灯笼12-15份
CN201110169	2011.06.22	一种清热	注发明	2012.12.26	丁小燕	丁小燕	不公告发	不公告发	A23F3/34	A23F3/34	本发明涉及一种清热解毒的金银花保健茶。它是将配方量的金银花
CN201030516	2010.09.15	包装盒（详外观设计	包装盒	2011.04.06	乌尔滨大	乌尔滨大	郝大平	郝大平	09-03	09-03	1.本外观设计后视图与主视图相同，省略后视图。2名称：包装盒(]
CN200810232	2008.11.25	一种清热	注发明	2009.06.17	洛阳市康	洛阳市康	井健华	井健华	A61K36/75	A61K36/75	一种清热解毒药物的组配方法，采用黄连1%-2%、黄芩2%-4%、
CN201510095	2015.03.02	清热解毒	注发明	2016.10.05	李德伟	李德伟	李德伟	李德伟	A23F3/34	A23F3/34	传统的凉茶只有清热解毒作用，没有滋补身体、强壮身体的作用。

Fig. 2. Miniature of the patents collected.

6.2 Collection of Chinese Herbal Medicine Names and Stop Words

The purpose of the collection of Chinese herbal names and stop words is to filter four character clauses and improve the accuracy of the four character effect phrase recognition. According to the writing habits of Chinese, the length of the general sentence is more than 6 Chinese characters. After splitting an abstract text into clauses with punctuations, four character clauses usually belong to the following several cases:

(1) Concise and comprehensive four character effect phrase. They are the objects we intend to collect in this paper.

(2) Name of Chinese herbal medicine with four characters. Such as "王不留行(emplanting)" "罗布麻叶(Apocynum)" "紫花杜鹃(Purple azalea)" etc.

(3) Name and proportion of Chinese herbal medicine. Such as "当归5克(Angelica 5 grams)", "桔梗3%(Platycodon 3%)" and so on.

(4) Proportion of Chinese herbal medicine. Such as "3-5克(3-5 grams)", "0.6%" and so on.

(5) Short clause containing stop words. For example, "通过比较(by comparison)", "简单易行(simple and easy to do)", "口感清爽(refreshing taste)".

We can see that case 2-4 can be filtered through the matching of Chinese herbal names and digits. Case 5 can be filtered by the matching of stop words.

In our experiment, we collected 84 stop words and 1117 Chinese herbal medicine names.

6.3 Initial Seed Words

In 100 patent abstract texts, we extract four character clauses and manually select 40 four character medicine effect phrases as initial seeds. Part of seeds is shown in Table 3.

Table 3. Part of seeds.

补精益髓(Replenishing marrow)
滋阴解毒(Nourishing yin and detoxifying)
消痈散结(Dispel carbuncle and disperse knot)
生津止渴(Produce saliva and slake thirst)
脓毒肿痛(Septic swelling and pain)
活络定通(Activating collaterals)
养肝明目(Nourishing the liver to improve visual acuity)
消除水肿(Elimination of edema)
驱风除湿(Dispelling wind and eliminating dampness)
利水消肿(Inducing diuresis to alleviate edema)
行气健脾(Regulating qi for strengthening spleen)
益气养血(Nourishing qi and nourishing blood)
消暑降温(Relieve summer heat and cool down)
散结消肿(Eliminate the mass and relieve swelling)
凉血解毒(Cool blood and detoxify)
活血散结(promote blood circulation and disperse knot)

6.4 Iterations

When the window size is set to 1, the number and accuracy of four character medicine effect phrases obtained by multiple iterations are shown in Table 4.

Table 4. Results of multiple iterations.

Iteration	Number of patents	Number of new seeds	Number of improper seeds	Precision	Number of all seeds
0	100	40	0	100%	40
1	1000	179	4	97.8%	215
2	2000	370	4	98.9%	581
3	3000	436	4	99.1%	1013
4	4000	402	2	99.5%	1413
5	5000	352	2	99.4%	1763

As we can see from Table 4, the accuracy of the method remains at a rather high level. It demonstrates the validity of our method.

7   Conclusions

As the first step of patent analysis and mining, the accuracy of Chinese word segmentation directly affects the results of analysis and mining. However, because of historical reasons, TCM patents often use four character phrases in classical Chinese to describe the effect of medicines. Such four word character phrases cannot be identified by mainstream Chinese word segmentation systems. Starting from a small number of annotated data, and taking advantage of the location characteristics of four character phrases, this paper proposes a co-training based method to collect four character effect phrases in TCM patents. In the future work, we will apply this method to the identification of other items in TCM patents, such as disease names and preparation methods.

Acknowledgments. This research is supported by National Key Research and Development Program of China under grant number 2017YFC1405403, National Natural Science Foundation of China under grant number 61075059, Green Industry Technology Leding Project (product development category) of Hubei University of Technology under grant number CPYF2017008, Natural Science Foundation of Anhui Province under grant number 1708085MF161, and Key Project of Natural Science Research of Universities in Anhui under grant number KJ2015A236.

References

1. Lupu, M., Fujii, A., Oard, D.W., Iwayama, M., Kando, N.: Patent-Related Tasks at NTCIR. Current Challenges in Patent Information Retrieval Series, vol. 37. Springer, Berlin, Heidelberg, New York (2017)
2. Roda, G., Tait, J., Piroi, F., Zenz, V.: CLEF-IP 2009: Retrieval Experiments in the Intellectual Property Domain. Lecture Notes in Computer Science, vol. 6241. Springer, Berlin, Heidelberg, New York (2009)
3. Sharma, P., Tripathi, R., Singh, V.K., Tripathi, R.C.: Automated patents search through semantic similarity. In: International Conference on Computer, Communication and Control (IC4). IEEE, Piscataway, NJ (2016)
4. Wang, F., Lin, L.: Domain lexicon-based query expansion for patent retrieval. In: International Conference on Natural Computation, Fuzzy Systems and Knowledge Discovery, pp. 1543–1547. IEEE, Piscataway, NJ (2016)
5. Zhang, L., Lei, L., Tao, L.: Patent mining: a survey. ACM Sigkdd Explor. Newsl. **16**(2), 1–19 (2015)
6. Magali, R.G.M., Juan, R.S., Zenilton, K.G., Paulo, E.M.: Automatic patent clustering using SOM and bibliographic coupling. Braz. J. Inf. Syst. **10**(1), 6–18 (2017)
7. Shanie, T., Suprijadi, J.: Text Grouping in Patent Analysis Using Adaptive K-means Clustering Algorithm. American Institute of Physics Conference Series, vol. 1827. AIP Publishing (2017) Article ID 020041
8. Shamsi, F.A., Aung, Z.: Automatic patent classification by a three-phase model with document frequency matrix and boosted tree. In: 5th International Conference on Electronic Devices, Systems and Applications, pp. 1–4. IEEE, Piscataway, NJ (2017)
9. Li, W.Q., Li, Y., Chen, J., Hou, C.Y.: Product Functional Information Based Automatic Patent Classification: Method and Experimental Studies, Information Systems, vol. 67, pp. 71–82. Elsevier, Amsterdam (2017)

10. Triulzi, G., Alstott, J., Magee, C.L.: Predicting technology performance improvement rates by mining patent data. In: SSRN Electronic Journal. SSRN, Rochester, NY (2017)
11. Fu, T., Lei, Z., Lee, W.C.: Patent citation recommendation for examiners. In: IEEE International Conference on Data Mining, pp. 751–756. IEEE, Piscataway, NJ (2016)
12. Wang, F., Lin, L. F., Yang, Z.: An ontology-based automatic semantic annotation approach for patent document retrieval in product innovation design. In: Applied Mechanics and Materials, vol. 446–447, pp. 1581–1590. Trans Tech Publications Inc, Switzerland (2013)
13. Okamoto, M., Shan, Z., Orihara, R.: Applying information extraction for patent structure analysis. In: International ACM SIGIR Conference on Research and Development in Information Retrieval, pp. 989–992. ACM, New York (2017)
14. Xu, M., Sun, F., Jiang, X.: Multi-label learning with co-training based on semi-supervised regression. In: 2014 International Conference on Security, Pattern Analysis, and Cybernetic, pp. 175–180. IEEE, Piscataway, NJ (2014)
15. Wang, W., Lee, X. D., Hu, A.L., Geng, G.G.: Co-training based Semi-supervised web spam detection. In: International Conference on Fuzzy Systems & Knowledge Discovery, pp. 789–793. IEEE, Piscataway, NJ (2013)
16. Iosifidis, V., Ntoutsi, E.: Large scale sentiment learning with limited labels. In: Acm Sigkdd International Conference on Knowledge Discovery & Data Mining, pp. 1823–1832. ACM, New York (2017)
17. Blum, A.: Combining labeled and unlabeled data with co-training. In: Conference on Computational Learning Theory, pp. 92–100. ACM, New York (1998)
18. Deng, N., Chen, X., Ruan, O., Wang, C., Ye, Z., & Tian, J.: The construction method of clue words thesaurus in Chinese patents based on iteration and self-filtering. In: International Conference on Emerging Internetworking. Springer, Berlin, Heidelberg, New York (2017)
19. Deng, N., Chen, X., Li, D.: Intelligent recommendation of Chinese traditional medicine patents supporting new medicine's R&D. J. Comput. Theor. Nanosci. **13**, 5907–5913 (2016)
20. Na, D., Xu, C.: Automatically generation and evaluation of stop words list for Chinese patents. Telkomnika **13**(4), 1414–1421 (2015)
21. Deng, N., Chen, X., Ruan, O., Wang, C., Ye, Z., Tian, J.: PaEffExtr: a method to extract effect statements automatically from patents. In: 11th International Conference on Complex, Intelligent and Software Intensive Systems. Springer, Berlin, Heidelberg, New York (2017)
22. Chen, X., Deng, N.: A semi-supervised machine learning method for Chinese patent effect annotation. In: 2015 International Conference on Cyber-Enabled Distributed Computing and Knowledge Discovery, 243–250. IEEE

Smart Fire-Alarm System for Home

Jihoon Kang, Shreya Basnet, and Sardar M. Farhad$^{(\boxtimes)}$

Victoria University, Sydney, Australia
sardar.farhad@vu.edu.au

Abstract. Internet of Things (IoT) is an emerging technology that can be used to design smart systems without incurring lots of money. Although, the legislation forces to install an analog fire alarm system to every household in modern era, the system is not smart enough to send immediate notifications to the concerned individual. The analog fire alarm cannot avoid the immediate risk or damage of fire as no one stays at home all the time. The earlier the incident is reported to the concerned individual the better to reduce the fire damage. In this paper, we have proposed a smart fire-alarm system that consists of a smoke sensor and a sound sensor that can detect the smoke and the noise of the analog fire alarm system (with a view to detect whether the analog fire alarm system is working) respectively, and can notify the user about the status immediately. We have implemented the proposed smart fire-alarm system and conducted experiments to determine the effectiveness of the system. The results show that the system is reliable under various conditions of smoke and alarm noises.

Keywords: Internet of everything · IoT · Sensor network · Smart home

1 Introduction

The Internet of Things (IoT) has been announced as one of the technology mega trends that will change the world according to Forbes [1]. The IoT enables the network of devices, vehicles, and home appliances containing electronics, software, actuators, and connectivity to connect, interact and exchange data [4]. The research on alarm system has recently got a lot of interests in the research community [6–9,12,14,15].

The IoT based fire alarm system that detects smoke and determines the location of the fire using camera and an intuitive algorithm has been illustrated in [9]. Raspberry Pi has been utilised to control different Arduino which are coordinated with a few sensors and a camera. A 360° transfer engine is integrated with the camera to snap the picture in whatever angle the fire is detected. An intuitive algorithm has been devised to identify the location of the fire. The system sends notification to the user in the event of fire with location information.

The temperature and humidity alert system to prevent food poisoning is illustrated in [8]. The system utilised temperature and humidity sensors and

© Springer Nature Switzerland AG 2020
L. Barolli et al. (Eds.): CISIS 2019, AISC 993, pp. 474–483, 2020.
https://doi.org/10.1007/978-3-030-22354-0_42

a Raspberry Pi. The data from the sensors are sent to an open source cloud system. The cloud system in turn sends notification when the sensor data reach a certain threshold.

New devices equipped with network function to become one of Internet of Things are emerging to develop smart homes and thereby smart cities. According to the regulation 76B under the Development Act 1993, the residences have to install battery powered or hard-wired (240 volt mains powered) fire alarms at home [5] in Australia and as well as in all modern societies. However, these alarms can be malfunctioned for any reason and could cause widespread damage of the property. Most importantly, these analog fire alarm systems are unable to send any notification to the user on the fire events. With the advent of IoT based cheap and convenient technologies, it is possible to install smart fire-alarm system that can not only detect fire but also notify the user about fire incident immediately. In this paper, we have proposed a smart fire-alarm system that consists of an Arduino, a smoke sensor and a sound sensor. As an enhancement, the proposed system can detect whether or not the existing analog fire alarm triggers in the fire event. In summary, the major contributions of this work are as follows:

- We have proposed an IoT based smart fire-alarm system.
- We have implemented the system and conducted experiments and the results show that the proposed system is accurate and reliable.
- We have determined the sensor settings such that it detects the smoke and analog fire alarm noise reliably and sends notifications accordingly. The experimental results ensured that system is not susceptible to any false environmental noise.

The rest of the paper is structured as follows: Sect. 2 illustrates the related works on IoT, Sect. 3 outlines the details of the smart fire-alarm system, Sect. 4 presents the experimental evaluations, finally we conclude in Sect. 5.

2   Related Works

Cloud based temperature and humidity alert system to prevent food poisoning is illustrated in [8]. The food safety standards specify that bakery foods must be stored at safe temperatures. The humidity creates the problem even worse. The proposed system of this paper incorporates the temperature and humidity sensors. The cloud framework collects the data from the sensors and send notification if the temperature and humidity reach a certain threshold.

Fire detection using CCTV technology and image processing is presented in [13]. To give early extinguishing of a fire disaster, extensive quantities of finders which intermittently measure smoke fixation or temperature are sent in structures. The system can be monitored in real time via Internet.

An MQTT (Message Queue Telemetry Transportation) broker using Amazon Web Service (AWS) has been illustrated in [10]. The broker is utilised to monitor and control room temperatures. The MQTT broker has been utilized as a platform to collect the data from the sensors and actuators via Wi-Fi channel.

A smart wireless home security system is proposed in [11] that sends notification to the owner in case of any trespass enters the home. The motion sensors and micro-controller have been used to design the system and WiFi enabled Internet has been used to send notification.

The system that detects the fire and identifies the location of the fire and can send a notification that includes the location information. They have incorporated Raspberry Pi, Arduino, smoke sensors, and a camera. An intelligent algorithm is developed that gets the data from the sensors and pictures from the camera and can send notification in the case of fire. The notification includes the location of the fire.

The IoT based fire detection framework that helps to recognise fire as quickly as time permits and thus save the valuable human lives has been illustrated in [9]. Raspberry Pi has been utilised to control different Arduino which are coordinated with a few sensors and a camera. A 360° transfer engine is integrated with the camera to snap the picture in whatever angle the fire is detected.

3   Proposed Smart Fire-Alarm System

The proposed smart fire-alarm system is used along side traditional analog fire alarm system. In this section, we explain the architecture of the system to explain the structural composition of the system. We explain the flow chart of the system to illustrate the functional decomposition of the system.

3.1   Architecture

The proposed smart fire-alarm system notifies the user in case of existence of fire at home and send the status of the analog fire alarm whether it is making noise or not. The smart system consists of an alarm-noise detector which is a sound sensor, and a smoke detector (gas sensor) as illustrated in Fig. 1. The system is connected to a router for Internet connection. The system periodically upload both the gas and noise sensor data to the open source cloud platform known as IFTTT [3].

The cloud platform checks the values and determines whether there is fire and the analog fire alarm is working. The cloud framework sends appropriate messages to a designated phone number based on identified situations as presented in Fig. 2. The noise sensor detects noise value of the analog fire alarm system. The gas sensor detects the presence of smoke. Based on status of noise and gas sensors, user will be notified accordingly. The related pseudo code of the flowchart is shown in Appendix A.

4   Experiments

We have implemented the proposed smart fire-alarm system using Arduino WeMo's D1 R2 V2.1 where micro-controller and ESP8266 Wi-Fi are integrated,

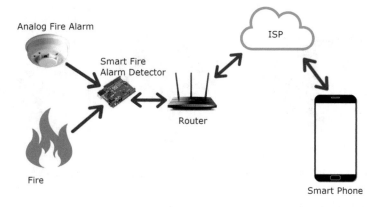

Fig. 1. Architecture of the proposed smart fire-alarm system.

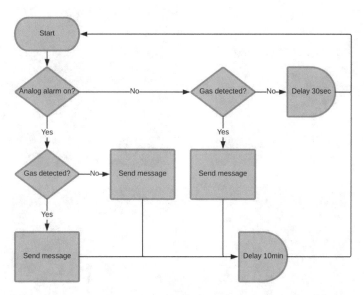

Fig. 2. Flowchart of the proposed smart fire-alarm system. Based on status of analog fire alarm and smoke sensor, user will be notified differently.

the MQ-2 gas or smoke sensor, and the XC4388 sound sensor as shown in Fig. 3. We program this device through the regular Arduino IDE.

4.1 Experimental Setup

Arduino, MQ-2 gas sensor and SC4388 noise sensor are connected each other. For this interconnection, the ground pin and Vcc pin of MQ-2 gas sensor and SC4388 noise sensor connect with ground pin and 3.3v pin of Arduino. The analog out pin of MQ-2 gas connects with A_0 pin of Arduino. The digital out pin

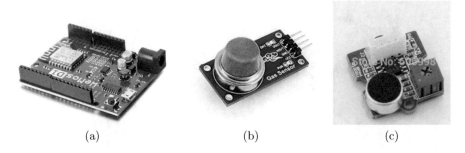

Fig. 3. (a) Arduino WeMo's D1 R2 V2.1 (b) MQ-2 gas or smoke sensor (c) XC4388 sound sensor.

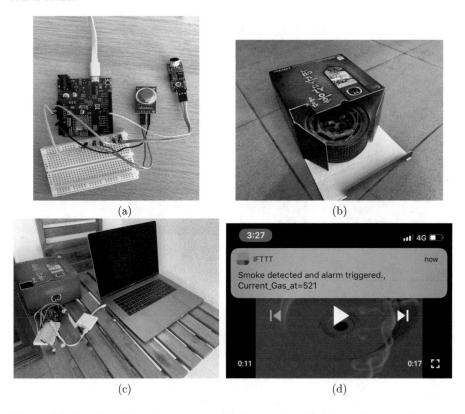

Fig. 4. (a) Circuit of fire-alarm system (b) Smoke box (c) Integrating circuit, smoke box, and Internet connection (d) Received notification on mobile.

of sc4388 connects with D8 pin of Arduino. Final assembled modules picture is shown in Fig. 4(a). We place assembled module near fire alarm and power point. We connect with USB power supplier with power in USB b type connector in Arduino. The application can be downloaded from the official website as shown in [2]. After installation, from the tools menu 'Board' and 'Upload speed' need to be set as 'WeMos D1 R1' and '115200' respectively.

Experimental setup has been conducted with the help of a smoke box to minimise the risk of fire incident as shown in Fig. 4(b). The smoke box has mosquito coils inside to make safe smoke. Arduino board with all sensing module assembled is connected to laptop. Arduino software is running with serial monitor on laptop. The laptop reads the data from the sensors via the Arduino and send the data to IFTTT which in turn generates notification. We play a loud analog fire alarm sound to trigger the sound sensor. Whole setup is placed outdoor environment to remove any possible incident with actual fire alarm as shown in Fig. 4(c). Example of notification on user's smart phone is shown in Fig. 4(d).

4.2 Experimental Results

The shortest response time setup can provide faster reaction because of minimised software coding. However, most of fire alarm sound has such a pattern that makes the sound sensor detects trigger point consistently.

Experiments have been conducted on four categories and observed the time taken to trigger the events: smoke only without loops, smoke only with loops, sound only with loops, and smoke and alarm both with loops. The loop is to remove any possible false positive scenario. We have conducted each experiments ten times and observed the time taken that are presented in Table 1. The time taken to detect the sound event without loops is the quickest, followed by smoke only without loops. Both smoke and sound detection experiment takes the longest time.

Table 1. Observation times required for four experimental evaluations.

Iteration	Smoke only without loops (s)	Smoke only with loops (s)	Sound only with loops (s)	Smoke and alarm both with loops (s)
1	6.55	7.78	5.25	10.27
2	7.43	7.23	4.13	8.95
3	6.26	8.11	6.11	9.33
4	6.99	7.98	4.87	11.32
5	6.32	7.39	5.13	10.87
6	7.86	8.22	5.09	9.14
7	5.46	6.80	4.96	8.78
8	6.09	7.22	4.87	10.92
9	4.13	5.87	5.03	10.19
10	7.09	8.19	5.32	9.47
Average	6.42	7.48	5.08	9.92

At the time when gas sensor detects enough amount of gas to trigger the event, if fire alarm sound volume is lower than minimum trigger volume because of inconsistency, smoke detection alarm system will think as Smoke only situation. Solution for this case is once gas or sound sensor detect high enough value to trigger the event, check both sensors again 30 loops repeatedly with delay of 0.05 s each time. This setup will make 1.5 s longer to send notification, but it can give more accurate information. Final version of codes that includes looping has been attached in this document as Appendix B.

5   Conclusion

This paper presents a smart fire-alarm system consisting smoke and sound sensors that can send immediate notification to the user in case of a fire at home. The system not only identifies smoke but also determines whether the builtin analog fire alarm is working or not. As the user gets an immediate notification by the system on a fire event, the user can take immediate prompt actions to minimise the loss. We have implemented the smart fire-alarm system and conducted experiments. The results show that the system is reliable under various smoke and alarm conditions.

A   Pseudo Code of the Proposed System

Pseudo code for smart fire-alarm system is shown as follows:

```
Procedure SmartAlarmDetector
    While true Do
        set noise to False;
        set gas to False;
        If NoiseSensor is True Then
            set noise to True;
        If GasSensor is True Then
            set gas to True;
        If noise is True Then
            If gas is True Then
                send message 1
            Else send message 2
            Delay
        Else If gas is True Then
            send a message 3
        End If
    End While
End
```

B   Arduino Code

```
    #include <Arduino.h>
#include <ESP8266WiFi.h>
#include <ESP8266WiFiMulti.h>
#include <ESP8266HTTPClient.h>
#define USE_SERIAL Serial

ESP8266WiFiMulti WiFiMulti;

void setup() {
  pinMode(A0, INPUT);
  pinMode(D8, INPUT);
//  pinMode(D11, OUTPUT);
  USE_SERIAL.begin(115200);
  // USE_SERIAL.setDebugOutput(true);

  USE_SERIAL.println();
  USE_SERIAL.println();
  USE_SERIAL.println();

  for (uint8_t t = 4; t > 0; t--) {
    USE_SERIAL.printf("[SETUP] WAIT %d...\n", t);
    USE_SERIAL.flush();
    delay(1000);
  }

  WiFi.mode(WIFI_STA);
  WiFiMulti.addAP("daniel", "0467697987");

}

void loop() {
  // wait for WiFi connection

  int soundValue = analogRead(A0);
  delay(10);
  int gasValue = digitalRead(D8);
  int soundSensitivity = 800;
  USE_SERIAL.print("sound = ");
  USE_SERIAL.println(soundValue);
  USE_SERIAL.print("gas = ");
  USE_SERIAL.println(gasValue);
  int gasSensitivity = 750;
  bool sound = false;
  bool gas = false;
  String msg = "?value1=";
//in url, characters after "key/" should be replaced with user's own key.
  String url = "http://maker.ifttt.com/trigger/cycle2/with/key/cdd3BA5g0Adb28BIDtAMVC";
  String finalurl = url;

  if (soundValue > soundSensitivity){
    sound = true;
  }
  if (gasValue > gasSensitivity){
    gas = true;
  }
//if space within msg not working? than put %20 instead of space
  if (sound == true){
    if (gas == true){
      msg += "Fire detected and alarm triggered.";
    }else{
      msg += "Fire not detected but alarm triggered.";
    }
  }else{
    if (gas == true){
      msg += "Fire detected but alarm not triggered.";
    }
```

```
}

  if ( msg != "?value1=") {

    if ((WiFiMulti.run() == WL_CONNECTED)) {

  HTTPClient http;

  finalurl += msg;
  finalurl += "&value2=";
  finalurl += "Current Gas at = ";
  finalurl += gas;

  USE_SERIAL.print("[HTTP] begin...\n");
  http.begin(finalurl); //HTTP

  USE_SERIAL.print("[HTTP] GET...\n");
  // start connection and send HTTP header
  int httpCode = http.GET();

  // httpCode will be negative on error
  if (httpCode > 0) {
    // HTTP header has been send and Server response header has been handled
    USE_SERIAL.printf("[HTTP] GET... code: %d\n", httpCode);

    // file found at server
    if (httpCode == HTTP_CODE_OK) {
      String payload = http.getString();
      USE_SERIAL.println(payload);
    }
  } else {
    USE_SERIAL.printf("[HTTP] GET... failed, error: %s\n", http.errorToString(httpCode).c_str());
  }

  http.end();

    }

  }
  delay(100);
}
```

References

1. The 9 technology mega trends that will change the world in 2018. www.forbes.com. Accessed 11 Mar 2019
2. Arduino software code. www.arduino.cc/en/Main/Software. Accessed 12 Mar 2019
3. IFTTT free cloud platform. ifttt.com. Accessed 12 Mar 2019
4. Internet of Things (IoT). en.wikipedia.org. Accessed 11 Mar 2019
5. Smoke alarms - what you need to know, regulation 76B under the development act 1993. http://www.mfs.sa.gov.au. Accessed 11 Mar 2019
6. Deshmukh, A.D., Shinde, U.B.: A low cost environment monitoring system using Raspberry Pi and Arduino with Zigbee. In: International Conference on Inventive Computation Technologies (ICICT), vol. 3, pp. 1–6. IEEE (2016)
7. Farhad, S.M., Alahi, I., Islam, M.M.: Internet of Things based free parking space management system. In: 2017 International Conference on Cloud Computing Research and Innovation (ICCCRI). pp. 1–6. IEEE (2017)
8. Hitendra, P., Sajan, S., Farhad, S.M.: Cloud based temperature and humidity alert system to prevent food poisoning. In: The 9th International Conference on Cyber Security and Communication Systems (ICCSCS 2018), pp. 1–12. Springer (2018)

9. Imteaj, A., Rahman, T., Hossain, M.K., Alam, M.S., Rahat, S.A.: An IoT based fire alarming and authentication system for workhouse using Raspberry Pi 3. In: 2017 International Conference on Electrical, Computer and Communication Engineering (ECCE), pp. 899–904. IEEE (2017)

10. Kang, D.H., Park, M.S., Kim, H.S., Kim, D., Kim, S.H., Son, H.J., Lee, S.G.: Room temperature control and fire alarm/suppression IoT service using MQTT on AWS. In: 2017 International Conference on Platform Technology and Service (PlatCon), pp. 1–5. IEEE (2017)

11. Kodali, R.K., Jain, V., Bose, S., Boppana, L.: IoT based smart security and home automation system. In: 2016 International Conference on Computing, Communication and Automation (ICCCA), pp. 1286–1289. IEEE (2016)

12. Liu, Q., Ma, Y., Alhussein, M., Zhang, Y., Peng, L.: Green data center with IoT sensing and cloud-assisted smart temperature control system. Comput. Netw. **101**(C), 104–112 (2016). https://doi.org/10.1016/j.comnet.2015.11.024

13. Shinde, R., Pardeshi, R., Vishwakarma, A., Barhate, N.: Need for wireless fire detection systems using IoT. Int. Res. J. Eng. Technol. (IRJET) **4**(1), 1078–1081 (2017)

14. Tang, J., Sun, D., Liu, S., Gaudiot, J.L.: Enabling deep learning on IoT devices. Computer **50**(10), 92–96 (2017)

15. Technologies, A.: Home automation and monitoring projects for Raspberry Pi. http://projects.privateeyepi.com/. Accessed 11 Mar 2019

A Framework for Early Detection of Antisocial Behavior on Twitter Using Natural Language Processing

Ravinder Singh[1(✉)], Jiahua Du[1], Yanchun Zhang[1], Hua Wang[1],
Yuan Miao[1], Omid Ameri Sianaki[2], and Anwaar Ulhaq[3]

[1] Victoria University, Footscray Park, VIC, Australia
{ravinder.singh,yanchun.zhang,Hua.Wang,
Yuan.Miao}@vu.edu.au, jiahua.du@live.vu.edu.au
[2] Victoria University, Sydney, NSW, Australia
omid.amerisianaki@vu.edu.au
[3] Charles Sturt University, Sydney, NSW, Australia
aulhaq@csu.edu.au

Abstract. Online antisocial behavior is a social problem and a public health threat. A manifestation of such behavior may be fun for a perpetrator, however, can drive a victim into depression, self-confinement, low self-esteem, anxiety, anger, and suicidal ideation. Online platforms such as Twitter and Facebook can sometimes become breeding grounds for such behavior. These platforms may have measures in place to deter online antisocial behavior, however, such behavior still prevails. Most of the measures rely on users reporting to platforms for intervention. In this paper, we advocate a more proactive approach based on natural language processing and machine learning that can enable online platforms to actively look for signs of antisocial behavior and intervene before it gets out of control. By actively searching for such behavior, social media sites can possibly prevent dire situations that can lead to someone committing suicide.

1 Introduction

Antisocial behavior is one of the ten personality disorders included in 'Diagnostic and Statistical Manual of Mental Disorders (DSM-5). These ten disorders are characterized into three different clusters and antisocial personality disorder falls in Cluster B, along with Borderline Personality Disorder, Histrionic Personality Disorder, and Narcissistic Personality Disorder [1]. It is a prevalent pattern of disregard for, and violation of the rights of others. A person with antisocial personality disorder fails to conform to social norms with respect to lawful behavior. The person can become irritable and aggressive and can be consistently irresponsible when it comes to dealing with other people. The person may lack remorse and mistreat others [1, 2]. There may be many elements that lead to a person developing antisocial behavior: genetic influences, maternal depression, parental rejection, physical neglect, poor nutrition intake, adverse socioeconomic or sociocultural factors are few of them [1, 3–7]. These factors can be categorized broadly into three main categories: Neural, Genetic and Environmental [8]. Antisocial

© Springer Nature Switzerland AG 2020
L. Barolli et al. (Eds.): CISIS 2019, AISC 993, pp. 484–495, 2020.
https://doi.org/10.1007/978-3-030-22354-0_43

personality disorder (ASPD) is one of the most reliably diagnosed conditions among all personality disorders. Many psychiatrists are reluctant to treat people who suffer from ASPD because there is a widespread belief that it is untreatable, however, there is increasing evidence that in certain cases it can be treated [9].

Online antisocial behavior is a widespread problem and threatens free discussions and user participations in many online communities. In many cases, it can be devastated for victims and deter them from using these platforms [10]. Online antisocial behavior appears to be an Internet manifestation of everyday sadism. An individual who possesses and display such behavior online seems to enjoy at the expense of others ignoring distress and harassment it may cause [11]. Apart from Sadism, attention-seeking, boredom, a desire to cause damage to the community and revenge are some of the motivations identified, that relate to antisocial behavior [12]. Antisocial behavior annoys and interferes with a person's ability to lawfully go about his business. Current measures in place to discourage antisocial behavior rely mainly on users reporting it directly to platforms [13]. In most cases, victims are reluctant to confront such behavior online as they are scared of retaliation. Therefore, most cases of antisocial behavior go unnoticed. The online platforms encourage freedom of speech but fail to draw a line between free speech and unacceptable behavior. Current measures in place do not seem to effectively prevent people from explicitly displaying antisocial behavior, exposing a lot of people, who may fall into a vulnerable group of people, to be on the receiving end of such behavior.

Twitter is one of the most popular social media platforms that encourages people to share views and content. A user contributes in the form of a tweet, which is a 280-word text, and may contain an image, a video, a link to an article etc. The platform encourages user participation in the form of discussions on topics of interests, however, this may bring along some undesirable behavior such as bullying, abuse, and harassment [2]. Online antisocial behavior is prevalent mainly among users of age 18–27 and has also been linked to excessive use of online platforms. Perpetrator seems to enjoy at the cost of others [11, 12, 14]. Ramifications of excessive use also lead to other psychological disorders, and to employ measures in place to curtail their impact on society is imperative [15]. Twitter and other platforms rely on users to report such behavior, in which case, platforms intervene and usually block the offensive user. The platform has systems in place that may prevent the distribution of illegal material, spam, nudity, pornography, etc. but nothing for antisocial behavior [16]. So on the one hand, the platform connects users to enable the exchange of information, ideas, and other useful resources and on the other hand it facilitates the spread of antisocial behavior and related issues, and put a large number of people at risk [17]. Online platforms need to have a system in place that automatically detects any antisocial activity and restrict or eliminate it without letting it affect other users.

2   Present Work

In this paper, we discuss a framework based on natural language processing and machine learning that can automatically detect antisocial behavior online and can enable platforms such as Twitter to proactively prevent it from spreading by having appropriate measures in place. Most of the research conducted on antisocial behavior has been qualitative in nature focusing mainly on deep case study analysis. Study groups are often chosen manually and are small in numbers. These studies are cumbersome in nature and may require a lot of resources and time. In today's world, we humans spend most of our time online. Access to the Internet has changed the way we live and do things in our daily life. We spend more time in front of screens today than we ever did. Most of our daily tasks such as work, social interactions, banking, shopping, and entertainment, etc. take place online. Since the way we live and do things has changed significantly, we need new ways to explore and study personality and behavior traits [17]. The research for this project has been conducted by collecting data from the social media site Twitter. Since this data is generated during our interactions with the outer world, it has a lot of information related to our human behavior and personalities. This research project aims to extract such information and use it to build a machine learning model to detect antisocial behavior online.

2.1   Natural Language Processing

Natural language processing is a field concerned with the ability of a computer to understand, analyze, manipulate, and potentially generate human language. By human language, we are simply referring to any language used for everyday communication. This can be English, Spanish, French, or Mandarin. A programming language such as Python, that we have used in this research, does not naturally know what any given word means. All it sees is a string of characters. For example, it has no idea what antisocial actually means. It sees that it is a ten-character long word, but the individual character doesn't mean anything to Python and certainly, the collection of those characters together does not mean anything either. Humans know what an 'A' and a 'S' means and together those 10 characters make up the world 'antisocial', and we know what that means. So natural language processing is the field of getting the computer to understand what 'antisocial' signifies, and from there we can get into the manipulation or potentially even generation of that human language. We probably experience natural language processing on a daily basis without even knowing. Natural language processing is a broad and evolving field that encompasses many topics and techniques. The core component of natural language processing is extracting all the information from a block of text that is relevant to a computer understanding the language. There are many techniques for natural language processing and machine learning methods are the most promising of all. Machine learning is a field of study that gives computers the ability to learn without being explicitly programmed. To conduct our experiments in this research, we used the Python programming language and NLTK package. The natural language toolkit is the most utilized package for handling natural language processing tasks in Python. Usually called NLTK for short, it is a suite of open-source tools originally created in 2001 at the University of Pennsylvania for the purpose of making natural language processing in Python easier. NLTK is great because it basically provides a jumpstart to building any natural language processing tasks by providing basic tools that can then be chained together rather than having to build all those tools from scratch.

2.2   Online Antisocial Behavior

To understand antisocial behavior online, we went through the diagnostic criteria explained in DSM-5 [1]. The term antisocial personality disorder is primarily used in a clinical setting and may be used to explain the behavior of a person who is against societal norms. To be antisocial may mean to be against rules, laws, norms, and acceptable behavior. Furthermore, against the rule and law may refer to failure to obey laws and legal system, engaging in criminal activities, arrest, etc. A person with antisocial personality may also lie, deceive people and manipulate others for self-amusement and profit. The person may get irritable and aggressive easily and is inclined to engage in fights. The person may also be impulsive, irresponsible and lack remorse for action. Not all psychological disorders can be diagnosed by a person's writing, but a few can, and antisocial behavior is one of them. Since it can be diagnosed by the way a person writes, we are able to detect such behavior online from tweets, online posts, reviews, and comments. In any text, antisocial behavior is expressed by using words, and context of use of those words. There are a number of rude words and short phrases that can be associated with antisocial behavior. It may seem easy for a human to pick up such behavior through text, however, may not be that easy for a machine. One reason is that some rude words can be used in humor or in sarcasm, which may not always be considered antisocial. Also, the context of a text plays an important role to classify it as an antisocial text. Use of slang, the order of words, local culture, etc. all play an important role in classifying a piece of text. Some words and phrases that are normal to use in one country may imply rude or antisocial behavior in other. An example of that is an experience shared by a friend from Australia, who was in a café in the US and asked for a 'White Coffee'. This is a normal way of getting a coffee with milk in Australia, however, in the US the guy at the café, who was a person of color, thought that my friend was rude and racist. My friend should have asked for 'coffee with milk' instead of 'white coffee'. Under certain circumstances, it is difficult for even humans to know the exact intentions of a person from his or her writings, and we can imagine how hard this could be for a machine. A machine or a computer relies on a set of rules and instructions to take any action, however, in the case of natural language processing, it is not so straight forward. There are few techniques used in natural language processing and we in this research project have used a machine learning approach. To train a machine learning model to detect antisocial behavior from a person's writing requires a lot of training and testing data, along with ground truth validation. In this research, we sought the help of a psychology graduate for labeling our dataset and ground truth validations. We, in this research, propose a framework that can be used to detect such behavior online with high accuracy and precision. To the best of our knowledge, this is one of the first studies in which machine learning has been used to detect psychological disorders online. The proposed framework in this study can be used for all type of online platforms including social networks.

2.3   Framework

Proposed framework consists of collecting the right tweets and labeling them. Once labeled, these tweets and labels need to be verified by a qualified person. The qualified person in our scenario is the person who has a thorough understanding of psychological disorders and can diagnose them in a clinical setting. Once the data, which is a set of tweets in our case, is properly labeled, we can use natural language processing techniques to clean and pre-processed it. These natural language processing techniques are discussed in detail in the following sections. Once the data is cleaned, it is used to train and test a machine learning model. For our model building, we tried five most popular machine learning algorithms: Logistic Regression, Support Vector Machine, Random Forest, Decision Tree, and Naïve Bayes. The results obtained using all these algorithms were very similar, while Support Vector Machine performed the best. Support Vector Machine is one of the most trusted algorithms in machine learning and worked well in our scenario as well. Once the model is built it can be integrated into any online platform, including social media platforms. We believe that the model will perform well with both stored and live stream data. In the case of live stream data, once a piece of text (tweet, post, news) is triggered to have antisocial semantics, it can either be removed by an algorithm or can be reviewed by a human staff member who can take further actions. A proactive approach like this can help reduce the prevalence of such behavior online and encourage healthy and clean online discussions.

3   Methodology

3.1   Data

We collected 55,810 tweets from Twitter between Oct 2018 and Feb 2019. Tweets are 280-words text that users share with others on the Twitter platform. When Twitter first started, the limit on the length of these tweets was 140 words and was later increased to 280 words as the popularity of the platform soared. We used various phrases such as "I do not care about the law", "I wish you die soon", "Go to hell" etc. to search and collect these tweets. Text data collected online is typically in a semi-structured or unstructured form. Our data was no different and was in semi-structured from when first collected. Therefore, some tweets were missing delimiters and had no indication of any punctuation. We used functions from the NLTK library to structure our data set. Once the dataset was in a structured form, we annotated the dataset manually with two categories: Tweets that conveyed antisocial behavior and tweets that did not. Once the dataset was annotated, we wanted to get it verified by someone from the area of Psychology. We hired a psychology graduate to do so. The person had a thorough understanding of all the personality disorders and could diagnose them in a clinical setting.

Psychological disorders can be classified into three categories: Personality disorders, Behavior disorders and State of mind disorders. Behavior and State of mind disorders fluctuate and usually cannot be detected from a person's writing. Behavior may change from time to time and so can State of mind. However, Personality disorders or traits do not fluctuate and stays with a person for a longer period of time. Since these traits stay with a person for a longer period they manifest through a person's speech and online writings. Antisocial behavior is one such personality disorder that can be reliably detected from online corpora. Our annotator was able to go through each and every tweet manually to see if it qualified as an antisocial tweet. If it did, it was labeled one. Once labeled, our dataset was ready to be explored further.

3.2 Data Pre-processing

This phase involved removing punctuation from our tweets, followed by tokenization, which means dividing a sentence into individual words. Once tokenization was done, the next step was to remove stop words. Stop words are the words which do not contribute much to the meaning of a sentence. Examples of such words are: the, is, are, etc. After removing the stop words, we used stemming to cut down words into their shortest form. This is done to reduce the work for our algorithm. All these steps are explained in the following paragraphs.

The first step in the pre-processing phase was to remove punctuation from the tweets. In order to remove punctuation, we had to actually have a way to show Python, the programming language we used, what punctuation even looked like. We accomplished by using the String package in Python. The reason that we care about removing punctuation is that period, parentheses, and other punctuations look like just another character to Python, but realistically, the period does not help pull the meaning out of a sentence. For instance, for us "I like research." with a period, is exactly the same as, "I like research". They mean the same thing for us, but when we give these sentences to a machine learning algorithm, the algorithm says those are not equivalent things. We wrote a function to cycle through each and every character, checked if it was punctuation, and discarded it if it was. This was done to reduce the workload of our algorithm. By removing punctuations, our algorithm had to deal with fewer characters in the learning process.

Now that we had removed punctuation, we could move on to tokenizing our text. Tokenization is spitting some string or sentence into a list of words by white spaces and special characters. For example, we could take a sentence "I am doing research" and split it into four words: 'I', 'am', 'doing', and 'Research'. Instead of seeing the whole sentences, our algorithm could see four distinct tokens, and it knew what to look at. Some of the words were more important than others. For instance, the words 'the', 'and', 'of', and 'or', appear frequently but offered little information about the sentence itself. These are what we call stop words. We removed these words to allow our algorithms to focus on the most pivotal words in our tweets. From the example above, if we remove 'I' and 'am', we are left with 'doing research'. This still gets the most important point of the sentence, but now our algorithm is looking at half the number of tokens.

The next step in the process was stemming. Stemming is the process of reducing inflected or derived words to their word stem or root. In other words, it means to chop off the end of a word, to leave only the base. This means taking words with various suffixes and condensing them under the same root word. For example, we can stem words such as 'connection', 'connected', and 'connective' to one word 'connect'. Stemming shoots for the same goal by reducing variations of the same root word and making our algorithm deal with fewer words. Without stemming, our algorithm will need to keep all three words: 'connection', 'connected', and 'connective' in memory, increasing the workload and making our machine learning model less efficient. In simple words, the whole idea of all these steps is to reduce the size of the corpus for our machine learning model to deal with. For stemming, we used the Porter stemmer from the NLTK package.

3.3   Vectorization

Now that we had a clean text data that we could use to build our machine learning model, we needed to convert it into a form that could easily be understood by our model. The process is called vectorization. This is defined as the process of encoding text as numbers to create feature vectors. A feature vector is an n-dimensional vector of numerical features that represent some object. In our context, that meant, we had to take individual tweets and convert them into a numeric vector that represented those tweets. The way we did this was by taking our dataset, that had one line per document, with the cell entry as the actual text message and converted it into a matrix that still had one line per document, but then we had every word used across all documents as the columns of our matrix. And then within each cell was counting, representing how many times that certain word appeared in that document. This is called a document-term matrix. Once we had the numeric representation of each tweet, we then carried down with our machine learning pipeline and fitted and trained our machine learning model. We vectorized text to create a matrix that only had numeric entries that the computer could understand. In our case, counting how many times each word appeared in each tweet. A machine learning model understands these counts. If it sees a one or a two or a three in a cell, then the model can start to correlate that with whatever we're trying to predict. In our case, that was antisocial behavior. We analyzed how frequently certain words appeared in a tweet in context to other words, to determine whether the tweet manifested antisocial behavior. In this research, we used both Word Frequency (WF) and Term Frequency-Inverse Document Frequency (TF-IDF) methods of vectorization. We did this to see a difference in the results with our machine learning model. The count vectorization created the document-term matrix and then simply counted the number of times each word appeared in that given document, or tweet in our case, and that is what stored in the given cell. The equation for this is:

$$wf(w,d) = \frac{number of occurrences of a word in tweet}{total number of all words in a tweet}$$

Term Frequency-Inverse Document Frequency, which is often referred as TF-IDF, created a document-term matrix, where there was still one row per tweet and the column still represented single unique term, however, instead of the cells representing the count, the cell represents a weighting that was meant to identify how important a word was to an individual tweet. We started with the TF term, which is the number of times that a term occurred in a tweet divided by the number of all terms in the tweet. For example, if we use "I like research" and the word we are focused on is 'research' then this term would be 1 divided by 3 or 0.33. The second part of this equation measures how frequently this word occurs across all the tweets. We started by calculating the number of tweets in the dataset and divided that by the number of text messages that this word appeared in and then took the log of that equation. For example, if we had 20 tweets and only one had the word 'research' in it, then the inverse document frequency means log (20/1). Now we had two parts of the equation: Term frequency and inverse data frequency. The last step was to multiply both to get a weight for the word 'research' in the tweet. The equation is as follows:

$$W_{i,j} = tf_{i,j} \times \log(\frac{N}{df_i})$$

Both the matrices have the same shape and the only difference is the values in the cells. After vectorization, we had our data set that could be used by algorithms to build a machine learning model. Machine learning is the field of study that gives computers the ability to learn without explicitly programmed. We do that by training a model using data and an algorithm and then testing its accuracy using more data. To this end, we divided our dataset into different buckets to train and validate our model. In this project, we used the K-fold Cross Validation method to divide our data, in which we used tenfold Cross-Validation. The full data set was divided into 10 subsets and the holdout's method was repeated 10 times. Each time, 9 subsets were used to train the model and the tenth subset for testing it. Results were stored in an array and the method was repeated 10 times with different testing sets each time. In the end, the average of all test results was taken to come up with the final result.

While building our model, we tried five different machine learning algorithms and each algorithm was implemented twice using two different vectorization methods: Word Frequency and TF-IDF. As discussed earlier, not much research has been conducted to deter online antisocial behavior and therefore we felt the need to explore all the available machine learning algorithms to get an optimum result with our dataset. We were able to get high accuracy and precision using the following five algorithms: Logistic Regression, Support Vector Machine, Random Forest, Decision Tree, and Naïve Bayes. Results are discussed in the next section.

4   Results and Discussion

Online antisocial behavior is relatively a new area of research. When social media platforms such as Twitter and Facebook started getting traction, they bought in some of the issues along with them. Antisocial behavior is one of them. To the best to our knowledge, there hasn't been much work done to detect and prevent antisocial behavior online. There are studies on cyberbullying and trolling, which can fall under the umbrella term of anti-social, however, not much has been researched on the detection of other aspects of such behavior. By using natural language processing and machine learning techniques, we actually have done a reasonably good job at detecting all forms of antisocial behavior. Following are the results we got from trying five different classifiers and using count vectorization. The accuracy we got was quite similar with all the classifiers used. Precision, Recall, and F1 scores were also quite similar with all these algorithms (Table 1).

Table 1. Vectorization using word frequency feature method

Classifier	Feature	Accuracy	Precision	Recall	F1 Score
Logistic Regression	WF	99.76%	99.58%	99.66%	99.62%
Support Vector Machine	WF	99.82%	99.69%	99.73%	99.71%
Random Forest	WF	98.09%	99.20%	94.71%	96.90%
Decision Tree	WF	99.71%	99.51%	99.56%	99.54%
Naïve Bayes	WF	98.84%	98.88%	97.56%	99.04%

All five algorithms detected antisocial behavior with high accuracy and precision. A tweet that was classified as containing antisocial semantics was the one that contained some sort of swear and rude word in order to upset or annoy someone. Not all the tweets that were classified positive contained swear words. The sentiment, semantic, and context of the text was also taken into consideration in deciding whether the tweet represented antisocial behavior. While classifying, some of the tweets were at the borderline or represented more of sarcasm than antisocial behavior. Such tweets were eliminated and were not used. Since this is one of the first studies trying to detect online antisocial behavior in all its different forms, we wanted to keep the things simple for our algorithms and model. The tweets, on which we had doubt to whether to classify them as positive or negative, were eliminated from training and testing dataset. Since Most of our tweets were quite clearly positive or negative, it made the job of classifying algorithm much easier, as there were limited number of words and phrases that our model had to learn to distinguish between positive and negative tweets. This study can be further extended by adding text that is more complex to classify, even by some human standards. We assume that by adding those sorts of tweets will impact the accuracy and precision metrics, however, will enable our model to generalize better on any data set.

As mentioned above, we tried our classifiers with TF-IDF vectorization as well. The results are shown below and are very similar to the results that we got using the count vectorization method. Support Vector Machine, when used with count vectorization, showed the best result, however, Random Forest was better when TF-IDF vectorization method was used. Overall, we managed to get good results with both vectorization techniques and were able to detect antisocial behavior from Twitter with a high accuracy (Table 2).

Table 2. Vectorization using TF-IDF feature method

Classifier	Feature	Accuracy	Precision	Recall	F1 Score
Logistic Regression	TF-IDF	99.48%	99.64%	98.71%	99.17%
Support Vector Machine	TF-IDF	99.79%	99.77%	99.58%	99.67%
Random Forest	TF-IDF	97.76%	99.31%	94.14%	96.67%
Decision Tree	TF-IDF	99.64%	99.46%	99.40%	99.43%
Naïve Bayes	TF-IDF	93.97%	98.54%	81.55%	99.45%

The following charts show the similarities between accuracy, precision, recall and F1 score using two different vectorization techniques: Word Frequency and TF-IDF. In both cases the results are highly similar. Reasons for such similar results could be the size of the dataset and the pre-processing techniques that we used. In regards to the size of the data set, even though we had around 55,000 tweets, more tweets could have bought in more variations in the text data. In regards to the pre-processing techniques, we believe our stemmer did a good job truncating all the important words to their roots, assisting both the vectorizing techniques to perform well. As it can be seen from the charts below, Support Vector Machine and Logistic Regression performed the best and Naïve Bayes lagged behind in almost every measuring metrics. We propose the use of Support Vector Machine for our model based on its performance on our dataset and its overall credibility dealing with different types of datasets.

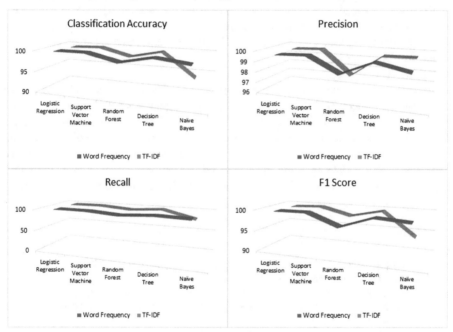

This study has some limitations and one of them is the size of our dataset. Our dataset contains around 55,000 tweets but a larger dataset could have bought in more diversity in regards to the feature words and phrases that our algorithms used to learn from. Secondly, we used around 30 different phrases to search for tweets. Once these tweets were collected, they were labeled as either antisocial or no antisocial. In our next study, we would like to increase the number of these phrases to at least 100. This is to bring in more diversity to words, phrases, contexts, semantics, and scenarios, used to train our classifier. The limited size of our dataset and the limited number of phrases used to search tweets may have contributed to such a high accuracy in classifying tweets. Nevertheless, the framework is a new approach and can be further refined and tuned to work with larger and diverse datasets and platforms.

5   Conclusions and Future Work

This research introduces a data-driven approach to detect and prevent antisocial behavior online. Social media platform Twitter has a responsibility to prevent its platform from becoming a breeding ground for antisocial behavior. Similarly, some other online platforms also enable the spread of antisocial semantics that plague the whole idea of freedom of speech online. It obstructs constructive discussion and leads to many users abandoning participation. At this stage, most of these platforms rely on users reporting such behavior. These platforms may have some measures in place to prevent antisocial behavior online, however, these are not as effective as they should be. We, in this research, proposed a framework based on natural language processing and machine learning techniques that can enable online platforms to proactively detect and restrict antisocial behavior. As can be seen by our results, our model can detect antisocial behavior on Twitter with high accuracy. The model can be integrated into an online system to depict such behavior on a live data stream. Once detected, appropriate action can be taken such as deleting the tweet or even blocking the user to prevent future incidents

In this study, we have explored data mainly from Twitter. Further studies can be conducted by collecting data from all sorts of online platforms. The diversity of data used will enable machine learning model to learn and perform better in real-world scenarios. Furthermore, we would like to explore other personality and behavior disorders that fall under the same category as antisocial behavior. Diagnostic criteria for these disorders overlaps in some instances and can present a challenge in training a model to classify and distinguish these disorders with high accuracy and precision.

References

1. A. P. Association: Diagnostic and statistical manual of mental disorders (DSM-5®). American Psychiatric Pub (2013)
2. Cheng, J., Danescu-Niculescu-Mizil, C., Leskovec, J.: Antisocial behavior in online discussion communities. In: Ninth International AAAI Conference on Web and Social Media, pp. 61–70 (2015)
3. Gard, A.M., Dotterer, H.L., Hyde, L.W.: Genetic influences on antisocial behavior: recent advances and future directions. Curr. Opini. Psychol. **27**, 46–55 (2018)
4. Flouri, E., Ioakeimidi, S.: Maternal depressive symptoms in childhood and risky behaviours in early adolescence. Eur. Child Adolesc. Psychiatry **27**(3), 301–308 (2018)
5. Woeckener, M., et al.: Parental rejection and antisocial behavior: the moderating role of testosterone. J. Crim. Psychol. **8**(4), 302–313 (2018)
6. McGuigan, W.M., Luchette, J.A., Atterholt, R.: Physical neglect in childhood as a predictor of violent behavior in adolescent males. Child Abuse Negl. **79**, 395–400 (2018)
7. Jackson, D.B.: The link between poor quality nutrition and childhood antisocial behavior: a genetically informative analysis. J. Crim. Justice **44**, 13–20 (2016)
8. Baskin-Sommers, A.R.: Dissecting antisocial behavior: the impact of neural, genetic, and environmental factors. Clin. Psychol. Sci. **4**(3), 500–510 (2016)
9. Meloy, J.R., Yakeley, A.J.: Antisocial personality disorder. A. A **301**(F60), 2 (2011)

10. Liu, P., Guberman, J., Hemphill, L., Culotta, A.: Forecasting the presence and intensity of hostility on Instagram using linguistic and social features. arXiv preprint arXiv:1804.06759 (2018)
11. Buckels, E.E., Trapnell, P.D., Paulhus, D.L.: Trolls just want to have fun. Pers. Individ. Differ. **67**, 97–102 (2014)
12. Shachaf, P., Hara, N.: Beyond vandalism: wikipedia trolls. J. Inf. Sci. **36**(3), 357–370 (2010)
13. Guberman, J., Hemphill, L.: Challenges in modifying existing scales for detecting harassment in individual tweets. In: Proceedings of the 50th Hawaii International Conference on System Sciences (2017)
14. Herring, S., Job-Sluder, K., Scheckler, R., Barab, S.: Searching for safety online: managing "trolling" in a feminist forum. Inf. Soc. **18**(5), 371–384 (2002)
15. Drouin, M., Miller, D.A.: Why do people record and post illegal material? Excessive social media use, psychological disorder, or both? Comput. Hum. Behav. **48**, 608–614 (2015)
16. Sest, N., March, E.: Constructing the cyber-troll: Psychopathy, sadism, and empathy. Pers. Individ. Differ. **119**, 69–72 (2017)
17. Singh, R., Zhang, Y., Wang, H.: Exploring human mobility patterns in Melbourne using social media data. In: Australasian Database Conference, pp. 328–335. Springer (2018)

Optimizing the Waiting Time for Airport Security Screening Using Multiple Queues and Servers

Mohamad Naji[1(✉)], Ali Braytee[2], Ali Anaissi[3], Omid Ameri Sianaki[1], and Ahmed Al-Ani[2]

[1] College of Engineering and Science, Victoria University, Sydney, Australia
mohamad.naji@vu.edu.au
[2] University of Technology Sydney, Ultimo, Australia
[3] The University of Sydney, Sydney, Australia

Abstract. Airport security screening processes are essential to ensure the safety of passengers and the aviation industry. Security at airports has improved noticeably in recent years through the utilisation of state-of-the-art technologies and highly trained security officers. However, maintaining a high level of security can be costly to operate and implement and can cause delays for passengers and airlines. In optimising a security process it is essential to strike a balance between time delays, security and reduced operation cost. This paper uses queueing theory as a method to study the impact of queue formation and the size of the security area on the average waiting time for the case of multi-lane parallel servers. An experiment is conducted to validate the proposed approach.

Keywords: Airport security screening process · Queueing theory · Queue formation

1 Introduction

The aviation industry has grown rapidly, as a preferred mode of passenger transportation. The number of passengers has been on a continuous increase in airports through the world. For example, in 2017 37 million passengers travelled through Sydney airport, and things will increase to approximately 74 million passengers by 2033. It estimated that the world wide number of passengers travelling by air will increase to reach 6 billion by 2033 [1]. This growth in the aviation industry and in numbers of passengers has made improving and tightening the security screening process essential to provide safety and comfort for passengers and airlines. Enhancing the security of the aviation industry and passengers has made the industry safer but is very costly [2]. Accordingly, it is important to strike a balance between increasing security and reducing processing time, cost and congestion. For this reason, researchers have used methodologies such as queueing theory and fuzzy reasoning to analyse, optimise and study the effects of congestion caused by security screening processes [1].

© Springer Nature Switzerland AG 2020
L. Barolli et al. (Eds.): CISIS 2019, AISC 993, pp. 496–507, 2020.
https://doi.org/10.1007/978-3-030-22354-0_44

Queueing theory was the first method used to optimise security screening process for a single server since 1970. [3] was the one of the first researchers to use and apply queueing theory to optimise the screening process of a single server. [2] proposed optimal static and dynamic policies for multilevel security systems to optimise the security process. [4, 5] introduced similar models and followed the concept of a Poisson input, exponential service time, and a single server, known as M/M/1, which followed the first-in-first-out (FIFO) service discipline for a multi-level single server. [6] sought to improve the security screening process by introducing a selectee lane and assigning to it passengers that may represent a threat.

Different authors have used fuzzy reasoning as a method of optimising the security screening process, as in the work of [7–9]. In the work of [10, 11], a model was introduced to enhance multi-criteria decision-making among a group of experts under critical circumstances to improve the decisions of security officers. [12] presented a fuzzy reasoning system to improve the efficiency of a baggage screening system with respect to uncertainty of machine (X-ray) and human factors.

Although the literature presents different methods of optimising the security process, to the best of our knowledge, no studies have investigated the impact of queue formation and the size of the security area on the average waiting time of passengers traversing security screening. Most of the aforementioned studies of airport security screening processes that used a queueing theory as a method have implemented approaches based on a single server, such as M/M/1 and M/G/1 (M/M/1 was introduced earlier and in M/G/1, where M stands for a system with Poisson arrivals, G is generally identically distributed, while 1 is the number of servers [13] and then multiplied the outcome by the total number of servers to determine the total average waiting time for the whole system. This assumption cannot be practically applied for many reasons, such as differences in human experience, knowledge of the security screening process and the existence of special needs passengers who require more processing time, passengers with slow movement and families with more passengers, as well as differences between machines such as X-ray machines and metal detectors. For all these reasons, we considered these issues in our model during the implementation phase in order to determine the average waiting time for a whole system with multiple servers. We consider an M/M/S/K queueing system with Poisson input, exponential service time, multiple servers and the buffer or area size (where M stands for a system with arrivals assumed to be a Poisson process, exponentially distributed service time, S number of servers, and K is the buffer size). To the best of our knowledge, this study represents the first attempt to explore the impact of queue formation on optimising the average waiting time for the airport security screening process. Furthermore, this study is considered to be the first attempt to use multiple parallel servers that are implemented in different scenarios to replicate airport security screening real-life settings.

The rest of this paper is structured as follows. Section 2 gives brief review of queueing theory. Section 3 illustrates the security screening procedure and the model description, and presents the implementation of the proposed model. Section 4 illustrates the results and discussion, followed by the conclusion.

2 Brief Review of Queueing Theory

Queueing theory is an approach based on mathematical formalisation and probability estimates. It was invented primarily to deal with computer system and call applications, but has also been used in other real-life queuing applications in contexts such as airports, trains, hospitals, banks and supermarkets. Queuing theory aims to analyse and determine the length of queues, service time, average waiting and sojourn time, as well as other system characteristics. Depending on the application, some queueing theory values must be predetermined, such as average arrival rate (denoted by λ), average service rate (denoted by μ), service and arrival type, number of servers or channels, buffer size and queue discipline [14].

The behaviour of a queueing system was described by Kendall in 1953, who described a queueing model that includes the service and arrival type, the number of servers, buffer capacity and service discipline, denoted as follows: A/B/C/D/E, where A is inter-arrival time distribution, B is service time distribution, C is number of servers, D is buffer size, E is the discipline, such as FIFO (first-in-first-out) or LIFO (last-in-first-out) [13, 15].

3 Methodology

In this work, we considered three different scenarios for Q (Q = 1, 2 and 3) and estimated the average waiting time per passenger for each scenario.

3.1 Security Screening Procedure

Passenger security screening process procedures vary from country to country, and may be determined by the airport administration or based on government laws. For example, in Sydney airport when passengers have checked in and completed the immigration process, they go through the security process, where they are allowed to carry only a small bag known as a carry-on bag or hand luggage. This bag should not exceed a certain weight, as decided by the airline regulations and should contain less than 100 ml of liquid. Different types of machine are used in various airports to ensure that no prohibited or illegal items are being carried. These include traditional X-ray machines, CT (computed tomography) machines which can display bags in 3-D and allow 360-degree rotations, and metal detector devices that are used to scan passengers' bodies [16–18].

Due to the variability of service distribution and number of servers, it should be clear that the assumptions of M/M/1 or M/G/1 for multiple-servers cannot be applied for the security screening process, so in our implementation we take into consideration multiple servers in parallel with different service distributions, i.e., the M/M/S queueing theory assumption (where M represents a system with Poisson arrivals, exponentially distributed service time and S is the number of servers), for applicability to real-life airport security settings.

In this model, queueing theory and M/M/S/K (where M represents a system with Poisson arrivals and exponentially distributed service time, S is the number of servers and K is the buffer size) discipline are used to build the system and study the impact of dividing the main buffer into multiple buffers and different buffer sizes for S parallel servers, to optimise throughput by reducing the average waiting time for passengers going through the security screening process.

In general, for M/M/S, the fraction of time a server is assumed to be busy can be expressed as:

$$C(s, \alpha) = \frac{\frac{\alpha^s}{s!\left(1-\frac{\alpha}{S}\right)}}{\sum_{k=0}^{s-1}\frac{\alpha^k}{k!} + \frac{\alpha^s}{s!\left(1-\frac{\alpha}{S}\right)}} \tag{1}$$

while the average waiting and response (sojourn) times respectively are:

$$E(W) = \frac{C(s, a) * \mu}{(1 - \rho)s} \tag{2}$$

$$E(R) = E(W) + \mu = \frac{C(s, a) * \mu}{(1 - \rho)s} + \mu \tag{3}$$

3.2   Queue Formation and Model Description

In general, most airports use a single passenger queue for the security screening process. In this study, however, we consider three queuing scenarios: single, two and three queues. Also, we investigate their impact on passenger waiting time. The scenarios of our queue formation are presented as follows:

The numbers of servers allocated for the queues are as follows, (i) one queue: all fourteen servers; (ii) two queues: each queue is allocated seven servers; (iii) three queues: the first two queues are allocated five servers each and the third queue is allocated four servers.

(i) First scenario: the total area is utilised to form a single queue and the passenger at the top of the queue is allocated to one of the S parallel servers, as shown in Fig. 1.

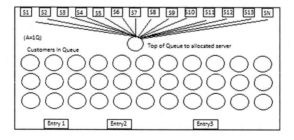

Fig. 1. First queueing scenario: One queue

(ii)   Second scenario: the total area is divided into two equal queues, i.e., Q1 = Q2 = A/2 and the number of servers allocated to each queue is S/2, as shown in Fig. 2.

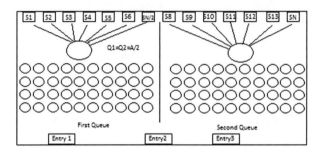

Fig. 2. Second queueing scenario: Two queues

(iii)   Third scenario: the total area is divided into three queues. The total area A is divided into three queues or buffers, and then the size of each buffer is based on the number of servers allocated to it, as shown in Fig. 3.

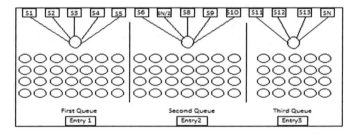

Fig. 3. Third queueing scenario: Three queues

Consider a multi-server parallel queueing system to service passengers according to the first-come-first-served (FCFS) concept. The arrival of passengers is assumed to follow a Poisson distribution, with the mean arrival rate of λ and successive mean time between arrivals of $1/\lambda$. Depending on the queue formation, a passenger is assigned by the security officers to join a certain queue (1, 2 or 3), based on available free space. Passengers are expected to be served by one of the S servers according to an exponential distribution with mean service rate of μ and service completion rate of $1/\mu$, while the capacity of the area is assumed to be finite.

In general, $P_N Q_1 S_i =$ number of passengers being served by server i in first queue. The total number of passengers served in the system (all queues) is

$$P_N Q_{total} = P_N Q_1 + P_N Q_2 + \cdots + P_N Q_N \tag{4}$$

The probability of blocking is defined as the chance that a customer will lose service due to high demand and lack of resources. For example, the probability of blocking of 0.01 means that 1% of customers will be denied or will lose service. The probability of blocking for different numbers of queues for M/M/S/Q (where M stands for a system with Poisson arrivals, exponentially distributed service time, S stands for number of servers and Q is the number of queues) is

$$P_Q = \sum_{q=1}^{Q} \sum_{r=1}^{S} \frac{\left(1 - \frac{\lambda\mu}{r}\right)\left(\frac{\lambda\mu}{r}\right)^q}{\left(1 - \frac{\lambda\mu}{r}\right)^{q+1}} \tag{5}$$

Equation (5) illustrates that the probability of blocking will decrease when the number of queues increases. In this model, queueing theory is used to build a system that follows the M/M/S/K concept. A modified Lindley equation (Eq. (7)) is used to determine the average waiting time of N passengers.

The Lindley Eq. (6) or Lindley process is based on a discrete time stochastic process, which can be used to describe the evolution of a queue length over time or to determine the average waiting time of passengers in a queue. For example, the first arriving passenger does not need to wait, so $W_n = 0$. Subsequent passengers will need to wait if they arrive at a time before the previous passengers are served [19].

The Lindley equation is used in this work due to its simplicity and suitability for the security area queue formation. The simplicity is manifested by the applicability of a single equation for determining the waiting and sojourn time of passengers. Also, Eqs. (1) and (2) do not take into consideration the difference in arrival times IA and the execution time of the previous passenger in determining the waiting time of passengers. The Lindley equation is

$$W_{n+1} = W_n + X_n - IA_n \tag{6}$$

where W_n is the waiting time of the previous passenger, X_n is the service time of the previous passenger; IA_n is the time between the previous and current arrival (where both passengers are in service or waiting mode). Figure 4 describes the concept where T_n is the arrival time of passenger n and T_{n+1} is the arrival time of passenger n + 1.

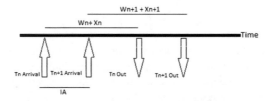

Fig. 4. The concept of the Lindley equation

To be more applicable for queueing processes such as airport security screening and to produce accurate results, another parameter such as walking time must be incorporated in the formula. Thus, the modified formula is

$$W_{n+1} = W_n + X_n + W_{walking} - IA_n \tag{7}$$

where $W_{walking}$ is the time required by a passenger to walk when the passenger comes to the top of the queue and is the first to be served.

4 Experimental Results and Discussion

The following section present the experimental setting follows by the result.

4.1 Experimental Settings

As in most implementations, some parameters must be predetermined or initialised before the simulation. These include service time (μ), number of arrivals, time between arrivals ($1/\lambda$), size of the security area, number of servers available (S) and finally walking time. According to [20] the time required for a passenger to complete the security screening process varies between 15 and 60 s and is 25 s on average, so a time of 25 s is chosen as the mean average of the exponentially distributed service time. The number of passengers is initially considered to start at 500 and then sequentially increased by 500 passengers to reach a maximum of 8000. The inter-arrival rate between two consecutive passengers is exponentially distributed at the rate of 1 s. Small, medium and large security areas are considered, i.e., A = 50, 75 and 100 m^2, and the total number of servers S is assumed to be 14 and they are available for all scenarios. Finally, the walking speed of a passenger with a carry-on bag who is at the top of the queue and the first to be served is assumed to be 1.5 seconds per metre.

In this implementation four different cases are used:

1. Normal variation: the service time is assumed to be exponentially distributed with average service mean of 25 s applied to all servers.
2. Extra variation and randomness: the service times for the various servers and passengers are completely different, which could be close to the real case in airport security areas. This case suits the M/M/S implementation that assumes iid (independent and identically distributed) service time. This case is implemented using the equation:

$$(1 + rand)(exprnd(service_{time})) \tag{8}$$

where the function rand is used to generate random numbers whose elements are uniformly distributed in the interval between 0 and 1, and exprnd generates random numbers from the exponential distribution with mean service_time which is 25 s, based on the random exponential distribution of service time.

3. The first three lanes are slow lanes (for special needs passengers), while the rest are normal: in most airports there are a few security lanes that are used by families with children and passengers with special needs. Such passengers may require triple the service time of other passengers. The service time of passengers in the first three lanes is expressed by the equation:

$$((1 + rand) * 2) * (exprnd(service\_time)) \qquad (9)$$

4. The first lane is an express lane (for crew and staff), the last three lanes are slow, and the remaining lanes (or lanes in between) are normal: The fast lane is implemented using the equation:

$$((1 + rand) * 0.1) * (exprnd(service\_time)) \qquad (10)$$

The middle 10 lanes (lanes 2 to 11) are as described in the first case and are considered normal lanes. Lanes 12 to 14 are considered to be slow and implemented as described in case 3.

The three different sized areas are applied in each case to study the impact of this implementation on the average waiting time. The aim of the four different cases in this model is to represent the variations in service time, and therefore the parameters of the model are chosen to reflect the cases discussed above. The variation in service time is applied in this implementation for several reasons:

Firstly, knowledge of the passenger screening process is different from one passenger to another, and is influenced by previous travel experience. According to [21], this affects the service time. Secondly, efficacy can vary from one machine to another [16, 22, 23]. Thirdly, security officers vary in their experience, which could affect the time needed to screen luggage (service time). Screeners should be able to interpret images from the scanners and distinguish between different items in a short period of time, to know the content of the bag without opening it, and to check whether the carry-on luggage contains any threat. Accordingly, we use an exponentially distributed service time with a mean of 25 s for all servers.

4.2   Results

We first consider the small area size of 50 m^2 and consider all three queuing scenarios of 1, 2 and 3 queues. Those three scenarios are applied for each of the four server cases described in the previous section. Figures 5, 6, 7 and 8 show the average waiting time when the number of passengers is varied between 500 and 8000. The figures show that the average waiting time per passenger for the three-queue formation is less than that of the two-queue, which in turn is less than that of the one-queue formation. This indicates that the average waiting time per passenger decreases when the number of queues increases, but the enhancement in waiting time between Q2 and Q3 is less than that between Q1 and Q2 in all cases.

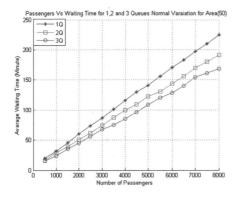

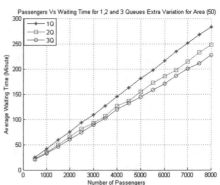

Fig. 5. Normal variation - Area 50 m^2

Fig. 6. Extra variation - Area 50 m^2

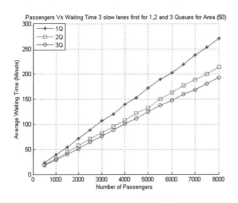

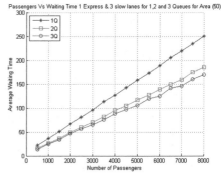

Fig. 7. Three slow lanes first - Area 50 m^2

Fig. 8. First lane express & last three lanes slow - Area 50 m^2

Figures 5, 6, 7 and 8 show that when the number of passengers increases from 500 to 8000 and the area size increases too, such as from 50 to 75 then 100 m^2, the average waiting time per passenger increases respectively. But if the size of the area is doubled from 50 to 100 m^2, the average waiting increases for the larger value only at the beginning, until the number of passengers reaches the maximum capacity of the area (saturation value), and then the increase becomes less for all scenarios. For example, for the first scenario, when the area size is 50 m^2 and the passengers numbers 500, 5000 and 8000, the values of the average waiting time per passenger for the one-queue formation (1Q) are 19.9, 140.8 and 224.4 respectively, whereas if the size of the area changes to 100 m^2, the values are 34.4, 145.2 and 228.3. This increment in average waiting time looks very obvious at the beginning but subsequently becomes slower. This finding is similar in all the different cases. The impact of area size on average waiting time is due to the following:

Firstly, the waiting time of $n + 1$ passenger is the cumulative sum of the previous n passengers, according to the Lindley equation or process (7).

Secondly, larger security areas can accommodate more passengers.

Thirdly, the walking time required for a passenger to reach the top of the queue is greater.

As shown in Figs. 5, 6, 7 and 8, there is a noticeable difference between the graph of one queue and that of two and three queues. This could be justified by the walking time that passengers need to spend to reach the service lanes from the top of the queue. This time could be noticeably higher in the case of one queue, as the distance to the furthest service lanes is much greater than in the case of two queues. However, the differences in distance (and consequently in walking time) begin to decrease as the number of queues increases, which justifies the smaller difference between the waiting times of two queues compared to three queues. From this result, we believe that there will be no noticeable improvement if we increase the number of queues from three to four, and this is applicable in all scenarios, depending on the size of the security area and the number of servers.

The graphs look linear, mainly because we use the same formula for the average service time, which is independent of the number of passengers, and we use a linear equation (Eq. (7)) to determine the average waiting time. Hence, when the number of passengers is doubled, the time is approximately doubled. This is applicable to all the graphs of 1, 2 and 3 queues.

Our results shows that, the number of passengers and average waiting time per queue for each individual case, each queue formation, area size of 50 m^2 and 8000 passengers. In all cases the number of passengers and the time required to be served for different queues differs due to varying service times.

For example in the first case and the two-queue formation, the total number of passengers served per queue is 4000 passengers, while the estimated service times for the first and second queues respectively are 95.29 and 94.4. For the three-queue formation, the numbers of passengers for each queue are 2672, 2671 and 2657 respectively, and the average service time per queue is 53.43, 54.52 and 62.07 min respectively.

However, in the fourth case and the two-queue formation, the two queues serve the same number of passengers in less time than that required by the first queue, whereas for the three-queue formation the first queue serves more people in less time. This is because the first lane is a fast lane (express lane) and it has been allocated to the first queue for both queue formations, which require less service time. Meanwhile, because the last three lanes are slow lanes, the second queue for the two-queue formation and the third queue for the three-queue formation require longer service times. This complexity adds value to this model as it reflects the real screening process scenario in most airports.

It is worth mentioning that we have chosen the model parameters that include service time, number of passengers and the time difference between arrivals to be as close as possible to real-life values. The service time formula will be verified when we collect real airport data.

The fourth case (one express lane at one end, three slow lanes at the other end and normal lanes in between) produces a noticeable improvement in average waiting time per passenger for the three-queue formation compared to that of one and two queues. It is worth mentioning that this case is practical for many airports, as it is necessary to have a fast lane for crew, staff and business passengers, slow lanes for passengers with special needs and normal lanes for other passengers.

5 Conclusion

Airport security screening is essential to provide safety for the aviation industry and passengers travelling by air. Experienced officers and advanced technological instruments help to enhance security. However, the delay caused by security screening remains a concern. Technical methods have been proposed to model and optimise this process. Despite the achieved improvements, delays still present a major concern. This paper used queueing theory to build a system with multi-lane parallel servers to study the impact of queue formation and area size on the average service time. Unlike existing methods which do not consider variations between the different servers, our proposed method successfully takes this issue into consideration. This implementation will be evaluated in the future with real data that we plan to collect. Further studies are required to develop new methods which address the limitations of existing methods and the theory behind them, by taking into consideration human and machine factors, to determine the optimal numbers of servers or required security officers, as well as the ideal area for the screening process.

References

1. Naji, M., Abdelhalim, S., Al-Ani, A., Al-Kilidar, H.: Airport security screening process: a review. In: CICTP 2017: Transportation Reform and Change-Equity, Inclusiveness, Sharing, and Innovation-Proceedings of the 17th COTA International Conference of Transportation Professionals, January 2018
2. Lee, A.J., Jacobson, S.H.: The impact of aviation checkpoint queues on optimizing security screening effectiveness. Reliab. Eng. Syst. Saf. **96**(8), 900–911 (2011)
3. Gilliam, R.R.: An application of queueing theory to airport passenger security screening. Interfaces **9**(4), 117–123 (1979)
4. Olapiriyakul, S., Das, S.: Design and analysis of a two-stage security screening and inspection system. J. Air Transp. Manage. **13**(2), 67–74 (2007)
5. Leone, K., Liu, R.R.: Improving airport security screening checkpoint operations in the US via paced system design. J. Air Transp. Manage. **17**(2), 62–67 (2011)
6. Nie, X., et al.: Simulation-based Selectee Lane queueing design for passenger checkpoint screening. Eur. J. Oper. Res. **219**(1), 146–155 (2012)
7. Skorupski, J., Uchroński, P.: A fuzzy model for evaluating airport security screeners' work. J. Air Transp. Manage. **48**, 42–51 (2015)
8. Skorupski, J., Uchroński, P.: Fuzzy inference system for the efficiency assessment of hold baggage security control at the airport. Saf. Sci. **79**, 314–323 (2015)
9. Skorupski, J., Uchroński, P.: A fuzzy system to support the configuration of baggage screening devices at an airport. Expert Syst. Appl. **44**, 114–125 (2016)
10. Skorupski, J.: Multi-criteria group decision making under uncertainty with application to air traffic safety. Expert Syst. Appl. **41**(16), 7406–7414 (2014)

11. Skorupski, J., Uchroński, P.: Managing the process of passenger security control at an airport using the fuzzy inference system. Expert Syst. Appl. **54**, 284–293 (2016)
12. Skorupski, J., Uchroński, P.: A fuzzy reasoning system for evaluating the efficiency of cabin baggage screening at airports. Transp. Res. Part C: Emerg. Technol. **54**, 157–175 (2015)
13. Cooper, R.B.: Introduction to Queueing Theory. Amsterdam, North Holland (1981)
14. Avi-Itzhak, B., Levy, H., Raz, D.: Quantifying fairness in queueing systems: principles and applications. Preprint (2004)
15. Sztrik, J.: Basic Queueing Theory, p. 193. University of Debrecen, Faculty of Informatics (2012)
16. Almazroui, S., Wang, W., Zhang, G.: Imaging technologies in aviation security. Adv. Image Video Process. **3**(4), 12 (2015)
17. Popovic, V., Kraal, B.J., Kirk, P.J.: Passenger experience in an airport: an activity-centred approach. In: IASDR 2009 Proceedings, pp. 1–10 (2009)
18. Blalock, G., Kadiyali, V., Simon, D.H.: The impact of post-9/11 airport security measures on the demand for air travel. J. Law Econ. **50**(4), 731–755 (2007)
19. Asmussen, S.: Applied Probability and Queues, vol. 51. Springer Science & Business Media, New York (2008)
20. Kirschenbaum, A.A.: The cost of airport security: the passenger dilemma. J. Air Trans. Manage. **30**, 39–45 (2013)
21. Babu, V.L.L., Batta, R., Lin, L.: Passenger grouping under constant threat probability in an airport security system. Eur. J. Oper. Res. **168**(2), 633–644 (2006)
22. Wang, X., Zhuang, J.: Balancing congestion and security in the presence of strategic applicants with private information. Eur. J. Oper. Res. **212**(1), 100–111 (2011)
23. Martín-Cejas, R.R.: Tourism service quality begins at the airport. Tour. Manag. **27**(5), 874–877 (2006)

Video Classification Using Deep Autoencoder Network

Farshid Hajati[1,2(✉)] and Mohammad Tavakolian[3]

[1] College of Engineering and Science, Victoria University Sydney,
Sydney, Australia
farshid.hajati@vu.edu.au
[2] School of Information Technology and Engineering, MIT Sydney,
Sydney, Australia
fhajati@mit.edu.au
[3] Center for Machine Vision and Signal Analysis (CMVS),
University of Oulu, Oulu, Finland
mohammad.tavakolian@oulu.fi

Abstract. We present a deep learning framework for video classification applicable to face recognition and dynamic texture recognition. A Deep Autoencoder Network Template (DANT) is designed whose weights are initialized by conducting unsupervised pre-training in a layer-wise fashion using Gaussian Restricted Boltzmann Machines. In order to obtain a class specific network and fine tune the weights for each class, the pre-initialized DANT is trained for each class of video sequences, separately. A majority voting technique based on the reconstruction error is employed for the classification task. The extensive evaluation and comparisons with state-of-the-art approaches on Honda/UCSD, DynTex, and YUPPEN databases demonstrate that the proposed method significantly improves the performance of dynamic texture classification.

1 Introduction

Classification and recognition tasks in different applications have been one of the interesting topics in recent years [1–8]. Videos contain dynamic textures which can be described as a visual process including a group of elements with random motions. Video dynamics widely exist in real-world video data, e.g. regular rigid motion like windmill, chaotic motion such as smoke and water turbulences, and sophisticated motion caused by camera panning and zooming. The modeling of video dynamics is challenging but very important for subsequent vision tasks such as video classification, dynamic texture synthesis, motion segmentation, and so on.

Despite all challenges, great efforts have been devoted to find a robust and powerful solution for video-based recognition. Furthermore, it is commonly substantiated that effective representation of the video content is a crucial step towards resolving the problem of dynamic texture classification. In the past decade, a large number of approaches for video representation have been proposed, e.g. Linear Dynamic System (LDS) based methods [9], Local Binary Pattern (LBP) based methods [10], and Wavelet based methods [11]. Unfortunately, the current methods are sensitive to

© Springer Nature Switzerland AG 2020
L. Barolli et al. (Eds.): CISIS 2019, AISC 993, pp. 508–518, 2020.
https://doi.org/10.1007/978-3-030-22354-0_45

undesirable external phenomena. Coupled with these drawbacks, other methods frequently model the video information within consecutive frames on a geometric surface so that it is represented by a subspace [12], a combination of subspaces [13], a point on the Grassmann manifold [14], or Lie Group of Riemannian manifold [15]. This requires prior assumptions regarding specific category of the geometric surface on which the samples of the video are believed to lie on.

A growing body of literature has investigated numerous approaches for video classification [11, 16]. Among them, Local Binary Pattern (LBP) based methods [10] have been widely used in texture analysis. Zhao et al. [17] extended the LBP to both space and time domains and proposed two new variants. Volume Local Binary Pattern (VLBP) [17] is an extension which combines both spatial and temporal variations of the video. Also, Local Binary Pattern on Three Orthogonal Planes (LBP-TOP) [17] computes LBP of three individual $x - y$, $x - t$, and $y - t$ planes to describe the video.

Recently, there is a huge growing research interest in deep learning methods in different areas of computer vision [18, 19]. Deep learning methods set up numerous recognition records in image classification [20], object detection [21], face recognition and verification [22]. Deep models have much more expressive power than traditional shallow models and can be effectively trained with layer-wise pre-training and fine-tuning [18]. Xie et al. [23] represented relationship between noisy and clean images using stacked denoising autoencoders. However, deep autoencoders are rarely used to model time series data. Despite this, there have been some works on using variants of Restricted Boltzmann Machine (RBM) [24] for specific time series data such as human motion [25]. Some other deep models have been put forward to address video data with convolutional learning of spatiotemporal features [26]. Needless to say that learning deep frameworks require huge amount of training data and is quite costly in computational demand. As a result, the deprivation of training data is indeed an obstacle to deploy a deep model for video classification tasks.

This paper presents a novel deep learning framework which makes no prior assumptions with respect to the underlying geometry and explore automatically the structure of the complex non-linear surface on which samples of video are present. The proposed method (see the block diagram in Fig. 1) first defines a Deep Autoencoder Network Template (DANT) whose weights are initialized with an unsupervised layer-wise pre-training using Gaussian Restricted Boltzmann Machines (GRBM). In order to learn class specific Deep Autoencoder Network (DAN), the initialized DANT is then separately trained for each class using all videos of that class. Therefore, DANs can represent videos of each class based on the learnt structure of the corresponding class. A query video is represented using the learnt class specific DANs for classification purpose. The representation errors from the respective DANs are then computed and a voting technique is used to decide which class the query video shall belong to.

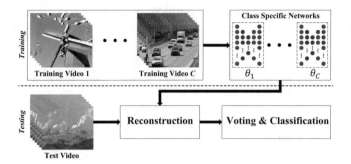

Fig. 1. The block diagram of the proposed method.

2   Deep Autoencoder Network

We first define a Deep Autoencoder Network Template (DANT) which will be used to learn the underlying structure of the data. The architecture of the DANT is shown in Fig. 2. For such a deep network, an appropriate parameter initialization is mandatory to achieve a good performance. Therefore, we initialize the parameters of DANT by performing a pre-training in a greedy layer-wise fashion using Gaussian Restricted Boltzmann Machines. The DANT with initialized parameters is then separately fine-tuned for each of the C classes of the training videos. We therefore end up with a total of C fine-tuned class-specific Deep Autoencoder Networks (DANs). The fined-tuned models are then used for video classification.

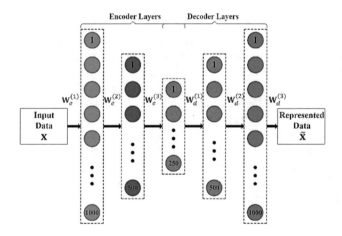

Fig. 2. The structure of the proposed Deep Autoencoder Network (DAN).

2.1 The Deep Autoencoder Network Template

As can be seen in Fig. 2, the proposed DANT is based on an autoencoder which comprises an encoder and a decoder. Both the encoder and the decoder have three hidden layers, with a shared third layer. The encoder calculates a compact low dimensional representation of the input data. We can formulate the encoder as a combination of non-linear functions $s(\cdot)$ to map the input data $\mathbf{x}$ to a representation $\mathbf{h}$ given by

$$
\begin{aligned}
\mathbf{h}_1 &= s\left(\mathbf{W}_e^{(1)}\mathbf{x} + \mathbf{b}_e^{(1)}\right) \\
\mathbf{h}_2 &= s\left(\mathbf{W}_e^{(2)}\mathbf{h}_1 + \mathbf{b}_e^{(2)}\right) \\
\mathbf{h} &= s\left(\mathbf{W}_e^{(3)}\mathbf{h}_2 + \mathbf{b}_e^{(3)}\right)
\end{aligned}
\tag{1}
$$

where $\mathbf{W}_e^{(i)} \in \mathbb{R}^{n_i \times n_{i-1}}$ is the encoder weight matrix for layer i having n_i nodes, $\mathbf{b}_e^{(i)} \in \mathbb{R}^{n_i}$ is the bias vector and $s(\cdot)$ is a non-linear sigmoid activation function.

The encoder parameters are learnt by combining the encoder with the decoder and jointly training the encoder-decoder structure to represent the input data by optimizing a cost function. So the decoder can be defined as a series of non-linear functions which calculate an approximation of the input $\mathbf{x}$ from the encoder output $\mathbf{h}$. The approximated output $\tilde{\mathbf{x}}$ of the decoder is obtained by

$$
\begin{aligned}
\mathbf{x}_1 &= s\left(\mathbf{W}_d^{(1)}\mathbf{h} + \mathbf{b}_d^{(1)}\right) \\
\mathbf{x}_2 &= s\left(\mathbf{W}_d^{(2)}\mathbf{x}_1 + \mathbf{b}_d^{(2)}\right) \\
\tilde{\mathbf{x}} &= s\left(\mathbf{W}_d^{(3)}\mathbf{x}_2 + \mathbf{b}_d^{(3)}\right)
\end{aligned}
\tag{2}
$$

We represent the complete encoder-decoder structure by its parameters $\theta_{DANT} = \{\theta_{\mathbf{W}}, \theta_{\mathbf{b}}\}$, where $\theta_{\mathbf{W}} = \left\{\mathbf{W}_e^{(i)}, \mathbf{W}_d^{(i)}\right\}_{i=1}^{3}$ and $\theta_{\mathbf{b}} = \left\{\mathbf{b}_e^{(i)}, \mathbf{b}_d^{(i)}\right\}_{i=1}^{3}$.

2.2 DANT Parameter Initialization

The above defined DANT is used to learn class specific networks. This is done by individual training of the DANT for videos of each class in the training set. The training is performed through stochastic gradient descent with back propagation [27]. The training may fail if the DANT is initialized by inappropriate weights. Thus, the parameters of the template are firstly set up by performing unsupervised pre-training. For this purpose, a greedy layer-wise approach is adopted and Gaussian RBMs are used.

RBMs [24] are generative undirected graphical models with a bipartite structure of two sets of binary stochastic nodes called the visible ($\{v_i\}_{i=1}^{N_v}, v_i \in \{0, 1\}$) and the hidden layer nodes ($\left\{h_j\right\}_{j=1}^{N_h}, h_j \in \{0, 1\}$). The visible layer's nodes are symmetrically connected with the nodes of the hidden layer through a weight matrix $\mathbf{W} \in \mathbb{R}^{N_h \times N_v}$, but there are no intra-layer node connections. The joint probability $p(\mathbf{v}, \mathbf{h})$ of the RBM is given by

$$p(\mathbf{v}, \mathbf{h}) = \frac{1}{Z} \exp(-E(\mathbf{v}, \mathbf{h})) \tag{3}$$

where Z is the partition function used as a normalization constant. $E(\mathbf{v}, \mathbf{h})$ is the energy function of the model defined as

$$E(\mathbf{v}, \mathbf{h}) = \sum_i b_i v_i - \sum_j c_j h_j - \sum_{ij} w_{ij} v_i h_j \tag{4}$$

where $\mathbf{b}$ and $\mathbf{c}$ are the biases of the visible and the hidden layers, respectively. In order to learn the model parameter $\{\mathbf{W}, \mathbf{b}, \mathbf{c}\}$ of the RBM, the training is performed by the numerical technique of Contrastive Divergence (CD) [28].

We can extend the standard RBM, which is used for binary stochastic data, to the real value data by appropriate modifications in its energy function. Gaussian RBM (GRBM) is one of such popular extensions whose energy function is defined by changing the bias term of the visible layer.

$$E_{GRBM}(\mathbf{v}, \mathbf{h}) = \sum_i \frac{(v_i - b_i)^2}{2\sigma_i^2} - \sum_j c_j h_j - \sum_{ij} w_{ij} \frac{v_i}{\sigma_i} h_j \tag{5}$$

where σ_i is the standard deviation of the real valued Gaussian distributed inputs to the visible node v_i. It is possible to learn σ_i for each visible unit but it become staggering when using CD for GRBM parameter learning. Alternatively, we use another approach and fix σ_i to a constant value in the data pre-processing phase.

Inasmuch as there are no intra-layer node connections, inference becomes readily tractable for the RBM to the contrary of most directed graphical models. The probability distributions for GRBM are given by

$$p(h_j|v) = s\left(\sum_i w_{ij} v_i + c_j\right) \tag{6}$$

$$p(v_i|h) = \frac{1}{\sigma_i \sqrt{2\pi}} \exp\left(\frac{-(v_i - u_i)^2}{2\sigma_i^2}\right) \tag{7}$$

where

$$u_i = b_i + \sigma_i^2 \sum_j w_{ij} h_j \tag{8}$$

Since our data are represented by real values, we use GRBMs to initialize the parameters of the proposed DANT. Two layers are considered at a time and the GRBM parameters are leant. At first, we assume the nodes of the input layer as the visible units

v. Thus, the nodes of the first hidden layer are considered as the hidden unit **h** of the first GRBM and the parameters are tuned. The activations of the first GRBM's hidden units are then used as an input to train the second GRBM. The process is repeated for all three hidden layers of the encoder. The weights learnt for the encoder layers are then tied to the corresponding decoder layers.

3   Video Classification

In this section, we first introduce the formulation and describe how to classify query videos using the representation error. Assume there are C training videos $\{\mathbf{X}_c\}_{c=1}^{C}$ with the corresponding class labels $y_c \in \{1, 2, \cdots, C\}$. Notice that a video sequence is denoted by $\mathbf{X}_c = \{\mathbf{x}^{(t)}\}_{t=1}^{T}$, where $\mathbf{x}^{(t)}$ contains raw pixel values of the frame at time t. The problem is assigning class y_q to the query video sequence $\mathbf{X}_q$.

3.1   Learning Class Specific Network

In order to initialize the parameters of the DANT using GRBMs, we randomly shuffle a small fraction of the training video sequences such that there are video sequences from all classes in this subset. We use this subset for layer-wise GRBM training of all encoder's layers. The parameters of the decoder layers are then configured with their corresponding tied parameters of the encoder layers. This process assures that the proposed network rarely gets stuck in a local minimum point.

At this point, the DANT structure with the initialized weights is trained to learn class specific DANs. Here, the training of a DAN, θ_c, is carried out by minimization of the representation error over all frames $\mathbf{x}^{(t)}$ of the video $\mathbf{X}_c$.

$$J\left(\theta_{DANT}|\mathbf{x}^{(t)} \in \mathbf{X}_c\right) = \sum_t \left\|\mathbf{x}^{(t)} - \tilde{\mathbf{x}}^{(t)}\right\|^2 \tag{9}$$

where $\tilde{\mathbf{x}}^{(t)}$ is the t-th reconstructed frame of the video $\mathbf{X}_c$.

In order to avoid over-fitting and enhance generalization of the learnt model to unknown test data, the regularization terms are added to the cost function of DANT. A weight decay penalty term J_{wd} and a sparsity constraint J_{sp} are added.

$$J_{reg}\left(\theta_{DANT}; \mathbf{x}^{(t)} \in \mathbf{X}_c\right) = \sum_t \left\|\mathbf{x}^{(t)} - \tilde{\mathbf{x}}^{(t)}\right\|^2 + \lambda_{wd}J_{wd} + \lambda_{sp}J_{sp} \tag{10}$$

where λ_{wd} and λ_{sp} are regularization parameters. J_{wd} ensures small values of weights for all hidden units. It is defined as the summation of the Frobenius norm of all weight matrices.

$$J_{wd} = \sum_{i=1}^{3} \left\|\mathbf{W}_e^{(i)}\right\|_F^2 + \sum_{i=1}^{3} \left\|\mathbf{W}_d^{(i)}\right\|_F^2 \tag{11}$$

where $\mathbf{W}_e^{(i)}$ and $\mathbf{W}_d^{(i)}$ are the weight matrices of the i-th layer of the encoder and the decoder, respectively.

Moreover, J_{sp} enforces that the mean activation $\bar{\rho}_j^{(i)}$ (over all training samples) of the j-th unit of the i-th hidden layer is as close as possible to a spartisty target ρ which is a very small value constant. J_{sp} is further defined regarding the KL divergence.

$$J_{sp} = \sum_{i=1}^{5}\sum_{j} \rho \log \frac{\rho}{\bar{\rho}_j^{(i)}} + (1 - \rho) \log \frac{1 - \rho}{1 - \bar{\rho}_j^{(i)}} \tag{12}$$

So, a class specific model θ_c is obtained by optimizing the regularized cost function J_{reg} over all frames of the class X_c. Since the sigmoid activation functions are non-linear and a number of layers are joined together, the autoencoder structure is capable of learning very intricate non-linear structures.

$$\theta_c = \arg\min_{\theta_{DANT}} J_{reg}\left(\theta_{DANT}|\mathbf{x}^{(t)} \in X_c\right) \tag{13}$$

3.2   Classification

Given a query video sequence $\mathbf{X}_q = \left\{\mathbf{x}^{(t)}\right\}_{t=1}^{T_q}$, we separately reconstruct it from all class specific DANs θ_c, $c = 1, \cdots, C$, using Eqs. (1) and (2). Suppose $\tilde{\mathbf{x}}_c^{(t)}$ is the t-th frame of the reconstructed query video sequence $\tilde{\mathbf{x}}_{q_c}$ based on the c-th class model θ_c. After computing the reconstruction errors for all C classes, the vote $v^{(t)}$ is assigned to the class whose network has reconstructed the frame $\mathbf{x}^{(t)}$ with the minimum reconstruction error.

$$v^{(t)} = \arg\min_{c}\left\|\mathbf{x}^{(t)} - \tilde{\mathbf{x}}_c^{(t)}\right\|_2 \tag{14}$$

The votes casted by all frames of $\mathbf{X}_q$ are then counted and the candidate class which achieves the maximum number of votes is declared as the class y_q of the query video sequence $\mathbf{X}_q$.

4   Experimental Results

The performance of the proposed method is evaluated on three databases for the tasks of video-based face recognition and dynamic texture classifications. For video-based face recognition, the performance evaluation is conducted on the Honda/UCSD [29] database. The DynTex [16] and the YUPPEN [30] databases are used for dynamic texture classification. For the Honda/UCSD the faces are detected, cropped, and resized using the same procedure as in [31].

4.1 Results on the Honda/UCSD Database

The Honda/UCSD database [29] contains 59 video sequences of 20 different subjects. The video sequences are recorded in an indoor environment for at least 15 s at 15 frame per second. Similar to [29], we use 20 video sequences for training and the remaining 39 for testing. We repeat our experiments ten times with different random selections of the training and testing sets.

The experimental results in terms of average recognition rates and the standard deviations of DAN and the state-of-the-art benchmark methods are summarized in Table 1. The results demonstrate that the proposed method achieves 100% classification on the Honda/UCSD dataset.

Table 1. Comparison of average recognition rates ± standard deviations (%) and equal error rates (%) on the Honad/UCSD dataset [29].

Method	Avg. Recognition Rate (%) ± Standard Deviation	EER (%)
VLBP [17]	77.94 ± 1.00	23.15
VLBP + AdaBoost [32]	88.80 ± 0.60	4.02
MMD [13]	95.55 ± 1.84	3.51
MDA [33]	96.44 ± 1.37	2.74
CDL [15]	98.97 ± 1.32	0.63
SANP [31]	99.36 ± 0.10	0.22
RNP [34]	95.90 ± 2.16	3.08
DAN	**100.00 ± 0.00**	**0.00**

4.2 Results on the DynTex Database

The DynTex database [16] is a standard database for dynamic texture analysis containing high-quality dynamic texture videos such as windmill, waterfall, smoke, etc. It contains over 650 videos recorded in PAL in different conditions. Each video has 250 frames length with the frame rate of 25 frame/sec. Table 2 compares the rank-1 recognition rates of the proposed DAN and the benchmark approaches. Following the standard protocol, we use Leave-One-Out (LOO) cross validation in the next experiments on the dynamic texture data.

Table 2. Comparison of the rank-1 recognition rate (%) of the proposed method with benchmark approaches on the DynTex database [16].

Method	Recognition Rate (%)
VLBP [17]	95.71
LBP-TOP [17]	97.14
DFS [35]	97.63
BoS Tree [36]	98.86
MBSIF-TOP [37]	98.61
st-TCoF [19]	98.20
DAN	**99.07**

4.3   Results on the YUPPEN Database

The YUPPEN database [30] is a stabilized dynamic scene dataset. This dataset was introduced to emphasize scene-specific temporal information. YUPPEN consists of 14 dynamic scene categories with 30 videos per category. The sequences in the YUPPEN dataset have significant variations, such as frame rate, scene appearance, scaling, illumination, and camera viewpoint. We present the experimental results on this database in Table 3. It can be observed that the DAN outperforms the state-of-the-art benchmark methods. The results confirm that the proposed DAN is effective for dynamic scene data in a stabilized setting.

Table 3. Comparison of the rank-1 recognition rate (%) of the proposed method to benchmark approaches on the YUPPEN database [30].

Method	Recognition Rate (%)
LBP-TOP [17]	84.29
BoSE [11]	96.19
SOE [11]	80.71
SFA [38]	85.48
CSO [39]	85.95
st-TCoF [19]	99.05
DAN	**99.16**

5   Conclusion

In this paper, we presented a novel deep learning framework for video classification. A multi-layer deep autoencoder network was designed which was first pre-trained for appropriate weight initialization and then used for learning class specific networks. The class specific network is capable to capture the underlying non-linear complex structure of videos. In order to classify a given query video, we adopt a voting strategy based on the minimum reconstruction error. The proposed Deep Autoencoder Network (DAN) are evaluated on three standard video datasets and achieved the best performance among the competing methods.

References

1. Hajati, F., et al.: Dynamic texture comparison using derivative sparse representation: application to video-based face recognition. IEEE Trans. Hum.-Mach. Syst. **47**(6), 970–982 (2017)
2. Hajati, F., Faez, K., Pakazad, S.K.: An efficient method for face localization and recognition in color images, in 2006 IEEE International Conference on Systems, Man and Cybernetics (2006)
3. Hajati, F., et al.: Surface geodesic pattern for 3D deformable texture matching. Pattern Recogn. **62**, 21–32 (2017)

4. Barzamini, R., et al.: Short term load forecasting using multi-layer perception and fuzzy inference systems for Islamic Countries. J. Appl. Sci. **12**(1), 40–47 (2012)
5. Shojaiee, F., Hajati, F.: Local composition derivative pattern for palmprint recognition, in 2014 22nd Iranian Conference on Electrical Engineering (ICEE) (2014)
6. Pakazad, S.K., Faez, K., Hajati, F.: Face detection based on central geometrical moments of face components, in 2006 IEEE International Conference on Systems, Man and Cybernetics (2006)
7. Ayatollahi, F., Raie, A.A., Hajati, F.: Expression-invariant face recognition using depth and intensity dual-tree complex wavelet transform features, vol. 24: SPIE. 1–13, 13 (2015)
8. Abdoli, S., Hajati, F.: Offline signature verification using geodesic derivative pattern, in 2014 22nd Iranian Conference on Electrical Engineering (ICEE) (2014)
9. Ravichandran, A., Chaudhry, R., Vidal, R.: Categorizing dynamic textures using a bag of dynamical systems. IEEE Trans. PAMI **35**(2), 342–353 (2013)
10. Ojala, T., Pietikainen, M., Maenpaa, T.: Multiresolution gray-scale and rotation invariant texture classification with local binary patterns. IEEE Trans. PAMI **24**(7), 971–987 (2002)
11. Feichtenhofer, C., Pinz, A., Wildes, R.P.: Bags of spacetime energies for dynamic scene recognition, in Proc. IEEE CVPR, pp. 2681–2688 (2014)
12. Kim, T.K., Kittler, J., Cipolla, R.: Discriminative learning and recognition of image set classes using canonical correlations. IEEE Trans. PAMI **29**(6), 1005–1018 (2007)
13. Wang, R., et al.: Manifold-manifold distance and its application to face recognition with image sets. IEEE Trans. Image Process. **21**(10), 4466–4479 (2012)
14. Harandi, M., et al.: Graph embedding discriminant analysis on Grassmannian manifolds for improved image set matching, in Proc. IEEE CVPR, pp. 2705–2712 (2011)
15. Wang, R., et al.: Covariance discriminative learning: a natural and efficient approach to image set classification, in Proc. IEEE CVPR, pp. 2496–2503 (2012)
16. Péteri, R., Fazekas, S., Huiskes, M.J.: DynTex: a comprehensive database of dynamic textures. Pattern Recogn. Lett. **31**(12), 1627–1632 (2010)
17. Zhao, G., Pietikainen, M.: Dynamic texture recognition using local binary patterns with an application to facial expressions. IEEE Trans. PAMI **29**(6), 915–928 (2007)
18. Bengio, Y.: Learning deep architectures for AI. Found. Trends Mach. Learn. **2**(1), 1–127 (2009)
19. Qi, X., et al.: Dynamic texture and scene classification by transferring deep image features. Neurocomputing **171**, 1230–1241 (2016)
20. Azizpour, H., et al.: From generic to specific deep representations for visual recognition, in Proc. IEEE CVPR, pp. 36–45 (2015)
21. Sermanet, P., et al.: Overfeat: integrated recognition, localization and detection using convolutional networks. arXiv preprint arXiv:1312.6229 (2013)
22. Sun, Y., Wang, X., Tang, X.: Deep learning face representation from predicting 10,000 classes, in Proc. IEEE CVPR, pp. 1891–1898 (2014)
23. Xie, J., Xu, L., Chen, E.: Image denoising and inpainting with deep neural networks, in advances in neural information processing systems, pp. 350–358 (2012)
24. Smolensky, P.: Information processing in dynamical systems: foundations of harmony theory. In: Parallel Distributed Processing: Explorations in the Microstructure of Cognition, pp. 194–281. MIT Press (1986)
25. Taylor, G.W., Hinton, G.E., Roweis, S.: Modeling human motion using binary latent variables, in advances in neural information processing systems, pp. 1345–1352 (2007)
26. Taylor, G.W., et al.: Convolutional learning of spatio-temporal features, in Proc. ECCV, pp. 140–153 (2010)
27. Hinton, G.E., Osindero, S., Teh, Y.W.: A fast learning algorithm for deep belief nets. Neural Comput. **18**(7), 1527–1554 (2006)

28. Hinton, G., et al.: Unsupervised discovery of nonlinear structure using contrastive backpropagation. Cogn. Sci. **30**(4), 725–731 (2006)
29. Lee, K.C., et al.: Video-based face recognition using probabilistic appearance manifolds, in Proc. IEEE CVPR, pp. 313–320 (2003)
30. Derpanis, K.G., et al.: Dynamic scene understanding: the role of orientation features in space and time in scene classification, in Proc. IEEE CVPR, pp. 1306–1313 (2012)
31. Hu, Y., Mian, A.S., Owens, R.: Face recognition using sparse approximated nearest points between image sets. IEEE Trans. PAMI **34**(10), 1992–2004 (2012)
32. Hadid, A., Pietikainen, M.: Combining appearance and motion for face and gender recognition from videos. Pattern Recogn. **42**(11), 2818–2827 (2009)
33. Wang, R., Chen, X.: Manifold discriminant analysis, in Proc. IEEE CVPR, pp. 429–436 (2009)
34. Yang, M., et al.: Face recognition based on regularized nearest points between image sets, in Proc. IEEE Int. Conf. and Workshops on Automatic Face and Gesture Recognition (FG), pp. 1–7 (2013)
35. Yong, X., et al.: Dynamic texture classification using dynamic fractal analysis, in Proc. IEEE ICCV, pp. 1219–1226 (2011)
36. Coviello, E., et al.: Growing a bag of systems tree for fast and accurate classification, in Proc. IEEE CVPR, pp. 1979–1986 (2012)
37. Arashloo, S.R., Kittler, J.: Dynamic texture recognition using multiscale binarized statistical image features. IEEE Trans. Multimedia **16**(8), 2099–2109 (2014)
38. Thériault, C., Thome, N., Cord, M.: Dynamic scene classification: learning motion descriptors with slow features analysis, in Proc. IEEE CVPR, pp. 2603–2610 (2013)
39. Feichtenhofer, C., Pinz, A., Wildes, R.P.: Spacetime forests with complementary features for dynamic scene recognition, in Proc. BMVC, pp. 1–12 (2013)

The 13th International Workshop on Engineering Complex Distributed Systems (ECDS-2013)

An Approximate Forecasting of Electricity Load and Price of a Smart Home Using Nearest Neighbor

Muhammad Nawaz[1], Nadeem Javaid[1(✉)], Fakhar Ullah Mangla[2],
Maria Munir[2], Farwa Ihsan[2], Atia Javaid[1], and Muhammad Asif[3]

[1] COMSATS University Islamabad, Islamabad 44000, Pakistan
nawazkhan.cui2018@gmail.com, nadeemjavaidqau@gmail.com
[2] University of Sargodha, Sargodha 40100, Pakistan
fakhar.mangla@uos.edu.pk, mariamunir.uos2016@gmail.com,
farwaihsan.uos2016@gmail.com
[3] The Islamia University of Bahawalpur, Bahawalpur 63100, Pakistan
muhammadasifkhancs@gmail.com
http://www.njavaid.com/

Abstract. In Smart Grid, electricity demand and price forecasting literature has focused on Industrial, Buildings, and Residential sector demand, but this paper focuses on short term electricity demand and price forecasting for residential customer. Here we take smart meter data of hourly based from a smart home. First standardize and selected important features by using Recursive Feature Elimination with Linear Support Vector Classifier (RFE-LSVC). Second, do forecasting through K-Nearest Neighbors (KNN), Decision Tree (DT), Random Forest (RF) and Support Vector Regression (SVR) models and perform comparative analysis among models against four scenarios and provided best solution among all for individual scenario. This work proposed best solution of smart home's load and price forecasting for smart grid to manage demand response efficiently. We evaluated every Models with Mean Absolute Percentage Error (MAPE).

1 Introduction

Electricity demand is increasing by the ever increasing global population. Developing countries governments are shifting their Traditional Grid (TG) to Smart Grid (SG) for efficient demand response of electricity. The smart grid is modern electric grid that intelligently and efficiently manages the power generation, distribution and consumption of electricity by introducing such type of technologies which enables two way communication between utility and consumer of electricity and also focus on satisfaction of consumer. By 2020, the EU aims to take the place of 80% traditional electric meters with smart meters to support the objectives of control increasing electricity prices and for comfort of consumer [3]. The Advanced Metering Infrastructures (AMIs) consists of Smart-Meters (SMs)

© Springer Nature Switzerland AG 2020
L. Barolli et al. (Eds.): CISIS 2019, AISC 993, pp. 521–533, 2020.
https://doi.org/10.1007/978-3-030-22354-0_46

is main functional technological innovation in SG domain which function for Demand-Response (DR) model in which energy demand through individual SM is based on consumer usage demand and in parallel provides response for energy off peak hours or on peak hours from utilities. In SG, utility manage electricity demand from SM of Smart-Homes (SHs) so efficiently as remain minimum burden on the SG [6].

To manage the DR of the SH, Electricity markets get benefits form load forecasting on the base of consumer usage pattern data gathered from their SM. This load forecasting becomes base for several important decisions such as price for specific hours, power generation schedules to fulfill demand and using source of power generations. Now price forecasting is also crucial and beneficial for consumers and energy market participants for auction strategies formulation and speculation planning. From price forecasting, consumer get benefit of low electricity price as to shift their load from on peak hours to off peak hours. Through price forecasting, electricity producers can maximize their profit and consumers can minimize the cost of their electricity usage.

AMIs main function of dual connection between utility and consumer bring much benefits for both consumers and utilities. Consumer is now aware of his load used by individual appliance and contribute to shift load from on peak hours to off peak hours for getting compensations from utility. Utility on the other hand also get benefits from the goal of AMIs mentioned as:

- Electricity buying decisions on consumer's hand.
- Customer satisfaction in selecting best service provider.
- Easiness of electricity theft detection.
- Minimize electricity price cost.
- Minimum error in electricity bills due to accurate calculation and two way involvement.

In SHs installed AMI, consumer is well known of every appliance's consumed electricity in every moment. Utility provide options to consumer for different prices against specific hours and this is possible only through precise prediction of load and price against it. Consumer take decisions for usage of electricity and totally aware of expected bill for that month.

1.1   Motivation

After reviewing different forecasting techniques for different scenario in literature, the following is motivation of our work.

- Day-ahead load and price forecasting for individual SH is not taken in consideration. Literature take into account mostly residential or market data for short term load forecasting. Very Short Term Load and Price Forecasting (VSTLPF) and Short Term Load and Price Forecasting (STLPF) for individual SH capable SG to check individual's patterns [3]. With VSTLPF and STLPF, SG will be able to know one-hour, twelve-hours, one-day and one-week ahead load demand from SH and efficiently manage DR.

- Classifiers like Artificial Neural Networks (ANN), Support Vector Machine (SVM) and Wavelet Transform (WT) etc. have low generalization potential therefore have an over-fitting problems when they are used.

- Data preprocessing as noise removal, Feature Importance (FI), Feature Selection (FS) and Feature Extraction (FE) through Decision Tree (DT), Recursive Feature Elimination (RFE) etc. used on being classified data, become base for good accuracy of forecasting classifier.

1.2   Related Work

For load and price forecasting in SG, there are several techniques discussed in literature. Electricity consumption and forecasting is divided into four main categories [3], as: (1) Very Short Term Load and Price Forecasting (VSTLPF), (2) Short Term Load and Price Forecasting (STLPF), (3) Medium Term Load and Price Forecasting (MTLPF) and (4) Long Term Load and Price Forecasting (LTLPF). These categories are based on their time differences (e.g. minute-to-minute, hour-to-hour, day-to-day, week-to-week, year-to-year, etc.) for calculation and forecasting of load. There is also a category of forecasting techniques for these type of load data. In different scenario researcher use different techniques as Random Forest (RF), Neural Networks (NNs) and Particle Swam Optimizations (PSO).

In paper [17], using EMD-LSTM NN with XGboost as feature importance, performs better in short term load forecasting than without XGboost. Author implemented Stacked Denoising Auto encoders (SDA) model for day-ahead hourly forecasting in [16] and get better result with extended SDA. For continuously day-ahead price forecasting is also done by consideration of market integration in [7] by using deep neural network and novel feature selection algorithm. They focus only two markets and increase 3.2% accuracy. [12] predicted day-ahead price as hourly base for market.Author predict 24 h individual price and reduce Root Mean Squared Error 16%. Feature Selection (FS) is important process for accurate forecasting. In [1], Author proposed new techniques of feature selection for accurate forecasting of load and price in SG. Author reduce redundancy using information theoretic criteria and hybrid filter-wrapper approach. Short term load forecasting on five year data from university campus is done by using moving average method and random forest method in [11] to get better accuracy than using only random forest as classifier. Author in [4] calculate short-term load forecasting but in large data by using combination of convolutional neural network with K-means algorithm. They perform forecasting as clustered large data into subsets using K-means and training convolutional neural network on that subsets.

SG consists of a large data and for forecasting many excellent classifier cannot perform well due to redundancy in features. So it is most important to be vary clean and preprocessed a data which is going to use for classification. Author in [15] use Random Forest (RF) and Relief-F based on Gray Correlation Analysis (GCA) for feature selection and Kernel Function (KF) and Principle Component

Analysis (PCA) for feature Extraction to get better forecasting performance of classifier. Short term load forecasting by using Integrated Intelligent Energy Management process is proposed in [2] where 56 different scenarios are checked with compared their results. Author in [8] forecast load of electricity of building consumer with Sliding Window Empirical Mode Decomposition (SWEMD) feature selection and Improved Elman Neural Network (IENN) for classification. In [10] for short term residential load forecasting, Author compared result of Regression Trees (RT), Neural Networks (NN) and Support Vector Regression (SVR) and showed that RT is better than others. One important factor as lifestyle of consumer from the load pattern is mentioned in [14] where Author forecasted short term load using private aggregate data from AMI Applications to Enhance Privacy Technologies (EPT) to protect private data. In SG prosumers are those who buy and sale energy and [9] proposed distributed electricity trading system to facilitate the peer-to-peer electricity sharing among prosumers. Here through agent coalition system, prosumers became able to form coalitions and negotiate electricity trading. Sometime there occurs security problems in SG, so [5] proposed a model where consumer can monitor and send their data over SG network without any data compromisation. We see forecasting techniques with different scenarios in literature and notice that involvement of big data, data normalization and classification increase DR efficiency of SG (Table 1).

1.3   Contributions

In this paper, we emphasize short term electricity load and price forecasting for smart home. Our aim is to highlight the problem as one classifier is not suitable for different scenario of forecasting in SG. Here this problem with some solutions is discussed very precisely. To obtain our goal, we take into consideration four forecasting models and proposed four forecasting scenarios and check performance of each model at each scenario with similar parameter. We provide following contribution in this paper.

- **Model's Comparative Analysis:** We implemented K-Nearest Neighbors (K-NN), Random Forest (RF), Decision Tree (DT), Support Vector Regression (SVR) models for load and price forecasting for SH. After the comparative analysis among these four models, we find the best solution model for load and price forecasting for small data which will benefit much to SH and SG. Actually literature shows many forecasting models for different scenarios but we perform forecasting with four models and compare them with respect to their accuracy, implementation complexity, time consumption for implementation. One model can never be used for all different forecasting scenarios like one-hour, one-day, one-week, one-month and one-year ahead forecasting. This paper showed that there is one best model for specific forecasting scenario but that model could behave bad in other scenarios. This model's analysis proposes best forecasting model for specific scenario over small data and these models are best solution for SG while dealing with different scenarios. This work actually benefit SG much for accurate load and price forecasting for a SH.

Table 1. Related work

Scheme	Techniques used	Consumer satisfaction	Objectives
[1]	Hybrid Filter Wrapper Approach	×	New feature selection method is proposed for better feature selection during forecasting in energy market
[2]	ANN, PSO, MAACPSO, ANFIS	×	Short-term load forecasting with integrated intelligent energy management process for smart grid
[3]	Deep Neural Network	✓	Short-Term Appliance level energy profiling and forecasting for residential household
[4]	CNN, K-Means	×	Short-term load forecasting using 1.4 million of load records as big data
[7]	Deep Neural Network	✓	Forecasting day-ahead electricity price in Europe with market integration and novel feature selection algorithm
[8]	Improved Elman Neural Network	×	Short term load forecasting and proposed enhanced version of empirical model decomposition for feature selection
[10]	Regression Tree, SVR, NN	×	Short-term residential load forecasting with three models and compare their results
[11]	Moving Average, RF	✓	Short-term load forecasting with 2-stage predictive analytics using data of private university Seoul, Korea
[12]	Multivariate Model	×	Hourly based day-ahead price forecasting for market with Root Mean Squared Error 16%
[13]	Artificial Neural Network	✓	Day-ahead average electricity load forecasting using smart meter data
[15]	DE-SVM	✓	Get efficient result of electricity price forecasting by using feature selection, extraction and dimensionality reduction techniques
[16]	Stacked Denoising Autoencoders	✓	Day-ahead electricity price forecasting and model comparison with others
[17]	EMD-LSTM NN	×	Short-term load forecasting with Xgboost-based k-means framework for Feature Importance Evaluation

- **VSTLPF:** From a SH, we perform forecasting in SM's minutes and hourly-based data. We done very short term electricity load and price forecasting for SH with the objective of enabling utilities and consumer to precisely know the electricity load and price. In very short term load and price forecasting we implemented and compare forecasting models for one-hour and twelve-hours ahead forecasting at SH data. This capable SG to know about one-hour and twelve-hours ahead load demand from SH and take proper decision to response their upcoming demand.

- **STLPF:** Short term load and price forecasting consist of one-day and one-week ahead load and price forecasting for SH. Utilities became able to know daily and week ahead demand of load from consumer and decided energy generation on this base. Consumer remains updated about their demand and price against energy from utilities so that he can shift their load from on peak hours to off peak hours in order to save money. Ahead load demand and price knowledge make consumer more satisfied and involvement for energy saving, shifting load through on peak hours to off peak hours in SG.

We addressed these issues in SG which are much benefited for both utility and consumer for electricity generation and consumption. SG decide energy generation resources and show energy price detail of specific hours to consumer so that they became able to decide energy consumption and transfer their load from on peak hours to off peak hours.

The rest of paper is organized as follows: Sect. 2 describes proposed system model and Sect. 3 is simulation results of system model. In Sect. 4, there described Performance Evaluation. Paper's conclusions and future work is described in Sect. 5.

2   System Model

Proposed system model is based on following steps (Fig. 1).

2.1   Input Data Configuration

The dataset used in this work is Appliance-level home dataset procured from the Smart* project[1] in the year 2016. There are four SMs against HomeG, so we chose common appliances as shown in Table 2. Here February month 2016 data of HomeG is taken for model's performance analysis. For models training and testing, we divided dataset into 80% and 20% respectively. In order to achieve higher accuracy, we normalize data as first convert data into watt per hour and then remove outliers from it. Price against load is calculated by multiplication of load (Use) data with specific value and then use this price feature to forecast price against load according our scenarios.

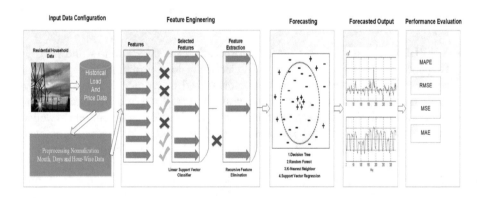

Fig. 1. Proposed system model

---

[1] Smart* Dataset is taken from UMass Smart Repository. The Goal of Smart* project is to optimize home energy consumption. http://traces.cs.umass.edu/index.php/Smart/Smart.

Table 2. Dataset main features

Features			
Date & Time	Use	Price	Water Pump
Kitchen Outlets1	Ejector Pump	Range Oven	PGR Outlets
Dishwasher	Refrigerator	Wall Oven	Kitchen Lights
Basement Lights	Wilo Pum	HVAC Air Handler	

2.2   Feature Engineering

Here Feature Engineering process is performed on dataset. This step is crucial for better performance of models so here Important features are selected by RFE-LSVC. RFE-LSVC is feature selection method that fits the model and remove the features which have weak impact on prediction target. Table 3 show the selected features which have higher impact on prediction accuracy. Another important thing is dataset splitting ratio which impact much in forecasting accuracy so we checked it with different ratio but results are with 80% at training and 20% at testing.

Table 3. Selected features

Features			
Date & Time	Use	Price	Water Pump
Ejector Pump	Range Oven	PGR Outlets	Refrigerator
Wilo Pum	HVAC Air Handler		

2.3   Forecasting

Accurate load and price forecasting of SH is main issue of literature and major concern of this article, here we explained how each forecasting model work. After feature engineering, selected features became base for forecasting through models as discussed below.

1. *Decision Tree (DT):*
   DT raise regression model in tree structure form. Simultaneously it breaks down dataset into small subsets and incrementally develop decision tree. At the end their is a tree with leaf nodes and decision nodes. Here Standard Deviation (SD) is used to homogeneity of numerical sample and build branches of tree. If numerical sample is homogeneous then its answer will be zero.

$$SD = \sqrt{\frac{\Sigma(x - \bar{x})^2}{n}} \tag{1}$$

Coefficient of Variance (CV) decide when to stop branching of tree.

$$CV = \frac{S}{\bar{x}} * 100\% \tag{2}$$

$$SDR(T, X) = S(T) - S(T, X) \tag{3}$$

Our dataset is splitted in different attributes and SD of each branch is calculated. The resulting SD is then subtraction from SD of before the split to form Standard Deviation Reduction (SDR) as described in equation. For decision node, attribute with high value of SDR is chosen this process run recursively and terminate when CV for branch become smaller than threshold or few branches e.g. 3. So when data points for all branches is equal to 3 then algorithm stop creating branches and assign average of each branch to its leaf node. If number of instances at leaf node are more than one then it calculate average of them which is final value for target.

2. **Random Forest (RF):**
   A supervised machine learning model which create a forest of trees and then makes it random. RF creates multiple decision trees from dataset and then merge them to get more accurate prediction results. During tree generation, it add randomness to model and search best feature among random subsets of features instead of most important feature while splitting a node. RF randomly select features to generate decision trees and average the result.

3. **Support Vector Regression (SVR):**
   SVR work with the principles of Support Vector Machine with few minor differences. But the main idea is individualizing hyperplane which should maximize margin, minimize error and part of error is tolerated. In SVR, first input 'x' is mapped onto k-dimensional feature with nonlinear mapping then linear model $f(x, w)$ is created.

$$f(x, w) = \sum_{i=1}^{k} w_i g_i(x) + b \tag{4}$$

Here $g_i(x), i = 1, ..., k$ are a set of nonlinear transformations and b is 'bias'. SVR use ε-insensitive loss function $L(y, f(x, w))$ which is used to measure quality of estimation.

4. **K-Nearest Neighbors (K-NN):** K-NN model is supervised machine learning algorithm that store available cases and predict on the base of distance function 'd'. For prediction, it calculate 'k' number of neighbors with distance function and then assign a class or value of its nearest neighbor which is output value. s

$$d = \sqrt{\sum_{i=1}^{k}(x_i - y_i)^2} \tag{5}$$

2.4   Performance Evaluation

After getting output from classifiers, here we check the performance accuracy by using classifier performance evaluation technique. Here we calculate accuracy of classifier with the calculation of Mean Absolute Error (MAE), Mean Squared Error (MSE), Root Mean Square Error (RMSE) and Mean Absolute Percentage Error (MAPE). And then calculate accuracy in percentage. Also evaluated models with respect to their training time and accuracy with each forecasting scenario.

3   Simulations

For simulation of proposed work, we performed it with python simulator running on Intel core i3 system, 8 GB RAM and 320 GB storage. We use four forecasting models with four different scenarios. We considered K-NN as best forecasting model and name it 'Actual' then compare it with others in all scenarios. Simulation results are discussed below.

3.1   Hour-Ahead Forecasting

After training at 80% of dataset, here perform one and twelve-hours ahead forecasting through four models and get their results. Simulation results of models are given bellow. Figure 2 is the load forecasting for 12 h ahead load demand. Figure 3 show the model's price forecasting comparison against 12 h ahead load.

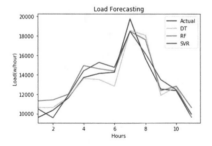

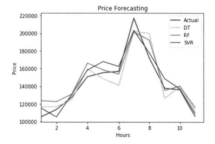

Fig. 2. 12 h ahead load forecasting comparison of models

Fig. 3. Price forecasting for 12 h ahead load

3.2   Day-Ahead Forecasting

This section is simulations of day-ahead load and price forecasting. All models are trained with same parameters but they behave differently in forecasting as their results are shown here. Figure 4 is model's result against one-day ahead load forecasting. And Fig. 5 show the price forecasting results for one-day ahead.

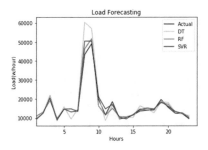

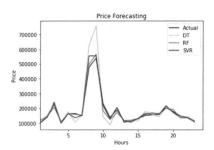

Fig. 4. Model's results for One-day ahead load forecasting

Fig. 5. Price forecasting for One-day ahead load

3.3   Week-Ahead Forecasting

Here simulations of one week-ahead load and price forecasting from all models are shown. Figure 6 is graph of week-ahead load forecasting. Figure 7 show the results of one-week ahead price forecasting by models. But Fig. 8 show the MAPE of each models for day-ahead forecasting and Fig. 9 is MAPE of models for week-ahead forecasting.

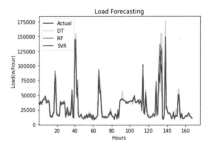

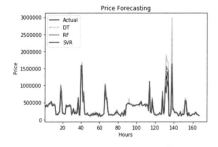

Fig. 6. One-week ahead load forecasting comparison

Fig. 7. One-week ahead price forecasting

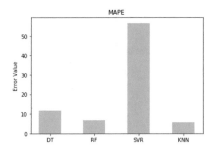

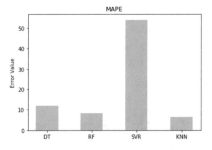

Fig. 8. Performance evaluation at One-day ahead forecasting

Fig. 9. Performance evaluation at One-week ahead forecasting

4   Performance Evaluation

To evaluate performance of classifiers, we used MAE, MSE, RMSE and MAPE. MAPE is percentage deviation therefore we use it for calculation of classifier's forecasting accuracy in percentage. Table 4 is comparison of classifiers for forecasting accuracy.

Classifier's forecasting results against four different scenarios is shown in table where 'Time (ms)' column show their training and forecasting time in millisecond. From this table we can analyze that in first scenario, K-NN perform well and RF is better than DT and SVR in accuracy but it is time consuming. If there would be large dataset then RF will perform badly and DT should be considered as good.

In second scenario for twelve-hours ahead forecasting, K-NN is good but here DT's is better than RF and SVR. Also, DT's and RF's accuracy increased here as compared to their accuracy in previous scenario. In third and fourth scenario, after K-NN, RF gave good accuracy as compared to others and it show its highest accuracy in third scenario as compared to its accuracy in other scenarios. Overall K-NN with k value 3 is good for all scenarios mentioned in this paper. Here small dataset is considered due to lazy learner this classifier don't perform good at large dataset.

Table 4. Model's comparative analysis

Model	Type	Time (ms)	MAPE	Accuracy (%)
DT	One-hour ahead	36	10.452	89.548
RF	One-hour ahead	360	9.649	90.351
SVR	One-hour ahead	203	69.642	30.358
K-NN	One-hour ahead	11	3.047	96.952
DT	12-hours ahead	36	10.452	89.548
RF	12-hours ahead	360	9.649	90.351
SVR	12-hours ahead	203	69.642	30.358
K-NN	12-hours ahead	11	3.047	96.952
DT	One-day ahead	36	10.452	89.548
RF	One-day ahead	360	9.649	90.351
SVR	One-day ahead	203	69.642	30.358
K-NN	One-day ahead	11	3.047	96.952
DT	One-week ahead	36	10.452	89.548
RF	One-week ahead	360	9.649	90.351
SVR	One-week ahead	203	69.642	30.358
K-NN	One-week ahead	11	3.047	96.952

5   Conclusion and Future Work

In this paper, big data is used for short term load and price forecasting only for a SH by SG. For electric load and price forecasting, after preprocessing of data we perform two further process on it, One feature engineering and other forecasting models on data. RFE with LSVC are used for selecting more important features from data. Load and price forecasting are performed by four models with four different forecasting scenarios as one-hour, twelve-hours, one-day and one-weak ahead forecasting. We notice that our models showed different forecasting accuracy at different scenarios and we compare their accuracy against each scenario. At the end there is model's comparisons Analysis table which explain each models accuracy rate against forecasting scenarios. Load and price forecasting for SH by SG, this paper is best solution for SG to forecast accurate load and price with best performing model against each scenario because one model can never gave good results in every scenario. In future, we will do medium and long term load and price forecasting with best forecasting models for these scenarios which will make SG efficient to predict and fulfill SH demand.

References

1. Abedinia, O., Amjady, N., Zareipour, H.: A new feature selection technique for load and price forecast of electrical power systems. IEEE Trans. Power Syst. **32**(1), 62–74 (2017)
2. Aydarous, A., Elshahed, M.A., Hassan, M.M.: Short term load forecasting as a base core of smart grid integrated intelligent energy management system. In: 2017 International Conference on Modern Electrical and Energy Systems (MEES), pp. 192–195. IEEE, November 2017
3. Din, G.M.U., Mauthe, A.U., Marnerides, A.K.: Appliance-level short-term load forecasting using deep neural networks. In: 2018 International Conference on Computing, Networking and Communications (ICNC), pp. 53–57. IEEE, March 2018
4. Dong, X., Qian, L., Huang, L.: Short-term load forecasting in smart grid: a combined CNN and K-means clustering approach. In: 2017 IEEE International Conference on Big Data and Smart Computing (BigComp), pp. 119–125. IEEE, February 2017
5. Gao, J., Asamoah, K.O., Sifah, E.B., Smahi, A., Xia, Q., Xia, H., Zhang, X., Dong, G.: Gridmonitoring: secured sovereign blockchain based monitoring on smart grid. IEEE Access **6**, 9917–9925 (2018)
6. Jindal, A., Singh, M., Kumar, N.: Consumption-aware data analytical demand response scheme for peak load reduction in smart grid. IEEE Trans. Ind. Electron. **65**(11), 8993–9004 (2018)
7. Lago, J., De Ridder, F., Vrancx, P., De Schutter, B.: Forecasting day-ahead electricity prices in Europe: the importance of considering market integration. Appl. Energy **211**, 890–903 (2018)
8. Liu, Y., Wang, W., Ghadimi, N.: Electricity load forecasting by an improved forecast engine for building level consumers. Energy **139**, 18–30 (2017)
9. Luo, F., Dong, Z.Y., Liang, G., Murata, J., Xu, Z.: A distributed electricity trading system in active distribution networks based on multi-agent coalition and blockchain. IEEE Trans. Power Syst. (2018)

10. Lusis, P., Khalilpour, K.R., Andrew, L., Liebman, A.: Short-term residential load forecasting: impact of calendar effects and forecast granularity. Appl. Energy **205**, 654–669 (2017)
11. Moon, J., Kim, K.H., Kim, Y., Hwang, E.: A short-term electric load forecasting scheme using 2-stage predictive analytics. In: 2018 IEEE International Conference on Big Data and Smart Computing (BigComp), pp. 219–226. IEEE, January 2018
12. Raviv, E., Bouwman, K.E., Van Dijk, D.: Forecasting day-ahead electricity prices: utilizing hourly prices. Energy Econ. **50**, 227–239 (2015)
13. Sulaiman, S.M., Jeyanthy, P.A., Devaraj, D.: Artificial neural network based day ahead load forecasting using smart meter data. In: 2016 Biennial International Conference on Power and Energy Systems: Towards Sustainable Energy (PESTSE), pp. 1–6. IEEE, January 2016
14. Tudor, V., Almgren, M., Papatriantafilou, M.: Employing private data in AMI applications: short term load forecasting using differentially private aggregated data. In: 2016 Intl IEEE Conferences on Ubiquitous Intelligence & Computing, Advanced and Trusted Computing, Scalable Computing and Communications, Cloud and Big Data Computing, Internet of People, and Smart World Congress (UIC/ATC/ScalCom/CBDCom/IoP/SmartWorld), pp. 404–413. IEEE, July 2016
15. Wang, K., Xu, C., Zhang, Y., Guo, S., Zomaya, A.: Robust big data analytics for electricity price forecasting in the smart grid. IEEE Trans. Big Data **5**, 34–45 (2017)
16. Wang, L., Zhang, Z., Chen, J.: Short-term electricity price forecasting with stacked denoising autoencoders. IEEE Trans. Power Syst. **32**(4), 2673–2681 (2017)
17. Zheng, H., Yuan, J., Chen, L.: Short-term load forecasting using EMD-LSTM neural networks with a Xgboost algorithm for feature importance evaluation. Energies **10**(8), 1168 (2017)

Survey on Intelligent Chatbots: State-of-the-Art and Future Research Directions

Ebtesam H. Almansor[1,2(✉)] and Farookh Khadeer Hussain[1]

[1] Faculty of Engineering and Information Technology,
University of Technology Sydney, Sydney, Australia
EbtesamHussain.Almansor@student.uts.edu.au, farookh.hussain@uts.edu.au
[2] Community College, Najran University, Najran, Saudi Arabia

Abstract. Human-computer interaction (HCI) is an area of interest which plays a major role in understanding the interaction between humans and machines. Dialogue systems or conversational systems including chatbots, voice control interfaces and personal assistants are examples of HCI application that have been developed to interact with users using natural language. Chatbots can help customers find useful information for their needs. Thus, numerous organizations are using chatbots to automate their customer service. Thus, the needs for using artificial intelligence has been increasing due to the needs of automated services. However, devolving smart bots that can respond at the human level is challenging. In this paper, we survey the state-of-art chatbot approaches from based on the ability to generate appropriate responses perspective. After summarizing the review from this aspect, we identify the research issues and challenges in chatbots. The findings of this research will highlight directions for future work.

Keywords: Conversation system · Chatbot ·
Responses generating approach

1 Introduction

Human-computer interaction (HCI) is a technology that allows communication between users and computers using natural language [1]. An automated conversation system (chatbot) is one human-machine conversation approach that has been designed to convince humans they are conversing with a human instead of a machine. Chatbots have been widely used in several domains, such as customer service, website help and education. Recent studies predict that 80% of businesses plan to implement chatbots by 2020 [2]. The main benefits of using chatbots for companies is that their customer service processes are automated as the chatbot can answer customers' questions about products or services. However, building a smart chatbot is challenging as requires contextual understanding, text entailment and language-understanding technology [3]. Therefore, various

© Springer Nature Switzerland AG 2020
L. Barolli et al. (Eds.): CISIS 2019, AISC 993, pp. 534–543, 2020.
https://doi.org/10.1007/978-3-030-22354-0_47

forms of artificial intelligence (AI) and natural language processing (NLP) are required.

AI aims to make communication between humans and computers easier by using natural language [4]. However, the complexity of human language has resulted in the need for AI scientists to provide models that can understand human language using the NLP approach. NLP is the area of research that explores the capability of computers to understand the human language [5].

The main focus of this literature review is to survey the existing literature and find the challenges and issues related to a chatbot. Therefore, in this paper, we adopt a comprehensive survey of the conversation system.

This paper is organized as follows: the first section presents an overview of chatbots; the second section presents a detailed background about chatbots; the third section details the classification framework for different types of chatbots; the fourth section presents the conclusion and future work.

2 Background of Chatbots

The idea of a chatbot comes from the "imitation" game or the Turing test which was created by Alan Turing (1950) [1]. This game aimed to determine whether a computer could imitate human behavior. The first chatbot, called ELIZA, was developed in 1966 [11]. This system used keyword matching and minimal context identification; however, this bot is a primitive system that lacked the ability to maintain a conversation between humans and bots. In the 1980s, the ALICE (Artificial Linguistic Internet Computer Entity) chatbot was created. This bot was considered to be significant due to the use of the Artificial Intelligent Markup Language (AIML) [12]. The idea behind AIML was to declare the pattern-matching rules which connect user-submitted words and phrases. The Jabberwack chatbot was built to simulate natural human language to learn from previous conversations and then the contextual patterns were used to select the most relevant response [13]. Additionally, commercial chatbots called Lingubots were developed to customize the template to analyze the word structure and grammar of the user's input [14].

2.1 Chatbot Applications in Public Sectors

Recently, the importance of chatbot in the public sector has taken place for example, chatbot was used for political purposes to inspire public opinion and intervene any discussion in social media about politics [15]. Another chatbot has been proposed as a digital channel of communication between citizens and the government [16]. In the education sector, a chatbot has been used to enhance critical thinking and support learners in learning a new language as the user can learn from the chatbot through their conversations [17]. An educational bot combining an intelligent tutoring system and learner modeling was designed to support learners [18]. Another chatbot was proposed for medical students for educational purposes [19]. In the health care sector, Your.MD chatbot was developed to provide relevant health information for patients [5]. Shawar and Atwell developed an algorithm for retraining a chatbot in a specific domain about a specific topic in any language [20]. Their algorithm was applied on two different languages, Arabic and Afrikaans, using the different corpus, the Qu'ran to compute frequently asked questions and the corpus of Spoken Afrikaans, respectively.

In the past few years, chatbots have been increasingly used by several organizations to increase the response time to customers in answering their questions and also reduce operational costs. Chatbot applications have been used in both the private sector, including the virtual assistants that are powered by voice (e.g. Siri, Alexa, Google now, Cortana) and public sector gaming agencies, telecommunications, banking (implementing transactions), tourism (booking hotels or tickets), media (news provision), retail, stock market and insurance companies [16]. Additionally, governments have used chatbots on social media platforms such as Twitter as a new form of political communication [21].

3   Proposed Classification for Chatbot Approaches

With the growth of AI technology, NLP researchers aim to build automatic conversation agents that respond to humans in a suitable time. There are two categories of conversation agents, namely task-oriented and non-task-oriented chatbots. Figure 1 shows our proposed classification of the chatbot categories.

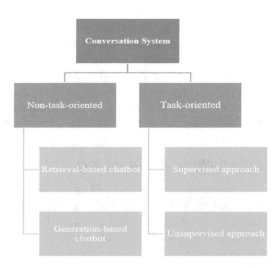

Fig. 1. Proposed classification for chatbot

3.1   Task-Oriented Dialogue System

A task-oriented system is one which is designed based on hand-crafted rules to help users achieve their goal or complete a specific task in a specific domain such as making a booking, traveling, shopping or ordering food [3,22]. This system has been widely used in both industry (Apple Siri, Microsoft Cortana, Facebook Messenger and Google Now) and academia [23]. They are focusing on developing natural language understanding (NLU) methods which parse the utterance from the user into predefined semantic slots. In the next paragraph, we discuss some of the task-oriented systems which have been implemented. There are two approaches, the supervised approach and the non-supervised approach.

3.1.1   Supervised Approach

The supervised approach depends on handcrafted feature extraction and annotated datasets [24]. As an example of this approach, Roy et al. [25] proposed a model that parsed the user's input into a semantic representation; then these representations are exploited by the dialogue manager to decide on the appropriate response. The dialogue manager tracks the conversation to determine who has spoken last, whether the information is private or shared; what plan should be followed and to what degree does the system understand the user's input. Then the dialogue system will respond to the user's input using NLP. There are also several approaches that can be used by dialogue systems to respond to the user, such as the slot fill approach [6]. Sequence-to-sequence (seq2seq) was used in a task-oriented dialogue system in [26]. However, this model cannot map the correct entities to the generated input. Thus, a copy-augmented Seq2Seq model was used to copy the relevant information from the knowledge base (KB) in [27]. The supervised approach has several limitations, such as the data annotation process and handcrafted feature extraction incur high costs and has poor scalability [24]. Thus, the unsupervised approach has been investigated.

3.1.2   Unsupervised Approach

The unsupervised approach learns features automatically from unlabeled datasets [24]. One of the most effective approaches is a deep learning approach which trained on an end-end neural network [28]. A convolution network model has been developed to capture the interaction between message and response [29]. An end-to-end-oriented dialogue system has been developed that uses the pipe-lined Wizard-of-Oz framework to collect the dialogue datasets. The main advantage of this model is the conversation developed without any need to use handcraft features to make assumptions [28]. Wang et al. [30] developed a system that leverages syntax features to measure text similarity.

3.2   Non-task-Oriented Dialogue System

A non-task-oriented dialogue system provides users with the means to participate in different domains such as a game, chitchat or entertainment, without providing the user with any help to complete any task in a specific job [31]. An example of this system is the chatbot which chats with the user in a similar way to a human and provides reasonable and relevant responses [3,31,32]. ELIZA is a chatbot that uses a combination of rules and patterns. In 1981, another chatbot was developed using simple text parsing rules to construct the dialogue system, PARRY [33]. Both systems did not use data for learning purposes. Therefore, a data-driven approach has been proposed to overcome some of the limitations of the non-task-oriented dialogue system. This approach enables chatbots to learn from the massive amount of available conversations on social media or the Web2.0, which enhances the communication between humans and chatbots. The developed methods are either retrieval-based or generation-based. Retrieval-based models can obtain response candidates from a pre-built index, rank the candidates then chose the response from the top-ranked ones [3,32].On the other hand, generation-based methods use natural language generation (NLG) to select the response [34].

3.2.1   Retrieval-Based Chatbot

A retrieval-based chatbot uses a selection of responses from ranking data rather than generating a new response. A dialogue system with a knowledge base that contains a considerable number of question-answer pairs was developed based on a statistical language model to select suitable responses [14,35]. The informal response interactive system (IRIS) is a chatbot that based on a vector space model. This bot searches for candidate responses to users' input using cosine similarity metrics [31]. Another example of a retrieval-based chatbot is the chatbot that proposed in [32] which can obtain response candidates with a research engine by utilizing the text similarity between the message and the response to select a proper response. This approach combines several text similarity features including topic similarity, cosine similarity and translation score. These features are used to rank the candidate responses [32]. Another approach is based on capturing the interaction between message and response using a convolution network [29]. Measuring text similarity using leverage syntax features was proposed in [30]. Considering the Using conversation history to support multi-turn conversations was studied in [22,36]. The authors used a neural learning architecture to select the best response and the model was tested on the Ubuntu Dialogue Corpus. Another retrieval model which considers conversation history and topic information to select a response in a single turn was proposed in [3], where the authors used an attention mechanism to enhance message-response matching. This chatbot can solve this problem by providing immediate responses. Thomas [37] built a chatbot that supports e-business and gives an immediate response to customers based on the frequently asked questions (FAQs). The author used a combination of artificial intelligent markup language (AIML) to answer general questions and latent semantic analysis (LSA) to answer questions related to services [37]. Another chatbot that focuses on healthcare was built to support people who have busy lives. This bot uses deep learning to address the problem of predefined rules as it focuses on the quality of the interaction between humans and computers. This is considered to be a self-adapting approach as it uses previous interactions with the chatbot [38].

3.2.2 Generation-Based Chatbot

A generation-based chatbot generates responses instead of selecting them from the underlying model. Knight et al. stated that using data-driven generation which depends on the utterance might be useful in a knowledge system [39]. A hybrid generation algorithm that combines finite state machine (FSM) grammars and a corpus-based language model has been developed in [40]. The FSM grammar is controlled by the concept of n-gram and words that take terminal and non-terminal co-occurrences into account. Due to the limited number of derivations, this approach achieved faster performance than real-time performance. Statistical machine translation has been investigated in translating an internal dialogue state into natural language [9]. Employing a statistical machine translation approach as a generic approach was proposed in [8,41]. Also, a sequence to sequence (S2S) framework was developed to generate responses to users' input [34,42–44]. Shang et al. [45] proposed a sequence to sequence framework which encodes a message with a neural network and generates responses using another recurrent neural network with an attention mechanism. A conventional S2S generation model was utilized in [46]. The authors modified the loss function to promote the diversity of responses.

4   Conclusion and Future Work

In this paper, we briefly introduced the history of chatbots and its use in public sectors. Then the proposed classification of the existing approaches and the types of chatbots was explained. In this research, some findings have been founded and need to be investigated. As most of the existing research focuses on improving the responses in chatbot other linguistic features need to be investigated such as using emotional or sentiment analysis. Also, the importance of user interface as an inter- active aspect of chatbots need more attention. There is a need for standard a framework to measure the quality of chatbot which is lacking in the existing literature. Also, using artificial intelligence to improve current chatbots and services could be another research direction. Deep learning is a promising approach that could be applied to develop more effective chatbots.

References

1. Shawar, B.A., Atwell, E.: Different measurements metrics to evaluate a chatbot system. In: Proceedings of the Workshop on Bridging the Gap: Academic and Industrial Research in Dialog Technologies, Association for Computational Linguistics, pp. 89–96 (2007)
2. Sandbank, T., Shmueli-Scheuer, M., Herzig, J., Konopnicki, D., Richards, J., Piorkowski, D.: Detecting egregious conversations between customers and virtual agents. arXiv preprint arXiv:1711.05780 (2017)

3. Wu, Y., Li, Z., Wu, W., Zhou, M.: Response selection with topic clues for retrieval-based chatbots. Neurocomputing **316**, 251–261 (2018)
4. Radziwill, N.M., Benton, M.C.: Evaluating quality of chatbots and intelligent conversational agents. arXiv preprint arXiv:1704.04579 (2017)
5. de Haan, H., Snijder, J., van Nimwegen, C., Beun, R.J.: Chatbot personality and customer satisfaction (2018)
6. Mesnil, G., Dauphin, Y., Yao, K., Bengio, Y., Deng, L., Hakkani-Tur, D., He, X., Heck, L., Tur, G., Yu, D.: Using recurrent neural networks for slot filling in spoken language understanding. IEEE/ACM Trans. Audio Speech Lang. Process. **23**(3), 530–539 (2015)
7. Li, J., Monroe, W., Ritter, A., Galley, M., Gao, J., Jurafsky, D.: Deep reinforcement learning for dialogue generation. arXiv preprint arXiv:1606.01541 (2016)
8. Ritter, A., Cherry, C., Dolan, B.: Unsupervised modeling of Twitter conversations. In: Human Language Technologies: The 2010 Annual Conference of the North American Chapter of the Association for Computational Linguistics, pp. 172–180. Association for Computational Linguistics (2010)
9. Langner, B., Vogel, S., Black, A.W.: Evaluating a dialog language generation system: comparing the mountain system to other NLG approaches. In: Eleventh Annual Conference of the International Speech Communication Association (2010)
10. Venkatesh, A., Khatri, C., Ram, A., Guo, F., Gabriel, R., Nagar, A., Prasad, R., Cheng, M., Hedayatnia, B., Metallinou, A., et al.: On evaluating and comparing conversational agents. arXiv preprint arXiv:1801.03625 **4**, 60–68 (2018)
11. Weizenbaum, J.: ELIZA—a computer program for the study of natural language communication between man and machine. Commun. ACM **9**(1), 36–45 (1966)
12. Wallace, R.: The elements of AIML style. Alice AI Foundation (2003)
13. Carpenter, R.: Jabberwocky. Jabberwocky.com (1997–2006). www.jabberwocky.com. Accessed 16 May 2006
14. Leuski, A., Traum, D.: NPCeditor: creating virtual human dialogue using information retrieval techniques. Ai Mag. **32**(2), 42–56 (2011)
15. Suárez-Serrato, P., Roberts, M.E., Davis, C., Menczer, F.: On the influence of social bots in online protests. In: International Conference on Social Informatics, pp. 269–278. Springer (2016)
16. Androutsopoulou, A., Karacapilidis, N., Loukis, E., Charalabidis, Y.: Transforming the communication between citizens and government through AI-guided chatbots. Gov. Inf. Quart. **36**, 358–367 (2018)
17. Goda, Y., Yamada, M., Matsukawa, H., Hata, K., Yasunami, S.: Conversation with a chatbot before an online EFL group discussion and the effects on critical thinking. J. Inf. Syst. Educ. **13**(1), 1–7 (2014)
18. Kerly, A., Hall, P., Bull, S.: Bringing chatbots into education: towards natural language negotiation of open learner models. Knowl. Based Syst. **20**(2), 177–185 (2007)
19. Kerfoot, B.P., Baker, H., Jackson, T.L., Hulbert, W.C., Federman, D.D., Oates, R.D., DeWolf, W.C.: A multi-institutional randomized controlled trial of adjuvant Web-based teaching to medical students. Acad. Med. **81**(3), 224–230 (2006)
20. Shawar, B.A., Atwell, E.: Fostering language learner autonomy through adaptive conversation tutors. In: Proceedings of the The fourth Corpus Linguistics Conference
21. Forelle, M., Howard, P., Monroy-Hernández, A., Savage, S.: Political bots and the manipulation of public opinion in Venezuela. arXiv preprint arXiv:1507.07109 (2015)

22. Yan, R., Song, Y., Wu, H.: Learning to respond with deep neural networks for retrieval-based human-computer conversation system. In: Proceedings of the 39th International ACM SIGIR Conference on Research and Development in Information Retrieval, pp. 55–64. ACM (2016)
23. Henderson, M., Thomson, B., Young, S.: Deep neural network approach for the dialog state tracking challenge. In: Proceedings of the SIGDIAL 2013 Conference, pp. 467–471 (2013)
24. Chen, Y.N.: Unsupervised learning and modeling of knowledge and intent for spoken dialogue systems. In: Proceedings of the ACL-IJCNLP 2015 Student Research Workshop, pp. 1–7 (2015)
25. Roy, N., Pineau, J., Thrun, S.: Spoken dialogue management using probabilistic reasoning. In: Proceedings of the 38th Annual Meeting on Association for Computational Linguistics, pp. 93–100. Association for Computational Linguistics (2000)
26. Zhao, T., Lu, A., Lee, K., Eskenazi, M.: Generative encoder-decoder models for task-oriented spoken dialog systems with chatting capability. arXiv preprint arXiv:1706.08476 (2017)
27. Eric, M., Manning, C.D.: A copy-augmented sequence-to-sequence architecture gives good performance on task-oriented dialogue. arXiv preprint arXiv:1701.04024 (2017)
28. Wen, T.H., Vandyke, D., Mrksic, N., Gasic, M., Rojas-Barahona, L.M., Su, P.H., Ultes, S., Young, S.: A network-based end-to-end trainable task-oriented dialogue system. arXiv preprint arXiv:1604.04562 (2016)
29. Hu, B., Lu, Z., Li, H., Chen, Q.: Convolutional neural network architectures for matching natural language sentences. In: Advances in Neural Information Processing Systems, pp. 2042–2050 (2014)
30. Wang, M., Lu, Z., Li, H., Liu, Q.: Syntax-based deep matching of short texts. arXiv preprint arXiv:1503.02427 (2015)
31. Banchs, R.E., Li, H.: IRIS: a chat-oriented dialogue system based on the vector space model. In: Proceedings of the ACL 2012 System Demonstrations, pp. 37–42. Association for Computational Linguistics (2012)
32. Ji, Z., Lu, Z., Li, H.: An information retrieval approach to short text conversation. arXiv preprint arXiv:1408.6988 (2014)
33. Colby, K.M.: Modeling a paranoid mind. Behav. Brain Sci. 4(4), 515–534 (1981)
34. Shang, L., Lu, Z., Li, H.: Neural responding machine for short-text conversation. arXiv preprint arXiv:1503.02364 (2015)
35. Leuski, A., Patel, R., Traum, D., Kennedy, B.: Building effective question answering characters. In: Proceedings of the 7th SIGdial Workshop on Discourse and Dialogue, pp. 18–27. Association for Computational Linguistics (2009)
36. Lowe, R., Pow, N., Serban, I., Pineau, J.: The Ubuntu dialogue corpus: a large dataset for research in unstructured multi-turn dialogue systems. arXiv preprint arXiv:1506.08909 (2015)
37. Thomas, N.: An e-business chatbot using AIML and LSA. In: 2016 International Conference on Advances in Computing, Communications and Informatics (ICACCI), pp. 2740–2742. IEEE (2016)
38. Rai, S., Raut, A., Savaliya, A., Shankarmani, R.: Darwin: convolutional neural network based intelligent health assistant. In: 2018 Second International Conference on Electronics, Communication and Aerospace Technology (ICECA), pp. 1367–1371. IEEE (2018)
39. Knight, K., Hatzivassiloglou, V.: Two-level, many-paths generation. In: Proceedings of the 33rd Annual Meeting on Association for Computational Linguistics, pp. 252–260. Association for Computational Linguistics (1995)

40. Galley, M., Fosler-Lussier, E., Potamianos, A.: Hybrid natural language generation for spoken dialogue systems. In: Seventh European Conference on Speech Communication and Technology (2001)
41. Ritter, A., Cherry, C., Dolan, W.B.: Data-driven response generation in social media. In: Proceedings of the Conference on Empirical Methods in Natural Language Processing, pp. 583–593. Association for Computational Linguistics (2011)
42. Serban, I.V., Sordoni, A., Bengio, Y., Courville, A., Pineau, J.: Building end-to-end dialogue systems using generative hierarchical neural network models. In: 30th AAAI Conference on Artificial Intelligence, AAAI 2016, pp. 3776–3783 (2016)
43. Vinyals, O., Le, Q.: A neural conversational model. arXiv preprint arXiv:1506.05869 (2015)
44. Zhang, W.N., Zhu, Q., Wang, Y., Zhao, Y., Liu, T.: Neural personalized response generation as domain adaptation. World Wide Web, pp. 1–20 (2017)
45. Shang, L., Sakai, T., Lu, Z., Li, H., Higashinaka, R., Miyao, Y.: Overview of the NTCIR-12 short text conversation task. In: NTCIR (2016)
46. Li, J., Galley, M., Brockett, C., Gao, J., Dolan, B.: A diversity-promoting objective function for neural conversation models. arXiv preprint arXiv:1510.03055 (2015)
47. Engelbrech, K.P., Gödde, F., Hartard, F., Ketabdar, H., Möller, S.: Modeling user satisfaction with hidden markov model. In: Proceedings of the SIGDIAL 2009 Conference: the 10th Annual Meeting of the Special Interest Group on Discourse and Dialogue, pp. 170–177. Association for Computational Linguistics (2009)
48. Ultes, S., Schmitt, A., Minker, W.: On quality ratings for spoken dialogue systems-experts vs. users. In: Proceedings of the 2013 Conference of the North American Chapter of the Association for Computational Linguistics: Human Language Technologies, pp. 569–578 (2013)
49. Hone, K.S., Graham, R.: Towards a tool for the subjective assessment of speech system interfaces (SASSI). Nat. Lang. Eng. **6**(3–4), 287–303 (2000)
50. Möller, S.: Subjective quality evaluation of telephone services based on spoken dialogue systems. ITU-T Recommendation, p. 851 (2003)
51. Higashinaka, R., Minami, Y., Dohsaka, K., Meguro, T.: Issues in predicting user satisfaction transitions in dialogues: individual differences, evaluation criteria, and prediction models, pp. 48–60. Springer (2010a)
52. Higashinaka, R., Minami, Y., Dohsaka, K., Meguro, T.: Modeling user satisfaction transitions in dialogues from overall ratings. In: Proceedings of the 11th Annual Meeting of the Special Interest Group on Discourse and Dialogue, pp. 18–27. Association for Computational Linguistics (2010b)
53. Schmitt, A., Schatz, B., Minker, W.: Modeling and predicting quality in spoken human-computer interaction. In: Proceedings of the SIGDIAL 2011 Conference, pp. 173–184. Association for Computational Linguistics (2011)
54. Hara, S., Kitaoka, N., Takeda, K.: Estimation method of user satisfaction using N-gram-based dialog history model for spoken dialog system. In: LREC (2010)
55. Schmitt, A., Schatz, B., Minker, W.: A statistical approach for estimating user satisfaction in spoken human-machine interaction. In: 2011 IEEE Jordan Conference on Applied Electrical Engineering and Computing Technologies (AEECT), pp. 1–6. IEEE (2011b)
56. Martinez, F.F., Blázquez, J., Ferreiros, J., Barra, R., Macias-Guarasa, J., Lucas-Cuesta, J.M.: Evaluation of a spoken dialogue system for controlling a HiFi audio system. In: Spoken Language Technology Workshop 2008, SLT 2008 IEEE, pp. 137–140. IEEE (2008)
57. Machinery, C.: Computing machinery and intelligence-AM turing. Mind **59**(236), 433 (1950)

58. Liu, C.W., Lowe, R., Serban, I.V., Noseworthy, M., Charlin, L., Pineau, J.: How not to evaluate your dialogue system: an empirical study of unsupervised evaluation metrics for dialogue response generation. arXiv preprint arXiv:1603.08023 (2016)
59. Yuwono, S.K., Biao, W., D'Haro, L.F.: Automated scoring of chatbot responses in conversational dialogue
60. Lowe, R., Noseworthy, M., Serban, I.V., Angelard-Gontier, N., Bengio, Y., Pineau, J.: Towards an automatic turing test: learning to evaluate dialogue responses. arXiv preprint arXiv:1708.07149 (2017)
61. DeVault, D., Leuski, A., Sagae, K.: Toward learning and evaluation of dialogue policies with text examples. In: Proceedings of the SIGDIAL 2011 Conference, pp. 39–48. Association for Computational Linguistics (2011)
62. Gandhe, S., Traum, D.: A semi-automated evaluation metric for dialogue model coherence, pp. 217–225. Springer (2016)
63. Guo, F., Metallinou, A., Khatri, C., Raju, A., Venkatesh, A., Ram, A.: Topic-based evaluation for conversational bots. arXiv preprint arXiv:1801.03622 (2018)
64. Yu, Z., Xu, Z., Black, A.W., Rudnicky, A.: Strategy and policy learning for non-task-oriented conversational systems. In: Proceedings of the 17th Annual Meeting of the Special Interest Group on Discourse and Dialogue, pp. 404–412 (2016)
65. Bowman, S.R., Vilnis, L., Vinyals, O., Dai, A.M., Jozefowicz, R., Bengio, S.: Generating sentences from a continuous space. arXiv preprint arXiv:1511.06349 (2015)
66. Li, J., Monroe, W., Shi, T., Jean, S., Ritter, A., Jurafsky, D.: Adversarial learning for neural dialogue generation. arXiv preprint arXiv:1701.06547 (2017)
67. Chakrabarti, C., Luger, G.F.: Artificial conversations for customer service chatter bots: architecture, algorithms, and evaluation metrics. Expert Syst. Appl. **42**(20), 6878–6897 (2015)

A Novel Approach to Extend KM Models with Object Knowledge Model (OKM) and Kafka for Big Data and Semantic Web with Greater Semantics

CSR Prabhu[1,2]([✉]), R. Venkateswara Gandhi[3,4], Ajeet K. Jain[4,5],
Vaibhav Sanjay Lalka[6], Sree Ganesh Thottempudi[7],
and PVRD Prasad Rao[8]

[1] Quantum University, Roorkee, Uttarakhand, India
csrprabhu@kmit.in
[2] KMIT, Hyderabad, Telangana, India
[3] UOT, Jaipur, Rajasthan, India
ravegag@gmail.com
[4] CSE, KMIT, Hyderabad, Telangana, India
ravegag@gmail.com, jainajeet123@gmail.com
[5] KLU, Vijayawada, Andhra Pradesh, India
[6] Philips R&D, Bengaluru, India
vaibhavsanjaylalka@gmail.com
[7] University of Heidelberg, Heidelberg, Germany
sganeshhcu@gmail.com
[8] CSE, KL University, Vijayawada, India
pvrdprasad@kluniverisity.in

Abstract. In knowledge management (KM), everything is identified by an object and the world is a collection of objects. In the context of NLP, identifying grammar is a tedious task. Every sentence is being identified as a triplet: *subject, predicate* and *object*. The circumstantial detail of different tenses representing past and future of the grammar are very difficult to analyze. To identify such diversity of data - including streaming data, we propose a model, called Object Knowledge Model (OKM). Current scenario of RDF(S) defines metadata at every stage which is static in nature and lacks flexibility. Mainly OKM intends to provide a common framework for expressing machine-processable information with greater semantics than RDF—being modeled to identify the metadata in streams built on Kafka. This paper primarily discusses OKM grammar and how metadata can be analyzed implementing OKM. With known Stanford NLP technique plus others, the proposed OKM grammar is more flexible and demonstrates to be superior. The grammar checks genuinity of framing sentences (in various languages) as each language has its roots and connotations intrigued, and we claim that this proposed model is a better fitter for NLP understanding and projecting the merits of using them in KM models.

Dr.CSR Prabhu—Former DG (NIC), currently Director (R & D).

L. Barolli et al. (Eds.): CISIS 2019, AISC 993, pp. 544–554, 2020.
https://doi.org/10.1007/978-3-030-22354-0_48

Keywords: Knowledge management · Big data · Hadoop eco system ·
Data Lakes · Kafka · RDF(S) · Web semantics · OKM

1 Introduction

A new form of Web content that is meaningful to computers will unleash a revolution
of new possibilities [1]. Some papers discusses query processing [7], and uses RDF
(Resource Descriptor Format) [3] for search engine tools [4, 5]. In this paper we
propose Object Knowledge Model (OKM) for semantic web instead of RDF and RDF
Schema. The metadata for the data is to be defined such that the stream data is
understood and processed in a better manner. Introducing a new model for assigning
proper semantics to Big Data, we discuss the OKM appropriateness for adding
semantics in Big Data. The presented OKM approach can be applied in various
knowledge representation domains like in semantic web, cross lingual and multilingual
search and also in information extraction of named entities. In the past literature there
have been some models for semantic search [6, 8–10] which can be considered as
standard models for developing semantic search tools.

This implies that Big Data is failing in its characteristics while handling streaming
data. For overcoming this, Kafka can be added in the Apache Hadoop framework. This
can help Big Data to some extent but Kafka fails at adding natural language semantics.
Internet has become a huge and updating information warehouse [2], another storage
repository that holds a vast amount of raw data in its native format is the Data Lake.
Data Lakes is another alternative of handling huge data. Data Lake uses a flat architecture to store data. Data Lakes comprises of huge data which is collected from
different sources of data. Such data which is stored in Big Data or Data Lakes are to be
properly identified by adding semantics. In this paper we discuss a new model called
Object Knowledge Model (OKM) for designing metadata for the web resources. The
different data models for this purpose are also described in this paper.

1.1 Data Lakes

The data lake concept has been well received by enterprises to help capture and store
raw data of many different types at scale and low cost to perform data management
transformations, processing and analytics based on specific use cases. The first phase of
the data lake growth was in consumer web-based companies. The data lake strategy has
already shown positive results for these consumer businesses by helping increase speed
and quality of web search, web advertising (click stream data) and improved customer
interaction and behavior analysis (cross-channel analysis). This led to the next phase
for data lakes, which was to augment enterprise data warehousing strategies.

2 Present Work

OKM (Object Knowledge Model) provides meaning to metadata in the form of objects. The triplets i.e. subject, predicate and object are the main components of sentences where OKM mainly targets on adding many attributes to the object which is important to define the metadata of the data present in Big Data or Data Lakes, in addition to the greater semantics of Natural Language which OKM provides.

In this paper we discuss a new model called Object Knowledge Model or OKM for designing metadata for the web resources. Extracting the greater semantics from the sentence given in any language with OKM, a frame based knowledge representation framework for enabling semantic/knowledge/context based search. OKM aims to enhance the semantic/context/knowledge based search of not only web pages but also research papers, equity reports etc. OKM which is a frame based knowledge representation model for the sentences. The heart of the OKM is its grammar which is proposed in this paper which will parse given any sentence in English. English sentence consists of set of descriptors, noun phrase and verb phrase. The OKM grammar is written in such a way that given any sentence it will generate the parse tree with non-leaf elements as set of descriptors (d), noun (n) and verb (v) and leaf elements as attributed words of the sentence ex: Ram (111). OKM provides for a generic intermediate frame based knowledge representation framework for representing greater semantics of natural language text. It also has modeling capability for metadata of databases comprising stream data. OKM can be used as a basis for object identification for streams and Data lakes. OKM can be implemented by converting OKM frames of Ontologies into metadata of data streams in Kafka. OKM has also been implemented for natural language text using NLTK POS tagging favorably and a better comparison with Stanford Universal clustered Dependencies.

3 Design and Implementation of Object Knowledge Model for Text Data

Simple English sentence
First of all let's see the structure of a simple sentence in English grammar consisting of
Set of descriptors (Adjective, Adverb, Articles) + Noun phrase + Verb phrase
Ex: Ram went to school by bus.
Brother, a letter for you.
Word Attributes
The attributes which are assigned are divided into two parts:

- Noun Attributes
- Verb Attributes

Noun Attributes: Given a sentence all the nouns and pronouns in the sentence are attributed by some of features such as gender 'g' (1: male, 2: female, 3: neutral), number 'n' (1: singular, 2: dual, 3: plural) and case 'c' (1–8) ordered as per Sanskrit Grammar cases.

Cases in order are as follows:

1. Subjective
2. Objective
3. Instrumental
4. Dative (indirect object)
5. Ablative
6. Genitive (possessive)
7. Locative
8. Vocative

Attributed noun word will look like: Noun word (g n c) ex: Ram (1 1 1). Verb Attributes: Given a sentence all the verbs in the sentence are attributed by some of the features such as person 'p' (1: first person, 2: second person, 3: third person), number 'n' (1: singular, 2: dual, 3: plural) and tense 't' (−1: past tense, 0: present tense, 1: future tense). Attributed verb word will look like:

Verb word (p n t) ex: went (3 1-1). Now, given a sentence let's say:

Ram went to school by bus.

When parsed through OKM converted into (((Ram 111) (went 31-1)) ((school 312)) ((bus 313))).

Prepositions such as the, an, and are neglected in OKM framework.

The parentheses which are assigned to the sentence are nothing but segmenting a sentence using OKM grammar.

3.1 OKM Grammar

The OKM grammar is given below:

K ::   = (d o)*
d ::   = (D)*
D ::   = set of descriptors (adjectives, adverbs) o :: = on* ov*
on ::   = nr na
na ::   = (i, t, n, u, g, n, c, u)*
i ::   = object id
t ::   = object type (stream data)
g ::   = 1..3;; gender
n ::   = 1..3;; number (singular, dual, plural)
c: = 1...7 case (seven cases)
u: = units
nr ::   = set of noun roots
ov ::   = vr va
vr ::   = set of verb roots/procedure/process/web service (with Oni as input and Oni + 1 as output) va ::= (p n t)*
va :   = (i, t, u, p, n, te)*
p ::   = 1..3;; person (first, second, third)
n ::   = 1..3;; number (singular, dual, plural)
t_e ::   = 1....9;; tenses and moods. An asterisk (*) denotes zero or more repetitions of the preceding element.

4   Results for OKM Implementation of Text Data

Now, given a sentence we have used NLTK POS tagging (Parts of speech tagging) to get the parts of speech of each word in sentence according to the sentence structure [15]. Then we have extracted the gender 'g' and number 'n' attributes of noun words and number 'n' and tense 't' attributes of verb word from POS-tagging and common NLTK libraries. But for case 'c' which is noun attribute and person 'p' which is verb attribute we have considered the dependencies [13] (noun and verb) by sentence segmentation using OKM grammar and Stanford universal dependencies. We have parsed the sentence and generated the parse tree using the OKM framework. In the tree the non leaf element as descriptor 'd', noun 'n' and verb 'v' and leaf elements as attributed words of the sentence and is compared with the parse tree generated by the grammar of Stanford.

The output generated is shown in the figure below: For sentences: Note: In OKM, 999 is the default value given to the D in all the figures (Fig. 1).

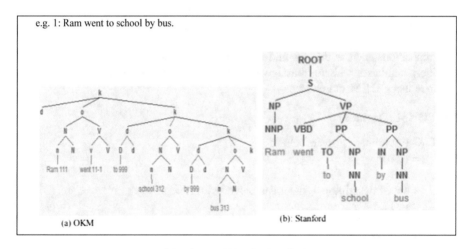

Fig. 1. (a) OkM (b) Stanford

OKM is easy to understand, as seen in Fig. 3(b), there are some non-terminals used which were not used in any above figures of Stanford and there are many non-terminals to remember when compare to OKM grammar. It is due to the reason that OKM has only three prime non-terminals i.e. set of descriptors 'd', noun 'n' and verb 'v' whereas Stanford has many non-terminals which is hard to remember. There are certain cases where the sentence segmentation is same in both OKM and Stanford as seen in examples in the Fig. 4 and 6. The sentences "here my pen is" and "Jai Mata Di" are segmented as (here) (my pen) and (Jai Mata Di) respectively. Whereas in OKM each and every noun and verb is attributed with the semantic knowledge which is not there in the Stanford. Further this work can be extended for any language. Given a sentence in any language first we convert it into English using Python library Goslate and then apply the above procedure [16–25].

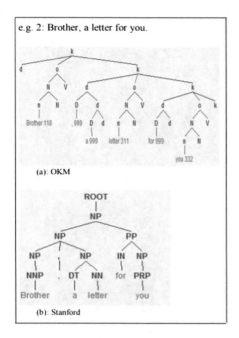

Fig. 2. (a) OkM (b) Stanford

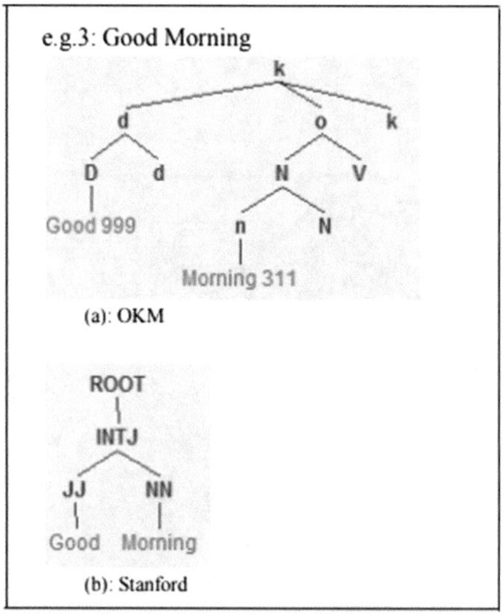

Fig. 3. (a) OkM (b) Stanford

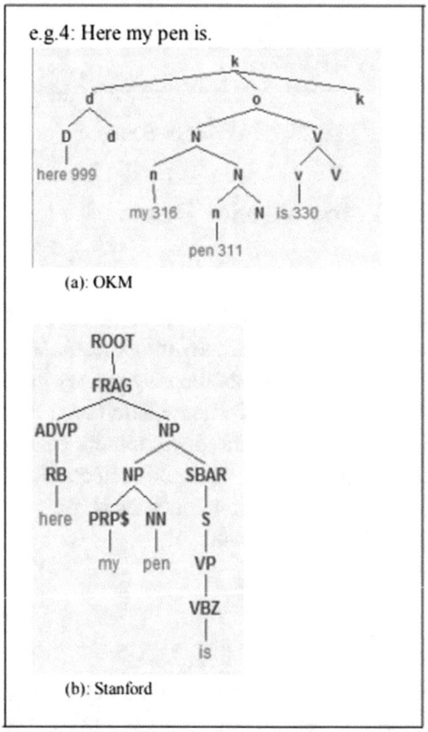

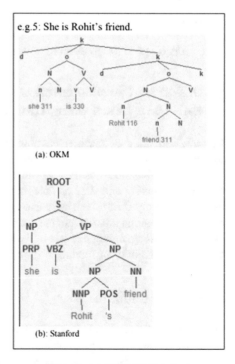

Fig. 4. (a) OkM (b) Stanford

Fig. 5. (a) OkM (b) Stanford

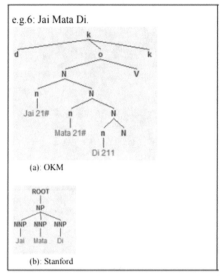

e.g. 7: The leaf is falling from the tree.

(a): OKM

(b): Stanford

e.g.6: Jai Mata Di.

(a): OKM

(b): Stanford

Fig. 6. (a) OkM (b) Stanford   **Fig. 7.** (a) OkM (b) Stanford

5   Design and Implementation of OKM for Stream Data with Data Lakes Using Kafka

Kafka is a stream data management platform using metadata OKM schema or ontology of metadata will be translated (by human or any tool) for stream into metadata in Kafka. Streaming analytics means doing analytics in real time as the data comes in as opposed to running analytics on data that is permanently stored somewhere (such as a data lake). Many data-driven organizations that are pursuing the development of use cases like recommendation engines, predictive maintenance, or fraud detection are moving toward streaming analytics. It's very likely that streaming analytics is, or soon will be, in a majority of roadmaps for companies large and small.

6   Kafka Metadata

This section gives examples on Kafka Metadata. For any given object stream we can define its schema in OKM implementation of so defined schema is in the following example. Here we present a scenario of input steam (type: = text). We have to create metadata in Kafka based on the overall schema defined using OKM as above. In the example a single object with type of stream data as text is defined as Metadata in Kafka (Similarly any other objects with other types can be defined in OKM first and then actual metadata for the same can be produced).

Example1

```
// pertaining to Stream(string)data type

import akka.actor.ActorRef;
import akka.kafka.ConsumerSettings;
import akka.kafka.KafkaConsumerActor;
import akka.kafka.KafkaPorts;
import akka.kafka.Metadata;
import akka.pattern.Patterns;
import java.time.Duration;
import java.util.List;
import java.util.Optional;
import java.util.concurrent.CompletionStage;
import java.util.stream.Collectors;
import org.apache.kafka.common.PartitionInfo;

Duration timeout = Duration.ofSeconds(2);
ConsumerSettings<String, String> settings =
consumerSettings.withMetadataRequestTimeout(timeout);
ActorRef consumer = system().actorOf((KafkaConsumerActor.props(settings)));

CompletionStage<Metadata.Topics> topicsStage =
    Patterns.ask(consumer, Metadata.createListTopics(), timeout)
        .thenApply(reply -> ((Metadata.Topics) reply));

  // convert response
  CompletionStage<Optional<List<String>>> response =
    topicsStage
      .thenApply(Metadata.Topics::getResponse)
      .thenApply(
        responseOptional ->
          responseOptional.map(
            map ->
              map.entrySet()
                .stream()
                .flatMap(
                  entry -> {
                    String topic = entry.getKey();
                    List<PartitionInfo>partitionInfos =entry.getValue();
return partitionInfos.stream().map(info -> topic + ": " + info.toString());
                  })
                .collect(Collectors.toList()))));
```

Example2:

```
Private Metadata getMetadata(Collection<TopicPartition> partitions)
{
Node node0 = new Node(0, "localhost", 100, "rack0");
Node node1 = new Node(1, "localhost", 100, "rack1");
Node[] nodes = {node0, node1};
Set<Node> allNodes = new HashSet<>();
allNodes.add(node0);
allNodes.add(node1);
Set<PartitionInfo> parts = new HashSet<>();
for (TopicPartition tp : partitions) {
  parts.add(new PartitionInfo(tp.topic(), tp.partition(), node0, nodes, nodes));
}
Cluster cluster = new Cluster("cluster-id", allNodes, parts, Collections.emptySet(), Collections.emptySet());
Metadata metadata = new Metadata();
metadata.update(cluster, Collections.emptySet(), 0);
return metadata;
}
```

7   Conclusion

(A) OKM for text

At present metadata of Data Lakes and Streams can be defined in Kafka as objects with objected are to be defined for ontology purposes, it is possible to define and describe the same in OKM. After that OKM statements can be converted (either manually or by preprocessor software) into metadata in Kafka which gets reflected in its implementation. Given any sentences, OKM grammar is able to parse sentence and segment the sentence into features but assigning the attributes to the word depends on the accuracy of the Stanford POS Tagging and universal dependencies. There are certain conclusions which were drawn from the results. OKM is more powerful than Stanford in sentence segmenting which can be seen in some of the cases for ex in the Figs. 2, 5 and 7.

(i)   For the sentence in Fig. 2 the sentence is segmented as:
      Stanford: (Brother, a letter) (for you) OKM: (Brother) (a letter for you)
(ii) For the sentence "She is Rohit's friend" in Fig. 5, the sentence is segmented as:
      Stanford: (She) (is Rohit's friend). OKM: (She is) (Rohit's friend).
(iii) For the sentence "The leaf is falling from the tree" in Fig. 7 the sentence is
      segmented as:
      Stanford: (The leaf) (is falling from the tree)
      OKM: (The) (leaf is falling) (from the tree).

In the above three examples one can clearly see the difference in sentence segmenting where OKM is more intelligent when compared to Stanford grammar.

(B) OKM for Data Lakes, Streams in Kafka:

Kafka goes beyond Hadoop with a schema registry, self-compressing storage that understands the concept of a "key" and other characteristics that assume data will change. Dozens of "writers" and "consumers" build to this open standard, empowering integration of transaction and other rapidly changing information with enterprise data stores, processing platforms and more. We have to first use OKM for making a design of a schema or ontology which can be subsequently be translated into Metadata of Kafka.

8   Future Work and Applications

- OKM can be applied in Semantic Web as a tool for better knowledge representation in KM models.
- It can be applied in Search Engine technology for cross lingual, multilingual search.
- The proposed OKM can be used for scoring the code mixing [14] in the Facebook test chatbots.
- It can be applied in language based learning by deploying OKM as a Knowledge Representation methodology with learning parameters as input and modification/ extension of the knowledge represented.
- OKM can be deployed for stream data with Kafka.
- The StanfordCoreNLP (Stanford tool) takes lesser time than the proposed OKM, but our work expresses the meaning of the sentence in a better way than the StanfordCoreNLP (Stanford tool), the decision score to compare the OKM and StanfordCoreNLP (Stanford tool) can be produced as future work.

References

1. Berners-Lee, T., Hendler, J., Lassila, O.: The semantic web. Sci. Am. **284**, 34–43 (2001)
2. Dai, W., You, Y., Wang, W., Sun, Y., Li, T.: Search engine system based on ontology of technological resources. JSW. **6**(9), 1729–1736 (2011)
3. Marzano, G.: Using resource description framework (RDF) for description and modeling place identity. Procedia Comput. Sci. **77**, 135–140 (2015)
4. Kim, K., Moon, B., Kim, H.: R3F: RDF triple filtering method for efficient SPARQL query processing. World Wide Web **18**, 317–357 (2013)
5. Jeon, M., Hong, J., Park, Y.: SPARQL query processing system over scalable triple data using SparkSQL framework. J. KIISE. **43**, 450–459 (2016)
6. Segaran, T., Evans, C., Taylor, J.: Programming the Semantic Web. O'Reilly, Beijing (2009)
7. Suryanarayana, D., et al.: Stepping towards a semantic web search engine for accurate outcomes in favour of user queries: using RDF and ontology technologies. In: 2015 IEEE International Conference on Computational Intelligence and Computing Research (ICCIC). IEEE (2015)
8. Hussain, S.M., et al.: Palazzo matrix model: an approach to simulate the efficient semantic results in search engines. In: 2015 IEEE International Conference on Electrical, Computer and Communication Technologies (ICECCT). IEEE (2015)
9. Suryanarayana, D., et al.: Cognitive analytic task based on search query logs for semantic identification. IJCTA **9**(21), 273–280 (2016)
10. Lewis, D.: Feature selection and feature extraction for text categorization (2000). https://doi.org/10.3115/1075527.1075574

11. Pakhale, K.D., Pawar, S.S. Focused retrieval of e-books using text learning and semantic search
12. Ding, L., Finin, T., Joshi, A., Pan, R., Cost, R., Peng, Y., Reddivari, P., Doshi, V., Sachs, J.: Swoogle: a search and metadata engine for the semantic web. In: International Conference on Information and Knowledge Management, Proceedings, pp. 652–659 (2004). https://doi.org/10.1145/1031171.1031289
13. de Marnee, M.-C., Manning, C.: Stanford typed dependencies manual (2008)
14. Vyas, Y., Gella, S., Sharma, J., Bali, K., Choudhury, M.: POS tagging of English–Hindi code-mixed social media content, pp. 974–979 (2014). https://doi.org/10.3115/v1/d14-1105
15. Kaur, R., Singh Garcha, L., Garag, M., Singh, S.: Parts of speech tagging for Indian languages review and scope for Punjabi language. Int. J. Adv. Res. Comput. Sci. Softw. Eng. **7**, 214–217. https://doi.org/10.23956/ijarcsse/v7i4/0140
16. Wagner, W.: Lang. Resour. Eval. **44**, 421 (2010). https://doi.org/10.1007/s10579-010-9124-x
17. Stanford parser. http://nlp.stanford.edu:8080/parser/
18. Sanskrit grammar. https://en.wikipedia.org/wiki/Sanskrit_grammar
19. What is the nominative case? definition, examples of nominative pronouns. https://writingexplained.org/grammar-dictionary/nominative-case
20. English for students. http://www.english-for-students.com/Noun-Cases.html
21. The Stanford Natural Language Processing Group. https://nlp.stanford.edu/software/tagger.shtml
22. Wikipedia (Grammar). https://en.wikipedia.org/wiki/Grammatical_case
23. Karak (Case). http://hindigrammar.in/case.html
24. English Grammar Guide. https://www.learnenglish.de/grammar/
25. Learning English Online. https://www.grammarbank.com/

Evaluation of Performance on N-Wavelength V2X Wireless Network in Actual Road Environment

Akira Sakuraba[1]([⊠]), Yoshitaka Shibata[1], Noriki Uchida[2],
Goshi Sato[3], and Kazuaki Ozeki[4]

[1] Regional Cooperative Research Division, Iwate Prefectural University,
Takizawa, Iwate-Pref, Japan
{a_saku, shibata}@iwate-pu.ac.jp
[2] Faculty of Information Engineering, Fukuoka Institute of Technology,
Fukuoka, Fukuoka-Pref, Japan
n-uchida@fit.ac.jp
[3] Resilient ICT Research Center, National Institute of Information
and Communications Technology, Sendai, Miyagi-Pref, Japan
sato_g@nict.go.jp
[4] P&A Technologies Inc., Morioka, Iwate-Pref, Japan
k_oozeki@pa-tec.com

Abstract. Road state sensing is one of essential technology to realize autonomous driving. High-level autonomous vehicle requires road state information with highly realtime and granularity in order to drive itself safely in any condition of road even snowy or icy road in winter. There is several methods to obtain current road state of vehicle running point, however vehicle requires some methods to exchange road state information among other vehicles or roadside units. This work proposes a Vehicle-to-Everything (V2X) wireless communication system which is oriented to exchange road state information. We designed the system for mutual exchanging road state information with other vehicles. In our design, proposed system switches the best wireless link according to signal or type of data. We evaluate our basic system design for V2V wireless communication in real road environment. We have measured basic network performance between roadside unit and vehicle which moves actual velocity of vehicle. The result suggests our method has enough network performance even in realistic road traffic.

1 Introduction

In winter season of cold weather district, there is a volatile road surface condition which varies even short distance and within a short times. Drivers in the region can easily encountered with the complicated road surface condition which contains dried, icy, wet, or compacted snow, this could be challenge for realizing high-level autonomous vehicle such as SAE 5 [1] not only today's manual driving vehicle. Hence, road condition sensing is a necessary elements of the future driving.

© Springer Nature Switzerland AG 2020
L. Barolli et al. (Eds.): CISIS 2019, AISC 993, pp. 555–565, 2020.
https://doi.org/10.1007/978-3-030-22354-0_49

We can find a lot of road sensing methods in order to understand forward road surface condition with sensors. For example, dynamics sensor such as accelerometer is a popular equipment to determine road surface condition in civil engineering, this type of method collects oscillation of the probe vehicle body with onboard dynamics sensors. However, these sensing methods is designed for offline processing which is analyzed, in other words, analyzing will be progressed after probe vehicle have completed to record road surface condition. Thus we have to consider a wireless communication in effort to exchange with other vehicle road state information which is determined onboard in realtime.

In this work, we propose a Vehicle-to-Everything (V2X) wireless communication system which is oriented to exchange road state information. We designed the system to deliver road state information and onboard sensor data between vehicles which is conducted by vehicle-to-vehicle (V2V) network, and between vehicle and roadside unit (RSU) which is delivered over vehicle-to-road (V2R) communication. Both wireless communications are configured without any public packet wireless network. Our communication system has cognitive wireless network which has multiple wireless standard and switches the best link to deliver data in response to network condition and types of data. We focus to describe V2V wireless communication in this paper.

We had a field experiment of V2V communication in actual public road and vehicle speed, in order to measure how our method and configuration is reasonable. The result suggests our proposed method has reasonable performance to exchange road state information in actual road environment.

2   Related Works

Road sensing technique processes in order to determine what road condition is corresponded to by analyzing of sensor output which is placed on moving vehicle. These methods had been introduced in field of pavement engineering with non-contact onboard sensor to collect pavement condition while probing vehicle is running. In this method, dynamics sensors especially acceleration sensor is one of popularly sensor to obtain road condition such pothole or bump of pavement. Du et al. proposed a measurement method in order to estimate the International Roughness Index (IRI) using Z-axis acceleration sensor [2]. Casselgren et al. introduced a road surface condition determination method with near infrared (NIR) sensor [3]. This method uses three different wavelength of NIR laser sensors in order to determine paved road condition into dry, wet, icy, and snowy.

Floating car data (FCD) is a vehicular telemetry method which provides to collect status of the remote place such as road traffic flow, the system contains public wireless network such 3G/3.5G/4G cellular network for data delivering. In other hand, CarTel [4] has V2R communication on private wireless network which is short-range vehicular adhoc network (VANET) based on 802.11b WLAN, it was possible to deliver data which is capable about 30 KiB/s end-to-end throughput from vehicle to internet.

We can cite a number of VANET proposal methods which is based on DSRC or 802.11p, and 802.11abgn for the popular wireless standard using for V2R or V2V communication device. WAVE system [5] is a prototype of DSRC operated on

5.8 GHz band and designed to notify traffic signs for driver with V2R communication. RSUs which equips the improved DSRC based wireless device are now in operation at actual expressway of Japan.

V2X applications especially require wireless communication short latency and less packet loss ratio not only high efficiency data delivering, there is several implementations of cognitive radio for V2X communication which consists of multiple wireless link and switches it situationally. Ito et al. developed 2-wavelength wireless communication consists of 920 MHz and 802.11bgn 2.4 GHz WLAN device to reduce latency which is caused time of authentication and association which is performed before delivering on WLAN band, they confirmed that the system allows to increase amount of data delivering between vehicle and RSU [6].

3   Proposed System and Configuration

This section describes road condition sharing process of our proposal system which can obtain sensor value and determine road condition on the vehicle, and then exchanging with other vehicle and/or RSU and share among neighbor vehicles.

3.1   Overview

Figure 1 illustrates an overview of our system. Our proposed system can share with highly real-time road states for drivers in manual driving.

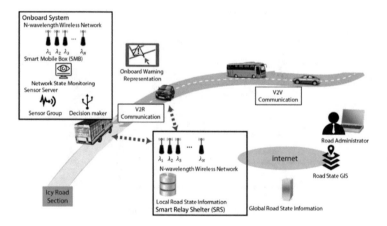

Fig. 1. System overview of N-wavelength V2X wireless communication system

Onboard unit is loaded on each vehicle which consists of Sensor Server and SMB (Smart Mobile Box). Sensor Server deals collecting of multiple onboard sensor output firstly. SMB analyzes for decision road states of up to current travelling point which describes as road state corresponding to dried/wet/snowy/icy/whiteout etc. with granularity. Next, SMB associates road state and their location. Then SMB delivers it with raw sensor data to RSU called Smart Relay Shelter (SRS) on N-wavelength V2R cognitive wireless network. On other side, roadside server provides road data and states of wide area is collected by other vehicle via the same network. This cognitive network also takes V2V communication directly with a vehicle which is heading to the opposite direction from own vehicle. At the same time, SRS also perform to deliver regional road states and acquires other area road surface states with cloudlet on cloud computing resources, therefore our approach allows to provide service is public cellular network independent.

Finally, the system provides visualization of road states based GIS (Geographic Information System) for users of road such drivers or operation managers of public transportation system such bus or taxi. For operation managers or driver who is not in driving, our proposal system represents road states as generic GIS designed for using on smart device and PC. Each vehicle equips road surface warning system in order to represent and warn for driver who is in driving, if ahead road condition would get worse which is described as icy, whiteout, etc. system will warn it in visually on display device and playout of warning message by human voice.

Continually acquiring of road states information is required to provide precise information, it could be efficient to load onboard unit on frequently running vehicles along the same route such bus.

These functions allows acquiring, sharing, and representing of road condition, it could realize safer road traffic even winter season.

3.2   Configuration of System

System configuration is shown in Fig. 2.

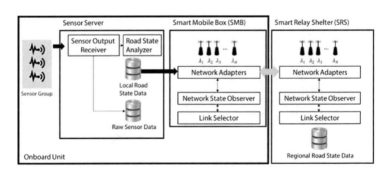

Fig. 2. System Configuration

Sensor server monitors output value from sensor group which is composed of various sensors which includes acceleration, gyro, geomagnetic, temperature, humidity, and atmospheric pressure, quasi electrostatic fields (QEF), NIR sensor, RGB image sensor, and positioning of vehicle in order to determine road surface state where vehicle is running. Sensor server is based on edge computing architecture and processes acquired sensor values to identify road state by machine learning algorithm. After road state decision has been completed, the result of analyze is sent to SMB.

SMB has several wireless networks with different wavelength wireless network device and decision maker which wireless device should be delivered to other SMB or SRS.

3.3   Cognitive Wireless Network for V2V Communication

The system consists of multiple network devices which have different wavelength wireless links. In implementation at this time, we install three different wireless standard based devices which is composed of 2.4/5.6 GHz band WLAN and 920 MHz band LPWA. Our V2V wireless communication is based on following two wireless standard to exchange road state information between vehicles.

2.4GHz Band WLAN. 802.11bgn is the most popular wireless standard and used in many field today, as well as VANET platform. 2.4 GHz band is a license-free band both indoor and outdoor. However, as this wireless band would be encountered with terrible interference due to carriers of other traffics, therefore 2.4 GHz band has concerned of system could not expect much performance especially in urban area. In current implementation, we designed the system to deliver road state information over 2.4 GHz WLAN which could be composed of less amount of data.

920 MHz Band LPWA. 920 MHz band wireless network, as known as LPWA, is one of license-free long distance radio communication technology. Network performance of this type wireless technology is designed to perform hundred to several kilo bps throughput with several to several ten kilometers range distance, it is difficult to transmit large amount of data as primary data transmission link but it is suitable to utilize it as controlling message link. Our system delivers network configuration information of WLAN before vehicle enters into WLAN communication range. This procedure will be described in Sect. 3.4.

3.4   Message Exchanging Sequence in V2V Communication

Figure 3 describes data transmission V2V communication using 2-wavelength network devices from initiation to be connected to WLAN. Firstly destination vehicle which

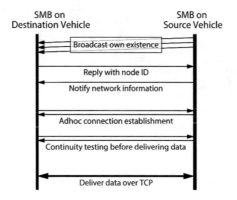

Fig. 3. Vehicle-to-vehicle wireless communication

requires road state information from other vehicle, waits broadcasts own existence information from source vehicle which has already road state information, via 920 MHz band. If SMB on destination vehicle received broadcast message, sends node ID in response. In V2V communication, source vehicle answers IP address related information, SSID, and wireless passphrase to destination vehicle. After destination vehicle received these information, it activates WLAN adapter and attempts to connect in managed mode on WLAN until both vehicles in range of it. After WLAN connection has been established, destination vehicle measures for continuity, finally road condition exchange will be taken over TCP between both vehicles after continuity test has been completed.

4   Prototype Environment

We built three-wavelength wireless communication device which uses Buffalo UI-U2-300D for 802.11bgn WLAN USB device and Oi Electric OiNET-928 or OiNET-923 for 920 MHz band LWPA wireless network, these devices are connected to SMB.

Controlling of SMBs are based on x86-64 architecture desktop system, we implemented them on Intel NUC Kit NUC5i5RYH barebone PC kit with Ubuntu 16.04 and 18.04 Linux system. Application on SRS and SMB is written by C, Ruby, and Bash Script.

5   Evaluation of V2V Communication

Figure 4 illustrates outline of evaluation for V2V wireless communication which consists of 920 MHz band LPWA and 2.4 GHz band WLAN in actual road environment. We performed to measure performance on TCP networking on WLAN which is conducted by onboard SMB which is placed on two vehicles moving to another direction.

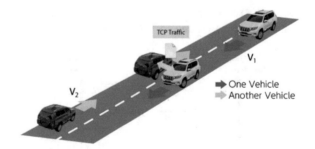

Fig. 4. Overview of both vehicle direction trials and installation of SMB wireless units

Fig. 5. Placement of wireless unit and antenna on both vehicle

5.1 Configuration and Network Setup

Firstly, we prepared pair of SMB which consists of 920 MHz LoRa standard LPWA unit and 802.11bgn WLAN adapter, both wireless devices is connected to a PC based SMB host. In the experiment, TCP traffic was generated with iPerf3 [7] and it has single-direction from one vehicle to another vehicle. In order to do, we configured SMB on another vehicle will be performed as AP, and one vehicle's one will be STA consistently.

Generically LPWA realizes very long distance communication range, however, it could not provide such wide area range in case of antenna is placed on rooftop of the vehicle due to lower feeding point of LPWA antenna height. Hence we configured spreading factor as known as SF value of LPWA to higher value in order to secure robust communication even both vehicles are moving in non-line-of-sight place each other. We configured LPWA unit in advance, the parameter was center frequency 920.6 MHz with mode SF11 which can transmit 537 bps of designed throughput with receive sensitivity of -136 dBm. This configuration is composed of balanced between robustness and practical performance for V2V communication.

5.2 Environment for in-the-Field Experiment

Secondly, we loaded LPWA antennas on rooftop of the one and another vehicle, and installed USB WLAN adaptor on left door mirror of both vehicle, the WLAN adaptor faces to the sidewalk. Figure 5 indicates placement of Tx/Rx antenna of wireless unit on the vehicles.

We planned this evaluation in actual public two-lane road which is illustrated in Fig. 6. The length of evaluation section on road was 1.1 km and road alignment was almost straight and almost smooth terrain but there was several grade section which includes 3.3% grade in 90 m length. There is non-line-of-sight between both vehicles at the initial due to the terrain, buildings, and rows of trees. Initial distance of both vehicle was 1.08 km.

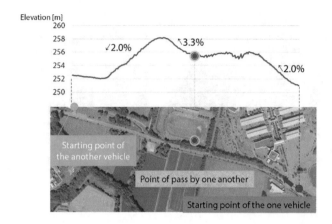

Fig. 6. Road alignment and grade of the experimental section on public road

We performed the experiment with various absolute vehicle velocities at 20, 30, 40, and 50 km/h. We have conducted 10 one-way trials in each velocity scenario. The condition of moving velocity of the one vehicle and the velocity of another vehicle were almost equally and vehicle stayed at a constant speed in every scenario.

The condition of road traffic density was slightly higher and weather was clear during until we had completed whole trials.

5.3   Measuring Target

We build an evaluation system before performing before running on actual road. While our evaluation, system records received signal strength index (RSSI), throughput and delivered data amount on WLAN link. SMB recorded the log for this evaluation which is installed on one vehicle behaved as STA.

5.4   Result

Figure 7 indicates averaged RSSI and averaged throughput of WLAN link which is both plotted in every second, the result recorded as of the time after the link established. The result is separately plotted in each velocities. We have observed that succeeded rate of TCP connection establishing was 92.5% of whole scenarios.

These graphs denote that both the peak value of RSSI and throughput appeared earlier elapsed time when the velocity was faster. Generically, RSSI has a strong correlation with throughput, we can consider that this result was reasonable.

Throughput on TCP traffic was affected vehicle velocity. The result shows that it was available in 38 s after establishing when the vehicle was moving at 20 km/h, it was decreased by 15 s in the scenario at 50 km/h. The peak of throughput was not drastically changed by varying vehicle's velocity.

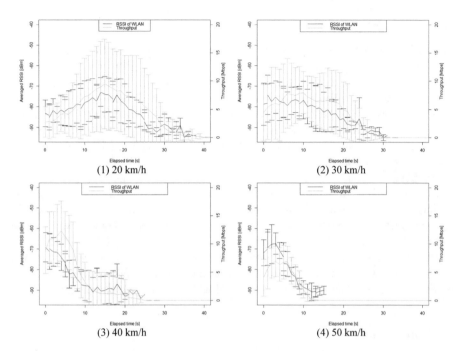

(1) 20 km/h

(2) 30 km/h

(3) 40 km/h

(4) 50 km/h

Fig. 7. Change of averaged RSSI and throughput after WLAN connection has been established in V2V communication. Each graphs is correspond to 20, 30, 40, and 50 km/h of velocities. Error bar describes standard deviation.

We have also investigated total amount of delivered data while the vehicle is moving in experiment scenario.

Figure 8 illustrates this result corresponding to vehicle velocity. Total data amount was almost linear decreased depending on increasing of vehicle velocity, it was varied from 24.2 MiB at 20 km/h to 7.03 MiB at 50 km/h scenario.

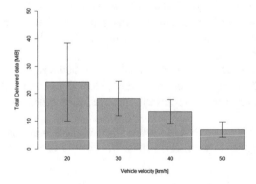

Fig. 8. Averaged total delivered amount of data which is corresponding to vehicle velocity. Error bar describes standard deviation.

5.5   Discussion

Surprisingly the result shows that the system has some capability for realistic vehicle velocity up to 50 km/h which is possible in actual road traffic. However, total delivered data amount was varied even the same condition by environmental factors, such as terrain, existing of oncoming or overtaking vehicle, or distance of the following vehicle, etc. Especially, terrain of the road influents line-of-sight (LOS) or difference height of Tx/Rx antenna on the vehicle which is occurred by terrain, therefore these factors could make significant effects to RSSI. Thus we should consider the position of WLAN antenna on the vehicle.

In higher vehicle velocity scenario, our prototype recorded higher RSSI after connection establishing. We can analyze this result is caused with distance between vehicle was closer while WLAN adapter is establishing with 4-way handshake, the distance when it completed to establishing would be shorter than slower velocity scenario. To obtain much amount data with exchanging over V2V communication, we should implement a new WLAN connection procedure which conducts shorter time compared with conventional method, for instance Fast Initial Link Setup (FILS) which is standardized as IEEE 802.11ai [8], it could realize as one of solution for it.

6   Conclusion

This work proposed a design for N-wavelength wireless communication system for V2X environment to exchange road surface conditions. This system has three of fundamental functions; observing of physically state of road surface by multiple sensors on the vehicle, analyze these sensor output for decision of road surface conditions, and transmit calculated road condition to roadside device or other vehicle to share road condition which the vehicle travelled until now. Design of our communication subsystem utilizes cognitive wireless network which consists of WLAN for data delivering and LPWA as the control message link.

We have evaluated our proposed cognitive wireless system on actual road environment. The evaluation results indicated the system has capability for delivering several to several ten MiB road state information over V2V communication even vehicles are moving in realistic speed on actual public road. It is a possibility to realize a public packet network independent system which would be worked in rural or mountainous area road.

In our future work we will implement actual data delivering application which uses in the inter-urban highway. We are also now planning to carry out the social experiment for road state sharing system using improved prototype.

Acknowledgments. This research was supported by Strategic Information and Communications R&D Promotion Program (SCOPE) No. 181502003, Ministry of Internal Affairs and Communications, Japan.

References

1. Taxonomy and Definitions for Terms Related to Driving Automation Systems for On-Road Motor Vehicles J3016_201806, SAE International (2018)
2. Du, Y., Liu, C., Wu, D., Jiang, S.: Measurement of international roughness index by using - axis accelerometers and GPS. Math. Probl. Eng. **2014**, 1–10 (2014)
3. Casselgren, J., Rosendahl, S., Eliasson, J.: Road surface information system. In: Proceedings of the 16th SIRWEC Conference (2013)
4. Hull, B., et al.: CarTel: a distributed mobile sensor computing system. In: Proceedings of the 4th Conference on Embedded Networked Sensor Systems (2006)
5. Tsuboi, T., Yamada, J., Yamauchi, N., Hayashi, M.: Dual receiver communication system for DSRC. In: Proceedings of 2008 2nd International Conference on Future Generation Communication and Networking, pp. 459–464 (2008)
6. Ito, K., Hashimoto, K., Shibata, Y.: V2X communication system for sharing road alert information using cognitive network. In: Proceedings of 8th International Conference on Awareness Science and Technology, pp. 533–538 (2017)
7. iPerf: Download iPerf3 and original iPerf pre-compiled binaries. https://iperf.fr/iperf-download.php. Accessed 22 April 2019
8. IEEE 802.11ai-2016: IEEE Standard for Information technology - Telecommunications and information exchange between systems - Local and metropolitan area networks - Specific requirements Part 11: Wireless LAN Medium Access Control (MAC) and Physical Layer (PHY) Specifications Amendment 1: Fast Initial Link Setup (2016)

The 12th International Workshop on Intelligent Informatics and Natural Inspired Computing (IINIC-2019)

Evaluation of Mobile Health Services in Health Organizations

Alsaleh Saad[(⊠)]

Hail Health Affairs, Ministry of Health, Riyadh, Kingdom of Saudi Arabia
saabalsaleh@moh.gov.sa

Abstract. Mobile health (mHealth) has a crucial role to play in the health sector as it can improve communication and enhance the integration of healthcare processes. The aim of this study is to evaluate and enhance the mHealth services currently available through Saudi Ministry of Health (SMOH) mobile applications. To achieve this aim, the study adopted an applied (evaluation) research method and the data were collected through direct observation. Although the SMOH has developed five mHealth applications to deliver its mHealth services, the study found that they did not include several important health services such as appointment reservation, opening and updating of medical records, patient referee and consultation, physician directory, request for medical reports, and health risk assessments. Implications drawn from the study findings could expand an area of mHealth services in the Kingdom of Saudi Arabia (KSA) which has yet to be adequately explored and evaluated. Additionally, the study findings may provide valuable insights for health professionals and administrators.

1 Introduction

Adoption of electronic health technology by end users, particularly patients, is an essential element in the area of Information Systems (IS). Recently, a new technology which has attracted the attention of health professionals is Internet-connected mobile technology (mHealth). It exploits mobile telecommunication and multimedia technologies and integrates them into new mHealth care delivery systems. With the use of modern smart mobile devices, mHealth is an emerging area in telemedical and telecare systems that offers media-rich as well as context-aware features that are extremely useful for electronic-health applications [1, 2]. The smartphone technology (as a telecommunication technology) has accelerated the implementation of telemedical platforms to distribute the awareness of health. Since the advent of smartphones, mHealth has gained more attention from participants in the healthcare system as it could completely change the way healthcare is viewed, managed and delivered [3].

Currently, there are more than 40,000 health and medical applications offered in the Apple App store [4] and this number is expected to grow annually by about 23% over the next few years [5]. A study in the KSA recently found that a large number of patients use their personal smartphone to surf the internet instead of using a personal

© Springer Nature Switzerland AG 2020
L. Barolli et al. (Eds.): CISIS 2019, AISC 993, pp. 569–577, 2020.
https://doi.org/10.1007/978-3-030-22354-0_50

desktop computer, as they did previously [6]. The Saudi Ministry of Health (SMOH) currently has five mHealth applications and they are available in three main stores include Apple App, Google Play, and Windows. This study has evaluated the health services provided by the SMOH through these mHealth applications and makes some recommendations to enhance the service efficacy of these applications.

2   Literature Review

2.1   The mHealth Concept and Importance

The concept of mHealth involves various types of mobile technologies that are used for health purposes. According to [6], the term "mHealth" refers to public and medical health practices which obtain support from a mobile device such as mobile phones and other wireless devices. It includes several mobile technologies such as mobile phones, personal digital assistants, smartphones, patient monitoring devices, mobile telemedicine/telecare devices, MP3 players, and mobile computing [7]. For this the purposes of this study, mHealth means using a health application via a smartphone for health and medical-related matters.

According to [8], smartphones offer the of promise improved clinical efficiency, medical quality, care coordination and reduction of healthcare cost. MHealth is important as it has a crucial role to play in healthcare field, since it can improve communication and enhance the integration of care processes [9, 28, 36], support public healthcare [10, 32], improve the performance of healthcare workers and health systems [29], and enhance health access and health delivery [4, 32, 35, 38]. A majority of the global population has access to real-time communication and information services via networks of mobile phones [12] and such services are currently part of people's daily activities [13, 33]. More specifically, mHealth applications enable individuals to accomplish most of their healthcare services conveniently [13]. In addition, mHealth technologies can provide a platform to deliver messages to patients for many reasons such as supporting them to adhere to their treatment, increasing their awareness about a particular medical issue, reminding them about their next appointment or regular check-ups with their provider, or even informing them about some important upcoming health event. Researchers have suggested that mHealth technologies gain extended capabilities when they are used in conjunction with each other in relation to consumer health communication [2]. In the KSA, there is considerable support for the adoption of mHealth technology to improve the management and health awareness of medical problems [5].

2.2   Enhancement of mHealth Services Quality

Any successful health organization must continue to improve and enhance its services quality. The importance of mHealth quality has been demonstrated in many studies [13, 15, 30]. According to [11] "the role of service quality in fostering the growth of mHealth services has gained much attention in the academic and practitioner communities" (182). Researchers [40] emphasized that the health service "is considered as the major element of fourth-generation health systems and the key to their success and future evolution" (547). Poor quality leads to complications and the need for additional care, which raises costs substantially [16]. Three parameters of service quality in mHealth settings have been suggested by [12] which are: knowledge and competence of the provider (training and orientation for healthcare professionals); capacity of access and monitoring devices (aware of the limitation of mobile devices); operational compatibility among multiple platforms and interoperability of information systems. A prior study found that there is an association between service quality and the perceived quality of health [17]. In another study [18], it was found that the service quality provided via a mobile platform is influenced by IS, the mobile network, or the information itself. This study focused on only one element, namely; the information provided.

2.3 Mobile Health Applications

The mHealth applications are run on various mobile devices. According to [1], "Many e-health application developers have chosen Apple's iOS mobile devices such as iPad, iPhone, or iPod Touch as the target device to provide more convenient and richer user experience, as evidenced by the rapidly increasing number of m-health apps in Apple's App Store" (2022). The mHealth applications available in the App store have been classified into seven types which are: drug information, medical information reference, medical practitioners' decision support (which includes physicians, surgeons, and nurses), medical educational tools, disease tracking tools, blood pressure tools, medical calculators, and others health tools including eye charts, medical images, color test tools, and timers reminding users to take medicine [1].

The mHealth applications have been developed and used in numerous health fields such as diabetes, obesity, depression treatment [23–25], patient monitoring, emergency response, emergency management [37, 41], preventive care, and remote diagnosis [42]. However, it has been argued that most mHealth applications have a simple functionality and do little more than provide information [31]. More than half of mobile health applications received fewer than 500 downloads for reasons that include poor quality, the absence or lack of guidance on their benefits, as well as lack of support from health professionals [39].

3 Research Methodology and Data Analysis

3.1 Research Methodology

Although many methods can be used for exploring, describing or evaluating mobile technologies, the present study opted for the evaluative (applied) research method, because this method concentrates on evaluating and enhancing mHealth application products. According to [43], one goal of applied research is to evaluate software and system products. Furthermore, applied research is one of the research methods being used in mobile human-computer interaction research; it is aim-directed and results in some kind of product being produced [44]. Moreover, applied research is "relevant in relation to design and implementation of systems, interfaces and techniques, which meet certain requirements for performance, user interaction and user satisfaction" [44, p. 321]. The data were collected through direct observation and survey of SMOH mobile applications which are available and run through both IOS and Android operating systems. The direct observation (including taking of notes, photographs, and video recordings) has historically been favored by user-centered design since it a method that situates researchers in the context in which technology use occurs [45]. After collecting the data, the researcher evaluated the services available based on his work experience. In IS research, the "data collection almost always involves the researcher's direct engagement in the setting studied…, thus, the researcher is the instrument for collecting and analyzing data; the researcher's impressions, observations, thoughts, and ideas are also data sources" [45, p. 39]. Due to space constraints in this article, the research stages are illustrated in Fig. 1.

Fig. 1. Research stages flowchart

3.2   Data Analysis

The SMOH currently has five mobile applications, which are: Main App, E-Directory App, Citizen's Voice App, the Children's Vaccination Reminder App, and Evidence-Based Healthcare (EBHC) Mobile App. The main interfaces of these applications are shown in Fig. 2.

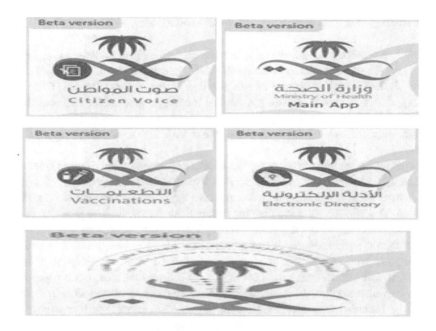

Fig. 2. Interfaces of SMOH mHealth applications

According to [47], the Main App was developed to get an overview of the SMOH, including its vision, mission and values, apart from news about from news about SMOH and world health days. Users can add events to the calendar, view what is new on the SMOH Portal, and obtain contact information. Users can also benefit from several e-services, such as the transactions inquiry service, which enables a user to inquire about e-transactions, and the payment orders inquiry service. The second application aims to enable users to search the website and locate SMOH facilities. In addition, the application provides information about the number of beds in a given hospital and the contact number of the facility, so that users can contact the facility directly via their smartphone. The "Driving" feature in this application allows users to navigate their way to the facility from wherever they are, by drawing a path on the interactive maps available on their smartphones, showing how to reach the hospital or health center and guiding users through the most efficient way. The third application (Citizen's Voice) enables citizens and residents to make their voices heard by ministry officials. It includes opportunities for making complaints, inquiries, and suggestions. The Children's Vaccination Reminder App is designed to remind parents of the dates of basic vaccinations against a range of diseases targeted by vaccination, according to the latest ministry vaccination schedule. This app also enables parents to manage their children's information, to view the vaccination history of each child, as well as the schedule for future vaccinations. The last application (EBHC), is intended to help clinicians implement guidelines in all hospitals and healthcare centers and provide standardized healthcare for patients.

4 Finding, Discussion and Recommendations

Although the SMOH has developed five mHealth applications to assist in the delivery of its E-Health services, the study found that the apps currently available do not include several essential and important health services such as appointments reservation, opening a medical record, updating a medical record, patient referee and consultation, physician directory, health risk assessments, request for medical reports. These essential services are available in many other mHealth applications used in other countries.

Based on the finding of the study, the researcher presents some recommendations to enhance the SMOH health mobile service which is provided through its mHealth applications. Firstly, it is recommended that an appointment reservation service should be added. This service will save time for both patients and healthcare professionals and it will decrease the effort required. Secondly, an app to open and update medical records should be introduced. When patients are able to open a medical record electronically, it will reduce the waiting time compared with achieving it manually. Besides, it will also decrease the cost as it is an electronic-based, not a paper-based, record. It is also suggested that an electronic medical consultation and referee service should be included into the SMOH application. This service will save physicians valuable time. The study recommends providing the application by means of a physicians' directory in which physicians could be classified based on their specialization and their locations in order to facilitate ease of communication between patients (or potential patients) and relevant physicians.

Health risk assessments should also be added to the application, as these are important for many of the SMOH patients. The health risk assessments may include obesity, mental health, diabetes, and heart disease. The facility to request a medical report is one of the main services which it is recommended should be available in the SMOH application. Adding such a service will save patients time as it will eliminate the time and effort needed to physically travel to the hospital to obtain the report, especially when the hospital is located in another province or city. The mHealth technologies have several benefits and advantages in the healthcare area. According to several studies, these advantages include: a reduction in human errors [14], and reduction in costs [1, 19]. Furthermore, the technologies "improve portability, immediacy, convenience, comparatively low unit efficiency, facilitate communication between patients and therapists…and leverage widely in chronic illness care, and care of elderly and disabled people" [1, p. 2024] as well as increasing health self-care and autonomous behaviors [34]. Additionally, they encourage healthy behaviors that may address several potential health problems and reduce the number of physical patient visits to health service providers [2, 21, 22, 26].

An additional advantage for health service providers (health organizations) of developing mHealth applications is that less physical infrastructure will be necessary [27] thereby resulting in lower costs.

5   Conclusion

This research is the first study to evaluate mHealth apps currently provided by the Saudi Ministry of Health. It has made several recommendations to enhance the SMOH mHealth services. The study could serve as a guide for health administrators as well as health practitioners interested in developing and producing mHealth applications for their potential patients, particularly from the perspective of patient services.

Acknowledgment. The author of this study wishes to record his appreciation to the Saudi Ministry of Health for its moral support.

References

1. Liu, C., Zhu, Q., Holroyd, K.A., Seng, E.K.: Status and trends of mobile-health applications for iOS devices: a developer's perspective. J. Syst. Softw. **84**(11), 2022–2033 (2011)
2. Istepanian, R.S., Lacal, J.C.: Emerging mobile communication technologies for health: some imperative notes on m-health. In: Engineering in Medicine and Biology Society, 2003. Proceedings of the 25th Annual International Conference of the IEEE, vol. 2, pp. 1414–1416. IEEE (2003)
3. Nasi, G., Cucciniello, M., Guerrazzi, C.: The performance of mHealth in cancer supportive care: a research agenda. J. Med. Internet Res. **17**(1), e9 (2015)
4. IMS Institute for Healthcare Informatics: Patient apps for improved healthcare: from novelty to mainstream. http://www.imshealth.com/deployedfiles/imshealth/Global/Content/Corporate/IMS%20Health%20Institute/Reports/Patient_Apps/IIHI_Patient_Apps_Report.pdf. Accessed 10 Dec 2015
5. Molina-Recio, G., García-Hernández, L., Castilla-Melero, A., Palomo-Romero, J.M., Molina-Luque, R., Sánchez-Muñoz, A.A., … Salas-Morera, L.: Impact of health apps in health and computer science publications. A systematic review from 2010 to 2014. In: Bioinformatics and Biomedical Engineering, pp. 24–34. Springer International Publishing (2015)
6. Alotaibi, M.M.: A study on the needs for educating and managing fasting diabetes through mobile health in kingdom of Saudi Arabia. Int. J. Comput. Syst. **2**, 4 (2015)
7. Mechael, P., Sloninsky, D.: Earth Institute at Columbia University. (2008). Towards the development of an mHealth strategy: a literature review. Retrieved from: http://www.who.int/goe/mobile_health/mHealthReview_Aug09.pdf
8. Putzer, G.J., Park, Y.: Are physicians likely to adopt emerging mobile technologies? Attitudes and innovation factors affecting smartphone use in the Southeastern United States. Perspectives in Health Information Management/AHIMA, American Health Information Management Association, 9(Spring) (2012)
9. Almaiman, A., Bahkali, S., Alfrih, S., Househ, M., El Metwally, A.: The use of health information technology in Saudi primary healthcare centers. Integrating Inf. Technol. Manage. Qual. Care **202**, 209 (2014)
10. Kahn, J.G., Yang, J.S., Kahn, J.S.: 'Mobile' health needs and opportunities in developing countries. Health Aff. **29**(2), 252–258 (2010)
11. Akter, S., D'Ambra, J., Ray, P.: Development and validation of an instrument to measure user perceived service quality of mHealth. Inf. Manag. **50**(4), 181–195 (2013)

12. Akter, S., Ray, P.: mHealth-an ultimate platform to serve the unserved. IMIA Yearb. Med. Inf. 75–81 (2010)
13. Shields, T., Chetley, A., et al.: ICT in the health sector: summary of the online consultation. InfoDev Working paper 2010. www.infodev.org/en/Document.84.aspx(2010). Accessed 10 May 2010
14. Franco, M., Tursunbayeva, A.: Mobile technology and public health organisational system. Symphonya. Emerg. Issues Manage. **1**, 81–90 (2014)
15. Mechael, P.: The case for mHealth in developing countries. Innovations: Technol. Governance, Globalization **4**(1), 103–118 (2009)
16. Porter, M.E., Teisberg, E.O.: Redefining healthcare: creating value based competition on results. Harvard Business School Press, Boston, Massachusetts (2005)
17. Choi, H., Lee, M., Lm, K.S., Kim, J.: Contribution to quality of life: a new outcome variable for mobile data service. J. Assoc. Inf. Syst. **8**(12), 598–618 (2007)
18. Koivisto, M.: Development of quality expectations in mobile information systems. In: International Joint Conference on Computer, Information, and Systems Sciences and Engineering (2007)
19. Kumar, S., Nilsen, W.J., Abernethy, A., Atienza, A., Patrick, K., Pavel, M., ... Hedeker, D.: Mobile health technology evaluation: the mHealth evidence workshop. Am. J. Prev. Med. **45**(2), 228–236 (2013)
20. Riley, W.T., Rivera, D., Atienza, A., Nilsen, W., Allison, S.M., Mermelstein, R.: Health behavior models in the age of mobile interventions: are our theories up to the task? Transl. Behav. Med. **1**, 53–71 (2010). PubMed: 21796270
21. Krishna, S., Boren, S.A., Bales, E.A.: Healthcare via cell phones: a systematic review. Telemed. J E Health. **15**(3), 231–240 (2009)
22. Quinn, C.C., Shardell, M., Terrin, M., Barr, E., Ballew, S., Gruber-Baldini, A.L., Cluster, A.: Randomized trial of a mobile phone personalized behavioral intervention for blood glucose control. Diabetes Care **34**, 1934–1942 (2011)
23. Bexelius, C., Lof, M., Sandin, S., Lagerros, Y.T., Forsum, E., Litton, J.E.: Measures of physical activity using cell phones: validation using criterion methods. J Med Internet Res. **12**(1), e2 (2010)
24. Burns, M.N., Begale, M., Duffecy, J., et al.: Harnessing context sensing to develop a mobile intervention for depression. J Med Internet Res **13**(3), e55 (2011)
25. Steinhubl, S.R., Muse, E.D., Topol, E.J.: Can mobile health technologies transform health care? JAMA **310**(22), 2395–2396 (2013)
26. Curioso, W.H.: New technologies and public health in developing countries: the Cell PREVEN project. In: Murero, M., Rice, R. (eds.) The Internet and Health Care: Theory, Research and Practice, pp. 375–393. Lawrence Erlbaum Associates, Mahwah (NJ) (2006)
27. Aranda-Jan, C.B., Mohutsiwa-Dibe, N., Loukanova, S.: Systematic review on what works, what does not work and why of implementation of mobile health (mHealth) projects in Africa. BMC Public Health **14**(1), 188 (2014)
28. Medhanyie, A.A., Little, A., Yebyo, H., Spigt, M., Tadesse, K., Blanco, R., Dinant, G.J.: Health workers' experiences, barriers, preferences and motivating factors in using mHealth forms in Ethiopia. Hum. Resour. Health **13**(1), 2 (2015)
29. Lewis, T.L., Wyatt, J.C.: mHealth and mobile medical apps: a framework to assess risk and promote safer use. J. Med. Internet Res. **16**(9), e210 (2014)
30. Becker, S., Miron-Shatz, T., Schumacher, N., Krocza, J., Diamantidis, C., Albrecht, U.V.: mHealth 2.0: experiences, possibilities, and perspectives. JMIR mHealth and uHealth **2**(2), e24 (2014)

31. Castaño Labajo, V., Xiao, J.: (2015). Market entry, strategy and business development in mobile health (mHealth) industry
32. Gagnon, M.P., Ngangue, P., Payne-Gagnon, J., Desmartis, M.: m-Health adoption by healthcare professionals: a systematic review. J. Am. Med. Inform. Assoc. ocv052 (2015)
33. Vincent, C.J., Niezen, G., O'Kane, A.A., Stawarz, K.: Can standards and regulations keep up with health technology?. JMIR mHealth and uHealth 3(2), e64 (2015)
34. Kaphle, S., Chaturvedi, S., Chaudhuri, I., Krishnan, R., Lesh, N.: Adoption and usage of mHealth technology on quality and experience of care provided by frontline workers: observations from rural india. JMIR mHealth and uHealth 3(2), e61 (2015)
35. Hoque, M.R., Karim, M.R., Amin, M.B.: Factors affecting the adoption of mHealth services among young citizen: a structural equation modeling (SEM) approach. Asian Bus. Rev. 5(2), 60–65 (2015)
36. Azfar, A., Choo, K.K.R., Liu, L.: Forensic Taxonomy of Popular Android mHealth Apps. arXiv preprint arXiv:1505.02905. (2015)
37. Beratarrechea, A., Kanter, R., Diez-Canseco, F., Fernandez, A., Ramirez-Zea, M., Miranda, J., … Rubinstein, A.: Challenges of implementing mHealth interventions for lifestyle modification in prehypertensive subjects in Argentina, Guatemala, and Peru. In: Technologies for Development, pp. 119–127. Springer International Publishing, Chicago (2015)
38. Johnston, M., Mobasheri, M., King, D., Darzi, A.: The imperial clarify, design and evaluate (CDE) approach to mHealth app development. BMJ Innovations, bmjinnov-2014 (2015)
39. Lahoti, A.A., Ramteke, P.L.: Advanced healthcare system using e-health & m-health in cloud & mobile environments. Int. J. Eng. Sci. Res. Technol. (2015) ISSN: 2277-9655
40. Hamine, S., Gerth-Guyette, E., Faulx, D., Green, B.B., Ginsburg, A.S.: Impact of mHealth chronic disease management on treatment adherence and patient outcomes: a systematic review. J. Med. Internet Res. 17(2), e52 (2015)
41. Panayides, A.S., Antoniou, Z.C., Constantinides, A.G.: An overview of mHealth medical video communication systems. In: Mobile Health, pp. 609–633. Springer International Publishing (2015)
42. Wynekoop, J.L., Conger, S.A.: A review of computer aided software engineering research methods. In: Proceedings of the IFIP TC8 WG 8.2 Working Conference on the Information Systems Research Arena of the 90's, Copenhagen (1990)
43. Kjeldskov, J., Graham, C.: A review of mobile HCI research methods. In: Human-Computer Interaction with Mobile Devices and Services, pp. 317–335. Springer, Berlin, Heidelberg (2003)
44. Hagen, P., Robertson, T., Kan, M., Sadler, K.: Emerging research methods for understanding mobile technology use. In: Proceedings of the 17th Australia conference on Computer-Human Interaction: Citizens Online: Considerations for Today and the Future, pp. 1–10. Computer-Human Interaction Special Interest Group (CHISIG) of Australia
45. Kaplan, B., Maxwell, J.A.: Qualitative research methods for evaluating computer information systems. In: Evaluating the Organizational Impact of Healthcare Information Systems, pp. 30–55. Springer, New York (2005)
46. MOH Apps for Smartphones: (2015). Retrieved from http://www.moh.gov.sa/en/Support/Pages/MobileApp.aspx

Intelligent Student Attendance Management System Based on RFID Technology

Bektassov Dias[1], Asif Mohammad[1], He Xu[1,2(✉)], and Ping Tan[2]

[1] School of Computer Science and Technology,
Nanjing University of Posts and Telecommunications, Nanjing 210003, China
dias_95_95@mail.ru, asifnupt@outlook.com,
xuhe@njupt.edu.cn
[2] Tongda College of Nanjing University of Posts and Telecommunications,
Yangzhou 225127, China
{xuhe,tanping5.20}@njupt.edu.cn

Abstract. Nowadays Radio Frequency Identification (RFID) based applications are rising in many areas such as industry, health-care, transportation. During the class time, manual collection of attendance consumes considerable amount of time and resources. Having many students in classroom, taking attendance is a difficult task and how to automatic get the attendance rate is very important for school management. In this paper, we proposed an easier way of using the RFID technology on how a school teacher is capable of automatically record the attendance instead of taking attendance manually. An RFID-based attendance management system is designed and implemented, and information service system including programmable devices and web-based applications are used. The test results show that the presented system is practical.

1 Introduction

Presently, there are two methods of recording student's attendance: calling its name or taking its signature [1]. It takes about 10 min to call out the name of each student in a class which is over 100 students. This time is waste and could be used in studies. Radio Frequency Identification (RFID) based systems take records at the beginning and the end of class [2]. But this kind of system requires placing a card on card-reader at the beginning and taking it out at the end of each class. This method saves time, forces students to attend classes, keeps data in safe and reduces work for teachers. A teacher can record student's activity during class time (such as playing mobile, sleeping), and can keep valuation of students [3]. With the help of smart attendance management system using RFID, the authority can get the students attendance information's and activities in different classes individually. The system performance includes data management, tracking student's attendance, sending reports, monitoring records, maintenance records, providing statistics, student registrations, attendance checking and teacher accounts.

Various kind of techniques for biometrics are used for verifying identification such as face recognition, irises, barcode face recognition, signatures, fingerprint, voice, Bluetooth and others [3, 4]. These systems are costly and needs to be maintained by programmers whereas our proposed system is fully automated, cheap and more accurate. In 2014, Kurniali designed an RFID student's attendance system, where the system can collect the data of those who attend classes by RFID-reader [5]. Eliminating or reducing the manual labor activities were our main purpose. The proposed system is more advanced and the main goal is to encourage students to attend their classes. Many traditional systems for student's attendance management take time consumption, increasing workforce requirements and double efforts respectively. These procedures improved educational and teaching level and corrected the teaching skills in academic sectors like schools, colleges, universities or any other educational institutes. In this paper we are proposing an attendance management and storing information service system for students, teachers and school managers. With web-based application and by using RFID technology, the proposed system manages student's attendance records and provides the capabilities of tracking student attendance. The RFID-based system of students is analyzed and focused on system features and core tests.

2 System Architecture

Students are required to come to the desired classes at least one minute before it starts. There is "on time" and "late" function that indicates whether student came late or on time. Whenever the RFID tag is placed on the RFID reader, the reader sends tag ID and current time to a designed PHP file. The PHP file adds those information and makes SQL request to database. Then it takes the attendance with a sign of "on time" or "late" depending on the arrival time. But students are not allowed to take their ID cards out of a card reader before class is over. Once it is taken out, it will be considered as leaving time. This system have sign up and log in functions. Teachers can create their accounts and store class information and monitor their students.

This Smart attendance management system consists of various electronics, Hardware and software components. Block diagrams are shown in Figs. 1 and 2:

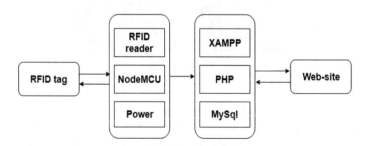

Fig. 1. Block-diagram of the proposed system

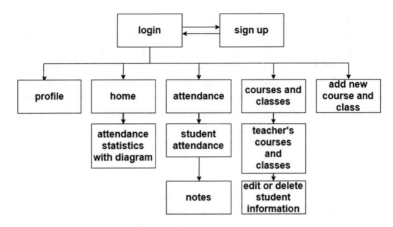

Fig. 2. Block-diagram of the web-site.

Both hardware and software components are needed to make an implementation process. It has been selected based on the following criteria: accessibility, cost effective, easy usage. For hardware, NodeMCU, RFID-RC522 and wires (see Fig. 3) are used. NodeMCU is developed under C language and works on the Espressif NON-OS SDK. The firmware was initially developed as a companion project to the ESP8266-based NodeMCU development modules, but the project is now community-supported, and the firmware can be run on any ESP module. The RC522 is a highly integrated reader/writer IC for contactless communication at 13.56 MHz. The MFRC522 reader supports ISO/IEC 14443 A/MIFARE and NTAG.

Fig. 3. NodeMCU and RFID-RC522 wire connection.

The RFID reader connected to NodeMCU microcontroller device which contains WiFi module and connected by pins (see Table 1). The NodeMCU sends the data to a local XAMPP server via using WiFi; PHP and MySQL are used to record and keep data of the attendance information. All of the records are available on Web-site.

Table 1. Pin wiring.

NodeMCU	RC-522
D4 (GPIO2)	SDA (SS)
D5 (GPIO14)	SCK
MOSI	MOSI
D6 (GPIO12)	MISO (SCL)
	IRQ
GND	GND
D3 (GPIO0)	RST (FLASH)
3.3v	3.3v

This system contains a teacher's login side where a teacher can sign-in or if he/she does not have an account they can sign-up (see Fig. 4). Each teacher is required to have its own account. If they do not have an account they should sign up on "sign up" page. Once teacher's profile has been created, they can login to system. At the home page they can see the feature that the system automatically generates the attendance statistics in the form of bar graph representing the number of on time and late of each class (see Fig. 5). Profile page is made for them to edit their profile information such as surname, last name, password, photo, e-mail and so on.

Attendance page contain each student attendance information (see Fig. 6). "Courses and classes" feature are for teachers to keep their courses and classes information in their own page. By "Add new course and class" function teacher's will store their course and class name (see Fig. 7). There is "student registration" page for students to register. They can check their attendance information on "student attendance" page.

Fig. 4. Login page   **Fig. 5.** Home page.

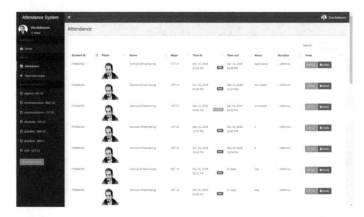

Fig. 6. Attendance page.

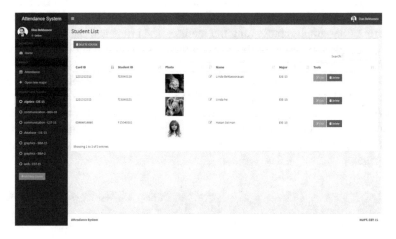

Fig. 7. Courses and classes page.

When the RFID tag is placed on reader – the WiFi based RFID reader circuit sends card ID to the database. When the card is taken out the circuit sends card ID again for the second time. PHP file will make record in database with the ID and its time. By this way we can record the duration of a student staying in class.

To get serial ID of a tag, RC522 library is used. This library will write and read, and various kind of RFID cards on Arduino or NodeMCU using RC522 based reader connected using the Serial Peripheral Interface (SPI) interface. Devices that connect to WiFi network are called stations (STA). Connection to WiFi are provided by an AP (access point), that turns to a hub mode for one or more connections. The AP is also connected to a wired network. The AP is generally integrated with a NodeMCU to provide access from WiFi network to the Internet. Each access point has its unique SSID (Service Set IDentifier), that essentially is the name of network when connecting a device (station) to the WiFi. Each ESP8266 module can operate as a station, so we can connect it to the WiFi network. It can also operate as a soft access point (soft-AP), to establish its own WiFi network. Therefore, we can connect other stations to such modules. In addition, ESP8266 is also able to operate both in station and soft access point mode at the same time. This offers the possibility of building mesh networks. To send data over WIFI we used the next libraries. They are available in Arduino library management.

Head files:

```
#include <ESP8266WiFi.h>
#include <WiFiClient.h>
#include <ESP8266WebServer.h>
```

The following code is written to check if a tag is on device or not. Once a tag is found on reader, the RFID reader terminal sends output message "Tag on". Once it had been taken out the message "Tag out" will be sent. The system is made as much as simple in order to be flexible and reliable.

Code:

```
rfid_tag_in = rfid_tag_out;
_rfid_error_counter += 1;
if(_rfid_error_counter > 2){
_tag_found = false;
}
byte bufferATQA[2];
byte bufferSize = sizeof(bufferATQA);
if(result == rc522.STATUS_OK){
if ( ! rc522.PICC_ReadCardSerial()) {
return;
}
_rfid_error_counter = 0;
_tag_found = true;
}
rfid_tag_present = _tag_found;
if (rfid_tag_in && !rfid_tag_out){
Serial.println("Tag on");
}
if (!rfid_tag_in && rfid_tag_out){
Serial.println("Tag out");
}
```

3   Results and Analysis

The system has two aims: the first one is by using RFID tag the system will automatically record, manage and register students' attendances, and the second aim is to encourage students to attend classes. Table 2 shows an analysis based on various criterion between traditional attendance systems with our proposed method. Automatically attendance record is more convenient for those who used to it. In one hand, barcode, magnetic stripe, biometrics are worse than RFID based individual staff statistics for less information accuracy and artificial identification of assets, traditional manual management and attendance management records.

Table 2. Comparisons of usual and suggested systems.

System	Method	Time	Need do quickly	Rapidity	Reliability	Fallibility	Easy to use
Usual system	Human work	>10 min	Yes	Slowly (human)	No (not safe)	High (paper)	No
Proposed system	Machine work	Instant	No	High (machine)	Yes (safe)	Low (database)	Yes

Table 3. Comparisons of attendance registration systems.

Different methods	Bar-code	Magstripe	Finger scanner	Proposed
Fallibility	Yes	Yes	Yes	No
Easy to use	No	No	Yes	Yes
Price	Cheap	Expensive	Expensive	Cheap
Secure	Yes	Yes	Yes	Yes
Influence of covering the data reader	System failure	System failure	System failure	No influence
Accuracy	Low	Low	High	High
Maintenance cost	Low	Low	High	Low
Energy use	High	High	Low	Low
Distance between reader and device or finger to register	Direct	Direct	Direct	<50 cm (wireless)

Typically, quick feedback redeems codes (QR codes) takes a long time to scan, barcodes and swipe cards are not reliable, impressive low data accuracy, artificial detection, traditional manual handling and separate employee statistical attendance records and banknotes attendance cards and reasoning documentation are not favorable. Whereas, our system which works with RFID technology can reach several advantages such as user-friendliness, affordability, security, adaptability, safety and reliable, can work without human interaction, providing attendance status and generating the attendance statistics on its own; there is no need to spend extra time on attendance record. The comparison of current generation attendance systems [6–9] with our system are shown on Table 3.

4   Advantages

The proposed system is totally automated and it has no need human interference. In many ways, it can help colleges, universities and many educations institutes to easily deploy attendance system. Depend on the results, attendance and data are time-efficient and system-wise efforts, and it limits the cost of high power as well. It will save time and money, and eliminate manual process. It has no confusions that this attendance management is a great deal of manual activities, involving in presence and leave entry and count the hours. The system has good web interface for students and teachers. Using this presence system teachers can track students entering and leaving time during the class room more accurately and quickly. Teachers can make notes for each student's behavior in class time. Thus, how the teachers can evaluate the students properly. This system can avoid proxy or false attendance.

5   Conclusion

To sum up, the purpose of building an intelligent student attendance management system based on RFID technology is successfully gained in this paper. Compared to the traditional method of attendance system, showing efficiency performance, the proposed system has obvious advantageous, and comfortable and acceptable for teachers and students. By using this smart attendance system, the student's attendance information with RFID tag can be get in time. This system can reduce manpower. In this paper, the system we described is convenient for schools, college, university and higher educational levels and easy to use. Attendance systems based on RFID eliminates paper works and avoids human errors that could happen while filling a paper with attendance manually. Using this system, the teachers can make notes on every student, it is a great expectation that the system can replace usual way of recording attendance. This system works properly and meets user requirements.

Acknowledgements. This work was supported in part by the CERNET Innovation Project under Grant NGII20180605, in part by the National Natural Science Foundation of China under Grant 61602261, in part by the STITP of the Nanjing University of Posts and Telecommunications (NUPT) under Grant SZDG2018014, in part by the Postgraduate Research and Practice Innovation Program of Jiangsu Province under Grant SJKY19_0825.

References

1. Yuru, Z., Delong, C., Liping, T., et al.: The research and application of college student attendance system based on RFID technology. Int. J. Control Autom. **6**(2), 273–282 (2013)
2. Patel, R., Patel, N., Gajjar, M., et al.: Online students' attendance monitoring system in classroom using radio frequency identification technology: a proposed system framework. Int. J. Emerg. Technol. Adv. Eng. **2**(2), 61–66 (2012)
3. Walia, H., Jain, N., et al.: Fingerprint based attendance systems—a review. Int. Res. J. Eng. Technol. **3**(5), 1166–1171 (2016)

4. Patel, U.A., Swaminarayan Priya, R., et al.: Development of a student attendance management system using RFID and face recognition: a review. Int. J. Adv. Res. Comput. Sci. Manage. Stud. **2**(8), 109–119 (2014)
5. Kurniali, S., et al. The development of a web-based attendance system with RFID for higher education institution in Binus University. In: EPJ Web of Conferences, vol. 68. EDP Sciences (2014)
6. Rjeib, H.D., Ali, N.S., Farawn, A.A., et al.: Attendance and information system using RFID and web-based application for academic sector. Int. J. Adv. Comput. Sci. Appl. **9**(1), 266–274 (2018)
7. Kumar, J., Kumar, A., et al.: Automatic attendance monitoring and tracking system using Bluetooth and face identification. Int. J. Adv. Res. Electron. Commun. Eng. **5**(4), 1166–1170 (2016)
8. Sayanekar, P., Rajiwate, A., Qazi, L., Kulkarni, A.: Customized NFC enabled ID card for attendance and transaction using face recognition. Int. Res. J. Eng. Technol. **3**(9), 1366–1368 (2016)
9. Jacob, J., Jha, K., Kotak, P., Puthran, S., et al.: Mobile attendance using near field communication and one-time password. In: IEEE 2015 International Conference, pp. 1298–1303 (2015)

An Agent-Based Intelligent Data Presentation Mechanism for Multifaceted Analysis

Kazuto Sasai[1]([✉]), Hiroshi Matsumura[2], Ryota Fukutani[2], Gen Kitagata[2,3], and Tetsuo Kinoshita[2,3]

[1] Graduate School of Science and Engineering, Ibaraki University,
4-12-1, Nakanarusawa, Hitachi, Ibaraki 316-8511, Japan
`kazuto.sasai.z@vc.ibaraki.ac.jp`
[2] Graduate School of Information Science, Tohoku University,
2-1-1, Katahira, Aoba-ku, Sendai, Miyagi 980-8577, Japan
[3] Research Institute of Electrical Communication, Tohoku University,
2-1-1, Katahira, Aoba-ku, Sendai, Miyagi 980-8577, Japan

Abstract. Network management tasks are increasing its complexity because of the development of the IoT environment, virtualization technologies. To keep the safety and security of the network systems, the advances of analysis method for various network data is required for the practical problem-solving. Although many IT tools to support the data analysis works in network management, it is not enough for adaptation to developing environment. Also, the solutions do not consider the knowledge variety of users. In this paper, we propose an agent-based approach for data presentation to support the analysis from multiple aspects in network management works. Concretely, we design the self-organized analysis procedure based on the cooperation among agents and the interactive requirement estimation mechanism to improve the support functions for the diverse of the users' knowledge. The evaluation experiments show the effectiveness of the prototypical system designed according to the proposed approach.

1 Introduction

Cybersecurity and network systems management increase the significance according to the development of advanced internet technology [1]. Due to the growing complexity of network structures, automated monitoring, analyzing, and the decision is necessitated functions of the network management systems (NMSs) for the inhibition of the administration burdens [2]. In recent years, several solutions of autonomous management have been proposed under the projects based on the keywords of big data and AI. Precisely, the technologies of big data improve the capability of data processing [3], and the studies around AI corresponding to machine learning enhance the precision of recognition functions [4,5]. However, the composition of AI and big data did not cover another surface

© Springer Nature Switzerland AG 2020
L. Barolli et al. (Eds.): CISIS 2019, AISC 993, pp. 587–595, 2020.
https://doi.org/10.1007/978-3-030-22354-0_52

of the big data. Because the machine learning requires pre-collected training data, under the unpredictable situations requiring the high-level heuristics, the autonomic functions of the AI capability has a limitation and needs a human hand to address the difficult tasks.

The other perspective of big data is an exploration of the data indicating a new explanatory variable for particular concepts. To obtain the view, we have to examine not only the existing dataset but also the possible combinations with the applicable analysis methods, and therefore the choice of the significant factors is not always natural. Although professional humans can find out the explanatory variables by using their domain knowledge and experience, it is difficult for beginners. In the field of multi-agent systems, the studies of human-agent interaction pay attention to the empowerment technologies for the collaboration between humans and agents [6]. An issue discussed by the researchers is the process of the cooperation requires accountability which reveals internal information flow in the autonomous multi-agent collaboration because the presentation of this information strongly supports the humans to make heuristic decisions. Therefore, the requirement of accountability expects another type of autonomy which establishes the functionality to present data flows inside the autonomous network systems.

On the other hand, there are several studies to automize the synthesis of data analysis flows. An example of automatic construction approach is for the data stream processing that means a computation flow consists of the unit function for a shot piece of data sequence generated continuously [7,8]. Further, the construction method of the user interfaces which align the graphs, texts, and other visualizations are also proposed [9]. Although the technics for automatic construction of analytic data flows are considered to be useful for the burden reduction of the humans, however, it is not enough to adopt multi-agent systems to address the accountability. We have developed an autonomous monitoring and decision-making system for network management tasks based on the concept of Active Information Resource (AIR), which is an information resource with the functions to interact each other actively and support the users [10]. We also discussed agent-based data analysis tools [11] to assemble the evolutional functionality to AIR-based network management. However, the feature is still closed to the system inside; therefore, it is necessary to consider the presentation and interaction capability of the autonomous data analytics systems to improve the accountability of multi-agent systems.

In this paper, we propose an intelligent data presentation method by using a multi-agent approach. The advantage of our proposal is not only autonomic construction of data analysis flows but also expand perspective to address the understandability and the heuristic problem-solving. The next section provides the proposed method and the prototypical system. Section 3 shows the result of the evaluation. We summarize and conclude this paper in Sect. 4.

2   Agent-Based Data Presentation Method

In this section, we propose a data presentation method based on a multi-agent approach to improving the data utilization for network and systems management tasks. The proposal indicates to build the data processing tools as the autonomous agents and to construct the agent organizations as network data analysis flows. Eventually, the agent organizations present their visualized analysis results for the users according to the requirement. Figure 1 shows an overview of this research. Firstly, we define a Data Processing Agent (DPA) as an encapsulated Data Processing Tool (DPT). Secondly, the DPAs are stored in the agent repository which is a particular feature of the agent framework, ADIPS/DASH [12] and are instantiated according to the request of the user. Finally, the visualized analysis results are presented from the organized analytics flows.

2.1 Data Processing Agents (DPA)

As mentioned in Sect. 1, in this paper, we define DPT as a generic term includes the tools for data collection, store, data processing, and visualization for the network and systems management works. The ADIPS/DASH [12] framework provides a schema to encapsulate the existing software and services. According to the design paradigm of the ADIPS/DASH framework, we define DPA as shown in Fig. 2. An agent is defined by the three parts, the base process including DPT, knowledge for utilization, and inference engine. The base process provides an interface with the actual DPT such as IO, calling the method, and so on, and the inference engine adopts knowledge to control the base process. In the working memory of the agent, the knowledge to utilize the DPT and stored data. Here, we define the four types of described knowledge, *workplace*, *repository*, *target*, *data*, and *relatedword*. *data* is a piece of knowledge about stored data, and *relatedword* is a piece of knowledge for interconnected other data flows (agent-organizations).

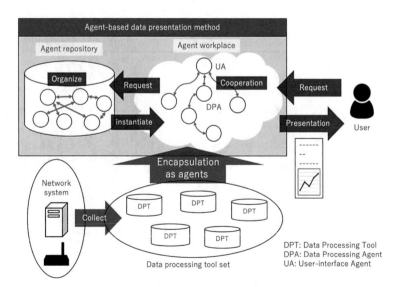

Fig. 1. Schematic diagram of the intelligent data presentation method

2.2   Organization Mechanism of DPA

In this section, we describe the organization mechanism for the proposed inventive data presentation method. The design for organization procedure is according to the organization protocol of the ADIPS/DASH framework. Figure 3 shows a schematic diagram of the message exchanged among the agents. When a request message is received by the UA, DPAs exchange messages to associate and connect each other and send related word to the others in parallel. After the iterated association of requested data flow information, multiple data presentation functionalities is established and provided to the user. Our idea of this organization scheme is the parallel execution of convergence to choose the optimal data flow construction together with the association of the related other data flow (alternative perspectives).

2.3   Prototypical System

Based on the design scheme of the DPAs and organization protocols of them, we build a prototypical system to test the proposed method. Concretely, we implemented the following DPAs:

- DPAs
    - Data store agent:
        - · stores system logs and statistical data of performance.
        - · has 64 *data* objects in the knowledge.
    - Search engine agent:
        - · provides the search engine for local pages.

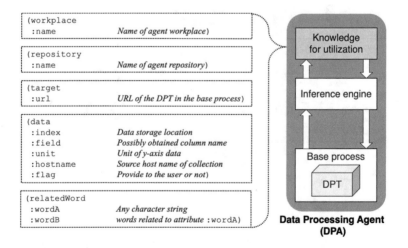

Fig. 2. Knowledge for utilization written in working memory of the ADIPS/DASH agent development framework. The structure is according to the JSON format, which is actually implemented as such format

      · has 3 *relatedword* objects in the knowledge.
- Graph agent:
  - · receives parameters and creates 2-dimensional charts such as bar charts and line charts.
- UA
  - Web service agent:
    - · sends a request from the Web server to other agents and returns search results from agents to the Web server.

For the prototypical system, we have readied three data processing tools, data store, search engine, and graph generation. The reason to choose the three DPTs merely is that is used for monitoring network and services. Additionally, for the user interface, we implemented the UI-Agent (UA) by web service framework to provide the user interface. In this paper, we introduce the natural language interface as the request made. The interface is provided as a web service which has merely input form and provided the analysis results aligned sequentially. The order if returned results are determined by the time, which means it displays from the result returned earlier. In the next section, we evaluate the proposed system using the prototypical system.

3   Case Study and Experiment

To evaluate the effectiveness and feasibility of the proposed intelligent data presentation method, here, we conduct the two case studies and a performance test experiment.

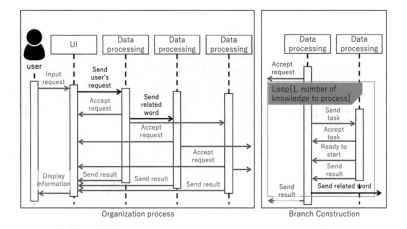

Fig. 3. Schematic diagram of message exchange in the organization process according to the individually implemented knowledge. The message "send related word" expand the analysis perspective parallely

3.1   Case 1: Request of "Mail Server"

Figure 4 shows the case when a request "mail server" into the input form. After receiving the request by UA, the agent organization protocol runs with cascading into the three data flows. Finally, The prototypical system provides three processed results as shown in the righthand side of Fig. 4. In this case, firstly, the system presents a search result of a search engine server, which retrieves the web page archived local knowledge base such as Wiki site. Secondly, the system provides the result of the same service, Fess with the search word "hostname of the mail server." This result comes from the word association in the knowledge of the DPA as installed with the "related-word." Thirdly, the system provides the result from Elasticsearch with the keyword "hostname of the mail server." This result indicates that the organization protocol can induce the other DPA has generated the different types of analysis tools for the same requested word through the word retrieved here.

From the case study, it results that a simple request by the word "mail server" can provide not only the indicated word but also related word and multifaced analytic result from different analysis service. Therefore, the user can compose the provided result information and possibly aware of the new ideas.

3.2   Case 2: Request of "IP Address of Web Server"

In addition to Case 1, Fig. 5 shows the case when the IP address of the mail server is inputted as a request of Case 2. The result of Case 2 is different from the result of Case 1. Case 1 when the word "mail server" is used for the analysis request provides the expanded term "Hostname of XX.XX.XX.XX" is used for the search

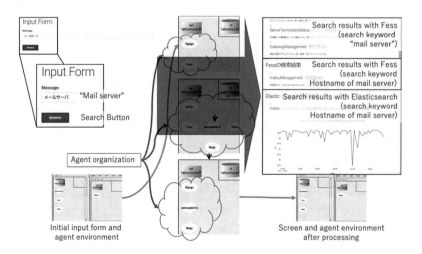

Fig. 4. Screenshot of the user interface and agent repository of dedicated development tool in Case 1. The organized data flows are cascaded into the several perspectives represented by the data processing chain of DPAs

keyword modified through the knowledge of DPA, "related-word." Firstly, the system provides the result of Fess by searching with the keyword and indicates the Wiki page concerned with the role of the server which has the target IP address. Secondly, the system also provides the result of Elasticsearch with the keyword. The presented visualized result is a temporal count of generated logs. If the user is not a professional administrator, the term of IP address is not familiar and difficult to imagine the target server from the IP address. However, in this case, the domain knowledge of the association between the IP address and the role in the network system is presented with the status information.

The result implies that multi-faced and intelligent data presentation method can support the administrators to aware the information resulted from multiple perspectives.

3.3 Evaluation Experiment

In addition to the previous case study, we experimented with estimating the performance advantage of the proposal organization functionality. Figure 6(a) showed the result of processing time which means the consumption time to make the results (search from the local pages) with the increase of the number of items. It is easy to see that even if the number of cases increased, the average processing time did not improve. Further, we have the result that the processing is finished within 1 s. It is considered that the searching the web pages is not such a big load for the DPA because the data is only text, and the number of entity is not so large.

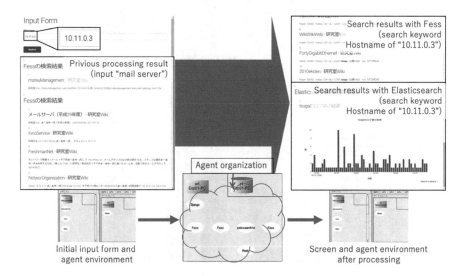

Fig. 5. Screenshot of the user interface and agent repository of dedicated development tool in Case 2. The organized data flows are cascaded into the several perspectives represented by the data processing chain of DPAs

Figure 6(b) shows the results of the processing time versus the number of items (the graph numbers to be organized). It is easily seen that the processing time increases if the number of elements increases. From the result, the average processing time per query (item) is calculated as around 6 sec. The implication of this result is parallelism of the data flow has a limitation of its scalability, and thus, the inhibition mechanism to select and pick up the processing flows should be implemented into the organization procedure.

Note that this result only evaluates processing performance but do not include objective analysis for the quality of data processing. Since there is a possibility to evaluate the support capability (quality) for heuristic analytics of network data, however, we remain this experiment as the future work. Additionally, the prototypical system does not include the re-organization function for failure DPAs. If the mechanism might be installed to the system, the result of Fig. 6 will be more robust.

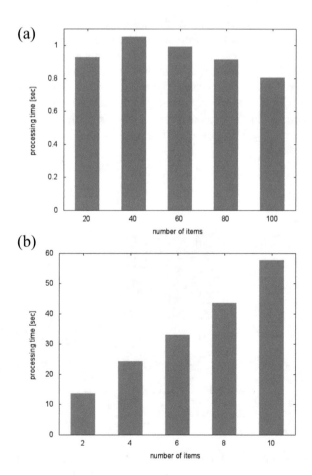

Fig. 6. Processing time for provide the graphs from the queried data. (a) Search engine agent retrieves from web sites such as wiki sites. (b) Data store agent and graph generation agent

4   Conclusion

In this paper, we have proposed an intelligent data presentation method for network and systems management works. The approach to constructing the data analysis flow is based on multi-agent organization procedure, and the encapsulation scheme of DPT is provided to define DPA. Additionally, we represent the sequence of the organization includes not only construct data processing flows but also connect another stream to expand the possibility to understand the actual phenomena. To evaluate the proposal, we conducted two case studies and an experiment, and the results support the advantage of our scheme. As the future work, we will make a variety of case studies and evaluate the support capability of the prototypical system.

References

1. Jing, X., Yan, Z., Pedrycz, W.: Security data collection and data analytics in the internet: a survey. IEEE Commun. Surv. Tutorials **21**, 586–618 (2019)
2. Samaan, N., Karmouch, A.: Towards autonomic network management: an analysis of current and future research directions. IEEE Commun. Surv. Tutorials **11**, 22–36 (2009)
3. Kambatla, K., Kollias, G., Kumar, V., Grama, A.: Trends in big data analytics. J. Parallel Distrib. Comput. **74**, 2561–2573 (2014)
4. Li, J., Liu, H.: Challenges of feature selection for big data analytics. IEEE Intell. Syst. **32**, 9–15 (2017)
5. Marjani, M., Nasaruddin, F., Gani, A., Karim, A., Hashem, I.A.T., Siddiqa, A., Yaqoob, I.: Big IoT data analytics: architecture, opportunities, and open research challenges. IEEE Access **5**, 5247–5261 (2017)
6. Jennings, N.R., Moreau, L., Nicholson, D., Ramchurn, S., Roberts, S., Rodden, T., Rogers, A.: Human-agent collectives. Commun. ACM **57**, 80–88 (2014)
7. Jabbar, S., Malik, K.R., Ahmad, M., Aldabbas, O., Asif, M., Khalid, S., Han, K., Ahmed, S.H.: A methodology of real-time data fusion for localized big data analytics. IEEE Access **6**, 24510–24520 (2018)
8. Jiao, Y., Wang, P., Feng, S., Niyato, D.: Profit maximization mechanism and data management for data analytics services. IEEE Internet Things J. **5**, 2001–2014 (2018)
9. Chen, X., Self, J.Z., House, L., Wenskovitch, J., Sun, M., Wycoff, N., Evia, J.R., Leman, S., North, C.: Be the data: embodied visual analytics. IEEE Trans. Learn. Technol. **11**, 81–95 (2018)
10. Sasai, K., Sveholm, J., Kitagata, G., Kinoshita, T.: A practical design and implementation of active information resource based network management system. Int. J. Energy Inf. Commun. **2**, 67–86 (2011)
11. Sasai, K., Tanimura, Y., Takahashi, H., Kitagata, G., Kinoshita, T.: An agent-based data analytics support tool for network management intelligence. Int. J. Energy Inf. Commun. **8**, 51–64 (2017)
12. Uchiya, T., Maemura, T., Xiaolu, L., Kinoshita, T.: Design and implementation of interactive design environment of agent system. In: IEA/AIE 2007. LNAI, vol. 4570, pp. 1088–1097. AAAI/ACM (2007)

Application of Improved Ant Colony Algorithm in Path Planning

Zhe Li[1(✉)], Ruilian Tan[2], and Baoxiang Ren[1]

[1] Air Force Engineering University, Xi'an 710051, China
kongyanshi@126.com
[2] College of Information Engineering, Armed Police Engineering University,
Xi'an 710086, China
madamtan@126.com

Abstract. Aiming at the problem of path planning in topography, this paper studies the influence of relevant factors on speed and path planning, and proposes an intelligent algorithm for terrain path planning based on slope factor and ant colony algorithm. Firstly, the slope factors affecting walking speed are analyzed, then the terrain data are pretreated, and the ant colony algorithm is improved according to the walking requirements. Finally, the optimal walking path is obtained. The simulation results show that the algorithm can achieve better terrain path planning according to walking requirements.

Keywords: Topography · Ant colony algorithm · Path planning

1 Introduction

Ant colony algorithm (ACO) is an intelligent optimization algorithm which has been widely studied in recent years [1–3]. It simulates the foraging behavior of ant colony in nature and is widely used in path planning. These studies mainly focus on people's daily life, such as the application to solve scheduling problem, assignment problem, traveling salesman problem [4, 5]. Ant colony algorithm has the advantages of positive feedback, robustness, parallelism and so on. Ant colony algorithm can calculate the shortest path problem better.

In this paper, aiming at the path planning problem in topography, the application of ant colony algorithm in topographic path planning is discussed under the background of the influence of slope factors. Firstly, according to the terrain data preprocessing, excluding inappropriate sites, improving ant colony algorithm according to the terrain characteristics to meet the actual needs, and finally according to the improved ant colony algorithm to solve the appropriate path, and analyze the algorithm parameters to set the optimal parameters. The simulation results show that, after the ant colony algorithm is improved and applied according to the requirements of terrain path planning, the most suitable traveling path can be calculated quickly and the problem of terrain path planning can be solved well [6–8].

© Springer Nature Switzerland AG 2020
L. Barolli et al. (Eds.): CISIS 2019, AISC 993, pp. 596–603, 2020.
https://doi.org/10.1007/978-3-030-22354-0_53

2 Ant Colony Algorithm

The basic idea of ant colony algorithm application is that the feasible solution of the problem to be optimized is represented by the path of ants. All paths of the whole ant colony constitute the solution space of the problem to be optimized. The shorter the path, the more pheromones released by the shorter the path, the more pheromones accumulated on the shorter path, the more ants choose the path. Ultimately, the whole ant will concentrate on the optimal path under the action of positive feedback [9, 10].

For route selection, our goal is to arrive at the designated location in the shortest time, and to make the path through the shortest and shortest time-saving. According to the principle of ant colony algorithm, Let the number of ants in the whole ant colony be m, number of sites to be reached is n. The distance between i and j is d_{ij} $(i, j = 1, 2, \cdots n)$. The pheromone concentration on the connection path is $\tau_{ij}(t)$. The concentration of pheromones on the connection paths is the same at the initial time named τ_0. Assuming the probability is p_{ij}^k which ant k arriving at the site i to j according to the concentration of pheromones between the sites, and the heuristic function is η_{ij} $(\eta_{ij} = 1/d_{ij})$, then there are:

$$
p_{ij}^k = \begin{cases} \dfrac{[\tau_{ij}(t)]^\alpha \cdot [\eta_{ij}]^\beta}{\sum\limits_{s \in A} [\tau_{is}(t)]^\alpha \cdot [\eta_{is}]^\beta} & s \in A \\ 0, & s \notin A \end{cases} \tag{1}
$$

According to ant colony algorithm, pheromones between different locations gradually disappear while ants release pheromones. The degree of volatilization of this pheromone is indicated by ρ $(0 \leq \rho \leq 1)$. Its pheromone updates can be expressed by the following formula:

$$
\begin{cases} \tau_{ij}(t+1) = (1 - \rho)\tau_{ij}(t) + \Delta\tau_{ij} \\ \Delta\tau_{ij} = \sum\limits_{k=1}^{n} \Delta\tau_{ij}^k \end{cases} \tag{2}
$$

Among them, $\Delta\tau_{ij}^k$ denotes the pheromone concentration of the ant from point i to point j, and $\Delta\tau_{ij}$ denotes the sum of the pheromone concentration of all ants from point i to point j. The pheromone release model adopts ant cycle system model, namely:

$$
\Delta\tau_{ij}^k = \begin{cases} Q/L_k, & i \text{ to } j \\ 0, & \text{others} \end{cases} \tag{3}
$$

Among them, Q is a constant, L_k represents the total amount of pheromones released by ants traveling a distance, and the length of the path passed by the first k ant.

3 Application of Improved Ant Colony Algorithm in Path Planning

Slope is one of the important factors affecting motion in terrain. Under the altitude of 2 km, the flatter the terrain, the faster the walking speed and the faster the arrival of the destination. Uphill not only needs to move horizontally but also climb up, which consumes physical energy. Generally speaking, the average speed of climbing decreases by about 5 km per hour for every increase of slope degree. Therefore, when choosing the route, if we can choose the place with lower slope as the route, we can reach the destination in a faster time. Based on the above principles, this paper proposes an improved ant colony algorithm for path planning, that is, before applying the ant colony algorithm to path planning, add the slope information of the location, select the more advantageous location, and combine the ant colony algorithm to plan the route, so as to save time and physical strength. The specific process is as follows (Fig. 1):

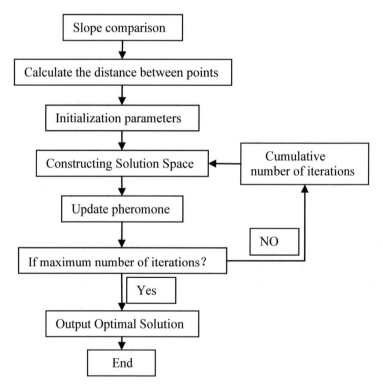

Fig. 1. Basic steps of ant colony algorithm for path planning

4 Algorithm Implementation

Steps:

Step 1: Compare the gradients of different places according to the requirements of observation points, and use MATLAB software to make traversal judgments and select the appropriate travel sites.

Step 2: According to the rectangular coordinate position of each point, the distance between each point is calculated, and then the symmetrical distance matrix is obtained.

Step 3: Initialize parameters and construct path solution space.

Step 4: According to the transfer probability formula (1), all eligible locations are visited, the length of the path is calculated, and the pheromone concentration is updated after each iteration according to formula (2).

Step 5: Cyclic iteration is performed to record the optimal path and its length for each iteration, and to determine whether the termination condition is satisfied. If it is satisfied, the algorithm is terminated and the optimal solution is output. Otherwise, it is transferred to step 3 until the maximum number of iterations is reached.

5 Simulation Analysis

5.1 Testing the Effectiveness of the Algorithms

In order to verify the effectiveness of the algorithm proposed in this paper, the algorithm is simulated by using MATLAB software. There are 20 sites covered by the target area. The experimental data of each point are shown in Table 2. In order to save time and energy, visits to places with slopes greater than or equal to $30°$ are cancelled. The plane rectangular coordinates and corresponding slopes are shown in Table 1. The number of ants m = 50, $\alpha = 1$, $\beta = 5$, $\rho = 0.1$, $Q = 1$, 300 iterations, starting point random selection, through simulation can get the three-dimensional coordinates of the song location as shown in Fig. 2, the optimal path planning as shown in Fig. 3.

Table 1. Experimental data from various locations

Place	Plane straight angular coordinate	Slope degree	Place	Plane straight angular coordinate	Slope degree
A	(26,13)	25°	K	(38,22)	18°
B	(31,22)	15°	L	(26,25)	32°
C	(47,13)	31°	M	(40,28)	10°
D	(34,35)	18°	N	(32,29)	10°
E	(32,12)	30°	O	(34,19)	15°
F	(30,49)	8°	P	(45,23)	18°
G	(25,17)	15°	Q	(33,26)	28°
H	(43,16)	16°	R	(34,32)	22°
I	(39,21)	15°	S	(25,23)	15°
J	(40,23)	33°	T	(37,14)	12°

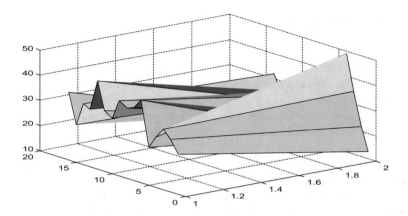

Fig. 2. Stereogram of coordinates of points

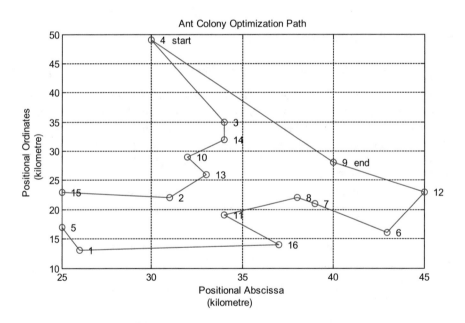

Fig. 3. Optimal path planning roadmap

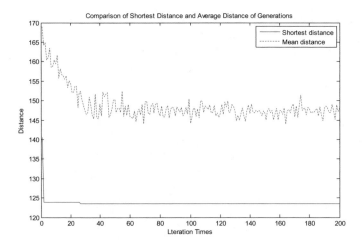

Fig. 4. Iterative change diagram of algorithms

From Fig. 2 and 3, we can see that by traversing the sites that meet the requirements, the sites with higher gradient C, E, L and J are excluded, and the sites that enter the ant colony algorithm are 16 sites with smaller gradient. After planning by the ant colony algorithm, we can get a shorter roundtrip route, starting from location 4 and returning from location 9. At the same time, we can see from Fig. 4 that, with the development of the ant colony algorithm, a shorter roundtrip route can be obtained. With the increase of iteration times, the shortest and average distances of each generation reach stability after 28 iterations, that is to say, the best path has been found at this time. The average value of the 20 times running in MATLAB shows that the total length of the shortest path is 111.7644 (km), which completes the task path planning well.

5.2 Examine the Influence of α and β on Path Planning in Ant Colony Algorithm

In standard ant colony algorithm, pheromone importance factor α and heuristic function importance factor β are important parameters affecting the optimization effect of the algorithm. Using the experimental data in Table 1, we first investigate the influence of pheromone importance factor α on path planning. We run 15 sets of data 300 times to get the average value. After sorting out, we can get Table 2.

Table 2. α's impact on the optimal path (m $= 50$, $\beta = 5$, $\rho = 0.1$, $Q = 1$)

Distance α	Average value	Distance α	Average value	Distance α	Average value
1	111.7438	6	111.8644	11	111.8787
2	111.7481	7	111.8644	12	111.8678
3	111.7544	8	111.7744	13	111.7777
4	111.7589	9	111.7776	14	111.8787
5	111.7643	10	111.7544	15	111.8812

As can be seen from Table 2, for the experimental data, when $\alpha = 5$, the minimum average path is 111.7643 km. Therefore, $\alpha = 5$ should be chosen in this experiment. Next, we examine the influence of heuristic function importance factor β on path planning when $\alpha = 5$ is used. We take an average of 300 times for 10 groups of data, and we can get Table 3 after sorting out.

Table 3. β's impact on the optimal path (m = 50, $\alpha = 5$, $\rho = 0.1$, $Q = 1$)

Distance α	Average value	Distance α	Average value
1	111.5643	6	111.7142
2	111.5663	7	111.7442
3	111.5693	8	111.7643
4	111.6433	9	111.7843
5	111.6943	10	111.8043

As can be seen from Table 3, when $\beta = 5$ and $\beta = 8$, the average shortest path is 111.6943 km. In summary, $\alpha = 5$, $\beta = 5$ or $\beta = 8$ are the most suitable parameter settings in the experimental data.

6 Conclusion

This paper focuses on the topographic route planning problem after the introduction of ant colony algorithm. Taking the slope factor as the research background, according to the actual travel demand and ant colony algorithm, a new algorithm suitable for terrain path planning is deduced, and the influence of algorithm parameters on the optimization performance is further studied, so as to determine the more appropriate algorithm parameters. The experimental results show that, after introducing ant colony algorithm, the terrain path planning can be completed well according to the needs. However, there are still many factors affecting the topographic path planning. The next research focus is how to adjust the algorithm parameters adaptively according to the topographic conditions, and judge the impact of various factors comprehensively, so that the function of the path planning is more in line with the actual needs of the path planning.

References

1. Xu, L., Zhang, S.: Study of path planning in obstacle environment based on an improved ant algorithm. Mach. Electron. **7**, 61–64 (2013)
2. Zhu, S., Xu, F., Teng, Z.: Application of improvement ants algorithm in solving shortest path. Comput. Technol. Dev. **21**(7), 202–205 (2011)
3. Wang, Y., Ye, Q.: Improved strategies of ant colony algorithm for solving shortest path problem. Comput. Eng. Appl. **48**(13), 35–38 (2012)
4. Zhang, Y., Chen, X.: General ant colony algorithm and its applications in robot formation. Pattern Recognit. Artif. Intell. **19**, 20(3), 3–8 (2007)

5. Clornei, I., Kyriakides, E.: Hybrid ant colony—genetic algorithm (GAAPI) for global continuous optimization. IEEE Trans. Syst. Man Cybern. Part B, Cybern. **42**(1), 234–245 (2012)
6. Shah, S., Kothari, R., Chandra, S.: Debugging ants: how ants find the shortest routs. In: 8th International Conference on Information, Communications and Signal Processing, pp. 1–5. IEEE (2011)
7. Colomi, A., Dorigo, M., Maniezzo, V.: Distributed optimization by ant colonies. In: Proceeding of the First European Conference of Artificial Life. Elsevier Publishing, Paris (1991)
8. Dorigo, M., Ganbardella, L.M.: Ant colony system: a cooperative learning approach to the traveling salesman problem. IEEE Trans. Evol. Comput. **1**(1), 53–66 (1997)
9. Jackson, D.E., Holcombe, M., Ratnieks, F.L.W.: Trail geometry gives polarity to ant foraging networks. Nature **432**(7019), 907–909 (2004)
10. Jia, Z., Siqing, B., Wang, H.: Path planning based on heuristic algorithm. Comput. Simul. **29**(1), 135–138 (2012)

An Innovative Model Based on FCRBM for Load Forecasting in the Smart Grid

Ghulam Hafeez[1,2], Nadeem Javaid[1(✉)], Muhammad Riaz[2], Khalid Umar[3], Zafar Iqbal[4], and Ammar Ali[1]

[1] COMSATS University Islamabad, Islamabad 44000, Pakistan
nadeemjavaidqau@gmail.com
[2] University of Engineering and Technology, Mardan 23200, Pakistan
[3] Bahria University Islambad, Islamabd 44000, Pakistan
[4] PMAS Agriculture University, Rawalpindi 46000, Pakistan
http://www.njavaid.com

Abstract. In this paper, an efficient model based on factored conditional restricted boltzmann machine (FCRBM) is proposed for electric load forecasting of in smart grid (SG). This FCRBM has deep layers structure and uses rectified linear unit (RELU) function and multivariate autoregressive algorithm for training. The proposed model predicts day ahead and week ahead electric load for decision making of the SG. The proposed model is a hybrid model having four modules i.e., data processing and features selection module, FCRBM based forecaster module, GWDO (genetic wind driven optimization) algorithm-based optimizer module, and utilization module. The proposed model is examined using FE grid data of USA. The proposed model provides more accurate results with affordable execution time than other load forecasting models, i.e., mutual information, modified enhanced differential evolution algorithm, and artificial neural network (ANN) based model (MI-mEDE-ANN), accurate fast converging short term load forecasting model (AFC-STLF), Bi-level model, and features selection and ANN-based model (FS-ANN).

1 Introduction

Electric load forecasting is an indispensable decision-making tool for energy management in both sectors of SG i.e., supply side and demand side. It also plays an important role in the secure and economic operations of SG [1]. Keeping aforesaid objectives the recent research in SG focus load scheduling based on optimization techniques [2,3]. However, the accuracy of electric load forecasting models is compromised due to their influence on stochastic factors such as climate change, human social activates, and country policies. Consequently, it is difficult to improve the forecast accuracy and hardly realistic to take all the influencing factors into account [4]. Thus, an intelligent model is required that intelligently take the key parameters to improve forecast accuracy.

Numerous models have been proposed and applied for an accurate load forecasting over the fast few decades such as legacy classical forecasting models

© Springer Nature Switzerland AG 2020
L. Barolli et al. (Eds.): CISIS 2019, AISC 993, pp. 604–617, 2020.
https://doi.org/10.1007/978-3-030-22354-0_54

including exponential smoothing, regression models, autoregressive integrated moving average (ARIMA) models, grey forecasting model (GM), and kalman filters [5]. The aforementioned forecasting models forecast the electric load but the accuracy is not up to the desired level due to their inherent limitations. The linear regression models depend on historical data and are not suitable to solve the non-linear problems. The ARIMA models taking into consideration previous and present data points while ignore other influencing factors. The GM models can only solve the problems with exponential growth trends. To overcome the aforementioned problems, in recent years, more effective models have been proposed to forecast electric load, such as an artificial neural network (ANN), multi-layer perceptron (MLP), radial basis fuzzy logic, machine learning, and intelligent system [6]. Though these effective methods outperform legacy methods, however, the provide accuracy is not satisfactory due to their limitations. The ANN-based models trapped into local minima and expert systems strongly rely on supervised learning. In this regard, integrated and hybrid models are developed [7], which are the combination of different individual models. The hybrid models outperform than individual models in terms of forecast accuracy.

In this paper, a novel FCRBM based electric load forecasting (FCRBM-ELF) model is proposed, which is a hybrid model. The major contributions are demonstrated as follows:

- The proposed model takes into account the exogenous influencing parameters in addition to historical electric load data for accuracy improvement.
- The new concept of candidates interaction is introduced for features selection. Also, the mutual information (MI) technique based features extraction criteria are extended to measure the candidate's interaction in addition to their relevancy and redundancy process.
- Due to better accuracy and fast convergence, RELU and is used with FCRBM which none of the existing models have used.
- The proposed GWDO algorithm is used in the optimizer to fine tune the adjustable parameters for feature selection technique to improve the forecast accuracy with affordable convergence rate.

The remaining paper is structured as follows: Sect. 2 demonstrates related work, Sect. 3 briefly describes the proposed system architecture, in Sect. 4, simulation results and discussions are described. Finally, Sect. 5 conclude the paper.

2   State of the Art Work

Electric load forecasting strategies are developed for many years in literature due to its importance in the decision making of SG. The forecasting strategies are categorized into four categories according to the forecasting period [8]. The first category is the very short-term forecasting [9] which corresponds to less than one day. The second category is the short-term forecasting which corresponds to the forecasting period of one day to one-week [10]. The third category is medium-term forecasting which corresponds to one week and a year ahead forecasting [11].

The fourth category is the long-term forecasting which corresponds to more than a year ahead forecasting [12]. Statistical tools and AI-based tools are commonly used for electric load forecasting. The recent and related work is summarized in Table 1.

3   The Proposed System Architecture

In literature, many authors used ANN based forecaster for load prediction due to its capability to predict the nonlinearity of consumers load. However, the performance of ANN-based models is not satisfactory in terms of accuracy. Thus, some authors integrated optimization module with ANN based forecaster, which improves significantly the forecast accuracy. However, the accuracy is improved at the cost of slow convergence rate. Moreover, the ANN-based models are suitable for small data size while their performance is degraded as the data size increases. Thus, we proposed a new electric load forecasting model based on FCRBM as shown in Fig. 1. The proposed model is subjected to accuracy, convergence rate, and scalability. The proposed system architecture comprises of four modules: (1) data processing and feature selection module, (2) FCRBM based forecaster module, (3) GWDO based optimizer module, and (4) utilization module. The detailed description is as follows:

3.1   Data Processing and Features Selection Module

The input data including historical load data and exogenous data (temperature, humidity, wind speed, and dew point) is fed into the data processing and features selection module. At first, the data cleansing is performed to recover the missing and defective values. Then, the clean data is normalized to remove the outliers and make the data within the limit of the activation function. The input data (X) includes electric load data $(P(h, d))$, temperature data $(T(h, d))$, humidity data $(H(h, d))$, dew point $(D(h, d))$, and wind speed $(W(h, d))$. The h shows particular hour and d shows particular day of historical data. The temperature, humidity, dew point, and wind speed are called exogenous variables. The normalized data is passed to through irrelevancy filter, redundancy filter, and candidate interaction phase subjected to removal of irrelevant, redundant, and nonconstructive information. The detailed description of relevancy, redundancy, and candidates interaction phases of features selection technique are as follows:

3.1.1   Relevancy Operation

The relevance of candidates input to the target variables is significant for abstractive features selection. For relevancy measurement in literature many techniques are used in which MI features selection technique is good. The MI measures the relevance between two variables x and y. The MI measurement is interpreted as observing y by on x and vice versa. The MI for continuous variables x and

Table 1. Recent and related work summary

Techniques	Objectives	Dataset	Remarks
ARIMA and exponential smoothing [9]	Forecast accuracy improvement for real-time scheduling of power generation	Great Britain grid	The accuracy is improved for univariate methods while the accuracy is low for multivariate methods
ANN and self-organizing map [13]	Decision support system to commercialize company bidding	Spanish grid	This model used meteorological and load data and ignored exogenous parameters which have a strong impact on the forecast accuracy
Differential polynomial neural network [14]	To reduce the generation cost and spinning reserve capacity	Canadian grid	This model has less accuracy and slow convergence which have a direct impact on spinning reserve and cost
ANN, ARIMA, and GM [15]	Accuracy improvement of the bulk power system	Fujian province of China	The accuracy is improved by incorporating large exogenous parameters at the expense of slow convergence rate and high complexity
Reglet and Elman neural network [16]	Improvement of the accuracy and capability for effective power system operation	AEMC	This model have large complexity that directly impacts the convergence rate
Support vector regression [17]	Accurate load forecasting to minimize energy imbalance and its associated cost	Irish CER	The parameters are optimally adjusted by the intelligent algorithm which improved the forecast accuracy at the expense of high execution time
ARMAHX and quasi-newton algorithm [18]	Forecast accuracy improvement for a market agent and system operators	Spanish and German energy market	The accuracy is improved by incorporating sigmoidal function, however, the execution time and complexity is increased
ANN, SVR, and fuzzy interaction regression [19]	Resilience improvement against data integrity attacks	GEFC 2012	The resiliency of the power system is improved at the cost of high modeling complexity
MI, ANN, and mEDE [21] and [20]	Accuracy and convergence rate improvement for EKPC and Daytown grid of USA	PJM market	This model is suitable for small data size and their performance are degraded for large data size
ANN-based hybrid models [22] and [23]	Accuracy improvement of microgrid	PJM market	The ANN-based models improved the forecast accuracy at the cost of high execution time

y is defined $I(x;\ y)$ for both individual $(p(x),\ and\ p(y))$ and joint probability distribution $(p(x,y))$. Assume that

$$S = \{x_1,\ x_2,\ x_3,....,x_M\},\tag{1}$$

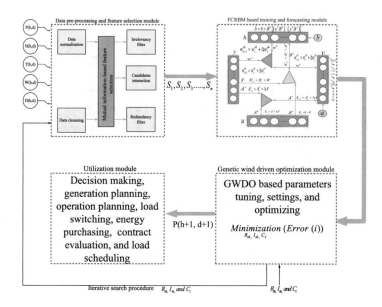

Fig. 1. The proposed system architecture

where S represents the set of candidate inputs and y is the target variable. The relevance of each candidate input with target variable y are checked. The relevance of candidate input x_i with target variable y is defined by the following Equation

$$D(x_i) = I(x_i;\, y), \tag{2}$$

where $D(x_i)$ represents candidate inputs to target variables.

3.1.2 Redundancy Operation

Many authors modeled the redundancy operation between the candidate inputs. The purpose is to remove the redundant information from the input data to improve convergence rate. The redundancy is evaluated in terms of the mutual information among the two candidate inputs. In literature, authors demonstrated that closely related candidate inputs reduce the performance of feature selection technique. The reason is that two candidate inputs have a large number of mutual information and less redundant information about the target variable. So, a variable with less redundant information about the target variable which may be incorrectly count as redundant and will be discarded, while it may be the key feature for forecaster. To overcome the aforementioned problem a redundancy measure based on interaction gain is modified as:

$$\begin{aligned} RM(x_i, x_s) &= Ig(x_i; x_s; y) \\ &= I[(x_i, x_s); y] - I(x_i; x_s) - I(x_s; y), \end{aligned} \tag{3}$$

where $RM(x_i, x_s)$ is the redundancy measure, x_i, x_s are candidate inputs, and y is the target variable. The Ig can be mathematically modeled in terms joint and individual entropy as:

$$Ig(x_i; x_s; y) = H(x_i, x_s) + H(x_i, y) + H(x_s, y) - H(x_i) - H(x_s) - H(y) - H(x_i, x_s, y), \tag{4}$$

where $H(x_i)$, $H(x_s)$, and $H(y)$ denote individual entropy and $H(x_i, x_s)$, $H(x_i, y)$, $H(x_s, y)$, and $H(x_i, x_s, y)$ denote joint entropy.

3.1.3 Interaction Session

In [22], used redundancy and irrelevancy filters for feature selection. However, the individual features may be irrelevant but become relevant when used together with other input candidates. Thus, the feature selection technique can be extended to interaction among the candidate inputs. If two candidate inputs x_i and x_s have redundant information about target y, then the joint *MI* of both candidates with y will be less than the sum of individual *MIs*. Thus, the result will be negative according to Eq. 3, which indicates redundant features x_i and x_s for the forecaster. The absolute value of Eq. 3 shows the amount of redundancy. On the other hand, if x_i and x_s candidate inputs interact with target y their interaction causes joint (x_i and x_s) MI with target y greater than the sum of individual MIs. Thus, the positive value of Eq. 3 indicates interacting features and its absolute value shows the amount of interaction. Hence, for redundancy and interaction the Eq. 3 can be modified as:

$$RM(x_i, x_s) = \begin{cases} \{Ig(x_i; x_s; y), & \text{if } Ig(x_i; x_s; y) < 0 \\ 0 & \text{otherwise} \end{cases} \tag{5}$$

$$In(x_i, x_s) = \begin{cases} Ig(x_i; x_s; y), & \text{if } Ig(x_i; x_s; y) > 0 \\ 0 & \text{otherwise} \end{cases} \tag{6}$$

where Eq. 5 is modified equation for redundancy measure and Eq. 6 is for interaction measure. Thus, the abstractive features are selected and fed into the forecaster module based on FCRBM.

3.2 FCRBM Based Forecaster Module

The purpose of this module is to devise a framework which is enabled via learning to forecast the future electric load. From Sect. 2 it is concluded that all forecast models are capable to predict nonlinear electric load profile. Thus, we chose FCRBM for forecaster module due to two reasons: (a) it predict the nonlinear electric load with reasonable accuracy and convergence rate, (b) and its performance is improving with the scalability of data. FCRBM is a deep learning model. It has four layers i.e., hidden layer, visible layer, style layer, and history layer. Each layer has a particular number of the neuron. In the forecaster module, FCRBM is activated by rectified linear unit (RELU) activation function. The RELU is chosen among the activation function because it overcomes the problems of overfilling and vanishing gradient, and has fast convergence as compared to other activation functions. The mathematical model of RELU is mentioned in Eq. 7.

$$f(x) = \max(0, \ x)$$
$$f(x) \begin{cases} 1 & \text{if } x \geq 0 \\ 0 & \text{otherwise} \end{cases} \tag{7}$$

The training and learning procedure iterates for a number of epochs to forecast the future load. To update weight and biases during training processes authors used different algorithms, i.e., gradient descent and backpropagation, levenberg-marquardt algorithm [23], and multivariate autoregressive algorithm [21]. The levenberg-marquardt algorithm trains the network faster as compared to gradient descent and backpropagation. Thus, the multivariate autoregressive algorithm is used for network training due to its fast convergence and better performance. The selected feature of data processing module $S_1, S_2, S_3, \ldots S_n$ is fed into the forecaster module, where the forecaster constructs training and testing data samples. The first three years of data samples are used for network training. On the other hand, the last year data samples are used for testing. The purpose is to enabled FCRBM based forecaster module via training to forecast the future load. The forecaster module returns error signal and the weights and biases are adjusted as per multivariate autoregressive algorithm. This error signal is fed into the optimization module to improve the forecast accuracy.

3.3 GWDO Based Optimizer Module

The preceding module returns the future predicted load with some error, which is minimum as per the capability of FCRBM, RELU, and training algorithm. To further minimize the forecast error the output of the forecaster module is fed into the optimizer module. The purpose of the optimizer module is to minimize the forecast error. Thus, the error minimization becomes an objective function for the optimizer module and can be mathematically modeled as:

$$\underset{R_{th}, \ I_{th}, \ C_i}{Minimize} \ Error \ (x) \quad \forall \ x \in \{h, d\} \tag{8}$$

where R_{th} is redundancy threshold, I_{th} is irrelevancy threshold, and C_i is candidates interaction. The optimizer module is based on our proposed GWDO algorithm. The optimizer module optimizes R_{th}, I_{th}, and C_i and feedback these parameters to data processing module. In data processing module, feature selection technique use optimized values of R_{th}, I_{th} thresholds and C_i candidates interaction for optimal selection of features. The integration of optimizer module with the forecaster module increase forecast accuracy at the cost of high execution time. Usually, the integration of optimizer with the forecaster module is preferred for those applications where accuracy is of high importance compared to convergence rate. For optimization, various techniques are available like linear programming, non-linear programming, convex programming, quadratic programming, and heuristic techniques. Linear programming is avoided because the optimization problem is non-linear. The non-linear programming is applicable here and returns more accurate results at the cost of large execution time. The convex optimization and heuristic optimization suffers from slow and premature convergence, respectively. Similarly, the DE [22] and mEDE [21] are not adopted because of slow convergence, low precision, and trapped into optimum. To cure the aforementioned problems we proposed GWDO. In other words, GWDO algorithm is preferred because it provides an optimal solution with a fast convergence rate. The proposed GWDO algorithm is a hybrid of GA and WDO. The GA enables the diversity of population and WDO has fast convergence. The forecasted future load is utilized in the utilization module for planning, operation, and unit commitment.

3.4 Utilization Module

The forecasted load is utilized for long term planning that needed state permits financing, right of ways, transmission and generation equipment, power lines (transmission lines and distribution lines), and substation construction.

4 Simulations Results and Discussions

For the performance evaluation of the proposed FCRBM-ELF model simulations are conducted in Matlab 2016, which is installed on a laptop having specifications of Intel(R) Corei3-CPU @2.4 GHz and 6 GB RAM with Windows 10. The proposed FCRBM-ELF model is compared with existing models i.e., MI-mEDE-ANN [20], AFC-STLF [21], Bi-level [22], and FS-ANN [23]. The aforementioned models are chosen due to closer similarity with the proposed model. For testing the proposed model real time hourly load data of FE grid is used. The dataset is taken from publicly available pennsylvania jersey maryland (PJM) [24]. The dataset is also considered in [21]. The dataset is of four years from 2014 to 2017. The first three years of data is used for training the FCRBM and last year data is for testing. The parameters used in simulations can be justified in [21]. The parameters listed are kept constant for existing and proposed model subjected to a fair comparison. The proposed model is tested in terms of four performance metrics, i.e., MAPD, RMSD, correlation coefficient (R) and execution time.

The first three performance metrics correspond to accuracy, which is defined as:

- Forecast accuracy: accuracy $= 100$-Error().

The last performance metric (execution time) corresponds to convergence rate, which is defined as:

- Convergence rate: execution time, the time required for a forecast model to complete its execution. The forecast model which have small execution time converges fast and vice versa. The execution time in this paper is measured in seconds.

The detailed description as follows:

4.1   Hourly Electric Load Prediction

The evaluation of hourly forecasted electric load of FE grid for the proposed forecast model (FCRBM-ELF) vs existing models (MI-mEDE-ANN, AFC-STLF, Bi-level, and FS-ANN) is illustrated in Fig. 2. It is clear that the proposed FCRBM-ELF model effectively forecasts the future load of FE grid. Both ANN and FCRBM based forecasters are capable to capture the nonlinearities of historical load time series data. The nonlinear prediction capability is due to the use of nonlinear activation functions (AFs) i.e., sigmoidal, rectified linear unit (RELU), and tangent hyperbolic (Tanh). The existing models (MI-mEDE-ANN, AFC-STLF, Bi-level, and FS-ANN) used sigmoidal AF and our proposed model select RELU because it has a fast convergence rate and solves the problems of overfitting and vanishing gradient. Figure 2 depicts that the proposed FCRBM-ELF model profile closely follows the target load profile as compared to existing models (MI-mEDE-ANN, AFC-STLF, Bi-level, and FS-ANN). It is clearly seen that the percentage error of the proposed FCRBM-ELF model is 1.10%, MI-mEDE-ANN is 2.2%, AFC-STLF is 2.1%, Bi-level is 2.6%, and FS-ANN is 3.6%, respectively.

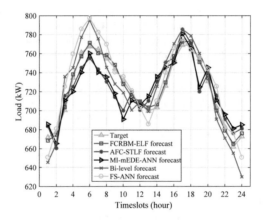

Fig. 2. Hourly load prediction of FE grid

4.2   Seasonal Electric Load Forecasting: Weekly Prediction with Hourly Resolution

The weekly electric load forecasting with hourly resolution is depicted in Fig. 3. This is the week ahead forecasted electric load of FE grid. It is worth mentioning that the proposed FCRBM-ELF model has better results as compared to the existing models (MI-mEDE-ANN, AFC-STLF, Bi-level, and FS-ANN). The proposed FCRBM-ELF model closely follows the target load which is clearly depicted in the zoomed box. The observation in terms of numerical values is that the percentage error of the proposed FCRBM-ELF model is 1.12%, MI-mEDE-ANN is 2.23%, AFC-STLF is 2.0%, Bi-level is 2.5%, and FS-ANN is 3.4%, respectively. The well-grounded reasons the for better performance of the proposed FCRBM-ELF model are the use of deep layer layout of FCRBM with RELU and integration of GWDO based optimization module.

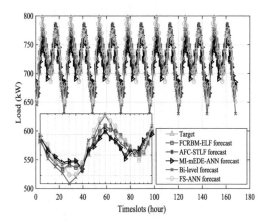

Fig. 3. Seasonal electric load forecasting for a week with an hourly resolution of FE grid

4.3 Performance Evaluation in Terms of Error and Convergence Rate

The performance analysis in terms of accuracy (error) and convergence rate (execution time) is illustrated in Figs. 4 and 5. The error indicates how much the forecasted value deviates from the target value. The smaller value of error results in high accuracy and vice versa. The error performance in terms of numerical values for both day and week ahead forecast is shown in Fig. 4a and b, respectively. The percentage error of FCRBM-ELF, MI-mEDE-ANN, AFC-STLF, Bi-level, and FS-ANN, 1.10%, 2.2%, 2.23%, 2.6%, and 3.6%, respectively. From the above discussion, it is concluded that Bi-level strategy is better than FS-ANN strategy in terms of error performance. The reason for this better performance is that forecast error is minimized by the integration of EDE based optimization module. However, this percent error is minimized at the cost of more execution time as depicted in Fig. 5. This Figure shows that the execution increases from 20 s to 95 s as the optimization module is integrated. Thus, it is concluded that there exists a tradeoff between accuracy and convergence rate. The proposed FCRBM-ELF model reduces this execution time due to the following reasons: (i) GWDO Sect. 3 is used in the optimization module instead of EDE and mEDE due to

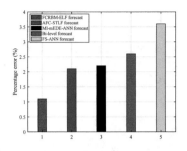

(a) Day ahead

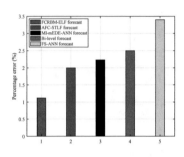

(b) Week ahead

Fig. 4. FE grid Electric load forecast: accuracy analysis in terms of percentage error

faster convergence, (ii) RELU is used instead of sigmoidal AF and multivariate autoregressive algorithm, (iii) FCRBM is used which performs better than ANN, (iv) for data pre-processing data cleansing and normalization are used, and (v) for features selection redundancy, irrelevancy, and candidate interaction process are used, while the existing models only use redundancy and irrelevancy. The aforementioned modifications in the existing models (MI-mEDE-ANN, AFC-STLF, and Bi-level) leads to reduce the execution time of 38 s. On the other hand, the proposed FCRBM-ELF model accuracy is improved as compared to existing models (MI-mEDE-ANN, AFC-STLF, Bi-level, and FS-ANN) [refer to Fig. 4]. However, the execution time of the proposed FCRBM-ELF model is more as compared to FS-ANN because with FS-ANN model no optimization module is used [refer to Fig. 5]. Thus, it is concluded from the above discussion that the proposed FCRBM-ELF model outperforms the existing models in terms of convergence rate and accuracy.

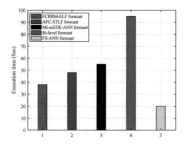

(a) Day ahead

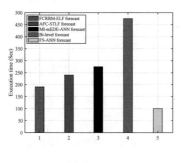

(b) Week ahead

Fig. 5. FE grid Electric load forecast: accuracy analysis in terms of Convergence rate

5   Conclusion

In this paper, the electric load forecasting problem is described. This problem is very complex due to the nonlinear behavior of consumers and influencing factors. Thus, an efficient electric load forecasting model based on FCRBM is proposed to provide accurate load forecast with affordable execution time. The proposed model is examined on FE grid data of USA. The obtained results are compared with other load forecasting models (MI-mEDE-ANN, AFC-STLF, Bi-level, and FS-ANN) in terms of both accuracy and convergence rate. It is validated that our proposed FCRBM-ELF model outperforms the other models in terms of forecast accuracy and convergence rate.

References

1. Zhang, X., Wang, J., Zhang, K.: Short-term electric load forecasting based on singular spectrum analysis and support vector machine optimized by Cuckoo search algorithm. Electr. Power Syst. Res. **146**, 270–285 (2017)
2. Javaid, N., Hafeez, G., Iqbal, S., Alrajeh, N., Alabed, M.S., Guizani, M.: Energy efficient integration of renewable energy sources in the smart grid for demand side management. IEEE Access **6**, 77077–77096 (2018)

3. Hafeez, G., Javaid, N., Iqbal, S., Khan, F.: Optimal residential load scheduling under utility and rooftop photovoltaic units. Energies **11**(3), 611 (2018)
4. Lin, C.-T., Chou, L.-D.: A novel economy reflecting short-term load forecasting approach. Energy Convers. Manag. **65**, 331–342 (2013)
5. Ryu, S., Noh, J., Kim, H.: Deep neural network based demand side short term load forecasting. Energies **10**(1), 3 (2016)
6. Li, H.-Z., Guo, S., Li, C.-J., Sun, J.-Q.: A hybrid annual power load forecasting model based on generalized regression neural network with fruit fly optimization algorithm. Knowl.-Based Syst. **37**, 378–387 (2013)
7. Chen, Y., Yang, Y., Liu, C., Li, C., Li, L.: A hybrid application algorithm based on the support vector machine and artificial intelligence: an example of electric load forecasting. Appl. Math. Model. **39**, 2617–2632 (2015)
8. Hahn, H., Meyer-Nieberg, S., Pickl, S.: Electric load forecasting methods: tools for decision making. Eur. J. Oper. Res. **199**(3), 902–907 (2009)
9. Taylor, J.W.: An evaluation of methods for very short-term load forecasting using minute-by-minute British data. Int. J. Forecast. **24**(4), 645–658 (2008)
10. De Felice, M., Yao, X.: Short-term load forecasting with neural network ensembles: a comparative study [application notes]. IEEE Comput. Intell. Mag. **6**(3), 47–56 (2011)
11. Pedregal, D.J., Trapero, J.R.: Mid-term hourly electricity forecasting based on a multi-rate approach. Energy Convers. Manag. **51**(1), 105–111 (2010)
12. Filik, Ü.B., Gerek, Ö.N., Kurban, M.: A novel modeling approach for hourly forecasting of long-term electric energy demand. Energy Convers. Manag. **52**(1), 199–211 (2011)
13. López, M., Valero, S., Senabre, C., Aparicio, J., Gabaldon, A.: Application of SOM neural networks to short-term load forecasting: the Spanish electricity market case study. Electric Power Syst. Res. **91**, 18–27 (2012)
14. Zjavka, L., Snášel, V.: Short-term power load forecasting with ordinary differential equation substitutions of polynomial networks. Electric Power Syst. Res. **137**, 113–123 (2016)
15. Liu, D., Zeng, L., Li, C., Ma, K., Chen, Y., Cao, Y.: A distributed short-term load forecasting method based on local weather information. IEEE Syst. J. **12**(1), 208–215 (2018)
16. Ghadimi, N., Akbarimajd, A., Shayeghi, H., Abedinia, O.: Two stage forecast engine with feature selection technique and improved meta-heuristic algorithm for electricity load forecasting. Energy **161**, 130–142 (2018)
17. Vrablecova, P., Ezzeddine, A.B., Rozinajová, V., Šárik, S., Sangaiah, A.K.: Smart grid load forecasting using online support vector regression. Comput. Electr. Eng. **65**, 102–117 (2018)
18. González, J.P., San Roque, A.M., Perez, E.A.: Forecasting functional time series with a new Hilbertian ARMAX model: application to electricity price forecasting. IEEE Trans. Power Syst. **33**(1), 545–556 (2018)
19. Luo, J., Hong, T., Fang, S.-C.: Benchmarking robustness of load forecasting models under data integrity attacks. Int. J. Forecast. **34**(1), 89–104 (2018)
20. Ahmad, A., Javaid, N., Mateen, A., Awais, M., Khan, Z.: Short-term load forecasting in smart grids: an intelligent modular approach. Energies **12**(1), 164 (2019)
21. Ahmad, A., Javaid, N., Guizani, M., Alrajeh, N., Khan, Z.A.: An accurate and fast converging short-term load forecasting model for industrial applications in a smart grid. IEEE Trans. Ind. Inform. **13**(5), 2587–2596 (2017)
22. Amjady, N., Keynia, F., Zareipour, H.: Short-term load forecast of microgrids by a new bilevel prediction strategy. IEEE Trans. Smart Grid **1**(3), 286–294 (2010)

23. Amjady, N., Keynia, F.: Day-ahead price forecasting of electricity markets by mutual information technique and cascaded neuro-evolutionary algorithm. IEEE Trans. Power Syst. **24**(1), 306–318 (2009)
24. htttps://www.pjm.com/. Accessed 8 Mar 2018

Proposal of Recollection Support System After Motorcycle Touring with Eye Tracking

Takahiro Uchiya$^{(\boxtimes)}$, Shotaro Sugiura, and Ichi Takumi

Nagoya Institute of Technology, Nagoya, Japan
{t-uchiya, takumi}@nitech.ac.jp,
s.sugiura@uchiya.nitech.ac.jp

Abstract. A lifelog is a digital record of human life and behavior. In recent years, with the development of wearable devices, it has become possible to record personal experiences for long periods. Future devices might record long lifelogs without imposing a heavy burden. Nevertheless, an important difficulty is that reviewing data recorded over a long period takes much time. Therefore, for this study, we aim to create a system that summarizes a lifelog video automatically for easy review. As described herein, we specifically emphasized eye tracking and examined which of three methods is most effective for video summarization: pupil diameter, saccade, or saliency map.

1 Introduction

A lifelog is a digital record of human life and behavior that includes various data such as video, audio, position information, heart rate, and pulse wave. Along with marketing and health management, such lifelogs are used to reflect upon memories as. When recalling memorable events, retrospectives using videos and pictures are easy to understand. In recent years, with the development of wearable devices, it has become possible to record personal experiences for long periods. Future devices might support the recording of long lifelogs without imposing any heavy burden.

By attaching a wearable device used for recording video and images to clothes and belongings, one can perform automatic photography without requiring any special operation. However, when recording for a long time by automatic shooting, a difficulty exists that the images become lengthy and numerous. When these videos and images are used to recall memorable events, they must be classified and searched manually, which hinders the review process.

To resolve this difficulty, numerous studies have examined systems to review lifelogs. Ohnishi et al. [1] found that such a system is valuable in recalling a scene that cannot be recorded manually. It turns out that looking back on a scene that cannot be recorded actively is effective. Therefore, we assume for the present study that a user is riding a motorcycle under circumstances in which recording cannot be done actively.

Earlier studies have specifically examined summarization of video recording while riding a motorcycle [2]. In the present study, the system records a scene of interest triggered by a button press, but an important difficulty is that it is necessary to have direct awareness about the button pressing while driving, which is hazardous.

© Springer Nature Switzerland AG 2020
L. Barolli et al. (Eds.): CISIS 2019, AISC 993, pp. 618–625, 2020.
https://doi.org/10.1007/978-3-030-22354-0_55

The purpose of this study is to resolve difficulties of earlier studies and to ascertain an effective method for video summarization for recalling memories. To resolve the difficulties described above, a lifelog is recorded only while driving a motorcycle: no button pressing is performed. For an earlier study, we developed a system that uses image processing to extract scenes that are useful for recalling memories from image feature quantities [3]. We developed a system that extracts and presents images (images of a spacious sky) that are generally likely to be impressive, but another difficulty arises: individual differences in preferences are ignored.

Therefore, this study specifically assesses a method using biological signals as a method of scene extraction considering personal preferences. Especially in the case of biological signals, research on gaze information and preferences is conducted actively. This study is designed to extract preferences using gaze information.

2 Proposed Method

Figure 1 portrays an outline of the proposed system. It is assumed that automatic shooting of visibility based video and acquisition of gaze information are performed while driving a motorcycle. The system calculates pupil diameter, saccade, and saliency map scores from video and gaze information, and extracts scenes of interest.

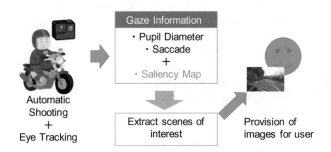

Fig. 1. Outline of the proposed system.

2.1 Saliency Map

Two main causes of eye movement exist. Bottom-up factors cause external stimuli. Top-down factors cause human thinking. The gaze movement caused by the bottom-up factor is a movement that responds to the color, brightness, shape, and movement of the object. Top-down factors include movement by human preference and search for a target. The saliency map is a map that presents the location at which gaze is likely to occur. The saliency map is a visualization of the gaze that is caused particularly by an external stimulus (bottom-up factor). Three factors of luminance, color, and direction are used especially for external stimuli. A feature map is created for each of the three elements and is added to create a saliency map. Each pixel has a value from 0 to 1. Figures 2 and 3 display examples of saliency maps.

A saliency map is said to be related to the ease of watching during driving [4]. The pupil diameter and saccade were effective only at the time of image viewing and were not considered during driving. We inferred that preferences can be extracted considering driving conditions by combination with a saliency map.

Fig. 2. Original image.   **Fig. 3.** Saliency map.

2.2   Pupil Diameter Score Calculation Method

The pupil diameter size is used as a score. Nevertheless, the pupil diameter must be corrected for light because it changes not only with interest but also according to the incidence of external light. Therefore, corrections were made according to a method used in an earlier study [5]: an equation was used for calculation to eliminate the influence of changes attributable to the external light of the pupil. This time, a video viewing experiment was conducted. It is a correction method used when a person watches a video. Equation 1 shows how to calculate the pupil area correction value.

$$y = 1.165 - 0.0016x \tag{1}$$

In that equation, y represents the pupil area correction value; x stands for the screen brightness value. The correction is performed by multiplying the calculated correction value by the actual pupil area.

2.3   Saccade Score Calculation Method

Saccade is a high-speed eye movement that occurs during information search. It is said to occur up to three times a second. Therefore, the velocity is calculated from the distance between samples. A threshold is set so that the maximum is three times per second. The eye movement beyond the velocity is regarded as a saccade.

2.4   Saliency Map Score Calculation Method

Scores are calculated using viewpoints and saliency maps. The score of the saliency map of the position of the viewpoint is taken as the score of the degree of interest.

3 Experiment

3.1 Outline of Experiment

We conducted simple experiment by watching a video presentation about motorcycle touring. An eye tracking "Pupil Headset" was used for gaze measurement. The experimental procedure is explained below.

1. Sit in front of the screen in a dark room.
2. Wear the Pupil Headset.
3. Align the screen height with the eyes so that the face–screen distance is 45 cm.
4. Calibrate the gaze.
5. Watch the video and conduct eye gaze measurement.

Subjects were five university students. No consideration was given to possession or non-possession of a motorcycle license. We recorded the scene of interest by having the subject press a button. The video is that of a motorcycle running in the mountains.

3.2 Method of Extraction and Evaluation

The scene of interest appeared 3.5 s before the button is pressed. Extraction was performed in descending order of score. The degree of matching between the interest scene of the correct answer obtained by pressing the button and the extraction by the system was evaluated using MAP: an index that gives a higher value for a correct answer (interesting scene) in a higher rank when all scenes are ranked. This time, we used random extraction as a basic method for evaluation by comparing them. Random sampling takes an average of 100 trials.

3.3 Results and Consideration

We describe results obtained when using three methods of pupil diameter, saccade, and saliency map. Table 1 presents results of calculating MAP for each index and for random extraction.

Table 1. MAP for respective methods

	Pupil diameter	Saccade	Saliency map	Random
Subject A	0.132971971	0.081217867	0.108933911	0.092316778
Subject B	0.082158398	0.080666269	0.151014947	0.101751424
Subject C	0.055990324	0.182981258	0.113048877	0.086915242
Subject D	0.211752108	0.090289701	0.135161116	0.081174413
Subject E	0.079375118	0.09306767	0.081819494	0.092316778
Average	0.112449584	0.105644553	0.117995669	0.090894927

Mann-Whitney's U test was applied with significance inferred for 5% per side to evaluate whether a significant difference was found between each method and a random result. Results showed no significant difference for Pupil diameter – Random or

Saccade – Random, but a significant difference was found for Saliency map – Random. We consider each one hereinafter.

3.3.1   Pupil Diameter – Random

We describe the point at which no significant difference was found between random and pupil diameter. First, for the normalized scores of pupil diameter of the respective subjects, the average was taken with a width of 500 samples before and 300 samples after based on the button press timing. Furthermore, Fig. 4 presents the average of all subjects. The vertical axis shows the normalized score. The horizontal axis shows the number of frames (time). The blue line at 500 frames is the button press timing. Other light blue lines show the normalized pupil diameter scores.

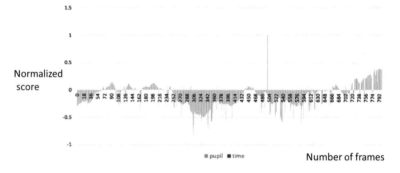

Fig. 4. Average scores of pupil diameter.

As the figure shows, the score is lower before and after pressing the button. A change in pupil diameter is thought to occur because of factors other than interest as the cause of ineffective pupil diameter. The pupil diameter is also adjusted in size with changes in distance [6]. Additionally, the phenomenon reportedly occurs when a viewer feels depth, even during video viewing [7]. The motorcycle driving video includes a sense of depth, as shown in Fig. 5. Therefore, it is thought that the image used in this experiment makes it easy to feel the sense of depth, and that the pupil diameter was adjusted for a sense of perspective.

Fig. 5. One scene of motorcycle running.

It was also considered that a difficulty arose in the correction method for external light of the pupil diameter. The correction method used this time uses the average of the luminance values of the screen. The line of sight cannot be considered. Results suggest that the correction could not be performed sufficiently and that the scene of interest could not be extracted sufficiently. The pupil change depends on the brightness of the central visual field compared to the peripheral visual field. For example, when a scene has only a bright part, but is mostly dark, the high luminance part is regarded as characteristic and gaze is likely to occur. However, in such a scene, the average value of the luminance is low, so the pupil diameter is corrected to be small. Especially in the case of a motorcycle, bright scenes such as scenes coming out of a tunnel are easy to see.

3.3.2   Saccade – Random

We reported that no significant difference was found between random and saccade results. As with pupil diameter, the result of averaging 500 samples before and 300 samples after button pressing can be shown. Figure 6 shows averages of all subjects.

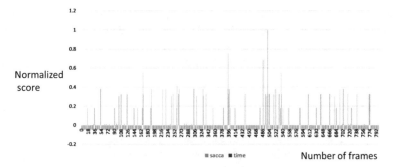

Fig. 6. Average scores of saccade.

The vertical axis shows the score of the normalized saccade. The horizontal axis shows the number of frames (time). The dark blue line at 500 frames is the timing of the button press. In this figure, a saccade occurs even before the button is pressed, but it appears that a saccade has occurred in other places as well. The cause of ineffective saccades might be high-speed eye movement caused by factors other than interest. It is possible that the movement image of the motorcycle has high-speed movement of the landscape and that the gaze movement that responds to the movement.

Furthermore, some difficulty might remain in the saccade identification method. Because saccades are classified as being above or below some set threshold, outliers might increase the threshold. Most are no longer identified as saccades. Results suggest that extraction using saccades was ineffective when assuming running on a motorcycle.

3.3.3   Saliency Map – Random

A significant difference was found between random results and the saliency map results, which proves the method as effective. Similarly to the pupil diameter and saccade figures, Fig. 7 shows the average of all subjects for the front 500 samples and the rear 300 samples based on the button press.

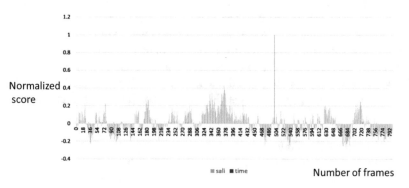

Fig. 7. Average scores of saliency map.

The vertical axis shows the score of the normalized saliency map. The horizontal axis shows the number of frames (time). The dark blue line at 500 frames shows the button press the timing. The graph shows that the score is high 3.5 s (210 samples) before the button is pressed. Therefore, the saliency map is regarded as effective.

4   Conclusion

This study was conducted to make an appropriate summary of life log images of motorcycle touring. To achieve this purpose, we examined which of three indices (pupil diameter, saccade, and saliency map) calculated from eye-gaze information obtained by eye tracking and visual field images was effective for image summarization. Results of comparison using the random method revealed only the saliency map as effective.

Future issues to be addressed in this research are presented below.

- Examination of pupil diameter correction method
- Consider how to identify saccades
- Use of a saliency map considering a time series

References

1. Ohnishi, Anna, et al.: Method to efficiently retrieve memorable scenes from video using automatically collected life log. IEICE Tech. Rep. **115**(27), 23–28 (2015)
2. Takahiro Sugiyama, Hidemitsu Maki, Yuichi Arai, Yoichi Takebayashi: A cruising-assist system of motorcycle based on video knowledge contents creation. The 20th Annual Conference of the Japanese Society for Artificial Intelligence (2006)

3. Shotaro Sugiura, Takahiro Uchiya, Ichi Takumi: Supporting recollection of memories after motorcycle touring by image selection. Proc. of IEEE Sixth Global Conference on Consumer Electronics (2017)
4. Mazda Motor Corporation.: Visual-field control device, 2018-016119 (2018)
5. Nakayama, Minoru, Yasuike, Ikki, Shimizu, Yasutaka: Controlling the effects for brightness in the measurement of pupil size as a means of evaluating mental activity. Japan Soc. Edu. Tech. **15**(1), 15–23 (1991)
6. Lee, E., Lee, J., Park, K.: Experimental investigations of pupil accommodation factors. Invest. Ophthalmol. Vis. Sci. **52**(9), 6478–6485 (2011)
7. Tetsuri Inoue: Eye movement and accommodation when viewing 2D and 3D images. J. Inst. Tel. Eng. Japan, pp. 423–428 (1996)

The 10th International Workshop on Frontiers on Complex, Intelligent and Software Intensive Systems (FCISIS-2019)

Design and Implementation of Cloud Service System Based on Face Recognition

Hao Wu, He Xu$^{(\boxtimes)}$, and Peng Li

School of Computer Science,
Nanjing University of Posts and Telecommunications, Nanjing 210023, China
{1217043220,xuhe,lipeng}@njupt.edu.cn

Abstract. Face recognition technology can be applied to many aspects in smart city, and the combination of face recognition and deep learning can bring new applications to the public security. The use of deep learning machine vision technology and video-based image retrieval technology can quickly and easily solve the current problem of quickly finding the missing children and arresting criminal suspects. The main purpose of this paper is to propose a novel face recognition method for population search and criminal pursuit in smart cities. In large and medium-sized security, the face pictures of the most similar face images can be accurately searched in tens of millions of photos. The storage requires a powerful information processing center for a variety of information storage and processing. To fundamentally support the safe operation of a large system, cloud-based network architecture is considered and a smart city cloud computing data center is built. In addition, this paper proposed a cloud server architecture for face recognition in smart city environments.

1 Introduction

With the development of smart cities and Internet of Things, face recognition technology can be applied to many aspects in smart cities. The combination of face recognition and deep learning can bring a brand new security experience to the public through a combination of hardware and software. Face recognition is one of the most challenging research directions in pattern recognition, image processing and artificial intelligence. The traditional face recognition technology consists of three main process: face image preprocessing, facial feature extraction and classifier design. Face image preprocessing is based on the results of face detection, image processing and ultimately serves the feature extraction process. The obtained original image cannot be directly used because of various restrictions and random disturbances. It must be preprocessed in the early stage of image processing, such as gray correction and noise filtering. Face feature extraction is the process of describing the face features of the original high-dimensional space in the low-dimensional feature space and modeling the features of the face. The purpose is to extract low-dimensional features which are good for classification. Feature extraction is the key of face recognition. An effective face feature extraction method not only helps to simplify subsequent classifier design, but also improves the recognition rate.

With the advent of deep learning, the development of the convolution neural network (CNN) has greatly improved the accuracy and time rate of image search

© Springer Nature Switzerland AG 2020
L. Barolli et al. (Eds.): CISIS 2019, AISC 993, pp. 629–636, 2020.
https://doi.org/10.1007/978-3-030-22354-0_56

technology. The use of deep learning machine vision technology can quickly and easily solve the current problem of missing children and quickly find the suspects. In medium and large-sized security, how to search, compare, analyze the photos of specific people and accurately search several face images are difficult because the face images may similar in tens of millions of photo documents. However, how to realize such complicated system and surveillance video storage requires a powerful information processing center for a variety of information storage and processing. Due to the large amount of traffic, a reasonable data processing platform and cloud server need to be built for population tracking system in the smart city. The purposes of this paper are to speed up the processing of image retrieval, enhance the accuracy of image recognition, reduce the pressure on cloud server backstage. So in this paper we design and implement a cloud server architecture for face recognition in smart city environment.

The following paper is organized as follows. Section 2 describes the related work. Section 3 describes the architecture proposed in this paper. Section 4 shows the experimental results. Finally, we summarize the paper and give comments on the future development.

2 Related Work

The traditional video-based image retrieval method is based on the text keywords of video retrieval and content-based video retrieval technology (Content-Based Video Retrieval, CBVR). Due to limited ability to describe, subjectivity and heavy workload, video retrieval based on text keywords cannot perform typical application and cannot meet the demand of video depth retrieval.

Deep learning simulates the structure of the human brain, uses the basic structure of the convolutional layer, the pooling layer and the fully connected layer of the convolutional neural network, so that the network structure can learn and extract related features. Therefore, deep learning has a strong learning ability on the data. In view of the traditional video retrieval feature extraction algorithm (color feature, texture feature and shape feature, etc.), deep learning feature extraction can provide a more accurate depiction of video images, so as to greatly reduce the retrieval range and achieve fast and accurate retrieval. Deep learning can be used to quickly search images. In the reference [1, 2], they introduce the relationship between deep learning technology and neural network, as well as deep learning and machine learning, and illustrate the application of deep learning in all aspects of life. At present, the commonly used image search algorithms are CNN (Convolutional Neural Network) algorithm, R-CNN (Regions with CNN features) [3]. However, for the capture of moving objects and the direct use of CNN algorithm, the desired effect cannot be achieved. MDNet (Multi-Domain Network) [4] includes an image model and a language model. An image model that enhances the multi-scale feature set and the utilization efficiency is proposed. The language model, combined with improved attention mechanisms, aims to read and explore discriminative image feature descriptions from the reports, learning direct mappings from sentences to image pixels [5]. The algorithm can be effectively used to achieve the goal of moving objects to map search technology, but considering the search at the same time, the data storage must be considered, when a single camera can produce 3.6 GB of data per hour, and the storage requirements of surveillance video data produced by these large number of cameras can reach PB or even EB level [6]. This leads to host a lot of storage space and cloud platform to carry the operation of the system, data compression and reasonable cloud platform [7, 8], which is very important for the smart city search system [9].

The purposes of this paper are to speed up the processing of image retrieval, enhance the accuracy of image recognition, reduce the pressure on cloud server backstage, and thus we build a cloud server architecture for face recognition in smart city environment.

3 The Approach

The entire face recognition tracking system consists of four main modules:

Hardware facilities: it includes sensors, projector, 3D printer model. The projectors project the images on the physical model of the platform to display thematic maps.

City big data: it collects all kinds of urban big data and provide basic data support for urban planning.

Calculation model: it provides the basic data processing and data support for urban planning through data mining model.

User Interface: it provides intuitive data presentation and rapid population tracking.

Face search system architecture for smart city (CityFusion System) is shown in Fig. 1:

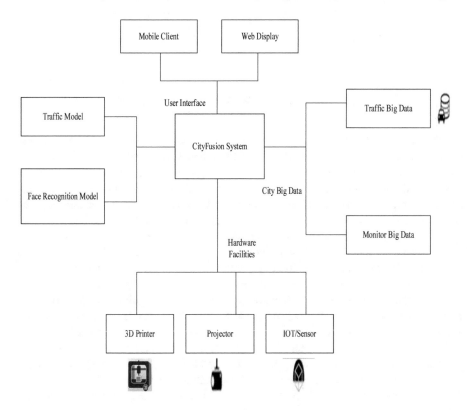

Fig. 1. The function architecture of face search system

The design of cloud structure of face search system is of great significance to the processing speed of face recognition system and the security of the system. The architecture is shown in Fig. 2:

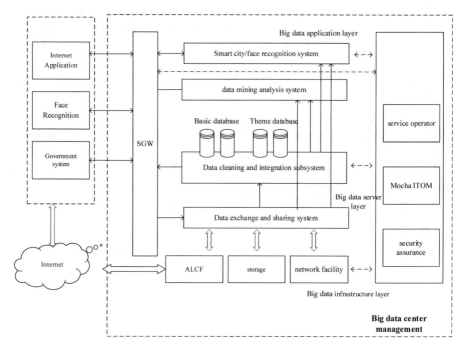

Fig. 2. The cloud-based face search system architecture

The big data infrastructure layer provides the operating environment for the uptime of big data service layers and big data application layers, including computing, storage, and networking facilities. Big data service layer provides various data services for government information systems and industry application systems, including data exchange and sharing services, data cleaning and integration services, and data analysis and mining services [5]. These data services can also be used by upper-level big data applications. The application service gateway provides a unified access point for all kinds of services and applications provided by users. Here, users outside the big data center include government departments, enterprises and public institutions, and the general public who access various services and applications provided by big data centers through government information systems, face recognition applications and Internet applications respectively. The support system provides necessary management and operation support functions for big data center management. According to the major functions of big data center in smart city, a big data center mainly has logical information exchange interfaces with the following three types of external systems. First, interface are connected with Internet applications. Big data center is connected through the Internet application portal, APP and other forms, and a variety of smart city applications are available to the general public. Second, interface with the industry application system. Big data center obtains important raw data from the industry application system, and for the application of the industry data exchange and sharing, data cleaning and data mining integration and other types of data mining services [10]. Third, interface with government information systems. Big data center obtains important original data from the government information system, and for the government information system to provide data exchange and sharing [11].

Due to the prevalence of smart cities and the increasing number of urban populations, previous requests from users for direct access to a single cloud server have been unable to meet user needs, resulting in slow access speeds [12]. A lot of web requests are blocked, and the server's CPU, disk IO, and bandwidth are under pressure. In this point, the processing method of the newly-designed cloud server needs to be redesigned, as shown in Fig. 3. On the one hand, the website images, JS, CSS, HTML, and application service-related files are stored in a distributed file system. On the other hand, the CDN will be static. Resource distributed cache implements "near access" at each node. By separating the dynamic request and the static request access ("separation"), the server's disk IO and bandwidth access pressure is effectively resolved [13]. As shown in the Fig. 3, the architecture adopts CDN + ESC + OSS + RDS, which accelerates the search efficiency of face recognition. Compared with the previous cloud architecture, the search speed is increased [14, 15].

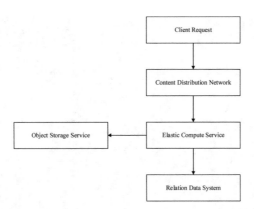

Fig. 3. Cloud server architecture

The information transmission between the web browser and the web server uses a secure socket. For the sake of data transmission security, HTTPS adds SSL are used to verify the identity of the server, and encrypts the communication between the browser and the server. Compared with the previous cloud architecture for smart city, the security of the proposed cloud architecture system has been greatly improved.

4 Experimental Results

In the hardware, further improvements are made by using the ordinary personal machine which has a 6 core and12 thread Intel Xeon CPU E5-2420 processor. In the million-scale standard data set, the recognition accuracy of the target is achieved is with more than 80% in 100 ms.

APP is implemented based on operator LBS(Location Based Service) and mobile GPS(Global Positioning System) positioning capabilities. Camera image recognition, voice reporting, video reporting and other functions are set in one of the operating results, which is shown in Fig. 4.

The user can upload the lost or captured image picture of the person on the client to obtain the approximate location of the person with the help of the camera of the person or the information of the lost population published by the user and is browsed by other users, facilitate common citizens and police to check and search lost people.

The system also can be used to look for criminals, the location of criminals can be viewed by watching the surveillance installed in the city. On the other hand, the police can use mobile phones or computers to upload criminal photos. The criminal's photos will be uploaded to the server's database. The city's camera captures the face photos in real time and matches the criminal's photos. If the match is successful, it returns the geography of the camera position that captured the suspect.

Fig. 4. App upload interface

5 Conclusion

In this paper, we devote ourselves to the current smart city environment, using the camera for video-based face recognition technology to track and find the lost people, which can quickly and easily find the missing children and the elderly, reduce crime rates. With the prevalence of smart city, the number of urban population accessing system is gradually increasing, the required storage data space is becoming larger and larger, resulting in the system efficiency becoming slower and slower. Therefore, we mainly solve the problem from the following aspects: (1) Reduce the storage capacity and speed up the search efficiency in smart city. (2) Establish a reasonable smart city platform and redeploy the cloud server architecture. Therefore, we should consider that the search efficiency is within a reasonable range and that the amount of the system storage is minimized. According to Moore's Law, with the development of economy and science, the storage and computing power of computers will be greatly improved in the next 10 years. Smart city will better serve all aspects of life. The search efficiency and accuracy of the face tracking system based on cloud platform will also be greatly improved with the passage of time. But the ubiquitous camera also violates our privacy at the same time. Urban surveillance systems need to consider privacy issues within the scope of the video surveillance and smart city tracking services to the community.

Acknowledgments. This work was supported in part by the National Natural Science Foundation of China under Grant 61672296, Grant 61602261, Grant 61762071 and Grant 61872194, in part by the Major Natural Science Research Projects in Colleges and Universities of Jiangsu Province under Grant 18KJA520008, and the 1311 Talent Plan of the Nanjing University of Posts and Telecommunications (NUPT).

References

1. Schmidhuber, J.: Deep learning in neural networks: an overview. Neural Netw. **61**, 85–117 (2014)
2. Zhu, X.X., Tuia, D., Mou, L., Xia, G.S., Zhang, L., Xu, F., et al.: Deep learning in remote sensing: a review. IEEE Geosci. Remote Sens. Mag. (2017, in press)
3. Razavian, A.S., Azizpour, H., Sullivan, J., Carlsson, S.: CNN features off-the-shelf: an astounding baseline for recognition, pp. 512–519 (2014)
4. Zhang, Z., Xie, Y., Xing, F., Mcgough, M., Yang, L.: MDNet: a semantically and visually interpretable medical image diagnosis network. In: 2017 IEEE Conference on Computer Vision and Pattern Recognition (CVPR), pp. 3549–3557 (2017)
5. Venugopal, M.P., Mishra, D., Subrahmanyam, G.R.K.S.: Computationally efficient deep tracker: guided MDNet. In: National Conference on Communications, pp. 1–6 (2017)
6. Li, D.R., Cao, J.J., Yuan, Y.: Big data in smart cities. Sci. China Inform. Sci. **58**(10), 1–12 (2015)
7. Hashem, I.A.T., Chang, V., Anuar, N.B., Adewole, K., Yaqoob, I., Gani, A., et al.: The role of big data in smart city. Int. J. Inf. Manag. **36**(5), 748–758 (2016)
8. Bilgaiyan, S., Sagnika, S., Das, M.: Workflow scheduling in cloud computing environment using Cat Swarm Optimization. In: IEEE Advance Computing Conference, pp. 680–685 (2014)
9. Lopez, V., Miñana, G., Sánchez, O., González, B., Valverde, G., Caro, R.: Big + Open data: some applications for a Smartcity. In: IEEE International Conference on Progress in Informatics and Computing, pp. 384–389 (2016)
10. Mastroianni, C., Cesario, E., Giordano, A.: Balancing speedup and accuracy in smart city parallel applications. In: European Conference on Parallel Processing, pp. 224–235. Springer, Cham (2016)
11. Zdraveski, V., Mishev, K., Trajanov, D., Kocarev, L.: Iso-standardized smart city platform architecture and dashboard. IEEE Pervasive Comput. **16**(2), 35–43 (2017)

12. Alworafi, M.A., Al-Hashmi, A., Dhari, A., Suresha, Darem, A.B.: Task-scheduling in cloud computing environment: cost priority approach (2018)
13. Shahdi-Pashaki, S., Teymourian, E., Tavakkoli-Moghaddam, R.: New approach based on group technology for the consolidation problem in cloud computing-mathematical model and genetic algorithm. Comput. Appl. Math. **37**(1), 1–26 (2018)
14. Salhi, H., Odeh, F., Nasser, R., Taweel, A.: Benchmarking and performance analysis for distributed cache systems: a comparative case study. In: Performance Evaluation and Benchmarking for the Analytics Era, pp. 147–163 (2018)
15. Xiong, L., Yang, L., Tao, Y., Xu, J., Zhao, L.: Replication strategy for spatiotemporal data based on distributed caching system. Sensors **18**(1), 1–14 (2018)

Analysis and Design of Group RFID Tag Security Authentication Protocol

Ke Zang, He Xu$^{(\boxtimes)}$, Feng Zhu, and Peng Li

School of Computer Science,
Nanjing University of Posts and Telecommunications, Nanjing 210023, China
{1317044710, xuhe, zhufeng, lipeng}@njupt.edu.cn

Abstract. Aiming at solving the defects of security and privacy of conventional RFID authentication protocols and low authentication efficiency, this paper proposes a group tag security authentication protocol. In the group authentication protocol, it needs to prove that certain tags belong to the same group and exist at the same time. The applicability of RFID authentication protocol is introduced in low-cost tag applications, and an improved scalable lightweight RFID authentication protocol is proposed. The tags only generate pseudorandom numbers and perform join ($\|$) and exclusive-OR (XOR) operations. The response message of the tag is identified by using an existing bit arbitration query tree based on grouping mechanism algorithm. Keeping update to the encrypted information in the tag is synchronized with the server throughout the group certification process. Based on these improvements, this protocol improves the efficiency and security of RFID system.

1 Introduction

The Internet of Things (IoT) is one of the most important components of modern technology systems, allowing devices to synchronize and connect to each other in real time. The use of Radio Frequency Identification (RFID) systems as node identification mechanisms is considered to be one of the necessary requirements for implementing IoT applications, and it helps to solve the problem of identifying things in a cost-effective manner. With the development of the IoT technology, the climax of RFID scale application projects has appeared. For example, large-scale clothing brands such as LA CHAPELLEUR, UNIQLO, Zara, H&M, and HLA have introduced RFID technology in factories, warehouses, and stores. Another noteworthy is the unmanned retail, Amazon Go unmanned offline supermarket, Alibaba at the second Taobao Creativity Festival launched Ali unmanned supermarket "Amoy coffee" and other unlicensed real cases have emerged, which let RFID technology become the focus of public and capital attention. However, because the IoT involves wireless channels for communication that are open to all types of malicious attackers, it is often plagued by various security and privacy issues [1, 2]. Security issues can be a problem that cannot be ignored in the development of RFID. The security of information exchange between a large number of tags and readers is mainly related to the security authentication protocol. Aiming at some existing security risks, the researchers proposed a series of improved security protocols for different scenarios.

© Springer Nature Switzerland AG 2020
L. Barolli et al. (Eds.): CISIS 2019, AISC 993, pp. 637–645, 2020.
https://doi.org/10.1007/978-3-030-22354-0_57

2   Research Works

As the key technology of the IoT sensing layer, RFID is widely used in warehousing, logistics, transportation and other fields, with the advantage of non-contact rapid identification [3]. The application of RFID technology brings convenience to the manufacturing industry and daily life. However, due to the insecure wireless channel which has privacy and security issues, mutual authentication is a way to improve security. For example, in medical information systems, the main problem is that unlicensed entities is not allowed, and authenticated entities can only access information within their own rights.

Literature [4] proposed an RFID security authentication protocol based on PRE-SENT algorithm, which can enhance the security of RFID system without occupying too much hardware resources, so it can be applied to low-cost RFID systems. Literature [5] proposes a protocol against denial of service (DoS) attacks, which uses a hybrid approach consisting of two complementary component protocols, Aloha Filtering (AF) and Poll & Listen (PL). The two are combined with the advantages of both protocols to quickly and completely identify blocked valid RFID tags. Literature [6] proposed an asymmetric design principle, and a lightweight technology was developed to generate dynamic tokens for anonymous authentication. The verification protocol does not need to implement complex and hardware-intensive cryptographic hash functions. Instead, a few simple and hardware-efficient operations are performed to ensure that the privacy of the tag is not compromised. Literature [7] proposed a security authentication protocol based on the Edwards curve for mobile RFID systems, which improves its ability to prevent side channel attacks, and applies elliptic curve to achieve security authentication. The protocol is more effective against a variety of attacks. Literature [8] proposed a new lightweight RFID tag ownership transfer protocol, and proved that the protocol safely implements the defined ideal function, namely, two-way authentication, which has the advantages of tag anonymity, anti-asynchronous attack, and backward privacy protection. The basic idea of the protocol is to solve the problem of random number exposure by using packet random numbers, but the weakness of the random number encryption algorithm leads to the protocol prototype being vulnerable to impersonation attacks, nonlinear attacks and DoS attacks. Literature [9] proposed an improved RFID authentication protocol, which effectively solved the nonlinear attack and DoS attack in [8]. However, the literature [10] pointed out that the unreliability of the key update operation in [9] which will led to forward and backward traceability attacks. So they proposed an improved key update algorithm to ensure that it could not be traced. Reference [11] proposes an ultra-lightweight authentication protocol (called ULRMAPC) that uses only ultra-lightweight displacement operations and XOR operations. The authors claim that their protocol is highly efficient and resistant to common problems with RFID attacks such as DoS, reader and tag analog attacks, but analysis in [12] shows that the ULRMAPC protocol is vulnerable to DoS, reader and tag simulation and desynchronization attacks. By applying elliptic curve, a protocol with forward security and non-traceability is proposed to RFID mutual authentication [13]. However, calculating the public key operation will impose a heavy burden on the tag which is not feasible in low-cost RFID tag.

With the development of mobile intelligent terminals, the demand for mobile RFID systems has become more and more important. Recently, a series of mobile RFID authentication protocols have been proposed. Literature [14] proposed an RFID authentication protocol using a mobile reader. The protocol implements EPC-C1G2 compliance on the tag using only pseudo-random function operations, but it is vulnerable to replay and synchronization attacks [15]. Literature [16] proposed a serverless RFID authentication protocol (ISLAP protocol) for mobile RFID systems, and then the authors [17] improved the shortcomings in the literature [16] in terms of forward untrackability and non-traceability. However, the improved ISLAP protocol cannot effectively resist replay, impersonation and desynchronization attacks. Reference [18] proposes a new lightweight RFID packet authentication protocol for multiple tags in a mobile environment, which allows partial tag loss. Unresponsive tags will not interfere with the entire authentication process, which can ensure that objects can be authenticated in a timely manner. The participating entities of the above protocol only include readers and tags. The reader needs to download an access list containing multiple tags, which are authorized for authentication, and therefore have higher requirements on the storage capacity of the reader. Aiming at solving the defects of security and privacy of conventional RFID authentication protocols and low authentication efficiency, this paper will present a group tag security authentication protocol.

3 Group Tag Based Security Authentication Protocol

3.1 Description of the Protocol Process

The protocol is divided into two processes: initialization and authentication. During the initialization process, the reader and tag are set according to the secret parameters. The authentication phase represents the server to verify the reader. The authentication process is divided into two phases: group proof generation and group certificate verification. In the group proof generation phase, the reader generates a group certificate based on part of the evidence in the tag response. In the group certificate verification phase, the backend server verifies the validity of the group certificate. Table 1 shows the protocol symbolic meaning. The protocol execution flow is shown in the Fig. 1.

Table 1. Symbolic table

Symbol	Meaning
BS	Backend server
R	Reader
T	Tag group
IDi	Tag id in group
Pi	Tag identifier
KTi	Tag groupkey password
RNq,RNv,RN1...,RNn	Temporary random number
IDR	Reader id

(*continued*)

Table 1. (*continued*)

Symbol	Meaning
PR	Reader identifier
EK	Encrypt with key password
KR	Reader key password
RA	Reader authorization tag
sk1,sk2	Key (used to generate the identifier PR, Pi)
g()	generator of pseudo random number
S	Check value
⊕	XOR operation
‖	Connection operation

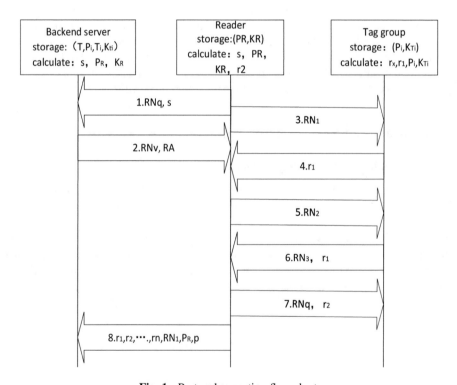

Fig. 1. Protocol execution flow chart

3.1.1 Initialization

The RFID certification protocol contains three parts: a tag group set T = (T1, T2,..., Tn), a reader (R) and a back-end server (BS). The BS shares the same password (KTi) in the tag group. Both the BS and the tag use the same pseudo-random number generator and the same key (sk1) for calculating the tag identifier (Pi). Tag identifier using formula (1) to calculate, where IDi is the tag ID. The server stores a tag information set (T, Pi, KTi, Ti), where T represents the group where the tag belongs to. Only collection (Pi, KTi) needs to be stored one time for each tag. The server and reader share the same key (KR). The server and each reader share a common pseudo-random generator and key (sk2) for computing the reader identifier. The reader identifier is using formula (2) to calculate, where IDR is the reader ID. Both the server and the reader maintain the collection (PR, KR).

$$P_i = g(ID_i || \; sk_1 \oplus K_{Ti}) \tag{1}$$

$$P_R = g(ID_R || \; sk_2 \oplus K_R) \tag{2}$$

3.1.2   Certification

(1) Generate a group proof

(a) It is assumed that each reader can be verified by the server when generating a group certificate for a group of tags. The reader first generates a temporary random number Nq by formula (3). Then set status to 1 where the identifiers PR and KR have been updated by calculate formula (4) and (5). Then the values of RNq and s are sent to the BS.

$$s = g(P_R || \; K_R \oplus RN_q) \tag{3}$$

$$P_R = g(P_R, RN_q) \tag{4}$$

$$K_R = g\,(K_R, RN_q) \tag{5}$$

(b) The BS calculates formula (6) based on the values of RNq and PR and KR. The result is stored in the database and determined if it is the same as the received value s. If the two values of s are the same, the reader will successfully be authenticated. Then an authorization (RA) is generated, where PR and KR also use the formula (4) and (5) to updated value. The BS stores the set (PR, KR) and the authorization set (R, T, RA) before and after the update process. This means that the reader has the right to generate a group certificate for the tag group (T) based on the identifier calculate (RA) by formula (6) containing the tag key password and the tag. The BS then generates a temporary random number RNs and sends it to the reader along with the RA, at the same time, both the tag identifier and the key password (Pi and KTi) are updated by formula (7) and (8).

$$RA = E_{KR}(P_i) \tag{6}$$

$$P_i = g(P_i, RN_q) \tag{7}$$

$$K_{Ti} = g(K_{Ti}, RN_q) \tag{8}$$

$$r_x = g(P_i, K_{Ti}, RN_1) \tag{9}$$

(c) The reader generates a temporary random number RN2 and passes it to the tag to initialize the protocol.

(d) The tag generates a temporary random number RN3, calculates the formula (10), and sets the state of the tag to 1.

$$r_1 = g(P_i\| RN_2 \oplus RN_3) \tag{10}$$

(e) Upon receiving the values of RN3 and r1, the reader calculates the value of r1 if the calculated value is the same as the received r1 value. If the two values are not the same, the protocol process ends. If the values are the same, the reader calculates Eq. (11) and sends the RNq and r2 values to the tag.

$$r_2 = g(P_i\| RN_3 \oplus RN_q) \tag{11}$$

(f) Upon receiving the values of RNq and r2, the tag calculates r2. If the values are the same, the tag will update its identifier (Pi) and the secret value (KTi) by using formula (7) and (8), and store the old and new values of Pi and KTi. And its state is set to 0.

(2) Verify grouping proof

(a) After authenticating all visible tags, the reader calculates the result of formula (12), based on the tag response, and then the reader sends it to the backend server: r1, r2,…, rn, RN1, PR, p.

$$p = MAC_{KR}(r_1 \oplus r_2 \cdots \oplus r_n, RN_1, P_R) \tag{12}$$

(b) The BS searches for the corresponding (PR, KR) based on the reader's identifier PR, and if it is no match, the verification fails. If a match is found, the authentication set (R, T, RA) of the search reader is executed. If no match is found, this means that the reader is not authenticated and the verification fails. If the validation set is found, the backend server calculates formula (12). If the calculated value is the same as the received value, the group certificate is considered to have been verified.

3.2 Protocol Security Analysis

The reader impersonates the attack: In the protocol, the reader needs to send a request to the backend server to initiate tag verification. The backend server is stored according to the tag identifier and the authentication secret key (s) values, and only the real reader can generate a valid s value. After the verification is successful, the backend server will send encrypted messy information to the reader. Since the encryption key is a secret shared only by the reader and the backend server, the attacker cannot intercept the message or disguise it as a reader.

$$s = g(P_R\| K_R \oplus RN_q) \tag{13}$$

Tag spoofing attack: The reader authenticates the tag based on the tag identifier and the pseudo-random generator before accepting some of the evidence from the tag. Therefore, the spoofing tag will not be completed.

Anonymity of readers and tags: Readers and tags use an identifier and a secret number during protocol execution. This allows the update identifier and key password to be synchronized. Therefore, an attacker cannot obtain information about a reader or tag based on the identifier and key password.

Mutual authentication: Mutual authentication between readers and tags is a security issue that is often occurred during broadcasts generated by packet verification. In the protocol, the reader first sends a temporary exchange random number to the tag to initialize the verification process. In this process, both the reader and the tag use the tag identifier which only knows the legal reader and the response tag, thereby achieving mutual authentication between the reader and the tag.

Replay attack: An intruder obtains communication messages by eavesdropping on communication among the server, reader and tag. When an intruder sends these messages to the reader, it can be easily found for duplicate messages because different one-time temporary exchange random numbers are used in each transmission.

Step-by-step attack: During protocol execution process, readers and back-end servers, readers and tags are always synchronized. The identifier and key are updated after successful authentication and state variables are used to store the status of the record update.

Proof the integrity of the group: The proof of parsing contains part of the evidence in the tag. These partial evidences are generated by pseudo-random number generators based on their respective tag identifiers and secrets. Since the intruder does not know the pseudo-number generator, identifier, and secret, it is impossible for the intruder to falsify part of the evidence of the tag. In addition, the group certification uses the public key through the MAC function between the reader and back-end server to verify part of the evidence from the tag. Since the MAC function is a secure one-way function, the intruder cannot derive the contents of the MAC from its value.

4 Protocol Efficiency Analysis

In the improved protocol, when the total number of tags is determined, the authentication time of the protocol will change as the number of groups changes. Here we simulate that when the total number of tags is different, the system authentication time trend changes with the number of groups. As shown in Table 2, regardless of the number of tags in the database, the trend of system authentication time will quickly drop to a minimum, and then slowly increase as the number of groups increases. If the risk of external attacks that may be experienced in actual applications is large, since the tags can be appropriately divided into more groups to achieve a higher security level. If the requirements for authentication efficiency are high and the attacker's capabilities are weak, then a smaller number of groups with higher system verification efficiency can be obtained.

Table 2. Minimum authentication time under the total number of different tags

Total number of tags	2^{10}	2^{12}	2^{16}
Number of tag groups	16	32	128
Authenticate time (ms)	0.87	4.20	10.47

5 Conclusion

This paper presents a scalable protocol for RFID systems. In the proposed protocol, readers and tags use identifiers to ensure their anonymity. The security analysis shows that the protocol is resistant to against impersonation of readers, forged tags, replay attacks, and desynchronization attacks. From the performance analysis, it can be observed that the protocol requires a relatively low number of clock cycles, making it suitable for application which require low-cost RFID tags for high level security.

Acknowledgements. This work was supported in part by the National Natural Science Foundation of China under Grant 61672296, Grant 61602261 and Grant 61762071, in part by the Major Natural Science Research Projects in Colleges and Universities of Jiangsu Province under Grant 18KJA520008, and the 1311 Talent Plan of the Nanjing University of Posts and Telecommunications (NUPT).

References

1. Fu, W., Qian, Z.H., Cheng, C., et al.: Anti-collision algorithm for bit arbitration query tree based on grouping mechanism[J]. J. Commun. **37**, 41–68 (2016)
2. Yu, L.L., Zhang, L., Li, J., et al.: A PSO clustering based RFID middleware[C], pp. 222–225. IEEE, New York (2018)
3. Akgün, M., Çağlayan, M.U.: Providing destructive privacy and scalability in RFID systems using PUFs. Ad Hoc Networks, pp. 32–42 (2015)
4. Zhang, X., Han, D., Cao, G., et al.: RFID security authentication protocol based on present algorithm[J]. J. Commun. **36**, 65–74 (2015)
5. Liu, X., Xie, X., Zhao, X., et al.: Fast identification of blocked RFID tags[J]. IEEE Trans. Mob. Comput. **17**, 2041–2054 (2018)
6. Chen, M., Chen, S., Fang, Y.: Lightweight anonymous authentication protocols for RFID systems[J]. IEEE-ACM Trans. on Networking **25**, 1475–1488 (2018)
7. Yang, Y.L., Peng, C.G., Zhou, Z., et al.: Mobile RFID security authentication protocol based on Edwards curve[J]. J. Commun. **35**1, 32–138 + 145 (2016)
8. Yuan, B.Q., Liu, J.Q.: Proof of security of RFID tag ownership transfer protocol[J]. J. Commun. **36**, 83–90 (2015)
9. Dehkordi, M.H., Farzaneh, Y.: Improvement of the hash-based RFID mutual authentication protocol. Wireless Pers. Commun. **75**, 219–232 (2014)
10. Alavi, S.M., Baghery, K., Abdolmaleki, B.: Traceability analysis of recent RFID authentication protocols. Wireles Pers. Commun. **83**, 1663–1682 (2015)
11. Fan, K., Gong, Y., Liang, C., et al.: Lightweight and ultralightweight RFID mutual authentication protocol with cache in the reader for IoT in 5G. Secur. Commun. Network, 3095–3104 (2016)
12. Aghili, S.F., Ashouri-Talouki, M., Mala, H.: DoS, impersonation and de-synchronization attacks against an ultra-lightweight RFID mutual authentication protocol for IoT. J. Super-computing, 509–525 (2018)
13. Farash, M.S., Nawaz, O., Mahmood, K., et al.: A provably secure RFID authentication protocol based on elliptic curve for healthcare environments. J. Med. Syst, 165–171 (2016)
14. Sundaresan, S., Doss, R., Piramuthu, S., et al.: Secure tag search in RFID systems using mobile readers. IEEE Trans. Dependable Secur Comput. 230–242 (2015)

15. Jannati, H., Bahrak, B.: Security analysis of an RFID tag search protocol. Inf. Process. Lett. 618–622 (2016)
16. Pourpouneh, M., Ramezanian, R., Salahi, F.: An improvement over a server-less RFID authentication protocol. Int. J. Comput. Network Inf. Secur. 65–171 (2014)
17. Shahrbabak, M.M., Abdolmaleky, S.: SRMAP and ISLAP authentication protocols: attack and improvements. IACR Cryptology ePrint Archive, 731–741 (2016)
18. Shen, J., Tan, H., Zhang, Y., et al.: A new lightweight RFID grouping authentication protocol for multiple tags in mobile environment. Multimedia Tools and Applications, 761–783 (2017)

Overcoming Data Security Challenges of Cloud of Things: An Architectural Perspective

Farhad Daneshgar[(✉)], Omid Ameri Sianaki, and Amnna Ilyas

College of Engineering and Science, Victoria University Sydney,
Sydney, Australia
{farhad.daneshgar, omid.amerisianaki}@vu.edu.au,
amnnailyas@gmail.com

Abstract. With the advancement in internet technology, the adoption of Internet of Things devices is no more a dream. By having millions of objects with the ability to communicate autonomously and having the limited sense to process and compute the data, the cloud can be leveraged as a potential platform for improving the rate of latency and data processing in data transmission. The integration of IoT and cloud computing gives birth to the phenomena called Cloud of Things (CoT) that despite being a feasible solution, it comes with some costs related to the data security. The main goal of this study is to develop a *security architecture of CoT*, called SACT for maintaining data security. The proposed framework provides a three-tier architecture for data transmission from one device to another.

Keywords: Data security · A Cloud of Things · Internet of Things · Architectural approach

1 Background of Study

The adoption of Wireless Senor Networks (WSN) has introduced a new era of the machine to machine communication that enabled a new phenomenon known as the Internet of Things (IoT). IoT is an association of devices (usually of small objects) that are restricted by the storage and processing capacities (Alessio et al. 2016). Cloud, on the contrary, provides its users with the unlimited scalable storage and processing facility, thus making many to ponder upon the integration of both concepts (Diaz et al. 2016). This gave birth to the concept termed as Cloud of Things (CoT). The CoT is a phenomenon which combines the resource benefits of IoT with the storage and analysis facilities provided by the cloud service (Alessio et al. 2016). Some of the benefits obtained by the above integration include expansion of application and scope of IoT as a service, improved storage for additional data, enhanced processing capabilities, and rapid communication among devices but little research has been performed on the drawbacks of this association (Diaz et al. 2016).

The primary objective of the current study is to identify the main challenges in data security within the CoT process. A preliminary investigation by the authors revealed existence of four categories of research themes that collectively constitutes the current literature on CoT data communication, these include *(i) securing services provided by*

L. Barolli et al. (Eds.): CISIS 2019, AISC 993, pp. 646–659, 2020.
https://doi.org/10.1007/978-3-030-22354-0_58

CoT, (ii) securing CoT architecture, (iii) securing the management of CoT service, and
(v) *current solutions available to overcome the above challenges.* Among the various
data security challenges of CoT, the current study focuses on the architectural challenge
of CoT.

2 Introduction

Due to the standardization of the TCP/IP over the span of thirty years the Internet has
grown to a stage where the usage of IoT devices has become a reality. This is due to the
wide availability of wireless networks coupled with the low-cost sensor-embedded
devices that allow smart objects communicate with each other (Fox and Kamburugamuve 2016). This information exchange is facilitated by the sensors or actuators who
respond to a certain situation based on the environment that they are communicating in
(Uckelmann et al. 2011). Also, as the IoT allows open data sharing among different
autonomous devices, real-time data integration by different objects possessing different
capabilities and intentions is one major drawback faced by the IoT.

Recently, the cloud has raised as the main complementary technology to the IoT for
serving as a platform for performing required data storage and processing, and allowing
independent data sharing among various devices (Cheng et al. 2014). Cloud can be
referred to as a model that enables both sharing of, and access to, pools of computing
resources in the form of applications, storage, servers, networks, or services on an on
demand basis, providing the user with the opportunity to scale resources based on their
needs (Mell and Grance 2011). Table 1 below demonstrates complementary aspects of
IoT and cloud computing.

Table 1. Complementary aspects of IoT and cloud (adopted from Donato et al. 2016)

	IOT	Cloud
Displacement	Pervasive	Centralized
Reachability	Limited	Ubiquitous
Components	Real world things	Virtual resources
Computational capabilities	Limited	Virtually unlimited
Storage	Limited or none	Virtually unlimited
Role of Internet	Point of convergence	Delivering services
Big Data	Source	Means to manage

According to Singh et al. (2015), the main reason for using cloud as a platform for
IoT is because of the following reasons:

- Wide coverage of cloud services allowing remote objects located in different parts
  of the world to communicate freely with each other
- Cloud allows a scalable solution for IoT enabled devices to adjust the storage and
  analysis facility based on the machine's requirement only

- Cloud services improve the performance of IoT devices by sharing the load in terms of data storage and processing hence, allowing better performance of the object services

With the collaboration of Cloud and IoT it was (and is) widely believed that for the IoT to continue providing intended functions, cloud must be used as a platform. Such integration however, revealed many issues and threats. A summary of these issues has already been expressed by Aazam and Hung (2014):

1. Identity management of devices including mobile devices, requiring the creation of an upgraded network each time a device enters network in absence of a secure algorithm
2. Location of data storage which if compromised may lead to various attacks and data loss
3. Privacy invasion by intruders by intercepting, stealing, or modifying data
4. Lack of security due to security standard limitations
5. Unrequired data being communicating which not only impacts processing capabilities but also leads to attacks in disguise
6. Security threats from network insiders and trust issues of sharing sensitive information with an unknown service provider

The above issues indicate a lack of security in CoT-enabled systems and networks. This in turn highlights the need for implementing strong security measures. Many authors suggested theories for countering the issues by focusing on one or more aspects related to the threats posed by the CoT implementation. These theories are briefly mentioned in the Literature Review section of this paper.

Among the challenges mentioned above, data communication security is a major issue due to the open data sharing among devices. The major objectives of the current study are summarized as below:

- To identify various data security issues of Cloud of Things, and to create themes that signify various category those challenges as an analytical tool used by future researchers.
- To investigate the possibility of proposing a high-level solution for the identified theme.

Based on the above objectives the following research questions have been derived for the study:

RQ1 - What is the role of data communication in the Cloud of Things structure?

RQ2 - What are the main challenges for providing secure data communication in Cloud of Things?

RQ3 - What is a suitable solution to counter the architectural challenges faced in a CoT environment?

3 Literature Review

The aim of this review is to provide a context for responding to the three research questions of the study. In doing so, the review is divided into four parts. In Sect. 3.1, data security issues within the CoT structure is discussed as a pre-cursor to answering the first research question of the study. Sections 3.2 and 3.3 provides reviews of security challenges that CoT environments are currently facing for data communication, as well as existing frameworks for overcoming those challenges. It is expected that the materials reviewed in Sect. 3 will lead to the proposal of a synthesized framework for the current study, the latter is discussed in Sect. 4.

3.1 Security of Data Communication in Cloud of Things

Security has been a major concern for CoT as it combines two major concepts of IoT and Cloud that each involve unresolved security threats for data communication taking place among various objects from both internal and external parties. Internal parties include cloud service providers and other tenants of the network who utilize services of the same provider, and can attack the network due to multi-tenancy feature allowing easy entrance to the network. The external parties include attackers who attack the network without being part of it by hacking of a compromised node or the gateway. Also, as the concept is relatively new, the theories suggested by many are yet incomplete, and fail to counter the above threats (Bhattasali et al. 2014). According to Botta et al. (2016), data collection and transmission services provided by the CoT take place among various objects, making CoT all about communications. Furthermore, many authors believe that security of communication remains a major problem in the CoT structure; the reason is that the information remains prone to many attacks conducted in the forms of SQL injections, eavesdropping, man-in-the-middle, and many other methods (Botta et al. 2016). With rising demand in the use of IoT enabled devices, it is estimated by Cisco that the number devices connected together will be more than 20 billion by the year 2020 (Baqer and Kalam 2017).

To resolve the above issues, the use of *over the edge data processing and searching model* has been suggested (Baqer and Kalam 2017). According to this model, it is essential that the data processing and storage is performed near the end device to lower the latency and time required for data related operations. For this purpose, they suggest the use of edge technology by having edge servers in the form of routers or mobile devices which help in improving the data access and data bandwidths while data sharing and also improve data security (Baqer and Kalam 2017).

According to Soldatos et al. (2014), with various heterogeneous devices communicating with each other, interoperability and communication becomes a major concern for CoT users. As each machine uses a different set of languages, protocols, and programs, it would be difficult for the devices to perform various communication functions. One solution to this issue is the use of open source systems allowing every device to communicate with others through the implementation of open sources system, no matter what language or program is supported by the machine. But it also poses a major security concern as an open source system itself may lure many malicious attacks due to allowing every machine to connect to the IoT network. To overcome challenges of opens source system, Soldatos et al. (2014) suggested the use of an open source IoT platform for semantic interoperability through the use of W3C semantic sensors, sensor middleware for secure data collections, and visual tools to deploy IoT applications.

Privacy intrusion is another major concern in CoT communications. This is not only due to the advanced ability of today's attackers but also due to the nature of the service that involves multi-tenancy by sharing common storage space with multiple people, and also relying on a third party for storage purposes. This calls for a need for implementation of a secure communications system that maintains security of the data being transferred and received from one machine to another. To avoid the breach of privacy, Henze et al. (2015) recommend the use of UPESCI model which requires the cloud users to create their own privacy spheres allowing them to customize their privacy settings as per their needs, requiring the cloud service providers to ensure that all the requirements of the sphere are being met (more details of this proposed model are provided in the next section). To ensure that the cloud providers provide desired requirements, a third party will be involved to check the compliance of each party with the terms stated in the service agreement (Henze et al. 2016).

3.2   Current Security Challenges of CoT

This study provides a classification of the CoT data security challenges with three categories; these are:

First Category: Threats to Networking Cloud and Edge Data Centers in IoT

Cloud offers a scalable and *pay as you go* structure with massive data storage for IoT enabled devices. The researchers recently highlighted the use of edge or fog assisted cloud that allows the communications to be processed, scrutinised, and filtered on the edge of the network. Fog-assisted cloud ensures maximum privacy and security protection by allowing only qualified communications entering the network. But this configuration also introduces several underlying security threats when implemented. The reason is that data analytics activities performed during the transfer of information to and from the cloud to the edge remain prone to multiple attacks in the form of perception layer attacks, network attacks, and application layer attacks. These attacks prevail as the traditional Cloud Data Centres (CDC) format for analysing information remains insufficient for analysing the complex data communications performed by various activities (Puthal et al. 2016).

When data is transferred from and to the Cloud Data Centre, it passes through 3 main layers namely perception layer, network layer, and application layer (Puthal et al. 2016). Each of these layer poses unique security threats to the data being transferred. Below is a list of a few of the threats at each layer:

Perception Layer Attacks

- Data tampering by the intruder through hacking the system, capturing a sensor device of the network, and entering the secure network through a fake device
- Attacker forging the ID of a single or more sensor devices to attack the network
- Lack of encrypted key protection or authentication failures

Network Layer Attacks

- Replay attacks by attacker capturing the data and using it multiple times to gain access or information from the network
- Timing attacks by gaining access to one secret key and then using that to gain information of all secret keys within the estimated time
- Denial of service attacks causing the network to jam and not allowing the authorized users to gain access to the system
- Routing attacks by posing threats to the routing structure
- Man in the middle attacks
- Spoofing attacks and wormhole attacks diverting information to the attackers

Application Layer Attacks

- Manipulation of system loopholes
- Data availability to unauthorized attackers
- Authentication weaknesses allowing attackers to break into the system

Second Category: Data Security Challenges in Integration of IoT and Cloud Computing

With the rising advancements in the field of Mobile Communications Cloud (MCC), many have suggested the use of IoT as a ground support for enhancing the services offered by MCC. The main aim of MCC is to provide data and information access services to the users from anywhere in the world. This is possible with the use of extensive cloud services allowing the users for storage, location, context awareness, and memory (Stergiou et al. 2018).

Though, cloud provides the users with additional facilities, the communications autonomy can be overwhelming for cloud and mobile service providers. IoT is suggested as a solution for allowing autonomous communications among machines. This however introduces the following security issues:

- The IoT enabled systems use an old operating system as a base to extend services while services required by the MCC are relatively new. This may result in some security threats as attackers in the past have been able to intrude secure IoT systems and increase the chances of both known and unknown attacks become relatively high.
- Inability of users to protect network by having a secure strong password.
- Insecure devices connecting to the network (Stergiou et al. 2018).

Third Category: Man-In-The Middle Attacks in a Fog Based Architecture

A fog paradigm is used to support cloud services during the IoT communications process. This is done by allowing over the edge communications for machines for secure and fast communications. This not only helps in providing a secure environment, but also enables the cloud to save resources, overcome the challenge of computation limitations, and ensure a steady network system along with accurate location services functions and high mobility support with low latency involved through allowing processing functions to take place at the edge of the network. However, Stojmenovic and Wen (2014) argue that it also comes with certain security threats introduced by hackers. This is due to the authentication verification of the users at different levels of gateways. The attacker can either tamper, steal, or spoof IP address for gaining an unauthorized access to the system. Hackers achieve this by intruding the network with a compromised gateway.

3.3   Existing Framework for Overcoming Data Security Issues in CoT

One main goal of the current study has been the development of a framework for overcoming security challenges of the CoT architectural; and we call it *security architecture of CoT (or SACT)* that is discussed in 4.2 in detail. In other words, we specifically focus on the architectural aspect of the data security in the CoT. To this end, various existing frameworks of CoT architecture that are relevant to the data security are reviewed in this section.

3.3.1   Gateway Design for CoT

The core function of CoT is to gather and deploy dispersed resources for performing its functions. Therefore, special focus has to be placed on how the data is transferred from one end to another, leading to resource gathering and distribution. Petrolo et al. (2016) proposed a model for the creation of gateways that are expected to generate less overhead with negligible impact on the performance of the CoT. This model is based on the use of lightweight technologies, the latter requiring the adoption of a container-based approach combined with a distributed cloud structure. The reason behind using distributed cloud is that the enormous amount of communication is exchanged among the intelligent nodes of the machines, and this creates a high amount of data that may be redundant and unsecure which require filtering and compression. Since the distributed cloud allows the processing to be conducted near the end device, the devices will be empowered to process the data based on their requirements along with reducing the layover time substantially (ibid.).

Furthermore, a container-based virtualized system in the above model plays the role of a lightweight substitute for the hypervisor-based gateway systems as it allows the operating system to reduce the workload by assign each container to process a fraction of data. Each container although works independently but is able to communicate with other containers working in the operating system. Each container working on the same OS has a separate interface and functions to support a certain.

The above arguments led to the development of another CoT gateway model that employs the use of smart gateways to enable recognition of the data before transmitting it to other devices (Aazam and Huh 2014). Such smart gateways perform multiple tasks before uploading data on the fog to ensure that only filtered data is stored and further processed.

3.3.2   Fog Computing and Its Role in the Internet of Things

Fog Computing provides enhanced geographic distribution, supports large number of devices, enhanced interactivity, heterogeneity, interoperability, and supports online analytics. As a result, Fog Computing can potentially improve the operation of the CoT and compliments the IoT more than Cloud does by supporting functionalities such as smart grids, wireless devices, and connected vehicles (Stojmenovic and Wen 2014).

More specifically, the use of fog computing can provide the following benefits to the CoT:

1. Providing an architecture with massive infrastructure of computers, storage, and networking devices.
2. Orchestration and resource management of the Fog nodes.
3. Innovative services and applications to be supported by the Fog (ibid.).

3.3.3 Architectural Characteristics of a CoT API

As various objects communicate with one another in a cloud environment, a need arises for interoperability among devices for object identification, management, and organization of objects. The reason is that if the object identification is compromised, the disguised attackers would be able to enter the network, compromising data security devices. In order to prevent this, Aazam and Huh (2014) proposed an API based solution including the following components:

1. IOTCloud Controller
2. Message Broker
3. Sensors
4. Applications

3.3.4 CloudThings: Common Architecture for Integrating the Internet of Things with Cloud Computing

Things or objects in an IoT architecture require high storage facility as the sensors generate a huge amount of data, as well as web-based facilities to connect from anywhere along with real-time processing for heterogeneous devices, interoperability and data security, and use of internet resources. To cater to these demands, authors of the paper proposed the use of Cloud-based IoT platform model. The suggested model is based on the idea that CloudThings architecture can be an online platform that allows system integrators and solution providers to leverage a complete Things application infrastructure for developing, deploying, operating, and composing Things applications and services. This is aimed to not only allow the optimum operations of the system but also allows the data to be securely communicated among devices through the use of a secure platform.

4 Proposed Secured Architectural Framework for CoT

The proposed architectural framework is a synthesis of the existing frameworks and theories, and is based on a three-tier model to resolve the security issues involved in data transmission from one device to another more effectively. We propose having the *smart gateways* for the tier one, a *layered based security architecture* in layer two, and *Fog services* at the tier three, all explained in the next section.

The use of smart gateway is aimed to conduct the preliminary examination of the data by identifying the resource usage and consumption patterns of a device, and comparing it with the past behavior, along with checking the priority tags for the data allowing important data to go further. The second tier will encrypt the data using onion routing, DTLS services, and AES facilitated encryptions. The last tier will then store those data with less priority while transferring the high priority data.

For details of the proposed security CoT architecture, we first provide a summary of the related work in the following section. This is followed by detailed description of our proposed framework in Sect. 4.2.

4.1 Related Work

The literature gathered proposes various network architecture suitable for having secure communications among multiple autonomous machines in a cloud of things structure. Each network architecture provides a unique set of characteristics and consists of multiple elements for making the network secure. Many of the existing frameworks suggest the use of smart gateways, Fog Computing, and layered-based encryptions for achieving a secure network architecture.

Smart gateways are considered the backbone of the network as they play the major role in communication process by being an intermediary between the cloud and the sensory networks. Kotis and Katasonov (2012) suggested that since entities in the CoT structure are interoperable and sensor-enabled, the use of a semantic gateway approach remains essential. The main reason is that smart gateway will allow the CoT-enabled devices to be interoperable and interconnected, and also allow 3rd party vendors to develop various software. This in turn will help the machines to be semi-automated requiring limited human intrusion and data to be handled by computer under a centralized configuration.

Gateways are deployed to act as a data exchange hub for both cloud and sensors. On this basis, Rahman et al. (2016) claim that the use of gateway can benefit the CoT-enabled devices in various ways including efficient power consumption, availability of bandwidths, and efficient processing as the gateway although are stationary in nature but yet provide a non-resource constrained empowerment to the users. Based on the above model, any geographically dispersed machine, when tries to communicate to the cloud using sensors, shall face a secure gateway structure that will perform the data preprocessing and filtering function to drop unnecessary data packets, and then send the filtered data to the cloud for storage and analysis.

Wu et al. (2012) suggested that gateways can be used to ensure that the integration of resources is performed with ease, and the usage of services becomes affordable for small enterprises. The model they proposed integrates various components including (i) restful gateway that connects sensors to the cloud and perform lightweight encryptions for data security, (ii) adaptors to meet the technical standard of each machine and communicate in a suitable language, and (iii) middleware for service composition for doing BPM (Business Process Management) and providing web services.

Fog Computing on the other hand is considered essential for smooth CoT operation because it allows the data to be transferred and received at a relatively high speed due to the pre-processing and data storage functions the FC provides to the cloud. Also, it significantly reduces the latency rate and improves CoT performance.

As stated before, one main challenge faced by a CoT-enabled structures is the data storage wastage and network energy wastage that are caused when the unnecessary communications are passed to the main cloud without any priority being assigned. To overcome these issues, Gupta et al. (2017) proposed an *ifogsim* framework that ensures quality of service in a CoT format by performing resource management functions. The *ifogsim* model has four components including (i) monitoring component used for keeping a check on sensors and actuators and ensuring their maximum utilization, (ii) resource component that is the backbone of the architecture used to manage limited resources in the optimal manner, (iii) power component to ensure minimum wastage, and (v) the application module component used to identify data dependencies in a graphical manner.

Today's world demands high mobility and requires the technology to offer consumer centric approach with reduced latency and faster services. To achieve these goals, Datta et al. (2015) suggested the use of Fog Computing to perform initial data analysis for resource intersection management and real time data processing. Their proposed model comprises of three stages where the first stage is application discovery that when a data packet is collected performs a check on application specifications and protocols and forwards the packet to either nearest node for processing; or if no response is gathered then passes it to a node dealing with similar applications. The second stage is to manage the resources by performing lightweight encryptions on outgoing and decoding incoming packets for security check. The final stage is to supply the processed data to the cloud for storage purposes.

4.2   Proposed Framework

Our proposed architectural framework is a synthesized version of some of the existing framework that use Fog Computing, smart gateways, and layered based network security systems. However, the added benefit of our proposed framework arises from the fact that slight improvements are still needed to leverage the benefits proposed by the above concepts. Most of the work contributed by many authors mainly focus on either of the three stated, and ignoring the combined benefits that can be gained through the implementation of all three. Our proposed framework provides a data security architecture for the CoT-enabled systems that maintains reliability, robustness, and security through the use of an intelligent system. It consists of a three-tier approach focusing on the use of smart gateways, fog computing, and layered based encryption systems. This is explained below.

Tier 1 - The smart gateway Tier: This Tier is responsible for establishing a connection between the outside network and the fog. Our proposed SACT model aims to utilize the limited data protection service provided by smart gateways to drop data packet that are from unreliable sources, in order to make the network secure for the users. For this reason, Tier 1 is proposed to allow the identification of priority data based on the information gathered from the applications environment requiring the gateway to perform resource management functions by prioritizing data and then creating a priority que for the data assigned the highest value in terms of priority.

Tier 2 – Layered-based security: This tier ensures that each of the following five steps, corresponding to five layers, involved in forwarding data to the fog and then to the cloud are executed. These five steps are shown in Fig. 1 below.

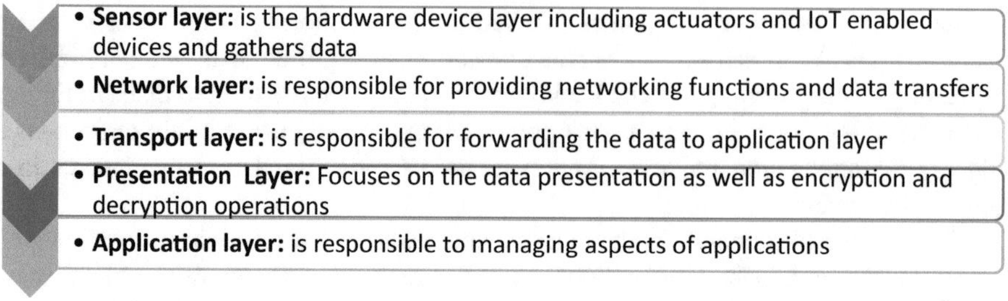

- **Sensor layer:** is the hardware device layer including actuators and IoT enabled devices and gathers data
- **Network layer:** is responsible for providing networking functions and data transfers
- **Transport layer:** is responsible for forwarding the data to application layer
- **Presentation Layer:** Focuses on the data presentation as well as encryption and decryption operations
- **Application layer:** is responsible to managing aspects of applications

Fig. 1. Proposed layered architecture for CoTs

Each layer at Tier 2 performs a significant duty in the data transfer and upload processes to the cloud. To ensure the maximum protection, the gateways will be used for performing a preliminary check on the activities, responses, power usage, and resource usage of the devices connected to the network, and to identify any suspicious behavior. The data that are considered less important will be stored temporarily in the smart gateway while the data considered important will be passed through the layers and the nodes. A set of user-supplied criteria will indicate the priority and importance of the data and this will be discussed later in more details. The above five layers are explained in more details below:

The gateway in Tier 1 passes the data to the second Tier where the data will pass through five layers as highlighted in Fig. 1. The first layer is the *sensor layer* which will capture data to be forwarded to the network layer. The network layer will use an *onion routing program* to ensure that the identity of the data sender and receiver remain uncompromised from attackers. On this basis, the latter either forwards or dispatches the data using a proxy identity. The network layer then implements the *IPsec protocol* for security and forwarded to the next layer.

The data at the transport layer will establish a security link with the application layer to ensure security maximization. At this stage it is possible to utilize the TCP; however, due to the overhead costs involved, the current paper suggests the use of DTLS (Datagram Transport Layer Security). The implementation of this protocol will not only be cheaper but will also be autonomous, requiring no human intervention as the protection link is automatically established. One more reason is that, as the data is passed through so many layers making it secure, the use of VPN or TCP remain mundane. The presentation layer will then employ encryption and decryption functions using the AES methods and the data representation is configured. Lastly, the application layer will check the data for what was sent or being sent to the user and will pass the data to the fog.

Tier 3 - Fog Computing Tier: this layer is responsible for processing the pre-processed significant data, and temporarily store the data that had been assigned lower priority, while allowing the data with high priority to pass through the nodes to the cloud for further analysis or storage. The above proposed three-Tiered CoT architecture is shown in Fig. 2 below:

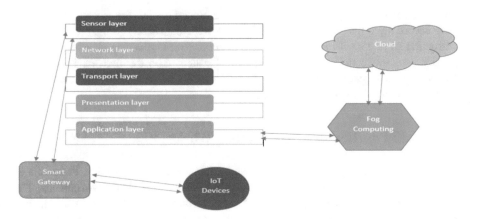

Fig. 2. Architecture of the proposed secure architectural framework for CoT (PSACT)

4.3 Functions of Proposed Framework

The proposed framework provides two main functions for maintaining data security in CoT. These are *data filtration* and *pre-processing functions*. With the help of the smart gateway and the use of fog at the second Tier, the data filtration can easily be performed. The reason for data filtration is to ensure the maximization of the security features offered by the proposed framework. Any packet which seems suspicious or is from an unidentified source will be recognized, as it may lack the priority tag attached to each data transferred from a genuine source and hence, that particular packet will be dropped by the gateway. Also, in the case of high load on gateway, any data deemed less critical will be automatically dropped to ensure that the data with high priority will reach the user on a timely manner.

For data prioritization, tags can be attached to each data packet; the gateway and fog will be able to identify the critical data and will be able to pass the data to the cloud based on the basis of the criticality of the information being carried.

4.4 Limitations of Proposed Framework

The following are the limitations that we have identified for the proposed SACT:

Time consuming: Although, the security system is made to ensure that the data remains unchanged and protected from any attacks but yet due to the multiple number of protocols implemented during the data being transferred from one place to another may result in high time being consumed for processing the data bearing lesser priority value which may not be the case in the time slot based system.

Use of DTLS may impact the security: The use of Datagram Transport Layer Security protocol is although cost and time efficient as it requires no human intervention and uses the IPV4 protocols along with UDP (User Datagram Protocol) but yet it remains prone to attacks as it is lesser secure than the TCP protocols which uses the TCP stack to divide the data into various packets and sends them to TCP program layer which reassembles the data and the sends it to the application layer.

Loss of data without priority tagging: As mentioned earlier, the smart gateway will work according to the priority tagging assigned to the data based on the pre-determined instruction so in case if the data although, being critical but having no match with the priority data instructions is transferred to the system then the data packets will be dropped by smart gateway due to lack of understanding as the gateway acts on the basis of the instructions stored in the system.

5   Concluding Remarks and Future Work

This study proposed a synthesized architectural framework for the CoT with the aim of addressing some of the data security challenges in the current CoT environments. The proposed framework offers a number of benefits including data prioritization, data delivery timeliness, resource usage efficiency, and most importantly, provides a high level of data security through various techniques suggested to be applied. The suggested framework is expected to ensure the maximum protection of the CoTs architecture to ensure the safety of communications taking place among various devices and with this thesis it expected that the security issues identified will be resolved and also future researchers can also have a pathway for further analysis.

Although, the proposed framework aims to encounter multiple issues highlighted in the research and provides a suitable solution for many of the problems that may arise due to architectural insecurity but yet, it remains prone to certain drawbacks. The two main limitations of the framework are identified in this study are tagging issues and time-consuming nature of the framework. One area of our future work will be focusing on removing the above identified limitations. The other area of our future work (in progress) is to adopt design science research methodology, and implement a knowledge artefact of the proposed SACT model to be tested and commented by experts in the field, as a proof of concept.

References

Aazam, M., Huh, E.-N.: Fog computing and smart gateway based communication for cloud of things. Int. Conf. Future Internet of Things and Cloud. **7**, 464–470. Research Gate (2014)

Alessio, B., Donato, W., Persico, V., Pescapé, A.: Integration of cloud computing and Internet of Things: a survey. Future Gen. Comp. Syst. **56**(4), 684–700 (2016)

Baqer, M., Kalam, M.A.: Secure data sharing and searching at the edge of cloud-assisted Internet of Things. IEEE Cloud Comp. **6**, 25–95 (2017)

Bhattasali, T., Chaki, N., Chaki, R.: Secure and trusted cloud of things. J. Netw. (2014)

Botta, A., Donato, W.D., Persico, V., Pescapé, A.: Integration of cloud computing and Internet of Things: a survey. Future Gen. Comp. Sys. **56**, 684–700 (2016)

Cheng, Y., Tao, F., Zhang, L., Lu, B.: CCIoT-CMfg: cloud computing and Internet of Things-based cloud manufacturing service system. IEEE Trans. Ind. Inform. **10**(2) (2014)

Datta, S.K., Bonnet, C., Harri, J.: Fog computing architecture to enable consumer centric Internet of Things services. Int. Sym. Cons. Elec. (ISCE) **10**, 1–7 (2015)

Diaz, M., Rubio, B., Martin, C.: State-of-the-art, challenges, and open issues in the integration of cloud and IoT. J. Netw. Comp. Appl. **67**(2), 99–117 (2016)

Donato, W., Alessio, B., Persico, V., Pescapé, A.: Integration of cloud computing and Internet of Things: a survey. Future Gen. Comp. Syst. **56**(3), 684–700 (2016)

Fox, G., Kamburugamuve, S.: Architecture and measured characteristics of a cloud based Internet of Things API. J. Comp. Netw., pp. 18–29 (2016)

Gupta, H., Dastjerdi, A., Ghoush, S., Buyya, R.: iFogSim: a toolkit for modeling and simulation of resource management techniques in the Internet of Things, edge and fog computing environments. Cloud Comput. Distrib. Syst. **10** (2017)

Henze, M., Hermerschmidt, L., H¨außling, R., Kerpen, D., Rumpe, B.: A comprehensive approach to privacy in the cloud-based Internet of Things. Future Gen. Comp. Syst. **8**(4), 701–718 (2015)

Henze, M., Hermerschmidt, L., Karpen, D., Häubling, R., Rampe, B., Wahlre, K.: A comprehensive approach to privacy in the cloud-based Internet of Things. Future Gen. Comp. Syst. **56**, 701–718 (2016)

Kotis, K., Katasonov, A.: Semantic interoperability on the web of things: the semantic smart gateway framework. In: Sixth International Conference on Complex, Intelligent, and Software Intensive Systems, Itlay: IEEE (2012)

Mell, P., Grance, T.: The NIST definition of cloud computing. Recom. Nat. Inst. Stand. Tech., pp. 1–7 (2011, September)

Petrolo, R., Morabito, R., Mitton, N.: The design of the gateway for the Cloud of Things. Ann. Telecommun. **3**, 1–19 (2016)

Puthal, D., Nepal, S., Rajiv, R., Chen, J.: Threats to networking cloud and edge datacenters in the Internet of Things. IEEE Cloud Comput. **3**(3), 64–71 (2016)

Rahman, A., Nguyen, T., Negash, B., Anzanpour, A., Azimi, I., Jiang, M., Liljeberg, P.: Exploiting smart e-Health gateways at the edge of healthcare Internet-of-Things: a fog computing approach. Elsevier Future Gen. Comp. Syst. J., pp. 2–46 (2016, November)

Singh, J., Pasquier, T., Ko, H., Bacon, J., Eyers, D.: Twenty security considerations for cloud-supported Internet of Things. Internet of Things J. **3**, 1–16 (2015)

Soldatos, J., Kefalakis, N., Hauswirth, Calbimonte, Aberer, M., Zaslavsky, Skorin-Kapov, L.: OpenIoT: open source internet-of-things in the cloud. In: Proceedings of the International Workshop, FP7 OpenIoT Project (pp. 13–25). Croatia, Springer (2014)

Stergiou, C., Psannis, K., Kim, B.-G.: Secure integration of IoT and cloud computing. Future Gen. Comp. Syst. **78**, 964–975 (2018)

Stojmenovic, I., Wen, S.: The fog computing paradigm: scenarios and security issues. Proc. 2014 Federated Conf. Comp. Sci. Inform. Syst. **2**, 1–8. IEEE (2014)

Uckelmann, D., Harrison, M., Michahelles, F. An architectural approach towards the future internet of things. Archit. Internet of Things, **2**(1), 156–185 (2011).

Wu, Z., Itälä, T., Tang, T., Zhang, C., Hämäläinen, M., Liu, Y. Gateway as a service: a cloud computing framework for web of things. 9th Int. Conf. Telecomm. (ICT) **3**, 1–6. Research Gate (2012)

Construction and Parallel Implementation of Homomorphic Arithmetic Unit Based on NuFHE

Liu Wenchao, Pan Feng$^{(\boxtimes)}$, Wang Xu'an, Tu Guangsheng,
and Zhong Weidong

Engineering University of People's Armed Police, Xi'an, Shaanxi 710086, China
646589015@qq.com, wangxazjd@163.com, wiqj027@126.com

Abstract. The existing homomorphic operator is not efficient and flexible, two parallel multi-bit homomorphic operators are designed for this problem in this paper. Firstly, the homomorphic 1-bit adders and 1-bit comparators are implemented based on CGGI scheme. Secondly, using the parallel technology, based on the GPU hardware, a homomorphic adder and comparator that can calculate an arbitrary bit length are implemented. Finally, the nuFHE homomorphic algorithm library is called to program the homomorphic adder and comparator. The results of experiment show that the parallel homomorphic adder and comparator improve the efficiency and flexibility of homomorphic operations and can meet the needs of some scenarios.

1 Introduction

Fully homomorphic encryption (FHE) is a kind of encryption scheme that performs the same operation on the ciphertext and is equivalent to the encryption scheme that performs the same operation on the plaintext. It satisfies both confidentiality and Can perform operations on ciphertext. It plays an important role in privacy protection and data security in the cloud computing environment.

1.1 The Development of Fully Homomorphic Encryption

In 2009, Gentry constructed the first CPA-security FHE scheme [1], marking the emergence of the first generation of FHE algorithms, which attracted the attention of the cryptologist. In 2011, Brakerski and Vaikuntanathan constructed a homomorphic encryption scheme BV11a [2] based on Ring learning with errors (RLWE), which gave birth to the second-generation FHE scheme. In 2013, Gentry, Sahai and Waters proposed an approximate eigenvector method and constructed an FHE scheme GSW [3] based on the LWE problem, marking the beginning of the third-generation FHE scheme. In 2016, Chillotti et al. based on the GSW variant on the ring, TGSW, constructed the efficient double-layer FHE scheme CGGI [4] with its bootstrapping process in less than 0.1 s, and proposed its optimized version in 2017 [5]. It is a kind of more efficient FHE schemes.

© Springer Nature Switzerland AG 2020
L. Barolli et al. (Eds.): CISIS 2019, AISC 993, pp. 660–668, 2020.
https://doi.org/10.1007/978-3-030-22354-0_59

At present, the efficiency of the homomorphic encryption scheme is constantly improving, and it is in the stage of moving from theory to application, but there is still a certain distance from practical use. Therefore, designing an efficient homomorphic encryption algorithm and optimization on its programming libraries is the key to promoting the practicality of the homomorphic encryption scheme.

1.2 Using Hardware to Accelerate Full Homomorphic Passwords

GPU (Graphics Processing Unit), also known as graphics card, is a processor originally used for image computing work, and is currently used in parallel computing [6]. Compared with the CPU, the GPU has powerful floating-point calculation and parallel computing capabilities, and is more suitable for computation-intensive and parallel-oriented programs, which can provide good computational support for the fast implementation of the algorithm. Considering that the bootstrapping of the FHE algorithm involves a large number of matrix multiplications and NTT operations, these computationally intensive operations can be accelerated by the GPU.

NuFHE [9] implements the FHE algorithm from TFHE [10] using CUDA and OpenCL and it refers to paper [4–8]. However, nuFHE only provides basic gates interface and does not provide such as adders, comparators and multipliers. Therefore, the development of a practical homomorphic circuits using the interface provided by nuFHE is one of the basic works of practical application of the CGGI scheme.

1.3 Our Work

Based on the nuFHE fully homomorphic encryption library, this paper implemented an efficient multi-bit homomorphic adder and comparator that can calculate plaintext in arbitrary lengths by CPU multi-threading technology and GPU hardware. The experimental results show that the parallel multi-bit plaintext homomorphic adder and comparator designed in this paper can calculate the plaintext of arbitrary bit length, and the operation speed is faster. Compared with the existing homomorphic adder and comparator, the efficiency and flexibility are improved. It can meet practical requirements in some scenarios.

2 Preliminary

This paper is based on the nuFHE software library. The underlying solution of nuFHE library is the CGGI scheme. This section mainly introduces the CGGI scheme.

Search problem: given access to polynomially many fresh random homogeneous TLWE samples, find their key $s \in \mathbb{B}_N[X]^k$.

Decision problem: distinguish between fresh random homogeneous TLWE samples from uniformly random samples from $\mathbb{T}_N[X]^{k+1}$

Definition 2-1 (TLWE) [4]. Let $n \geq 1$ be an integer, N is the power of 2, noise parameter $\alpha \geqslant 0$, randomly select the private key $s \in \mathbb{B}_N[X]^k$. Massage $\mu \in \mathbb{T}_N[X]$ is encrypted as $c = (a, b) \in \mathbb{T}_N[X]^k \times \mathbb{T}_N[X], b \in \mathbb{T}_N[X]$ has Gaussian distribution $D_{\mathbb{T}_N[X], \alpha, s \cdot a + \mu}$ around $\mu + s \cdot a$. The sample $c = (a, b)$ is random *iff* its left member a (also called mask) is uniformly random When the vector a is taken from a uniform distribution.

Definition 2-2 (Phase) [4]. Let $c = (a, b) \in \mathbb{T}_N[X]^k \times \mathbb{T}_N[X]$ and $s \in \mathbb{B}_N[X]^k$, define phase $\varphi_s(c) \triangleq b - s \cdot a$ of TLWE sample. Phase $\varphi_s(c)$ is linear over $\mathbb{T}_N[X]^{k+1}$ and is $(kN + 1) - lipschitzian$ for the l_∞ distance: $\forall x, y \in \mathbb{T}_N[X]^{k+1}$, $\|\varphi_s(x) - \varphi_s(y)\|_\infty \leq (kN + 1)\|x - y\|_\infty$.

Definition 2-3 [4]. Let c be a random variable $\in \mathbb{T}_N[X]^{k+1}$, which we'll interpret as a TLWE sample. All probabilities are on the Ω-space. We say that c is a valid TLWE sample *iff* there exists a key $s \in \mathbb{B}_N[X]^k$ such that the distribution of the phase $\varphi_s(c)$ is concentrated. If c is trivial, all keys s are equivalent, else the mask of c is uniformly random, so s is unique.

The message of c denoted as $b \in \mathbb{T}_N[X]$ is the expectation of $\varphi_s(c)$.

The bootstrapping process in the CGGI scheme is described in Algorithm 1.

The homomorphic operation provided in the CGGI scheme: input TLWE ciphertext c_1, c_2 and call the bootstrapping algorithm to implement the homomorphic gate circuit.

$$\text{HomNAND}(c_1, c_2) = \text{Bootstrapping}((\mathbf{0}, 5/8) - c_1 - c_2)$$
$$\text{HomXOR}(c_1, c_2) = \text{Bootstrapping}(2(c_1 - c_2))$$
$$\text{HomAND}(c_1, c_2) = \text{Bootstrapping}((\mathbf{0}, -1/8) + c_1 + c_2)$$
$$\text{HomOR}(c_1, c_2) = \text{Bootstrapping}((\mathbf{0}, 1/8) + c_1 + c_2)$$
$$\text{HomNOT}(c) = (\mathbf{0}, 1/4) - c$$

where the "0" in bold indicates the 0 vector of the n dimension and the HomNAND, HomXOR, HomAND, HomOR, and HomNOT refer to the homomorphic NAND gate, the homomorphic XOR gate, the homomorphic AND gate, Homomorphic OR gate, homomorphic NOT gate.

Algorithm 1[4] Bootstrapping TLWE- to - TLWE

$Input : (\boldsymbol{a}, b) \in LWE_{s,\eta}(\mu), BK_{s \to s'',\alpha}, KS_{s' \to s,\gamma}, s' = KeyExtract(s'') \in \mathbb{Z}^{kN},$

$msg = \mu_0, \mu_1 \in \mathbb{T}.$

$Output : LWE_{s,\nu} (\{ \begin{array}{ll} \mu_0 & \varphi_s(\boldsymbol{a}, b) \in (-\frac{1}{4}, \frac{1}{4}] \\ \mu_1 & else \end{array})$

1 : $\overline{\mu} \triangleq \frac{\mu_0 + \mu_1}{2}, \overline{\mu}' \triangleq \frac{\mu_0 - \mu_1}{2}.$

2 : $\overline{b} \triangleq \lfloor 2Nb \rceil, \overline{a}_i \triangleq \lfloor 2Na_i \rceil, i \in 1, n$.

3 : $testv = (1 + X + ... + X^{N-1}) \times X^{-\frac{2N}{4}} \cdot \overline{\mu}' \in \mathbb{T}_N[X].$

4 : $ACC \leftarrow (X^{\overline{b}} \cdot (\boldsymbol{0}, testv)) \in trivialTLWE_{a=0}(\pm \overline{\mu}' + \overline{\mu}' X^1 ... - \overline{\mu}' X^{N-1}).$

5 : $for\ i = 1\ to\ \frac{n}{2}$

6 : $Keybundle_i = X^{-\overline{a}_{2i-1} - \overline{a}_{2i}} BK_{i,1} - X^{-\overline{a}_{2i-1}} \cdot BK_{i,2}$

$$- X^{-\overline{a}_{2i}} \cdot BK_{i,3} + BK_{i,4}.$$

7 : $ACC \leftarrow Keybundle_i \boxdot ACC.$

8 : $\boldsymbol{u} = (\boldsymbol{0}, \overline{\mu}) + SampleExtract(ACC).$

9 : $return\ KeySwitch_{KS_{s' \to s,\gamma}} (\boldsymbol{u}).$

---

3   Homomorphic Circuits

This chapter mainly introduces the design of the homomorphic adder and comparator. It is mainly divided into two parts: First, the CGGI scheme on GPU is used to construct an efficient 1-bit homomorphic adder and comparator. Secondly, the parallel multi-bit homomorphic adder and comparator are designed and implemented by using CPU multi-threading technology and GPU hardware.

For the 1-bit adder and comparator in the ciphertext state, the operation can be realized by the homomorphic operation described below: plaintext $m_0, m_1 \in Z_2$, defining the operations **AND, XNOR, XOR, OR** on the ciphertext, and this operations satisfying the following homomorphic properties:

$$\textbf{dec}(\textbf{AND}(\textbf{enc}(m_0), \textbf{enc}(m_1))) = \textbf{AND}(m_0, m_1)$$
$$\textbf{dec}(\textbf{XNOR}(\textbf{enc}(m_0), \textbf{enc}(m_1))) = \textbf{XNOR}(m_0, m_1)$$
$$\textbf{dec}(\textbf{XOR}(\textbf{enc}(m_0), \textbf{enc}(m_1))) = \textbf{XOR}(m_0, m_1)$$
$$\textbf{dec}(\textbf{NOT}(\textbf{enc}(m))) = \textbf{NOT}(m)$$

Combined with the above method, a 1-bit homomorphic numerical circuits can be designed with reference to the implemented plaintext 1-bit numerical circuits.

3.1   1 Bit Homomorphic Adder

This subsection mainly introduces the design process of 1-bit adder in plaintext as shown in Fig. 1. The designs of 1-bit homomorphic adder can be constructed accordingly as programmed in Algorithm 2.

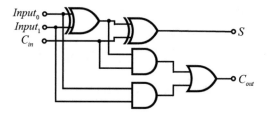

Fig. 1.   1-bit adder using ALU

Algorithm 2 1-bit homomorphic adder

Input : $Input_0$, $Input_1$, C_{in}, pk , sk

Output : S, C_{out}

1: $C_0 = \textbf{enc}(Input_0, pk)$, $C_1 = \textbf{enc}(Input_1, pk)$
    $C_2 = \textbf{enc}(C_{in}, pk)$

2: $C_S = \textbf{XOR}(\textbf{XOR}(C_0, C_1), C_2)$

3: $C_{c_{out}} = \textbf{OR}(\textbf{AND}(\textbf{XOR}(C_0, C_1), C_2), \textbf{AND}(C_0, C_1))$

4: $S = \textbf{dec}(C_s, sk)$, $C_{out} = \textbf{dec}(C_{c_{out}}, sk)$

5: **return :** S , C_{out}

3.2   1 Bit Homomorphic Comparator

This subsection mainly introduces the design process of 1-bit comparator in plaintext as shown in Fig. 2. The designs of 1-bit homomorphic comparator can be constructed accordingly as programmed in Algorithm 3.

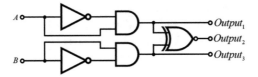

Fig. 2. 1-bit comparator using ALU

3.3   Parallel Multi-bit Homomorphic Circuits

Traditional multi-bit circuits are implemented by calling 1-bit circuits bit by bit and sending the results back to the client one by one, involving excessive decryption operations and communication, and without taking advantage of hardware parallelism. This method has high computational complexity and low efficiency.

Algorithm 3 1-bit homomorphic comparator

Input : A , B , pk , sk

Output : $Output_0, Output_1, Output_2$

1:    $C_0 = \mathbf{enc}(A, pk)$, $C_1 = \mathbf{enc}(B, pk)$

2:    $OUTPUT_1 = \mathbf{AND}(\mathbf{NOT}(C_0), C_1)$

3:    $OUTPUT_2 = \mathbf{XOR}(\mathbf{AND}(\mathbf{NOT}(C_0), C_1), \mathbf{AND}(C_0, \mathbf{NOT}(C_1)))$

4:    $OUTPUT_3 = \mathbf{AND}(C_0, \mathbf{NOT}(C_1))$

5:    $Output_0 = \mathbf{dec}(OUTPUT_0), Output_1 = \mathbf{dec}(OUTPUT_1),$
$Output_2 = \mathbf{dec}(OUTPUT_2)$

6:    **return :** $Output_0, Output_1, Output_2$

Based on the 1-bit homomorphic comparator designed in Subsect. 3.1 and 3.2, this subsection uses the multi-threading technology of the CPU and the GPU-accelerated 1-bit homomorphic circuits to construct a multi-bit numerical circuits that can add and compare arbitrary bit lengths. The implementation process is:

First select the appropriate plaintext space, because only the 1-bit plaintext is encrypted here, so the plaintext space is selected as Z_2.

The method of design parallel multi-bit circuits is as follows:

(1)   Select two binary plaintexts $m = (x_0, \cdots, x_n)$ and $n = (y_0, \cdots, y_n)$. The two plaintexts are encrypted respectively by using the FHE scheme CGGI, to obtain the corresponding ciphertext $M = enc(m, pk)$, $N = enc(n, pk)$.

(2)   Using the multi-threading technique for the two ciphertexts, and in parallel calling the 1-bit homomorphic adder and comparator described above to perform the homomorphic addition and comparison operation and temporarily storing the result.

(3)   Perform the corresponding homomorphic gate operation on the result of the temporary storaged to obtain the final result.

For two binary strings whose input length is n bit, the 1-bit homomorphic circuits are sequentially obtained from the high-order to the low-order, and after decryption is completed, the program ends and outputs the result, as shown in Algorithm 4 and 5.

4   Experiment and Analysis

Based on the nuFHE full homomorphic encryption library, this paper writes a 1-bit homomorphic adder and comparator, and uses the CPU multi-threading technology and GPU hardware to design and implement a parallel multi-bit homomorphic adder and comparator. During the experiment, the running time of the homomorphic adder and comparator on the GPU (CUDA platform) were measured.

Algorithm 4 multi-bit homomorphic adder
Input : $m = (x_0, \cdots, x_n)$, $n = (y_0, \cdots, y_n)$, pk, sk
Output : $Out = (out_0, \cdots, out_n)$
1:   **parallel for** $i = 0$ *to* n
$M_i = \mathbf{enc}(x_i, pk)$, $N_i = \mathbf{enc}(y_i, pk)$
$FLevel_i = \mathbf{AND}\,(M_i, N_i)$
$SLevel_i = \mathbf{OR}(M_i, N_i)$
2:   $OUT_i = \mathbf{OR}(FLevel_i, SLevel_i)$
3:   **for** $i = 1$ *to* n
$OUT_i = \mathbf{OR}(FLevel_i,$
$\mathbf{AND}\,(SLevel_i, OUT_{i-1}))$
4:   **parallel for** $i = 0$ *to* n
$Out_i = \mathbf{dec}(OUT_i, sk)$
5:   **return :** Out

Algorithm 5 multi-bit homomorphic comparator
Input : $m = (x_0, \cdots, x_n)$, $n = (y_0, \cdots, y_n)$, pk, sk
Output : out
1:   **parallel for** $i = 0$ *to* n
2:      $M_i = enc(x_i, pk)$, $N_i = enc(y_i, pk)$
3:      $FLevel_{i,1} = \mathbf{AND}\,(\mathbf{NOT}(M_i), N_i)$
4:      $FLevel_{i,2} = \mathbf{XNOR}(\mathbf{AND}\,(\mathbf{NOT}(M_i), N_i),$
$\mathbf{AND}\,(M_i, \mathbf{NOT}(N_i)))$
5:   $SLevel_0 = FLevel_{i,2}$, $OUT = FLevel_{i,1}$
6:   **for** $i = 1$ *to* n
7:      $SLevel_i = \mathbf{AND}\,(SLevel_{i-1}, FLevel_{i,1})$
8:      $OUT = \mathbf{XOR}(OUT, SLevel_i)$
9:   $out = \mathbf{dec}(OUT, sk)$
10: **return :** out

Experimental environment: hardware uses Intel(R) Core(TM) i7-8700 3.20 GHz CPU, 8G memory, NVIDIA GTX1080 8G graphics card, software system runs Ubuntu 16.04 LTS 64-bit operating system (Linux kernel 4.13), test program based on CUDA 9.2 platform, compiled with the gcc 7.2 compiler. For the binary plaintext bit strings of different lengths, the time required for the homomorphic adder and comparator are shown in Tables 1 and 2.

Table 1. Multi-bit homomorphic adder run time test

Plaintext length (bit)	10	20	30	40	50
Time (ms)	158.9	217.1	387.2	463.5	505.1
Plaintext length (bit)	60	70	80	90	100
Time (ms)	571.6	698.3	812.0	904.7	1083.4

Table 2. Multi-bit homomorphic comparator run time test

Plaintext length (bit)	10	20	30	40	50
Time (ms)	142.0	185.5	331.4	400.1	513.2
Plaintext length (bit)	60	70	80	90	100
Time (ms)	641.1	678.4	741.7	873.1	984.7

5 Conclusion

Based on the nuFHE fully homomorphic encryption library, this paper establishes an efficient multi-bit homomorphic adder and comparator. Firstly, a 1-bit homomorphic adder and comparator is programmed, and parallel multi-bit plaintext homomorphic adder and comparator is realized by CPU multi-threading technology and GPU hardware. The experimental results show that the parallel multi-bit homomorphic adder and comparator designed in this paper can calculate the plaintext of arbitrary bit length, and the operation speed is faster. Compared with the existing homomorphic comparator, the efficiency and flexibility are improved. Ability to meet practical requirements in some scenarios.

Acknowledgments. This work is supported by National Cryptography Development Fund of China Under Grants No. MMJJ20170112, National Natural Science Foundation of China (Grant Nos. 61772550, U1636114, 61572521), The National Key Research and Development Program of China Under Grants No. 2017YFB0802000, Natural Science Basic Research Plan in Shaanxi Province of china (Grant Nos.2018JM6028). This work is also supported by Innovation Team Research Fund No. KYTD201805 of Engineering University of People's Armed Police.

References

1. Gentry, C.: Fully homomorphic encryption using ideal lattices. In: Proceedings of the 41st Annual ACM Symposium on Symposium on Theory of Computing—STOC 2009, pp. 169–178. ACM (2009)
2. Brakerski, Z., Vaikuntanathan, V.: Fully homomorphic encryption from ring-LWE and security for key dependent messages. In: *Annual Cryptology Conference* (pp. 505–524). Springer, Heidelberg (2011)
3. Gentry, C., Sahai, A., Waters, B.: Homomorphic encryption from learning with errors: conceptually-simpler, asymptotically-faster, attribute-based. In: Advances in Cryptology—CRYPTO 2013, pp. 75–92. Springer, Heidelberg (2013)
4. Chillotti, I., Gama, N., Georgieva, M., et al. Faster fully homomorphic encryption: bootstrapping in less than 0.1 seconds. In: International Conference on the Theory and Application of Cryptology and Information Security—ASIACRYPT 2016, pp. 3–33. Springer, Heidelberg (2016)
5. Chillotti, I., Gama, N., Georgieva, M., et al. Faster packed homomorphic operations and efficient circuit bootstrapping for TFHE. In: International Conference on the Theory and Application of Cryptology and Information Security—ASIACRYPT 2016, pp. 377–408. Springer, Cham (2017)

6. Owens, J.D., Houston, M., Luebke, D., Green, S., Stone, J.E., Phillips, J.C.: GPU Computing. Proc. IEEE **5**(96), 879–899 (2008)
7. Gentry, C., Sahai, A., Waters, B.: Homomorphic encryption from learning with errors: conceptually-simpler, asymptotically-faster, attribute-based. In: Annual Cryptology Conference, pp. 75–92. Springer, Heidelberg (2013)
8. Ducas, L., Micciancio, D.: FHEW: bootstrapping homomorphic encryption in less than a second. In: Annual International Conference on the Theory and Applications of Cryptographic Techniques, pp. 617–640. Springer, Heidelberg (2015)
9. GPU-powered Torus FHE implementation: https://github.com/nucypher/nufhe
10. Fast Fully Homomorphic Encryption Library over the Torus: https://github.com/tfhe/tfhe

The 10th International Workshop on Virtual Environment and Network-Oriented Applications (VENOA-2019)

A Proposal of Code Correction Problem for Java Programming Learning Assistant System

Nobuo Funabiki[1(✉)], Sai He[1], Htoo Htoo Sandi Kyaw[1], and Wen-Chun Kao[2]

[1] Department of Electrical and Communication Engineering,
Okayama University, Okayama, Japan
funabiki@okayama-u.ac.jp
[2] Department of Electrical Engineering,
National Taiwan Normal University, Taipei, Taiwan
jungkao@ntnu.edu.tw

Abstract. To advance Java programming educations, we have developed a Web-based *Java Programming Learning Assistant System (JPLAS)*. JPLAS offers several types of programming exercises to cover various learning stages of Java programming, where any answer can be marked automatically on the server. The *code writing problem* requests a student to write a source code satisfying the specifications described in the *test code*, such as names and data types of classes, methods, and variables, so that the answer code is marked by running the test code on *JUnit*. However, it appears challenging for a novice student to extract the specifications from the test code properly. In this paper, we propose a *code correction problem* as a new type programming exercise for reading and debugging codes, which gives a source code with errors called a *problem code* with the corresponding test code, and requests students to correct the problem code to pass the test code. We also propose the *error generation algorithm* to generate the problem code from a sample source code automatically. We verify the effectiveness of our proposal through applications of seven assignments of the code correction problem to 45 students in the Java programing class in Okayama University.

1 Introduction

Recently, *Java* has been widely used in various practical application systems in societies and industries due to the high reliability, portability, and scalability. For instance, Java was selected as the most popular object-oriented programming language in 2015 [1]. Strong demands have emerged from industries in expanding Java programming educations. Correspondingly, a plenty of universities and professional schools are currently offering Java programming courses to meet this challenge. A typical Java programming course consists of grammar instructions in classes and programming exercises in computer operations.

© Springer Nature Switzerland AG 2020
L. Barolli et al. (Eds.): CISIS 2019, AISC 993, pp. 671–680, 2020.
https://doi.org/10.1007/978-3-030-22354-0_60

To advance Java programming educations, we have developed a Web-based *Java Programming Learning Assistant System (JPLAS)* [2]. As the server platform, *Linux* is adopted for the operating system, *Tomcat* is for the Web application server, and *MySQL* is for the database. The applications in JPLAS are implemented based on the *MVC model*, where *Java* is used for *model (M)*, *HTML/CSS/JavaScript* is for *view (V)*, and *JSP* is for *control (C)* [3]. JPLAS offers several types of programming exercises to cover various learning stages of Java programming, where any answer can be marked automatically on the server. Currently, JPLAS provides the *element fill-in-blank problem* [4], the *code completion problem* [5], the *value trace problem* [6], the *statement element fill-in-blank problem* [7], and the *code writing problem* [8], to support self-studies of Java programming.

Among them, the *code writing problem* has been designed for a student to write a source code from scratch. Any answer code from a student is automatically marked on the server by running the *test code* on *JUnit*, so that this system allows students to repeat the cycle of writing, testing, modifying, and resubmitting a code on their own. Nevertheless, each student needs to write a source code satisfying the specifications that are described in the *test code*, such as names and data types of classes, methods, and variables, which can be difficult for a novice at Java programming. As shown in [2], the performance of students is rapidly dropped when they start solving programming exercises in JPLAS using test codes.

Here is an example test code. In **source code 1** for Math class, plus method returns the summation of two integer arguments. In **test code 1** for MyMath class, testPlus method tests plus method by comparing the result for 1 and 4 and its expected result 5. The test code imports *JUnit* packages containing test methods at lines 1 and 2, and declares MathTest at line 3. @Test at line 4 indicates that the succeeding method represents the test method. Then, it describes the procedure for testing the output of plus method.

source code 1

```
1  public class Math {
2    public int plus(int a, int b) {
3      return( a + b );
4    }
5  }
```

test code 1

```
1   import static org.junit.Assert.*;
2   import org.junit.Test;
3   public class MathTest {
4     @Test
5     public void testPlus() {
6       Math ma = new Math();
7       int result = ma.plus(1, 4);
8       assertThat(5, is(result));
9     }
10  }
```

In this paper, we propose a *code correction problem* as a new type of exercise problem in JPLAS so that a student will study a test code and debug an erroneous source code by referencing the test code. Here, a source code with errors called a *problem code* and the corresponding *test code* are given to a student, where the problem code cannot pass the test code due to the errors. Then, the student needs to correct the problem code to pass it. Besides, we propose the *error generation algorithm* to generate the problem code from a sample source code automatically. For evaluations, we generated seven assignments of the code correction problem, and asked 45 students to solve them. The results and questionnaire verify the effectiveness of our proposal.

In [10], Ahmadzadeh et al. showed that even students with a sufficient understanding of programming are not adept at debugging codes. Then, through solving the code correction problem, it is expected that students will improve debugging skills. In [11], Griffin claimed that if students spend as much time on reading, tracing, and debugging codes as on writing codes, they will learn programming more effectively. To solve the code correction problem, students are required to read the problem code and the test code carefully, and debug the problem code to make it pass the test code.

The rest of this paper is organized as follows: Sect. 2 presents the code correction problem. Sect. 3 shows the evaluation results of the proposal. Finally, Sect. 4 concludes this paper with future works.

2 Proposal of Code Correction Problem

In this section, we present the code correction problem and its generation procedure.

2.1 Definition of Code Correction Problem

In the code correction problem, a pair of a *problem code* and a *test code* are given to a student. In the problem code, the name of a class, a method, and a variable, the data type, and the access modifier can be a candidate for error, because they can be described in the test code. Then, the student needs to fix all the errors of the problem code to produce the *answer code*, which will meet the requirements of the test code on *JUnit*. After the answer code is submitted to the JPLAS server, it will be evaluated by the test code. Through solving the code correction problem, it is expected that students will read the both codes carefully, and debug the problem code while extracting the specifications for the source code from the test code.

2.2 Example of Code Correction Problem

Figure 1 demonstrates an example of the code correction problem. That is, students will receive both the problem code and the test code. Then, students correct the problem code by changing `private` to `public`, `Sampling` to `Sample`, and `short` to `static int`.

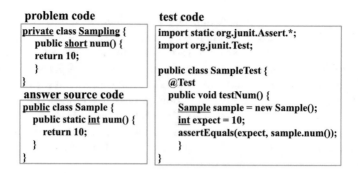

Fig. 1. Example of code correction problem.

2.3   Flow of Code Correction Problem

The following procedure indicates the flow of generating and solving a code correction problem:

1. Problem generation
   A teacher generates a new code correction problem on his/her computer by applying the *error generation algorithm* in the next subsection to a proper source code that has been found from a textbook or a Web site.
2. Assignment registration
   A teacher registers a new assignment that consists of the problem statement, the problem code, and the test code at the server using the Web browser.
3. Assignment solution
   A student selects one assignment, makes the answer code by correcting the errors in the problem code, and submits it to the server using the Web browser.
4. Answer code test
   The server tests the answer code by executing the test code on *JUnit*, and returns the result to the student.
5. Answer code modification and resubmission
   The student modifies the answer code and resubmits it to the server if necessary.

2.4   Error Generation Algorithm

The error generation algorithm generates errors in the source code by changing elements that can be extracted from the test code. To be specific, this algorithm injects the source code with errors by changing the access modifier of a class or a method, the data type of a method or a variable, and the name of a class, a method, or a variable. Besides, it injects errors into the behavior of the source code by changing the operator, the constant number, or the constant variable name in an equation or a conditional expression. The procedure of the algorithm is as below:

1. Access modifier
   An access modifier, `public`, `protected`, and `private`, of a class or a method is randomly changed to another one among them.
2. Data type
   A data type, `int`, `short`, `long`, `float`, and `double`, of a method or a variable is randomly changed to another one among them. Here, `void` is not changed.
3. Name
   A name of a class, a method, or a variable is switched to another name by applying the *error name generation method* in the next subsection.
4. Behavior
   An operator, a constant number, or a constant name in an equation or a conditional expression is changed:
   - an arithmetic operator, such as `+`, `*`, `-`, or `/`, is randomly changed to another one.
   - an conditional operator, such as `>`, `<`, `&&`, or `!=`, is randomly changed to another one.
   - a constant number is randomly changed to the similar number.
   - a constant name is changed to another name by applying the *error name generation method*.

2.5 Error Name Generation Method

The following procedure describes the *error name generation method*.

1. Error name generation using dictionary
   A set of candidates for error names that have similar meanings as the original name, are extracted from the dictionary, *WordNet*, using the similar word estimating function. Then, one candidate is randomly selected from this set for the error name.
2. Error name generation using word list
   If a proper error name is not found by 1, the word whose spelling is most similar to the original name among the word in the *word list* is selected for the error name. The *word list* needs to be prepared by the user of the algorithm.
3. Error name random generation
   If a proper error name is still not found by 2, the error name is generated by randomly adding or removing one character, or changing to another character, in the original name.

It is noted that a combined name using the *Camel method* is first divided into the set of individual names, and then, the above procedure is applied to each individual name. After that, the individual generated error names are combined into one name.

2.6 Verification of Problem Code

After generating the problem code, the validity of each injected error in the code is verified respectively by running the test code with the problem code that exists only one error. The following is the verification process:

1. Save all the errors detected by the error generation algorithm in the problem code in the *error list*.
2. Select one error in the *error list*, and apply this error into the source code.
3. Run the test code with this source code that has only one error.
   - If the test is failed, this error is valid because it can be identified by this test code.
   - Otherwise, this error is invalid because it cannot be found by this test code, and will be removed from the *error list*.
4. Generate the valid problem code by injecting the source code with the specified number of errors by randomly selecting them from the *error list*.

2.7   Problem Difficulty Adjustment

The difficulty of the generated problem code can be adjusted by injecting the source code with the adjusted number of errors. For a given parameter D ($0 < D \leq 1$) from the user, the number of errors N is calculated by $N = D \times L$ if $D \times L < M$, $N = M$ otherwise, where M represents the number of errors in the *error list* and L does the number of statements (lines) of the source code. It is noted that N becomes an integer value by rounded.

3   Evaluation

In this section, we discuss the preliminary evaluation of the code correction problem and the error generation algorithm in a Java programing class in Okayama University.

3.1   Assignments

To evaluate their effectiveness, we generated seven assignments for the code correction problem running the algorithm with $D = 0.5$. The generated assignments cover the following topics of Java programming, where it is expected that the correct solution rate decreases as the assignment ID increases:

1. fundamentals of class and constructor,
2. fundamentals of method,
3. four basic arithmetic operations and loops with `for`,
4. conditional operations with `if`, `switch`, and `operator`,
5. interface,
6. simple data structure with *Queue*, and
7. generics.

   Generics in 7. can give a general data type to a class or a method by using the symbols, `<>`. Then, a specific data type is defined when the class or the method is used. For example, `List<Integer>` defines `List` class of `Integer`. *Generics* is often used in libraries in Java.

At the evaluation, we first showed the overview of the assignments and explained the roles of test codes to the students who are taking the Java programming course. Then, we asked them to solve the assignments within 100 min. Before this evaluation, these students have completed the fundamental Java grammar and programming courses.

3.2   Solution Results

Table 1 reveals the average correct answer rate and the average number of answer submissions among the students for each assignment. In *assignment 1*, the average number of answer submissions appears high, with the high correct rate. Seeing that a great number of students were not familiar with the test code, they made mistakes in correcting the problem code in this easy assignment. Then, the average number of answer submissions decreases as they solve more assignments until *assignment 4*, where the average correct rate is also relatively high. However, the average correct rate becomes low both in *assignment 6* and *assignment 7*, because these assignments are difficult for them.

Table 1. Solution results.

Assignment ID	Ave. correct rate (%)	Ave. # of submissions
1	90.5	14.9
2	85.2	10.9
3	86.8	12.8
4	69.1	7.4
5	80.9	5.4
6	60.0	6.7
7	44.7	7.2

3.3   Questionnaire Results

After solving the seven assignments, we asked the students to answer a questionnaire with three questions, as shown in Table 2. As the first question, about 70% students replied positively, which indicates that they can study how to read and use the test code through solving the code correction problem. Similarly, about 70% students replied positively in the second question, which signifies that they can improve the Java programming ability by solving this problem. However, about 70% students replied negatively in the last question, which means that these assignments for the code correction problem are difficult for them. Therefore, it is a must to adjust the problem difficulty by properly setting D for each assignment, to avoid discouraging the novice in solving the code correction problem in JPLAS.

Table 2. Questionnaire questions and answers.

ID	Question	Yes	May yes	Neutral	May no	No
1	Do you understand how to read test code?	13	19	6	5	2
2	Do you improve Java programming?	14	18	8	1	4
3	Do you think the assignments are difficult?	16	17	8	4	0

3.4   Student Opinions

Several students replied the following negative opinions in the questionnaire:

1. It is hard to understand the test results by the test code in the interface.
2. There are a lot of spelling errors in the problem codes. It is better to contain other errors in them.

The reason for 1. Comes from the implementation of the test result interface to students in JPLAS. The test log message of *JUnit* is displayed directly as shown in Fig. 2. *JUnit* log message is usually difficult for them. The improvement of this interface with simple messages for the novice will be explored in future studies.

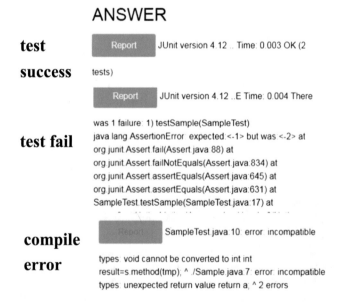

Fig. 2. Test result interface in JPLAS.

The reason for 2. Is that a source code generally has plenty of elements whose names can be changed by the *error generation algorithm*, which become spelling errors. When selecting the specified number of errors from the *error list*, various types of errors should be selected so that their numbers become balanced. In addition, it is necessary to include other types of errors in the algorithm for the diversity of the code correction problem. In [12], by analyzing programming errors from students who took introductory Java courses, they can be classified into three categories: syntax errors, semantic errors, and logic errors. They should be considered in error generations for the code correction problem, which will be examined in future studies.

4   Conclusion

This paper proposed the *code correction problem* in JPLAS for studying to read and debug Java source codes. It gives a source code with errors to students with the corresponding test code, and requests them to correct the source code to pass the test code. The *error generation algorithm* was also presented to generate a problem code from a sample source code automatically. The effectiveness was verified through applications of seven assignments to 45 students in our department. The future works include the improvement of the test result interface in JPLAS, the selection of errors for balancing error types in the problem code, and the consideration of other error types in the algorithm.

References

1. Cass, S.: The 2015 Top Ten Programming Languages. http://spectrum.ieee.org/computing/software/the-2015-top-ten-programming-languages/?utm_so
2. Ao, S., et al. (eds.): IAENG transactions on engineering sciences - special issue for the international association of engineers conferences 2016, vol. II. World Science Publishing, pp. 517–530 (2018)
3. Ishihara, N., Funabiki, N., Kuribayashi, M., Kao, W.-C.: A software architecture for Java programming learning assistant system. Int. J. Comput. Soft. Eng. **2**(1), 116 (2017)
4. Funabiki, N., Zaw, K.K.T., Ishihara, N., Kao, W.-C.: A graph-based blank element selection algorithm for fill-in-blank problems in Java programming learning assistant system. IAENG Int. J. Comput. Sci. **44**(2), 247–260 (2017)
5. Kyaw, H.H.S., Aung, S.T., Thant, H.A., Funabiki, N.: A proposal of code completion problem for Java programming learning assistant system. In: Proceedings of CISIS, pp. 855–864 (2018)
6. Zaw, K.K., Funabiki, N., Kao, W.-C.: A proposal of value trace problem for algorithm code reading in Java programming learning assistant system. Inf. Eng. Express. **1**(3), 9–18 (2015)
7. Ishihara, N., Funabiki, N., Kao, W.-C.: A proposal of statement fill-in-blank problem using program dependence graph in Java programming learning assistant system. Inf. Eng. Express. **1**(3), 19–28 (2015)
8. Funabiki, N., Matsushima, Y., Nakanishi, T., Amano, N.: A Java programming learning assistant system using test-driven development method. Int. J. Comput. Sci. **40**(1), 38–46 (2013)
9. JUnit. http://junit.org/

10. Ahmadzadeh, M., Elliman, D., Higgins, C.: An analysis of patterns of debugging among novice computer science students. In: Proceedings of ITiCSE, pp. 84–88 (2005)
11. Griffin, J.M.: Learning by taking apart: deconstructing code by reading, tracing, and debugging. In: Proceedings of SIGITE, pp. 148–153 (2016)
12. Hristova, M., Misra, A., Rutter, M., Mercuri, R.: Identifying and correcting Java programming errors for introductory computer science students. In: Proceedings of SIGCSE, pp. 153–156 (2003)

A Study for Investigating Driver's Condition by Radar

Toshiyuki Haramaki$^{(\boxtimes)}$ and Hiroaki Nishino

Division of Computer Science and Intelligent Systems,
Faculty of Science and Technology, Oita University, Oita, Japan
{haramaki,hn}@oita-u.ac.jp

Abstract. In this paper, we show how to decrease the occurrence of car accidents and to instantly detect sudden health problems that occur to the driver. In order to carry out this system, we aim to calculate the risk of safe driving in real time by constructing a system that monitors drivers using radar and observes driver behavior and biological information. Conventionally, observation based on optical sensing has been taken as a method of watching the driver's situation. In order to realize observation by that method, a sensing device is mounted in the vehicle toward the driver, and the driver's situation is determined based on the input sensor value. However, the method we propose uses radio waves as sensors. This system determines whether the driver can concentrate on driving by continuing to observe the driver using a radar sensor. By utilizing the Doppler radar, the driver who is driving without contact is always observed. In addition to facial expressions and body movements of the driver, we are intended to acquire heart rate, breathing, sweating on the body surface and so on.

1 Introduction

In recent years, the demand for private cars in the world is increasing year by year. At the same time, the number of traffic accidents is further increased. Reducing the amount of traffic accidents benefits society. One of the means to attain that is to drive a car automatically. So far, many researchers have been actively working on the development and realization of automatic driving technology. However, some issues have yet to provide a fully automated operation. For example, if the auto brake system detects a pedestrian or object in front of a moving vehicle, the vehicle will decelerate or stop automatically. It contributes to reducing rear-end collisions. The adaptive cruise control system reduces the driver's burden on accelerator and brake operations. However, these techniques may reduce the driver's awareness of caring for safe driving. Automotive control functions are under an obligation to ensure driver safety, even if the driver is not perfect. After all, the best way to decrease the risk of causing a traffic accident is to have a human driver drive with responsibility and drive safely after having acquired the skills of safe driving. To date, the ability to release the driver from the physical operation is currently being implemented with some limitations. Advanced Safety Vehicles (ASVs) and Advanced Driving Support Systems (ADASs) are becoming practical applications to support driving safety. However, the driver is still obliged to drive safely even if the car has these features.

© Springer Nature Switzerland AG 2020
L. Barolli et al. (Eds.): CISIS 2019, AISC 993, pp. 681–690, 2020.
https://doi.org/10.1007/978-3-030-22354-0_61

Accumulation and distraction of the driver's fatigue prevent the driver from driving safely. To reduce traffic accidents. It is necessary to detect their condition. Until now, various research and development have been carried out to detect the driver's condition. Some are in everyday use. However, I think that a system that can really be trusted has not appeared yet. Therefore, there is necessary to have a way to support the driver's driving behavior in some way so that the driver can continue driving safely. There are for two main ways to support safe driving. One of them is directly driving support. Precise driving support means adding another driving operation to the driver's driving to prevent an abnormal situation such as an accident. A specific example is ABS (Antilock Brake System). Another is joint driving support. With joint driving support, all driving operations are left to the driver, and by providing appropriate information and advice to the driver while driving. The driver is supported to drive safely.

Numerous research projects are actively being carried out to implement automated driving technology. The ability to release the driver from physical operations is currently implemented under some limitations. Examples of its implementation include ASV and ADAS. These systems are becoming practical applications to support driving safety. However, unless the car is a fully automated car of Level 5 defined by the Society of Automotive Engineers (SAE) International, the driver is required to drive safely even if these driver assistance functions are installed. There are various types of research activities to advise the driver to drive safely for supporting it. Kang shows that there is several ways to monitor the condition of drivers and vehicles [1]. It's imperative to use hybrid means based on multiple sensor information to detect drowsiness of the driver and to correctly recognize the driving situation for supporting safe driving. And It's necessary to realize safe driving assistance by combining and using the appropriate parts of these functions.

In this paper, we propose a new approach to realize an safe driving support system. In our previous studies, we always monitor the driver's situation directly related to the occurrence of traffic accidents with multiple sensors, and at the same time monitor the behavior of the vehicle with multiple sensors. By analyzing the sensor data of the driver's situation and the vehicle behavior made, as a result, the relationship between the driver's state and the vehicle is derived, and to keep the vehicle safer, how is the driver? It provides the driver with advice on how to lead. An approach that is newly introduced this time is to use an FMCW-UWB (Frequency Modulated Continuous Wave Ultra-Wide Band) radar as a sensor for observing the driver's situation. The realization method is reproduced below. First, the radar system is included with the car. The driver's body is irradiated with radar while driving. By observing the reflected wave, it becomes possible to grasp the driver's physical condition. Previous studies have used optical cameras, infrared sensors, heart rate sensors, GSR sensors, etc. to observe the driver. By using a modern radar sensor in addition to these sensors, it is expected that the driver's condition can be observed with higher accuracy. This time, the performance observation was conducted utilizing the radar sensor for evaluation, and the actual observation data of the sensor was confirmed. Based on the results of this experiment, we will discuss the points of functional enhancement when a radar sensor is added to the efforts we have made to date.

2 Related Work

Many research projects are actively being carried out to implement automated driving technology. The ability to release the driver from physical operation is currently being implemented under some limitations. ASV and ADAS, which support driving safety, are being put to practical use. However, the driver is still obliged to drive safely even if the car has these features. There are various types of research activities to advise the driver to drive the vehicle safely.

Leem et al. show a method for monitoring a robust vital sign of a driver while driving by using an IR-UWB (Impulse Radio – Ultra Wide Band) radar [2]. As a feature of the method, they propose a new algorithm that can estimate the vital signs of the driver even if the driver is moving by driving activity. It is proposed to use the algorithm to grasp the driver's condition in real time and to help prevent traffic accidents. Akhlaq et al. show a mechanism that uses a combination of a CMOS image sensor and a sensor that acquires the behavior of a car to realize a low-cost and effective ADAS (Advanced Driver Assistance System) [3]. Kim et al. analyze the detection probability of the FMCW-UWB radar for oil reservoir water level gauge and show that the detection performance can be sufficiently exhibited by appropriately setting the margin of the ADC bit and the margin of the frequency [4]. Vesselenyi et al. show an algorithm for estimating the driver's level of consciousness and drowsiness by measuring and analyzing the driver's EEG (Electroencephalogram) and EMG (Electromyography) while driving to prevent traffic accidents [5]. Diraco et al. show show how to acquire vital data contactlessly by sensing using radar [6].

Previous systems provide advice based solely on quantitative assessment of observed sensor data. As an extension, the driver's behavior and important information are further adopted to find out the appropriate advice for the driver. By capturing these data, the system can provide more effective warning information to the driver. Various presentation methods have been studied to communicate dangerous situations by observing human conditions with multiple sensors [7, 8].

These enhancements use the knowledge gained from previous studies to investigate. Figure 1 shows an experiment where a device equipped with many sensors is installed in a car, and research is continued to observe the driver's condition and the behavior of the car, when a prototype is created and installed in the car and experimented. At this time, the driver's condition was continuously observed using a heart rate sensor, a GSR sensor, an ultrasonic sensor, and a temperature sensor. Data on vehicle operation was acquired using an IMU sensor. Those data were processed by a controller in the car via wireless communication, and the obtained results were used to give advice to the driver for safe driving.

We are developing a safe driving training system that dynamically improves the algorithm of advice by further developing these approaches and learning the relationship between the advice to the driver and the resulting behavior [9]. Figure 2 shows a new safe driving training system to a driver. The system aims to provide appropriate advice to a driver by measuring the driving behaviors while continuing to make advice to the driver. It detects the distance between the front and the own vehicles, and the driver's vital sign. Next, the system analyzes the data and estimates the risk of causing

Fig. 1. The sensors and controller mounted on the car

a traffic accident in real time. Then, the system advises the driver to avoid possible accident by making longer inter-vehicle distance based on the estimation. Finally, the system measures whether the driver changes his/her driving behaviors after the advice. The system gives better advice by learning outcomes for evaluation.

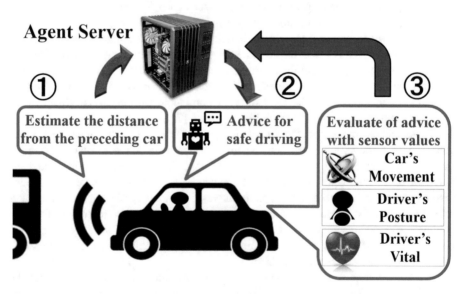

Fig. 2. The sensors and controller mounted on the car

3   System Design

In this chapter, we introduce the design for observing the driver with radar and the design for safe driving of the driver using the observation results. As an application example of a radar sensor, a method of emitting a radar to the outside of a vehicle to grasp the presence of people and objects around has already been put to practical use. In our research, we use radar sensors to keep track of the drivers in the car. In order to accurately grasp the situation of F driver, we use the FMCW broadband radar. By using it, the distance between the sensor and the observation target, the positional relationship, and the relative velocity can be acquired [10]. Furthermore, in recent years, research has been advanced on vital sensing that measures heart rate and respiration of a person using a radar sensor [11].

3.1   Driver Monitoring Method by Radar

Figure 3 shows a diagram showing components for estimating a driver's condition using a radar sensor and a processing flow for realizing it. The components that make up this system are a radar sensor installed in the car, a Processor unit that extracts the driver's observation results from the radar sensor, and an Estimation unit that determines the state from the driver's observation results. When using this system, a radar sensor, which is a transmitter and receiver of radio waves, is installed in the car beforehand. The radar sensor emits radio waves to the driver while driving. And a sensor receives a radio wave reflected by a driver and its circumference. Received data will continue to be sent to the processor unit at any time. The processor unit calculates the distance, direction and movement to the driver by analyzing the data received by the sensor. The estimation result is used to estimate the driver's condition based on the calculation result and comparison with past learning results. In order to estimate the state from the driver's observation result, it is necessary to know in advance what kind of data observation result is.

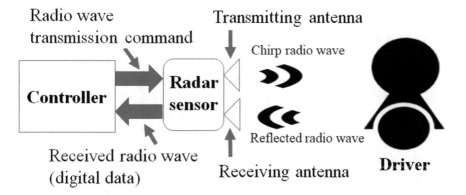

Fig. 3. Monitoring target for current network topology visualizer.

3.2   Safe Driving Support Including Driver Monitoring by Radar

Figure 4 shows the flow of components of a system that provides advice for safe driving by observing the driver with a radar sensor. This system includes a plurality of sensors installed in a car, an agent server for selecting advice information based on observation data of the driver, and an interface device for presenting advice to the driver. By using a radar sensor for driver's observation, we think that it is possible to acquire vital information such as driver's breathing and heartbeat. The sensor module has one edge computing node for observing the distance to the preceding vehicle, and a radar center for observing the driver's condition to evaluate the effects of the vehicle behavior and advice. The agent server provides two processing routines for realizing advice for safe driving. The advice routine allows the driver to maintain the distance between vehicles. The feedback learning routine evaluates the behavior of the car and the condition of the driver. We improve the effectiveness of the advice through these processes.

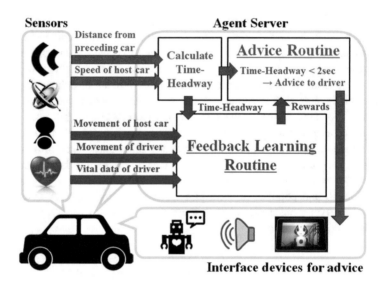

Fig. 4.   Safe driving support system by observing the driver and the car.

The advice routine uses the monitoring data provided by the sensor node to calculate the inter-vehicle time to the preceding vehicle. If the time is less than 2 s, the system advises the driver to increase the distance. The system then uses several means to present this advice to the driver, including voice, video and robot movements. After the agent server advises the driver, the feedback learning routine processes the vehicle behavior and driver state obtained from the edge node. At that time, it is determined whether the driver has followed the instruction. When the driver operates according to the system's advice, the system determines that the advice has been applied to the driver. If the driver does not act based on the advice, it is judged whether the advice is inappropriate for the driver or whether the driver is losing concentration. In the subsequent process, advice methods appropriate to the situation are used. By accumulating this learning activity, the system will gradually provide better advice.

4 Preliminary Experiments

This preliminary experiment was performed in a large radio wave darkroom. The reason is to prevent transmission radio waves from leaking and to conduct experiments in a pure radio environment. The radar sensor used in this experiment is a millimeter wave radar sensor evaluation kit manufactured by TI (Texas Instruments). This evaluation kit is equipped with a single-chip millimeter wave radar sensor and a chip for processing acquired observation data at high speed. Furthermore, the real-time data capture adapter used with this kit has the function of continuing to transmit the millimeter wave reception data acquired by the sensor evaluation kit reception antenna by streaming. In this experiment, this data was continuously sent to a notebook computer, and the experiment was conducted while confirming the observation results in real time by visualizing the data.

As software for visualizing the observation results, we used mmWave SDK and mmWave Studio provided by TI. Using this software, it has a function to graphically display the position and direction of the object and the speed of movement based on the reception result of the reflected wave. In addition, it is also possible to design a chirp, which is a transmission waveform of the millimeter wave radar, on the GUI screen, and to evaluate the observation performance by the chirp. In this experiment, as objects to be observed, in addition to human beings, objects of various sizes and various materials were tried. The experiment was continued while confirming the graphical representation of the observation results. Because of the experiment, in the case of observing a human and the case of observing an object, a clear difference was observed in the observation result.

Figure 5 shows a radar sensor evaluation kit that is in the center of the photograph on the left side of the experimental scene in the radio anechoic chamber. By operating this from an external computer, the photograph on the right side where the object was observed shows that the position of a person or a thing is changed, and the radar is emitted and observed. In this preliminary experiment, it was found that the radar observation results are different between things and people, and that the observation results are different when people are moving and when they are stopped. When a stationary object was observed, the observation result did not change with the passage of time, but when a person was observed, the observation result did not change with the passage of time, even if it was stopped or walking. It turned out that it keeps changing finely.

Figure 6 shows an example of observation results of object detection using a radar sensor. This chart is the display screen of mmWave Demo Visualizer, which is one of the functions of the mmWave Studio software. In this table, points are plotted at positions where it is determined that an object exists because of analysis of reflected waves observed by a radar sensor. The horizontal axis represents the distance between the left and right. Some of the points on the upper right place large objects such as 1 m square boxes and small 20 cm square objects. These objects are shown at one point

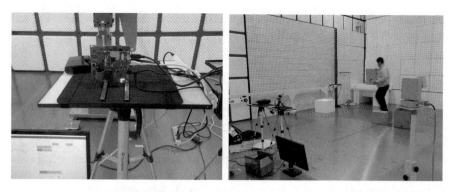

Fig. 5. Server status visualization for IT infrastructure visualizer.

regardless of their size. The point cloud at the bottom left of the table is the result of observing the state of human movement. Thus, the observation results are completely different between the stationary object and the moving person. Furthermore, even when a human being is at rest within the observation range, the human observation results are represented by a plurality of points, and the drawing of the points turned on and off with the passage of time.

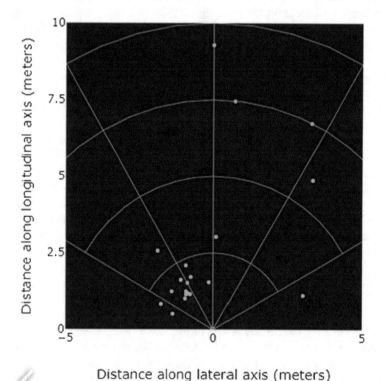

Fig. 6. Plot of observation results on mmWave demo visualizer.

5   Conclusions

In this paper, to reduce traffic accidents. We proposed a safe driving support system by monitoring the driver while driving and providing appropriate support according to the situation. To realize that, in this research, by using the radar as a driver monitoring method, more accurate state identification is possible compared with the conventional monitoring by sensors and cameras, and the detailed statement by using the radar It showed that identification was possible. In addition, as a preliminary experiment to demonstrate the theory, when observing the driver using a radar installed in a car, the observation results on the difference between the observation results of routine radar and abnormal radar were introduced. Based on these findings, it is possible to identify the driver's condition and provide appropriate assistance by continuing to observe the driver using a radar. Furthermore, the ability to use radar to obtain driver's condition means that both results and vehicle driving condition data are correlated, improving the driver's ability to drive safely under all conditions Let Or, conversely, I think that it can be used to analyze whether the possibility of an accident occurring increases.

In the future, we aim to realize the function of proposing a driving plan such as the timing of a break by predicting in advance the driver's drowsiness and concentration decline. In addition, when the system detects that the driver's condition has fallen into a very perilous situation, a mechanism has been devised to provide the driver with a function to safely stop the vehicle in cooperation with the partial automatic driving system. Want to. Furthermore, by linking with the fully automated driving system, the driver and the passenger whose condition has suddenly changed while driving can realize the function of safely transporting it to the hospital and public facilities, etc. and carry out experiments to establish the method.

References

1. Hang-Bong Kang: Various approaches for driver and driving behavior monitoring: a review. Proc. of 2013 IEEE Int. Conf. on Comp. Vis. Workshops, 616–623 (2013)
2. Leem, S.K., Khan, F., Cho, S.H.: Vital sign monitoring and mobile phone usage detection using IR-UWB radar for intended use in car crash prevention. J. Sens. MDPI 17(6), 1240 (2017)
3. Akhlaq, M., Sheltami, T.R., Helgeson, B., Shakshuki, E.M.: Designing an integrated driver assistance system using image sensors. J. Intell. Manuf. 23(6), 2109–2132 (2012)
4. Kim, S.D., Lee J.H.: Performance analysis of FMCW-UWB radar for oil tank level gauge. Proc. Int. Conf. 2011 First ACIS/JNU Comp. Netw. Syst. Ind. Eng. (2011)
5. Vesselenyi, T., Rus, A., Mitran, T., Tataru, B.,Moldovan, O.: Vehicle driver drowsiness monitoring and warning system. Proc. of Int. Conf. 2016 Automot. Trans. Eng. pp. 873–880 (2016)
6. Diraco, G., Leone, A., Siciliano, P.: Radar sensing of vital signs in assisted living applications. Proc. of 2017 Italian Forum of Ambient Assisted Living, pp. 3–22 (2017)
7. Haramaki, T., Yatsuda, A., Nishino, H.: A robot assistant in an edge-computing-based safe driving support system. Proc. 21st Int. Conf. NBiS 2018, pp. 144–155 (2018)
8. Okazaki, S., Haramaki, T., Nishino, H.: A safe driving support method using olfactory stimuli. Proc. 12th Int. Conf. CISIS 2018, pp. 958–967 (2018)

9. Haramaki, T., Nishino, H.: An improved safe driving training system based on learning of driving behaviors. Proc. IEEE Int. Conf. 2019 ICCE-TW (2019)

10. C lovescu, Sandeep Rao: The fundamentals of millimeter wave. Texas Instruments. http://www.ti.com/lit/wp/spyy005/spyy005.pdf Accessed 25 April 2019

11. Driver vital sign detection demonstration using mmWave radar sensors. https://training.ti.com/jp/driver-vital-sign-detection-demonstration-using-mmwave-radar-sensors Accessed 25 April 2019

Evaluation of Searching Method for Various Contents Using Cache Routers with Transmission Records in Unstructured Peer-to-Peer Networks

Daiki Takeda[1] and Shinji Sugawara[2(✉)]

[1] Nagoya Institute of Technology, Nagoya 466-8555, Japan
daiki@sugawara-lab.org
[2] Chiba Institute of Technology, Narashino 275-0016, Japan
shinji.sugawara@it-chiba.ac.jp

Abstract. In the previous research, we proposed a method to guide queries using query transmission records in each relay peer and router under the condition where content cache routers available for efficient content searching in unstructured peer-to-peer networks. However, there are many kinds of contents exchanged over the network, and the actual system needs to search the targets from the mixture of those kinds of contents. The efficiency of the proposed system was not evaluated sufficiently when different content types are mixed in the research. Then in this paper, we evaluate the effectiveness of the method in the environment where some different kinds of contents are mixed and the target contents are searched from that mixture.

1 Introduction

Recently, with the advancement of terminals and broadbandization of communication networks, various contents such as music and moving pictures have been actively exchanged among users on the network. Such contents are increasing in capacity and diversification, and content sharing system using Peer-to-Peer (P2P) network is attracting attention as a method for efficiently searching for and sharing the contents desired by the user on the network [1]. A system on P2P network is a system configuration in which terminals called "peers" communicate with each other on an equal basis and provide services, and has advantages on load distribution, scalability, and fault tolerance, over client-server systems.

Content sharing systems using P2P networks can be classified into hybrid type and pure type systems according to the presence or absence of servers that manage the meta data of entire system. In hybrid P2P system, the location information of each peer and the list of the contents held by that peer are collectively managed by the index server. When a peer requests a content, the content can be easily searched by making an inquiry to the index server, however as in the client-server systems, the problems arise on the index server, concerning scalability due to the enlargement of the network scale, and the fault tolerance. On

L. Barolli et al. (Eds.): CISIS 2019, AISC 993, pp. 691–701, 2020.
https://doi.org/10.1007/978-3-030-22354-0_62

the other hand, in pure P2P network systems, peers share equally the retention of index information and various processing, the problems seen with hybrid P2P are unlikely to occur. However, when a peer is searching for its target content, since there is no index server, it is necessary for the peer to forward a query to the peers connected with itself (its adjacent peers) on the overlay network. The peer that has received the query can further search the peer holding the target content by repeating transmission of the query to its adjacent peers, but redundant forwarding of the query increases the amount of traffic in the network.

Therefore, the solution is examined for both structured and unstructured pure P2P systems. Structured P2P is a system that allow a predetermined peer to hold content index information, and Chord [2] and CAN [3] use DHT (Distributed Hash Table) and so on. In DHT, both content and peer identifiers are converted into hash values and then linked to each other, and each content index information is stored in its corresponding peer. This enables the user to search his/her target content at high speed and with high accuracy from the hash value of the identifier of the content. However, in the networks where the topology changes dynamically due to joining and leaving peers, problems such as increase in operation cost and unsuitability for partial match search due to the use of hash functions occur. On the other hand, unstructured P2P system like Gnutella [4] and Freenet [5] does not impose structural constraints on peers and contents so it is possible to construct more flexible systems than structured P2P systems. However, peers holding target contents cannot be easily identified, and since flooding is usually used to search for the contents, the increase in the number of queries and the decrease in the search accuracy become problems. From the above, there is a trade-off between system flexibility and ease of content search in P2P system operations.

In the previous research, we proposed a content sharing system on unstructured P2P network which can cope with the rapid increase of the number of users and enable stable and flexible system operation even when peers are relatively frequently joining in and leaving the network [6]. In this system, in order to solve the problem that the amount of traffic is increased by the increase of the number of queries, assuming the use of content cache routers, trajectories of the queries that have reached the peers holding the target contents are stored as Transmission Records (TRs) [7] in each relay peer and router, and when the contents requested in the past is requested again, the queries are directed only to the appropriate peers for the contents' discovery according to the information of TRs. In addition, even in an environment where the network topology changes dynamically due to the peers' joining and leaving the network, TRs are continuously updated to maintain their consistency and this causes high search accuracy is achieved and the amount of the network traffic is suppressed.

However, there are many kinds of contents exchanged over the network, such as texts, music, pictures, videos and so on, and the actual system needs to search the targets from the mixture of those kinds of contents. In the previous research, the efficiency of the proposed system was not evaluated sufficiently when different content types are mixed. Then in this paper, we evaluate the effectiveness of the method in the environment where some different kinds of contents are mixed and the target contents are searched from that mixture.

The rest of this paper is organized as follows. Firstly, Sect. 2 describes the Related Works and Sect. 3 explains the searching method proposed in the previous research. Section 4 discusses the effectiveness of the method in the case of searching for multiple types of contents, after describing its evaluation method and evaluation results. Finally, Sect. 5 concludes this paper.

2   Related Works

Although various approaches for content searching have been researched, here we introduce Breadcrumbs [8] and Adaptive Probabilistic Search (APS) [9]. Both of them have a profound connection with our proposal.

2.1   Breadcrumbs

Breadcrumbs is intended to improve the content delivery efficiency by each node (i.e., router) caching some content items in a client-server content delivery system. The load concentration to a server can be suppressed by setting query guiding information in the routers located on the content transmission path between the server and the requesting user. In this method, five entries are recorded in the query guiding information (Breadcrumbs: BC) as shown in Table 1.

Table 1. Breadcrumbs.

BC entries	Details
Content ID	ID of the requested content
Upstream	ID of the upstream node
Downstream	ID of the downstream node
Download time	Time when the last content delivered
Request time	Time when the last query arrived

To demonstrate Breadcrumbs concretely, let consider the example shown in Fig. 1. On the content delivery path from the server to the user (i.e., requesting node), nodes located on the server side are defined as upstream nodes, and nodes located on the user side are defined as downstream nodes. When an upstream (downstream) node does not exist, it means that the upstream (downstream) node is the server (user).

In the Fig. 1, node A has already acquired the content item from the server, and nodes D, C, and B record the query guiding information as illustrated by the downstream arrows. In this situation, when node F requests the same content item as node A, a query is sent from node F to the server. As the query is forwarded, it encounters the query guiding information at node C and reaches

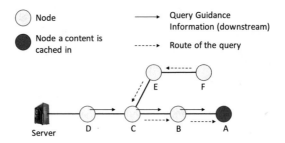

Fig. 1. Example of Breadcrumbs' behavior.

node A with the guidance of the downstream arrows. Finally node F downloads the content item from node A.

If the query reaches node A, but the content item is not cached there, the query is guided to an upstream node by tracing back the route according to BC. Then the query guiding information on nodes B and C is disabled. Accordingly, the query successfully reaches the server and node F downloads the content item.

However, the system has the problems of Single Point Of Failure (SPOF) because the system is based on the client-sever model. Therefore, we remove the SPOF by employing an unstructured P2P network. In addition, our proposed method is able to search the content items effectively by setting the query guidance information in the peers as well as the routers.

2.2   APS

Tsoumakos et al. proposed a content searching method called Adaptive Probabilistic Search (APS). In APS, each peer possesses a table of indices and index values to the adjacent peers for the requested content items. Indices are identifiers of adjacent peers that are the candidates for forwarding a query. Each index value indicates the degree that a peer should be select as a forwarding peer.

When a search is initiated, the peer determines the forwarding peer by k-walker random walk, and the peer with the higher index value is typically chosen as the forwarding peer. As a query is transmitted, the index value of the adjacent peer in the peer is decreased by 10. However, when the query reaches the requested content item, the query retraces its route and increases only the requested content's index value by 20 in each peer along the route.

In the content sharing system employing an unstructured P2P network, the users cannot sometimes get their requesting content items because the content items vanish from the network because of peers' dropping out. In addition, because communication traffic for searching content items increases when the search range is enlarged to improve the content acquisition rate, we use content cache routers which are able to cache some content items in order to find the requesting content items near the requesting peers.

3 Proposed Method

As mentioned above, we proposed a method using the concept of Breadcrumbs and content cache routers in an unstructured P2P network in the previous research [6]. Specifically, TRs, which are trails of queries that reach the peer or the router possessing the requested content item, are kept in each of relay peer and router. When the same content is requested again, the query can be guided appropriately by TRs, allowing for an efficient search and content retrieval. TRs consist of 4 entries as shown in Table 2. Content ID is the ID of the requested content item. Upstream is the ID of the upstream peer or router to which the peer or router sends the content item. Downstream is the ID of the downstream peer or router from which the peer or router receives the content item. Reference Time is the time when the TR was referred lastly. Because peers can join or leave the network, the content has to be prevented from vanishing as peers drop out. Therefore, a utility content item is allocated in the requested peer as well as in the owner replica allocation method [10] when the peers receive a content item. In this case, if a query is guided to the downstream peer but the content item has been removed, the content item can be retrieved by sending the query back to the most upstream peer since this peer plays the role as the server. Modifying the upstream and downstream directions in TRs keeps the query guiding route, and a query can be guided to an appropriate peer in the next retrieval. However, when peers possessing TRs leave the network, TR needs to be controlled to maintain consistency. In the proposed method, the peers possessing TRs periodically check the connection to its adjacent peers to ensure TRs consistency, and TRs are updated or deleted as necessary. In addition, if the storage capacity of a peer is deficient in replica placement of the content item, the replica is allocated by removing other content items with the lowest utility. Then TRs related to the removed content item are updated or deleted.

Table 2. Transmission record.

TR entries	Details
Content ID	ID of the requested content
Upstream	ID of the upstream peer or router
Downstream	ID of the downstream peer or router
Reference time	Time when the TR was lastly referred

Using this control method, we can suppress the network load and improve the accuracy of content searching in a network with a dynamic topology.

Unfortunately, due to limitation of space, we omit the concrete descriptions of the proposed method's algorithms which was shown in [6]. However, note the following things. TTL (Time To Live) of queries is set to an appropriate value. Query guiding information possessed by a peer based on a logical link is called trp (transmission record in a peer) while query guidance information possessed by a router based on a physical link is called trr (transmission record in a router).

4 Evaluation

4.1 Evaluation Method

In this paper, the efficiency of the proposed method is evaluated from the following three view points by using computer simulations.

1. Content Acquisition Rate
   The ratio of the number of successful contents acquisitions over the number of total contents requests.
2. Content Searching Load
   This is the total load on the network for content searching. Concretely, it is a summation of the capacity of queries for content searching, confirmations of the contents transmissions, making, updating, and discarding TRs, multiplied by the number of hops on the network as the transmission distances.
3. Content Transmission Load
   This is the total load on the network for content transmissions. Concretely, it is a summation of the capacity of contents multiplied by the number of hops on the network as the transmission distances.

As the methods for comparison, three methods such as flooding with TTL constraint using cache routers, the method which is the same with proposal except caching and trr functions, and the method which is the same with proposal except trr function are prepared.

On top of that, the contents assumed to be shared have three categories, such as texts, music, and videos in order to consider the variety of the contents.

4.2 Evaluation Conditions

The parameters of the computer simulations for the evaluations is shown in Table 3. To make the network topology in the simulations, BA model [11] network consisting of 1,000 routers is used. The initial number of peers on the network is set to 1,000, and each peer is assumed to be connected directly with a router selected randomly in the network. In the initial state of each simulation run, each peer has 50 randomly selected content items, and logically connected to some (from one to four) randomly selected peers to make a overlay network which is maintained not to be broken apart.

Peers' join, leave the network, and content requests occur in randomly selected 1% of the whole peers per unit time. Categories of the requested contents such as texts, music, and videos are selected randomly with a probability proportional to the ratio of the categories of the contents existing in the network. Each title of the contents is selected according to Zipf distribution from the selected category.

Table 3. Simulation parameters.

Parameters	Values
Number of simulation runs	100 [times]
Period of a simulation run	5,000 [unit times]
Number of routers	1,000
Initial number of peers	1,000
Maximum number of peers	2,000
Initial number of deployed contents in each peer	50
Total number of content titles	30,000
Capacity of cache in a router	10 [MB]
Initial TTL value	6
Threshold H_{th}	3
Interval of repeating processes	5 [unit times]

When a peer newly joins, the peer connects to a router selected randomly from all the routers in the whole network, with holding 50 content items, and at the same time, the peer logically connects to another peer selected randomly from all the active peers in the network. When a peer which was once joining in the network in the past joins again, the peer connects again to the same router with the lastly connected one, with holding the contents possessed at the time of lastly leaving the network, and at the same time, the peer logically connects to another peer newly and randomly selected from all the active peers in the network.

When a peer leaves the network, if the peer is logically connecting to multiple peers, a peer is selected from the peers and the rest of the peers are all logically connected again to the selected one, in order to avoid the logical network's disjunction.

The storage capacity of each peer is given according to a normal distribution the average and the standard deviation of which is 5 [GB] and 1 [GB] respectively. Caching policy of each router is LRU (Least Recently Used), and its storage capacity is set to 10 [MB]. Total number of contents titles is set to 30,000, and the capacity of each category of the contents is given according to normal distributions showing in the Table 4. The capacity of texts is given based on the web sites' capacities gathered and shown in [12], the capacities of music and videos are assumed to be five and ten minutes volumes, respectively. The ratio of the three categories in content searchings is given as four scenarios. The capacity of a query is set to 200 [B], and initial TTL value is given to be 6. Furthermore, threshold H_{th} is set to be 3, and the regularly repeating processes are executed in every 5 unit times.

Table 4. Capacity distributions for contents category.

Category	Capacity distribution
Texts	Normal distribution (Ave.: 1,800 [kB], S.D.: 200 [kB])
Music	Normal distribution (Ave.: 8,000 [kB], S.D.: 200 [kB])
Videos	Normal distribution (Ave.: 100 [MB], S.D.: 20 [MB])

4.3 Evaluation Results and Discussions

The results of evaluations using computer simulations are shown in Figs. 2, 3 and 4. Each value of the results is the average of 100 simulation runs. Three figures show the relationship between the mixture ratio of three content types (texts, music, and videos) and evaluation indexes such as content acquisition rate, searching load, and content transition load.

From Fig. 2, we can see that the acquisition rate becomes worse in every method's result when the ratio of videos is large. This is because the existence of many videos with large capacity reduces the number of content replicas held in each peer, and this makes it easy for the contents to be lost when the contents reallocation and peers' leaving the network occur. On the other hand, when the ratio of texts is large, the acquisition rate becomes the highest in every method's result. This is because the number of content replicas held in each peer increases, and on top of that, since the number of requests for music contents is reduced, the number of cache replacements in routers is reduced, and the contents that are disappeared from the peers can be retrieved from the cache in routers. Figure 2 also shows that the acquisition rates of the proposed method are always larger than those of any other method. This is because the video files which cannot be cached in routers can be efficiently retrieved from the peers possessing them by the appropriate query guidance with trp, and other files can be frequently retrieved from routers' cache by trr, in the proposed method.

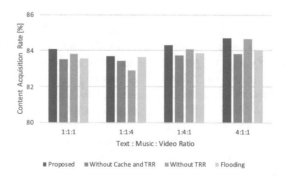

Fig. 2. Content acquisition rate.

Figure 3 shows that the content searching load becomes large in any method when the ratio of videos is large. This is because the cache usage reduces since the number of requests for video contents increases, and this makes it difficult for the number of query transmissions to be reduced by routers' cache. On the other hand, when the ratio of the texts is large, the load becomes small in any method. The reason is that the number of cache replacements reduces and usage of cache increases since the number of requests for contents to be cached in routers is large and the number of music contents is small, when the ratio of the texts is large. Figure 3 also shows that the proposed method reduces the content searching load better than any other method in all the conditions with combinations of content ratios we prepared. The reason of this is that the proposed method confines the range of query delivery and at the same time, guides the queries to the peers or routers holding the target contents.

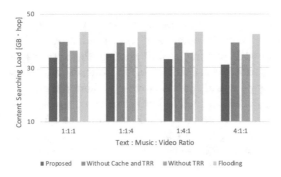

Fig. 3. Content searching load.

Figure 4 illustrates that the content transmission load becomes large when the ratio of the video is large in any method. This is simply because the number of requests for video contents with large capacity becomes large. On the other hand, when the ratio of the text is large, the load becomes the smallest in any method. This is also simply because the number of requests for text contents with small capacity becomes large. Figure 4 also presents that the proposed method reduces the content transmission load better than any other method in all the conditions with combinations of content ratios we prepared. As mentioned in the discussions on the searching cost above, this is because the proposed method can find the target contents with small number of query transmissions by using query guidance, and this causes that the total number of hops for query transmissions reduces.

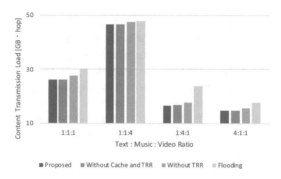

Fig. 4. Content transmission load.

5   Conclusion

In the previous research, we proposed a content sharing system on unstructured P2P network which enables stable and flexible system operation even when peers are relatively frequently joining in and leaving the network, assuming the use of content cache routers, trajectories of the queries as Transmission Records (TRs) in each relay peer and router. However, there are many kinds of contents exchanged over the network, such as texts, music, pictures, videos and so on, and the actual system needs to search the targets from the mixture of those kinds of contents. Then in this paper, we evaluate the effectiveness of the method in the environment where some different kinds of contents are mixed and the target contents are searched from that mixture.

From three view points, such as content acquisition rate, content searching load, and content transition load, the result of evaluations by computer simulations says that the proposed method shows the effectiveness of the proposed method better than any other compared method, even in the cases where a variety of contents are mixed and shared in the network. The reason of the effectiveness is that the proposed method confines the range of query delivery and at the same time, successfully guides the queries to the peers or routers holding the target contents by using TRs.

As a future work, we need to make a further evaluation of the proposed method with more variety of content types and larger number of content mixture combinations.

Acknowledgements. This work was partially supported by JSPS KAKENHI Grant Number JP17K00134.

References

1. Sunaga, H., Hoshiai, T., Kamei, S., Kimura, S.: Technical trends in P2P-based communications. IEICE Trans. Commun. **E87-B**(10), 2831–2846 (2004)
2. Stoica, I., Morris, R., Karger, D., Frans Kaashoek, M., Balakrishnan, H.: Chord: a scalable peer-to-peer lookup service for Internet applications. In: ACM SIGCOMM 2001, pp. 149–160, August 2001

3. Ratnasamy, S., Francis, P., Handley, M., Karp, R., Shenker, S.: A scalable content-addressable network. In: ACM SIGCOMM 2001, pp. 161–172, August 2001
4. Gnutella. http://gunutella.wego.com
5. Freenet. https://freenetproject.org
6. Takeda, D., Sugawara, S., Fukushima, N., Ishibashi, Y.: An efficient content searching method using query transmission records with content cache routers in unstructured peer-to-peer networks. In: Proceedings of the Third International Symposium on Computing and Networking - Across Practical Development and Theoretical Research - (CANDAR), pp. 200–206, December 2015
7. Tomimatsu, T., Sugawara, S., Ishibashi, Y.: Query guidance with transmission records for efficient content searching in unstructured peer-to-peer networks. In: Proceedings of IEEE 2013 International Communications Quality and Reliability Workshop (CQR), May 2013
8. Rosensweig, E.J., Kurose, J.: Breadcrumbs: efficient, best-effort content location in cache networks. In: Proceedings of IEEE INFOCOM 2009, pp. 2631–2635 (2009)
9. Tsoumakos, D., Roussopoulos, N.: Adaptive probabilistic search for peer-to-peer networks. In: Proceedings of ICS 2003, pp. 102–110, September 2003
10. Lv, Q., Cao, P., Cohen, E., Li, K., Shenker, S.: Search and replication in unstructured peer-to-peer networks. In: Proceedings of ICS 2002, pp. 84–95, June 2002
11. Barabási, A., Albert, R.: Emergence of scaling in random networks. Science **286**, 509–512 (1999)
12. HTTP Archive. http://httparchive.org/

A Piano Performance Training System Based on Visual and Tactile Guidance

Hokuto Tsutsumi[1] and Hiroaki Nishino[2(✉)]

[1] Graduate School of Engineering, Oita University, Oita, Japan
v17e3007@oita-u.ac.jp
[2] Faculty of Science and Technology, Oita University, Oita, Japan
hn@oita-u.ac.jp

Abstract. Learning and improving piano play performance supported by an ICT-based training system is a difficult and time consuming task. It is especially difficult for beginners because existing technologies force users the continuation of excessive burdens on their vision and concentration. We propose a piano performance training system where the users can easily focus on the acquisition of piano performance skills with lighter cognitive loads. The proposed system provides a training environment based on visual and tactile guidance. It enables the users to keep their eyes on a screen for checking the performance instructions presented on it and concentrate on typing keys with correct fingers according to the instructions. It also uses the tactile stimuli for intuitively prompting the users to hit the keys with correct fingers on a real keyboard. The system recognizes the users' fingering, judging whether they hit the right keys with correct fingers, and feedback the results via tactile sensation. It enables even beginners to easily practice piano performance with correct fingering by self-study.

1 Introduction

Obtaining piano performance skills by self-study is always a challenging task. Taking one-to-one lessons with an expert is the best way for a learner, especially a beginner, but relying on an ICT-based training system is another choice. Such a system graphically displays the order and timings of keystrokes and the learner needs to exactly replicate the keystrokes on a real keyboard.

Figure 1 shows a typical training environment using an existing piano performance learning system. The system presents a graphical panel called piano roll score on a system monitor as shown in the figure. It guides the learner for which key on the keyboard to type with which finger as an animation. A set of red rectangle marks are scrolling down according to the progress of a song and the learner needs to hit a specific key when the lower end of a mark contacts its target key. Each mark corresponds to a specific musical note and its length presents the sound length. Therefore, the learner should carefully watch the animation and hit the corresponding key on the real keyboard at the exact same time as the contact happened. The system also instructs the learner about the finger to hit the key by showing the fixed number assigned to each finger on the graphical keyboard as shown in the left side of the figure.

© Springer Nature Switzerland AG 2020
L. Barolli et al. (Eds.): CISIS 2019, AISC 993, pp. 702–712, 2020.
https://doi.org/10.1007/978-3-030-22354-0_63

This process requires the learner to visually confirm the keying position and fingering information on the screen, moving his/her line of sight onto the real keyboard, and hitting the designated key with the correct finger on the keyboard. While iterating these operations is an essential practice for mastering the "fingering" technique to hit a right key with a correct finger, it imposes a heavy burden on the learner's vision. It is a very difficult hand-eye coordination task.

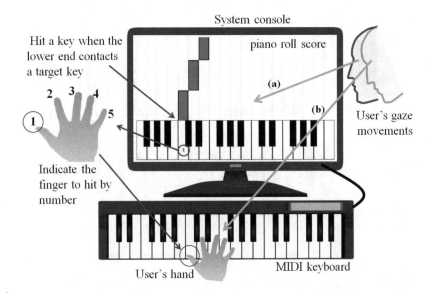

A user should iterate the following two gaze movements during the training process:
(a) check the keystroke position and finger to hit on the screen, and
(b) hit the key by the designated finger on the real keyboard.

Fig. 1. Example training environment in an existing piano performance learning system.

The authors devised and implemented the system to solve the above-mentioned problems as shown in Fig. 2. It tracks the learner's hand movements with a camera and displays each finger's position by its number on the graphics keyboard as shown on the left side of the figure. This function enables the learner to check both of the target key to hit and his/her current finger positions only by watching the console screen. It allows the learner skip the step of quickly comparing the screen and the actual keyboard every time he/she hits a key. He/she can concentrate on accurately reproduce the song indicated by the piano roll score with correct fingering while focusing on the screen. Additionally, the system presents vibratory stimuli to the finger to hit the target key, so the learner can intuitively perceive the finger with less visual load. The system provides a self-study environment for intuitively practicing the piano playing skills through a better hand-eye coordination task with light cognitive loads.

The rest of the paper is organized as follows. After we describe related research projects in Sect. 2, we show the proposed system overview and elaborate its implementation method in Sect. 3. After that, we account for some experimental results and observations found in evaluations in Sect. 4. Finally, we conclude the paper with some issues for further improvements in Sect. 5.

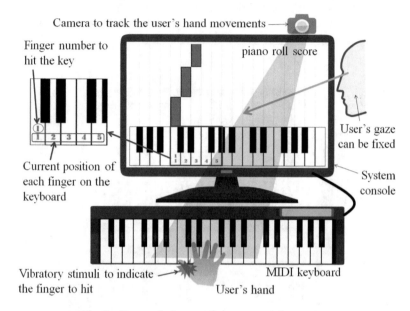

Fig. 2. Proposed piano performance training system.

2   Related Work

There are some precedent research activities for developing piano performance learning systems for beginners. Takegawa et al. proposed a method for directly projecting performance information such as a key and a finger number to hit the key on a real keyboard [1]. A user performs exercise by looking at the contents projected on the keyboard. The system recognizes the user's fingering by tracking colored markers attached to his/her finger and suppresses the projection of subsequent performance information if he/she mistakes the fingering as a penalty. The system provides the penalty function for motivating the user to improve his/her playing skills.

Rogers et al. developed a system enabling beginners who cannot read musical note to learn piano playing skills [2]. Their system projects the piano roll score on the surface placed adjacent to a real keyboard and interactively guides the user to follow the contents. They compared the system with other setups using a traditional paper note and an existing tool, and verified the system was effective in reducing the user's cognitive load during practice.

These systems are designed and implemented based on the AR (augmented reality) technology for presenting support information around the user's hands in a real space. In contrast, our approach collects and presents all guidance information on a computer screen. We consider this approach provides a more natural hand-eye coordination task environment for young learners who can touch-type on PCs.

3 System Implementation

3.1 Functional Overview

Figure 3 shows the appearance and software organization of the proposed system. The system consists of four software components as shown in the figure.

A user firstly selects a specific song to practice and the MDM (Music Data Manager) reads the sound source (MIDI) file and the fingering data (csv) of the corresponding song from the database. The SDM (System Display Manager) illustrates the playing method by sequentially displaying the piano roll score and the finger number to hit the key on the console screen based on the music data selected by the user. The FRM (Fingering Recognition Manager) tracks the user's finger movements captured by a camera mounted on the console and visualizes the movements by showing each finger's position with its number on a graphical keyboard. The user can grasp his/her finger positions by the visualized information on the screen without looking at his/her hands on the real keyboard. The user can concentrate on hitting the keyboard according to the displayed instructions by only watching the screen. Furthermore, the TDM (Tactile Display Manager) directly gives vibratory stimuli to the user's finger to hit the key through the tactile display worn by the user in parallel with the piano roll information.

The system provides an environment in which the user can acquire the piano performance skills by synergistically utilizing his/her visual and tactile sensations.

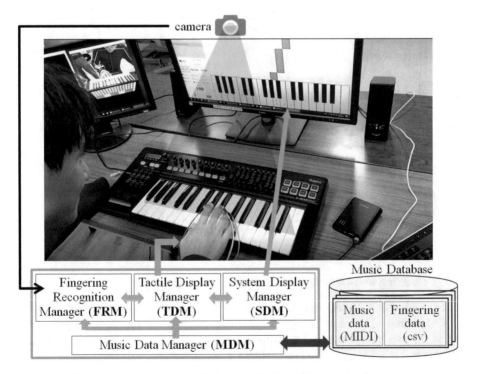

Fig. 3. Appearance and software organization of the proposed system.

3.2   System GUI Interface

Figure 4 shows a captured image of the system console managed by the SDM module. The screen layout is roughly composed of four regions. All functions for controlling the system are grouped together on the left side area numbered as 1 in the figure for the user's convenience. They are GUI widgets for selecting a MIDI device to use, selecting a set of music and fingering data to practice, starting and pausing the exercise function, and setting optional functions. As regards the optional functions, the user can control his/her exercise pace (*auto* mode to activate the system at a preset speed or *manual* mode to perform by the user's own pace), and turn on/off the finger movements tracking and the tactile display functions. The system displays a score computed based on the number of user's typing and fingering errors for each exercise in the lower left region numbered as 2 in Fig. 4. This function allows the user to quantitatively checking the progress of his/her performance.

The regions numbered as 3 and 4 in the figure are the music window for drawing the piano roll score according to the progress of the exercise, and the graphical keyboard window for indicating the finger number on the target key to hit. When the user enables the finger movements tracking function, the system displays the user's finger positions with their corresponding numbers in the graphical keyboard. Therefore, the user can concentrate on the exercise by only looking at the screen without visually checking his/her own hands status as explained in Fig. 2.

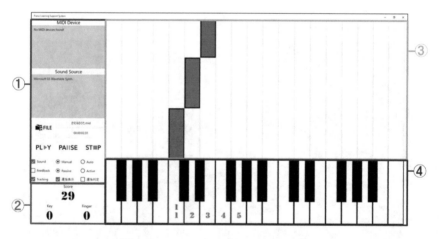

Fig. 4. Layout of system console screen.

3.3   Tracking and Detection of Finger Movements

The system uses a camera mounted at the top of the display monitor for detecting the keyboard area and the user's fingertip positions as shown in Fig. 3. Since the system tracks the user's fingering with a normal web camera without attaching any markers or sensors to his/her fingers, the user can easily set up the system.

In order for the system to correctly track the user's fingering on the keyboard, the system should identify the keyboard position. The system firstly takes a keyboard image by the camera as shown in Fig. 5(a). Next, it uses the keyboard image for accurately detecting the number of keys and the layout of white and black keys using binarization and labeling functions supported by OpenCV as shown in Fig. 5(b).

After that, the system continuously tracks and detects the user's fingering using a pair of the live camera image and the keyboard layout image as shown in Fig. 5(c). Both images have the same resolution and the key positions on the keyboard are completely matched at the pixel level between the two images. Then, the system obtains the user's finger positions captured in the live image using an open source motion tracker called OpenPose [3]. It estimates a human pose with a monocular camera using a deep learning technique. It estimates the joint positions of the user's arms and fingers, and returns their two-dimensional coordinate values in the live image. The system determines the key positions on which the user's fingers are placed by finding keys that contain his/her finger coordinate values in the keyboard image. For example, the system finds that the user's middle fingertip is placed on the eleventh white key in the keyboard layout image as illustrated in Fig. 5(c). Finally, the system draws each finger number on the detected key position in the graphical keyboard as shown in Fig. 4.

The system performs the above-mentioned process in every screen update frame. We adopt a GPU (NVIDIA's GeForce GTX 1070) to track the user's fingering in real time with OpenPose. As a result, the system can detect the fingertip positions at about 18 fps for one-handed keystrokes and at about 12 fps for two-handed operations. The system maintains this performance even in fast hand and finger movements. The tracking accuracy can further be improved by increasing the saturation value of the live image to clarify the boundaries between the hand and the background.

3.4 Tactile Display Function

In the previous system, we designed and implemented a tactile display for directly indicating a finger to hit the key via vibratory stimuli [4]. While the sensitivity of finger pad is better than back of finger, attaching a vibrator on the finger pad interferes with the user's keying operations. Therefore, we developed a tactile display for stimulating the back of finger in our previous system as shown in Fig. 6(a) [5]. Since this display stimulates the portion close to the base of finger, the user occasionally confuses a stimulation part between index and middle fingers.

In order to solve the above-mentioned problem, we newly developed another display for giving vibratory stimuli to a fingernail as shown in Fig. 6(b). This makes use of the result found in a previous research that the vibratory stimuli to the fingernail induce effective tactile sensations on the finger pad side [6]. The user wears the controller (Raspberry Pi) on his/her wrist with a wristband and attaches five wired disk-shaped vibration motors to his/her fingernails. The mobile battery powers the controller.

The system sends a finger number to the controller via a Bluetooth connection and then the controller actuates the corresponding motor by PWM (pulse width modulation) control. It can also present chord sounds by simultaneously activating multiple motors in addition to single notes. Since the user can change vibration strength by adjusting the parameter of the controller, he/she can set his/her optimum strength that doesn't interfere with his/her performance. As a result, the user quickly judges and reproduces the fingering on the real keyboard by sensing the stimulus presented to the fingertip.

(a) keyboard image captured by a web camera   (b) Keyboard layout detected by the system

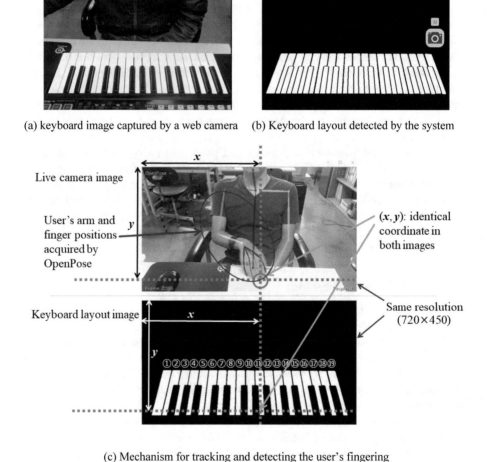

(c) Mechanism for tracking and detecting the user's fingering

Fig. 5. Functions for tracking and detecting the user's fingering on the keyboard.

4 Experiments

4.1 Experiment 1: Accuracy of Fingering Recognition Function

We conducted some experiments for verifying the effectiveness of the system. Since the accuracy of the fingering recognition function is a crucial issue for both of the visual and tactile guidance, we measured the fingering recognition rate when a subject is playing a real song. We selected a song titled "Minuet BWV Anh.114" in this experiment. It is an exercise song usually played by beginners but it requires some practical skills such as crossing fingers and jumping hands on the keyboard.

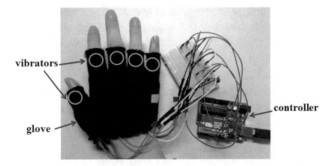

(a) Tactile display for stimulating the back of each finger (previous version).

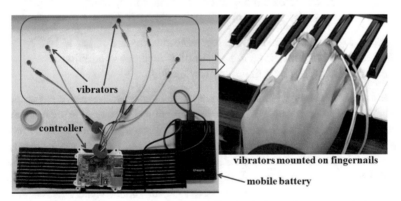

(b) Tactile display for stimulating the nail of each finger (current version).

Fig. 6. Tactile displays used in the system.

We registered a set of music and fingering files of the song and asked an experienced player who could play it without fail. We asked him to play in two different tempos, normal (quarter note = 100) and slow (quarter note = 60) tempos, to check the recognition rate at different playing speeds. We also asked him to play the song five times at each tempo and measured the number of recognition errors (the number of times the system judged the correct keying as incorrect one) appeared in the part played by his right hand. Table 1 shows the average number of errors and the average recognition rates measured at two tempos. The song has a total of 192 keystrokes.

Table 1. Results in experiment 1.

Playing speed	Number of errors	Recognition rate
Normal tempo	53.4	72.2%
Slow tempo	7.4	96.1%

Because the system spends about 300 to 350 ms to get a hand posture from a captured live image, it is considered that the delay time has a substantial impact on degradation in the recognition rate at the normal tempo. The effect of this delay time is particularly apparent when the player jumps his hands on the keyboard. We found another case to cause the degradation when the player crosses his fingers. Since the thumb is temporarily hidden under the index finger when hitting a key in such case, the thumb is misrecognized as the index finger. In the slow tempo, on the other hand, the system stably achieves the high recognition rate throughout the experimental tasks.

From the above results, the fingering recognition function is quite effective when beginners train by confirming their fingering at relatively slow tempos, but further performance improvement is required for users training at normal or faster tempos to improve their skills.

4.2 Experiment 2: Usefulness of Visual and Tactile Guidance Functions

In this experiment, we asked ten beginners for using the system in the following three operation modes to verify the usefulness of the visual and tactile guidance functions:

- Mode A: to present the piano roll score and the finger number to hit,
- Mode B: to activate the fingering recognition function in addition to mode A, and
- Mode C: to activate the tactile guidance in addition to mode B.

We select the same song with its right hand part as used in the above experiment and set the system for matching the tempo to the subject's pace (the manual mode as explained in Sect. 3.2). We asked each subject to practice three times in one mode and then play full in mode A. The number of subjects who practice in modes A, B, and C are 4, 3, and 3, respectively. We take a video of their final full play performance and measure the number of typing and fingering mistakes and playing times as factors to verify the effectiveness on the training session in each mode. Table 2 shows the results in average. We also conducted a questionnaire survey on the effectiveness and comfort of fingering recognition and tactile stimulation after the experimental task.

Table 2. Results in experiment 2.

Playing mode	Average number of mistakes		Average playing time (mm:ss)
	Typing	Fingering	
A (piano roll + finger)	5.57	9.00	02:28
B (A + fingering recog.)	6.00	5.67	02:49
C (B + tactile)	5.67	5.00	03:00

From Table 2, we found that the subjects trained in mode B and C made less fingering mistakes than the subjects trained in mode A while they spent longer time in the full play trial. Mode A subjects tended to make mistakes by rushing their performance and some of them played with their own incorrect fingering. On the other hand, Mode B and C subjects tried to play slowly and prevent keying and fingering mistakes in the final performance. Based on these observations, we verified that the implemented system functions are quite useful to the subjects for acquiring and improving their performance in the training.

In the post-experimental questionnaire, we found both of the fingering visualization and the error detection functions are highly rated. In particular, some subjects stated they could easily perceive error-prone parts in the song because the system immediately pointed out a mistake at the moment it happened. Other subjects replied they can pay close attention to the information displayed on a screen throughout the training session and it was quite efficient to focus on the exercise. Another positive comment was the system enables users to improve their playing techniques by just like enjoying video games even if they cannot read a music note. While many subjects positively supported the tactile guidance function, some subjects felt that the controller was heavy and the wiring between the vibration motors and the controller was bothered. We, therefore, need to further improve the tactile display design.

5   Conclusions and Future Work

We proposed a piano performance training system effectively assisted by both visual and tactile sensations. It enables beginners to focus on easy-to-understand guidance information and acquire skills for piano performance by self-study. The system tracks user's hand movements by using a camera, presenting his/her current finger positions on a screen, and noting if he/she incorrectly performs fingering. The system conveys notifications as vibratory stimuli in conjunction with visual feedback without disturbing the user's concentration on exercise. We designed and implemented the system and verified its effectiveness through experiments. We asked subjects for exercising a real song in the experiment and observed some gains for using the implemented visual and tactile guidance functions. We also found some issues to further improve such as the performance of fingering recognition and the comfort of tactile display.

In the near future, we would like to conduct more experiments to verify the applicability for training with practical and challenging songs. Trial use of the system in actual piano learning environment is another important challenge.

References

1. Takegawa, Y., Terada, T., Tsukamoto, M.: A piano learning support system considering rhythm. Proc. International Computer Music Conference, pp. 325–332 (2018)
2. Rogers, K., Röhlig, A., Weing, M., Gugenheimer, J., Könings, B., Klepsch, M., Schaub, F., Rukzio, E., Seufert, T., Weber, M.: P.I.A.N.O.: Faster piano learning with interactive projection. Proc. 9th ACM Int'l Conf. on Interactive Tabletops and Surfaces, pp. 149–158 (2014)

3. Cao, Z., Simon, T., Wei, S.-E., Sheikh, Y.: Realtime multi-person 2D pose estimation using part affinity fields. Proc. IEEE Int'l Conf. on Computer Vision and Pattern Recognition, pp. 1302–1310 (2017)
4. Tsutsumi, H., Nishino, H., Kagawa, T.: A piano performance trainer with tactile guidance. Proc. IEEE Int'l Conf. on Consumer Electronics-Taiwan, pp. 19–20 (2017)
5. Tsutsumi, H., Nishino, H., Kagawa, T.: A tactile assistance for improving fingering skill in piano performance. Proc. IEEE Int'l Conf. on Consumer Electronics-Taiwan, pp. 209–210 (2018)
6. Ando, H., Kusachi, E., Watanabe, J.: Nail-mounted tactile display for boundary/texture augmentation. Proc. of Int'l Conf. on advances in computer entertainment technology, pp. 292–293 (2016)

Study on Autonomous Outing Support Service for the Visually Impaired

Eiji Aoki[1(✉)], Shinji Otsuka[1], Takeshi Ikenaga[2], Hideaki Kawano[2], and Masaaki Yatsuzuka[3]

[1] Institute for Hyper Network Society, Oita, Japan
{blue,otsuka}@hyper.or.jp
[2] Kyushu Institute of Technology, Kitakyushu, Japan
{ike,kawano}@ecs.kyutech.ac.jp
[3] Autobacs Seven Co., Ltd, Tokyo, Japan
yatsuzuka@autobacs.com

Abstract. When the visually impaired people go out alone, they rely on white canes and Braille blocks. However, there are problems on the route, such as obstacles on the sidewalk Braille blocks. As a result, a traffic accident or a fall accident from the station's platform has occurred. That is the cause many cases where people go out without worrying about troubles. In this research and development, those obstacles are detected using a stereo camera linked to a smartphone, and the content is guided to the user by a wearable speaker. In addition, the location of the user is captured by GPS and guided to the destination. There are two stages, one for "Mobility Support Service" and "Watch Over Service".

1 Introduction

In Japan, where population decline continues, rationalization measures are adopted in various fields of industry, and the impact is also spreading to the lives of general citizens. Taking railway as an example, the effect is remarkable in local areas. For example, the railway company announced that 8 out of 11 stations in Oita city will be unmanned. It says "The smart support station will be introduced where cameras and interphones will replace station staff". Therefore, voices of anxiety spread to people with disabilities and dissenting opinions erupted at the briefing sessions. In the unmanned station, there are no plans to install platform doors, and there are also stations that are not equipped with even Braille blocks. The governor of Oita also complained about this reality, and the railway company postponed the unmannedization except for one station that had already compatible with barrier-free. The occurrence of an accident in which the visually impaired people falls from the station's platform has been widely reported in recent years. Based on this situation, we conducted preliminary survey if there could be support by utilizing ICT for the visually impaired people. It is because to consider that recent technological development related to IoT and AI will be more vital than the inconvenient life in local areas: [1–3].

As a result of survey, it was found that not only the accident at the station but also the visually impaired people tend to worry about trouble and refrain from going out

© Springer Nature Switzerland AG 2020
L. Barolli et al. (Eds.): CISIS 2019, AISC 993, pp. 713–722, 2020.
https://doi.org/10.1007/978-3-030-22354-0_64

when they go out alone relying on the white cane. The trouble is collision with a bicycle or pedestrians placed on Braille blocks. It was also found that the number of guide dogs available to the visually impaired people is surprisingly low. In the welfare of persons with disabilities by administrative agencies, there is a system of "companion support" as a public service. The administrative burden of a system is increasing year by year in the country, prefectures and municipalities. The purpose of this research and development is to provide the visually impaired people with the freedom to go out and to reduce the public expense burden associated with the support of the agency. It aims at the support service which can go out autonomously by combining existing technology and products. First of all, it is a service to eliminate anxiety and inconvenience and to go out with confidence. Moreover it is necessary when going out, such as providing information for moving and activities, eliminating anxiety for meals, which means appropriate and effective assistance. It not only takes the user to the place they want to go, but also roles such as reading and writing for the visually impaired people. Among them, the two cases to be developed this time are the "Mobility Support Service" for providing information and mobility support, and the "Watch Over Service" that family members can feel relieved. We use stereo cameras, wearable speakers, smartphones, and software for linking their data. Specifically, the camera detects an object and guides the user to the destination by voice guidance of obstacles and routes from the speaker as needed. We also developed a device with a built-in GPS and emergency button and attached it to the white cane. This makes it possible to transmit location information and an emergency signal to the call center, and in the event of an emergency, prepares a mechanism for watching over contacting relations like a family: [4, 5] (Fig. 1).

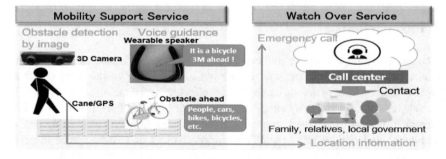

Fig. 1. Overall service outline image

2   Equipment Configuration

The visually impaired people usually go out using white canes or guide dogs. In addition to that, they go out with wearing a stereo camera and a mobile assistance tool consisting of a processing unit Jetson, a mobile battery, a wearable speaker and a smartphone, with the assumption of this equipment configuration. And built-in GPS and emergency button is attached to a normal white cane. At the time of departure, firstly the destination is set to Google map by voice from the wearable speaker and microphone connected with the smartphone. When a stereo camera recognizes an obstacle in forward direction on the way, the distance, direction, and object name of the obstacle are alerted by voice via a wearable speaker. The location information is always acquired by a white cane, and uses it for specifying the place so that the location of the visually impaired people is not unknown. If you set the area in advance, an alert email will be sent to your family. At the time of disaster, it is also possible to collectively contact the local government with the location of the visually impaired people who is using it.

As shown Fig. 2, Jetson, which recognizes image data from a stereo camera, is an AI computing platform that accelerates the parallel processing of NVIDIA's mobile embedded system with GPU. With high performance, low power deep learning computing and computer vision. Jetson is an ideal platform for embedded projects that require intensive computing. The Jetson TX2 development kit, which is a standard product, is rich in various interfaces, but the size (170 mm × 170 mm) is large and not suitable for portable use. Therefore, replacing the GPU on the Jetson TX1/2 Carrier Board made by Macnica Implementation (87 mm × 50 mm). It is housed in a commercially available plastic case for use.

Fig. 2. The development kit for the standard Jetson TX2 and macnica Jetson TX1/2 Carrier Board, and the GPU has been miniaturized.

As shown Fig. 3, a stereo camera is ZED Mini of STEREO LABS, which is possible to easily perform space grasping and space mapping. The data can be recognized up to a maximum distance of 0.7–7 m, angles up to 30° horizontally and 45° vertically. It is fixed on the outerwear with the clip for commercially outdoor camera. The wearable speaker for transmitting voice uses a microphone also equipped to instruct the smartphone. In the habit of the visually impaired people who judge the situation by sound, the type that completely occludes the ear is dangerous, so the shoulder type design (SHARP AQUOS sound partner AN-SS1) and the sound conduit design (SONY Xperia Ear Duo XEA 20) that the surrounding sound can be heard were used.

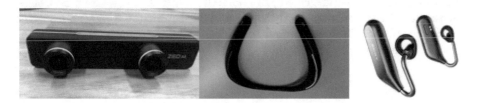

Fig. 3. ZED Mini, SHARP AQUOS and SONY Xperia Ear Duo

We use a trained model called YOLO (You Only Look Once) written using the Darknet machine learning framework. YOLO is a real time object detection algorithm, which can learn a huge number of images by using a representative deep learning technique "CNN (Convolutional Neural Network)" often used in image recognition, and it is possible to identify an object of about 80. As shown recognition image Fig. 4. In this research and development, YOLO compatible with ROS (Robot Operating System) was adopted, and object recognition was performed as fast as 14 to 16 frames per second. The library used by ROS is adopted because it is effective in the field of image recognition and 3D measurement. Instructs data acquisition, processing, and output on this ROS.

Fig. 4. Recognition image of learned object

Look at the flow of the internal processing of obstacle detection in the mobility support tool. First, a color image is acquired from the two lenses of ZED Mini, and a distance image is generated in ZED. Next, transfer the ZED color image and distance image to Jetson's ROS, and insert the color image into YOLO for object recognition and YOLO for traffic light detection. Each determination result is transmitted to a program that makes various determinations. The determination program also obtains the ZED distance image, calculates the distance at the object coordinates, and determines whether to output the object distance or object type (such as a person or a car) with probability. With respect to an object judged to be a pass, the value of several frames is held, and it is judged whether the object is approaching the user. At this point, it is limited to objects in the 12 o'clock direction, with the exception of crossing objects. We transfer this result to smartphone with Wi-Fi, and, furthermore, upload to the cloud of Google assistant via MQTT. The result is output by voice from the Google assistant on the smartphone. The guidance to the destination uses Google map. The guidance to the destination is started by talking to the microphone of the voice AI system, "Please route by walk to 00".

3 Demonstration Experiment

In this research and development, demonstration experiments were conducted in Oita city and Beppu city, Oita prefecture. Oita prefecture is one of 47 prefectures in Japan, located in the east side of Kyushu. As a commemorative event for the International Year of Disabled Persons held in Oita in 1981, it held the world's first "Oita International Wheelchair Marathon" as a wheelchair-only marathon. It has become the world's largest and highest-rated event, which will be held every year for about 30 years. Oita city is a prefectural office location and a core city. It has a population about 477,000 people and an area of 502 km^2. Designated as a new industrial city, it has developed as an industrial city such as steel and petrochemical, and in recent years there are also many electronic parts and device production. Beppu city is the second largest in the prefecture, about 119,000 people and an area of 125 km^2. It is a hot spring sightseeing spot boasting the largest number of springs in the world and there are many hotels and recreation facilities. Many medical and research institutes are located that use hot springs, and because of this relationship, the facility for supporting people with disabilities, "The House of the Sun" has been established, and now more than 400 disabled people working. There is a facility in the neighboring Hiji town too, and more than 320 people with disabilities work. We show the number of visually impaired people and population ratio targeted for this project. The ratio of visually impaired people in Beppu city is higher than the population ratio of the whole country because there are many hotels because it is a tourist resort, and there are many massagist and needlework that become the visually impaired people's work (Table 1).

Table 1. The ratio of population and visually impaired

Item	Japan	Oita prefecture	Oita city	Beppu city
Population	126,933,000	1,143,976	476,868	119,020
Visually impaired	337,997	3,692	1,191	497
Ratio	0.27%	0.32%	0.25%	0.42%

3.1 Task to Solve

With the development of ICT, the means by which the visually impaired people can obtain information have been dramatically improved. The influence is also improving communication skills. In particular, with the spread of smartphones in recent years, it has been developed voice-reading applications and destination guidance applications, and become indispensable to daily life for the visually impaired people. In urban areas, services such as destination guidance and shopping support are being developed, but in Oita prefecture and other regions with a small population, it is hardly widespread. For out-of-office support, public services such as companion support and guide dogs are the mainstream. Actually, there are the following three issues regarding going out, and we aim to solve by this research and development.

Burden amount on public services is increasing.

The government has implemented subsidy programs of companion support and grants to guide dogs as a public service for the visually impaired people. There is no expense and the upper limit by the income, but the personal cost is 10%, the remaining 1/2 to the country, 1/4 to prefectures, 1/4 to municipalities. The purpose use is limited, commuting to school is unavailable, and the use of entertainment is limited depending on municipalities, so the scope of application is wide. This is a relatively new public service launched in October 2011, its use frequency is increasing year by year. Because this system has been widely recognized, it has been found that this system can be used even in the case of not being used at first, and the provider side is gradually expanding the service usage range. Guide dogs are owned by the Guide Dog Association, and will be lent free to the visually impaired people. The user can use it only by carrying out the initial expenses of dog gear and the expenses necessary for daily life (dog food and toilet seat). However, training of guide dogs requires a large amount of expenses such as about 4 million yen to 8 million yen for facilities, feeding and trainers. So they are covered by donations and support costs from local governments. Oita prefecture subsidizes approximately 2 million yen for loaning guide dogs and about 40 to 50,000 yen for medical expenses annually. The guide dogs can active approximately eight years. As of December 2018, 13 dogs are active in Oita prefecture. On the other hand, as a negative aspect, guide dogs live their lives at the user's home, they are sometimes avoided from the trouble of feeding and excreting.

Reduce going out and tend to stay home.

The visual impaired people go out relying on white canes or guide dogs and Braille blocks, but places that go for the first time or places that do not have Braille blocks tend to have a high hurdle to go out and limit the range of activity. They often collide with obstacles parked on Braille blocks, get out on a driveway to avoid them, and often encounter dangerous cases. It may be troubled with a walking people that is not noticed because of pedestrians and smartphone operations that are stopped on Braille blocks. In the case of pedestrian crossings, if it is not equipped with an acoustic device, it is not possible to judge blue and red. But it crosses according to surrounding sounds and signs, there is a danger that it changes to red light on the way. Then they avoid going out under these circumstances. People who have been visually impaired since childhood have a tendency to go out relatively in order to perform walking training in blind schools. However, those who become adult and who have lost their eyesight tend to refrain from going out. According to the Ministry of Land, Infrastructure, Transport and Tourism's "Study on Transportation Facility Development Plans Contributing to Barrier-Free City Planning", the average number of times go out weekly is 3.92 times for healthy people and 2.27 for the visually impaired people.

Feeling uneasy about evacuation at the time of disaster.

At the time of disaster such as an earthquake, a typhoon, or a heavy rain, it is difficult for the visually impaired people to move to evacuation sites alone even in an emergency response because they do not know the surrounding situation. If they go out, it will be more difficult, and if they encounter disaster, they will get stuck without being able to evacuate. When locked in a building, it is difficult for them to escape by themselves and it is difficult for them to be rescued. In Oita city disaster prevention plan, if you register as well as elderly people and need-to-care persons, the local welfare officer performs safety confirmation and evacuation guidance at the time of disaster. However, no special measures have been taken for the visually impaired people. Mirairo Co., Ltd. surveyed people with disabilities throughout the country about their fears against the disaster and their evacuation situation, triggered by the damage from the heavy rain in West Japan in July 2018. According to the survey results, about 90% of people with disabilities say that they felt uneasy about disasters. In addition, about 80% of those who received evacuation instructions and recommendations answered that "we did not immediately secure security". The reasons are: "It is difficult to walk far in the evacuation area", "I thought that I would disturb the people around me if I have visual impairment", and "I was not able to go to the evacuation center because both couples are blind and there are guide dogs" are listed.

3.2 Experiment Contents

In Oita city and Beppu city, which are verification areas, we conducted questionnaire survey on the following nine items for the visually impaired people. Three types of verification were conducted: "staff verification", "accompanying verification", and "non-accompanying verification" (Table 2).

Table 2. Questionnaire survey.

Usability of equipment	Detection status of obstacles	Response time
Ease of hearing voice	Relevance of induction	Effectiveness in rainy weather
Accuracy of location information acquisition	Emergency call button effectiveness	Increase or decrease in feeling of security when going out

Staff verification 2 days.
The blindness association staff of all blind verify from the standpoint of the administrator whether there is a defect in the equipment configuration or it can withstand the use of the persons with disabilities, prior to accompany verification of the visually impaired people. In fact, two staff members participated.

Accompany verification 6 days.
Select one from the multiple walking courses set in the verification area, and perform verification while walking as a group of the visually impaired people and two verification personnel. One of the verification personnel mainly check the verification items and the other ensure the safety of the visually impaired people. We also carried out research with umbrellas and guide dogs in rainy weather. At the destination, we interviewed the feeling of use. A total of 69 visually impaired people were able to participate.

Non-accompany verification 7 days.
We have improved the equipment configuration in order to cope with the issues that came up during accompany verification. After that, by having a visually impaired

person use it for a week, we checked about the increase and decrease in the number of outings. In the verification, our staff ask and confirm the improvement for using the system. But basically, a visually impaired person is supposed to go out and act freely for a week (Fig. 5).

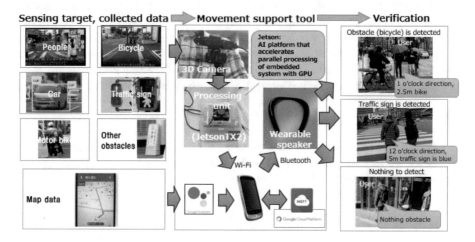

Fig. 5. Contents of demonstration experiment

4 Conclusion

4.1 Effect Verification

As for effect verification, the questionnaire survey was conducted through interviews at the time of accompany verification. There was 74 participation due to the planned number of participants, but the number of questionnaires that could be obtained was 69. Because canceled due to rainy weather, and there was a sudden occurrence of business.

4.2 Effect for Task to Solve

The effect of reducing the burden on public services was to estimate the cost reduction for those that met the conditions that could substitute for companion support in this effect verification. We calculate the target as among those who have used the service within 2 h of companion support. According to the questionnaire results after verification, 70% of users can go out with confidence, and 71% of users answered that the time to go out will increase, so 70% of the use within 2 h will be replaced by the system. The amount of burden within 2 h in Oita City in March 2018 is 91 for 2,792,000 yen, equal 30.700 yen per person. It will be 30.700 yen multiple 70%, equal 21.500 yen. Therefore, the reduction effect of the same amount of supporting contribution per person is 21.500 yen per month. On the other hand, the public service cost burden amount per person in Oita city is 54.400 yen per month at the whole 2018 fiscal year, so it is estimated that the reduction effect can suppose a little larger.

The effect of the tendency to stay home is that the survey results after the verification conducted indicate that 71% of the users spend more time going out. They are answering on a fill-in basis, "I want to go to places I have never been to", "I want to go to various places", "More opportunities for walks", "More to feel comfortable, so more", "Don't feel good even when I feel bad it's easy to go out". One week of non-accompany verification investigated increase in the number of outings.

About the dispelling effect about the anxiety against the response at the time of disaster, according to the questionnaire result after the verification, 70% of users can go out with confidence by providing safety confirmation by watch over service, emergency button and call center. In addition, the accuracy of guidance to the destination was evaluated from the route history of the watch over service at the time of accompany verification. Of the 31 data acquisitions, 29 data used safe and shortest moving courses. Therefore 93.5% was accurate induction.

In non-accompany verification, data were acquired about the change in walking speed. The same visually impaired person measured changes in walking time with and without the mobility support service at the verification. As a result, with the service one way average 26 min 06 s, without the service one way average 30 min 39 s, it was speed improvement of 15%.

As Braille blocks are individually managed by the country, prefectures, and municipalities for each road jurisdiction. There has been no information summarized centrally in a wide area. Therefore, in this research and development, open data using the format of "walking space network data" of the Ministry of Land, Infrastructure, Transport and Tourism, for Braille blocks in JR Oita station area and JR Beppu station area created as a format. The data is assumed to be published at the G-Spatial Information Center and used in the "Navigation application dedicated to the visually impaired people" (Table 3).

Table 3. The result of experiment

Cost reduction	Tend to go out	Dispelling anxiety	Accurate guidance	Walking speed
21,500/month	71% up	70% down	93.5%	15% up

5 Future Work

Although new hardware development was not performed, a mobility support tool was created by combining commercial off-the-shelf products, but it is necessary to put it in a backpack to carry around 2 kg for cameras, processors, batteries, and smartphones. According to the verification questionnaire, young people and men were evaluated not to feel the weight, but some women pointed out that it was heavy. Also, the presence of a cable connecting the camera and the processing unit makes it difficult for the visually impaired people to wear it alone. It is also troublesome that the devices need to be powered on /off and recharged individually. Power consumption was large due to Wi-Fi,

Bluetooth communication module activation, parallel operation of image recognition program and a heavy large-capacity battery was required. In the future, we will improve the device to an easy-to-carry, easy-to-operate device by using an integrated hardware configuration based on the processing device. In addition, we will eliminate unnecessary communication modules, promote energy saving by improving the efficiency of image processing programs, and make the battery smaller.

Improve the speed from image recognition to voice guidance. When it is necessary to issue a warning in an instant such as a sudden jump out, processing in the cloud may take time and the warning may not be in time. Therefore, we added a function so that the alert could be issued only by the processing device, and conducted the non-compliance verification to sound the alert sound with the external speaker. In the future, sound generator such as a buzzer will be integrated into the processor.

Obstacles are read out in order from the closest one at the time of one detection. Sidewalks with a lot of traffic are often read out as obstacles, and for the visually impaired people there is too much information. It took a long time to read it all, and it was found that it was not in time to read out the approaching obstacle. Instead of reading out all the obstacles in the detection range, it is necessary not to read out what is expected to be an obstacle. It is improved so that it can be judged whether it is an obstacle or not by predicting the timing when it deviates from the detection range and the next movement.

Acknowledgments. This work was supported in part by Ministry of Internal Affairs and Communications (MIC) in Japan, IoT service creation support project in FY2018.

References

1. Aoki, E., Otsuka, S., Ikenaga, T., Kawano, H., Yatsuzuka, M.: Study on regional transportation linkage system that enables efficient and safe movement utilizing LPWA. In: Proceedings of the 12th International Conference on Complex, Intelligent, and Software Intensive Systems (CISIS-2018), pp. 968–977, 4–6 July 2018
2. Aoki, E., Oba, Z., Watanabe, R.: Study on data utilization of regional industry in cross-cutting and systematic regional community networks. In: Proceedings of the 11th International Conference on Complex, Intelligent, and Software Intensive Systems (CISIS-2017), pp. 795–804, 10–11 July 2017
3. Aoki, E., Oba, Z., Watanabe, R.: Study on cross-cutting and systematic regional community networks. In: Proceedings of the 10th International Conference on Complex, Intelligent, and Software Intensive Systems (CISIS-2016), pp. 599–604, 6–8 July 2016
4. Aoki, E., Yoshitake, S., Kubo, M.: Study on sensor networks for elderly people living alone at home. In: Proceedings of 2015 International Conference on Consumer Electronics-Taiwan (ICCE-TW), pp. 132–133, 6–8 June 2015
5. Aoki, E., Yoshitake, S., Kubo, M.: Study on a nursing system using information communication technology. In: Proceedings of the 8th International Conference on Complex, Intelligent, and Software Intensive Systems (CISIS-2014), pp. 631–636, 2–4 July 2014

Classification of Arteriovenous Fistula Stenosis Using Shunt Murmur Analysis and Random Forest

Fumiya Noda$^{(\boxtimes)}$, Daisuke Higashi, Keisuke Nishijima, and Ken'ichi Furuya

Oita University, 700, Dannoharu, Oita-shi, Oita, Japan
{v1553046,v1453054,k-nisijima,furuya-kenichi}@oita-u.ac.jp

Abstract. Although patients undergoing hemodialysis generally have shunts implanted within a body part, problems such as blood vessel stenosis can occur. Patients undergoing hemodialysis can conveniently check their own shunt function by listening to shunt murmurs. However, manually judging the shunt function is difficult and requires experience. In this study, we propose a method to classify shunt stenoses using Random Forest (RF). The resistance index (RI) obtained from the ultrasound system is used as a class label. The normalized cross-correlation coefficient, the ratio of frequency power to mel frequency cepstrum coefficient (MFCC), was used as a feature to train in an RF classifier. As a result, the classification accuracy of RI by RF was found to be higher than that achieved by a support vector machine (SVM).

1 Introduction

Some patients, such as those with renal disease, undergo hemodialysis to eliminate waste products and excess water from their blood. An arteriovenous fistula (AVF), known as a shunt, can be implanted within these patients by anastomosing their arteries and veins. However, problems such as stenosis and occlusion may occur due to the age of the patients. If the functioning of a shunt is obstructed, the shunt must be implanted in another place for the patient to be able to continue undergoing hemodialysis. Therefore, patients can conveniently check their own shunt function by listening to shunt murmurs. However, the analysis of shunt murmurs requires experience as they are difficult to examine. Thus, a system that automatically examines shunt function is required.

In previous research [1], SVM has been used to analyze the shunt function. However, it cannot be used in a real environment because of low identification accuracy. Therefore, I thought that I would like to improve the identification rate by changing the classifier. Resistance index (RI) is a measure that indicates the difficulty of blood flow to the distal end. In Murakami's survey [2], it was confirmed that when the RI value was 0.6 or higher, the group of patients in which the shunt malfunction occurs tends to grow. It is nearly impossible to confirm the change in shunt function over time. However, if RI can be identified, it is expected to assist in the detection of changes over time.

© Springer Nature Switzerland AG 2020
L. Barolli et al. (Eds.): CISIS 2019, AISC 993, pp. 723–732, 2020.
https://doi.org/10.1007/978-3-030-22354-0_65

In this study, we propose a method to automatically classify shunt stenosis using Random Forest (RF), features obtained from frequency analysis, and RI. The random forest has a feature that it is likely to be highly accurate because it creates a plurality of decision trees and identifies them by majority decision. To confirm the effectiveness of this method, we performed cross-validation and evaluated the method based on accuracy and F-measure. Note that recording data and feature amount used here is the same as that used in the previous research.

2   Collection of Shunt Murmurs

Shunt murmurs were recorded using the recording equipment attained by connecting AT9903, which is an Audio Technica microphone, with FC-200, which is a FOCAL CORP chest piece. The microphone with the stethoscope chest piece is shown in Fig. 1. In addition to the microphone with the stethoscope chest piece to record shunt murmurs, a DR-05 IC recorder from TASCAM was used. Using these instruments, shunt murmurs from the anastomotic area of patients implanted with AVF were recorded before hemodialysis.

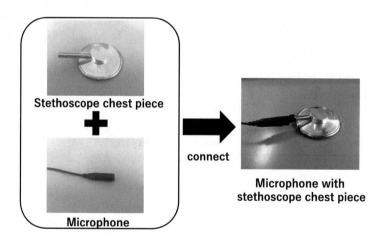

Fig. 1. Recording equipment

3   Feature Extraction

3.1   Mel-Frequency Cepstral Coefficients

Mel-frequency cepstral coefficients (MFCCs) is a low-dimensional spectrum feature defined in the quefrency region. To obtain MFCCs, we defined a filter object called Mel filter bank and obtained the general form of the spectrum by multiplying it with the amplitude spectrum. The Mel filter bank is a filter whose operation is fine in the low-frequency band and rough in the high frequency band on the Mel scale. The Mel scale [3] is shown in Eq. 1.

$$Mel(f) = 2595 \log 10(1 + \frac{f}{700})$$ (1)

By transforming the general form of the spectrum using discrete cosine transforms, we obtained the quefrency region. The low-dimensional features of the obtained quefrency region are called MFCCs. We used 16 dimensions of the MFCCs as a feature.

3.2 Ratio of Frequency Power

Features of the stenosis shunt murmurs appear in the frequency spectrum. Therefore, we attempted extraction of these features using Fourier transform. Fourier transform is one of the methods used in the frequency analysis of signals, and a frequency spectrum can be obtained by Fourier transforming the corresponding time signal. First, we obtained a frequency spectrum by applying Fourier transform on the shunt murmurs signal. Next, we divided the band of 1–2000 Hz into four sections of 500 Hz each and calculated the frequency power for each band.

$$p_1 = \sum_{f=1}^{500} \{20log10(abs(X(f)))\}$$ (2)

$$p_2 = \sum_{f=500}^{1000} \{20log10(abs(X(f)))\}$$ (3)

$$p_3 = \sum_{f=1000}^{1500} \{20log10(abs(X(f)))\}$$ (4)

$$p_4 = \sum_{f=1500}^{2000} \{20log10(abs(X(f)))\}$$ (5)

Later, we calculated the sum of frequency powers of the 1—2000 Hz band by adding the values obtained from Eqs. 2–5:

$$P_{total} = p_1 + p_2 + p_3 + p_4$$ (6)

Finally, for each sum calculated by Eq. 6, we determined the ratio for the band.

$$P_1 = \frac{p_1}{P_{total}}, P_2 = \frac{p_2}{P_{total}}, P_3 = \frac{p_3}{P_{total}}, P_4 = \frac{p_4}{P_{total}}$$ (7)

This process is shown in Fig. 2.

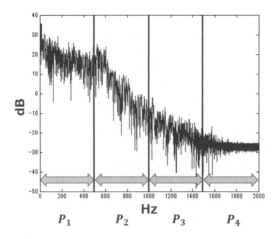

Fig. 2. Image of the ratio of the frequency power

3.3   Normalized Cross-Correlation Coefficient

The normalized cross-correlation coefficient [4] is obtained using normalized cross-correlation analysis and shows the correlation between two images or the similarity between systems. It is an effective means to investigate the correlation and similarity. It is obtained by treating the results of a wavelet transform as an image. Wavelet transformation [5] is a method of time-frequency analysis that estimates a signal by shifting and scaling a small wave called a wavelet. The function $\psi(t)$ that exists around the origin t = 0 with an average value of 0 is an example of a wavelet. By shifting and scaling this $\psi(t)$ on the t-axis, a basis $\psi_{a,b}(t)$ is generated.

$$\psi_{a,b} = \frac{1}{\sqrt{a}}\psi(\frac{t-b}{a}) \tag{8}$$

where a is a scaling parameter, called a scale, and b is a shift parameter. The inner product of $\psi_{a,b}$ and signal $f(t)$ is a wavelet transform. In this study, normalized cross-correlation coefficients were obtained for the shunt murmurs of two patients with low RI. The normalized cross-correlation coefficient is calculated using Eq. 9:

$$R = \frac{f(t,\omega) - \overline{f(t,\omega)} \times (g(t,\omega) - \overline{g(t,\omega)})}{\sqrt{(f(t,\omega) - \overline{f(t,\omega)})^2 \times (g(t,\omega) - \overline{g(t,\omega)})^2}} \tag{9}$$

where t denotes the time, ω denotes the frequency, and $\overline{f(t,\omega)}$ and $\overline{g(t,\omega)}$ denote the average brightnesses. $f(t,\omega)$ and $g(t,\omega)$ are calculated by Eqs. 10 and 11:

$$f(t,\omega) = \frac{1}{\sqrt{a}} \int f_1(t)\overline{\psi(\frac{t-b}{a})}dt \tag{10}$$

$$g(t,\omega) = \frac{1}{\sqrt{a}} \int f_2(t)\overline{\psi(\frac{t-b}{a})}dt \tag{11}$$

where $\overline{\psi(-)}$ is the complex conjugate of $\psi(-)$.

4 Proposed Method

In this study, we propose a method to automatically classify shunt stenoses using Random Forest, frequency analysis of shunt murmurs, and RI. When the RI value is 0.6 or higher, the group of patients in which the shunt malfunctions occur grows. Therefore, it is desirable that RI be classified into two groups: lower than 0.6 and 0.6 or higher. First, we classified the recording data into two classes based on their RI value: lower than 0.6 and 0.6 or higher. Next, data with a length of 0.8 seconds was cut out from the learning data of each class, and feature quantities learned by Random Forest were extracted. Further, we extracted features from the test data in the same manner as above and classified them using the learned Random Forest. As a result, the test data was categorized into two classes, one with an RI value lower than 0.6 and another with an RI value of 0.6 or higher. The flow of these processes is shown in Fig. 3.

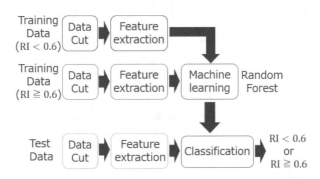

Fig. 3. Classification using Random Forest

4.1 Random Forest

Random Forest is a machine learning algorithm that estimates classes by combining the plurality of decision trees. Its name is derived from the introduction of random numbers in the construction of decision trees such that every decision tree is unique. The problem of decision tree is that it overfits the learning data. Using Random Forest, reducing the degree of excess adaptation by over-fitting multiple decision trees in different directions and considering the average of the results is possible. When a decision tree is created, bootstrap sampling is performed. This is done to randomly extract data from the learning data, allowing duplication, for creating a learning dataset. As a result, each decision tree of Random Forest is constructed for slightly different datasets. Moreover, as feature quantities are randomly extracted to create decision trees, it is possible to investigate their importance. Example of a random forest is shown in Fig. 4.

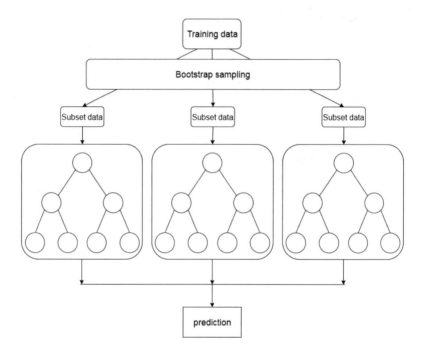

Fig. 4. Example of Random Forest

5   Experiments

It is aimed at confirming the effectiveness of the proposed method by comparing the identification by Random Forest, the identification by SVM, and the accuracy of human judgment.

5.1   Experimental Condition

Shunt murmurs used for classification were recorded using the microphone with stethoscope chest piece. We used the shunt murmurs at the anastomosis of 60

AVF patients (30 patients with an RI value lower than 0.6 and 30 patients with an RI value of 0.6 or higher) for classification. We used 300 sets of data in total. As a method of evaluation, we performed fifth-order cross-validation and evaluated it based on the accuracy and F-measure. While performing cross-validation, we used 80% of the total data for training and the remaining 20% for testing. In this study, we used data from different people for training and testing. We utilized the normalized cross-correlation coefficient, ratio of frequency power, and MFCCs as learning features. The experimental conditions are shown in Table 1.

Among the 62 patients used for this experiment (2 patients used for calculating the normalized cross-correlation coefficient and 60 patients used for the experiment), 32 patients demonstrated an RI value lower than 0.6 and 30 patients demonstrated an RI value of 0.6 or higher. The relationship between the judgment of the medical staff and RI for 61 people (excluding the one who was not judged by the staff) is shown in Table 2. If the RI value is lower than 0.6 and the judgment of the staff is undoubted, or if the RI value is 0.6 or higher and the judgment of the staff is doubted, the decisions are regarded as correct. In this survey, the accuracy of staff judgment is observed to be 59%. This is used as a performance indicator. The parameters considered for RF are shown in Table 3. We are examining the number of decision trees and the depth of decision trees.

Table 1. Experimental conditions

Sampling frequency	48 kHz (microphone)
Data length	0.8 s
Number of test data	60
Number of training data	240
Feature	MFCCs (16 dimensions) Ratio of frequency power (4 dimensions) Normalized cross-correlation coefficient (2 dimensions)

Table 2. Relationship between staff judgment and RI (Number of patients)

		Staff judge	
		No doubt of stenosis	Doubt of stenosis
Measure using ultrasonic diagnostic equipment	Less than 0.6	28	4
	More than 0.6	21	8

Table 3. RF parameters

Parameters	Range
Number of decision trees	1–21
Depth of decision tree	1–21

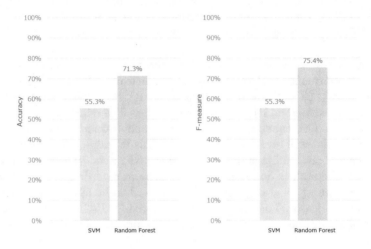

Fig. 5. Result of calculating accuracy and F-measure

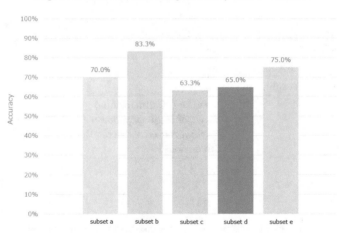

Fig. 6. Result of cross-validation of random forests

5.2   Results and Discussion

The classification results of the experiment are summarized in Table 4. Figure 5 shows the result of calculating accuracy and F-measure from Table 4. RF is the result when the number of decision trees is 1 and the depth of decision trees is 3. In recognition of shunt murmur using SVM, the correct answer rate

Table 4. Result of SVM and Random Forest (2)

| | | Measure using ultrasonic diagnostic equipment | |
		Less than 0.6	More than 0.6
Classification of SVM	Less than 0.6	83	67
	More than 0.6	67	83
		Measure using ultrasonic diagnostic equipment	
		Less than 0.6	More than 0.6
Classification of Random Forest	Less than 0.6	132	68
	More than 0.6	18	82

was 55.3% and the F-measure was 55.3%. In the analysis of shunt sound using Random Forest, the accuracy rate was 71.3% and the F-measure was 75.4%. The correct answer rate and F value were higher in the random forest than in the SVM. The results of cross-validation of random forest are shown in Fig. 6. It is probable that the correct answer rate will change if the training data changes as there is a difference in the accuracy rate of cross-validation. Furthermore, to use it in a real environment, the identification accuracy is inadequate. Therefore, investigating future ways to improve the identification accuracy of the shunt murmurs is necessary.

6   Conclusion

Some patients, such as those with renal disease, check their own shunt function by listening to shunt murmurs. However, analysis of shunt murmurs requires experience as they are difficult to examine. Thus, a system that automatically examines the shunt function is required. In this study, we proposed a method of classifying shunt stenoses using Random Forest. As a result of the identification experiment, it was confirmed that the identification rate of the Random Forest is higher than that of the SVM. Also, the discrimination rate is higher than the staff's judgment, although it can not be compared simply. To realize the objective, it is necessary to examine feature quantities and classifiers for analyzing shunt murmurs as the present identification accuracy is insufficient.

Acknowledgments. This work was supported by JSPS KAKENHI grant numbers 18K11377, 16K00245, and 15H02728.

References

1. Higashi, D.: Classification of arteriovenous fistula stenosis using shunt murmurs analysis and support vector machine. VENOA, 4 July 2018
2. Murakami, K., et al.: Effectiveness of ultrasonic pulse Doppler method in shunt Management'. Kidney and Dialysis, pp. 39–43 (2003)
3. Shikano, S., et al.: Speech recognition system, Ohmsha (2001)

4. Sasaki, K., et al.: Functional assessment of vascular access based on time-frequency analysis of shunt murmurs, IEICE, IEICE Technical report, vol. 114, no. 54, US2014-12, pp. 25–30 (2014)
5. Nakano, H., et al.: Signal and Image processing by the wavelet. Kyouritsu Publisher (1999)

Effects of Equivalent Sources Arrangement on Spatial Interpolation of Impulse Responses

Masashi Yoshino[1]([✉]) and Ken'ichi Furuya[2]

[1] Graduate School of Engineering, Oita University,
700, Dannoharu, Oita-shi, Oita, Japan
v18e3028@oita-u.ac.jp
[2] Faculty of Engineering, Oita University,
700, Dannoharu, Oita-shi, Oita, Japan

Abstract. In this paper, we focus on spatial interpolation of impulse responses. In conventional studies, the amplitude value of the interpolated impulse responses at the initial part is small. To solve this problem, we propose a method of changing the arrangement of equivalent sources used to reconstruct the impulse responses. From the experiment result, we confirmed that the error of the impulse responses at the initial part decreased.

1 Introduction

With the recent development of virtual reality (VR) and the like, much research on acoustic high presence reproduction is underway. To reproduce a high presence sound field, it is necessary to install many speakers and measure many impulse responses as acoustic characteristics of rooms. By analyzing the impulse response, various acoustic characteristics, such as the frequency transfer characteristic and the reverberation curve, can be obtained. However, measuring impulse responses of whole rooms is expensive and time consuming because a large number of microphones have to prepared and installed. To solve this problem, we consider spatial interpolation of impulse responses.

In the conventional study [1], two interpolation methods, namely the plane wave decomposition method (PWDM) and the time-domain equivalent source method (TESM), are used to interpolate the impulse responses. In the two interpolation methods, equivalent sources are arranged around the region to be interpolated, and plane waves or spherical waves are generated from the equivalent sources to reconstruct impulse responses. In solving the inverse problem, regularization is used to stabilize and improve the interpolation accuracy. However, the amplitude value of the interpolated impulse responses become small in the initial part.

© Springer Nature Switzerland AG 2020
L. Barolli et al. (Eds.): CISIS 2019, AISC 993, pp. 733–742, 2020.
https://doi.org/10.1007/978-3-030-22354-0_66

In this paper, we analyze the effects of equivalent sources arrangement and interpolation accuracy by simulation and real environment experiments using two conventional methods.

2   Model and Interpolation of the Impulse Response

In this section, the model equations of two conventional impulse responses interpolation methods are presented.

Then, an interpolation method using the presented model equations is described.

2.1   Plane Wave Decomposition Method

The PWDM can be represented by a finite weighted sum of plane waves, as shown in Fig. 1. The square represents the microphone to be interpolated, the triangles represent the training microphones to interpolate the impulse responses, and the black circle represents the speaker measuring the impulse responses. Plane waves arrive from directions of equivalent sources arranged to surround the microphone array. First, plane waves $\hat{\phi}_{f,l}$ at positions in three-dimensional coordinates x, are defined by the following equation:

$$\hat{\phi}_{f,l}(x) = e^{ik_{f,l}^T x} \tag{1}$$

where $f = 0 \dots N_f - 1$ is the frequency, x is the spatial variable, and $k_{f,l}$ is the wave number representing the direction of the lth plane wave. The symbol $\hat{A}$ indicates that A is a frequency domain representation. The PWDM model equation of impulse responses in the frequency domain is defined by the following equation, using the plane wave of (1):

$$\hat{p}_f(x) = \sum_{l=0}^{N_\omega - 1} \hat{\phi}_{f,l,x} \hat{\omega}_{f,l} \tag{2}$$

where $\hat{p}_f(x)$ is the sound pressure at position x and frequency f, N_ω is the number of equivalent sources, $\hat{\phi}_{f,l,x}$ is the (f,l)th plane wave at position x, and $\hat{\omega}_{f,l}$ is the (unknown) weight of the (f,l)th plane wave. If we want to predict the sound pressure at N_m positions x_m, (2) can be generalized, as follows:

$$\hat{P} = D_{p\omega}(\hat{W}) \tag{3}$$

where $\hat{P}$ is a $N_f \times N_m$ matrix containing the complex sound pressure, and $\hat{W}$ is a $N_f \times N_\omega$ matrix containing the weight. The linear operator $D_{p\omega}$ maps these weights to the complex sound pressures and essentially represents a dictionary of N_ω plane waves with N_f frequencies.

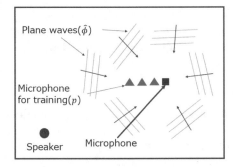

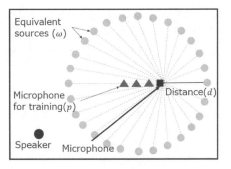

Fig. 1. Plane wave model used for PWDM. (The square represents the microphone to be interpolated, the triangle represent the training microphones to interpolate the impulse responses, and the black circle represents the speaker measuring the impulse responses.)

Fig. 2. Spherical wave model used for TESM. (The green circles represent equivalent sources that generates the spherical waves ϕ.)

2.2 Time-Domain Equivalent Source Method

TESM can be represented by a finite weighted sum of spherical waves from equivalent sources in time domain, as shown in Fig. 2. In this figure green circles represent equivalent sources that generates the spherical waves ϕ.

First, spherical waves ϕ_l at positions in three-dimensional coordinates x and time $t = 0 \dots N_t$ are defined by the following equation:

$$\phi_l(x, t) = \frac{1}{4\pi d_{l,x}} \delta(t - \frac{d_l}{c}) \tag{4}$$

where d is the distance between the sound source and receiver, δ is the Dirac delta function, and c is the sound velocity. The TESM model equation of impulse responses in the time domain using the spherical wave of (4) is the following:

$$p(x, t) = \sum_{l=0}^{N_\omega - 1} \delta(t) * \phi_{l,x}(t) * \omega_l(t) \tag{5}$$

where $p(x, t)$ is the sound pressure at position x and time t, $\phi_{l,x}(t)$ is the spherical wave, and $\omega_l(t)$ is the (unknown) weight of the spherical wave. The symbol $*$ indicates the linear convolution operation. Similar to (3), if we want to predict the sound pressure at N_m positions x_m, (5) can be generalized as follows.

$$P = D_{(s)t}(W) \tag{6}$$

where P is a $N_t \times N_m$ matrix containing the sound pressure, and W is a $N_t \times N_\omega$ matrix containing the weight. The linear operator $D_{(s)t}$ maps these weights to the sound pressures and essentially represents a dictionary of spherical waves.

2.3   Method of Interpolation

Interpolation is performed using the above described impulse response model equations.

First, $D_{p\omega}$ in (3) and $D_{(s)t}$ in (6) can be found from the experimental conditions and the coordinates x of the position we want to interpolate at.

Next, we estimate the weight W of the unknown equivalent sources. The estimation methods are described below.

Finally, by substituting the estimated weights into the two model Eqs. (3) and (6), we interpolate the impulse responses at the position we want to interpolate at.

2.4   Formulation of Inverse Problem

From the measured impulse responses, we estimate the optimal weights of the impulse response model equations. The inverse problem can be formulated as an optimization problem as follows:

$$W^\star = \arg \min_{W} f(W) = \frac{1}{2}\|D(W) - \tilde{P}\|_F^2 \tag{7}$$

where $D(W)$ is the model expression of the impulse response, $\tilde{P}$ are the measured impulse responses, and $\|\cdot\|_F$ is the Frobenius norm. As this optimization problem will in general lead to over-fitting, we consider $l1$-norm regularization and $l2$-norm regularization as follows:

$$W_1^\star = \arg \min_{W} f(W) + \lambda \|vec(W)\|_1 \tag{8}$$

$$W_2^\star = \arg \min_{W} f(W) + \lambda \sum_{l=0}^{N_\omega - 1} \|W_{:,l}\|_2 \tag{9}$$

where λ is a variable that controls the level of regularization and is adjusted using K-fold Cross Validation [1].

2.5   Optimization Algorithm

As an optimization algorithm, the forward-backward splitting (FBS) algorithm will be employed. The FBS algorithm can solve the optimization problem using the following structure:

$$W^\star = \arg \min_{W} f(W) + g(W) \tag{10}$$

where $f(W)$ is the function in (7) and $g(W)$ is a regularization term like $\lambda \|vec(W)\|_1$ in (8). It starts with the first guess W^0 and then iterates the following equation to get the optimal solution.

$$W^{k+1} = T_\gamma(W^k) = prox_{\gamma g}(W^k - \gamma \nabla f(W^k)) \tag{11}$$

where $T_\gamma(\cdot)$ is the forward-backward operator, $\nabla f(\cdot)$ is the Jacobian operator of $f(\cdot)$, and γ is the step-size. $prox_{\gamma g}(\cdot)$ is a proximal mapping of the function $g(\cdot)$ and is defined by the following equation:

$$prox_{\gamma g}(W) = \arg \min_U \frac{1}{2\gamma} \|U - W\|_F^2 + g(U) \tag{12}$$

The acceleration of the FBS algorithm using the quasi-Newton method can be expressed as follows, where the corrective direction S^k was added to accelerate the convergence of the algorithm.

$$W^{k+1} = T_\gamma(W^k) + \tau S^k \tag{13}$$

where τ is the step-size. The corrective direction S^k is calculated using the curvature information of the fixed-point residual R_γ, obtained from the quasi-Newton Limited memory BFGS [2].

$$R_\gamma(W^\star) = W^\star - T_\gamma(W^\star) = 0 \tag{14}$$

The step-sizes γ and τ are selected adaptively with two separate line-search procedures.

3   Proposed Method

In the conventional methods, equivalent sources were arranged to surround the microphone array. However, it is obvious that direct waves (spherical waves) are emitted from the position of the sound source (speaker) to the microphones. The small amplitude value of the impulse responses in the initial part is attributed to the fact that the direct waves cannot be well reproduced from this arrangement.

As shown in Fig. 3, equivalent sources are also arranged around the speaker. By doing so, we will attempt to better reproduce the direct waves from the speaker. Furthermore, at the first guess W^0 of the iterative equation (13), the initial value was set only for equivalent sources around the speaker. This was done to raise the influence of equivalent sources around the speaker, because the number is different between equivalent sources around the speaker and the microphone array. For the same reason, in the iterative equation (13), only the weight W^k of equivalent sources around the speaker are multiplied by the value.

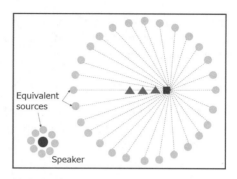

Fig. 3. Arrangement of equivalent sources in the proposed method.

4   Simulation Experiment

In the simulation experiment, we analyzed the influence of interpolation accuracy, given by the arrangement of the equivalent sources. We compared the conventional method, whereby equivalent sources surround the microphone array, and the proposed method, whereby equivalent sources are arranged around the speaker as well as around the microphone. We performed interpolation by applying, respectively, $l1$-norm regularization and $l2$-norm regularization to the two models.

4.1   Simulation Experimental Conditions

Impulse responses were generated by the method of images [3]. A bandpass 2nd order Butterworth filtering process from 20.0 Hz to 3.2 kHz was performed at a sampling frequency of 8 kHz. The reverberant acoustic environment consists of a $7.34 \times 8.09 \times 2.87\,\mathrm{m}^3$ box-shaped room. The speaker was placed at [2.00, 6.50, 1.43] m. The sound velocity was 343 m/s, the reflection coefficient of the wall was 0.55, and the reverberation time was set to $T_{20} = 0.097\,\mathrm{s}$. For the conventional method, 700 equivalent sources were arranged on the Fibonacci lattice [4] with a radius of 2.87 m around the microphone array. For the proposed method, in addition to those of the conventional method, 100 equivalent sources were arranged on the Fibonacci lattice with a radius of 0.5 m around the speaker. The radii and number of equivalent sources arranged around the speaker were determined by preliminary experiments. Three microphone arrays were used, as shown in Fig. 4. In all three experiments, the number of microphones is 21, of which 15 are used as training microphones, while the remaining 6 microphones are used to compute the NMSE. In the orthogonal microphone array, the microphone arrays are orthogonal to each other with the origin at [4.4, 3.1, 1.5] m. The linear microphone array is arranged with the central microphone at [4.6, 3.3, 1.7] m. In the spherical microphone array, microphones were placed at the vertices of a dodecahedron whose center was at [4.4, 3.1, 1.5] m and the radius of the circumscribed sphere was 15 cm. The other one was properly placed on

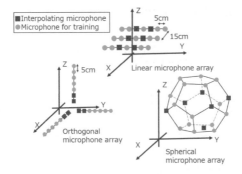

Fig. 4. Microphone arrays of the simulation experiment.

the lower side. As a measure of interpolation accuracy, we used the normalized mean square error (NMSE). The lower is the value of NMSE, the higher is the interpolation accuracy.

4.2   Simulation Experimental Result

Table 1 shows the interpolation NMSE (dB) of each microphone array in the conventional and proposed methods. From Table 1, TESM-L1 has the best interpolation accuracy for all three microphone arrays. It can be seen that the linear microphone array exhibits only a small difference between the two models. As the spacing between microphones is smaller than that of other microphone arrays, the amplitude value at the initial part could be interpolated with good precision. Therefore, arranging equivalent sources around the speaker did not have much influence on the impulse response.

With both the orthogonal and spherical microphone arrays, the interpolation accuracy improved the most with the PWDM-L1. Figure 5 shows the impulse and frequency responses of the microphone with the best accuracy among the PWDM-L1 in orthogonal microphone arrays, and the blue and red lines show the original and interpolated impulse responses, respectively. Figure 6 shows a comparison between the amplitude error in the initial part of the impulse response interpolated by PWDM-L1 in orthogonal microphone arrays of the conventional (blue line) and proposed (red line) methods. In the conventional method, the amplitude value of the interpolated impulse responses at the initial part was small. By arranging equivalent sources around the speaker, the amplitude value at the initial part became large. Thus the interpolation accuracy was improved. In fact, Fig. 6 shows the amplitude error of the proposed method is smaller than that of the conventional method at 0.011 s, reached by the direct wave from the speaker. From these results, we consider this method to be effective when the microphones interval is wide, and the initial part accuracy is low.

Table 1. NMSE (dB) of the simulation experiment.

	Orthogonal		Linear		Spherical	
	Conventional	Proposed	Conventional	Proposed	Conventional	Proposed
PWDM-L1	−5.26	−6.72	−9.51	−9.55	−4.98	−5.66
PWDM-L2	−7.05	−7.65	−12.24	−12.27	−7.22	−7.34
TESM-L1	−13.72	−13.86	−18.55	−18.52	−15.03	−15.59
TESM-L2	−7.68	−7.99	−12.08	−12.44	−7.38	−7.67

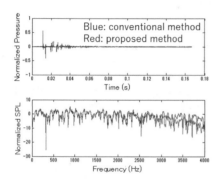

Fig. 5. Comparison of impulse and frequency responses of the simulation experiment.

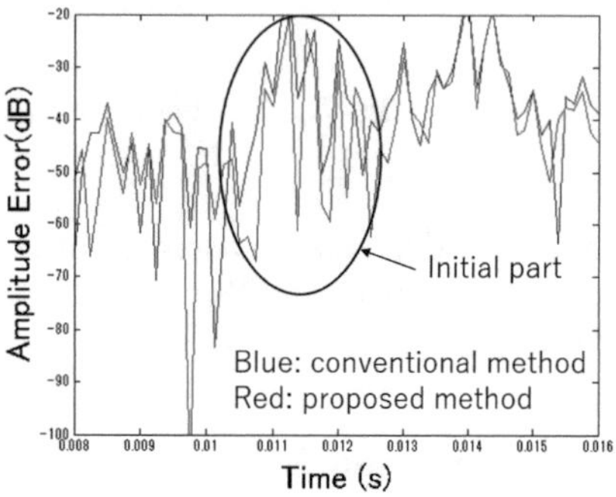

Fig. 6. Comparison of the amplitude error of the impulse response at the initial part of the simulation experiment.

5   Real Environment Experiment

In the real environment experiment, we analyzed the influence of the interpolation accuracy, given by the arrangement of equivalent sources. Similar to the simulation experiments, we performed the interpolation by respectively applying the $l1$ and $l2$-norm regularizations to the two models. We then, compared the conventional the proposed methods.

5.1   Real Environment Experimental Conditions

For the measurement signal of the impulse responses, the configuration 1002 of single-and multichannel audio recordings database (SMARD) [5] was used. The sampling frequency, room dimensions, arrangement of speaker, and equivalent sources are the same as in the simulation experiment. The microphone array used in the real environment experiment was only the orthogonal microphone array.

Table 2. NMSE (dB) of the real environment experiment.

	Conventional	Proposed
PWDM-L1	−5.47	−5.89
PWDM-L2	−7.26	−7.46
TESM-L1	−8.42	−9.19
TESM-L2	−7.66	−7.92

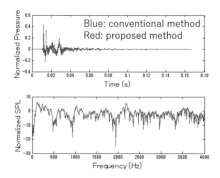

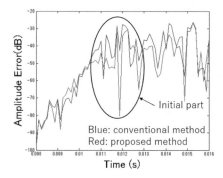

Fig. 7. Comparison of impulse and frequency responses of the real environment experiment.

Fig. 8. Comparison of the amplitude error of the impulse response at the initial part of the real environment experiment.

5.2 Real Environment Experimental Result

Table 2 shows the interpolation NMSE (dB) of the conventional and the proposed methods. It can be seen from Table 2, that TESM-L1 has the highest accuracy improvement, while PWDM experienced only a slight accuracy improvement with two regularizations. As equivalent sources of the PWDM represents the direction of the plane waves, the equivalent sources around the speaker overlaps with the equivalent sources arranged around the microphone array. Therefore, the impulse responses were not significantly influence.

In both models, the l1-norm regularization improved the accuracy more than the l2-norm regularization. The l2-norm tends to bring weights closer to zero, and, unlike the l1-norm, it does not eliminate unnecessary weights by setting them to zero. Thus, the l2-norm does not exclude weights of equivalent sources surrounding the microphone array in the initial part of impulse responses, which are unnecessary in generating direct waves from the speaker. Therefore, equivalent sources around the speaker did not have much influence on the direct wave generation. In fact, with TESM-L1, the ratio between the nonzero number of elements of the weight matrix W of the equivalent sources surrounding the microphone array to the ones around the speaker is 4 : 1, while, with TESM-L2, it is 9 : 1.

Figure 7 shows the impulse and frequency responses of the microphone with the best accuracy among the TESM-L1 interpolated microphones. The blue and red lines show the original and interpolated impulse responses, respectively. From the frequency response in Fig. 7, the low frequency band can be interpolated more accurately than the high frequency band. Figure 8 shows a comparison between the amplitude error in the initial part of the impulse response interpolated by TESM-L1 of the conventional (blue line) and proposed (red line) methods. As shown in Fig. 8, the amplitude error of the proposed method is smaller than that of the conventional method at 0.012 s, reached by the direct wave from the speaker. From this, it was possible to interpolate the direct wave from the speaker accurately by arranging equivalent sources around the speaker.

6 Summary

In this paper, equivalent sources are arranged not only around the microphone array but also around the speaker, which is the actual sound source, and the influence of the arrangement is analyzed. No significant improvement in interpolation accuracy was observed, but the amplitude value at the initial part could be interpolated with good accuracy. From regularization, this method also confirmed that the l1-norm performs more accurately than the l2-norm.

Acknowledgements. This work was supported by JSPS KAKENHI grant number 16K00245, 19K12044.

References

1. Antonello, N., et al.: Room impulse response interpolation using a sparse spatio-temporal representation of the sound field. IEEE/ACM Trans. Audio, Speech, and Language Processing **25**(10), 1929–1941 (2017)
2. Stella, L., et al.: Forward-backward quasi-Newton methods for nonsmooth optimization problems. Comput. Optim. Appl. **67**(3), 443–487 (2017)
3. Allen, J.B., et al.: Image method for efficiently simulating small-room acoustics. J. Acoust. Soc. Am. **65**(4), 943–950 (1979)
4. Gonzalez, A.: Measurement of areas on a sphere using fibonacci and latitude-longitude lattices. Math. Geosci. **42**(1), 49–64 (2010)
5. Nielsen, J.K., et al.: The single- and multichannel audio recordings database (SMARD). In: Proceedings of International Workshop Acoustic Signal Enhancement, pp. 40–44 (2014). http://www.smard.es.aau.dk/

Automatic Determination of the Optimum Number of Updates in Synchronized Joint Diagonalization

Taiki Izumi[1]([✉]), Yuuki Tachioka[2], Shingo Uenohara[3], and Ken'ichi Furuya[3]

[1] Graduate School of Engineering, Oita University,
700, Dannoharu, Oita-shi, Oita, Japan
v18e3001@oita-u.ac.jp
[2] Denso IT Laboratory, Shibuya cross tower 28F, Shibuya-ku, Tokyo, Japan
[3] Faculty of Engineering, Oita University, 700, Dannoharu, Oita-shi, Oita, Japan

Abstract. This study focuses on Synchronized Joint Diagonalization (SJD), which is a newly proposed sound source separation (BSS) method. SJD performs iterative updates of parameters for source separation. For its practical use, it is necessary to determine the optimum number of the iterations. We proposed to optimize it by observing the differences of the estimated activation matrix before and after updates during each iteration. We confirmed the effectiveness of this approach by BSS experiments.

1 Introduction

Currently, various blind sound source separation (BSS) methods have been proposed. There are various types, e.g., Independent Component Analysis (ICA) [1], Independent Vector Analysis (IVA) [2], non-negative matrix factorization (NMF) [3], Multichannel Non-negative Matrix Factorization (MNMF) [4], and Independent Low-Rank Matrix Analysis (ILRMA) [5]. ICA performs separation by assuming that each sound source is independent from each other. IVA uses the non-Gaussian nature of sound source signals. NMF performs separation for each commonly occurring frequency feature. MNMF, which is a multi-channel extension of NMF, performs high performance separation by using spatial information in addition to frequency information. ILRMA combines IVA and MNMF. A newly proposed BSS method that uses the non-stationarity of the sound source signal, synchronized joint diagonalization (SJD) [6], has been proposed. SJD simultaneously diagonalizes the correlation matrix among several time sections after dividing each utterance into several time sections.

SJD is similar to IVA. These BSS techniques above achieve source separation by an iterative algorithm. However, because SJD is a new BSS method, it is under how many times this iterative algorithm needs to obtain its good values.

In this paper, we proposed to optimize the number of iterations by observing the differences of the estimated activation matrix before and after updates at each iteration. We confirmed the effectiveness of this approach by BSS experiments.

L. Barolli et al. (Eds.): CISIS 2019, AISC 993, pp. 743–751, 2020.
https://doi.org/10.1007/978-3-030-22354-0_67

2 Blind Sound Source Separation (BSS) by SJD

2.1 Overview

Joint Diagonalization (JD) [7] of a correlation matrix among several time frames has been proposed as a BSS method utilizing non-stationarity of signals. SJD is a BSS method that is temporally synchronized the diagonal components corresponding to the same signal source while JD is used to solve multiple simultaneous diagonalization problems. Figure 1 shows the principle of BSS algorithm by SJD.

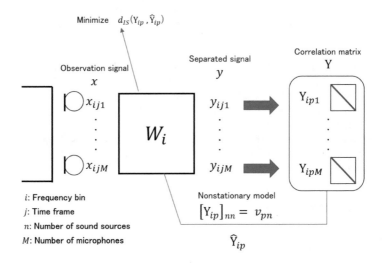

Fig. 1. BSS algorithm by SJD.

2.2 Formulation of SJD

Short-time Fourier transforms were performed on the observation signals from each microphone channel $m = 1, \cdots, M$ to obtain a time-frequency representation x_{ijm}. where $i = 1, \cdots, I$ represents a frequency bin and $j = 1, \cdots, J$ represents a time frame. Assuming that the observed signal is a linear mixture of independent sound source signals $s_{ijn} = [\boldsymbol{s}_{ij}]_n, n = 1, \cdots, N$, an $M \times N$ mixing matrix $\boldsymbol{A}_i$ can be defined, and the observation signal can be defined by the following equation.

$$\boldsymbol{x}_{ij} = \boldsymbol{A}_i \boldsymbol{s}_{ij}. \tag{1}$$

The purpose of the BSS is to obtain an $N \times M$ separation matrix $\boldsymbol{W}_i$ for each frequency bin $i = 1, \cdots, I$ only from the observation signal and to estimate the separated signal $y_{ijn} = [\boldsymbol{y}_{ij}]_n, n = 1, \cdots, N$ by the following equation.

$$\boldsymbol{y}_{ij} = \boldsymbol{W}_i \boldsymbol{x}_{ij}. \tag{2}$$

2.3 Joint Diagonalization (JD)

The simultaneous diagonalization of the correlation matrix is performed for each frequency bin i. J time frames are divided into P time sections $\mathscr{J}_p(p = 1, \cdots, P)$ and the correlation matrix $\boldsymbol{X}_{ip}$ of the observation signal and the correlation matrix $\boldsymbol{Y}_{ip}$ of the separation signal are obtained by the following Eq. (4) in each time section p.

$$\boldsymbol{X}_{ip} = \frac{1}{P} \sum_{j \in \mathscr{J}_p} \boldsymbol{x}_{ij} \boldsymbol{x}_{ij}^H, \tag{3}$$

$$\boldsymbol{Y}_{ip} = \frac{1}{P} \sum_{j \in \mathscr{J}_p} \boldsymbol{y}_{ij} \boldsymbol{y}_{ij}^H = \boldsymbol{W}_i \boldsymbol{X}_{ip} \boldsymbol{W}_i^H, \tag{4}$$

where H is the Hermitian transpose. Then, a separating matrix $\boldsymbol{W}_i$ is obtained for simultaneously diagonalizing the P correlation matrices $\boldsymbol{Y}_{ip}$ of the separated signal. In the case of $P = 2$, it can be strictly diagonalized, but when $P \geq 3$, it is generally impossible to obtain an exact solution.

2.4 Synchronized Joint Diagonalization (SJD)

To model the non-stationarity of the signal source, the diagonal matrix $\hat{\boldsymbol{Y}}_{ip}$ is defined by the following equation.

$$[\hat{\boldsymbol{Y}}_{ip}]_{nn'} = \begin{cases} v_{pn} & \text{if } n = n' \\ 0 & \text{if } n \neq n', \end{cases} \tag{5}$$

SJD minimizes multichannel Itakura-Saito divergence between $\boldsymbol{Y}_{ip}$ and $\hat{\boldsymbol{Y}}_{ip}$.

$$d_{IS}(\boldsymbol{Y}_{ip}, \hat{\boldsymbol{Y}}_{ip}) = \text{tr}(\boldsymbol{Y}_{ip} \hat{\boldsymbol{Y}}_{ip}^{-1}) - \log \left[\det \boldsymbol{Y}_{ip} \hat{\boldsymbol{Y}}_{ip}^{-1} \right] - N. \tag{6}$$

where the right hand side depends only on the time section p and the signal n, and does not depend on the frequency i. Let $\boldsymbol{V}$ be a matrix of size $P \times N$ and define it as $v_{pn} = [\boldsymbol{V}]_{pn}$ for formulation. SJD works by minimizing a cost function C that is defined as the sum total of all the following frequencies.

$$C = \sum_{i=1}^{I} \sum_{p=1}^{P} \left[\sum_{n=1}^{N} \left(\frac{[\boldsymbol{Y}_{ip}]_{nn}}{v_{pn}} + \log v_{pn} \right) - 2 \log |\det \boldsymbol{W}_i| \right]. \tag{7}$$

2.5 BSS Algorithm

The separating matrix $\boldsymbol{W}$ and the activation matrix $\boldsymbol{V}$ are alternately updated. By differentiating Eq. (7) with respect to v_{pn} and setting it to zero, the following updating Eq. (8) concerning the activation matrix $\boldsymbol{V}$ is derived.

$$v_{pn} = \frac{1}{I} \sum_{i=1}^{I} [\boldsymbol{Y}_{ip}]_{nn}. \tag{8}$$

Next, the separation matrix $\boldsymbol{W}_i$ for each frequency is updated by the following procedure.

$$U_{in} = \frac{1}{P} \sum_{p=1}^{P} \frac{1}{v_{pn}} \boldsymbol{X}_{ip}. \tag{9}$$

From Eq. (9), $\boldsymbol{W}_i$ is updated as a matrix for hybrid simultaneous diagonalization of N matrices $\boldsymbol{U}_{in}$ as follows.

$$\boldsymbol{w}_{in} = (\boldsymbol{W}_i\boldsymbol{U}_{in})^{-1} \boldsymbol{e}^n, \tag{10}$$

where $\boldsymbol{e}^n$ is an N dimensional vector in which only the nth row is unity. Then, normalization is performed using the following equation.

$$\boldsymbol{w}_{in} \leftarrow \frac{\boldsymbol{w}_{in}}{\sqrt{\boldsymbol{w}_{in}^{H}\boldsymbol{U}_{in}\boldsymbol{w}_{in}}}, \tag{11}$$

3   Proposed Method

3.1   Relationship Between Number of Updates and SDR

Here, we investigate the relationship between the number of updates and separation performance. The music data listed in Table 1 was used for the experiment. Separation performance was evaluated in terms of SDR the experimental result is shown in Fig. 2. It shows the number of updates and the result of the average signal-to-distortion ratio (SDR). It can be confirmed that separation performance saturated until 30 updates. However, the number of updates must be fixed before processing when it is unknown how many updates are required for each sound source.

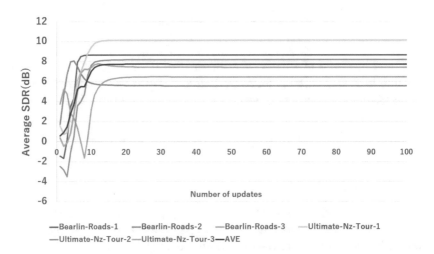

Fig. 2. Separation performance at each number of updates (P = 404).

3.2   Viewpoints

Figure 3 shows the difference between SDR and V at each number of updates from one to 30. As a result, it was confirmed that there was no the difference in V after updating about 10 times. Additionally, it was confirmed that the difference between the activation matrices V and the SDR had a negative correlation. Based on this fact, it is possible to reduce useless updates by terminating the updates at the time when the activation matrix V no longer changes. It can be possible to optimize the number of updates for each sound source.

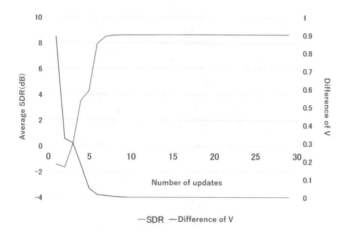

Fig. 3. Relationship between difference in V and SDR.

3.3   Optimization of Number of Updates Using Activation Matrix V

To optimize the number of updates, the difference between activation matrices V before and after update is used. Let V^q be the qth iteration in the following equations.

$$\left|V^{q-1} - V^q\right| < \varepsilon\left|V^1 - V^2\right|, \tag{12}$$

$$\left|\frac{V^{q-1}}{max(V^{q-1})} - \frac{V^q}{max(V^q)}\right| < \varepsilon\left|\frac{V^1}{max(V^1)} - \frac{V^2}{max(V^2)}\right|, \tag{13}$$

$$\left|\frac{V^{q-1}}{norm(V^{q-1})} - \frac{V^q}{norm(V^q)}\right| < \varepsilon\left|\frac{V^1}{norm(V^1)} - \frac{V^2}{norm(V^2)}\right|. \tag{14}$$

When the difference becomes less than ε of the difference between the first time and the second time, the iteration ends.

As a result of the preliminary experiment, we hardly observed any difference between these Eqs. (12), (13) and (14) criteria. In this experiment, the difference is judged by using the Eq. (12) formula.

4   BSS Experiment

4.1   Experimental Conditions

The mixed signals were music data and were composed of two sound sources. These signals were observed by two microphones ($M = 2$). This is shown in Table 1. I used the data prepared in the database [9]. In Fig. 4, the microphones are sequentially numbered from 1 to 14 from the right. The microphone numbers

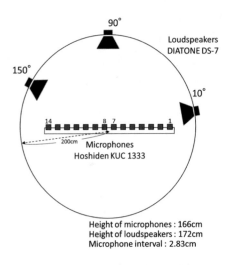

Fig. 4. Arrangement of sound sources and microphones.

Table 1. Music data composed of two parts.

ID	Author/Song	Part
B-1	Bearlin-Roads-1	Piano(90°)
		Vocal(10°)
B-2	Bearlin-Roads-2	Vocal(10°)
		Ambient(150°)
B-3	Bearlin-Roads-3	Piano(90°)
		Ambient(150°)
U-1	Ultimate-Nz-Tour-1	Guitar(90°)
		Synth(10°)
U-2	Ultimate-Nz-Tour-2	Synth(10°)
		Drum(150°)
U-3	Ultimate-Nz-Tour-3	Guitar(90°)
		Drum(150°)

Table 2. Parameters of SJD.

Reverberation time	300 ms
Sampling frequency	16 kHz
Frame size	1024
Shift size	256
Number of sources	2
Number of microphones	2
Number of time sections P	404202

used in this experiment are 6 and 8. The parameters of SJD are listed in Table 2. The number of time sections P was set to the maximum $P = 404$ and 202. The program was implemented in MATLAB and executed using an Intel Core i7-4770 3.4 GHz processor. The calculation time was evaluated using the real time factor (RTF) representing the ratio of the processing time to the length of the sound source, and the separation performance was evaluated with the SDR [10]. In Eq. (12), ε was set to 0.001.

4.2   Experimental Results

Tables 3 and 4 show the number of updates optimized by the proposed method for each piece of music with the number of time sections P of 404 and 202, SDR, and RTF, accordingly. The average SDR in the tables shows the average of the six SDR from B-1 to B-3 and from U-1 to U-3 at each number of updates. From these two tables, it can be confirmed that the performance of the proposed method using the optimized number of updates is equivalent to that of the conventional method using about 30 updates. If the number of updates was set to be 10, the performance was insufficient. These show that number of updates can be optimized by the proposed method using the difference of the activation

Table 3. Experimental results in case of $P = 404$ (Bold is the number of updates by the proposed method).

ID	Number of updates	SDR	RTF	Average SDR
B-1	30 times	8.64 dB	0.73	
	10 times	8.64 dB	0.092	
	12 times	8.64 dB	0.10	30 times
B-2	30 times	5.57 dB	0.73	7.73 dB
	10 times	5.86 dB	0.092	
	13 times	5.69 dB	0.11	
B-3	30 times	6.45 dB	0.73	
	10 times	3.22 dB	0.092	
	18 times	6.36 dB	0.14	10 times
U-1	30 times	10.12 dB	0.73	7.02 dB
	10 times	9.41 dB	0.092	
	16 times	10.11 dB	0.13	
U-2	30 times	8.19 dB	0.73	
	10 times	7.56 dB	0.092	
	15 times	8.11 dB	0.12	Proposed method
U-3	30 times	7.42 dB	0.73	7.75 dB
	10 times	7.42 dB	0.092	
	16 times	7.56 dB	0.13	

Table 4. Experimental results in case of $P = 202$ (Bold is the number of updates by the proposed method).

ID	Number of updates	SDR	RTF	Average SDR
B-1	30 times	8.65 dB	0.17	
	10 times	8.65 dB	0.063	
	12 times	8.65 dB	0.075	30 times
B-2	30 times	5.71 dB	0.17	7.67 dB
	10 times	6.02 dB	0.063	
	13 times	5.83 dB	0.08	
B-3	30 times	6.60 dB	0.17	
	10 times	5.40 dB	0.063	
	16 times	6.45 dB	0.091	10 times
U-1	30 times	10.17 dB	0.17	7.36 dB
	10 times	9.48 dB	0.063	
	17 times	10.16 dB	0.097	
U-2	30 times	7.92 dB	0.17	
	10 times	7.45 dB	0.063	
	16 times	7.95 dB	0.094	Proposed method
U-3	30 times	6.97 dB	0.17	7.69 dB
	10 times	7.18 dB	0.063	
	17 times	7.10 dB	0.1	

matrix V. In addition, RTF of the proposed methods was smaller than when 30 updates. Therefore, useless updates were avoided by the proposed method, and as a result, RTF was improved.

5 Summary

In this study, we focused on SJD, which is a new BSS method. We experimentally confirmed that there is a negative correlation between the difference in the activation matrix V before an after update and the SDR along with the iterations. Then, we proposed a method to optimize the number of updates by observating the difference of the activation matrix V. The proposed method can avoid unnecessary updating with the optimal number of updates for each song and reduce the processing time. In the future, it is necessary to confirm the effectiveness of the proposed method for other types of sound source, e.g., speech.

References

1. Lee, T.-W.: Independent Component Analysis-Theory and Applications. Kluwer, Norwell (1998)
2. Lee, I., et al.: Fast fixedpoint independent vector analysis algorithms for convolutive blind source separation. Sig. Process. **87**(8), 1859–1871 (2007)
3. Lee, D.D., et al.: Learning the parts of objects with nonnegative matrix factorization. Nature **401**, 788–791 (1999)
4. Sawada, H., et al.: Multichannel extensions of non-negative matrix factorization with complex-valued data. IEEE Trans. ASLP **21**(5), 971–982 (2013)
5. Kitamura, D., et al.: Dtermined blind source separation unifying independent vector analysis and nennegative matrix factorization. IEEE/ACM Trans. Audio Speech Lang. Process. **24**(9), 1626–1641 (2016)
6. Sawada, H.: Blind signal separation by synchronized joint diagonalization. In: 32nd SIP Symposium, pp. 332–337 (2017)
7. Ziehe, A., et al.: A fast algorithm for joint diagonalization with non-orthogonal transformations and its application to blind source separation. J. Mach. Learn. Res. **5**, 777–800 (2004)
8. Araki, S., et al.: The 2011 signal separation evaluation campaign (SiSEC2011): -audio source separation. In: Latent Variable Analysis and Signal Separation, pp. 414–422. Springer, Berlin (2012)
9. RWCP: Sound Scene Database in Real Acoustic Enviroment (RWCP-SSD), Speech Resources Consortium. http://research.nii.ac.jp/src/RWCP-SSD.html. Accessed 21 Aug 2018
10. Vincent, E., et al.: First stereo audio source separation evaluation campaign: data algorithm and results. In: Independent Component Analysis and Signal Separation, pp. 552–559. Springer, Berlin (2007)

Practice of Programming Education Using Scratch and NekoBoard2 for High School Student

Kazuaki Yoshihara[✉] and Kenzi Watanabe

Graduate School of Education, Hiroshima University, 1-1-1 Kagamiyama, Higashihiroshima, Hiroshima 739-8524, Japan
{dl73863,wtnbk}@hiroshima-u.ac.jp

Abstract. We have had practice of programming education for high school students using scratch, which is a visual programming environment, and NekoBoard2, which is a microcomputer board with multiple sensors. First, the students learned basics of programming and how to treat values that sensors measured and how to use mesh function in Scratch. Then, they produced original cooperative or competition games while giving ideas each other. Finally, they introduced their own game to everyone.

1 Introduction

The Japanese government announced a new course of study in 2017 and a Course of Study for high school in 2018 [1, 2]. According to the Course of Study, for technology education in junior high school, new learning content which is programming of interactive content using network is added to current programming by measurement control. According to the Course of Study for information education in senior high school, students have to learn programming. These new learning contents are difficult for beginners in programming such as junior and high school students.

In this paper, we have had practice of programming education for high school students using Scratch, which is a visual programming environment, and NekoBoard2, which is a microcomputer board with multiple sensors. Scratch has a mesh function to create interactive content. Students created cooperative or competition game as interactive contents using network by Scratch's mesh function.

2 Scratch and NekoBoard2

We used Scratch as programming environment and NekoBoard2 as microcomputer of measurement and control in this practice.

2.1 Scratch

Scratch is a programming environment developed by MIT media Lab [3]. Figure 1 shows the Scratch. By using Scratch, we can easily make a program by combining prepared blocks.

© Springer Nature Switzerland AG 2020
L. Barolli et al. (Eds.): CISIS 2019, AISC 993, pp. 752–756, 2020.
https://doi.org/10.1007/978-3-030-22354-0_68

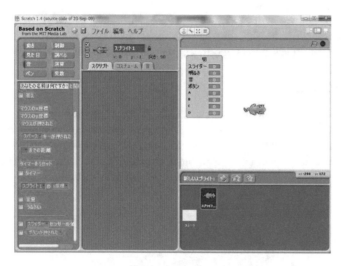

Fig. 1. Scratch

In this practice, we used mesh function in Scratch 1.4. Mesh is a method of having multiple Scratch projects interact, even if they are on different computers. We can use mesh by editing of Scratch's System Browser. Figure 2 shows a Scratch's System Browser.

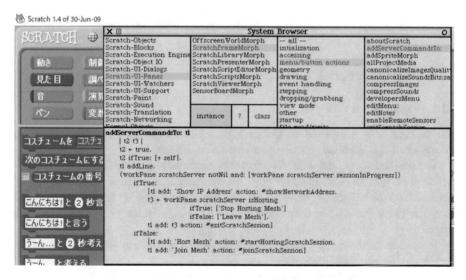

Fig. 2. Scratch's system browser

A Scratch has mesh available looks like an un-edited Scratch. However, Shift-clicking" Share" will reveal two new buttons: "Host Mesh" and "Join Mesh". "Host Mesh" begins a mesh session. "Join Mesh" joins a mesh session. The hosting computer's IP address will be shown when host user selects "Host Mesh". When the join user selects "Join Mesh", they enter in the IP address of the hosting computer that is hosting the mesh session. A mesh session can contain unlimited Scratch programs.

2.2   NekoBoard2

NekoBoard2 is a sensor board enable to physical operation with Scratch [4]. NekoBoard2 can read the value of the brightness sensor, a volume sensor, a slider sensor and a button sensor. In addition, NekoBoard2 can expand that four terminals to measuring resistance level connect to other sensors. Figure 3 shows NekoBoard2.

Fig. 3.   NekoBoard2

3   Teaching Practice

3.1   Practice Overview

Our practice was conducted at Eishin high school, and twice for three hours on December, 2018. Applied students participated in the practice.

The practice was divided into two parts. Students pair in pairs. One is host side and the other is join side. In the first half, the students learned how to use the values measured by NekoBoard2. Then, The join users input the IP address of the host computer to use mesh. And they learned how to share variables by writing the sample program.

In the second half, they created cooperative or competition games as interactive contents using network in pairs.

Finally, they published each completed games to all, and played the games freely which other students created.

3.2 Sample Game Program

The sample game program in Scratch included sequential processing, conditional branch processing, and loop processing. The Scratch programs are two programs using NekoBoard2 and mesh. One is a program for host and the other is a program for join. The game is a battle game which the characters hit back reflected ball. The characters move in conjunction with a NekoBoard2 slider. Coordinates of sprites are shared with each other using mesh function. The Students wrote the sample programs that shared coordinate values of own character and ball by mesh. Figure 4 shows Sample Game Program.

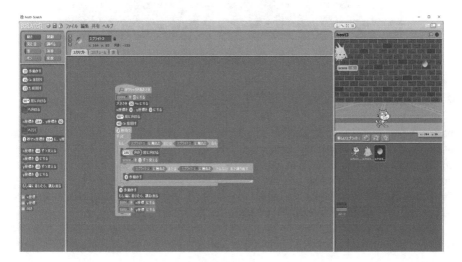

Fig. 4. Sample game program

3.3 Create Original Game

The students created games at the second half. They wrote the original programs by trial and error. They brought ideas while having a talk with the partner of the pair. They were considering what kind of information they should send to do the ideal action. Some students created game such as a cooperation shooting game, which unlike the sample game. Other students created game which added a new element to the sample game.

Student's comments after the practice show that "I hardly knew about sensor control and programming, but the explanation was easy to understand so that I was able to create a program while having fun." and "It was good to be able to create a program in cooperation with the partner of the pair". Figure 5 shows the state of the practice.

Fig. 5. The state of the practice

4   Conclusions

In this paper, we have had practice of programming education through the production of interactive programs using Scratch and Scratch's mesh function as a programming environment and NekoBoard2 as a sensor board. Although it is short time for six hours, the students repeated logical thinking through programming eagerly while creating of the game. In addition, some students were able to feel the goodness of making collaboratively by creating interactive content.

However, some students only modified the sample game program. Thus, it is necessary to improve this practice plan so that students can produce programs creatively.

Acknowledgements. We express our sincerely thanks to Mr. Yokota and the Eishin Gakuen giving us opportunities of the practice.

References

1. The MEXT, Junior High School Teaching Guide for the Japanese Course of Study. Technology and Home Economics. http://www.mext.go.jp/component/a_menu/education/micro_detail/__icsFiles/afieldfile/2018/05/07/1384661_5_4.pdf. Accessed 20 Apr 2019 (in Japanese)
2. The MEXT, Senior High School Teaching Guide for the Japanese Course of Study. Information. http://www.mext.go.jp/component/a_menu/education/micro_detail/__icsFiles/afieldfile/2018/07/11/1384661_6_1_2.pdf. Accessed 20 Apr 2019 (in Japanese)
3. MIT media Lab, scratch-imagine, program, share. https://scratch.mit.edu/. Accessed 20 Apr 2019
4. Switchscience: Nekoboard2. https://www.switch-scienece.com/catlog/2700/. Accessed 20 Apr 2019 (in Japanese)

IntelligentBox for Web: A Constructive Visual Development System for Interactive Web 3D Graphics Applications

Kohei Noguchi[1] and Yoshihiro Okada[1,2(✉)]

[1] Graduate School of Information Science and Electrical Engineering,
Kyushu University, Fukuoka, Japan
okada@inf.kyushu-u.ac.jp
[2] Innovation Center for Educational Resources (ICER),
Kyushu University Library, Kyushu University, Fukuoka, Japan

Abstract. This paper proposes Web version of IntelligentBox called IntelligentBox for Web. Originally, IntelligentBox is "a Constructive Visual Software Development System for Interactive 3D Graphic Applications", invented by Okada, el al. It was made for helping non-skilled users to develop interactive 3D graphics applications, particularly to create educational materials. IntelligentBox for Web provides equivalent functionalities of original IntelligentBox with much affordability for developing Web applications besides standard desktop applications. In IntelligentBox for Web, the authors used Reactive Programming for message passing of the communication among software components, and WebGL for displaying and interacting with 3D objects on a web-browser. Using Web technologies, users will not be bothered setting a development environment up when installing 3D graphics applications, and instead will only have to access a corresponding web page on a web-browser.

1 Introduction

3D technology has been through such a progress, that its capability is utilized in many different realms, and various 3D graphics applications have been emerged. Previously, to develop such applications, they needed dedicated programmers, and a complicated development environment. For its sheer cost, it was difficult to install applications for rather unrelated, non-technical realms of industry, for example, educations.

To solve this issue, Okada et al. invented IntelligentBox [1] that is a development system for 3D interactive contents, enabling intuitive development of interactive application for non-skilled users. With this, school teachers can make 3D educational materials to teach what was once difficult to learn using traditional paper textbooks, such as the structure of car shaft, or an internal combustion engine. For example, Fig. 1 shows how IntelligentBox works on a Windows system. It represents a car shaft model, and a user can interact with these parts like doors, a steering handle and tires.

There are, however, several problems with this development software. As described in [1], IntelligentBox is built onto an MVC framework [2], therefore programmers can

© Springer Nature Switzerland AG 2020
L. Barolli et al. (Eds.): CISIS 2019, AISC 993, pp. 757–767, 2020.
https://doi.org/10.1007/978-3-030-22354-0_69

easily add new Boxes or extend previous Boxes. However, nowadays the internet and Web technology have been progressed, and platform for the development of interactive contents have shifted from standalones to Web applications.

Fig. 1. A screen image of one of the applications developed using IntelligentBox.

In this paper, we propose new version of IntelligentBox as a Web application, and justify that this new version is much simpler and affordable. It can provide equivalent functionalities of the original IntelligentBox. The remarkable point of implementing as a Web application is that it is much easy to use compared to a standalone desktop application. Standalone desktop applications require installation, with many other requirements, to execute in a single computer. Users have to repeat this tedious process in each individual computer, which is troublesome. Web applications, on the other hand, require nothing but a single Web browser, and users are ready to go. Another remarkable point is that it can utilize latest technologies using Web or the Internet easily. Such technologies include WebVR, P2P, Blockchain, and many more Java-Script libraries. These capabilities extend IntelligentBox in various ways, allowing it to develop much diverse applications.

The remainder of this paper is organized as follows: we introduce related works in Sect. 2. Then we provide a detailed explanation of IntelligentBox in Sect. 3. Section 5 describes the actual implementation of IntelligentBox for Web. After that, we show its demonstration in Sect. 5. Finally, we conclude the paper and describe future works in Sect. 6.

2   Related Works

In 1980s, by the emerge of more powerful CPUs rather than ever and GPUs, 3D graphics and more rich graphical experiences became to be obtainable with much cheaper cost. However, the mechanism of 3D drawing was quite complicated, and at the time when 3D was new, developing 3D graphics applications was still difficult. At such time, OpenGL API [3] was released by Silicon Graphics (SGI), and provided abstract layer of 3D programming. Its simple use of 3D drawing API reduced developer's cost of understanding complex hardware-level programming, and therefore made 3D programming much easier.

The problem was, although 3D became much easier with OpenGL, their instruction set was quite abstract, and it required many instructions to implement rather simple tasks, such as "draw a cube". IRIS Inventor [4] was then developed to help these tasks, and provided simpler ways of displaying and organizing 3D models. IRIS Inventor, its current version is called OpenInventor, is a similar product compared to IntelligentBox. It has various 3D components, and each has a functionality and an actual 3D shape. Developers can combine those components to create 3D scenes, and users can directly manipulate those objects on the screen with mouse or keyboard, and their predefined mutual relationships invoke objects to act together.

The key difference between the two is that IRIS Inventor is a software development kit (SDK), and the individual component and its relationships have to be programmed beforehand in IRIS Inventor, whereas IntelligentBox does not need any programming. In IntelligentBox users can dynamically change its behavior and structure on-the-fly enabling much easier development of complicated 3D objects.

The other related works are Virtual Reality construction toolkit systems including VREAM [5], SUPERSCAPE VRT [5], REND386 [5], MR Toolkit [6], MER [7, 8], and so on. Most of them provide a script language that allows developers to define the behavior of each object existing in a 3D virtual world. Some of them also provide facilities that enable developers to build distributed 3D virtual worlds. However, these facilities are low-level functionalities so that developers have to make programs when building distributed applications. As well, there are some distributed Virtual Reality systems such as DIVE [9, 10], MASSIVE [11], and VLNET [12]. Although these are very powerful systems, using their essential mechanisms to construct distributed 3D graphics applications is not easy. RoomBox [13] concept of IntelligentBox [1] is expected to have an approach similar to Division's dVS/dVISE system [14], whose agent process migrates the objects-database changes to remote computers through P2P connection. However, dVS/dVISE system's agent is neither a visible nor manually operable object on the computer screen.

For the development of 3D graphics games, there are several popular game engines, e.g., Unreal Engine [15] and Unity [16]. These game engines have very powerful functionalities, however, they request developers to make any text-based programs.

3   IntelligentBox

IntelligentBox is an interactive 3D graphics software development system developed by Okada et al. It enables untrained users to assemble complicated interactive 3D graphics applications using many building blocks and simple manipulations. Those building blocks are called "Boxes", and each Box has its own shape and functionality, providing some interactivity. Their function varies from some basic ones like buttons, knobs or sliders, to complex ones like for a physical simulation or network communications. Boxes can be combined into a composite Box, with some capabilities of communicating their information among the Boxes. This allows Boxes to synchronize their states and work in synergy.

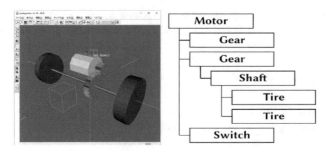

Fig. 2. An example of IntelligentBox (left) and its internal structure (right).

For example, the left part and right part of Fig. 2 show a screen image of an simple composite box in IntelligentBox and its internal structure, respectively. This example represents a typical gear box system in a toy car, and tells how kinetic energy transfers through gears and shafts. The motor is a timer, counting a value up if the activation switch is on. Each two gears are connected to the motor, and they rotate at different rotation ratios as motor counts. The shaft and tires are connected to the lower gear, and they follow its movement. These rotational components also accept dragging with a mouse device and can be rotated manually.

3D graphics applications developed using IntelligentBox are treated as composite Boxes, and each Box has its own functionality. Some Boxes act like normal GUIs such as buttons, knobs or sliders, and other Boxes provide animations according to user's interactive input, i.e., direct manipulation. Developers combine those Boxes, and make complicated Boxes treated as 3D graphics applications. IntelligentBox has four core mechanisms. These are MVC model, a parent-child relationship, a message passing protocol, and Model sharing.

3.1   MVC Model

MVC model is one of the design patterns we use today in regards of programming as shown in the left part of Fig. 3. This MVC model breaks the program code of an instance such as GUI component down into its three parts, which are Model, View and Controller. Model represents internal state of the instance. When someone changes the internal state of the instance by manipulating its Model, the Model notify the change to the other two components of Controller and View. View has the appearance of the instance and it takes a role of an output to the user through displaying the state of Model by visual representations. Controller receives and interprets an input from the user to change the state of Model by interactive operations onto the instance. In this way, each component is semantically distinguished, making them intuitive and easy to understand. Also, they are sparsely connected, code changes do not affect the other components, therefore easy to maintain.

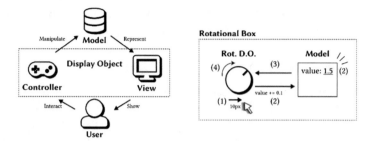

Fig. 3. Structure of MVC (MD) pattern (left) and message flows in Rotation Box between user's input and appearance change (right).

IntelligentBox provide many software components called Boxes that have this MVC structure. Actually, IntelligentBox provides user controls on each Box using a mouse device that directly affect the appearance of the Box, so there is not much merit separating Controller and View. Therefore, in IntelligentBox, Controller and View are combined into one component called Display Object, and the structure is called an MD (Model-Display Object) structure. In a Box, Model has essentially an associative array those values mean the state of the Box, and its entity is called a "slot". Each slot can contain a primitive data, such as a number, a string or a structure of those, and reference types are not allowed. Practically, those slots are treated as input/output of the Box, or just a simple storage. Model can be edited though dedicated method, and notifies its change to the corresponding Display Object. Display Object is what characterizes the Box it belongs. There are many types of Display Object, and each has its own policy of how to interpret user's input, edit slots of its Model, and to represent Model as a 3D object.

For example, Rotation Box is one of the Boxes which acts like a knob shown in the right part of Fig. 3. It consists of a generic Model and Rotation Display Object. The behavior is as follows.

(1) Rotation Display Object receives user's mouse drag operation.
(2) The interpretation policy of Rotation Display Object is "take x-axis of drag movement and change "value" slot according to it", and it acts so.
(3) Model then realizes its change, and notifies others, in this case, only the Display Object.
(4) The representation policy of Rotation Display Object is "change rotational attitude of the 3D object according to "value" slot", and it acts so after the notification.

3.2 A Parent-Child Relationship

When composing two Boxes, a parent-child relationship is automatically defined between them. That means, one Box becomes a child of the other Box. When this relationship is established, a child follows its position and orientation to the parent just like forward kinematics.

If a Box has its parent, one of its slots can be "connected" to one of the parent's slots. The two connected slots then become synchronized to have the same value, therefore a Box can send its data to another Box. The way of synchronizing is defined by a message passing protocol, explained in the next subsection.

3.3   A Message-Passing Protocol Called Slot Connection

Between two related Boxes, they can communicate with each other using a message passing protocol. In IntelligentBox a Model has slots, and each slot holds one type of data, and user can select one slot of each of related Boxes, then connect those. Data of connected slots then become always synced, and thus you can transfer one data to the other.

To accomplish this behavior, three messages are defined. Those are "Set", "Gimme", and "Update". Set is called from a child Box, telling the parent to set data to the connected slot. Gimme is also called from a child, asking the parent a data in the connected slot. Update is called from a parent Box, notifying all of its children that the Model has somehow updated, asking them to also update if they need.

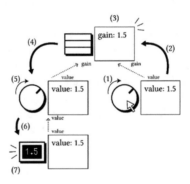

Fig. 4. An example of the message passing. Input on the right knob traverses throughout the entire hierarchy.

Those messages are invoked as follows. If one Box received user's input, it edits its Model and immediately issues Update, and Set if it has a parent. The parent which received Set, also edits its Model, sending Update, calling Set to its parent. This continues all the way up to its root. When children hear the Update from its parent, they call Gimme to the parent for fetching data according to its slot connection, and issue Update if needed. This also propagates to its ends. Note that Update will be issued only if Model changed its data. If Model was edited but no changes occurred, Update will not be issued. Therefore, there will be no infinite loops among Boxes if they are properly connected.

There are some Boxes that do not need any Set/Gimme, such as a simple display unit or a microphone unit, or you may want to temporarily disable some of Set/Gimme/Update features to test something. To achieve this, each Box has a set of flags to control its feature. If the flag were not set, respective features will be disabled entirely.

Figure 4 shows an example of set of Boxes. The sequence is as follows.

(1) User drags right knob, changing its slot "value".

(2) The Box has its parent and the slot "value" is connected to parent's slot "gain", so it calls Set to send data.

(3) The parent Box receives the Set, changing its slot "gain", invoking Update.

(4) Left knob hears the Update, calling Gimme to its parent to get the data of "gain", storing its slot "value".

(5) Left knob then changes its appearance to match the data, and simultaneously issues Update.

(6) A child of left knob, a display, hears parent's Update, and calls Gimme. It asks parent's slot "value", and stores the data to their slot "value".

(7) A display then refreshes its screen, matching the data.

3.4 Model Sharing

There is another way to sync data among Boxes. The above method only enables two connected Boxes to communicate, and since it is impossible to have multiple parents, there is no way to communicate between two Boxes that are hierarchically separated. To solve this, IntelligentBox has a feature called Model sharing, which enables a Box to treat its Model as the Model of another Box. Unlike slot connections, Model sharing copies the entire behavior to another Box, not a single slot. Note that Model sharing can only be done with the same type of Boxes, as Box behavior is dependent of the Model.

4 Web Implementations

This section explains how we implemented mechanism we explained earlier as a Web application. The original IntelligentBox is written in C++ standalone application, and our goal is to write equal application in JavaScript, as a frontend application. We introduce a new design pattern of programming, reactive programming, to handle event-driven application like this more fluently and WebGL [17]. Actually, we employ Three.js [18], one of the popular libraries for WebGL, to enable to display and interact with 3D graphics contents. We also describe how we improvised a new file format, replacing the old format IntelligentBox uses.

4.1 Reactive Programming

In IntelligentBox, there are an inter-Box messaging and an intra-Box messaging. In an inter-Box messaging, an Update of one Box causes other Boxes to issue Update messages, and what Box issues an Update changes depending on the environment, i.e. relationships. In an intra-Box messaging, Display Object catches user's input, and edit its Model. Model then issue its update message to related one, or possibly two Display Objects, invoking them to refresh its appearance, causing issues of inter-Box Update message. In the original IntelligentBox, this messaging was implemented as a function call, that is a sender calls each receiver's function, embedding data as its parameter.

One of the problems is that the sender needs to know, and manage all of its receivers. If some receivers are added or removed, the sender have to treat all of those miscellaneous processing. Another problem is that each receiver cannot much perform a receiver-specific conditional branching. You have to either include all of them in the called function, or let the sender do them beforehand. It causes the function swollen, or the sender to do unrelated things. The sender just wants to issue messages, and the receivers want to freely catch and manipulate messages.

To solve this problem, we introduce reactive programming [19]. Reactive programming is one of declarative programming, which is a contrast to the traditional imperative programming. In reactive programming, data can form a data stream, and any data change flows in the stream. you can then subscribe to the stream, that is to link some functions to the stream, invoking function calls automatically when data change happens. you also can attach filters or conditional branches to a stream, enabling the receiver to alter stream independently. In other words, receivers observe what a sender sends, so a stream is also called an observable, and this whole design is also called an observable pattern. Figure 5 shows an example of a reactive system.

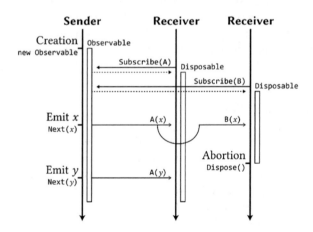

Fig. 5. An example sequence diagram of reactive system.

This design solves the problems we showed in previous design. The sender only has to create an observable and to issue message through the observable, letting receivers do whatever they want. This fact enables to drastically reduce the cost of the sender, and simplifies the code. Also, receivers can not only change its data value or branch, but also change timings, multiply or combine data in time, or even create a new stream based on data. This fundamentally differentiates it from event delegates or callbacks, as in reactive programming, specifically ReactiveX [20], a library which we use in this system, those temporal functionalities are built-in.

4.2   File Format

In order to save/load what you have created, IntelligentBox has an import/export feature which enables you to port and share boxes. However, the file format IntelligentBox uses is not based on any existing file formats other than it is text-based, and it is not documented either. It is extremely hard to parse and there is no reason to stick with it, therefore we improvised a new format. The new file format is based on JavaScript Object Notation (JSON) [21]. Boxes and their peripheral data are stored as JSON objects, with some additional information to recreate when it is loaded. The reason why we use JSON and not other formats is that it works seamlessly with JavaScript.

5   Demonstration

We made some simple examples shown in Fig. 6. In the left part, we made a simplified 2WD car, with four tires and one steering. The two back tires are directly connected to the body, and the front tires are indirectly connected via steering box. You can steer this car by dragging the steering box. In the right part, we made a simple monoplane with a handle to manipulate. The handle is a 2-DOF controller, which can handle 2-dimensional control by dragging the box. In this example, the gray handle next to the plane substitutes slots of the body part, allowing wings to rotate accordingly.

Figure 7 show IntelligentBox for Web working on mobile devices. In these devices, the viewpoint control for rendering 3D contents and the box operation are done by tapping and dragging instead of operating a mouse device. IntelligentBox for Web is based on the HTML5 standard including WebGL, and these Web browsers perform basically the same performance as desktop browser regardless of a desktop PC or a mobile phone.

However, there are some problems with them. Compared to desktop PCs, mobile devices have less computing power for the sake of their compactness and limited battery power. Because of that, on those devices, IntelligentBox for Web performs slightly worse than on PCs, as it lags or stutters a bit. Currently the system is not optimized for performance yet, and we have already found that the stutters are from dynamic UI elements, not from the actual system, so we do not think it is much of a problem, and we are pretty sure that we can solve this in the near future.

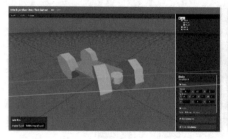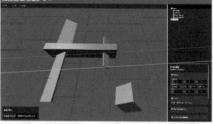

Fig. 6. Examples of IntelligentBox Web. Left: A real life example of a simplified car. Right: An example of a monoplane.

Fig. 7. IntelligentBox for web is available on mobile devices. Left: Google Nexus 7 (2012) on Android 5.1.1. Right: iPhone 5S on iOS 10.3.1.

6 Conclusions

In this paper, we introduced IntelligentBox for Web, Web version of IntelligentBox: A Constructive Visual Software Development System of Interactive 3D Graphics Applications, and explained its technology, and the implementation as a Web application. We used JavaScript for frontend logics, reactive programming for the message passing system, and WebGL for 3D contents. In conclusions, despite some lack of capabilities, we managed to run a clone system of IntelligentBox on a Web browser, enabling it to run on wide variety of platforms, increasing its portability.

As future work, we will solve several problems remaining in IntelligentBox for Web those are discussed in the previous section.

References

1. Okada, Y., Tanaka, Y.: IntelligentBox: a constructive visual software development system for interactive 3D graphic applications. In: Proceedings Computer Animation '95, pp. 114–125. IEEE, Geneva, Switzerland, April 1995
2. Goldberg, A., Robson, D.: Smalltalk-80: The Language and Its Implementation. Addison-Wesley (1993)
3. Open GL. https://www.opengl.org/
4. Strauss, P.S., Carey, R.: An object-oriented 3D graphics toolkit. In: Proceedings of the 19th Annual Conference on Computer Graphics and Interactive Techniques, pp. 341–349. ACM, Chicago, Illinois, USA, July 1992
5. Stampe, D., Roehl, B., Eagan J.: VIRTUAL REALITY CREATIONS. Waite Group PressTM, 200 Tamal Plaza, Corte Madera, CA 94925 (1993)
6. Shaw, C., Green, M., Liang, J., Sun, Y.: Decoupled simulation in virtual reality with the MR toolkit. ACM Trans. Inf. Syst. **11**(3), 287–317 (1993)
7. Anderson, D.B., Barrus, J.W., Howard, J.H., Rich, C., Shen, C., Waters, R.C.: Building multiuser interactive multimedia environments at MERL. IEEE Multimed. **2**(4), 77–82 (1995)
8. Barrus, J.W., Waters, R.C., Anderson, D.B.: Locals and beacons: efficient and precise support for large multi-user virtual environments. In: Proceedings of IEEE Virtual Reality Annual International Symposium, pp. 204–213. Santa clara, CA (1996)
9. Fahlen, L.E., Stahl, O., Brown, C.G., Carlsson, C.: A space based model for user interaction in shared synthetic environments. ACM INTERCHI '93 New York:43–48 (1993)

10. Hagsand, O.: Interactive multiuser VEs in the DIVE system. IEEE Multimed. **3**(1), 30–39 (1996)
11. Greenhalgh, C., Benford, S.: MASSIVE: a distributed virtual reality system incorporating spatial trading. In: Proceedings of IEEE 15th International Conference on Distributed Computing Systems (DCS'95),Vancouver, Canada:27–34 (1995)
12. Pandzic, I., Capin, T., Magnenat-Thalmann, N., Thalmann, D.: VLNET: a networked multimedia 3D environment with virtual humans. In: Proceedings of Multi-Media Modeling MMM '95, Singapore:21–32 (1995)
13. Okada, Y., Tanaka, Y.: Collaborative environments of IntelligentBox for distributed 3D graphics applications. Vis. Comput. **14**(4), 140–152 (1998)
14. Grimsdale, C.: dVS-distributed virtual environment system. In: Proceedings of Computer Graphics '91, pp. 163–170. Bleinheim Online, London, UK (1991)
15. Unreal Engine. https://www.unrealengine.com/ja/features
16. Unity. https://unity.com/ja
17. Marrin, C.: Webgl specification. Khronos WegGL Working Group (2011)
18. Danchilla, B.: Three.js framework. Beginning WebGL for HTML5, pp. 173–203. Apress, Berkeley, CA, USA (2012)
19. Wan, Z., Hudak, P.: Functional reactive programming from first principles. In: Proceedings of the ACM SIGPLAN 2000 Conference on Programming Language Design and Implementation, pp. 242–252. ACM, Vancouver, British Columbia, Canada, June 2000
20. Maglie, A.: ReactiveX and RxJava. Reactive Java Programming, pp. 1–9. Apress, Berkely, CA, USA (2016)
21. Bray, T. (ed.): The JavaScript Object Notation (JSON) data interchange format. RFC 8259, December 2017

Web-Based Collaborative VR Training System for Operation of Radiation Therapy Devices

Kotaro Kuroda[1], Kosuke Kaneko[2], Toshioh Fujibuchi[3],
and Yoshihiro Okada[1,2,4(✉)]

[1] Graduate School of ISEE, Kyushu University, Fukuoka, Japan
okada@inf.kyushu-u.ac.jp
[2] Cybersecurity Center, Kyushu University, Fukuoka, Japan
[3] Faculty of Medical Sciences, Kyushu University, Fukuoka, Japan
[4] ICER (Innovation Center for Educational Resources)
of Kyushu University Library, Kyushu University, Fukuoka, Japan

Abstract. In this paper, the authors propose a web-based collaborative VR training system for the operation of radiation therapy devices. Medical students have to train as a therapist if they want to be so. However, radiation therapy devices are very expensive and dangerous, so any training systems without using the devices are needed. The authors have already proposed the web-based VR training system for the operation of medical therapy devices, and made preliminary tests to confirm which kinds of VR devices are suitable for the proposed system. From the test results, the authors chose one training system configuration of VR devices and implemented a more practical web-based collaborative VR training system.

1 Introduction

Recently, VR applications have become popular and then, we can easily try them on various types of VR devices such as HTC VIVE [1], Oculus Rift [2], and even standard smartphones. In addition, by combining VR applications and real-time bi-directional communication, developers can provide users with collaborative VR experiences.

VR simulation systems are available as education tools [3–6] in various fields including physics, medicine, and architecture. Their immersive VR experiences help students to understand complex phenomena in the real world or the usage of certain specific tools. Particularly, in the field of medicine, VR simulation systems can help students to improve their operation skills for medical therapy devices.

Our research purpose is to propose a collaborative VR training system [7] for the operation of radiation therapy devices. Currently, our target is a radiation therapy device called TrueBeam [8] shown in Fig. 1. Several medical procedures including the operation of TrueBeam sometimes need to be performed by multiple persons, so training systems should support such collaborations. In addition, they should support also immersive operations such that operators often touch patients themselves in medical procedures. Moreover, we should consider their costs. As previous work [9], we have tested a couple of training system configurations of different VR devices for our purpose. From the test results, we chose one training system configuration of VR

L. Barolli et al. (Eds.): CISIS 2019, AISC 993, pp. 768–778, 2020.
https://doi.org/10.1007/978-3-030-22354-0_70

devices and implemented a more practical web-based collaborative VR training system. In this paper, we introduce it.

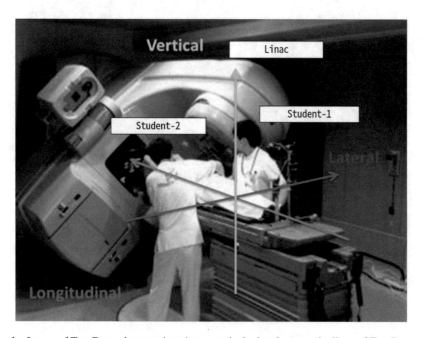

Fig. 1. Image of TrueBeam, here student-1 moves the bed and rotates the linac of TrueBeam via its controller and student-2 adjusts the orientation of a patient.

The remainder of this paper is organized as follows: in next Sect. 2, we describe related work and research objective. In Sect. 3, we explain the architecture of the proposed system and the details of its implementation. Next, we discuss the usefulness and problems of the proposed system in Sect. 4. Finally we conclude the paper in Sect. 5.

2 Related Work and Research Objective

Imura et al. [10] have already developed a VR training system for TrueBeam. In this system, the operation of TrueBeam and the setup of a patient are performed in the real world, and they react in a virtual world as shown in Fig. 2. The main purpose of this system is to help students to understand how to operate TrueBeam and how to adjust the position and orientation of a patient. To achieve this, a smartphone with a dedicated application is used to operate TrueBeam in the virtual world and a chest model of a patient with a gyroscope sensor device is used to realize the patient setup. As a development tool, Unity [11], which is one of the most popular game engines, is used to generate the virtual world of this system.

Virtual device and a patient represented as 3DCG model

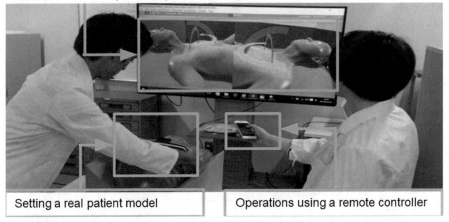

| Setting a real patient model | Operations using a remote controller |

Fig. 2. Image of operation training of TrueBeam on the system of Imura et al.

Figure 3 shows the components of Imura's system. This system consists of a PC for displaying the virtual world in a treatment room, a chest model of a patient with a gyroscope sensor device, a smartphone which displays TrueBeam controller, and a server which synchronizes the status of these three components by real-time bi-directional communication.

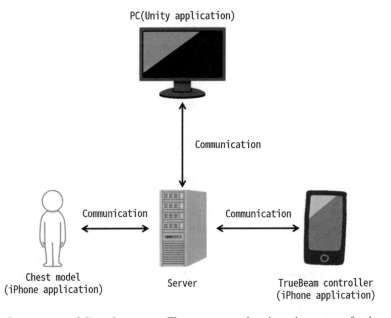

Fig. 3. Components of Imura's system. The server synchronizes the status of other three components using real-time bi-directional communication.

We have already modified Imura's system to become a web-based one available with VR/AR devices [7, 9] in order to provide students with more immersive training experience. Then this time, in our proposed system, students are supposed to see the virtual world through a standard smartphone with a HMD like Vox+ Z3 [12] because it is more affordable than HTC VIVE, Oculus Rift, and so on. However, when students wear a HMD, they cannot recognize exactly where the chest model and TrueBeam controller are. To avoid this problem, we use a gamepad device like SONY DUALSHOCK 4 [13] for TrueBeam operation instead of TrueBeam controller, and a motion tracking device like Leap Motion Controller [14] for the patient setup instead of the chest model. By making our previous system support these devices, it can be said that the system has become a web-based, collaborative, intuitive and affordable system.

3 System Architecture and Implementation

3.1 Overview of the System Architecture

For rendering 3D contents on a browser, we use Three.js [15], which is one of the most popular JavaScript libraries for WebGL API [16]. Figure 4 shows the system architecture of our proposed system. Each student uses a smartphone and a VR goggle as an HMD to see a virtual world as shown in Fig. 5. The direction of the viewpoint for rendering 3D contents is synchronized with the orientation of a smartphone, and the

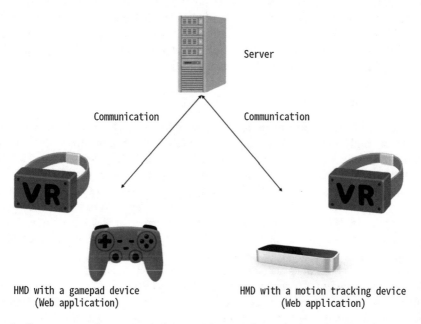

Fig. 4. Components of our proposed system, here students check a virtual world through a smartphone with an HMD. They also use a gamepad device or a motion tracking device to train the medical procedures of TrueBeam.

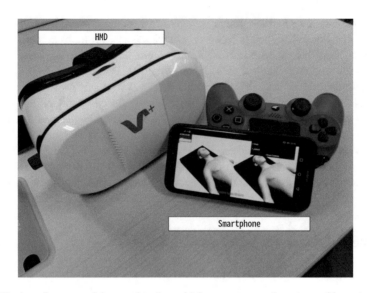

Fig. 5. Devices for recognizing a virtual world in our proposed system with a stereo view support.

stereo view is supported to be suitable for the HMD. Student-1, who is an operator of TrueBeam, uses a gamepad device to change the bed position and the rotation of the linac. Student-2, who treats a patient, uses a motion tracking device that senses his/her hands' gesture to change the orientation of the patient 3D model. These operation data of Student-1 and Student-2 are always sent to the server via web socket communication, and shared with each other. In our proposed system, Socket.IO [17], which is a Node.js [18] module for web socket communication, is used for that.

3.2   Gamepad Device

As shown in Fig. 6, TrueBeam has five degrees of freedom. The linac has one, which is its rotation angle around its local x-axis called GNT. The bed has four, three of them are its position in a 3D scene called LNG(x-direction), VRT(y-direction) and LAT(z-direction), the other is its rotation angle around the world y-axis called ANG.

Student-1 uses a gamepad device to exercise TrueBeam operation. This time, student-1 uses SONY DUALSHOCK 4, as shown in Fig. 7. When student-1 presses the R2 key, a ray will be casted from the center of a screen to pick the 3D object. With this key pressed, when student-1 presses the cross keys, the operation linked to the picked 3D object will be executed. Then the name of picked 3D object and the value of pressed key will be sent to the server as the 'operation' event of Socket.IO to synchronize the status of TrueBeam with student-2. The status of a gamepad device, whether each key is pressed or not, can be given via Gamepad API [19]. Figure 8 shows how to exercise TrueBeam operation on our proposed system.

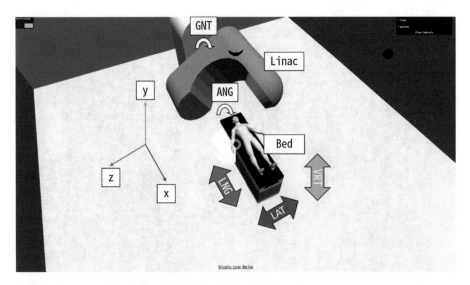

Fig. 6. Degrees of freedom of TrueBeam. Student-1 can exercise the operation of TrueBeam related to these five degrees of freedom via a gamepad device.

Fig. 7. Keys of SONY DUALSHOCK 4 available on our proposed system. Student-1 is supposed to press the cross keys with the R2 key pressed to exercise TrueBeam operation.

3.3 Motion Tracking Device

Figure 9 shows how to exercise the patient setup on our proposed system. We have to prepare a PC to connect to Leap Motion Controller, because it does not support a web application on smartphones. We use LeapJS [20], which is the official JavaScript SDK, to get hand data from Leap Motion Controller. We use the palm position, palm direction, and finger direction to display two 3D hand models in a 3D scene as shown in Fig. 10.

Once student-2 clenches his/her both hands, the patient setup mode will start, a ray will be casted from the center of a screen to pick the 3D patient object, and the direction from the left hand to the right hand will be synchronized to the rotation angle of the 3D

Fig. 8. Image of training of TrueBeam operation on our proposed system.

Fig. 9. Image of the patient setup on our proposed system. Student-2 needs a PC to connect to Leap Motion Controller.

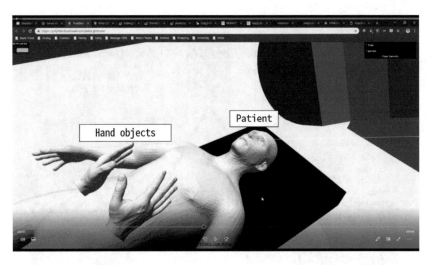

Fig. 10. Interaction between the 3D hand objects and the 3D patient object.

patient object. When student-2 clenches his/her hands again, the patient setup mode will stop. Then the name and the rotation angle of the 3D patient object will be sent to the server as the 'orientation' event of Socket.IO to synchronize the status of it with student-1.

3.4   Server Process

The source code of JavaScript below shows functions executed on the server. When the 'operation' event or 'orientation' event is issued, a callback function will be invoked. Its argument stores the name of the picked 3D object of TrueBeam and the key value of a gamepad device or rotation angle of the 3D patient object. In each callback function, its argument will be broadcasted to the other student to synchronize the 3D scene. Each function executes TrueBeam operation and the patient setup, and it is linked to the name of 3D object, so students can share the exactly same common state in the 3D scene as shown in Fig. 11.

Source Code.

```
operationNameSpace.on('connect', (socket) => {
   socket.on('operation', (object) => {
      socket.broadcast.emit('operation', object);
   });
});

orientationNameSpace.on('connect', (socket) => {
socket.on('orientation', (object) => {
      socket.broadcast.emit('orientation', object);
   });
});
```

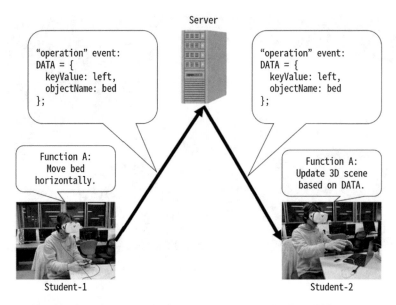

Fig. 11. Smaple communication in status synchronization of 3D scene.

4   Discussion

4.1   Features as a Collaborative System

In this research, we have developed a web-based collaborative VR training system for the operation of radiation therapy devices by applying web socket communication. Our proposed system enables students to share the exactly same common 3D environment with each other. Some medical procedures including TrueBeam operation and the patient setup are often necessary to be performed by multiple persons, so our proposed system is suitable for such a case.

4.2   Clarity of Interaction

In our proposed system, students are supposed to recognize a virtual world via a smartphone inside a VR goggle as a HMD. However, when they wear the HMD, they cannot see the other devices necessary for the training. To avoid this problem, we employ a gamepad device for TrueBeam operation and a motion tracking device for the patient setup through hands' gesture.

Regarding a motion tracking device, students are supposed to use it on a desk, however, there is still the same problem as mentioned above, we are looking for a way to fix it on a HMD. In addition, students cannot feel the haptic feedback such as the touch of a patient's body, so we have to find the solution for that.

4.3 Completeness of the System

Now, our proposed system is ready to use for the collaborative training for the operation of radiation therapy devices. However, it lacks the functionality of the learning feedback to analyze the improvement for students' operation. We are going to implement that functionality.

5 Conclusion

In this research, we have developed a web-based collaborative training system for the operation of radiation therapy devices. Medical students have to train medical procedures in order to become a radiation therapist. However, radiation therapy devices are expensive and dangerous. So, any affordable training systems are needed. In addition, some medical procedures are often necessary to be performed by multiple persons, training systems must be collaborative. Moreover, interactions between students in the real world and 3D objects in a virtual world must be immersive such that students touch a patient directly. By combining real-time bi-directional communication with some affordable extra devices, we have developed our proposed system that satisfy these requirements.

We still have several problems as mentioned in Sect. 4, so we will solve them as soon as possible. In addition, as future work, we have to evaluate the usefulness of our proposed system by asking medical students to actually use it for their training and to give us some feedbacks.

Acknowledgements. This research was partially supported by the project grant from the Kyushu University Education Innovation Initiative.

References

1. HTC VIVE. https://www.vive.com/us/product/vive-virtual-reality-system/. Accessed 15 Mar 2019
2. Oculus Rift. https://www.oculus.com/. Accessed 15 Mar 2019
3. Okada, Y., Ogata, T., Matsuguma, H.: Component-based approach for prototyping of Tai Chi-based physical therapy game and its performance evaluations. ACM Comput. Entertain. **14**(1), 4:1–20 (2016)
4. Akase, R., Okada, Y.: Automatic 3D furniture layout based on interactive evolutionary computation. In: Proceeding of the 5th International Workshop on Virtual Environment and Network Oriented Applications (VENOA-2013) of CISIS-2013, pp. 726–731. IEEE CS Press (2013)
5. Kosuki, Y., Okada, Y.: 3D visual component based development system for medical training systems supporting haptic devices and their collaborative environments. In: Proceeding of the 4th International Workshop on Virtual Environment and Network Oriented Applications (VENOA-2012) of CISIS-2012, pp. 687–692. IEEE CS Press (2012)
6. Miyahara, K., Okada, Y.: Collada-based file format for various attributes of realistic objects in networked VR applications supporting various peripherals. J. Mob. Multimed., Rinton Press **6**(2), 128–144 (2010)

7. Kuroda, K., Kaneko, K., Fujibuchi, T., Okada, Y.: Web-based VR system for operation training of medical therapy devices, VENOA 2018. In: Proceeding of the 12th International Conference on Complex, Intelligent, and Software Intensive Systems, pp. 768–777. CISIS (2018)

8. TrueBeam. https://www.varian.com/ja/oncology/products/treatment-delivery/truebeam-radio therapy-system. Accessed 8 Apr 2018

9. Kuroda, K., Kaneko, K., Fujibuchi, T., Okada, Y.: Web-based operation training system of medical therapy devices using VR/AR devices, 2018 In: The 13-th International Conference on Broadband and Wireless Computing, Communication and Applications (BWCCA-2018)

10. Imura, K., Fujibuchi, T., Kaneko, K., Hamada, E. and Hirata, H.: Evaluation of the Normal Tissues Dose and Exposure Efficiency in Lung-stereotactic Body Radiation Therapy, the 44th Autum Scientific Congress, Japanese Society of Radiological Technology, 13–15 October 2016 (in Japanese)

11. Unity. https://unity3d.com/jp. Accessed 8 Apr 2018

12. Vox+ Z3. http://www.vox-vr.com/vox-z3-vr-virtual-reality-headset-vr-glasses.html. Accessed 15 Mar 2019

13. SONY DUALSHOCK 4 https://www.playstation.com/en-us/explore/accessories/gaming-controllers/dualshock-4/. Accessed 15 Mar 2019

14. Leap Motion Controller. https://www.leapmotion.com/. Accessed 15 Mar 2019

15. Three.js. https://threejs.org/. Accessed 15 Mar 2019

16. WebGL API. https://developer.mozilla.org/en-US/docs/Web/API/WebGL_API. Accessed 15 Mar 2019

17. Socket.IO. https://socket.io/. Accessed 15 Mar 2019

18. Node.js. https://nodejs.org/en/. Accessed 15 Mar 2019

19. Gamepad API. https://developer.mozilla.org/en-US/docs/Web/API/Gamepad_API. Accessed 15 Mar 2019

20. LeapJS. https://developer-archive.leapmotion.com/getting-started/javascript. Accessed 15 Mar 2019

Designing a Simplified User Interface System for Smartphone Natives to Facilitate PC Operations

Keiji Urawaki and Makoto Nakashima$^{(\boxtimes)}$

Oita University, 700 Dannoharu, Oita-shi 870-1192, Japan
{v19e3005,nakasima}@oita-u.ac.jp

Abstract. A novel simplified interface system for smartphone natives, which allows users to operate a PC through their own smartphones, is proposed. Smartphones have overtaken PCs and are currently the most familiar devices for interacting with the world. We refers to those who grew up with such devices as "smartphone natives." Although they are quite adept at using such devices, they have little PC experience and limited operation skills. According to an interview with expert users, including university students and staff in the computer science department, we identified 55 effective shortcut keys that they use regularly and that once mastered will speed up the use of Windows OS and MS Word. The proposed interface system has the soft keys corresponding to these shortcuts, each of which is laid out on the smartphone screen according to its importance by applying a kind of heat map, the thumb zone, representing the comfortable area on a smartphone screen for touch with one-handed use. We have conducted a user study of smartphone natives, each of which is a novice user of a PC, and revealed the effectiveness of the proposed system in text entry and a future research direction to further investigate the effect of our system when operating a PC without the use of a keyboard/mouse.

1 Introduction

Smartphones have spread rapidly throughout this decade and have replaced PCs in terms of interacting with the world. For example, in Japan, according to the Communications Usage Trend Survey, Japanese smartphone penetration is now approaching 80% since 2010, the rate has gone from 9.7% to 75.1% in 2017 [5], whereas the rate of PCs has decreased from 83.4% to 72.5%. Meanwhile, more than 75% of the 13 to 19 years olds use smartphones for Internet use in 2017 and the percentage of smartphone-only users in the 13 to 39 demographic more than any other age demographic in the U.S. [2]. We refers to those who grew up with such devices as "smartphone natives" [6]. Although they are quite adept at using such devices, they have low use opportunities and limited operation skills of PCs. On the other hand, they know little about effective keyboard shortcuts (which hereafter we call simply shortcuts) to access frequently used commands when operating a PC and many applications. Thus, it is important to consider developing an effective system enables the smartphone natives to easily operate a PC as they increase in number year upon year.

© Springer Nature Switzerland AG 2020
L. Barolli et al. (Eds.): CISIS 2019, AISC 993, pp. 779–788, 2020.
https://doi.org/10.1007/978-3-030-22354-0_71

In this paper, we propose a simplified user interface system to facilitate PC operations by using a smartphone. In our pre-experiments of 30 smartphone natives, we confirmed that the text input speed of them when using a smartphone is significantly higher than when using a QWERTY keyboard. Based on an interview with skilled PC users, we chose 55 effective shortcuts that they use regularly and that once mastered will speed up the use of Windows OS and MS Word. The proposed interface system provides the soft keys corresponding to those shortcuts, each of which is laid out on the smartphone screen according to its importance by applying a kind of heat map, the thumb zone [3, 7], representing the comfortable area on a smartphone screen for touch with one-handed use.

We conducted a user study of smartphone natives, 9 university students, each of which is a novice user of a PC. They were asked to complete a document correction task on the MS Word file using the proposed system and then to answer the questionnaire about the usability of the system. The results revealed the effectiveness of the proposed system in text entry and a future research direction to further investigate the effect of our system when operating a PC without the use of a keyboard/mouse.

2   Related Work

There are several applications available that assist in controlling a PC remotely when using a smartphone. Remote Mouse [9] turns a smartphone into a wireless remote control for a PC. This application allows the users to input text through the software keyboard, perform mouse operations, control a media player and switch application windows by using a smartphone. FlickTyper [4] is a kind of software keyboard on a smartphone, which allows the users to input Japanese text for a PC using the flick input method. The user can swipe from a key in a certain direction to produce the desired character. Similar to Remote Mouse, this application can control a PC with mouse operations through a smartphone.

Many shortcuts for operating a PC are able to be simulated on the above applications. However, they are difficult to master for the novice users who have very little PC experience. FlickTyper requires pre-setting custom keys for the specific shortcuts by the user themselves on its software keyboard. Even though Remote Mouse can display all the keys needed for executing the shortcuts on its software keyboard, initially, the novice users are unfamiliar with the key combinations. As a result, these applications are not designed for the novice PC users.

3   Smartphone Natives and PC Operations

Before developing the proposed system, we conducted a preliminary survey in order to clarify the following questions about the practical skills of the smartphone natives in using a PC.

(i) How much faster is a smartphone native on inputting text on the smartphone than that on a PC with its QWERTY keyboard?

(ii) How many shortcuts for the OS, i.e., Windows OS, and applications on the PC does a smartphone native know?

3.1 Text Input Speed in Smartphone Natives

The thirty students aged 15 to 22 (11 university students and 19 high-school students) participated in the first preliminary survey. According to the pre-interview, almost all the students use their smartphones an average of over 2 h in a day, compared with little or no use of PCs. Each participant was asked to type in some Japanese text in the text editor application of his/her own smartphone and in MS-Word with the QWERTY keyboard for the PC in the author's research room within a specified period of time (university and high-school students had 3 and 2 min, respectively). When using the smartphone, all the participants used the flick input method. As the text for this survey, we used a past test of Certificated Speed Typist given by the National Association of Commercial High Schools [8] that was partially modified so that the frequency of the alphabetic letters and numerals would be the same.

The average text input speed when using a smartphone was 56.4 characters per minute, whereas the speed when using a QWERTY keyboard was 35.7 characters per minute. The former speed was statistically faster than the latter one (a Wilcoxon's sign rank test, $p < 0.05$). As a result, it could be said that the smartphone natives have the ability to input characters faster when using the smartphone than when using the keyboard.

3.2 Visibility of Shortcut Keys with the Smartphone Natives

To clarify the second question mentioned above, we selected 36 commands with shortcuts which we can use on the desktop PC with Windows 10 components and built-in tools, and 102 commands with shortcuts for MS-Word. Then we requested that skilled PC users and smartphone natives (who are not skilled PC users) filled in the questionnaire shown in Table 1. The questions were prepared to see which shortcuts are considered effective by skilled users in using a PC and how many such effective shortcut keys are recognized by the smartphone natives.

Table 1. Questions and choices

Questions	Choices
Q1. Do you know this command?	Yes, No
Q2. Do you know the shortcut for this command?	Yes, No
Q3. How often do you use this command?	3 = frequently use, 2 = sometimes use, 1 = rarely use
Q4. Would you like to use this command in future?	A 1–7 range Likert scale was used. (1 = very acceptable, 4 = neutral, 7 = very unacceptable)

First, 6 graduate students and 6 university staff majoring in Computer Science as the skilled users were asked to answer the questions for the 36 Windows shortcuts. The average percentage of the skilled PC users giving positive answers (i.e., yes) to Q2 was only 24.5% compared with 54.8% for Q1. Next, for 102 MS Word shortcuts, the same questionnaire was conducted to 7 graduate students majoring in Computer Science, which were different from the above mentioned students. Similar to the result of the previous questionnaire, the average percentage of the respondents was only 23.5% to Q2 compared with 42.0% for Q1. However, the answers of more than half of respondents to Q4 in the former and latter questionnaire were positive (i.e., answers were 1, 2, and 3) for 16 out of 36 and 39 out of 102 shortcuts, respectively. The average point for Q3 regarding these 55 shortcuts was 2.2. According to the those results, it could be said that the 16 shortcuts for Windows OS and 39 shortcuts for MS Word are recognized to be highly effective in operating a PC by the skilled users.

For these 55 shortcuts, we conducted the aforementioned questionnaire for 10 smartphone natives which are university students and have little or no experience with PCs. The average percentages of positive answers for Q1 and Q2 were less than 26% and 8%, respectively. On the other hand, the answers of more than half of respondents to Q4 were positive for 48 out of 55 shortcuts.

4   A Simplified User Interface System

From the results of the preliminary survey, we found that the smartphone natives know little about the effective commands and their corresponding shortcuts, whereas the text input speed on the smartphone was faster than that on the keyboard. Therefore, we designed the proposed system to meet the following two requirements.

- The system enables work on any type of smartphone.
- The system enables the user to easily utilize highly effective shortcuts in operating a PC even for a novice user.
- The system allows the user to input characters on the smartphone to the PC.

4.1   Function and Design

To fulfill the first requirement, this system is developed as a web application which can work through the web browser on a smartphone. The problem is getting the proposed system on the web page to work with PCs, and to enable the proposed system to send character inputs and/or command sequences to any PC's operating system and any PC application the user desires. To deal with this problem the CollaboTray technology [1] for application-sharing, which enables network access to any application, is utilized.

For the second and third requirements, the system allows the executions of shortcuts and text input by assuming that an ordinal user usually holds the mouse with his/her one hand and then holds the smartphone with the other hand. The problem is the layout of soft keys for prompting the smartphone natives to use the effective shortcuts and allowing them to input text for a PC. To design the soft key layout, we introduced the Thumb Zone [3, 7], which is the most comfortable area on the smartphone for touch with one-handed use shown in the left side of Fig. 1. The figure shows the thumb zones when the dominant hand is the left one, where the green area on the smartphone screen is naturally touchable, the yellow area is touchable by stretching the finger, and the red area is hard to touch. This difference in comfortability must be taken into consideration especially when arranging soft keys for the proposed system.

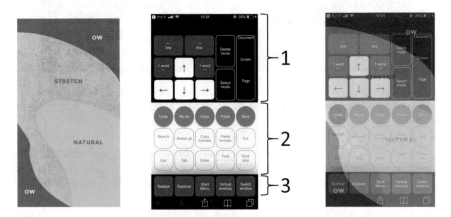

Fig. 1. The interface design of the proposed system on the smartphone for operating a PC. **Left**: The Thumb Zone [7] when the dominant hand is left. **Middle**: The initial state of the system. There is three areas for arranging soft keys on the screen. **Right**: The Thumb Zone [7] and our interface design.

4.1.1   Shortcut Key Arrangement

According to the survey result mentioned in Sect. 3.2, our system has 54 soft keys corresponding to the shortcuts (hereafter these soft keys are written as shortcut keys for simplicity) which in the survey are judged to be of highly effective. (Note: One shortcut for maintaining the macros for MS Word is omitted since it is not for novice users). We also selected 4 more convenient shortcuts for using a PC. We have two primary policies for arranging those effective shortcuts: (1) The shortcut keys of both caret operations and edit operations should be able to be used simultaneously. (2) The shortcut keys are arranged on the appropriate area in terms of their usage rate based on the questionnaire results in Sect. 3.2.

To meet the above policy, the interface design of our system is organized into three areas as shown in the middle of Fig. 1. The shortcuts for caret operations are placed in the upper-side, area 1, and the shortcuts for MS Word are placed in the area 2. The most frequently used shortcuts for Windows OS are placed in the lowest area. The following describes in more detail the ways in which the arrangement is obtained.

Area 1: Caret Operations
The left half of Fig. 2 shows the initial state of the area 1 for the shortcut keys corresponding to caret operations and arrow keys for moving the caret in the respective direction. Since this system assumes that the user holds the smartphone in the left hand, the arrow keys which are likely to be used frequently, are arranged on the bottom-left

side of this area. Four other basic shortcut keys for moving the caret one word to the right/left and to the beginning/end of the current line, are arranged around the arrow keys, which are black/white reverse ones with arrows.

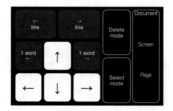 

Fig. 2. The arrangement of the shortcut keys for caret operations in Area1. **Left**: The initial state of Area1. **Right**: The right-side menu of the shortcuts for moving carets a long distance is expanded.

The arrow keys are also used for two modes of deleting and selecting characters. Tapping the soft keys, "Delete mode" and "Select mode," once activates the delete mode or select modes, respectively. The select mode is equivalent to the shift-lock state of the keyboard.

The other shortcut keys for moving the caret further away are stored in a kind of menu on the right-side of this area to secure a space for aforementioned keys. The right half of Fig. 2 shows the expanded menu in which all the keys are shown from the top: Six keys of shortcuts for moving the caret to the beginning/end of the document, up/down one screen, and up/down one page. The menus are expanded by touching any item name on the menu.

Area 2: Edit Operations for a Word Processor
As shown in Fig. 3, the shortcut keys for MS Word are placed on two panels A and B which are switched visible and invisible in turns by swiping up and down on this area. In order to save the user time and energy, the shortcut keys are categorized into two groups according to the usage rate of the skilled PC users in the survey result mentioned in Sect. 3.2. The panel A consists of shortcut keys which are frequently used more than ones for the panel B. To highlight this difference in usage rate, the panel A has a higher brightness compared to the panel B. Furthermore, in each panel, the arrangement of shortcut keys depends on touch comfortability with the Thumb Zone. For example, the most frequently used shortcut keys are placed on the area of 'naturally touchable.' In order to see more details of our arrangement policy, in the right side of Fig. 1, the Thumb Zone is superimposed over the initial state of our system. As a result, almost all the shortcut keys for MS word are arranged in the green and yellow areas on Thumb Zone.

Area 3: Operations on OS
The 16 highly effective shortcuts for Windows OS and built-in tools are arranged in three rows. These shortcut keys are arranged on the basis of their utilization rate found from the preliminary survey. As shown in the middle of Fig. 1, only the top row containing the most frequently used shortcut keys are initially shown on this area. The other rows appear as shown in the left half of Fig. 3 when the user swipes up on this area.

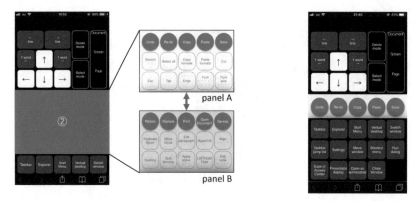

Fig. 3. The arrangement of the shortcut keys for MS Word in the area 2. **Left**: Two panels for storing the shortcut keys for MS Word. **Right**: The shortcut keys for Windows OS and built-in tools are expanded into the three rows.

4.1.2 Text Input

For text entry through the smartphone to a PC, the system allows the user to type in text using the inherent manner, e.g., the flick input method for Japanese, as they like. The smartphones' keyboard with the empty text box appears when the user swipes from right to left. The text box is for confirmation of the characters which are being inputted to a PC using this keyboard.

4.2 System Architecture and Flow

The architecture of the proposed system is depicted in Fig. 4. Data flows are also shown. The system consists of mainly two mechanisms of Interface system control and PC control. The former is responsible for handing all information needed to render the interface system as described above on the smartphone's web browser and any information about the operations on the smartphone to the PC control mechanism. The latter one receives the information about the operations occurred on the smartphone from the former mechanism and generates the events on the basis of the received data for the operating system to perform the specified functions of the shortcuts.

5 Experimental Results

We conducted an experiment to evaluate the usability of the proposed system in operating a PC in which the 9 university students aged 19 to 22 (5 females and 4 males) participated. Each participant is the smartphone native and uses the smartphone more frequently than the PC.

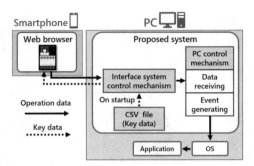

Fig. 4. The system architecture

5.1   Conditions and Apparatus

We prepared two conditions using two types of input devices: Keyboard & Mouse (**KM**) and the Proposed System & mouse (**PS**) on a smartphone. **KM** is the base condition that uses an ordinal keyboard and a mouse to operate a desktop PC. In **PS** condition, each participant used his/her own smartphone to operate the PC through the proposed system.

5.2   Tasks and Metrics

Each participant was asked to perform the task of polishing up the document using MS Word according to the specific instructions in the two conditions. In this experiment, we used two sets A and B of the document and the instruction note for polishing up the document. The content of each document is a draft of an invitation letter containing some deficiencies. The instruction note had 14 instructions on how to polish up the invitation letter, where 3 commands for the Windows OS built-in tools and 20 commands for MS Word were considered necessary to complete the task and the shortcut keys corresponding to these commands are implemented on the proposed system. Before staring the task, we explained to the participants how to use the proposed system and the shortcuts that were supposed to be used in this task. Then the participants experimented with a few minutes' trails to get used to the system. Note that, in the condition **PS**, we imposed a limitation on using the mouse for the instructions except some, e.g., using the dialog, in order to offer more opportunities to use the shortcuts.

The total time it took to complete the polishing up of each document was recorded. Furthermore, the time it took to input text in the two instruction (60 and 10 characters on the instruction 4 and 13, respectively) was also recorded. After having completed the tasks, each participant was asked to answer questionnaire about the proposed system.

5.3   Results

Figure 5 shows the total time taken to complete the task. Although, in condition **PS,** all the participants completed the task in a longer time than in condition **KM,** there is statistical difference (t-test, p < 0.05). This results implies that the participants needed more practice to get used to this novel interface system. The recorded times for instruction 4 and 13 were also shown in this figure. Although there is no statistical difference in the average time between the conditions PS and KM in instruction 4, the participants could input text in a shorter time using the proposed system than when using the keyboard in instruction 13 (t-test, p < 0.05).

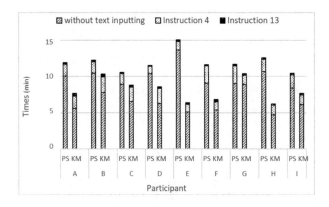

Fig. 5. The time it took to complete the task.

A 7-point Likert-scale based questionnaire was used to evaluate the proposed systems. Figure 6 shows the statements of the questionnaire and the statistics of the responses. When the responses 1–3 are considered as cons and responses 5–7 as pros, a binomial test found no statistical difference between pros and cons (p > 0.05) for any statement. In S2, S3 and S6, the proportion of pros tended to be high. Some participants

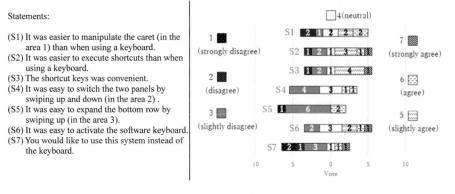

Statements:

(S1) It was easier to manipulate the caret (in the area 1) than when using a keyboard.
(S2) It was easier to execute shortcuts than when using a keyboard.
(S3) The shortcut keys was convenient.
(S4) It was easy to switch the two panels by swiping up and down (in the area 2).
(S5) It was easy to expand the bottom row by swiping up (in the area 3).
(S6) It was easy to activate the software keyboard.
(S7) You would like to use this system instead of the keyboard.

1 (strongly disagree)
2 (disagree)
3 (slightly disagree)
4 (neutral)
5 (slightly agree)
6 (agree)
7 (strongly agree)

Fig. 6. The result of the questionnaire

commented positively such as "It was convenient to be able to use the shortcuts with just one tap operation" and "I think it would be more convenient if I got used to the operation." The above results show that, from the view point of an interface design, the proposed system provides the advantage of being able to turn the smartphone into an effective remote control for a PC if the smartphone users are given more time to get used to it.

6   Conclusions

A simplified user interface system for the smartphone was proposed to support the smartphone natives unfamiliar with PC operations. From the result of the evaluation experiment, the proposed system is expected to be effective partially when processing tasks such as inputting text. In order to clarify how our interface design affects the PC operations of the smartphone natives, a trial on a long-term basis should be performed.

Acknowledgements. This work was partially supported by JSPS KAKENHI Grant Number 17K00274.

References

1. Abe, Y., Matsusako, K., Kirimura, K., Nakashima, M., Ito, T.: Tolerant sharing of a single-user application among multiple users in collaborative work. In Companion Proceedings of the ACM Conference on Computer-Supported Cooperative Work (CSCW'10), pp. 555–556 (2010)
2. DigitalGov. https://digital.gov/2016/04/26/trends-on-tuesday-smartphone-ownership-reaching-saturation-fueling-media-consumption/
3. Hoober, S., Berkman, E.: Designing Mobile Interfaces. O'Reilly Media, CA (2011)
4. InterLab FlickTyper. http://flicktyper.com/
5. Ministry of Internal Affairs and Communications, Japan. http://www.soumu.go.jp/johotsusintokei/statistics/statistics05a.html
6. Reed, R.: From digital natives to smartphone natives. In MomentFeedBlog. https://momentfeed.com/blog/from-digital-natives-to-smartphone-natives/
7. Scott, H.: How to design for thumbs in the era of huge screens. http://scotthurff.com/posts/how-to-design-for-thumbs-in-the-era-of-huge-screens
8. The National Association of Commercial High Schools. http://www.zensho.or.jp/puf/examination/pastexams/pc.html
9. Yao, R.: Remote mouse. https://www.remotemouse.net/

Development of Teaching Materials for Routing of Network Using a Full Color LED Tape as Physical Interface

Arisa Ishikawa[✉], Kazuaki Yoshihara, and Kenzi Watanabe

Graduate School of Education, Hiroshima University,
Higashi-Hiroshima, Hiroshima, Japan
{m196554,d173863,wtnbk}@hiroshima-u.ac.jp

Abstract. In Japan, the new course of study for technology education of junior high schools and for Information education of high schools included the contents of network. However, there are little teaching materials for these studies. Thus it is difficult students image network systems and technologies, because students have few realities for network systems.

So we have been developing teaching materials for experimental studying of routing technologies based on concept of physical visualizations.

1 Introduction

In the new course of study for technology education of junior high schools in Japan, "interactive contents" and "contents of network" are newly included. In addition, the new course of study for information technology of high schools included the contents of learning the basic of the network technologies. However, there are little teaching materials for these studies. In addition, network technologies can not see how it works. Thus it is difficult students image network systems and technologies, because students have few realities for network systems.

So We have been developing teaching materials for experimental studying of routing technologies based on concept of physical visualizations. The teaching materials can visualize physically the process of selecting a path for traffic in a network using full color serial LED tapes.

2 Development of Teaching Materials

2.1 Development Environment

The development environment of this teaching material is shown in Table 1.

L. Barolli et al. (Eds.): CISIS 2019, AISC 993, pp. 789–797, 2020.
https://doi.org/10.1007/978-3-030-22354-0_72

Table 1. Development environment

Classification	Details
Computer	Raspberry Pi 2 Model B V1.1
OS	Raspbian Stretch
Programming Language	Python
Equipment for physical visualization	Full Color LED tape

Figure 1 shows Raspberry Pi 2 Model B V1.1.

The Raspberry Pi is a low cost, credit-card sized computer. It is developed by Raspberry Pi Foundation for the purpose of teaching basic computer science at schools.

Fig. 1. Raspberry Pi 2 model B V1.1

Raspberry Pi is full of input and output ports. For example, it has HDMI ports, LAN port, USB port, and 40 I/O pins. Figure 2 shows the role of each pin.

pin number

3.3V	1	2	5V
GPIO 2	3	4	5V
GPIO 3	5	6	GND
GPIO 4	7	8	GPIO 14
GND	9	10	GPIO 15
GPIO 17	11	12	GPIO 18
GPIO 27	13	14	GND
GPIO 22	15	16	GPIO 23
3.3V	17	18	GPIO25
GPIO 10	19	20	GND
GPIO 10	21	22	GPIO 25
GPIO 9	23	24	GPIO 8
GND	25	26	GPIO 7
ID_SD	27	28	ID_SD
GPIO 5	29	30	GND
GPIO 6	31	32	GPIO 12
GPIO 13	33	34	GND
GPIO 19	35	36	GPIO 16
GPIO 26	37	38	GPIO 20
GND	39	40	GPIO 21

Fig. 2. The role of pins

Figure 3 shows Full color LED tape for physical visualization.

It is a tape in which a plurality of LEDs are connected. It has 3 terminals: 5V, GND, input terminal of PWM signal. It can be used by connecting 5V and GND terminals to power supply and input terminal of the PWM signal to the microcomputer.

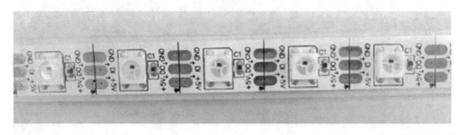

Fig. 3. Full color LED tape

In development of teaching materials for routing of network, the arrangement of equipment and allocation of IP address were as shown in Fig. 4.

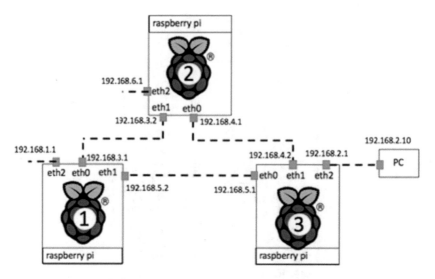

Fig. 4. Arrangement of equipment and allocation of IP address

The broken lines represent the ethernet cable. In this paper, we explain using this arrangement.

2.2   Packet Capture

Packet capture is to capture packets on the network. This teaching material has a packet capture function in python programming. Figure 5 shows a example of packet capture. We can specify what data to display on the monitor.

```
get icmp packet
rp3_eth0 > rp1_eth1, length 42, icmp type: 8
get icmp packet
rp1_eth1 > rp3_eth0, length 42, icmp type: 0
get icmp packet
rp3_eth0 > rp1_eth1, length 42, icmp type: 8
get icmp packet
rp1_eth1 > rp3_eth0, length 42, icmp type: 0
```

Fig. 5. A example of packet capture

This teaching material analyzes the captured packet, and it discards non-icmp packets. In addition, it discards non-type 8 packets. That is, it uses only ping echo request packets.

2.3   Socket Communication

In this teaching material, we use socket communication to adjust the lighting timing of the LED tape.

It is an instant that ICMP packets reaches their destination and is captured by each Raspberry Pi. If it is processed to light up the LED tape after capturing, all the LED tapes start to light at almost the same time. It is necessary to adjust the timing in order to light the packet path in order.

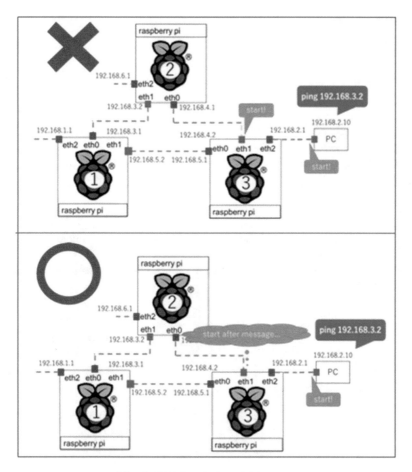

Fig. 6. Lighting timing of LED tape

In socket communication, TCP or UDP is mainly used. This time, we chose UDP, which has high real-time capability and simple processing.

We explain the process of timing adjustment with Fig. 6.

First, when Raspberry pi captures the packet, it refers to the source IP address. Second, baced on the result, it determined whether the LED tape shoud light first.

ex.) In Fig. 6, the LED tape between PC and Raspberry Pi(3) should light first.

And, LED tapes that should not light first will stand by without lighting. Third, once the first LED finish lighting, a message is sent to the Raspberry Pi which controls next LED tape.

ex.) In Fig. 6, Raspberry Pi(3) sends a message to Raspberry Pi(2).

Raspberry Pi which received the message makes the second LED tape light.

ex.) In Fig. 6, the LED tape between Raspberry Pi(3) and Raspberry Pi(2) start lighting.

3 How to Use

3.1 Preparation

What is required in this teaching material is the following for one Raspberry Pi (the number)

- Raspberry Pi (1)
- Full color LED tape (2)
- Jumper Wire M/M (6)
- Jumper Wire M/F (3)
- Condenser 1000μF (2)
- 330 Ω resistance (2)
- USB ethernet adapter (2)

Assign an IP address to each interface of Raspberry pi. And solder the LED tape and Jumper Wire M/M.

And connect Raspberry Pi and LED tape. Figure 7 is a schematic of the LED tape and the Raspberry Pi.

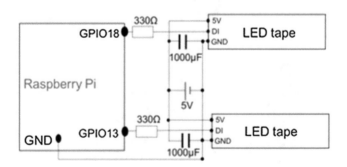

Fig. 7. A schematic of the LED tape and the Raspberry Pi

Figure 8 shows the arrangement of actual equpment and a state of the experiment.

This arrangement is only an example. It works even if you add more Raspberry Pi or change the position of the PC.

In Fig. 8, we send a ping from PC to 192.168.3.1. The LED tape between PC and Raspberry Pi(3) has finished lighting.

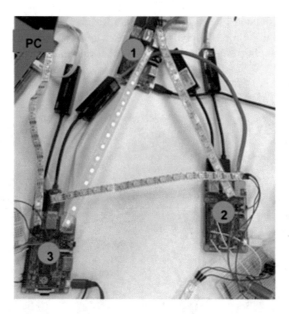

Fig. 8. The arrangement of actual equipment and a state of the experiment

Also, this teaching material acquires the route table of Raspberry Pi, and works properly even if the route is changed. There is no need to stop the program.

Figure 9 shows the state when sent a ping to 192.168.3.1 from PC in the shortest path.

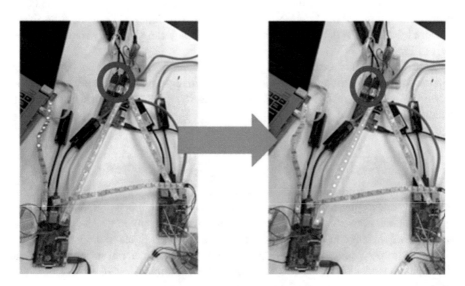

Fig. 9. Experiment of shortest path

The place surrounded by red circle is where Raspberry Pi(3) and Raspberry Pi(1) are connected. At that time, Raspberry Pi was set to perform routing control dynamically. So, if we unplug the cable there, the packet automatically reroutes. Figure 10 shows it.

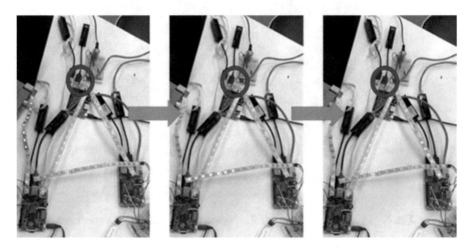

Fig. 10. Experiment of rerouting

4   Discussions

This teaching material is possible to physically visualize how packets are transferred from the terminal and reach the destination. This makes it possible to understand the structure of the network. In addition, student can experience the construction of the network by setting the devices and IP address. And they can physically see the result of their settings.

Also, this teaching material can physically visualize the flow of packets between different networks. This makes it possible to intuitively understand the routing of the network. The routing table can also be set statically, so student gain a better understanding the routing of the network.

5   Conclusion

In this study, we developed a teaching material for routing of the network using the full color LED tape as physical interface. By using this marerial, it is possible to experience learning about networks that is currently difficult. Also, by physically visualizing packet flow, student can intuitively understand about routing technology.

As future prospects, we want to practice this teaching material in junior high and high school students' class and to confilm the usefulness. Also, in this teaching material, students have to set IP address and to addition routing tables on the console screen. However, it is difficult for junior high and high school students to operate console screen. Therefore, we want to be able to set easily by them using the dials. Also, in case of dynamic routing, we want to see on the display how computer has chosen that route. In addition, we want to physically visualize technologies other than routing, such as NAT and NAPT.

Acknowledgements. This work was supported by JSPS KAKENHI Grant Number JP18K11570.

References

1. The MEXT, The course of study for junior high school. http://www.mext.go.jp/component/a_menu/education/micro_detail/__icsFiles/afieldfile/2019/03/18/1387018_009.pdf. Accessed 20 April 2019, in Japanese
2. The MEXT, Points of high school course of study revision. http://www.mext.go.jp/a_menu/shotou/new-cs/__icsFiles/afieldfile/2018/04/18/1384662_3.pdf. Accessed 20 April 2019, in Japanese
3. Raspberry Pi. https://www.raspberrypi.org/. Accessed 20 April 2019, in English
4. SWITCHSCIENCE, Serial LED tape with protective tube 1 reel (4 m). https://www.switch-science.com/catalog/2819/. Accessed 20 April 2019, in Japanese

Research of the Harvesting Date Prediction Method Using Deep Learning

Yukikazu Murakami[✉], Kengo Miyoshi, and Kazuhiro Shigeta

National Institute of Technology, Kagawa College,
355 Chokushi-cho, Takamatsu, Kagawa, Japan
{murakami,shigeta}@t.kagawa-nct.ac.jp

Abstract. Contract farming has a managerial advantage that farmers can directly negotiate prices with business partners. And it is necessary to predict harvesting date of agricultural crops precisely for contract farming. Conventionally, this precondition has been solved by experience rule of experienced farmers or some mathematical model and simple regression models. However, for new farmers who do not have experiential rules, contract farming is a difficult management method. And previous prediction models cannot consider complex relationship between environmental parameters. In order to solve this problem, this research proposed automatic harvesting date prediction method using statistical model by deep learning. This research confirmed that deep learning models exceed the accuracy of non-deep learning models. And this research proposed a practical system using our models and one existing Web application for management of farming tasks.

1 Introduction

The agricultural working population in 2017 had decreased by 790,000 compared to 2010. This number is 30.3% of the whole agricultural working population in 2010. In addition, the number of people aged 65 years or older during the same period, the percentage of workers has risen by about 4.9%. The reduction of the agricultural industry due to the decline of the farming population and the aging of the population have many problems. As a method of solving the problem, various work in productivity and agriculture have been considered to improve the efficiency more than before and to make agriculture more attractive industry [1]. This research focus on a "Contract cultivation" as one of the measures to improve the efficiency of agricultural management. We research on harvest date forecasting methods using deep learning to support contract cultivation.

2 Theory

This study proposes a harvest date prediction method for agricultural products using Deep neural network (DNN). DNN is one of a statistical models for realizing machine learning on a computer. It has made the middle layer of the

© Springer Nature Switzerland AG 2020
L. Barolli et al. (Eds.): CISIS 2019, AISC 993, pp. 798–803, 2020.
https://doi.org/10.1007/978-3-030-22354-0_73

hierarchical neural network multi-layer. It is possible to obtain the essential features more effectively than non-layered neural networks. However, deepening layers causes gradient loss problems in most cases [2]. Deepening of DNN can not achieve the expected effect unless gradient losses can be avoided appropriately. Therefore, it is important to select hyperparameters for appropriate learning (Fig. 1).

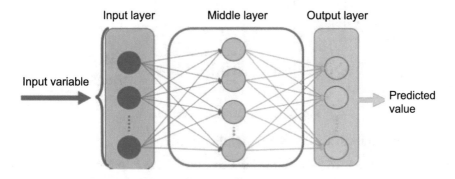

Fig. 1. Deep neural network

3   Overview of Harvest Date Prediction Method and WEB Implementation

3.1   Harvest Date Forecast Model

It shows a model for forecasting harvest dates at Fig. 2. We explain the behavior of the model. First, normalized input parameters in the data set are input to the input layer of the model. The input signal is outputted from the output layer as the harvest date through the middle layer. The middle layer has 4 layers. It uses the ReLU (Rectified Linear Unit) function in the first to third layers, and the tanh function is used in the fourth layer. In the output layer, the sum of the signals input from the fourth intermediate layer is used as the output.

3.2   Reliability Judgment Model

It is considered that the forecast by the harvest date forecast model may not fall within an allowable error (maximum 1 week). Therefore, when using the method, We use a reliability of the outputted forecast harvest date. In this study, we have developed a reliability judgment model that predicts the probability that the predicted harvest date is outputted within 3 days. It shows a model for forecasting harvest dates at Fig. 3.

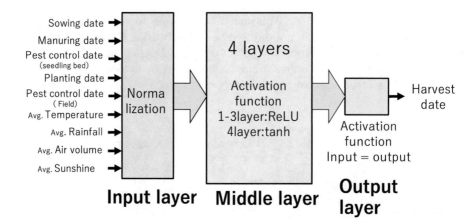

Fig. 2. Harvest date forecast model

3.3   Implementation to WEB Application

We have implemented the two models described above into the web application "iFarm" [3] for agricultural work management. By this implementation, users can execute this prediction method using only a WEB browser.

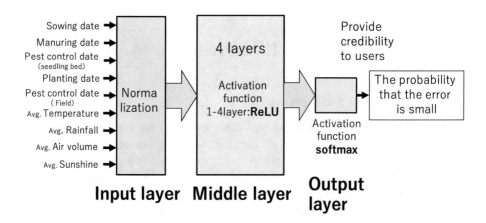

Fig. 3. Reliability judgment model

4   Accuracy Verification of Harvest Date Prediction Method

4.1   Data Set

We show the data set used for learning and evaluation of the model in Fig. 4. The data set consists of the work schedule in the cultivation cycle shown in Fig. 4 and the average temperature, precipitation, air volume and sunshine in the cultivation period.

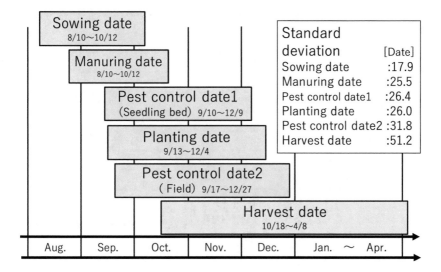

Fig. 4. Data set

4.2 Accuracy Verification of Harvest Date Prediction Model

For each data cycle in the data set, We verify what kind of accuracy the harvest date prediction model has. The accuracy of the model is evaluated using leave-one-out cross validation. We have used the "Absolute error" to the indicator of verification.

4.3 Accuracy Verification of Reliability Determination Model

We have used leave-one-out cross validation as well as the "harvest date prediction model" to evaluate its accuracy. As the evaluation method, the reliability of the probability of model output is rounded off, and the predicted value is converted to binary data similar to the value of the label. The result is collated with the label of the data set to obtain the correct answer rate.

5   Verification Result

Figure 5 shows the prediction accuracy for each data cycle possessed by the harvest date prediction model, and Fig. 6 shows the results for the correct answer rate of the reliability judgment model. It was shown that the prediction with the harvest date prediction model has 27 in 35 data cycles which predicted with an error within 3 dates from Fig. 5 and from Fig. 6, the reliability judgment model has a 82.9% maximum correct rate. Because of this, it is possible to judge that the reliability judgment model can successfully learn the magnitude relationship the harvest date prediction error using the work schedule data and the weather

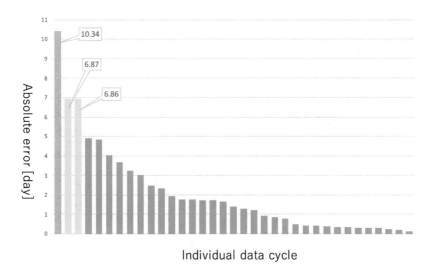

Fig. 5. Prediction accuracy for each individual data cycle

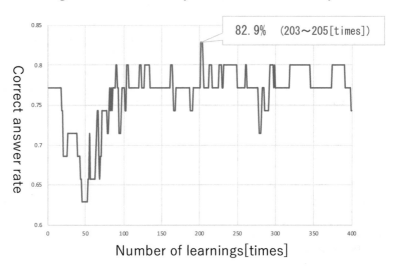

Fig. 6. Prediction accuracy of the reliability judgment model

data. However, in this data set, we also confirmed that there is data that does not fit within the allowable range of errors. In other words, the user can select the prediction result by using the reliability judgment model.

6   Conclusion

In this study, we aimed to develop an automatic forecasting system for harvest date with accuracy that can withstand practical use in contract cultivation. For that purpose, we have proposed a harvest date forecast model and reliability judgment model for lettuce using deep learning. From the verification results, it was confirmed that the prediction accuracy can withstand the actual use with linking the harvest date prediction model and the reliability determination model. As future prospects, we think that it is need to continuously collect data sets in cooperation with iFarm. Because to collect the cultivation cycles data requires a lot of time and effort, I think that cooperation with iFarm, which can easily collect data using IoT devices, is important for improving the accuracy of the system.

References

1. Ministry of Agriculture, Forestry and Fisheries, Present state and future of ICT agriculture (mainly AI agriculture) (2015). http://www.maff.go.jp/j/shokusan/sosyutu/sosyutu/aisystem/pdf/ictai.pdf. Accessed 18 Jan 2018
2. Japan Society for Artificial Intelligence: Deep Learning: The Modern Science Company, pp. 267 (2015)
3. Murakami, Y.: iFarm: development of web-based system of cultivation and cost management for agriculture. In: CISIS 2014, Birmingham, UK, 2nd–4th July 2014

The 9th Semantic Web/Cloud Information and Services Discovery and Management (SWISM-2019)

Enabling Accountable Collaboration in Distributed, Autonomous Systems by Intelligent Agents

Flora Amato[1]([✉]), Pasquale Femia[2], and Francesco Moscato[2]

[1] DIETI, University of Naples "Federico II", Naples, Italy
flora.amato@unina.it

[2] DiSciPol, University of Campania "Luigi Vanvitelli", Caserta, Italy
{pasquale.femia,francesco.moscato}@unicampania.it

Abstract. High degree of *distribution* is one of the leading features in present computer systems. Cloud Computing, Internet of Things and Blockchains are nowadays very hot research topics and we are going to use these architectures in many critical domains, like e-health, conservation of documents and acts in accordance with the law, economic transactions and contracts etc. When dealing with such heterogeneous systems, it is really hard to understand if a distributed collaboration among agents fulfils requirements, rules and current laws. In addition, if something goes wrong during collaboration, assigning accountability is even more complicated. This work aims at introducing a novel methodology and a formal framework able to attribute liability of failures or incorrect design and implementation both to humans and software agents in autonomous, distributed systems.

1 Introduction

Nowadays new distributed architectures are widespread and more and more domains exploit their features in order to make activities faster, reliable, transparent and more automated. In particular, they have introduced software (sometimes *intelligent*) agents in interactions and during collaborations where once only humans took part. Everyone relies on distributed systems as aid to everyday activities.

In particular, when dealing with national and international laws and rules, collaborations and distributed transactions among agents (both human and software) must follow *contracts* requirements, as well as all laws that can be applied to the domain where agents stipulate the contract. In particular, when cooperation fails in reaching contract goals, or when agents are not able to enact needed actions, liability of the failure should be assigned to some agent, but dealing with automated software as agents make liability difficult to assign.

In this context, many questions arise: are software agents acting in compliance with requirements, contract rules and, of course, with existing laws and rules? In the case of concurrency with humans and software agents, is the use of automated

ⓒ Springer Nature Switzerland AG 2020
L. Barolli et al. (Eds.): CISIS 2019, AISC 993, pp. 807–816, 2020.
https://doi.org/10.1007/978-3-030-22354-0_74

procedures fair? If something goes wrong (due both to malicious and incautious actions), are we able to pick out the agent responsible for failure or crime? What does it matter in a contract if a failure, or a damage, is caused by a software element ? Where Liability is?

Many are the cases where human and software agents are involved in distributed interactions. For example, trade markets sees High Frequency Trading as a problem for human agents [1], but this technology attracts more and more investments [2]. Public administration and construction industry have appointed Building Information Model [3]. Always more often National cybersecurity focuses on critical infrastructures [4] (like hospitals, railway systems etc.), where software malicious interventions cause economical or even human losses.

Dealing with these technologies involves the introduction of "third-parties" actors, creating intricate chains of responsibility that are very hard to disentangle when solving legal disputes and issues. The problem is understanding where responsibility of "something" when both human and software actors interacts when enacting rules defined in a contract. The difficulties are hidden in the fact that other actors develop software, which in turn uses other software and so on.

In many cases actors are not aware of all legal issues. Complex interactions and the presence of many actors in smart contracts, create chains of responsibilities that are not easy to manage.

In addition, malicious interventions may happens, or even the whole contract can be designed and defined with malicious intents. Actors in a "normal" interaction usually have to deal with national laws, but distributed, smart contracts and in general distributed systems on Internet show trans-national characteristics and International Laws rule their requirements. At the moment, requirements related to Cyber Security is one of the hot topic in the scientific literature: again, understanding if a distributed interaction of actors fulfil national and international rules about cyber security is a really hard task, and determining "where" and "who" a multi-agent system violates these laws a is even more difficult task.

Here, we present a methodology able to analyse if a distributed, multi-agent system collaboration is compliant with pre-defined rules (generally addressed as *contracts*), national and international laws (i.e. is a contract "sound" or "correct"?). In addition, liability of failure in collaboration is addressed, considering *hidden* (third-parties agents described before) agents.

A framework, whose architecture is described too, enforces the methodology. It is based on an Artificial Intelligence system able to help lawyers and judges, as well as users and providers in understanding what are the features needed by each actor of a smart contract, as well as in finding responsibilities in the case of failures.

In our methodology, formal models here enable formal planning: we use and extend here a methodology inherited from Software Engineering field [5]. The methodology we will use extends the "Beliefs, Desires, Intentions" [6] logics. The application of this model for analysis of compliance of given MAS systems with international Law has been proved effective in previous works [7–9].

This paper addresses the issues of attribution of liability in case of failures of *contracts*. We propose a novel methodology, based on formal models, able to describe behaviours of agents in contracts and to analyse collaboration among contract actors in order to state if any requirement or rule is violated.

2 Methodology

The methodology we present here aims at providing a way to analyse collaborations by considering all (or at least the most of) the laws, rules and conventions related to stipulations among parts, taking into account both national and international laws. The proposed methodology explicitly addresses software agents into collaborations, dealing with the problem of analysing responsibilities when a *contract* uses automatic procedures and intelligent systems in order to fulfil its goals.

In order to fulfil these goals, the methodology consist of five main steps, as depicted in Fig. 1.

The first thing to do is to formalize both *Requirements* and *Behaviours* of agents.

Formal Definition of Laws and Rules: Requirements formalization requires a description of laws and rules of the contracts that agents are going to enact. In this step we use an ontological model able to define which action is allowed (and what it is not) in a given state. In addition, Liability is addressed by associating agents responsibility to failure (or success) of general actions. We model laws and rules, as well as actors characteristics by using Ontology Web Language (OWL) and/or Resource Description Framework (RDF).

Formal Definition of Agents' Behaviours and Interactions: In this step we formalize the agents structures and behaviours, as well as we describe their interaction. The model used here is based on the Beliefs, Desires and Intentions(BDI) model and on First Order Logic (FOL) STRIPS [10,11] formalisms. Agents have their knowledge of the whole system (modelled by beliefs). Actions agents are able to perform, are in turn defined by: (a)a list of preconditions; (b) a list of effects. Precondition and Effects are defined in a proper ontology.

Agents interactions effectively model the *contract* we are going to analyse. Case by case, contracts require the definition of a proper Sequence Diagram from Unified Modelling Language (UML) (more details are found in [12]) Many sequences are defined: each of them describes one use case of the contract execution. Hence, many cases may occur, both for correct and incorrect termination of the *contract*.

Creation of Joint Planning Model: From models defined in previous steps, we create a joint model, that merges all ontological information described before. Notice that national and international laws are modelled once and it is not necessary to re-define them for new contracts. The joint planning model will be described deeply later. We express the joint model as a planning problem: agents interacts to reach common goals, by using their BDI properties. External events

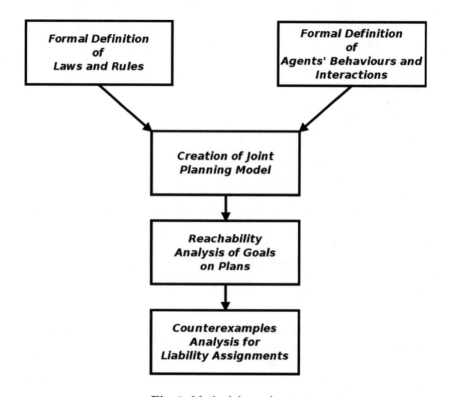

Fig. 1. Methodology phases

and agents are modelled as well. Plans are sequences of actions performed by agents in order to reach their own goals. Anyway during contracts enactments may have some actions enables, that can finally lead to a failure in contract execution. These conditions have to be analysed and they are the target of the next step.

Reachability Analysis of Goals on Plans: In this step, we apply model checking techniques in order to analyse reachability of the states where *contracts* executions terminates with success. This requires two steps, since we must consider both actions and their precondition and effects. We have to know, in any state of the *contract* execution, predicates that evaluates true in preconditions and effects. Hence, we need to use a *condition propagation* algorithm during all possible branches of contract execution. This algorithm too is described later.

Having provided analysable logic bases models for contract enactment, we can apply state search and model checking techniques. In this way, The problem of checking soundness of contracts, or the one of assigning liability in case of problems, will be translated into checks on the models of properties answering to questions like the following one: "Is it always true that a collaboration will ever reach contract goals respecting legal requirements?".

It is possible to prove that questions can be translated into a Computational Tree Logic (CTL) properties that can be easily checked on the proper model. Hence we provide in this phase an algorithm to generate and check these properties on models depicted by the users. Thanks to this methodology, we are able to understand if a contract is legally correct, and to assign liability in the case for any reason (including malicious, external interventions) the contract fails in its enactment.

Counterexample Analysis for Liability Assignments: The result of the previous step is a state transition model that we model-check in order to evaluate reachability of correct final states. If branches leading to incorrect termination (i.e. violating requirements, rules and laws) of the contract exist, the model checker generates counterexamples that can be used to understand who is the responsible of contract violation.

3   Multi-agent Joint Planning Model

Here we report a simplified version of the planning model introduced before. The model is based on a Multi-Agent System (MAS) that uses a representation of systems. We consider a variant of Beliefs, Desires, Intentions (BDI) logics [13], colored by temporal information that we do not report here for lack of space. A planing model PM is:

$$PM = (Agents, \mathcal{W}, \mathcal{TS}, \mathcal{F})$$

Where: $Agents$ is the set of agents we are considering in the contract enactment; $\mathcal{W}$ is the set of all beliefs, states, variables and predicates of the environment where the contract is enacted; $\mathcal{TS}$ is a Transition System that defines possible state transitions of agents in the environment and, finally, $\mathcal{F}$ is a set of formulas expressed in first order logics that characterize preconditions and effects in each state in $\mathcal{W}$.

The triple $\langle n, d, v \rangle$ to evaluates variables in $\mathcal{W}$, where n is the name of a variable, d is its domain and v the value assigned to the variable. A State $s \in \mathcal{W}$ is a set of variable evaluations.

For each state, the set *State Conditions* contains all formulas in $\mathcal{F}$ that holds in a given state s:

$$StateCondition(s) = \{\phi \in \mathcal{F}, s \in \mathcal{W} : s \models \phi\}$$

Notice that ϕ cannot be a sub-formula of other formulas holding in the state s.

In this work we consider $\mathcal{TS}$s and States with only one State Condition per state. If $s \models \psi$; $s \models \phi$ and $s \models \psi \wedge \phi$; then we consider only the last formula as State Condition in s.

Agents are then triples:

$$(Actions, Beliefs, Goals)$$

Where: *Actions* is the set of actions that an agent can execute (if its preconditions are met). Actions modify the environment changing $\mathscr{W}$ representation.
In particular, If preconditions of actions evaluate true, $\mathscr{W}$ changes such that all
predicates in the effects of the action evaluate true. Some actions may require
the execution of actions of other agent and may need *communication* as well.
Actions of *Reactive* agents also depend on external events or messages. Special
reactive agents are *Proactive* and the effects of the agents directly by the agent.
An agent that is neither reactive or proactive is addressed as *Resource*.

Beliefs include the knowledges the agents have about: the $\mathscr{W}$; the Agent
itself; other Agents.

Notice that an agent may have a belief about the $\mathscr{W}$ which in turn is *not*
true in the environment: in general, beliefs of each agents may not be exact.

Goals is a set of states in $\mathscr{TS}$ that represent states an agent want to reach
in order to terminate the contract with success. Abusing notation we call *Goal
Conditions* in $\mathscr{F}$ the formulas that are satisfied in goal states. Goal condition is
a State Condition for a goal state.

Beliefs are manages as Worlds variables and states.

Agents define the $\mathscr{TS}$ Transition System on $\mathscr{W}$ states by means of *Actions*.
An action $\alpha \in Actions$ is a triple:

$$(name, Precondition, Effects)$$

where *name* is trivially the name of the action; *Effects* is a set of formulas that
hold in the new state; *Precondition* is a formula that *evaluates* true in a state s
in order to *apply* (execute) the action and to produce a state transition:

$$s \overrightarrow{\alpha} s'$$

produces a transition from the state s to s'. If Precondition of α evaluates true
in s, s' will be the same of s, except for variables involved in Effects evaluation.
The values of these variables have to change in order to satisfy *all* effects in s':

$$\forall \phi \in Effects \; s' \models \phi$$

In addition we must consider that an agent executing an action can access
only to its local representation of $\mathscr{W}$, i.e. to its beliefs. Hence, if agent's beliefs
and World State are not synchronized (i.e., if the agents has a wrong belief about
the world), it is possible that Precondition is evaluated true on beliefs, but *not*
on $\mathscr{W}$ state.

In order to apply an action, we must execute the following two steps: (1) an
agent tries to apply Effects in a state s producing a transition from s to s' if
precondition is evaluated true on its *beliefs*; (2) if Precondition evaluates true in
World too, then $s \overrightarrow{\alpha} s'$ both in agent's Beliefs and in $\mathscr{W}$ too.

$\mathscr{TS}$ is then the Transition System defined by the application of all actions in
any state of $\mathscr{W}$, performed by *all* Agents in the model. The execution of an action
to build $\mathscr{TS}$ must follow the two steps previously defined. State transitions
apply both to agents *Belief* and to $\mathscr{W}$. Anyway Precondition control is enacted

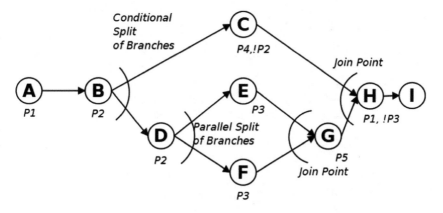

Fig. 2. Condition propagation

on Beliefs first. If evaluation fails on Beliefs, the action is not applied even if Precondition would evaluate true on $\mathcal{W}$. In this model, a **Plan** to reach a Goal $\mathcal{G}$ with given Goal Condition is a *path* from a starting state to a state where the Goal Condition holds. Notice that in a Multi Agent System, actions in a transition System can be executed by different agents, even concurrently. We consider here a path as a linear scheduling of concurrent applications of actions.

4   Condition Propagation

During Contract execution, actions of agents modify the truth values of predicates in $\mathcal{W}$. Hence, in order to analyse contract enactments, we need to know: (1) if execution of actions are enabled in a given state; (2) What are the effects of the action on $\mathcal{W}$. Hence, we need a way to describe Precondition and Effects of actions and, more important, how they propagate during contract enactment.

At this purpose, we use the idea of some Web Service languages and methods: in particular, we use the concepts related to the Ontology Web Language for Services (OWL-S) [14], thus juxtaposing actions to the concept of services and exploiting the ontological definition of first steps of the methodology we explained before. We use Precondition and Effects definition similar of the one discussed in [14].

One of the main problems to solve is that agents, while interacting during the enactment of a contract, may execute operations in parallel, of can choose their actions depending on some events or conditions. This leads to many possible schedules of actions during contracts enactment.

Notice that an action can be executed by an agent if $\mathcal{W}$ satisfies all its precondition in a given state. Anyway, before the enactment of the contract, $\mathcal{W}$ it is not known in a given state, because of the presence of parallel or conditional branches in contract definition.

Let us consider the planning model described by the graph in Fig. 2, where nodes represent actions (it is not important to know which is the agent executing

A:
$$\mathscr{P}^A = [((P_1, true))]$$
B:
$$\mathscr{P}^B_{before} = [((P_1, true))]$$
$$\mathscr{P}^B = [((P_1, true), (P_2, true))]$$
C:
$$\mathscr{P}^C_{before} = [((P_1, true), (P_2, true))]$$
$$\mathscr{P}^C = [((P_1, true), (P_2, false), (P4, true))]$$
D:
$$\mathscr{P}^D_{before} = [((P_1, true), (P_2, true))]$$
$$\mathscr{P}^D = [((P_1, true), (P_2, true))]$$
E:
$$\mathscr{P}^E_{before} = [((P_1, true), (P_2, true))]$$
$$\mathscr{P}^E = [((P_1, true), (P_2, true), (P_3, true))]$$
F:
$$\mathscr{P}^F_{before} = [((P_1, true), (P_2, true))]$$
$$\mathscr{P}^F = [((P_1, true), (P_2, true), (P_3, true))]$$
G:
$$\mathscr{P}^G_{before} = \mathscr{P}\mathscr{H}^E_{before} \cup \mathscr{P}^F_{before} =$$
$$[((P_1, true), (P_2, true), (P_3, true)),$$
$$((P_1, true), (P_2, true), (P_3, true))]$$
Collapsing Union:
$$\mathscr{P}^G_{before} = [((P_1, true), (P_2, true), (P_3, true))]$$
$$\mathscr{P}^G = [((P_1, true), (P_2, true), (P_3, true), (P_5, true))]$$
H:
$$\mathscr{P}^H_{before} = [((P_1, true), (P_2, false), (P_4, true)),$$
$$((P_1, true), (P_2, true), (P_3, true), (P_5, true)),$$
$$((P_1, true), (P_2, true), (P_3, true), (P_5, true))]$$
Collapsing Union:
$$\mathscr{P}^H_{before} = [((P_1, true), (P_2, false), (P_4, true)),$$
$$((P_1, true), (P_2, true), (P_3, true), (P_5, true))]$$
$$\mathscr{P}^H = [((P_1, true), (P_2, false), (P_3, false), (P_4, true)),$$
$$((P_1, true), (P_2, true), (P_3, false), (P_5, true))]$$

Fig. 3. Condition propagation

it), where edges introduce precedences of actions and where their Effects are listed below each edge.

For example, the execution of action **B** produces the effect that makes true the predicate *P2*, while execution of **C** makes *P4* true and *P2* false.

In addition, we consider parallel and conditional splits (we omit the management of split and join condition for brevity's sake).

Notice that condition propagation considers only Effects, because of Preconditions have only to be checked from $\mathscr{W}$ when executing the action.

By traversing the graph in Fig. 2, we build a set of sets of predicates the evaluates true. Inner sets contain predicates that evaluate true before and after the execution of an action.

Due to the presence of parallel and conditional branches, it is possible to have more the one inner set of predicates, and they are all possible before and after the execution of an action.

The formal definition of condition propagation is complex, but we exemplify it here for convenience. A deeper description of the algorithm is in [15].

Let $\mathscr{P}^X_{before}$ be the set of sets with evaluations of predicates before the execution of action X, and let $\mathscr{P}^X$ be the set after the execution.

Notice that new sets of predicates can be inserted in $\mathscr{P}^X$ if X follows a split point, and that some inner sets can collapse to one if X follows a join point.

Let us indicate with $(P_a, true\ or\ false)$ the evaluation of predicate P_a.

Finally, let us address sets of sets by including inner sets in square brackets. Condition propagation works as described in Fig. 3.

If preconditions of I include P_1 and P_4, it is clear that there is only one branch in contract execution that leads to the assertion of P_4 (i.e. only one of the sets in $\mathscr{P}^H$ contains P_4). Hence, there is a case when the contract cannot be executed, due to the choice of following the lower path in Fig. 3.

In this case, the liability of the failure of contract execution is in the cause of the choice of the lower path: it may be an external event, or a choice made by an agent.

5   Conclusions and Future Works

In this work we presented a methodology to analyse distributed collaboration among humans and software elements, making possible accountability of liability during the execution of a *contract*. We provided the definition of a formal model to enact analyses, which is based on BDI logics and MAS, and that can be used to generate a state transition model to analyse by means of model checking. Precondition and effects in the state transition model are managed by an algorithm called *Condition Propagation* we use here to support liability analysis.

Future works include the application of the methodology to blockchain [16] technologies and the definition of complex ontologies to define rules and laws to apply in the context of public and private contracts.

References

1. Baron, M., Brogaard, J., Hagströmer, B., Kirilenko, A.: Risk and return in high-frequency trading. J. Financ. Quant. Anal. 1–32 (2018)
2. Delaney, L.: Investment in high-frequency trading technology: a real options approach. Eur. J. Oper. Res. **270**(1), 375–385 (2018)
3. Porwal, A., Hewage, K.N.: Building information modeling (BIM) partnering framework for public construction projects. Autom. Constr. **31**, 204–214 (2013)
4. Paté-Cornell, M.-E., Kuypers, M., Smith, M., Keller, P.: Cyber risk management for critical infrastructure: a risk analysis model and three case studies. Risk Anal. **38**(2), 226–241 (2018)

5. Moscato, F., Amato, F., Amato, A., Aversa, R.: Model-driven engineering of cloud components in MetaMORP(h)OSY. Int. J. Grid Util. Comput. **5**(2), 107–122 (2014)
6. Wooldridge, M.: An Introduction to Multiagent Systems. Wiley, Hoboken (2009)
7. Amato, F., Moscato, F.: A planner for supporting countermeasures in large scale cyber attacks. In: Conference on Complex, Intelligent, and Software Intensive Systems, pp. 964–973. Springer (2017)
8. Amato, F., Moscato, F.: A model driven approach to data privacy verification in E-Health systems. Trans. Data Priv. **8**(3), 273–296 (2015)
9. Amato, F., Cozzolino, G., Mazzeo, A., Moscato, F.: An application of semantic techniques for forensic analysis. In: 2018 32nd International Conference on Advanced Information Networking and Applications Workshops (WAINA), pp. 380–385. IEEE (2018)
10. Fikes, R.E., Nilsson, N.J.: STRIPS: a new approach to the application of theorem proving to problem solving. Artif. Intell. **2**(3), 189–208 (1972)
11. Burfoot, D., Pineau, J., Dudek, G.: RRT-Plan: a randomized algorithm for strips planning. In: ICAPS, pp. 362–365 (2006)
12. Amato, F., Mazzocca, N., Moscato, F.: Model driven design and evaluation of security level in orchestrated cloud services. J. Netw. Comput. Appl. **106**, 78–89 (2018)
13. Wooldridge, M.: Agent-based software engineering. IEE Proc. Softw. Eng. **144**, 26–37 (1997)
14. Klein, M., Konig-Ries, B., Mussig, M.: What is needed for semantic service descriptions? A proposal for suitable language constructs. Int. J. Web Grid Serv. **1**(3–4), 328–364 (2005)
15. Amato, F., Moscato, F.: Pattern-based orchestration and automatic verification of composite cloud services. Comput. Electr. Eng. **56**, 842–853 (2016)
16. Swan, M.: Blockchain: Blueprint for a New Economy. O'Reilly Media Inc., Newton (2015)

Semantic Representation of Cloud Manufacturing Services and Processes for Industry 4.0

Beniamino Di Martino[✉], Valeria Di Traglia, and Ivan Orefice

Dipartimento Di Ingegneria, Università Della Campania "Luigi Vanvitelli", Caserta, Italy
beniamino.dimartino@unina.it,
valeriaditraglia@libero.it, ivanorefix@yahoo.com

Abstract. The manufacturing process has become increasingly complex, based on many new technologies and advanced networks, in response to changes in local, national and international markets. In particular, what companies are trying to adopt is the so-called Cloud Manufacturing, a business model based on the network sharing of some technologies that allow analysis and real-time management of data related to machinery, equipment, and all the resources that participate in the supply chain of a product. This work consists in having understood and developed a semantic representation of the services that Cloud Manufacturing offers in its implementation, and in particular the computer technologies on which it is based.

1 Introduction

One of the consequences happened with the phenomenon of globalization was competition in all sectors. In particular, industries are in competition for their survival, not only nationally but globally. The main element of an industry's life is innovation. Moreover, what companies try to adopt is the so-called Cloud Manufacturing, a business model based on the sharing of networks in some technologies and analysis of resource management participate in the assembly line of a product.

The first part consists of a brief introduction of the concept of cloud: the subjects that take advantage of this new production model that goes beyond all the problems of network management existed and the enabling technologies.

In the second part, a brief representation of what is the structural taxonomy of which the Cloud is composed. Then we move on to the "core" part of this work, the ontology, that, applied to the business process, becomes very useful for communication and information exchange between the different systems.

Finally, a brief conclusion regarding the potential for future development of Cloud Manufacturing.

1.1 State-of-the-Art of Cloud Manufacturing

Although manufacturers benefit from the implementation of cutting-edge networking technologies to find new solutions over the competitors, there are various problems in

© Springer Nature Switzerland AG 2020
L. Barolli et al. (Eds.): CISIS 2019, AISC 993, pp. 817–826, 2020.
https://doi.org/10.1007/978-3-030-22354-0_75

these existing network technologies that affect production within the manufacturing industry. Among these problems we can find: the sharing of production resources, because resources are centralized in the network and cannot be distributed through the latter due to the lack of management of production services in the network; the inability to access production resources (equipment, machinery) in the production network due to complications in transferring demanding resources to the network; difficulties in sharing knowledge between production units, such as suppliers, customers and partners, due to the geographical dimension, the regulations of the countries, the different operating systems and the amount of complex data and processes that are an integral part of manufacturing production.

To address these problems in the manufacturing sector, a new production model combines innovative technologies and existing production networks that have emerged in the last period to create a new model called "Cloud Manufacturing". This model can provide and share productive resources and capacities as services for users in corporations, which gives life to the so-called Industry 4.0, a concept that goes hand in hand with cloud manufacturing.

1.2 Cloud Manufacturing Definition

The fact that Cloud Manufacturing is considered an emerging concept and a living idea, which has not yet been defined, means that currently there is a variety of definitions for it. Borrowing the definition of NIST for Cloud Computing, Cloud Manufacturing can be defined as a method to enable, through the network, easy and on demand diffused access to a shared and configurable set of manufacturing resources (for example software to support the production, resources and production capacities) that can be acquired and released quickly and with minimum effort of management or interaction with the service provider. It can be seen as a production model that provides resources and production capacities and a knowledge base platform for collaboration between different users (consumers, producers, suppliers) to achieve their goals using the latest information technologies and advanced communication networks.

2   Taxonomy of Cloud Manufacturing

This chapter focuses on a description of everything related to Cloud Manufacturing, starting from a taxonomy. Particular attention will be given to the technologies that enable it, together with the services that these technologies offer. Capturing the requirements for cloud manufacturing and its types, its characteristics and attributes in the form of taxonomy can allow companies to understand and choose a system as much as best suited to their needs. The taxonomy analyzed, contained in the article "Taxonomy and uncertainties of cloud manufacturing" by the authors Yaser Yadekar, Essam Shehab and Jorn Mehnen, provides some macro categories that will be detailed below.

2.1 Deployment Models of Cloud Manufacturing

There are three types of deployment models in the cloud environment: public cloud, private cloud, and hybrid cloud. Each type is designed for a specific situation suitable for companies and to satisfy their requirements. A public cloud offers services and infrastructure from an external third-party service provider via the Internet. The advantage of this type of cloud is to reduce the cost of IT solutions in the company. However, security and privacy issues are the disadvantages of this type of implementation model, as the infrastructure is in the public domain. A private cloud offers companies the same services and the same infrastructure as the public cloud, but is managed internally, with the only business organization that exploits cloud services. Building and managing a private cloud can be an expensive option for organizations. Finally, a hybrid cloud consists of two types of clouds, a public cloud and a private cloud. This type of cloud is used by companies to determine how to distribute and share services, important information and the infrastructure inside or outside the company. Non-critical data is migrated to a public cloud while critical data is transferred to a private cloud.

2.2 Delivery Models of Cloud Manufacturing

There are two classifications delivery models: the first type depends on IT resources (storage, software, server and network), while the second type depends on resources and production capacity.

The type of IT resources includes three service delivery models, which can be considered as three levels of abstraction: Infrastructure as a Service (IaaS), Platform as a Service (PaaS) and Software as a Service (SaaS). In IaaS, all the hardware (servers, storage space and network components) needed to support any computing operation within the company is owned by the cloud service providers, and therefore controlled by them. PaaS provides the IT platform, which includes the operating system, the programming language and the database to the company as a service: everything is supported by SOAP and REST. The SaaS model provides users with software applications without the need to purchase, install and manage the application, in which application is performed via the Internet from the cloud. The other type includes all the production resources and capabilities involved in industrial aspects, which can be provided through a service model for cloud manufacturing users.

2.3 Stakeholders of Cloud Manufacturing

The main stakeholders in any typical information system environment are the suppliers, who sell, install, license, perform system maintenance, and consumers, who use, own, maintain and update the system. However, in a Cloud environment, new stakeholders appear and the role of providers and consumers changes. Interested parties in a cloud production can be classified into three main groups: cloud users, cloud providers and cloud operators.

Cloud providers are responsible for providing production resources to cloud users. Users are considered consumers or organizations, who subscribe to a cloud production service. Cloud operators own and manage production and are responsible for providing cloud services to users.

2.4   Resources in Cloud Manufacturing

The production resources can be divided into two groups: the first group concerns the soft resources, which includes software, analytics libraries and cloud services, while the second group, related to hard resources, includes production equipment, monitor devices control, materials, transport, storage and computational resources (server, platform).

2.5   IT Technologies of Cloud Manufacturing

Cloud manufacturing is supported by four main IT technologies: cloud computing, IoT, virtualization and artificial intelligence. All of these technologies are used in many areas of manufacturing factories and help workers in their job, from the design to the real production. These technologies are discussed further in this paper.

3   Ontology

In this chapter will be discussed the work environment with which the ontology of Cloud Manufacturing was developed, followed by the introduction of the ontology itself. An ontology is a formal, shared and explicit representation of a conceptualization of a domain of interest. In IT, they can serve a variety of purposes, including deductive reasoning, classification, different problem solving techniques.

3.1   Work Environment

The work environment used is Protégé. Protégé is a free and open source ontology editor, and also a knowledge management system. Provides a graphical user interface to define ontologies. Like Eclipse, Protégé is a framework for which various other projects suggest plugins. This tool uses the concept of classes as a starting point for the development of ontologies.

3.2   Cloud Manufacturing Ontology

As can be seen from image 1, the central point of the ontology developed is the Cloud Manufacturing, the main class (in addition to Thing) from which various "SubClassOf" relations start.

The six subclasses linked with the main class are: Deployment models, Models and Networks, Stakeholders, Manufacturing Resources, Delivery models and

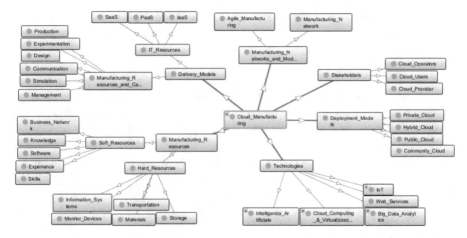

Fig. 1. Ontology of Cloud Manufacturing

Technologies. Each of them has additional subclasses, which have the same kind of relationship that the main Cloud Manufacturing class has with their daughters.

Figure 2 springs from a zoom of the ontology related to the class Technologies. Describes the technologies that enable cloud manufacturing, and that offer useful services in the corporate production process. It has five subclasses which are: IoT,

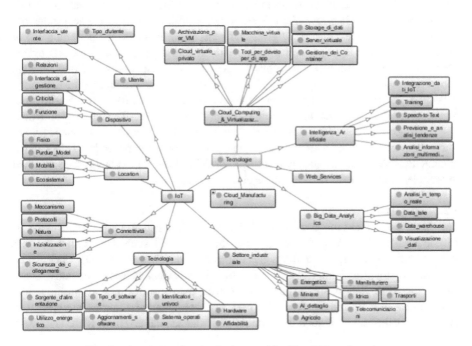

Fig. 2. Ontology of technologies used in Cloud Manufacturing

Artificial Intelligence, Web Services, Big Data Analytics and Cloud Computing & Virtualization.

3.2.1 Cloud Computing

This class, the Cloud computing, refers to a paradigm of providing services offered on demand from a supplier to an end customer through the Internet (such as data storage, processing or transmission), The resources are not fully configured and implemented by the supplier specifically for the user, but they are assigned, quickly and conveniently, thanks to automated procedures. When the user releases the resource, it is similarly reconfigured in the initial state and made available again in the shared pool of resources, with the same speed and economy for the supplier. The cloud computing architecture involves one or more real servers, generally in a highly reliable architecture (server cluster) and physically located in the data center of the service provider. The service provider exposes interfaces to list and manage its services. The administrator client uses these interfaces to select the requested service (for example a complete virtual server or just storage) and to administer it (configuration, activation, deactivation). The end customer uses the service configured by the administrator client.

3.2.2 IoT

Particular importance is assumed by the subclass IoT: this technology allows us to connect physical objects and automatically exchange data over the Internet using supporting technologies. In other words, it represents the possibility of connecting daily devices (coffee maker, oven, smartphone or machine tool) to the Internet to interact with other devices. Therefore, the IoT at an industrial level aims to connect industrial goods, such as engines, electricity grids and sensors, in the cloud thanks to the use of internet networks: it is a system that includes intelligent objects on the network, cyber-physical resources, information technologies associated and optional cloud or edge computing platforms, which enable intelligent and autonomous access in real time, collection, analysis, communications and exchange of processes and information regarding products and services, in order to optimize the overall value of production.

Some peculiar characteristics of IoT devices are: detection (radio frequency identification), communication technologies (wireless sensor network, embedded systems) and middleware. Radio frequency identification (RFID) is used to identify tags attached to an object and transfer data to the receiver wirelessly.

3.2.3 Big Data Analytics

The third classis about the myriad of data is generated, thanks to the widespread use of IoT devices in cloud manufacturing. First, the volume of such data is so large that RFID technology can generate thousands of pieces of data in a second. Data sets can be transmitted, stored and interrogated using traditional database tools. Secondly, every part of the RFID data is so abstract that it can be made up of different attributes like EPC, job ID, user ID and machine ID. These attributes should be further interpreted, so even the smallest parts of such data should be understandable. Third, RFID data is so complex that a large number of RFID events can involve workers, machines, jobs, places, time, production logic and materials.

3.2.4   Virtualization

An additional information technology studied and plotted into the ontology that is used by manufacturing companies thanks to cloud manufacturing is virtualization. Virtualization is an IT approach to create a multiple virtual version of a single physical resource or capacity, such as a server, storage device, network or even an operating system, to share it with other users or organizations on the network. It allows the sharing of resources among cloud users, which minimize the cost of using resources or physical capabilities for users. Moreover, another advantage of virtualization is related to the ability to operate and support legacy systems that require old libraries of operating systems, hardware and software.

3.2.5   Artificial Intelligence & Machine Learning

The last piece of IT technology studied and used in cloud manufacturing is the subclass related to artificial intelligence. Artificial intelligence used in industrial level refers to the application of artificial intelligence to industry. Unlike general artificial intelligence, which is a frontier research discipline for building computerized systems that perform tasks that require human intelligence, industrial AI is more interested in applying these technologies to tackle painful industrial creation points. and increasing customer value, improving productivity and information discovery. Although in a dystopian view of AI applications intelligent machines can take jobs away from humans and cause social and ethical problems, industry in general has a more positive view of AI and sees this transformation of the economy unstoppable and expects huge business opportunities in this process.

4   Use Case: Business Process of a Food Factory

It was analyzed the general case of a tomato-producing company that uses Cloud Manufacturing IT services. In the following graph (Fig. 3) we can observe the onto-logical representation the business process, as well as Fig. 4 which defines the set of activities needed to optimize and integrate the supply chain process in order to make the business of the company effective. The process is composed by four steps. In the first step, production is started. In this phase there is the activation of the drones and of all those IoT sensors positioned on the machinery within the tomato-producing industry. In the case of this industry, the use of drones gives the opportunity to photograph ground conditions and real-time environmental conditions. Since we are talking about Precision Farming, there are other instruments used, such as "smart" tractors, able to move without a person on board: they have the sensitivity to detect the environmental conditions through weather sensors and therefore, to prepare the start and end of a job or to anticipate their workflow or stop in case of adverse weather. Then, the data are sent to the Cloud platform that receives and processes them. The next step is the transformation. In this phase the quality control of the products is fundamental, which takes place through IoT sensors that are activated when required. In the case of the example in question, through barometric sensors it is possible to identify the optimal conditions of the land, in the event of sowing, maturation and harvest. In the transformation phase, in this case of the tomato, a series of smart and automation processes are activated which lead from the raw material to the finished product: juice, purée or sauces. Then his card is digitized in order to trace the entire journey of the product. After that, we move on to the third step, the distribution. Smart sensors, such as bar code readers, make it easier to identify lots of packaged products, so as to be able to improve the transport conditions by making the relative procedures cheaper: the paths, in fact, are studied through machine learning algorithms. At this point we move on to the fourth and last step, the one related to consumption. In this phase we find some smart technologies such as, for example, smart labels that include RFID technology, useful as they change color with the temperature and, therefore, alert the customer in the event that the food is left out of the coolers for too long. Additional smart sensors relate to the reduction of waste associated with consumption test the current state of deterioration of the product (in this case the tomato); there is even the phenomenon of smart packaging that gives information about the quality of the product, such as presence of microorganisms (with biological sensors) and of any gases (chemical sensors). Finally, an integral part of this level are also the meal kits, that are boxes with ingredients already dosed and partly prepared, useful for quickly cooking delicious dishes. In the case analyzed, we have products deriving from the use of tomatoes as sauces or various sauces.

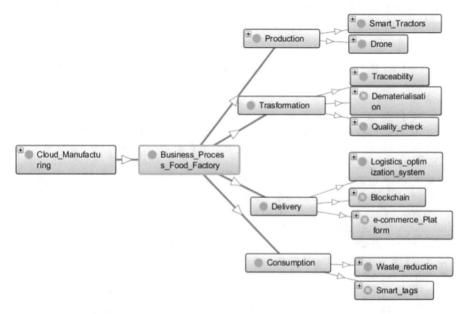

Fig. 3. Ontological representation of the Business process

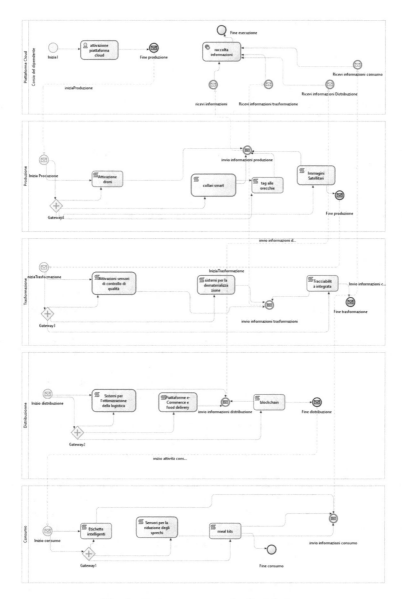

Fig. 4. Business process of a food factory

5 Conclusions

This work has the aim to analyze Cloud Manufacturing and the IT services that this model provides to manufacturing companies.

The rapid growth of advanced technologies in information systems and networks has allowed the manufacturing industry to apply new and complex production systems based on advanced networks. Cloud manufacturing is one of these emerging models and has a significant impact in the manufacturing sector. It is therefore essential to understand well the concept of Cloud Manufacturing, its capabilities and potential, but also its uncertainties, before adhering to it.

A possible future development proposed with this work could consist in dealing with additional issues related to Cloud Manufacturing such as: the Cloud Manufacturing life cycle; the benefits of adopting this model; and the role and responsibility of the stakeholders in this environment.

A further future development is certainly the improvement of the uncertainty factors of this cloud model, such as privacy, the security of stored and processed data and the factor relating to the specific techniques to possess in order to adopt Cloud Manufacturing. At the same time it would be interesting analyze the interoperability between different clouds and internal infrastructures, and at the same time to draw up guidelines for companies, for the choice of reliable Cloud Manufacturing systems.

References

1. Tao, Fei, Zuo, Ying, Xu, Li Da, Zhang, e Lin: IoT-based intelligent perception and access of manufacturing resource toward cloud manufacturing. IEEE Transactions on Industrial Informatics, vol. 10, no. 2, May 2014 (2014)
2. Boyes, Hugh, Hallaq, Bil, Cunningham, Joe, Watson, Tim: The industrial internet of things (IIoT): an analysis framework. In: Computers in Industry, vol. 101, pp. 1–12, ISSN 0166-3615 (2018)
3. Yadekar, Yaser, Shehab, Essam, Mehnen, Jorn: Taxonomy and uncertainties of cloud manufacturing. Int. J. Agile Syst. Manag. 9, 48–66 (2016). https://doi.org/10.1504/IJASM.2016.076577
4. Liu, N., Li, X.: A resource virtualization mechanism for cloud manufacturing systems. In: van Sinderen, M., Johnson, P., Xu, X., Doumeingts, G. (eds.) Enterprise Interoperability. IWEI 2012. Lecture Notes in Business Information Processing, vol. 122. Springer, Berlin, Heidelberg (2012)
5. Liu, Y., Xu, X.: Industry 4.0 and cloud manufacturing: a comparative analysis. In: ASME. J. Manuf. Sci. Eng. 139(3):034701-034701-8 (2016). https://doi.org/10.1115/1.4034667
6. Kubler, S., Holmström, J., Främling, K., Turkama, P.: Technological theory of cloud manufacturing. In: Borangiu, T., Trentesaux, D., Thomas, A., McFarlane, D. (eds.) Service Orientation in Holonic and Multi-Agent Manufacturing. Studies in Computational Intelligence, vol. 640. Springer, Cham (2016)

Optimization and Validation of eGovernment Business Processes with Support of Semantic Techniques

Beniamino Di Martino, Alfonso Marino, Massimiliano Rak,
and Paolo Pariso$^{(\boxtimes)}$

Engineering Department, University of Campania "L. Vanvitelli",
Aversa (CE), Italy
beniamino.dimartino@unina.it, {alfonso.marino,
massimiliano.rak,paolo.pariso}@unicampania.it

Abstract. E-government aspires to improve the quality of public services delivery, improving efficiency, effectivity and accessibly applying IT advanced methodologies and technologies. However, there are a lot of challenges that needs to be addressed in order to enable a full adoption of state of art solution in public administration. This preliminary study aims at proposing the design of semantic-based Decision Support System that helps some of the core actors in adoption of such solutions in e-government processes. The methodology relies on semantically annotated BPMN (Business Process Model Notation) and on inferential engines to support e-government processes re-engineering and optimization and e- government validation. The study will focus on the EU and Italian regulation and processes, in order to have as a reference real-world problems.

1 Introduction

E-government can be broadly defined as the use, on the part of public administrations, of information technologies combined with organizational changes in order to acquire new competencies to improve public services, ensure automated processes and support public policies. E-government consists in several activities as management of digital infrastructure, platforms, bandwidth, cyber security, health and school, plain of industrial investment.

The official inception of the e-government concept can be traced back to the European Commission document[1] entitled "Europe and the Global Information Society". In the document e-government is presented as one of the pillars of European economic development, conceived as the agile exchange of information which brings closer the center and the periphery of service provision and contributes to the creation of a more citizen and business friendly state. In order to keep track of the advances in the direction of a digitalized society, EU commission provides, since 2014, the Digital Economy and Society Index (DESI): a composite index that summarises some

---

[1] https://cordis.europa.eu/news/rcn/2730/en.

L. Barolli et al. (Eds.): CISIS 2019, AISC 993, pp. 827–836, 2020.
https://doi.org/10.1007/978-3-030-22354-0_76

30 relevant indicators on Europe's digital performance and tracks the evolution of EU Member States, across five main dimensions: Connectivity, Human Capital, Use of Internet, Integration of Digital Technology, Digital Public Services.

It is worth noting that, at state of art, in 2018 Italy ranks below 25th (out of 28) according to the DESI ranking.

One of the main inhibitors for the adoption of digitalization techniques is the difficulties of managing the flexibilities and continuous advances that IT technologies carry inside the process management of public administration.

Adoption of e-government approaches in public administration, as more deeply illustrated in Sect. 2 has a lot of limits, due to issues related to limits in capabilities from the public administration in clear definition and modeling of their processes, difficulties in modeling and automating the verification of law compliance and/or in manage the new (cyber-) security risks that adoption of new technologies involves.

According to such consideration we propose an innovative methodology in order to model and enforce e-government processes based on adoption of BPMN (Business Process Model Notation) semantically annotated. The methodology, through the semantic representation of e-government processes, aims at allowing solving problems like process change management (e.g. the continuous need to adapt PA processes due to digital economy), and interoperability among relevant processes (e.g. the need to automate interactions among different domains and/or different entities in PA processes). In line with this, the methodology can be considered strategic in order to reduce the tradeoff between these issues in every field of e-government activities.

The methodology aims at supporting Business processes modeling, optimization, validation, and the management of the e-government processes. It is forecasted that the present research will be of interest for business analysts who model the processes conceptually, technical developers responsible for implementing the technology for the processes, and moreover for the people who will manage and monitor the processes.

The remainder of this paper is organized as follows: next section summarises the state of art related to e-government in EU and Italy and on the adoption of semantic techniques for Business processes modeling, optimization and validation. Section 3 describes the E-government problem and summarize the open research challenges. Section 4 illustrate the methodology we propose. Section 5 summarises the conclusions and future works.

2 E-Government State of Art

E-government aspires to deliver public services more efficiently, effectively, accessibly by IT advanced methodologies and technologies. However, there are a lot of challenges that needs to be addressed in order to enable a full adoption of state of art solution in public administration. We identify the problems of e-Government in terms of three different goals: modeling, optimization and validation.

Modeling addresses the problem of describing the processes in a formal way, in order to support their automation. *Optimization* implies refining the processes in order to obtain the same results consuming less resources (where resources can be time, number of steps, personnel involved). Last, but no least, *Validation* aims at demonstrating that the process is compliant with actual regulation and the constraints that the processes is declared to be subject to.

It is worth noting that such challenges are an interdisciplinary problem that touches processes organization, regulations and IT technologies. Accordingly, here we briefly summarise the state of art of processes and regulations in Italy and then the existing techniques for processes modeling, optimization and validation.

2.1   E-Government in Italy

Italy e-government started to take shape at the end of the 90s with the administrative reform resulting from the "Bassanini law", n. 59 of the 15th of March 1997 which entrusted the government with the task of assigning functions and tasks to the Regions and Local Agencies to enact administrative simplification and Public Administration reform. Italy initiated the process later than other European countries, such as France, Germany and Belgium which had started the development of e-government in the 80s and, by the end of the decade, had managed to establish good European practices.

At state of art, in 2018 Italy ranks below 25th (out of 28) according to the DESI ranking (Digital Economy and Society Index) which refers to EU's research to measure and monitor Member States performance in implementing digital connectivity, use web services from the citizens and digital switchover of public services. Moreover, the index gotten in 2017 ONU E-Government Survey placed Italy to the 24th in the world ranking.

In Italy, by an organizational point of view, e-government can be implemented by overcoming macro limits related to some relevant business processes. Such complex processes, which have already been optimized by other European countries, can be pursued in Italy as well, provided that central and local administrations, especially Regions and municipalities, are fully and actively involved in the process.

Furthermore, to date the achievement of optimization and validation processes has proven quite challenging because it requires a rapid re-organization of the entire administrative apparatus and its modalities of service provision which entails revision of work flows, of planning priorities and strategies and of local public administration culture [1]. The pivot of the new system revolves around citizen and business centeredness and quality of service delivery. In this track a reengineering business processes is strategic in order to implement a change management process. The decentralization of power established by the reform laws calls for a radical reformulation of procedures, protocols, employee skill set, management competencies and of traditional hierarchies. These several bottlenecks highlight trade-offs between ITs opportunities, service management [2] within e-Government services.

2.2   Semantic-Based Business Processes Modeling, Optimization
and Validation

In order to meet the need for a uniform definition and representation of Business Processes, understandable and sharable among domain experts, and at the same time process able by software tools supporting their management (e.g. creation, modification, execution, simulation and optimization), many research efforts have been made to develop machine readable standard formalisms and notations. Remarkable standards were developed, such as the Business Process Execution Language (BPEL) [3] and the Business Process Model Notation (BPMN) [4]. BPMN is very adequate to describe a business process, and allows to represent many forms of information that must be integrated into a process model. The notation consists of graphical symbols to represent action, flow, or behavior of a process. BPMN provides a notation that can be readily understandable by all users:

- from the business analysts who model the processes conceptually;
- to the technical developers responsible for implementing the technology for the processes;
- to the people who will manage and monitor the processes.

Based on BPMN, many BPM - Business Process Management - tools have been developed, both open source and proprietary (among open sources ones, notable examples are Bonita, Camunda, Activity).

One limiting factor of almost all the BPMN based BPM tools existing today, is represented by a total lack of semantic information associated to the BPMN representation: this makes difficult to understand, compare and re-use Business Processes, even when they are developed by domain experts and process engineers working within the same organization, and furthermore strongly limit interoperability activities among different organizations. Hence, the need has raised to semantically annotate Business Processes with meaningful information which can give a clearer vision of the concepts involved in each of tasks, roles and rules defining them.

The practical benefits deriving from the application of semantics to business process modelling are presented in [5], in which the authors stress the important role played by semantic annotations in easing the design and development of new business processes. The European funded project SUPER [6] has provided a very complex framework for the modelling, management and semantic enrichment of business processes, focusing on the BPMN graphical notation and on the Web Service Modelling Ontology (WSMO) [7] to provide the semantics. The resulting framework is composed by a set of ontologies, organized according to a multi-layered structure, which deal with both structural aspects of the BPMN and domain specific concepts. Unfortunately, the WSMO representation has turned out to be very complex and cumbersome, and has been superseded by the Ontology Web Language - OWL [8] W3C standard representation.

The idea of applying patterns to business processes design and implementation is not new, as different research efforts have been carried out in the past to discover and define such patterns. The work presented in [13] presents an interesting set of patterns, not specifically related to business processes, but which can be easily represented in BPMN [14]. Such patterns describe both simple and complex combinations of basic control flow structures, identifying commonly adopted and re-usable solutions for workflow definitions. However, while such patterns can represent the building blocks for more complex business processes designs, they are too low level to be used stand-alone. Also, they miss semantic information which could be used to support the matching with semantically described cloud services. Higher level patterns, provided by the same authors in [15], provide representations of resource and data centered workflows which can be used in the definition of business processes. However, semantic- based representation are still missing, and in some cases a proper representation in BPMN could be not possible, due to the data and resource centric vision of such patterns, which is not fully compatible with the task-centric view of business processes. Activity patterns, representable both with UML activity diagrams and BPMN, are presented in [16]. Such patterns describe high level task workflows, which deal with basic activities in a process scenario, such as generic information requests, decision making, notifications and so on. One of the main drawbacks of such patterns is represented by their structural similarity: in most cases, the workflow is exactly the same, with the only difference being represented by task names. This can represent a problem since, without proper semantic annotations, it is impossible for an automatic framework to distinguish them.

3   E-Government Processes and Research Challenges

The optimization and validation of e-Government Business processes impact on the structure of the e- government system where there are some basic roles played by the actors in the same system:

- Politician who define a law;
- Process Analyst who define the process by an organizational point of view;
- Process Engineer who define process by an informatic point of view;
- Bureaucrat who implements the management process;
- Users who use e-government services based on the business process.

Adoption of e-government approaches in public administration has a lot of limits, due to issues related to limits in capabilities from the public administration in clear definition and modeling of their processes, difficulties in modeling and automating the verification of law compliance and/or in manage the new (cyber-) security risks that adoption of new technologies involves.

Particularly the main challenges are oriented to the optimization and validation of an e-Government BPM. This ambition, in a particular domain as the public administration and by an organizational point of view, clashes with the following matters for the Process Analyst:

- Challenge (egov-a) Process Analysis Is it possible to convert of the Government Business Processes from a documental paper process to an IT process?
- Challenge (egov-b) Change Management Process Are we able to implement, in an effective way, a Business Process in a Public Administration?

The Business Process Modelling Notation (BPMN) (see State of the Art section), the most used notation nowadays for process modelling, can describe how the tasks and activities are connected and to whom they are assigned, and define the orchestration of the entire process. However, it does not provide any information about the context in which the process is displaced. We think that this lack in expressiveness can be overcome by exploiting semantic-based technologies. In the following we propose five challenges a process engineer has to address when designing her solution.

- Challenge (BPMN-a) - Roles compatibility: Is there a way to make sure that a person or even an information system, to which a particular task is assigned, has the necessary authorization and capabilities to perform it?
- Challenge (BPMN-b) Process Correctness: Does a process meet the specified requirements? Is it structurally correct?
- Challenge (BPMN-c) Process re-engineering and optimization: Is it possible, according to some rules of thumb, to replace or to improve, wholly or partially, a process?
- Challenge (BPMN-d) Process re-use: Is it possible to reuse a process in other domains, adapting it to the right context?
- Challenge Process enhancement (BPMN-e): Is it possible to implement or substitute one or more of the defined automatic tasks by means of online (cloud) services, to enhance the process?

4 Methodology

The approach we propose relies on the idea of a methodology based on semantic techniques, which can support the Process Engineer in addressing and resolving the challenges stated above. The approach relies on (i) A Semantic, OWL based, model for representing Business Process Model will be employed and (ii) Business Process Patterns specifically tailored at eGovernment Domain semantically represented by means of a suitably defined semantic model.

The methodology relies on the results of the work presented in [9], which provides the definition of an OWL based ontology for the semantic description of a BPMN's structure: all the graphical elements of the standard are represented by OWL classes, with object properties marking their exact relationship. Such an ontology is then used to validate the process model and ensure that all of the constraints imposed by the notation are respected. The approach proposed in our work [10] is based on such a semantic representation, and extends it in order to support the recognition of specific process patterns and to provide a context-aware analysis of the process. However, such a semantic representation is too tied to the BPMN standard which, whilst being among the most used ones for business process representation, could be in the future replaced by more advanced ones. That is why in [11] we propose an OWL-S [12] based representation of business processes, which at the same time provides effective support for the mapping of tasks to cloud services and guarantees independence from the business process modelling language adopted.

The approach we propose address two key problems: e-government Process Optimization (i.e. correct organization of activities and minimization of resource consumption) and Process Validation (i.e. verification of process correctness respect to e-government processes, law compliance and risk analysis).

4.1   Tools Involved in the Methodology

In order to address such goals, our solution relies on (i) a tool, our Semantic Decision Support System (the SSS tool), that we aims at developing in the context of the project and (ii) on a methodology for e- government processes optimization and validation (the SSSeGov methodology).

The Semantic Decision Support System (SSS tool) is a solution that offers, in an integrated way, Semantic models, an Inference engine, and a graphical Annotator tool. The SSS tool, which is still under design process, include the following Components:

(a) Semantic based BPMN representation (S-BPMN), an OWL-based ontology devoted to describe and annotate BPMN processes.
(b) Knowledge Base, based on ontologies that collects the concptes related to e-government processes, law compliance and risk analysis for (cyber-)security issues. In practice it includes a (b.1) eGovernment Domain Ontology, a (b.2) Legal Domain Ontology and a (b.3) Security (threats) Ontology,
(c) Inferential Engine, to elaborate the process description for both optimization and validations, supported by Inference Rules' repositories specific for e-governemtn processes, law compliance and security evaluations. The repositories will include (c.1) Process Optimisation Rules, (c.2) Law Compliancy Rules, (c.3) Security (threats) Rules
(d) BPMN Graphical Semantic Annotator, devoted to help process analysts and processors to annotate the e-government processes in order to correctly perform process optimization and validation.

4.2   The Methodology

The SSS tool enable the implementation of our SSSeGov methodology for Optimization and Validation of e- Government Business Processes, which consists of the following Organizational Activities (described in detail in the following):

(1) Business Process Analysis and Definition
(2) BPMN Representation
(3) Semantic Enrichment of BPMN
(4) Reengineering and Optimization
(5) Law-aware Verification and Validation
(6) Security-aware Verification and Validation

Business Process Analysis and Definition: This activity involves the analysis and definition of the organizational process and it is developed as support of BPMN tools. The output of this step is the definition of the business process domain with the aim to point out the main strategic actors of the process, the inputs and outputs exchanged in the process, the activities and the procedures that settle the process, and the responsibilities that stem by the actions of the actors involved in the process.

BPMN Representation constitutes, as the the Business Process Analysis and Definition, a further activity to support BPMN tools. Particularly this activity is strictly propaedeutic to an effectiveness of a BPMN. This step includes actions as documenting the process in order to understand the organization's workflow, identification of the process tasks with the aim to improve performance and use of sources, and detection of the coordination and integration processes to enhance quality and efficiency.

Semantic Enrichment of BPMN is the critical step of the methodology, that relies on the eGov, LAw and Security ontologies and aims at enriching the BPMN descriptions with the concepts from the three target domains.

Semantic Annotation with Domain (eGov) Concepts, relying on the eGov ontology, will focus on enriching BPMN description with governmental services concepts. During this activity, the BPMN notation of the eGov Business Process is processed through a BPMN parser and a BPMN-to-OWL un-parser, and a - OWL compliant - ontology is produced, the "S-BPMN" (Structural BPMN) Ontology.

It is a representation based on a full scale semantic depiction of the Business Process notation. Once all the information regarding the process have been annotated, the produced S-BPMN ontology contains every and each element of the process: all the graphic components in the BPMN have their corresponding individual, with all its data and objects.

Definition of (eGov) Law Compliancy constraints and *Definition of Security constraints*, aims at enriching the description with ontologies related to actual law constraints and information needed to verify cyber-security risks.

SSS supported Reengineering and Optimization aims at analyzing an e-government process and re-engineerize it, in order to optimize the activities and resource consumption, typically reengineering and Optimization activities, performed by the process engineer, include: (i) Monitoring of the business process, (ii) Measurement of automed phases and not automed of the business process, (iii) Data collection of the business process bottlenecks, (iv) Definition of the expected performances from that business process, (v) Time Analysis of the each activities and procedures to settle the performance of the business process linked to the Reengineering process, (vi) Reengineering process on the base of the detected errors

In order to support Process Reengineering and Optimization, the SSS will enable three kinds of analysis: Structural Analysis. It uses the direct connections between BPMN and the S-BPMN semantic representation. Through this analysis it is possible to obtain information exclusively on the structure of the process: therefore, if taken individually, it may be used to make a structural verification of the process, then to locate any errors in the process model. Also, it can represent the base for the recognition of known configurations.

SSS supported Law-aware Verification and Validation, In the legal framework characterizing eGovernment activities [Constitution, European Treated, Rules (as the GDPR) and community directives, ordinary (among which the CAD) Laws, ministerial rules, circular, provisions of the Authorities of guarantee, only to quote some of it, is requie the role of interpreter (or rather of the one which must apply the right to the concrete case) to establish what dispositions are applicable to a specific phase of the digital administrative procedure.

SSS supported Security-aware Verification and Validation activity automates the risk analysis process and enable its continuous application in the e-government processes execution, thanks to the security-based semantic annotation of the BPMN. Applying the set of inferential rules and analyzing the annotated BPMN, it will be possible to: (i) identify applicable threats, (ii) analyze and evaluate related security risks, (iii) select the proper countermeasures to enforce in order to mitigate existing risk.

Applying the above illustrated approach we be able to (a) Optimize the processes, respecting law, security and resource-related constraints and (b) Validate the correctness of processes respect to expected process behavior. The research addresses, so, a lot of state of art challenges related to (i) state of art limits of the business process evaluations, (ii) the application of the BPMN to e-government context.

5 Conclusions

The optimization and validation of BPMN aims at solving several bottlenecks linked to the problems of mutual adaptation between the organization and the changing processes. Particularly, the business processes, realized in effective way with BPMN tools, have a high evolution speed that squeeze the bureaucracy of the public administration, in general very slow to receive the change of own external environment. The proposed architecture framework for e-government adoption will reduce confusion surrounding e-government infrastructure in the public sector through understanding the implementation processes, identifying requirements of information technology tools, highlighting the importance of the organizational management resources and the impact of barriers.

The application of the validation methodology allows to create E-government patterns able to:

- delivery added value, also in others public administration units;
- increase the automation of e-government processes;
- speed up the development process;
- enable the interoperability between different public administrations;
- incentive the reuse of e-government services as much as possible.

References

1. Schein, E.H.: Organizational culture. FB Editor, Ebbok (2017)
2. Normann, R.: Service Management: Strategy and Leadership in Service Business. John Wiley & Sons (1993)
3. Andrews, T., Curbera, F., Dholakia, H., Goland, Y., Klein, J., Leymann, F., Liu, K., Roller, D., Smith, D., Thatte, S. et al.: Business process execution language for web services (2003)
4. O.M.G. (OMG): Business process model and notation (bpmn) version 2.0. Technical report (2011)
5. Wetzstein, B., Ma, Z., Filipowska, A., Kaczmarek, M., Bhiri, S., Losada, S., Lopez-Cob, J.M., Cicurel, L.: A life cycle based requirements analysis. In: Proceedings of the 3rd workshop on semantic business process and product lifecycle management, Technical University of Aachen, Innsbruck (2007)
6. SUPER - Semantics Utilized for Process management within and between EnteRprises. http://projects.kmi.open.ac.uk/super/
7. Roman, D., et al.: Web service modeling ontology. Appl. ontology 1(1), 77–106 (2005)

8. McGuinness, D.L., Van Harmelen, F.: Owl web ontology language overview. W3C Recomm 10(10) (2004)

9. Rospocher, M., Ghidini, C., Serafini, L.: An ontology for the business process modelling notation. In: Proceedings of the Eighth International Conference (FOIS 2014) Formal Ontology in Information Systems, vol. 267. IOS Press (2014)

10. Di Martino, B., Esposito, A., Stefania, N., Maisto, S.A.: Semantic annotation of BPMN: current approaches and new methodologies. In: Proceedings of The 17th International Conference on Information Integration and Web-based Applications and Services (iiWAS2015). ACM Press (2015)

11. Di Martino, B., Esposito, A., Nacchia, S., Maisto, S.A.: A semantic model for business process patterns to support cloud deployment. COMPUTER SCIENCE, Springer 32(3–4), 257–267 (2017)

12. Burstein, M., Hobbs, J., Lassila, O., Mcdermott, D., Mcilraith, S., Narayanan, S., Paolucci, M., Parsia, B., Payne, T., Sirin, E., Srinivasan, N., Sycara, K.: OWL-s: semantic markup for web services(2004). http://www.w3.org/Submission/2004/SUBM-OWL-S-20041122/

13. van Der Aalst, W.M.P., Ter Hofstede, A.H.M., Kiepuszewski, B., Barros, A.P.: Workflow patterns. Distrib Parallel Databases 14(1), 5–51 (2003)

14. BPMI.org. Business process pattern examples http://www.workflowpatterns.com/vendors/documentation/BPMN-pat.pdf (2003). Accessed 10 March 2016

15. Russell, N., Ter Hofstede, A.H.M., Edmond, D., van der Aalst, W.M.P.: Workflow data patterns: identification, representation and tool support. In: Proceedings of the 24th International Conference on Conceptual Modeling, Klagenfurt, Austria. Springer, Heidelberg, 353–633 (2005). https://doi.org/10.1007/11568322_23

16. Thom, L.H., Reichert, M., Iochpe, C.: Activity patterns in process-aware information systems: basic concepts and empirical evidence. Int. J. Bus. Process. Integr. Manag. 4(2), 93–110 (2009)

Analysis of Existing Open Standard Framework and Ontologies in the Construction Sector for the Development of Inference Engines

B. Di Martino[1](✉), C. Mirarchi[2], S. Ciuffreda[2], and A. Pavan[2]

[1] Department of Engineering, Università della Campania "Luigi Vanvitelli",
Aversa, Italy
beniamino.dimartino@unina.it
[2] Department of Architecture, Built Environment and Construction Engineering,
Politecnico di Milano, Milan, Italy
{Claudio.mirarchi,alberto.pavan}@polimi.it,
simone.ciuffreda@mail.polimi.it

Abstract. The definition of shared dictionaries able to facilitate communication in the construction sector has been an open issue long since. The advent of ICT highlighted the existing issues imposing new challenges to allow the communication between systems. Nowadays, the digital transformation in the construction sector is led by Building Information Modelling. Several efforts have been spent both in research and industry developing open standards like the industry foundation classes and related ontologies to empower interoperability. However, this paper highlights the obstacles limiting use and exchange of information at the state of the art.

1 Introduction

The construction sector is a project-based industry characterized by informative-intensive processes that produce one-of-a-kind products that are characterise by high complexity. Project information is composed by several documents among which graphical drawings traditionally represent a fundamental component thanks to their ability to transfer articulated and complex information in technical and usually shared terms between technicians. However, the high variety of elements that can compose a building or an infrastructure, the geographical dislocation of the products and the increasing number of information requires to design a real estate posed several challenges in the communication between the subjects involved in a construction project. The introduction of Building Information Modelling (BIM) that is the use of object oriented software to develop information models able to transfer information in a structure and collaborative process, can limit the fragmentation of this information. The information models developed in BIM processes can be defined as a shared digital representation of physical and functional characteristics of built objects. However, the peculiarities of the construction sector, such as one-of-a-kind-product, long life-cycle of products, fragmented and spatially distributed chain are challenging the effective

© Springer Nature Switzerland AG 2020
L. Barolli et al. (Eds.): CISIS 2019, AISC 993, pp. 837–846, 2020.
https://doi.org/10.1007/978-3-030-22354-0_77

interoperability (i.e. the exchange and use of information between different systems [1]) in BIM as well as the possibility to effectively use the data contained in the information models. Since 1995, the BuildingSmart consortium is working on the Industry Foundation Classes (IFC), i.e. a common data model to represent and describe building processes [2]. However, the effective application and use of IFC in the construction industry processes demonstrated to be all but easy. The IFC schema comprehends only a small part of all the elements that can be included in a real estate and does not provide a standard classification for other elements or for sub-groups or aggregation of the same building components. Moreover, the study proposed in this paper highlighted several discrepancies between the standards that work together with IFC. This, can hamper the development of automated system able to use the data contained in information models dealing for example to automatic quality checking, code checking, computational design, etc. Thus, there is both a lack in the formal representation of objects to allow their correct use in information models and a lack in the correct semantic identification of these objects even when they are identified.

According to the existing limitation, around 2000 researchers start to explore the use of semantic web technologies to enhance the information exchange processes [3]. Following this trend, the Linked Data Working Group (LDWG) [4] developed the ifcOWL ontology based on the IFC standard and an IFC-to-RDF converter [5].

A common problem in the construction domain is the use of building information models to the automatic verification of the design according to standards, regulations, project requirements, etc. Usually, due to interoperability issues, building information models need to be prepared and improved to allow the application of automated verification processes. The preparation activity includes for example the definition of standard names for each building space, the inclusion of specific information in building object, etc. The automatic identification of both the building spaces destination of use and of real estate units (identified as aggregation of building spaces) can drastically reduce the time required to "prepare" the model for the check phase. The same situation can be extended to several cases of building information models use where the information has to be combined with external inputs and interoperability issues rise. A recent case study developed by the Authors highlighted the relation between the above mention ontological limitations and the difficulties in developing automated means to use building information models contents.

The objective of this work is to analyse the existing structure of the open standards related to IFC with the objective of clearly identify existing issues and relate these issues to the case study developed by the Authors. Moreover, starting from this last, this article will argue about the need of extending the existing ontological structures. This study paves the way for future analysis on the open standard structure that nowadays represent the reference point for the construction industry to promote its improvement according to the increasing needs of interoperability and data uses.

The rest of the paper is organized as follow. Section 2 presents the background related to the development of open standards in the construction sector. Section 3 highlights the open issues and the motivations to this study. Section 4 proposes a compared analysis between the main standards that constitute the semantic shared structure nowadays available in the construction sector. Section 5 deals with the implication of the analyzed issues to the development of automated processes to use data and information. Finally, Chapter 6 presents the conclusion and discussion of the article.

2 Background

IFC is nowadays recognized and defined in the construction sector as a standard common language. The IFC schema provides the building blocks for interoperability through its open and neutral data structure. It represents geometry, relations, processes and material, performance, fabrication, and other properties, using the EXPRESS language.

However, the use of IFC underlined several issues due to the unclear processes of import and export of IFC files. These processes are usually defined as black box embedded in the specific software used to export and/or import the IFC file. Lack of information in sharing building information models through IFC formats is registered in several cases, e.g. when models are exported and imported in the same BIM authoring tool, from architectural models to structural models, from BIM authoring tools to analysis software (e.g. for energy analysis) or from architectural models to facility management tools. Objects, relations, and attributes can be defined in several ways and their standard identification is defined at a specific level (e.g. ifcSpace) without considering sub-levels (e.g. ifcSpace: kitchen) and aggregation (e.g. ifcRealEstateUnit or IfcBuilding). Hence, data exchange usually needs the definition of ad hoc standards defined by organisation or by project with the consequent difficulties in using the information by means of automated processes.

IFC covers only the transport of data and information in the process. Processes, standard terminology and technical requirements are not included in the IFC schema. Indeed, Building Smart includes five basic standards, namely Information Delivery Manual (IDM), International Framework for Dictionaries (IFD), BIM Collaboration Format (BCF), Model View Definition (MVD), and Industry Foundation Classes (IFC) that all together constitute the standard for the interoperability in the construction sector (Table 1).

Table 1. IFC related standards from BuildingSmart [6]

Description	Name	Related standard
Describe processes	Information Delivery Manual IDM	ISO 29481-1 [7] ISO 29481-2 [8]
Transport information/data	Industry Foundation Class IFC	ISO 16739 [9]
Change Coordination	BIM Collaboration Format BCF	buildingSMART BCF
Mapping of Terms	International Framework for Dictionaries IFD	ISO 12006-3 [10] buildingSMART Data Dictionary
Translates processes into technical requirements	Model View Definitions MVD	buildingSMART MVD

IDM focuses on processes that require the exchange or share of information between project participants and the information required and resulting from the execution of these processes. It is strictly related to MVD that is the identification of the required model views, i.e. the *"subset of the IFC schema that is needed to satisfy one or many exchange requirements"*, needed for a specific use of the model. IFD provides a systematic collection of terms, vocabulary and attributes to establish a standard semantic in the construction sector. Nevertheless, the IFD is limited to the components defined in the IFC schema and suffer the same issue in the formal representation of building components described in the introduction. Figure 1 proposes the synthesis of the relations between the presented standards.

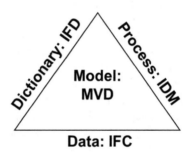

Fig. 1. Basic open standards from BuildingSmart [6]

From early 2000s researchers start to explore the use of semantic web technologies to enhance the information exchange processes [3]. Nowadays, there is an active research community working on this topic. The BuildingSMART Linked Data Working Group (LDWG) [4] developed the ifcOWL ontology based on the IFC standard. To automate the conversion process between IFC models and RDF models an IFC-to-RDF converter was developed [5]. Nevertheless, the direct use of the IFC into an RDF schema resulted in a complex data structure that can hinder the effective use of the information. Thus, in the literature can be found several studies that focuses on the development of ontologies that can integrate and/or simplify the ifcOWL ontology. The Linked Building Data Community Group at the W3C developed the Building Topology Ontology (BOT) [11]. Moreover, Pauwels and Roxin [12] explored different ways to simplify the ifcOWL ontology. Following the development of the BOT, an IFC to LBD converter was developed [13] while other studies focused on the extension of the BOT [14] and on the relation between existing ontologies and ontologies developed in the construction sector [15].

3   Motivation

Even if other studies revealed the need to extend the existing ifcOWL ontology, the proposed extensions are limited to a high level that does not allow the identification through a shared semantics of specific elements such as e.g. types of *IfcSpace*.

Moreover, the existing studies focuses on the IFC structure and its translation and/or relation to ontologies but without considering the dictionary used to define these objects. Thus, on the one hand there is the need to clarify the existing open standard configuration to understand the relations and possible issues between the existing standards to promote the development of ontologies that can overcome these problems. On the other hand, a recent study developed by the Authors demonstrated the need to integrate the ontologies nowadays developed to allow automated processes of data use. Hence, these paper proposes two analysis that together can be used to promote and support the development of improvements in the existing structure of open standards in the construction sector.

4 Comparison Between the Basics Open Standards

The main goal of the proposed analysis is to clarify the existing overall situation of definitions BuildingSMART uses in writing the different versions of IFC. The framework used in the model definition is defined in ISO 12006 *Part 3: Framework for object-oriented Information* [10], and it is intended as an EXPRESS-based taxonomy model able to describe relationships between concepts, groupings of elements and their relative properties. ISO 12006 *Part 2: Framework for classification* [16], instead, aims to define the main classes of objects of Architecture, Engineering and Construction (AEC) field, supported by the definitions of the objects introduced. In giving new definitions of classes, this document also gives an idea of relationship between the elements of construction sector (e.g. a *construction result* is the output of a *construction process*, starting from *construction resources*; a *work result* is the output of a *work process*, etc.).

The object-oriented framework proposed by ISO 12006 has been compared to the widely used by IFC and codified in ISO 16739. The analysis tries to match definitions and properties given in the two standards according to the proposed frameworks. The comparison has been held considering both IFC 2 $\times$ 3 and IFC 4 standards, as they are the most referenced by the industry in interoperability issues. Also, both ISO 12006:2007 and ISO 12006:2015 have been compared, as the updated version introduces several major changes in the definitions and in the framework in which definitions are applied.

Quantitatively, ISO 12006 has less definitions than IFC but, as it has already been sentenced, the first one aims to give narrower information about classes as framework for further application in AEC and civil/infrastructure domains.

Objects defined in ISO 12006 and IFC classes have been matched weighting the relationship between the terms on a five-step scale from 0% to 100% (0%, 25%, 50%, 75%, 100%) representing respectively the absence of a match up to a congruent match between the definitions.

4.1 ISO 12006:2007 to IFC Comparison

Since ISO 16739 defines the objects while introducing the framework of building objects, ISO 12006-2 terms were put in comparison with IFC definitions to find matches or mismatches. ISO 12006-2:2007 introduces a total of 20 terms, as a list, with no relation defined between the objects. The only exception is made in the definition of the terms *construction result, construction process, construction resource, work result* and *work process* (see Tables 2 and 3). According to the given definitions, both *construction resources* and *construction results* have an input-output relation with *construction process*, and the same works for work-related definitions.

As it is shown, the two codifications identify two different generalizations of the concept defined as *process* by IAI in ISO 16739 (a *process* is defined as object-occurrence located in time, indicating "when"). The same principles occur in the definition of *product* (defined as physical or conceptual object that occurs in space, as a specialization of *object*).

Tables 2 and 3 report an example of the comparison developed between the ISO 12006-2 and the IFC schema highlighting the existing lack of coherence between the two.

Table 2. Example of comparison between ISO 12006-2 and IFC

Item	UNI 12006-2		IFC			
	Term	Definition	Entity	Definition (from IAI)	Notes	%
1	Object	Any part of the perceivable world	IfcObject	Generalization of any semantically treated thing or process. Objects are things as they appear, tangible or non tangible	Supertype in IFC of Actors, controls, groups product and resources	100
2	Construction object	Object of importance to the construction industry	IfcProduct	Any object, or any aid to define, organize and annotate an object, that relates to a geometric or spatial context. Subtypes of IfcProduct usually hold a shape representation and a local placement within the project structure. In addition to physical products (covered by the subtype IfcElement) and spatial items (covered by the subtype IfcSpatialStructureElement) the IfcProduct also includes non-physical items, that relate to a geometric or spatial contexts, such as grid, port, annotation, structural actions, etc.	Object of importance may be represented both by IfcProduct or IfcResource	25

Table 3. *construction process* and *work process* comparison

Item	UNI 12006-2		IFC			
	Term	Definition	Entity	Definition (from IAI)	Notes	%
11	Construction process	Process which transforms construction resources into construction results	IfcProcess	IfcProcess is defined as one individual activity or event, that is ordered in time, that has sequence relationships with other processes, which transforms input in output, and may connect to other other processes through input output relationships. An IfcProcess can be an activity (or task), or an event. It takes usually place in building construction with the intent of designing, costing, acquiring, constructing, or maintaining products or other and similar tasks or procedures	IFC definition is more inclusive than the one of ISO 12006	100
13	Work process	Predominant construction process which results in a work result	IfcProcess		Narrower concept, possibly subtype of "construction process", or viceversa	50

4.2   ISO 12006:2015 to IFC Comparison

One of the innovations brought by the 2015 edition of the ISO 12006 is a new link between the part 2 and the part 3. While in previous editions the scope of bringing out a taxonomy model for the objects was discussed only in one part, in the updated edition ISO working groups have developed new and simpler terms with the goal of clarifying some collisions or misunderstanding in previous definitions. An example of this workflow is visible in Table 4, where the same objects have clearer and coded definition, linked one to each other with different kinds of relations. In this new edition, as a matter of fact, a basic process model is introduced, as the relation between resources, processes and product is clarified; type-of and part-of relationships are defined and introducing properties, still keeping definitions as broad as possible.

Table 4. ISO 12006-2:2015 to IFC comparison example

UNI 12006-2				IFC				
Item	Code	Term	Definition	2 × 3	4	Entity	Definition (IAI)	%
8	3.1.8	Space	Limited three-dimensional extent defined physically or notionally		x	IfcSpatialElement	A spatial element is the generalization of all spatial elements that might be used to define a spatial structure or to define spatial zones	50
9	3.1.9	Activity space	Space (3.1.8) defined by the spatial extension of an activity		x	IfcSpatialElement	A spatial element is the generalization of all spatial elements that might be used to define a spatial structure or to define spatial zones	50
10	3.2.1	Construction agent	Human construction resource (3.2.5) carrying out a construction process (3.3.2)	x	x	IfcActor	The IfcActor defines all actors or human agents involved in a project during its full life cycle. It facilitates the use of person and organization definitions in the resource part of the IFC object model. This includes name, address, telecommunication addresses, and roles	75

Misleading definitions and overlapping terms are still present in 2015 ISO document, referring to the definition of particular process stages or spaces with or without function. It is stated that all objects can be defined both as objects and also as systems of elements, which are objects themselves.

5   Automated Use of Information

In the scientific literature can be found some studies related to the automated recognition of building information models objects. For example, Belsky, Sacks and Brilakis [9] proposed a semantic enrichment engine to recognize building information modelling objects. An improved version of the system was proposed to classify and aggregate prefabricated elements in bridge projects [10]. And a recent experimentation proposes the comparison between the semantic enrichment engine and a machine learning application to recognize room types in information models [11]. Among these studies, the recognition of building spaces is of particular interest due to the related possibility to define automated process to control information models against regulations and/or design requirements. In this direction, the Authors developed an automated system based on the ifcOWL that can recognize the characteristics of spaces and identify their uses in the building such as kitchen, bathroom, bedroom, etc. However, the recognition of these elements required the introduction of ad hoc extensions in the ifcOWL ontology because both specific type of ifcSpace and the aggregation of these spaces to identify an apartment are not considered in the ifcOWL. The lack of a shared ontology able to comprehend these concepts leads to the use of ad hoc extensions that creates island of information and knowledge according to the specific application, thus limiting its diffusion and improvement. Hence, starting from the proposed study it is envisioned the need to extend the existing ifcOWL ontology according to the need of the real estate market that sees in the identification of spaces uses and in their aggregation according to the real estate units a crucial area of action (e.g. in selling and renting buildings, in manage maintenance activities, in checking the coherence according to existing regulations, etc.).

6   Discussion and Conclusions

Combining the result of the case study presented in Chapter 5 with the ones reported in the existing literature and with the analysis proposed in Chapter 4, it is clear the need to extend the existing ontology and to promote the integration of this ontology with an extended and shared dictionary that is able to identify also for the humans the means of the terms used in the ontological structure. In fact, one of the consequences of narrow definition reported in ISO 12006 is the misunderstanding between users and, most importantly, between systems that leads to the need of defining myriads of ad hoc processes. The possibility to mislead an object, as it is already coded how many information per stage of lifecycle are needed, grows with the increase of generic terms for the same concept (e.g. space, process, resource, etc.). On the other hand, strict frameworks do not ease interoperability as they may not include systems or innovative objects. In designing a framework for built environment models, it should be also taken into consideration that every object, needs to be defined in according to the specific need that are identified in the process – Level of Information Need (LOIN) [17].

References

1. IEEE, Standard Computer Dictionary. A Compilation of IEEE Standard Computer Glossaries (1990)
2. Laakso, M., Kiviniemi, A.: The IFC standard - a review of history, development, and standardization. J. Inf. Technol. Constr. **17**, 134–161 (2012)
3. Pauwels, P., Zhang, S., Lee, Y.C.: Semantic web technologies in AEC industry: A literature overview. Autom. Constr. **73**, 145–165 (2017). https://doi.org/10.1016/j.autcon.2016.10.003
4. Building Smart, Linked Data Working Group (2018). http://www.buildingsmart-tech.org/future/linked-data. Accessed 21 Jan 2019
5. Pauwels, P., Terkaj, W.: EXPRESS to OWL for construction industry: towards a recommendable and usable ifcOWL ontology. Autom. Constr. **63**, 100–133 (2016). https://doi.org/10.1016/j.autcon.2015.12.003

6. B. International, Open Standards - the basics (2014). https://www.buildingsmart.org/standards/technical-vision/open-standards/

7. ISO, ISO 29481-1 - Building information models – information delivery manual – Part 1: Methodology and format (2016)

8. ISO, ISO 29481-2 - Building information models – information delivery manual – Part 2: Interaction framework (2012)

9. ISO, ISO 16739: Industry Foundation Classes (IFC) for data sharing in the construction and facility management industries (2013)

10. ISO, ISO 12006-3: Building construction - organization of information about construction works - Part 3: Framework for object-oriented information (2007)

11. Linked Building Data Community Group, Building Topology Ontology (BOT) (2019). https://github.com/w3c-lbd-cg/bot. Accessed 21 Jan 2019

12. Pauwels, P., Roxin, A.: SimpleBIM: from full ifcOWL graphs to simplified building graphs. In: 11th European Conference on Product Process Modelling, pp. 11–18 (2016)

13. Bonduel, M., Oraskari, J., Pauwels, P., Vergauwen, M., Klein, R.: The IFC to linked building data converter - current status. In: Proceedings of the 6th Linked Data Architecture Construction Workshop Semantic, pp. 34–43 (2018)

14. Rasmussen, M.H., Pauwels, P., Lefrançois, M., Schneider, G.F., Hviid, C.A., Karshøj, J.: Recent changes in the building topology ontology. In: 5th Linked Data Architecture and Construction Workshop (2017). https://doi.org/10.13140/rg.2.2.32365.28647

15. Schneider, G.F.: Towards aligning domain ontologies with the building topology ontology. In: 5th LDAC Work., Dijon, France, 13–15 November 2017

16. ISO, ISO 12006-2: Building construction - organization of information about construction works - Part 2: Framework for classification (2015)

17. ISO, UNI EN ISO 19650-1 - Organization of information about construction works — Information management using building information modelling — Part 1: Concepts and Principles, 2018

Semantic Techniques for Validation of GDPR Compliance of Business Processes

Beniamino Di Martino, Michele Mastroianni(✉),
Massimo Campaiola, Giuseppe Morelli, and Ernesto Sparaco

University of Campania Luigi Vanvitelli, Via Roma, 29 81031 Aversa, CE, Italy
beniamino.dimartino@unina.it,
michele.mastroianni@unicampania.it,
{massimo.campaiola, giuseppe.morelli1,
ernesto.sparaco}@studenti.unicampania.it

Abstract. Starting from 25th May 2018, the EU Regulation 2016/679, known as GDPR (General Data Protection Regulation) - relating to the protection of individuals with regard to the processing and free circulation of personal data - is directly applicable in all Member States. The application of the GDPR caught a lot of companies and institutions unprepared and off guard. One of the most complex activities that the GDPR requires is the Data Protection Impact Assessment (DPIA). The aim of this paper is to provide an Expert System that is able to draw up the DPIA, and a prototype is shown.

1 Introduction

Privacy and data protection are at this time major issues both for enterprises, that require personal data of their customers to perform their IT services, and for public administrations, that manage personal data of citizens. In the European Union, the new regulation 2016/579 "General Data Protection Regulation" (GDPR), after May 25, 2018, has imposed many new constraints.

In particular, according to Article 35 of GDPR, the Data Protection Impact Assessment (DPIA) has become a mandatory analysis in order to identify and evaluate privacy threats in personal data management. A DPIA aims to conduct a systematic risk assessment, in order to identify privacy threats and impose technical and organizational controls to mitigate those threats.

The steps of a DPIA are [1]: (1) System characterization. (2) Specification of privacy targets. (3) Evaluation of degree of protection demand for privacy targets. (4) Identification of threats. (5) Identification of controls. (6) Implementation of controls. (7) Generation of PIA report. The DPIA is a complex analysis and must be written by experts from different branch of studies, i.e. privacy and domain experts, risk assessment and IT professionals too.

This paper aims to address this problem using artificial intelligence methods in order to draw up DPIA analysis. To do this, semantic techniques are applied in order to validate business processes for GDPR compliance.

© Springer Nature Switzerland AG 2020
L. Barolli et al. (Eds.): CISIS 2019, AISC 993, pp. 847–855, 2020.
https://doi.org/10.1007/978-3-030-22354-0_78

In this paper, a methodology for setting up an automatic rating of risk assessment due to personal data processing is proposed, and a prototype and some preliminary results are shown.

The paper is organised as follows: the basic methodology is shown in Sect. 2, the implementation details are presented in Sect. 3, and conclusions and future works are listed in Sect. 4.

2 Methodology

Description of processes to validate is done using BPMN (Business Process Model Notation) [2]. BPMN is an international standard for the representation of business and organizational processes. To describe data processing using BPMN has been used the "BPMN Ontology" developed by Bruno Kessler Foundation [3]. The ontology provides a complete description of the BPMN elements in Appendix B of the specific BPMN version January 2008. All elements of the BPMN are represented through classes according to a tree hierarchy.

Extraction of ontologies directly from the text of the GDPR is not trivial, since the regulation does not report any kind of classification. However, in Article 30, the GDPR requires that the controller (i.e. the chief of institution) writes and updates a document, called the *"Records of processing activities"*. That record shall contain all of the details of data attributes, purpose of processing, time limits for data erasure, and so on. It's possible and easy to extract ontologies from this document.

In our prototype we use a document named *"Linee guida in materia di privacy e protezione dei dati personali in ambito universitario"* (Guidelines on privacy and data protection issues in academic field) [4] and it was written by a meeting of the General Managers of the Academic Administrations (*"Convegno dei Direttori Generali delle Amministrazioni Universitarie"* - CODAU). This document contains a sample Record of processing activities and is used by all Italian universities (Table 1).

Table 1. Record of processing activities.

Record of processing activities	Details
Responsible	Name of responsible
Purpose & legal basis	Purpose of processing (e.g. Student career)
Data collection	Which data is collected? (personal, health…)
Data processing	How and where is data processed? (in house, cloud,…)
Data storage & transfer	Where is data stored? May data be transferred? (e.g. to foreign country)
Data deletion & retention	When the data may/must be deleted?
Security measures	Which are the technical and organisational security measures?

All these characteristisc are described using a semantic representation done by an ontology called "GDPR Ontology" realised in OWL [5]. OWL (Web Ontology Language) is a mark-up language to describe knowledge and a particular domain of application. OWL is an extension of RDF and they both belong to the semantic web project.

3 Implementation

In the first phase the Academic Record of processing activities has been analysed. From this document the main concepts have been extracted, and an OWL ontology called *University_GDPR_Ontology* was built in order to create an easily sharable and editable knowledge base. This ontology describes for every record of processing activity that includes the handled data and its type.

The type of a Datum can be Personal, Biometric, Health or Judicial. The following picture represents the structure of the above-stated ontology (Fig. 1):

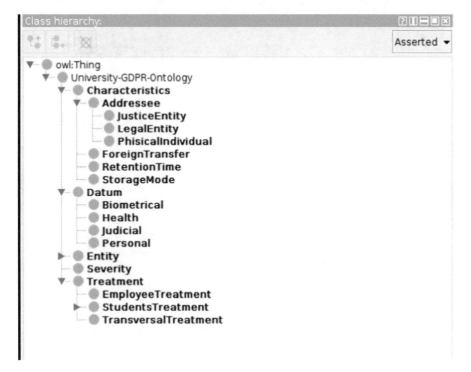

Fig. 1. "University_GDPR_Ontology" class hierarchy

For example, *Job Placement Processing* handles *Contact Data, Curriculum Vitae* and *Handicap Data* that are respectively two "Personal Data" and one "Health Datum":

Data in turn has some characteristics as:

- Retention time (less than 1 year, less than 10 years, unlimited)
- Datum transferred to foreign countries (yes or not)
- Storage digital security (encrypted or not)
- Addressee (law, public or private entity)

The up-mentioned Data Characteristics are related to data through object properties. This type of properties interconnect various classes. The following four object properties relate a *Datum* to its *Characteristics*:

- *Has_Addressee*
- *Has_StorageMaintenanceMode*
- *Has_RetentionTime*
- *Has_ForeignTransfer* (Fig. 2).

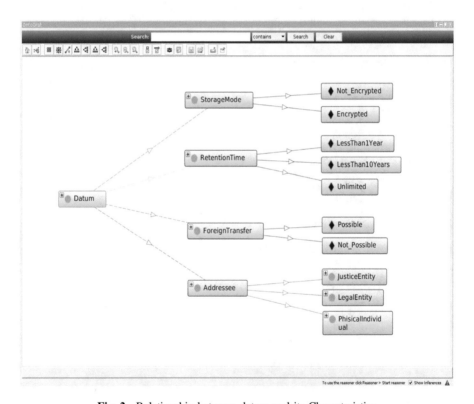

Fig. 2. Relationship between datum and its Characteristics

The property "*handles*" instead relates processing to data. This property is represented by the links between the Treatment *"Job Placement Processing"* and its handled Data.

Last but not least, the object property *Has_Severity* helps in the process of Data Protection Impact Assessment (DPIA). To assess the risk related to the fact that Data could be stolen or corrupted, a severity has been associated to either a *Data Type* and to *Data Characteristics*.

For the *Data Types* the following scores have been assigned:

Table 2. Data type risk factorfont sizes of headings. Table captions should always be positioned *above* the tables.

Data type	Score
Personal datum	1
Justice datum	2
Health datum	3
Biometric datum	4

Regarding the *Data Characteristics* the following scores have been assigned (Tables 3, 4 and 5, 6):

Table 3. Store maintenance mode score.

Data type	Score
Encrypted	0
Not encrypted	1

Table 4. Retention time score.

Retention time	Score
Less than 1 year	0
Less than 10 years	1
Unlimited	2

Table 5. Addressee score.

Addressee	Score
Justice entity	0
Government entity	1
Firm of physical Individual	2

Table 6. Foreign transfer score.

Foreign transfer	Score
No	0
Yes	1

This scoring system is arbitrary, and is only an example of one of the possible scoring systems that could be used. These scores can be easily modified editing them into the *"University-GDPR-Ontology"*. In this way, a domain expert who doesn't have knowledge of Prolog's logic programming, can easily customize the OWL ontology according to his necessities, because all of the domain information is stored only in the ontology and not in the expert system.

Once the ontology was created, its structure and individuals were converted in Prolog facts through the use of the Thea library developed by Vassilades, Wielemaker and Mungall [6]. Then the Expert System (implemented in Prolog) has been created.

It has two main purposes:

- To join BPMN Processes to Data Processing
- To estimate the Data Processing risk

The combination between BPMN Processes and Data Processing is based on BPMN topology and verbal pattern recognition. In fact, since BPMN diagrams only describe task's flow and interaction to compose processes, the only BPMN information that could be exploited to do this automatic classification was the task's name and its position in a certain pool. The expert system is in charge of doing this recognition. How the recognition is made will be clearer in the example at the end of this paragraph.

The second part involved the use of Prolog's built in predicates and list structures to detect the risk band of a given Treatment (described by a BPMN Process). The calculation of the risk is made starting from the Data that a Treatment handles.

Data has the *Characteristics* that are associated to a Severity (Score) as shown in the following figure (Fig. 3):

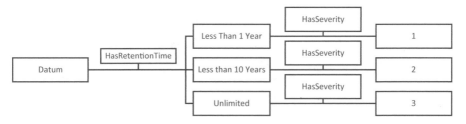

Fig. 3. Datum characteristic' severity

Each of the Characteristics of a Datum has a severity. The expert system sums the severities associated to Data Characteristics given by the scores in the tables above and multiplies this sum by a Risk factor associated to the Type of the Datum according to the scoring illustrated in Table 2.

Considered that a Datum can be associated to a risk score given by its type and characteristics, is becomes evident that a *Processing* can handle more than one *Datum*. In this case the choice that has been make is to assign the Processing a risk that corresponds to the *maximum* of its data risk scores. So for example if the *"Job Placement Processing"* handles three data with scores of 3, 5 and 18, the Treatment will have a risk score of 18. Obviously contact data and CV data are less risky than an Handicap datum.

It's important to notice that the maximum score that a Datum can reach is 24 (if the Datum is Biometrical, it is not encrypted, it's sent to a foreign country, it's stored for an unlimited time and it's destinated to a physical individual).

To do the DPIA, this range from 0 to 24 has been divided into 3 intervals:

- Low Risk → 0–8
- Medium Risk → 9–18
- High Risk → 19–24

In this way, the risk assessment related to a BPMN Process is classified in one of the up-mentioned intervals according to its score.

To show an example of the project's functionalities, the following BPMN Diagram is recognized to be a *"Job Placement Processing"* because the Expert System finds tasks that represent a communication between the Univerity's Job Placement Department and a Student and another communication between the same department and an external company (Fig. 4).

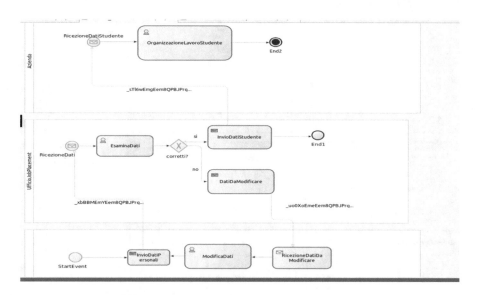

Fig. 4. JobPlacementProcessing – BPMN diagram

Once the BPMN process has been classified the second part of the Expert System plays its role, rating the process as a Medium Risk process, since its score is 18/24. In the following drawing is shown the call-tree that represents how Prolog rules are called and nested one in the other (Fig. 5).

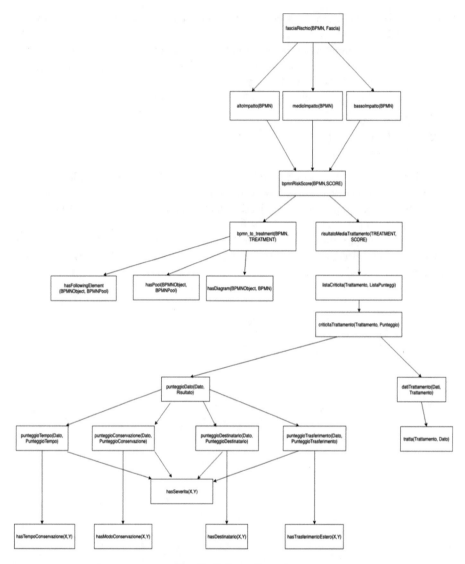

Fig. 5. Rules call graph

4   Conclusion and Future Works

In this paper we have discussed how to evaluate Business Processes with the new EU regulation GDPR. In particular, we explained how to make automatically one of the most critical activities required in the GDPR, the DPIA (Data Protection Impact Assessment). First of all, we focused on the development of the "University_GDPR_ Ontology", an ontology that represents the semantic description of academic treatments. Then, using Prolog, we matched BPMN diagrams to a particular academic data processing and we evaluate DPIA calculating the risk-score associated to a Business Process. Our study is still at a preliminary stage, so in the near future we intend to create an Ontology that describes also business, public and other type of data processing, and to generalize our Expert System to make DPIA for all kind of treatments described in the above-mentioned Ontology.

The scoring system represents an example of a possible scoring system. With the help of a domain expert it is possible to improve this system and make it more accurate.

References

1. Ahmadian, A.S., Strüber, D., Riediger, V., Jürjens, J.: Supporting privacy impact assessment by model-based privacy analysis. In: Proceedings of the 33rd Annual ACM Symposium on Applied Computing, pp. 1467–1474 (2018)
2. Shapiro, R., White; S.A., Bock, C., Palmer, N., Muhelen, M., Brambilla, M., Gagné, D. et al.: BPMN 2.0 Handbook 2nd Ed.: Methods, Concepts, Case Studies and Standards in Business Process Management Notation. Future Strategies Inc., Lighthouse Point (2011)
3. Rospocher, M., Ghidini, C., Serafini,L.: An ontology for the business process modelling notation. In: Formal Ontology in Information Systems: Proceedings of the Eighth International Conference, vol. 267, pp. 133–146. IOS Press, Amsterdam (2014)
4. Workgroup CODAU on Privacy and GDPR: Linee guida in materia di privacy e protezione dei dati personali in ambito universitario, Version 1.1 (2017). Online: http://www.codau.it/sites/default/files/verbali/all_3_lineeguida_privacy_gdpr_ravera.pdf
5. The W3C Consortium: OWL 2 Web Ontology Language Document Overview, Second Edition (2012). Online: https://www.w3.org/TR/owl2-overview/
6. Vassilades, V., Wielemaker, J., Mungall, C.: Processing OWL2 ontologies using Thea: An application of logic programming. In: Proceedings of the 5th International Workshop on OWL: Experiences and Directions, vol. 529. CEUR-WS.com (2009)

The 6th International Workshop on Hybrid/Cloud Computing Infrastructure for E-Science Application (HCCIEA-2019)

Low Power Wireless Networks
for Extremely Critical Environments

S. Ciccia$^{(\boxtimes)}$, A. Scionti, G. Giordanengo, L. Pilosu, and O. Terzo

Advanced Computing and Electromagnetics, LINKS Foundation,
10138 Torino, Italy
`simone.ciccia@linksfoundation.com`

Abstract. Navigation in the Arctic region is strongly compromised due to extreme environmental conditions. Among the others, correct positioning is the most critical aspect to manage in such environmental conditions. The main criticality for this can be ascribed to the low-orbit of satellites used for positioning, which implies very weak signals received at ground. To overcome such limitations, this work aims to investigate and develop a solution that provides highly accurate positioning by monitoring and mitigating the ionospheric impact on GNSS signals. Improving quality of the received signals allows for safer routes for Arctic shipping. To this end, this study foresees the deployment of an ad-hoc regional wireless network, where each node is a GNSS station. This paper describes the technological choices made to guarantee the operation of the GNSS stations network in such harsh environments. Specifically, the paper focuses on the description of the low power communication and computing subsystems.

1 Introduction

This work proposes a customized solution to support Arctic shippings, with specific focus on providing better positioning. The main goal of the project is the adoption of technologies that guarantee the GNSS ad-hoc networks operations in harsh environments. The setup of robust GNSS networks in Arctic and sub-Arctic regions is challenging, since the extreme environmental conditions and lack of land surface and power supply pose strong constraints. Also, connectivity among stations is critical, and the low elevation locking of the GNSS satellites makes this setup even more difficult. In this operational framework, we aims at enhancing the positioning accuracy integrating low-power systems able to handle, forecast and interpret ionospheric data useful to mitigate the effect of the ionosphere on high accuracy positioning systems [1]. The proposed solution provides information useful for correcting positioning to the GNSS service providers operating in the Arctic region. This information consists of different parameters concerning the level of ionospheric disruption on the signal coming from each GNSS satellite, and it will be included in the correction messages sent to the final users. The architecture of the system, sketched in Fig. 1a, has been conceived as a network of ground units, i.e.; NARWHALS System Units (NSU)s, with a

© Springer Nature Switzerland AG 2020
L. Barolli et al. (Eds.): CISIS 2019, AISC 993, pp. 859–868, 2020.
https://doi.org/10.1007/978-3-030-22354-0_79

data connection to a central controller, i.e.; the Central Controller Unit (CCU). The NSUs are fully-autonomous nodes deployed on the territory, on fixed reference points in the land. They continuously acquire GNSS data with a multi-constellation receiver, pre-process the received data on-board and send them to the CCU by means of long-range wireless communication. Finally, the CCU delivers the correction parameters to service provider, which will be responsible for forwarding them to the end users. The NSU is the core component of the such network, as depicted in Fig. 1a. The structure of the NSU has been designed taking into account modularity, and starting from three main building blocks: (i) a GNSS receiver; (ii) a low power computing stage; and (iii) a low-power communication infrastructure for interconnecting NSUs and the CCU. This modular approach is shown in Fig. 2a.

This paper describes the design choices for the low power computing and communication building blocks in different contexts of interest. These latter could be a network of NSUs deployed in a regional area, with a typical distance of tens of kilometres from the CCU, or a denser network of NSUs as in the harbour scenario illustrated Fig. 1b.

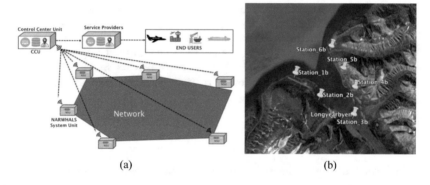

(a)   (b)

Fig. 1. Architecture of the proposed system (a), and harbour scenario (b) in the Svalbard Islands territory.

2   Low Power Computing

The low power computing subsystem is based on two different type of nodes, i.e., the NSU and the CCU, which are described in the following.

2.1   NARWHALS System Unit (NSU)

The NSU is a low-power computing module, and it is mainly devoted to run the GNSS software receiver. Beside receiving GNSS signals, the NSU is also responsible for pre-processing, filtering and compressing acquired data, which are then transmitted to the low-power communication module. Such latter module transfers pre-processed data to the CCU. The selection of the most suitable hardware that was able to satisfy the NSU operating requirements, mainly derive from being the system operating in a harsh environment. In addition, compatibility constrained the selection of the final hardware. To this purpose, the computing board is based on a X86_64 processor. This choice allows to run proprietary software application to run on top of a standard Linux OS keeping performance high. To this purpose the market offer many different solutions, ranging from those one targeting low-power applications (consuming very few watts) to high-end systems (i.e., multicore CPUs equipped with large amount of embedded cache memory and consuming up to tens of watts). Although ARM-based system with fair performance and consuming very few watts, they do not offer the right capabilities and compatibility guarantees to run the GNSS software receiver application. In addition, environmental conditions greatly reduce the choice for the hardware board. Indeed, most of the electronic components (including the CPU) on the board are designed to operate in a range of temperatures that goes from $0\,°C$ to $+40\,°C$. Conversely, in the Arctic environment the temperature range varies from $-35\,°C$ to $+8\,°C$.

A working prototype of such kind of low-power computing module (which satisfies all the previously mentioned constraints) has recently been validated and put into operation in the Antarctic region [2,3]. Such prototype was based on electronic components suited for operating in a large temperature range (industrial range), which is enough to satisfy hard temperature constraints. All the components, thus, have being selected to work on this range (from $-40\,°C$ to $+110\,°C$), including CPU, DRAM memory modules and disks. More specifically, the prototype was based on a low-power Intel Atom solution, sporting 2 physical cores running at $1.75\,GHz$ and equipped with $512\,KiB$ of internal cache each (no active cooling was necessary). Such capabilities are still enough to well support this applications.

During the validation phase of the prototype, different technical issues have been faced; all of them demonstrated to be crucial for setting up the proposed network. First of all, the hardware should be installed in a closed box, preventing water and dust to enter. This entails the absence of ventilation and, albeit being able to work at very low temperature, the module should be able to manage excessive warming in sunny days. A fundamental step between the implementation and the system deployment in field has been represented by the test of each building block of a NSU in an environmental chamber. Such test aimed at underlining possible behavioural anomalies due to extreme temperature and/or humidity levels excursion.

Besides the test on single devices adopted, an interesting point was the test of the overall system of Fig. 2a, whose conditions are not as straightforward to foresee as the sum of single elements working together. Figure 2b shows that the internal temperature of the box containing a NSU is quite higher (i.e., approximately $+30\,°C$) compared to the outside environmental temperature. Similar results have been achieved by switching on the NSU in a cold chamber, which has been kept to the lowest temperature for several hours. These preliminary tests were fundamental for avoiding any kind of problems due to the harsh environmental conditions (antarctic campaign) and thus before the real in-field system installation. This implies that the same design can be well suited for a system disposal in the Arctic context. To this end, the same experimentation will also be foreseen for this equipment, in order to get a preliminary feedback about the integrated solution and minimize issues while testing.

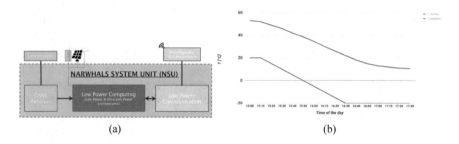

(a)   (b)

Fig. 2. Block diagram of the NSU (a); Measurements of the temperature inside the box (b).

2.2   Central Controller Unit (CCU)

The processing stage, hosted on a physical or virtual server, runs the algorithm used to forecast and calculate the mitigation parameters which are used to better determine positioning. Based on a sizing phase, the processing stage is instantiated as a single node or a cluster of nodes, each with the proper computing features needed to run the software in time for calculating and providing the positioning correction for a given period. The computing effort to which the CCU is subject to is also related to the reference scenario and, consequently, to the number of NSUs collecting data for a certain geographical area. Furthermore, moving this stage on a Cloud computing infrastructure allows to maximize the application scalability, enabling to manage a (possibly) growing number of connected nodes. Conversely, a physical implementation of such resource would need a careful planning to identify a balanced setup in order to achieve good performance at a reasonable price. It is worth of noting that, in case of a Cloud computing-based solution, the architecture still requires the presence of a central node, which is used to collect and centralize all the data coming from the connected NSUs.

As for the NSU, the CCU is equipped with a low-power communication module, in order to be able to receive messages from NSUs.

3   Low-Power Wireless Communication

In many application scenarios, sensors and actuators are distributed over a very wide area; Low-Power Wide-Area Networks (LPWAN), compared to short-range wireless standard based on Wi-Fi, Bluetooth and cellular connections, is more attractive since provides long-range and low-power connectivity [4–6]. In the following, an accurate research on LPWAN technologies is reported to find a possible candidate for the proposed network.

SigFox is a long-range and low-power standard providing a coverage up to 10 km and 50 km in urban and suburban area respectively. However, SigFox suffers of two main limitations. First the low-rate (100 bps); second, the transmit time restriction defined by ETSI regulation [7]. Despite the wide-coverage, these factors exclude it for the intended application.

LoRa is a more flexible low-power and long-range standard which claims a coverage of 2–5 km and 15 km in urban and suburban area respectively at 868 MHz. Someone tested 20 km with directional antennas [8]. Wider coverage can also be achieved at lower frequencies, however there is a limited number of usable channels. The achievable rate spans from 0.25 to 5.5 kbps depending on the link distance, i.e.; the higher the distance the lower the rate. LoRa differs from SigFox from the fact that some channels are not limited in transmit duty cycle and achieves higher rates.

DASH7 borns for wireless sensor and actuators with a protocol stack that reaches up to 2 km and data rate up to 167 kbps.

Weightless is a family of three open low-power wirelesss standards for Internet of Things (IoT) end-devices. Rates are very flexible, for example, Weightless W can reach up to 1 Mbps, however it covers up to 2 km with power consumption higher than other standards (e.g.; LoRa).

Ingenu is a robust protocol operating at 2.4 GHz while maintaining low-power operation. The main asset of Ingenu in comparison to alternative solutions is the high data rate (i.e.; up to 624 kbps uplink and 156 kbps downlink). On the contrary, the consumption is higher and the range shorter (up to 5–6 km) due to the high-band used.

According to this analysis, LoRa is the most flexible compromise between data rates, link distances and power consumption. Thus it is best candidate for this specific application, and in the following will be deeply analysed.

3.1   LoRa Standard Specifications

LoRa is interesting for the operation at 433/868 MHz. Table 1 illustrates the European regulation under the bandwidth of interest [9].

Table 1. Frequency bands, ERP and duty cycle for non-specific devices in Europe.

Op.Frequency [MHz]	ERP [dBm]	Duty Cycle %
433.05–434.79	+0	N/A
868.00–868.60	+14	<1.0
868.70–869.20	+14	<0.1
869.30–869.40	+10	N/A
869.40–869.65	+27	<10.0
869.70–870.00	+7	N/A

Table 2. Data rates, time overhead and typical Receiver sensitivity.

Spreading Factor	Rate [bps]	$T_{overhead}$ [ms]	$RX_{sense}(SF)$ [dBm]
SF12	250	1100	−136
SF11	440	570	−133
SF10	980	280	−132
SF9	1760	150	−129
SF8	3125	80	−126
SF7	5470	40	−123

The Effective Radiated Power (ERP) is related to the Equivalent Isotropic Radiated Power (EIRP) by the relation ERP = EIRP − 2.15 in *dBm*, where 2.15 dB is assumed to be the gain of a half-wave dipole antenna. The duty cycle is also specified and defines the amount of time the channel can be occupied. For instance, with a 10% duty cycle, in one hour the channel can be used for 360 s. Alternatively, the radio should repeatedly transmit for 36 s and shutdown for 3.6 s. It needs to be noticed that, when transmit duty cycle is reached in one channel, the same radio can use another channel to continue with the transmission of data.

LoRa is based on spread-spectrum technology and has an adaptive rate regulator. Possible rates are reported in Table 2. An end node tries and estimate the highest data rate it can use and be received correctly by the gateway. Starting with that estimation it initiates the transmissions. If no reply is received within the next expected downlink transmissions, the end node may try to establish connectivity by switching to the next lower data rate providing a more robust connectivity. The gateway is instead a complex receiver with a highly parallel architecture acting like (at least 8) different independent receivers, configurable on different channel frequencies (each of them can demodulate any Spreading Factor (SF) index from 7 to 12 in parallel) [10,11]. Finally, at protocol level, the LoRa packet of data consists of a preamble (i.e.; Physical layer (PHY) header, 16 bits CRC) and the payload. The preamble is used for detection and synchronization purposes, the PHY header describes the payload length which ranges from 13 to 255 bytes. The payload is transmitted with a (5,4) parity code channel encoding. In Table 2 the time overhead (i.e. the time used for the transmission of the preamble, the PHY header and the CRC) and the payload data rate (parity code included) are shown as function of the spreading factor.

4   An Example of Network Implementation

Figure 1b illustrates a possible implementation of a LoRa network based on this application. The nodes are distributed around the harbour with a maximum distance of 15 km with respect the gateway. Each node (after rates negotiation) will use a rate that depends on the position it is placed. For example, the farthest device will employ the minimum rate ($Rate\,(SF = 12) = 250$ bps). Each node communicates with the Gateway in a bidirectional manner.

4.1   Transmit Time and Link-Budget Analysis

LoRa networks need to be properly designed, especially in bandwidth with transmit duty cycle imposed. Thus, the first quantity of interest is the time to transmit a LoRa packet. This is described by Eq. (1):

$$t_{TX}\,(SF) = \frac{Pmax}{Rate\,(SF)} + T_{overhead}\,(SF)\ [s] \tag{1}$$

where $Pmax$ is the maximum payload in bits b that can be transmitted within a LoRa packet (i.e.; 255B, equivalent to 2048b), $Rate\,(SF)$ is the data rate in *bps* with which the payload is transmitted and $T_{overhead}$ is the time-overhead caused by the packet preamble. Then, we can compute the overall ON-time to transmit by Eq. (2):

$$tot_{TX}(SF) = t_{TX}(SF) \cdot N_{pack} \ [s] \tag{2}$$

where N_{pack} is the amount of data need to be transferred, that we call *Dsize*, divided by the maximum size of a LoRa packet *Pmax*.

To perform the link budget estimation, we account for the characteristics of a commercial device, i.e. the Adafruit LoRa module based on Semtech Radio transceiver [8,10]. The first parameter of interest is the receiver sensitivity, a function of the data rate. Table 2 reports an extract from the radio datasheet [11]. This information states that an incoming signal with a strength level lower than the sensitivity threshold is not perceived by the receiver circuitry. As a rule of thumb, the wireless link is designed with a safety margin, i.e.; the receive signal strength has to be stronger than the sensitivity threshold of at least +20 dB. The second is the receive power P_{RX}. Equation (3) describes the propagation by using the Friis formula [12];

$$P_{RX} = P_{TX} + G_{TX} + G_{RX} - 20 \log_{10} \left(\frac{4\pi d f_{ch}}{c} \right) \ [dBm] \tag{3}$$

where P_{TX} is the transmitted power in *dBm*, while G_{TX} and G_{RX} are the antenna gains in *dB* for the transmitter and the receiver respectively. The sum of the terms P_{TX} and G_{TX} gives the level of *EIRP* in *dBm*. Finally, the term to be subtracted is the attenuation due to the signal propagation path and is composed by the link distance d in *m*, the selected propagation channel f_{ch} in *Hz* and the speed of light c in *m/s* .

4.2   Network Simulation and Link-Budget Estimation

The required amount of data to transmit has established to be 1 kB each second. For a node placed at 15 km we suppose that it transmits at minimum rate, thus $Rate(SF = 12) = 250$ bps. From Eq. (2), $t_{TX}(SF = 12) = 9.36$ s and $N_{pack} = 4$ packets respectively. Then with Eq. (2) we found that $tot_{TX}(SF = 12) = 37.5$ s. This provides enough margin in term of time since such information is required by the network each second. While, in the best case the radio transmits at maximum rate, $Rate(SF = 7) = 5.5$ kbps. In this case $tot_{TX}(SF = 7) = 412$ ms. Concluding this Section, it can be stated that the design of a LoRa network strictly depends on the link distance of the farthest node and the quantity of data to be transmitted. Figure 3a shows the results of link budget simulation at 868 MHz. The computation has performed by considering an *EIRP* of 9 dBm (i.e.; under the law limit) and a common antenna of 5 dB in directivity for the receiver. The term Receive Signal Strength Indication (RSSI) is measure of the received signal as sum of direct and reflected path. The safety margin is indicated with a dashed line at +20 dB with respect the sensitivity threshold for the minimum and maximum rate (i.e.; $SF = 12$ and $SF = 7$ respectively). Theoretically, the maximum achievable link in line-of-sight is about 3.5 km with the maximum rate (i.e.; $SF = 7$) and 16.5 km with the lowest rate of (i.e.; $SF = 12$). The same computation is performed at 433 MHz, with the difference that, the $EIRP$ is set to 2.15 dBm (i.e.; $ERP = 0$ dBm) and $G_{RX} = 0$ dB. In this case, the maximum achievable link results between 8 km and 35 km with the highest and lowest rates respectively.

4.3   Advancement in Wireless Communication

Here, we provide an advanced discussion to reduce the consumption of the radio and/or increase the distance of a link by means of reconfigurable antennas. The concept of reconfigurable antenna for energy saving and/or interference reduction aims at maximizing radiation in the intended direction, while reducing the energy wasted in unwanted directions [13]. Reconfigurable directive antennas should be exploited on NSUs with the purpose of minimizing the energy consumption. With the sole purpose of giving an example, Fig. 4 shows the current absorption of the LoRa module during the transmission of a packet. The measurement is repeated for different transmit power level, i.e.; +5 dBm, +13 dBm and +20 dBm and for equal packet length and data rate. Remarkably, the absorbed current increases as the transmit power increases. Nodes equipped with such reconfigurable antennas would be able to direct the beam toward the direction of the gateway/concentrator. Thanks to the higher antenna gain, the nodes near the gateway can reduce the transmit power to the minimum, expecting a power saving according to Fig. 4 (e.g. +5 dBm). Instead, the farthest node will found a better signal quality and thus exploiting the adaptive rate regulator will find the best rate that can be used (less time to transmit data, less energy).

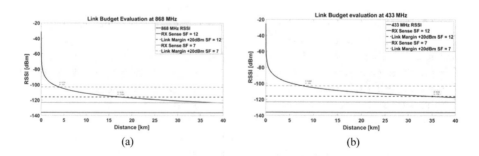

Fig. 3. Evaluation of the link budget at 868 MHz (a) and 433 MHz (b).

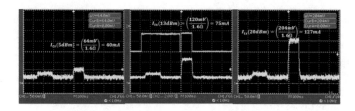

Fig. 4. Absorbed current when transmitting a LoRa packet with +5 dBm, +13 dBm, +20 dBm, from left to right, respectively

5 Conclusions and Future Perspectives

The paper addressed the feasibility study of a possible Arctic network, by selecting and analyzing a suitable low-power computing and communication solution. Via link-budget simulation we highlight that a wider coverage can be obtained at 433 MHz. However, channels in this band are limited and should be assigned to the farthest NSUs only. The amount of data to transfer judge on the decision of sharing these channels and this allows to overcome the duty-cycle limitation.

When there is no compromise between link-distance and channel assignments, two further options are valuable. The first is using pre-coding to minimize the amount of transferred data and favouring channel scheduling. The second solution is the involvement of a satellite operator to provide a connection that can not be covered by terrestrial standards, neither with directive antennas.

Acknowledgments. This work was supported by the NARWHALS project, which has received funding from the European Space Agency (ESA) under the Contract No.4000123171/18/NL/FG.

References

1. Aquino, M., Monico, J.F.G., Dodson, A.H., Marques, H., Franceschi, G.D., et al.: Improving the GNSS positioning stochastic model in the presence of ionospheric scintillation. J. Geodesy **83**(10), 953–966 (2009)
2. Mossucca, L., Pilosu, L., Ruiu, P., Giordanengo, G., Ciccia, S., et al.: GreenLab: autonomous low power system extending multi-constellation GNSS acquisition in Antarctica. In: Proceedings of the XXXII-nd International Union of Radio Science General Assembly and Scientific Symposium (URSI - GRASS), Montreal, Canada, August, pp. 19–26 (2017)
3. Pilosu, L., Mossucca, L., Scionti, A., Ciccia, S., et al.: Low power computing and communication system for critical environments. In: Proceedings of the 11th International Conference on P2P, Parallel, Grid, Cloud and Internet Computing (3PGCIC), Asan, Korea, October, pp. 221–232 (2016)
4. Centenaro, M., Vangelista, L., Zanella, A., Zorzi, M.: Long-range communications in unlicensed bands: the rising stars in the IoT and smart city scenarios. IEEE Wirel. Commun. **23**(5), 60–67 (2016)
5. Kim, M., Chang, S.: A consumer transceiver for long-range IoT communications in emergency environments. IEEE Trans. Consum. Electron. **62**(3), 226–234 (2016)

6. Adelantado, F., Vilajosana, X., Tuset-Peiro, P., Martinez, B., et al.: Understanding the limits of LoRaWAN. IEEE Commun. Mag. **55**(9), 34–40 (2017)
7. ETSI: ETSI EN 300 220-1 V2.4.1 (2012-01); Electromagnetic compatibility and radio spectrum matters. Accessed 12 February 2019. http://www.etsi.org/deliver/etsi_en/300200_300299/30022001/02.04.01_40/en_30022001v020401o.pdf
8. Adafruit: RFM95 LoRa Radio Feather M0. Accessed 22 February 2019. https://www.adafruit.com/product/3178
9. CEPT: ERC Recommendation 70-03: relating to the use of short range devices. Accessed February 2019. https://www.ecodocdb.dk/download/25c41779-cd6e/Rec7003e.pdf
10. Adafruit: RFM95/96/97/98 low-power long-range transceiver. Accessed March 2019. https://cdn-learn.adafruit.com/assets/assets/000/031/659/original/RFM95_96_97_98W.pdf?1460518717
11. IMST: WiMOD iC880A LoRaWAN Concentrator. Accessed on March 2019. https://www.wireless-solutions.de/downloads/Radio-Modules/iC880A/iC880A_Datasheet_V1_0.pdf
12. Gross, F.: Chapter 3: antenna fundamentals. In: Smart Antennas for Wireless Communications with MATLAB. McGraw-Hill (2005)
13. Ciccia, S., Giordanengo, G., Vecchi, G.: Reconfigurable antennas for ultra low-power radio platforms based on system-on-chip. In: Proceedings of the European Conference on Antennas and Propagation (EUCAP), April, London, UK (2018)

Unmanned Aerial Vehicle for the Inspection of Environmental Emissions

S. Ciccia[(⊠)], F. Bertone, G. Caragnano, G. Giordanengo, A. Scionti,
and O. Terzo

Advanced Computing and Electromagnetics, LINKS Foundation,
10138 Torino, Italy
simone.ciccia@linksfoundation.com

Abstract. These days, level of environmental pollution seriously concern people's health risk. For this reason, the demand of innovative techniques to determine pollution in sensible areas is ever increasing. This work presents a novel method for the control of environmental emissions. The proposed solution deploys an Unmanned Aerial Vehicle (UAV) to facilitates the responsible authorities to quickly act and real-time monitoring specific areas. The work encompasses both hardware and software integration to develop a new measurement tool that will be hosted by an UAV. The platform comprises digital optics sensors and toxic/pollutant gas concentration detectors along with low-power computing and communication capabilities.

1 Introduction

The reason UAVs are commonly employed these days is the easy-way of performing onerous tasks at minimal cost, as for instance the delivery of products [1]. Another good example is the search and rescue operation, where UAVs are able to real-time monitoring critical geographic areas (i.e.; high-altitude mountains) [2]. An important aspect of this application is the possibility of creates 3D-imaging to real-time monitoring situation and obtaining the exact position of the inspected location [3]. UAV applications are constantly growing. In [4] the authors achieved a traffic monitoring platform; while [5], UAVs are employed as Radio Frequency (RF) source localizer. They are also specifically involved in emergency situations as power suppliers and communication delivery systems [6,7]. Around these applications, this paper proposes a novel method to create 3D maps based on the level of the environmental emissions, which has never been done so far.

1.1 Structure of the Paper

The rest of the paper is organized as follows. In Sect. 2 the proposed approach is described, while Sect. 3 presents the overall architecture of the developed measurement platform. This section introduces the integrated sensors and

© Springer Nature Switzerland AG 2020
L. Barolli et al. (Eds.): CISIS 2019, AISC 993, pp. 869–875, 2020.
https://doi.org/10.1007/978-3-030-22354-0_80

describes the digitalization of the measurements (i.e.; Analog to Digital Converters (ADC) and Inter Integrated Circuit (I2C) conversion). Furthermore, Subsect. 3.3 presents the selected low-power computing and wireless communication module and then the integrated software to handle measurements is also described. Section 4 provides a description of the performed functional tests, and then some conclusions are drawn.

2   Proposed Approach

The deployment of UAVs allow rapid acquisition and map reconstruction of environmental data also in location not-accessible to humans, as for instance, due to the presence of toxic substance. The idea behind this work is illustrated in Fig. 1 and consists in:

1. Developing a multi-sensor platform for UAVs capable of real-time monitoring and recognising toxic gases and pollutant particles;
2. An on-board (local) processing of data incoming from sensors to elaborate a geo-location map with the related status of emissions. This information is transmitted to a ground gateway by means of wireless communication. The process is real-time, and the platform is able to communicates critical situations (with an alarm) to activate an immediate intervention.
3. The ground gateway acts as a internet-bridge for transferring the collected data to the cloud platform. The latter executes more complex algorithm (e.g.; data mining, computer vision) for a finer map elaboration and performs the data storage of the received data.
4. Finally, a Graphical User Interface (GUI) provides to the end-user a multilayer tool able to visualize geo-referenced 3D maps of toxic and pollutant substances.

3   Architecture of the Monitoring Platform

The proposed measurement prototype consists of a multi-sensor platform hosted by the UAV. This platform allows the sampling of environmental data, the pre-processing and the wireless transmission toward the ground gateway. The overall architecture is detailed in Fig. 2 and each module will be discussed.

3.1   Integrated Sensors

Two different types of sensors are used in our solution. For the detection of toxic gas substances, we use a set of electrochemical gas sensors. This kind of sensors estimate the concentration of a target substance by chemical reactions that occur in specific electrodes. The resulting current is proportional to the concentration of the specific gas and can be converted into a qualitative measure. The chemical reactions are not immediate and require a certain amount of time

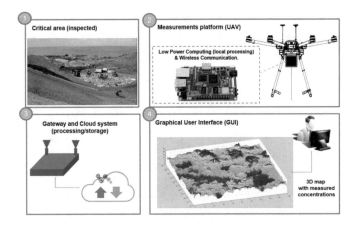

Fig. 1. Representation of the UAV and the installed measurement platform.

to give a valid output. Each sensor is characterized by a specific set-up time that have to be considered before accept the output as valid. Similarly, a certain amount of time have to pass between consecutive measurements. Those times are generally in the order of tenths of seconds up to a couple of minutes. The discharge of a sensor, when passing from air saturated with the target gas to air without the gas is generally lower than the set-up time. Experiments however shown that this is not always the case for all tested sensors. To preserve the accuracy of electrochemical sensors it is important to avoid saturating them for long periods of time, otherwise the chemical components could be damaged. Once opened after the first use, they are also characterized by an expiration time. After this time (in the order of months), the sensors are considered worn-out and no more reliable. The electric output of these sensors is sampled with an ADC and made available through I2C interface. A second class of sensors is optical, used to estimate particles of dust in suspension in air. A built-in fan pushes air between a laser ray and an optic receptor. Light get scattered by passing particles. The scattered light is converted into electric signals to be further processed. Analysis of signal waveforms is then automatically run to estimate the quantity and dimension of particles, since there is a direct relation between particles diameter and resulting light scattering. The used sensor is able to detect $PM10$, $PM2.5$ and $PM1.0$ particles. Output values are available through I2C interface. In the following the list of employed sensors:

- $H2S$: Hydrogen Sulphide sensor. At 100 ppm gives an output of 3 V, reaching a maximum of 4.8 V at 160 ppm.
- $NH3$: Ammonia sensor. At 100 ppm gives an output of 3 V, reaching a maximum of 4.8 V at 160 ppm.
- $CH3SH$: Mercaptan sensor. At 10 ppm gives an output of 3 V, reaching a maximum of 4.8 V at 16 ppm.

- *VOC*: Multi-gas and Volatile Organic Compound (VOC)s sensor calibrated to Carbon Monoxide (CO) equivalent. It is capable of a standard range of 0–1000 ppm CO equivalent and a maximum overload of 2000 ppm.
- *PM*10: Particle sensor with a measurement range from 0 to 500 $\frac{\mu g}{m^3}$ for *PM*10, *PM*2.5 and *PM*1.0, with a resolution of 1 $\frac{\mu g}{m^3}$.

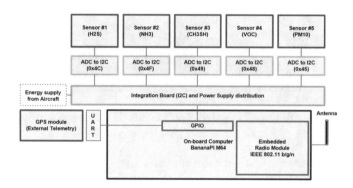

Fig. 2. Detailed flowgraph of the measurement platform.

3.2   Interface Board: Sensors to Computing

An intermediate module between a sensor and the I2C bus is necessary for the digitalization of the measured data. This task is performed by the integrated circuit MCP3221 [8]. This module is composed by a low-power ADC with 12-bit resolution and a minimal circuitry to provide compatibility with the I2C. The main reason we selected the MCP3221 is the possibility of uniquely addressing each sensor (up to 8). The proposed addressing is illustrated in the orange boxes of Fig. 2, which shows the connection to the I2C bus.

The electrical schema for the interface board is depicted in Fig. 3. Such board distributes the supply voltage to all sensors and to the Global Positioning System (GPS) module. At the input of each device a series of two capacitors, 1 µF and 10 ηF, are added to avoid over-voltages and under-voltages respectively. The I2C bus is shared from all sensors so that data flow into a single port (I2C0) on the computing device. This communication protocol has been selected since provides the communication link between many integrated circuits (i.e. Slaves) and a Master device with a bidirectional shared bus composed by two lines, the Serial Data Line (SDA) and the Serial Clock Line (SCL). Sensors are read in the following way:

- the Slaves (i.e.; sensors) are listening on the shared I2C bus;
- the Master put the address of the sensor he want to read and then release the control the bus;

- the Slaves check if the assigned address match with the received one. If this is the case, they send the sensed data on the shared bus.
- Finally, the master receives data and takes again the control of the bus.

An important aspect of this protocol is that data collision is avoided (each sensor has a univocal address).

The GPS module sends data by means of the Universal Asynchronous Receiver-Transmitter (UART) interface which consist of two separate data lines, one for the transmission and one for the reception.

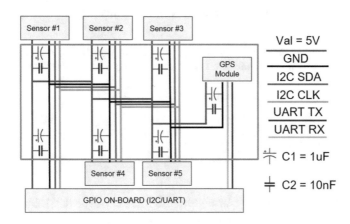

Fig. 3. Interface board: supply voltage distribution, I2C shared Bus and UART

3.3   Low-Power Computing and Communication

The sensing of data, the local processing and the related wireless transmissions are performed by the Banana Pi BPI-M64 single board computer [9]. The latter offers good computing performance (i.e.; it includes a 1.2 GHz Quad-Core ARM Cortex A53 64-Bit Processor) while ensuring low energy consumption. For the real-time transmission of the measured data we employed the on-board WiFi (AP6212); a 802.11 b/g/n commercial transceiver operating at 2.4 GHz [10]. Therefore, it can be connected to any Wi-Fi access point available (i.e.; which supports the 802.11 b/g/n standard). In our case, we setup an access point with another BPI-M64 board to receive data from the UAV.

The software managing the wireless communication has been written in ANSI-C and creates a socket interface with the AP6212 module to exchange packets in a bidirectional manner from the C application to the radio module. The latter takes care for the modulation and RF transmission. The same application run on the gateway in order to extract the received data.

4   Functional Tests

Two different test have been performed in order to evaluate the correct operation of the realized measurement tool.

First, we analysed the particulate matter. In this case we placed the acquisition board outdoor and we recorded the PM2.5 and PM10 every 20 min during the day. The mean values obtained from the relief have been compared with the value acquired by the nearest Arpa station, P.za Plouves, Aosta (AO) - Italy, in the same day [11]. Such comparison is reported in Table 1. The measured mean values are very close to those provided by the Arpa station. Slightly differences could be caused by the sensor circuitry and the calibration.

Table 1. Measurements of particulate matter: comparison between Arpa station and developed tool. The acquisitions are performed in data 21 March 2019.

Particulate	Arpa	Measured
Matter	$\frac{\mu g}{m^3}$	$\frac{\mu g}{m^3}$
PM2.5	15.0	16.5
PM10	21.0	24.8

A second test has been performed in order to check the detection of CO. The setup has been located in a cleaner camera and the VOC sensor is excited with a gas bottle. The result of the test is reported in Fig. 4 which clearly shows the presence of CO when this is released from the bottle.

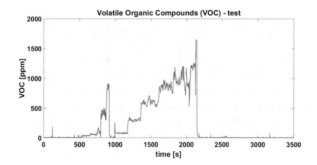

Fig. 4. Measurements of VOC in presence of CO.

5   Conclusions and Future Perspectives

In this paper we presented the general idea of a new method for the relief of environmental emissions. The sensor board that will be hosted by the UAV has been realized and tested to verify the accuracy of the acquired data. We are currently proving the functionality of the integrated GPS module and analysing the precision on the estimate position. We are developing a GUI for the final user. This service will show the levels of emissions in a 3D fashion to facilitate the responsible authorities in the determination of critical areas.

Acknowledgments. The authors would like to thank the companies Novasis Innovazione Srl and Aisico Srl of Pont-Saint-Martin, Aosta, Italy, for the valuable help in supporting this work.

References

1. Yang, N.K., San, K.T., Chang, Y.S.: A novel approach for real time monitoring system to manage UAV delivery. In: Proceedings of the 5th IIAI International Congress on Advanced Applied Informatics (IIAI-AAI), July, Kumamoto, Japan, pp. 1054–1057. IEEE (2016)
2. Naidoo, Y., Stopforth, R., Bright, G.: Development of an UAV for search and rescue applications. In: Proceedings of the IEEE Africon, September, Livingstone, Zambia, pp. 1–6. IEEE (2011)
3. Lauterbach, H.A., Nüchter, A.: Preliminary results on instantaneous UAV-based 3D mapping for rescue applications. In: Proceedings of the IEEE International Symposium on Safety, Security, and Rescue Robotics (SSRR), August, Philadelphia, PA, USA, pp. 1–2. IEEE (2018)
4. Elloumi, M., Dhaou, R., Escrig, B., Idoudi, H., Saidane, L.A.: Monitoring road traffic with a UAV-based system. In: Proceedings of the IEEE Wireless Communications and Networking Conference (WCNC), April, Barcelona, Spain, pp. 1–6. IEEE (2018)
5. Dehghan, S.M.M., Moradi, H.: A new approach for simultaneous localization of UAV and RF sources (SLUS). In: Proceedings of the International Conference on Unmanned Aircraft Systems (ICUAS), May, Orlando, FL, USA, pp. 744–749. IEEE (2018)
6. Zhang, Y., Ren, Z., Liu, L., Wei, C., Yin, C.: Design for a fast high precision UAV power emergency relief system. In: Proceedings of the 4th International Conference on Applied Robotics for the Power Industry (CARPI), October, Jinan, China, pp. 1–4. IEEE (2016)
7. Bupe, P., Haddad, R., Rios-Gutierrez, F.: Relief and emergency communication network based on an autonomous decentralized UAV clustering network. In: Proceedings of the SoutheastCon, April, Fort Lauderdale, FL, USA, pp. 1–8. IEEE (2015)
8. Microchip: Low-Power 12-Bit A/D Converter with I2C Interface. Accessed 20 Mar 2019
9. BPI: BPI Home, Banana Pi Single Board Computers. Accessed 29 October 2018
10. TheBIT co., Ltd.: AP6212 WI-FI/BT SIP MODULE User Manual WMDM-110G. Accessed 22 Mar 2019
11. Arpa Valle d'Aosta: Air Quality Measurements from Local Stations. Accessed 26 Mar 2019

A Classification of Distributed Ledger Technology Usages in the Context of Transactive Energy Control Operations

Fabrizio Bertone[(✉)], Giuseppe Caragnano, Mikhail Simonov,
Klodiana Goga, and Olivier Terzo

LINKS Foundation, Turin, Italy
{fabrizio.bertone,giuseppe.caragnano,mikhail.simonov,
klodiana.goga,olivier.terzo}@linksfoundation.com

Abstract. The Renewable Energy Sources (RES) have added large amounts of unpredictability in energy planning operations. The discrepancy between the forecast and the effective demand figures could led to dropouts and/or blackouts. To manage possible mismatch between demand and generation, a flexible behavior of energy users can be explored in the context of Transactive Energy Control (TEC). In the latest years, the potential of Distributed Ledger Technology (DLT) in various applications has been widely recognized by research literature. The main objective of this paper is to offer a structured exposition of different applicative scenarios in the view of using DLT in TEC context. This work explores the potential to diminish the unpredictability of RES by decentralized TEC operations in highly scaled context in which billions of small scale users could be efficiently aggregated into clusters capable to offer an equivalent of sufficient reserves.

1 Introduction

Traditional energy market is based on a highly hierarchical structure, where energy production, transmission and distribution are controlled by few subjects. This kind of infrastructure is transitioning towards new paradigms where institutional actors are joined by smaller operators in a more distributed and inclusive environment. The envisioned introduction of more and more Renewable Energy Sources (RES) enables a localized and diffuse production of energy. At the same time, RES are also introducing a higher volatility in energy availability and price, that can vary quickly and abruptly. Flat rate contracts are being dismissed and replaced by new models where energy will be sold by the minute.

This requires for the introduction of new enabling tools, allowing consumers and producers to be more actively involved and cooperate in energy exchanges. As already happened in stock markets, machines and algorithms will be pervasive and act on behalf of persons to timely trade energy. This requires especially robust and secure communication infrastructures, in order to assure the authenticity and trustworthiness of each message exchanged. Distributed Ledger Technology (DLT) is a set of promising technologies that could fit these needs and support new changing scenarios. DLT technologies are already being actively explored in energy sector by research

© Springer Nature Switzerland AG 2020
L. Barolli et al. (Eds.): CISIS 2019, AISC 993, pp. 876–885, 2020.
https://doi.org/10.1007/978-3-030-22354-0_81

community and enterprises, as demonstrated in this systematic review of projects from all over the World [1].

In this paper authors analyze the peculiarities of possible application of DLT technologies in particular to the new Transactive Energy context.

The document is structured as follow: Sect. 2 introduces specific aspects of the changing energy domain, with a focus on its economical side. Section 3 briefly describes Distributed Ledger Technologies with their current implementation details. In Sect. 4 authors propose a classification of DLT usages in the context of Transactive Energy operations. Final conclusions are drawn in Sect. 5.

2 Energy Domain Aspects

Energy can be considered to be both a product and a service. Electricity is by its nature complex to store for a later use, so it has to be readily be produced on request and consumed in (almost) real-time. Traditionally the energy market has been divided in two macro-groups: the wholesale generators and the retailer distributors.

Energy is primarily traded in the so-called Day-Ahead Market (DAM), where buyers (utilities) plan the amount of energy required to meet customers' needs for the following day -hour by hour- on the base of historical data (such as energy consumption on previous years) and additional information like weather forecasts. Sellers (owners of power plants) offer the availability of a certain amount of energy at a price that can vary based on production cost (due for example to the variability of oil and gas prices). The final price is then computed based on the totality of request and offer.

A second market, called Intraday Market (IM), is used to supplement the DAM and correct forecasts of demand in almost real-time. This is due to the uncertainty of the exact amount of energy that will be required the following day, that can be rarely foreseen precisely, and the real energy production, that can variate due to incidents or other unpredictable factors. This secondary market operates based on the first-come, first-served principle, where prices can greatly variate and be orders of magnitude higher than if bought in advance using DAM.

The electricity distribution system is physically bonded by Kirchhoff's laws and has very strict requirements in terms of stability and balance between production and consumption of power, where an incorrect management could easily lead to blackouts or disruption of infrastructures.

In the past, energy market was operating in a virtually unlimited scenario, due to the stability and inertia of mechanical generators. This is quickly moving into an increasingly complex and resource-limited scenario. The introduction of more and more Renewable Energy Sources (RES) shares is causing fluctuations and uncertainty in the availability of energy. Wind turbines can stop production both for lack of wind or in the opposite case of too much wind. Photovoltaic (PV) cells can produce energy only in presence of sun, so during night time or cloudy days their output is unavailable. Clouds can form quickly and unexpectedly, with the consequence of a sudden drop of power in the grid. As a reference of what is about to happen in following years, the

European EU 2030 Climate and Energy Policy set the goal of having around 32% of electricity generation from RES by 2030[1].

In a similar way, the Electric Vehicles (EV) recently introduced in the market, are a new source of uncertainty due to the high wattage nature of their batteries that can have a great impact on the grid during recharge.

A third category of energy market is then given by the balancing and Ancillary Services Market (ASM). These services are required in order to guarantee the reliability of the power system. In AC grids, the main indicator of grid (in)stability and (un) balance is given by the frequency, that must be always as close as possible to its nominal value (typically 50 or 60 Hz, depending on local standard) and has to be corrected as soon as possible in case of deviance.

The different markets and services are available and operates in different time slots, as shown in Fig. 1.

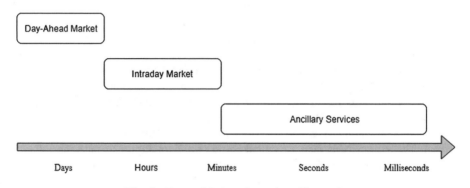

Fig. 1. Energy Markets Operations Timescale

In addition, the energy production is transitioning from a centralized model, where generation is led by few big producers, into a more scattered and diffused injection of power in the grid from medium and small RES producers and prosumers (a prosumer is an actor that is mainly an energy consumer but is also producing energy for self-consumption or to sell on the market).

The energy market is actively changing into a more inclusive and dynamic system, where availability can greatly vary by the minute, and prices as a consequence. New models and services are being developed in order to deal with this changing scenario.

2.1 Transactive Energy

The relatively quick introduction of a substantial number of distributed RES producers and small prosumers requires new economic and technological tools to improve energy market in the evolving scenario. A first solution to the problem of energy availability uncertainty caused by RES was given by the Demand-Response model (DR). In DR,

---

[1] https://ec.europa.eu/clima/policies/strategies/2030_en.

consumers are incentivized to adapt their energy consumption patterns in relation to the real-time capacity of the grid, by the mean of changing prices. However, DR only consider the consumption side of the energy market.

Since the initial introduction of DR programs, power systems became more and more complex, requiring further adaptations [2]. Transactive Energy Control (TEC) was then introduced to extend the DR model also to the production side, so that all resources behaves in a coordinated manner to optimize the system in a decentralized way [3]. This new framework empowers even smaller-scale actors to participate more actively in the energy market ecosystem [4].

3   Distributed Ledger Technology

Distributed Ledger Technology is the general and high-level name given to a set of technologies that allows the utilization of distributed databases independent from central controllers. The first example of such technology was introduced with the theorization and implementation of the Bitcoin cryptocurrency [5]. The aim of Bitcoin was the creation of a new class of currencies and transaction services that could be able to be self-managed with the aid of pre-determined algorithms. A fundamental characteristic of DLT is that information stored in the distributed ledger is immutable and globally verifiable by each node/user alone.

DLT systems can be divided in two complementary categories: permissionless or permissioned ledgers. Permissionless ledgers are "public", in the sense that anybody can participate in its operations both as a user and in the management of the service. Bitcoin is a kind of permissionless ledger, since anybody can join its protocol and participate as a currency user and in the transactions management. While the explicit lack of assigned administrative rights is a fundamental characteristic of the bitcoin protocol, other applications requires instead a more controlled environment. When the participation in a ledger is not open to the whole world we are in presence of a permissioned DLT.

3.1   DLT Models

While DLT is a generic name denoting some kind of distributed database (or ledger), the actual implementation can vary. Different models of distributed ledger have been proposed and demonstrated.

DLTs are data structures based on an innovative approach of integration between different technologies such as public key cryptography, distributed peer-to-peer networks and consensus mechanisms.

Typically, the ledger is implemented by a sequence of ordered blocks of transactions of agreed size. A block refers to a set of transactions that are grouped together and added to the ledger simultaneously. Each block contains a cryptographic hash relating to the previous block, which guarantees a close correlation between adjacent blocks. Generating blocks using hash functions and cryptographic algorithms results in an irreversible process that cannot be reverted.

Two different blocks may be offered as the next entry in the ledger. Addressing these problems requires the consent protocol.

A consensus protocol allows all participants (all the nodes of the DL), to agree on a single version of the truth, without the need for a trusted third party for validation.

The first implementation of a DLT, the Bitcoin, is based on the so called Blockchain (originally spelled block chain or just chain). While DLT and Blockchain are sometimes used interchangeably, the blockchain is only a specific kind of DLT, but other models also exists, the main alternative being the Directed Acyclic Graph (DAG). As discussed in this work [6], there are different implementations of DLT, of which some of the most used are: Blockchain, Directed Acyclic Graph, Hashgraph and Sidechain.

Blockchain. As the name suggests, ledgers based on blockchains store information inside chunks of data (blocks) chained together in an ordered sequence. In the Blockchain, the digital ledger of transactions is generated and modified by all the participants of a peer-to-peer network through a secure and shared protocol on the distributed network. All nodes in the network execute algorithms to evaluate, verify and match transaction information with the Blockchain transactions history. A new block is approved only if most of the nodes agree in favor of the transaction. The new block, once approved, is added to the existing chain [7].

Directed Acyclic Graph. The DLT technology could introduce innumerable advantages also to Internet of Things (IoT) infrastructures, increasing security and process automation. In order to overcome the technical limitations of the current blockchain architecture given the high computational cost and scalability issues, allowing the integration of low-power devices, validation complexity should be very minimal and confirmation times should be very short [8].

The Tangle represents an evolution for Distributed Ledger Technologies and is a technology behind the IOTA cryptocurrency [9]. Based on a concept of a Direct Acyclic Graph (DAG), Tangle is a scalable distributed database, without commissions and allows offline transactions. In DAG, blocks are not linked in a simple linear chained sequence. Each ancestor block, to be validated, needs to be confirmed by at least two other blocks corresponding to parallelized branches. This way blocks are distributed in a graph instead of a chain.

These features of the Tangle and its cryptocurrency make it possible to become the backbone of the emerging M2 M Economy.

Hashgraph. Hashgraph introduces a different model of distributed consensus, and tries to overcome the idea that the node that arrives first is the only one to validate the transaction.

As explained in [10], the hashgraph is compared to a tree that during its growth generates different branches but, unlike blockchain, instead of pruning new growth, each new branch is reinserted into the body.

In the Blockchain, if two miners create two blocks simultaneously, the community will choose one to continue and discard the other one. In hashgraph, every container is used and no one is discarded.

Sidechain. Sidechains are a separate blockchain connected to the main blockchain by a two-way peg. By creating an asset pegged, the sidechains remain independent of the main chain and remove its scalability limits. Thanks to this artifice, the sidechain developers could work and try new features without compromising the main chain.

The main idea [11] is to transfer assets by providing proofs of possession avoiding the need for nodes to track the sending chain. Therefore, it is possible to create a transaction while moving assets between different blockchains, creating a transaction on the first blockchain that locks the assets. Then, the second blockchain will contain a complementary transaction whose input is a cryptographic proof of previous locking.

3.2   Distributed Consensus Algorithms

In order to allow a distributed usage of the DLT, it is necessary to adopt an agreed and shared consensus algorithm. This allows third parties to trust the ledger information without the necessity to trust each other. Different consensus algorithms have been formulated and implemented, at the moment the two main alternatives are the Proof of Work and the Proof of Stake.

Proof-of-work (PoW), based on distributed consensus, is the system proposed by Bitcon to create and verify transactions on the network. PoW algorithms are designed so that they are quite energy intensive, because a significant amount of resources must be used to prove the work done to write the data in each block.

In order to avoid the high complexity and energy impact of PoW, an alternative algorithm called Proof of Stake (PoS), was proposed. In PoS, the validator of a new block is chosen in a deterministic way, according to his wealth (called stake), and not on his computational power.

3.3   Smart Contracts

Smart contracts are self-executing contracts whose terms of the agreement between the parties are written directly as an algorithm. The code and the agreements in the contract are not centrally recorded but belong to a distributed and decentralized blockchain network.

Szabo [12] has defined smart contracts as computerized transaction protocols that execute the terms of a contract. Many of the DLT platforms now available are based on -or support the execution of- smart contracts for managing rules and processes and generating transactions.

4   DLT Usages in TEC Operations

Distributed Ledger Technologies are already being studied in the context of energy exchanges. However the main focus of studies has been on the peer-to-peer (P2P) transaction between users in micro grids. In this section we try to instead define and categorize multiple different applications of DLT in the specific context of Transactive Energy operations.

4.1   Prices Signaling

One fundamental function of TEC systems is the price signaling, where information about instantaneous bids/asks, like in a stock market, are broadcasted to all participants. Since this information is the base for all following decisions about energy consumption, injection or accumulation, it is of the highest priority that this information in not falsifiable in any way. The immutability and verifiable characteristics of the DLT are well suited to support this data transmission layer.

4.2   Capacity Planning and Provisioning

With the increasing amount of Electric Vehicles connected to the grid, a flexible source of capacity is becoming available. The status and geographical position of EV is an information that can be somehow be known beforehand. This information could be integrated in the system to support the traditional markets like DAM and IM. In this case, the DLT could be used so that EV owners commit themselves to participate in the capacity provisioning for the following periods of time.

4.3   Agreement Compliance Enforcement

One important aspect to consider is how to enforce agreed contracts between participants in the new energy framework. Service Level Agreements (SLA), in order to be effective, have to consider specific performance metrics, corresponding to service objectives, and penalties in case those metrics are not satisfied. It is important to notice that a great amount of automatization is required in order to comply with real-time requirements of energy grid. DLT is well suited to be applied in different approaches. Both hardware and software mechanisms can be considered, as explained in the following.

Hardware Device. Hardware approach requires that a special flow control device is able to talk with a network of peers and get market signals in real time. It is required that the device can interact with the DLT and in particular understand and comply with the agreed smart contracts. The peculiar aspect of using an hardware device is that being an implementation of a Cyber-Physical System (CPS), it is able to enforce the compliance with the agreement. In particular, a flow controller has the ability to physically disconnect loads (or producing units) from the grid by means of relays or similar switches.

In a specular way, if the local node is provided with Battery Management Systems (BMS), power can also be injected in the grid on request or on convenience (if instantaneous energy price is high enough). An example of such kind of device is presented in [13].

Bonding Tokens. Another way to enforce agreement compliance is by using economic penalties or rewards managed completely by software. This allows more freedom of choice case-by-case to the user, that is able to break the agreement at the cost of losing some kind of "credit". In this model, a deposit of digital currency, or token, is used as a bonding mechanism to enforce compliance and "good behavior". In the same way, reward mechanisms can also be implemented in order to give a prize to users that act in ways that enhance the grid.

An important aspect to consider is the value given to the tokens. While in future energy scenarios the token value could very well follow energy prices and eventually be adjusted based on current grid status, a certain balance has to be maintained. If the required deposit is too high in value, users will be discouraged to participate in the system for fear of losing money. On the opposite, if the deposit is too low users could be not so obliged in following the rules. A proposition and study of such system is presented in [14].

4.4 Cost Optimization

Smart Contracts are algorithms that can support decision making using a set of pre-configured rules and inputs coming from the Ledger. Users of the TEC framework can then configure the local controller device in order to optimize the energy cost following custom preferences [15]. With the price signals coming from the data transmission layer, the controller can autonomously decide if at any given instant it is more convenient to consume or save energy by enabling/disabling smart appliances or connecting/disconnecting traditional "dumb" appliances.

4.5 Energy Accumulation and Sell-Back

The availability of Electric Vehicles or other Battery Management Systems (BMS) allows users to enhance the simple cost optimization process and implement a more sophisticate energy trading system. Using the same price signals, a user can decide to accumulate energy when it is more convenient, for example when there in a surplus of PV production, and sell it back to the network when prices are high, or even use local storage for local consumption in case of necessity.

4.6 Flexibility and Flexiramps as Ancillary Service

One of the functionalities provided by Balancing Ancillary Services, traditionally operated by specialized operators, is the creation of "ramps" that allows for smooth transitions between different levels of power flow and total capacity in the grid.

In addition to traditional operators, the production of ramps could also be enabled by the coordinated actions of multiple smaller actors. Ramps can be of the upward or downward type. Both kind of ramps can be produced by exploiting the flexibility of users. When an upward ramp is needed, users can use their flexibility to inject power in the system or alternatively reducing their consumption. Similarly, downward ramps can be created by increasing consumption (e.g. charging EV batteries) or dropping power injection to the grid (e.g. divert local PV production to local energy storage). The same, or a parallel, DLT infrastructure as the one used for price signaling could also be used for this kind of ancillary services coordination.

4.7   Energy Aggragators and Brokers

While single users could participate in the energy trading system by themselves, an aggregation of very small actors might be advisable. Users joining together in aggregated form could participate more actively and with a better bargaining leverage. Collective operation points can be established and shared between the participants creating a Virtual Power Plant [16] managed by a broker.

4.8   Energy Origin Tracking and Certification

A side advantage of DTL is the possibility of energy origin tracking. This way it is possible to know where and how each share of energy was produced. Some users (especially enterprises) could decide to buy preferably or exclusively energy produced by "clean" sources. This information could be eventually certified and shared using the public DLT information. Other users could decide to promote the local community and economy by paying for energy produced locally.

5   Conclusions

The energy market is a complex system undergoing a great amount of new challenges and modifications. The introduction of more and more Renewable Energy Sources and Distributed Energy Sources has a substantial impact on grid stability and operations. In order to overcome those new challenges, new trading and coordination frameworks are being introduced in the power management systems. Distributed Ledger Technologies are a set of new promising tools that can help in the distributed management of the grid management and energy exchanges. In this paper authors analyzed the current end future situation in energy systems and DLT implementations. A classification of different aspects that could be managed through DLT, in particular in the context of Transactive Energy has been proposed.

References

1. Andoni, M., Robu, V., Flynn, D., Abram, S., Geach, D., Jenkins, D., McCallum, P., Peacock, A.: Blockchain technology in the energy sector: a systematic review of challenges and opportunities. Renew. Sustain. Energy Rev. **100**, 143–174 (2019)
2. Chen, S., Liu, C.C.: From demand response to transactive energy: state of the art. Mod. Power Syst. Clean Energy **5**, 10 (2017)
3. Liu, Z., Wu, Q., Huang, S., Zhao, H.: Transactive energy: a review of state of the art and implementation. In: 2017 IEEE Manchester PowerTech, Manchester, pp. 1–6 (2017)
4. Simonov, M., Tibiletti, L.: Enabling small-scale actors to operate on markets of energy and ancillary services. In: Troussov, A., Maruev, S. (eds.) Techno-Social Systems for Modern Economical and Governmental Infrastructures, pp. 241–294 (2019)
5. Nakamoto, S.: Bitcoin: a peer-to-peer electronic cash system (2008)

6. Nabil, EI., Claus, P.: A review of distributed ledger technologies. In: Confederated International Conferences: CoopIS, C&TC, and ODBASE 2018, Valletta, Malt, Proceedings, Part II, 22–26 October 2018
7. Sachchidanand, S., Nirmala, S.: Blockchain: Future of financial and cyber security, pp. 463–467 (2016)
8. Strugar, D., Hussain, R., Mazzara, M., Rivera, V., Afanasyev, I., Lee, JY.: An architecture for distributed ledger-based M2 M auditing for electric autonomous vehicles. In: 33rd International Conference on. Advanced Information Networking and Applications (2019)
9. Popov, S.: The Tangle v1.6 (2016)
10. Baird, L., Harmon, M., Madsen, P.: Hedera: A governing council & public Hashgraph network (2018)
11. Back, S.A., Corallo, M., Dashjr, L., Friedenbach, M., Maxwell, G., Miller, A.K., Poelstra, A., & Timón, J. (2014). Enabling Blockchain Innovations with Pegged
12. Szabo, N.: The idea of smart contracts. In: Nick Szabo's papers and concise tutorials (1997)
13. Simonov, M., Bertone, F., Goga, K., Tibiletti, L.: Demand-side flexibility measurement and estimation For transactive energy control. In: 2018 IEEE International Telecommunications Energy Conference (INTELEC), Turin, 2018, pp. 1–6
14. Ferraro, P., King, C., Shorten, R.: Distributed ledger technology for smart cities, the sharing economy, and social compliance. IEEE Access 6, 62728–62746 (2018)
15. Thomas, L., Zhou, Y., Long, C., Wu, J., Jenkins, N.: A general form of smart contract for decentralized energy systems management. Nature Energy 4: 140–149 (2019)
16. Siano, P., De Marco, G., Rolán, A., Loia, V.: A Survey and Evaluation of the Potentials of Distributed Ledger Technology for Peer-to-Peer Transactive Energy Exchanges in Local Energy Markets. IEEE Sys. J. (2019)

Analysis of Job Scheduling Techniques in a HPC Cluster Deployed in a Public Cloud

Francesco Lubrano[1], Klodiana Goga[1(✉)], Olivier Terzo[1], Antonio Parodi[2], and Martina Lagasio[2]

[1] Fondazione LINKS, Torino, Italy
{francesco.lubrano,klodiana.goga,olivier.terzo}@linksfoundation.com
[2] CIMA Research Foundation, Savona, Italy
{antonio.parodi,martina.lagasio}@cimafoundation.org

Abstract. In this paper has been presented an analyses of different scheduling techniques on AWS ParallelCluster, an Amazon Web Services supported, open source cluster management tool for the deployment and management of High Performance Computing (HPC) clusters in the AWS cloud.

1 Introduction

In recent years, the Cloud environment has played a major role in running High-Performance Computing (HPC) applications, which are computationally intensive and data intensive in nature. Cloud providers offer highly customizable computing platforms, suitable, in terms of resource availability and configurability, for executing HPC applications. One of these application is the Weather Research and Forecasting (WRF) Model [1]. WRF model is a numerical weather prediction system commonly used for atmospheric research and operational forecasting. WRF model requires huge amount of computing resources and its software architecture makes it suitable for parallel computing; this implies that a HPC or cloud computing infrastructure is needed. Moreover is required an appropriate resource management system, in order to deal with parallel computation needs. AWS ParallelCluster [15] is one of the possible answers to the cited requirements. It allows to run HPC clusters in AWS Cloud Environment and offers a large range of configurations. Concerning the management of nodes, resources and parallel jobs, AWS ParallelCluster provides different batch schedulers. This paper presents a comparison among different Local Resource Management Systems (LRMS) and job scheduling techniques evaluated in the AWS Cloud environment.

2 Resource Management and Job Scheduling Techniques

In this section are described two of the most used software for resource management and job scheduling techniques in HPC environment.

L. Barolli et al. (Eds.): CISIS 2019, AISC 993, pp. 886–895, 2020.
https://doi.org/10.1007/978-3-030-22354-0_82

2.1 SGE

Sun Grid Engine [2] was an open-source batch-queuing system developed by Sun Microsystems and used for distributed resource management. Acquired at first by Oracle [3], it is now owned by Univa [4]. Starting from the original open source code, some communities forked SGE and continued to update and improve the code, such as Son of Grid Engine [5]. SGE acts as a job scheduler: it runs on top of a large number of computing nodes organized in clusters and it schedules, dispatches and manages parallel user jobs. For communications among processes, SGE supports Open MPI from version 1.2 [6]. Open MPI can automatically detect when it is running inside SGE system and will use automatically SGE to launch and kill processes. SGE is organized in daemons, computer programs running in background. Daemons can be grouped by host function. Figure 1 shows SGE architecture, and in particular it describe the positions, the roles and the interactions among daemons. The SGE architecture can be described identifying, according to their functions, three different host types: (*i*) **Submit host**, in charge of submitting, monitoring and managing jobs. It is the host directly controlled by the user; (*ii*) Master host, in charge of taking job-scheduling decisions, handling incoming user jobs and dispatching jobs to Execute hosts; (*iii*) Execution host, in charge of executing jobs. There is also a fourth host type, the administration host, able to make changes to the cluster's configuration. Each host type has a set of daemons running on it, in order to perform their own task. In particular, the **Master host** task is managed by three daemons: (1) *sge_qmaster* is responsible to manage the entire cluster and receives scheduling decisions about jobs from the sge_schedd daemon; (2) *sge_schedd* is responsible for deciding to which queue send a job and communicates its decisions to the sge_qmaster daemon; (3) *sge_commd* is the daemon who provides communication among SGE components. Instead, the **Execution host** is managed by the

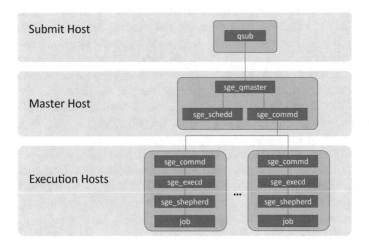

Fig. 1. SGE architecture

following daemons: (1) *sge_execd* is responsible for the execution of jobs and for the management of the queues in its host; (2) *sge_shepherd* is responsible for the control of running jobs; (3) *sge_commd* is the daemon who provides communication with the Master Host. Below are listed some common commands used to check and manage jobs and cluster status: (a) *qhost* displays status information about SGE execution hosts; (b) *qsub* is the user interface for submitting a job to SGE; (c) *qstat* displays job/queue status; (d) *qdel* removes a waiting or running job with the given id from the system.

2.2   SLURM

Simple Linux Utility for Resource Management (SLURM) [7] is an open source resource management and job scheduling system for large and small Linux cluster. The role of a resource management is to distribute computing power to user's jobs within a parallel computing infrastructure, with the target of efficiently assigning jobs to resources in order to satisfy user demands and achieve good performance in overall system's utilization. SLURM carries out this role ensuring fault-tolerance and high scalability and runs without kernel modifications. SLURM has three main functions: (1) it allocates exclusive and/or non-exclusive access to resources (compute nodes) to users for some duration of time; (2) it provides a framework with a defined set of commands for starting, executing, and monitoring work (normally a parallel job) on the set of allocated nodes; (3) it arbitrates contention for resources by managing a queue of pending work. Therefore, SLURM can manage resource allocation within a cluster providing the mean to execute parallel jobs in a parallel environment and, when there are more jobs than resources, can act as job scheduler, managing jobs' queue. Indeed, SLURM supports complex scheduling algorithms, resource limits and QoS policies. For communications among processes, SLURM supports Open-MPI through a dedicated plug-in [8]. It is possible to run OpenMPI using the srun command. According to Fig. 2, SLURM relies on a hierarchical structure where one node, which can be optionally redounded, acts as controller node to manage all the available resources. Slurmd is multi-threaded daemon running in each compute node, in order to provide node management, to read the common configuration file and to provide fault-tolerant hierarchical communications with the controller node. Slurmctld daemon runs in the controller node; it is in charge of controlling the entire cluster. Through cited daemons, SLURM manages different entities: (1) compute nodes, the principal resources in the cluster; (2) jobs, represent the resource allocation requests for a defined time frame; (3) partitions, job queues with constraint such as job size limit, job time limit, users permitted to use it, etc.; (4) job steps, sets of parallel tasks within a job. In order to ensure portability and customization, SLURM implement a general purpose plug-in mechanism. Plugins are typically loaded when the daemon or command starts, via configuration file or user options.

Below are listed some common commands used to check and manage jobs and cluster status. (a) *srun* submits a job for execution or initiate job steps in real time. srun allows users to requests arbitrary consumable resources;

(b) *salloc* allocates resources for a job in real time. This command spawn a shell, used to execute srun commands to launch parallel tasks; (c) *sbatch* submits a job in a form of bash script; (d) *sacct* reports job or job step accounting information about active or completed jobs; (e) *scancel* cancels a pending or running job or job step; (f) *sinfo* displays the state of partitions and nodes managed by SLURM; (g) *squeue* reports the state of running and pending jobs or job steps.

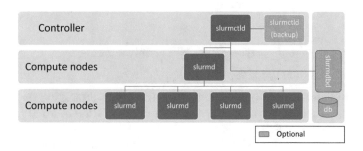

Fig. 2. SLURM architecture

3   Test-Bed Description

In this section the test-bed characteristics are described. We provided a complete description of the tool used for multiprocessing communications and a complete description of the framework used to launch the parallel cluster instance in AWS cloud environment. Finally, test results have been reported. For the purposes of this paper has been evaluated two schedulers SGE and SLURM which were already available in the AWS ParallelCluster template. The tests are based on simulating Weather Research and Forecasting (WRF) Model [1] installed with OpenMPI implementation.

3.1   Open MPI

Open MPI [9] is an open-source implementation of the Message-Passing Interface (MPI) Standard [10]. Open MPI was born to answer the request of an open-source software capable of managing parallel computer architectures. Parallel computing applications, such as MPI applications, can involve thousands of processors and this leads to face many issues related to scalability, such as process control, optimal resource management, latency, fault tolerance, application failure management, etc. These issues can impact on the overall performance of any parallel application and they have been taken into account in the design of the Open MPI architecture. Indeed, its architecture is based on component model, ensuring at the same time platform stability and the possibility to add

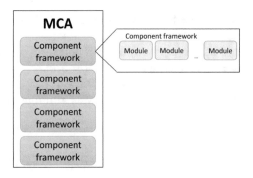

Fig. 3. Open MPI architecture

independent software add-ons. Open MPI is the result of merging the knowledge and the experience achieved by other three implementation of MPI: (1) FT-MPI from the University of Tennessee [11]; (2) LA-MPI from Los Alamos National Laboratory [12]; (3) LAM/MPI from Indiana University [13] Also contributions from PACX-MPI [14] team at the University of Stuttgart has been gathered in Open MPI. Therefore, Open MPI provides a combination of the best ideas and technologies coming from the other implementation listed above. In order to support a wide range of heterogeneous parallel machines, Open MPI provides features such as high performance drivers for interconnects computing resources even through multiple heterogeneous networks. It also provides check on data integrity and network failover, transparent to the running application.

MPI (Message-Passing Interface) is a communication protocol for programming parallel computers with the main goal of defining a portable, efficient, flexible and widely used standard for writing message-passing programs. The first version has been published in 1994 and today MPI is still a reference model in parallel computing environment. Open MPI is one of the implementations of MPI standard. Depicted in Fig. 3, the Open MPI architecture is a modular architecture, composed by three main functional areas: (1) the MPI Component Architecture (MCA);(2) the component frameworks; (3) the modules.

The MCA is the area that provides management services for all the others component. For example, MCA can accept and pass run-time parameters through the component framework and to each module. Component frameworks are components dedicated to a single task. Each component framework has different tasks and policies and it refers to a particular scenario, such as Point-to-point Transport Layer (PTL), a component framework in charge of managing network protocols. Each component framework can discover and manage modules. Modules are independent components that live in the component framework environment. In the case of PTL, each module of this framework corresponds to a particular network protocol and device.

3.2   Testbed Description

The tesbed is based on AWS ParallelCluster which is an open source cluster management tool officially supported by AWS. It was built on top of the Cnf-Cluster project [16] and it has the same aim: provide a simple and quick way to setup all the needed resources to build an HPC cluster in the AWS cloud environment. HPC clusters are sets of closely coupled compute, storage, and networking resources. They allow customers to run large scale scientific and engineering workloads.

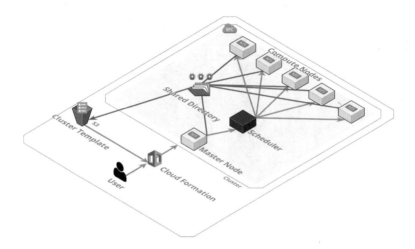

Fig. 4. Graphic representation of the testbed

Figure 4 represents the testbed used in this paper, based on the AWS Parallel Cluster template where EC2 instances, storage services and networking resources already configured to cooperate each others.

3.3   Job Scheduling Techniques Comparison

For this test have been used M5a [17] instances which are the latest generation of General Purpose Instances. This family provides a balance of compute, memory, and network resources, and is a good choice for many applications. All instances have the following specs: (1) 2.5 GHz AMD EPYC 7000 series processors; (2) EBS Optimized; (3) Enhanced Networking.

As well as WRF is a CPU intensive application [18] it has been decided to use a cluster of 80 cores in three different scenarios, but has been selected EC2 instances of the same family in order to have the same processor type as described in Table 1. In this paper the performance of WRF simulations have been run and modeling the same 24-h event. The WRF model setup consists of three nested domains. As well as the purpose of this paper is related to the

Table 1. Instance Types used for testing [17]

Instance Type	vCPU	Mem (GiB)	Storage	Dedicated EBS bandwidth (Mbps)	Network performa (Gbps)
m5a.xlarge	4	16	EBS-only	Up to 2,120	Up to 10
m5a.2xlarge	8	32	EBS-only	Up to 2,120	Up to 10
m5a.4xlarge	16	64	EBS-only	2,120	Up to 10

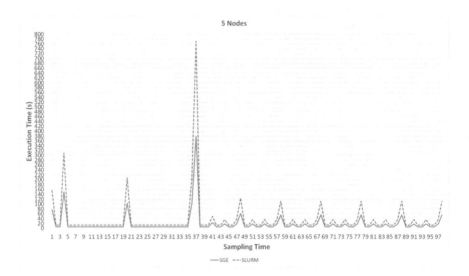

Fig. 5. Performance evaluation in the 5 nodes cluster

evaluation of the scheduling techniques used by the cluster, the focus will be on the execution time of each scheduler for each WRF sampled time. Since this type of simulation (three nested domains) requires more time to be executed the authors have extrapolated the execution time (compute time) during the first 1 min of WRF run, which include an interval of 98 sample times. Based on the execution time (compute time) during the first minute has been calculated the execution time of each scheduler to execute 6 h of WRF run Fig. 8. In the first scenario have been evaluated the performance of SLURM and SGE in a cluster with 5 Computing Nodes (m5a.4xlarge instances) which have 16 vCPU each. As represented in the Fig. 5 in this scenario the performance of both schedulers is quite similar, with SGE performing slightly better, with a lower execution time (computational time) for each WRF sampling time.

In the second scenario is used a cluster of 10 Compute Nodes (m5a.2xlarge instances) which have 8 vCPU each. In this scenario as depicted in Fig. 6 SLURM performs better, with a lower execution time (computational time) for each WRF sampling time.

In the third scenario is used a cluster of 20 Compute Nodes (m5a.xlarge instances) which have 4 vCPU each. From the Fig. 7 can be seen a clear difference

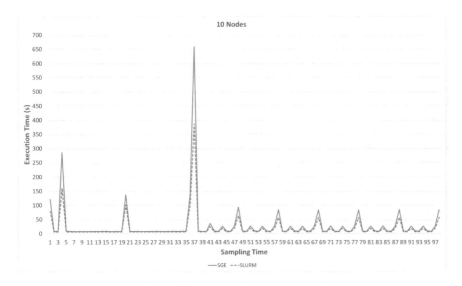

Fig. 6. Performance evaluation in the 10 nodes cluster

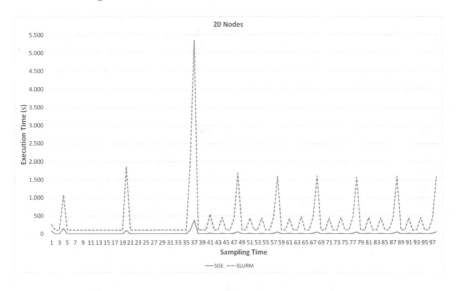

Fig. 7. Performance evaluation in the 20 nodes cluster

in performance between SGE and SLURM. The performance of SLURM in this scenario is very low compared with SGE (almost 15 times higher) but also compared to the other two scenarios.

In Fig. 8 has been represented the estimation time required to execute 6 h of WRF run in each scenario. In the 5 Compute Nodes Scenario SGE has a better performance, while in the 10 Compute Nodes Scenario SLURM performs better,

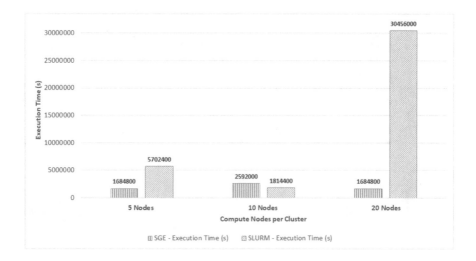

Fig. 8. SGE and Slurm comparison

instead it degrades in the third scenario with 20 compute nodes. This can be attributed to the fact that the compute nodes used in this scenario has lower performance although the total number of cores of the cluster remains the same.

4   Conclusions

This work has the purpose to highlight the differences between two different type of schedulers SGE and SLURM when run the same simulation and in different clusters, with the same total number of cores and with the same processor type, in order to evaluate the effective use of the parallel computational resources available. The tests have shown that increasing the number of cores initially SLURM performs better but its performance decreases strongly in the cluster with 20 compute nodes. Increasing the number of nodes doesn't imply that better performance will be achieved, there should a strict evaluation of the single compute node in terms of CPU and RAM and the scheduler to be used.

Acknowledgements. This research work is part of the STEAM Project, funded by the European Space Agency. ESA Contract No. 4000121670/17/NL/AF. The authors acknowledge all the partners of the project for their work and contribution.

References

1. The Weather Research & Forecasting Model. https://www.mmm.ucar.edu/weather-research-and-forecasting-model
2. Gentzsch, W.: Sun Microsystems: Sun grid engine: towards creating a compute power grid. In: Proceedings of the 1st International Symposium on Cluster Computing and the Grid (CCGRID 2001), p. 35. IEEE Computer Society, Washington, DC, USA (2001)

3. Oracle and Sun Microsystem. Strategic acquisition. https://www.oracle.com/sun/index.html
4. Univa completes acquisition of Grid Engine assets, becoming the Sole Commercial Provider of Grid Engine Software. http://www.univa.com/about/news/press_2013/10222013.php
5. Son of Grid Engine project. https://arc.liv.ac.uk/trac/SGE/
6. SGE support in Open MPI environment. https://www.open-mpi.org/faq/?category=sge
7. Jette, M., Grondona, M.: Slurm: simple Linux utility for resource management [PDF]. Proceedings of ClusterWorld Conference and Expo, San Jose, California, June 2003. https://slurm.schedmd.com/slurm_design.pdf
8. Slurm support in Open MPI environment. https://www.open-mpi.org/faq/?category=slurm
9. OpenMPI: Open source high performance message passing library. https://www.open-mpi.org
10. MPI: A Message-Passing Interface Standard version 3.1. https://www.mpi-forum.org/docs/mpi-3.1/mpi31-report.pdf
11. Fagg G.E., Dongarra J.J.: FT-MPI: fault tolerant MPI, supporting dynamic applications in a dynamic world. In: Dongarra, J., Kacsuk, P., Podhorszki, N. (eds.) Recent Advances in Parallel Virtual Machine and Message Passing Interface. EuroPVM/MPI. LNCS, vol. 1908. Springer, Heidelberg (2000)
12. Aulwes, R.T., Daniel, D.J., Desai, N.N., Graham, R.L., Risinger, L.D., Taylor, M.A., Woodall, T.S., Sukalski, M.W.: Architecture of LA-MPI, a network-fault-tolerant MPI. In: International Parallel and Distributed Processing Symposium (2004)
13. Squyres, J.M., Lumsdaine, A.: A component architecture for LAM/MPI. In: Dongarra, J., Laforenza, D., Orlando, S. (eds.) Recent Advances in Parallel Virtual Machine and Message Passing Interface. EuroPVM/MPI 2003. LNCS, vol. 2840. Springer, Berlin, Heidelberg (2003)
14. Gabriel, E., Resch, M., Beisel, T., Keller, R.: Distributed computing in a heterogeneous computing environment. In: Alexandrov, V., Dongarra, J. (eds.) Recent Advances in Parallel Virtual Machine and Message Passing Interface. EuroPVM/MPI 1998. LNCS, vol. 1497. Springer, Berlin, Heidelberg (1998)
15. AWS ParallelCluster project repository. https://github.com/aws/aws-parallelcluster
16. AWS CfnCluster. https://cfncluster.readthedocs.io/en/latest/
17. EC2 M5a instance type. https://aws.amazon.com/it/blogs/aws/new-lower-cost-amd-powered-ec2-instances/
18. Goga, K., Parodi, A., Ruiu, P., Terzo, O.: Performance analysis of WRF simulations in a public cloud and HPC environment. In: Complex, Intelligent, and Software Intensive Systems, pp. 384–396. Springer (2018)

The 1st International Workshop on Knowledge Creation and Innovation in Digital World (IKIDW-2019)

Collective Engagement and Spiritual Wellbeing in Knowledge Based Community: A Conceptual Model

Olivia Fachrunnisa[1], Ardian Adhiatma[1(✉)], and Heru Kurnianto Tjahjono[2]

[1] Department of Management, Faculty of Economics, Sultan Agung Islamic University, Semarang, Indonesia
{olivia.fachrunnisa,ardian}@unissula.ac.id
[2] Department of Management, Faculty of Economics and Business, Universitas Muhammadiyah Yogyakarta, Bantul, Indonesia
herukurnianto@umy.ac.id

Abstract. The purpose of this study is to develop a model of collective engagement on strengthening community spiritual welfare. In knowledge-based economy, community of practice (CoP) is created to support the dispersion of knowledge among community. However, the large area of expertise in knowledge-based community sometimes create little friction that will ultimately hard to solve. Therefore, we need a social engineering model to maximize the existence of the community towards the realization of community's welfare. Specific targets to be achieved is an understanding on the application of the concept of collective engagement, dimensions, forming factors and its impact on the community's spiritual wellbeing. A four propositions is build based on literature review. Future research agenda is provided to address validity of the proposed model.

Keywords: Collective engagement · Spiritual wellbeing · Knowledge-based community

1 Introduction

The problems that arise from the study of literature is the lack of studies on the concept of collective engagement at the level of community of practice. Community of practice (CoP) is a group of people who agree to develop expertise on the basis of knowledge or expertise that they have. Therefore, this study becomes important to draw up a description and basic concepts concerning collective involvement in an organization. Once the basic concept is found, then the dimension and scale of measurement will be developed to validate the new theory of collective community engagement.

Collaboration in a community should be increasingly emphasized to the members of CoP given the fact that each practitioner runs his own community, and even maybe just think about the welfare of people who become part of the community. Another fact is the imbalance of the spirit of competition and cooperate or collaborate among each group. The balance between cooperation and collaboration becomes an important factor

L. Barolli et al. (Eds.): CISIS 2019, AISC 993, pp. 899–906, 2020.
https://doi.org/10.1007/978-3-030-22354-0_83

to improve the community welfare. The next fact is that every community no longer seeks to develop core competencies but rather the willingness to master all the competence and autonomous or independent.

The results of this study will contribute to improving spiritual welfare of the community in terms of the theory of community wellbeing, especially in engagement theory that developed from the individual level to the organizational level following the measurement scale from the perspective of knowledge-based community. Engagement is a behavior that indicates the degree to which people moved to blend with the job in an organization. Entanglement of individuals with tasks and objectives of the group will create a positive effect on the level of innovation and creativity [1]. Collective engagement is more than the aggregate amount of the individual's involvement in the organization [2]. Engagement has two basic components. First, the individual must be aligned to a common goal and, secondly, they must be committed to mutually support each other's efforts [3]. When the members of each organization involved is focused on achieving the goal, then knowledge management practices such as sharing of information, shared values and shared vision will mutually reinforcing each other. In the end, when individual has full engagement and focus on the purpose of the organization's goals, then a relationship of mutual support and individual efforts will result in the group's energy, enthusiasm and focus on achieving common goals.

The concept of employee engagement has been discussed extensively by some previous researchers. In this case, employee engagement is measured at the individual level that indicates an individual engagement with the organization. Kahn [4] defines engagement as a more comprehensive description of the investment in the attitude of affective, behavioral and cognitive energy a person in the workplace. In a group or association, engagement will be measured at the level of organization involving all members of the organization's with other organizations such as, employers' associations, traders associations, business associations and other forums. Thus, collective engagement is a construct of the level of organizations and groups is an indicator of the presence of a motivational environment in organizations (the motivational aspect).

2 Literature Review

2.1 Value Congruence and Collective Engagement

Community is defined as a collection of citizens who are joined together because of geography proximity and interact with each other so that identity is created [5]. It is in line with Crane et al. [6] who define community as individual citizens or groups of citizens organized to represent their common interests. Whereas according to Fawcett et al. [7], community is a group-based approach including community members who might then have an impact/strategic on the company. Community engagement is a pattern of activities carried out by companies to work collaboratively with and through groups of people to overcome problems that affect the social welfare of these people [8, 9]. Hence, it can be concluded that community engagement is the attachment of community members to a community that works collaboratively for a common goal.

Research by Vogel et al. [10] states that job involvement will increase if the value congruence increases from low to high. Moreover, Rich et al. [11] stated that value congruence is positively related to engagement. This explains that the higher the value congruence will increase the level of engagement. The involvement of high work with a high level of conformity of values to the organization will create community engagement. Hence, it can be concluded that value congruence affects community engagement. In addition, Bhargava et al. [12] states that value congruence is positively related to the effectiveness of job performance. This will happen if someone has high self-esteem so that it results in more appropriate self-value with the value of the community which results in him becoming increasingly tied to his community.

P1: Value congruence (similarity values among community members) will produce a collective engagement.

2.2   Collective Engagement and Spiritual Welfare

In the research by Slåtten, Svensson, and Sværi [13], the condition that support employees to feel as "we" in company make the engage collectively, physically, cognitively, and emotionally to be creative and innovative in the role of their perfor-mances in achieving appraisal performance. Previous research by Albdour and Altar-awneh [14] also show that employee engagement influence positively towards organizational commitment. Slåtten and Lien [15] show that there is positive influence between collective engagement and company ability to create innovation. Cognitive collective engagement develops cognitive attitude, managing and sharing knowledge and experience, and involvement. Then, it can create emotional awareness to commit in the role of performance, increase the ability of individual to think more creative. This creativity is needed to increase innovation performance.

A positive effect on the level of innovation and creation is the result of individuals' engagement with the tasks and objectives of the group [1]. However, group leader must understand how to translate individual engagement to organizational engagement or collective engagement as suggested by Barrick, Thurgood, Smith, and Courthright [16]. In the organization, the aggregate amount of the individual's involvement can be called as collective engagement [17]. There are two basic components in engagement or involvement. First, organizational members should have a common goal and, sec-ondly, they must have an effort to support each other [18]. Members must be involved in achieving goal in order to reinforce a mutual value by sharing information, values and vision. In the end, when member has full engaged and focus on organizational objectives, the relationship of mutual support and individual effort will produce group's energy, enthusiasm and focus on achieving common goals.

However, despite the suggested link between employee engagement and organi-zational performance, there is no many researches examine engagement at the orga-nization level [19]. Moreover, Barrick et al. [16] has investigated the role of collective organizational engagement in mediating the relationship between firm resources and firm performance. However, they discuss collective organizational performance from all aspects which are physically, cognitively, and emotionally, while Zhang and Bartol [1] explain about the concept of psychological empowerment that will grow creative process engagement.

P2: Collective engagement (engagement collectively) will produce spiritual welfare of community members.

2.3   Engagement and Spiritual Wellbeing

Cooperation is an agreement between two or more mutual or cooperative as "two or more persons to carry out joint activities in an integrated manner which is geared toward a target or specific purpose." Thus, it can be concluded that by joining in a community is efforts to improve themselves, while the community needs to build a strong community association. One strategy to achieve the results and objectives of the organization, leaders is how to translate individual collective engagement. Kahn [4] defines engagement as a more comprehensive description of the investment in the attitude of affective, behavioral and cognitive energy a person in the workplace. Collective engagement is a construct of organizational levels and is an indicator of the presence of a motivational environment in organizations (the motivational aspect). Antecedent of collective organizational engagement is motivating work designs, HRM Practices and CEO Transformational Leadership Behaviors [20]. Three of the organization's resources are sufficient to meet the needs of meaningfulness, psychological safety and psychological availability. Meaningfulness influenced by the characteristics of the task and the job role, psychological safety is an individual comfortable feeling for his role in the organization, without fear of the consequences of negative self-image, status or career [4]. Operationally, motivating work design can be done by giving tasks and challenges that are meaningful to the individual.

A set of HR practices can be designed to improve psychological safety. HR practice-oriented investment and also inducement HR practices that increase the expectations of the individual in the organization is considered to be improving psychological safety. While psychological availability is how someone joins himself to a task or role to consider the adequacy of resources for the physical, emotional and psychological. It is influenced by the level of trust of one's feelings about his ability to work, as well as the relationship status of a person in the organization [20].

Collective Engagement will benefit a community in some way. First, when community members interact with each other, then they will share the element of positive behavioral elements such as affective, motivational and attributes that can improve performance attributes such as collective efficacy and the high potential of the group. Second, each member will be mutually comparing their input and output in the organization. This refers as a process of social comparison. Each will compete to customize their engagement results compare with contributions other members in the group. So, it is clear that collective engagement would improve organizational welfare.

Third, leaders are able to increase the level to which members feel connected and identify the destination itself with organizational objectives, which at a later stage will override his desire to achieve organizational goals that are more valuable [21]. Therefore, collective engagement increases organizational value through improved organizational welfare.

P3: Value congruence will generate high social identity community.

P4: The identity of a strong social community will generate the spiritual welfare of the community.

3   Research Method

This paper deals with the development of conceptual model on the relationship between value congruence, social identity, collective engagement and spiritual welfare. This paper is based on secondary data which has been collected from books, journals, newspapers, and internet. A conceptual framework has been developed describing the aspects of value congruence, collective engagement and social identity to create community spiritual welfare.

4   Conceptual Model

Based on literature review in the previous section, the conceptual models can be described in Fig. 1. Community spiritual welfare can be raised from collective engagement and social identity. Meanwhile, social identity and collective engagement would increase if members of the community have a high value congruence.

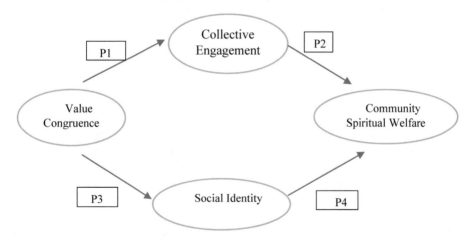

Fig. 1. Conceptual models of collective engagement and spiritual wellbeing in knowledge based community

In the complex business world, the main task of community of practices is to manage their members smoothly, with the objective of enhancing their psychological attachment to the community [22]. The link between HRM practices and organizational outcomes such as commitment, job satisfaction and performance have been discussed widely by scholars [22, 23]. With the development effort to concern on community welfare perspective, it is needed to perform a model to create social identity and member's engagement to achieve community welfare.

A social value is explained as an equality practice for all human beings as they are all community members. We have to respect their personality. A community which takes role as a pillar in economic activity, should enrich the welfare of its member. As a community member, s/he needs to engage with the community who has relationship with the institution. The institution, therefore, need to develop a creative process engagement. Social identity and collective engagement are hypothesized as requirements to create creative process engagement which will further enhance the community welfare.

Measurement of proposed concepts is provided in Table 1.

Table 1. Concepts measurement

No.	Variable	Definition	Measures
1.	Value congruence	Level of conformity between individual and organizations towards common goals' attainment	• Altruism • Relationship • Security • Authority • Prestige • Autonomy
2.	Collective engagement	Collective attachment of community members towards their community	• Problem identification • Involve in information searching • Ideas for problem solving • Involve in decision making process
3.	Social identity	Identity that represents a set of social relationships based on membership in a group or community	• Social categorization • Social identification • Social comparation
4.	Spiritual welfare	A positive feeling with life spirit to be a better man	• Calling • Membership • Happiness • Contribution • Satisfaction • Meaningful

5 Conclusion and Future Research

The concept of collective community engagement and its impact on spiritual community welfare has not been widely discussed in the current literature. This paper aims to build a conceptual model and hypotheses development for collective community engagement and spiritual community welfare. In the future, we plan to further validate the proposed models. Moreover, further research will be done in 4 stages of research which are:

1. Validate dimensions of all the proposed concepts.
2. Validate measurement of each variable (value congruence, collective engagement, social identity and community welfare). Items and concepts derived from the structural development and the theory of group engagement.
3. Implement several pilot studies to improve the variable measurement. Pilot studies will carry out to get the opinion of the general public or a layperson on the concept of collective engagement and spiritual well-being.
4. Conducting a survey to the respondents each CoP to test the validity of the instrument developed by structural and nomological.
5. Test the proposed model.

Acknowledgements. This research is funded by Directorate of Research and Higher Education, Indonesia, Fundamental Research Scheme, 2019.

References

1. Zhang, X., Bartol, K.M.: Linking empowering leadership and employee creativity: the influence of psychological empowerment, intrinsic motivation, engagement and creative process. Acad. Manage. J. **53**(1), 23 (2010)
2. Kline, R.B.: Principles and Practice of Structural Equation Modeling, 2nd edn. Guilford Press, New York (2005)
3. Christian, M.S., et al.: Work engagement: a quantitative review and test of its relations with task and contextual performance. Pers. Psychol. **64**, 89–136 (2011)
4. Kahn, W.A.: Psychological conditions of personal engagement and disengagement at work. Acad. Manage. J. **33**, 692–724 (1990)
5. Bakker, A.B., Xanthopoulou, D.: The crossover of daily work engagement: test of an actor-partner interdependence model. J. Appl. Psychol. **94**(6), 1562 (2009)
6. Crane, A., Matten, D., Moon, J.: Stakeholders as citizens? rethinking rights, participation, and democracy. J. Bus. Ethics **53**(1–2), 107–122 (2004). https://doi.org/10.1023/B:BUSI.0000039403.96150.b6
7. Fawcett, S.B., Paine-Andrews, A., Francisco, V.T., Schultz, J.A., Richter, K.P., Lewis, R., Williams, E.L., Harris, K.J., Berkley, J.Y., Fischer, J.L., Lopez, C.M.: Using empowerment theory in collaborative partnerships for community health and development. Am. J. Community Psychol. **23**(5), 677–697 (1995)
8. Scantlebury, M.: The Ownership Structures of Heritage Tourism Enterprises in Barbados and Their Institutional and Community Involvement. University of Waterloo, Waterloo, Canada (2003)
9. Vogel, R.M., Rodell, J.B., Lynch, J.W.: Engaged and productive misfits: how job crafting and leisure activity mitigate the negative effects of value incongruence. Acad. Manage. J. **59**(5), 1561–1584 (2016). https://doi.org/10.5465/amj.2014.0850
10. Rich, B.L., LePine, J.A., Crawford, E.R.: Job engagement: antecedents and effects on job performance. Acad. Manage. J. **53**, 617–635 (2010)

11. Bhargava, Shivganesh, Pradhan, Harsh: Work value congruence and job performance: buffering role of leader member exchange and core self evaluation. Asian Soc. Sci. **13**(1), 98 (2017)
12. Slåtten, T., Svensson, G., Sværi, S.: Empowering leadership and the influence of a humorous work climate on service employees' creativity and innovative behaviour in frontline service jobs. Int. J. Qual. Serv. Sci. **3**, 267–284 (2011)
13. Albdour, A.A., Altarawneh, I.I.: Employee engagement and organizational commitment: evidence from Jordan. Int. J. Bus. Manage. Invent. **19**(2), 192–212 (2014)
14. Slåtten, T., Lien, G.: Consequences of employees' collective engagement in knowledge-based service firms. J. Serv. Res. **2016**(8), 95–129 (2016). https://doi.org/10.1007/s12927-016-0006-7
15. Barrick, M.R., Thurgood, G.R., Smith, T.A., Courthright, S.H.: Collective organizational engagement: linking motivational antecedents, strategic implementation, and firm performance. Acad. Manage. J. **58**(1), 111–135 (2015)
16. Kim, H.J., Shin, K.H., Swanger, N.: Burnout and engagement: a competitive analysis using the big five personality dimensions. Int. J. Hosp. Manage. **28**, 96–104 (2009)
17. Bailey, C., Madden, A., Alfes, K., Fletcher, L.: The meaning, antecedents and outcomes of employee engagement: a narrative synthesis. Int. J. Manage. Rev. **19**, 31–53 (2017)
18. Fachrunnisa, O., Adhiatma, A., Mustafa.: Social identity, collective engagement, and communal patent for successful digital collaboration. Paper presented at the Conference 8th Knowledge Management International Conference (KMICe), Chiang Mai, Thailand (2016)
19. May, D.R., et al.: The psychological conditions of meaningfulness, safety and availability and the engagement of the human spirit at work. J. Occup. Organ. Psychol. **77**(11), 11–37 (2004)
20. Alfesa, K., et al.: The link between perceived human resource management practices, employee engagement and behavior: a moderated mediation models. Int. J. Hum. Resour. Manage. **24**, 23 (2013)
21. Rich, B.L., Lepine, J.A., et al.: Job engagement: antecedents and effects on job performance. Acad. Manage. J. **53**, 617–635 (2010)
22. Zeffane, R., Connell, J.: Trust and HRM in the new millennium. Int. J. Hum. Resour. Manage. **14**, 3–11 (2003)
23. Zangoueinezhad, A., Moshabaki, A.: Human resource management based on the index of Islamic human development. Int. J. Soc. Econ. **38**(12), 962–972 (2011). https://doi.org/10.1108/03068291111176329

Financial Technology and E-Corporate Governance Model for Small Medium Enterprises

Mutamimah[(⊠)]

Department of Management, Faculty of Economics,
Sultan Agung Islamic University, Semarang, Indonesia
mutamimah@unissula.ac.id

Abstract. This paper is designed to develop an E-Corporate Governance model for Small and Medium Enterprises that uses Fintech financing. The agency problem in the relationship between Fintech Corporation and SMEs is between creditor and debtor. In digital era today, most SMEs have accessed financing though Fintech Corporation. SMEs are included into high risk borrowers group, so in order to reduce moral hazard as well as credit risk E-Corporate Governance is needed. E-Corporate Governance is a system, structure, regulation and mechanism to control and monitor SMEs' behavior digitally, so moral hazard and credit risk can be reduced and SMEs' performance can increase.

Keyword: E-Corporate Governance · Small Medium Enterprises · Fintech · Credit risk

1 Introduction

SMEs have strategic role in Indonesia economy, however the development of SMEs are still slow due to the limit of access to financing (Haider 2018; Yoshino and Taghizadeh-Hesary 2016). Government has regulated the banks to allocate credits for SMEs in 2018 by 20%, although there are still some SMEs that are unable to access capital from financial institutions. It is caused by the condition where SMEs do not have sufficient collateral, high interest from banks, complicated procedures and insufficient financial reports (Haider 2018).

Along with digital advance, SMEs are also utilizing the development of information technology and according to Central Bureau of Statistic in 2017, recorded that 3.79 million of SMEs have used digital technology. It shows that SMEs have potential to carry out financial innovations called Fintech. Fintech can provide financial solutions (Arner et al. 2015). It is also as an application of digital technology for financial intermediation problems (Aaron et al. 2017). According to Bank Indonesia, one of Fintech forms is Peer to Peer Lending which is a debt-based transaction between individuals, Fintech Corporation, and business such as SMEs. Fintech with such model is suitable for SMEs financing because it has several advantages, including: (a) Fintech can reach the society without location limits, (b) It is unnecessary to provide physical collateral to get loans, this collateral is an obstruction for SMEs to access capital from

© Springer Nature Switzerland AG 2020
L. Barolli et al. (Eds.): CISIS 2019, AISC 993, pp. 907–913, 2020.
https://doi.org/10.1007/978-3-030-22354-0_84

non-digital financial institutions, (c) the procedures are simpler, faster and cheaper (Minerva 2016).

According to the Sharia Banking Outlook ("Karim Consulting Indonesia," 2017) there has been a shift in assessing whether someone or an institution deserves or not for financing from a Fintech company. Assessment is based on intangible components such as: personality, intelligence and integrity, social media use, online shopping application use as well as mobile use. The assessment is actually only limited to the requirements to get credit. But it is unknown either the risk of these funds will be returned on time or not, it is also called as credit risk.

Moreover, SMEs are included in the high risk borrower group (Zairani and Zaimah 2013) and non performing financing for SMEs is higher than non-SMEs (Mutamimah and Hendar 2017). If the requirements for obtaining credit are intangible, it does not guarantee that loan funds will be well managed by SMEs which will certainly harm Fintech Corporation, investors and other stakeholders. Borrowed capital must be monitored properly using information technology; this is the necessity of E-Corporate Governance for SMEs. E-Corporate Governance for SMEs is a system, structure, regulation and mechanism to manage, monitor and control the behavior of managers based on information technology, so there will be no misbehaving in the management of SMEs. Through E-Corporate Governance, managers of SMEs can reduce credit risk as well as manage SMEs effectively and efficiently in order to increase the performance of SMEs.

The fundamental differences in this research with the previous ones are: *First*, in previous phenomenon, the feasibility assessment to obtain credit approval from Fintech was intangible including personality, intelligence and integrity, social media use and mobile use. These criteria do not guarantee that the borrowed fund can be managed properly by SMEs. In other words, according to these criteria, SMEs as debtor still have high credit risk; hence it will harm depositors, Fintech Corporation, investor and other stakeholders. This research as concept paper proposes SMEs to apply E-Corporate Governance to reduce the high credit risk. *Second*, corporate governance in Fintech was only applied on Fintech Corporation as creditor so far. This is reflected in OJK Regulation number 15/SEOJK.05/2016 concerning the Report on the Implementation of Good and Not Good Corporate Governance for SMEs as debtor. Even though there are 3 parties involved in Fintech, namely the society as depositors, Fintech company as intermediary institutions and SMEs as debtors. Although loans given to SMEs are one source of income for Fintech corporation, they still must be responsible toward the depositors and shareholders (Zairani and Zaimah 2013).

Fintech Corporation as intermediary institution have 2 kinds of agency conflicts: (a) between depositors or investors and Fintech Corporation as manager has been arranged on OJK Regulation number 15/SEOJK.05/2016 concerning the Report on the Implementation of Corporate Governance; (b) between Fintech Corporation as creditors and SMEs as debtors. However, SMEs as debtors trusted by Fintech Corporation as principal must be able to carefully manage the funds borrowed from the company and reduce the risk. Thus it is necessary to arrange E-Corporate Governance. This is the gap that *has not been found* until now (Fig. 1).

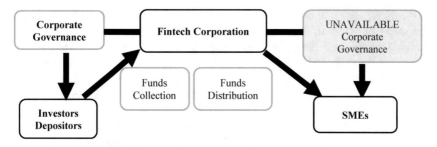

Fig. 1. Corporate Governance at Fintech Corporation

Henceforth Fintech Corporation must consider the quality of SMEs based on the implementation of E-Corporate Governance. This is supported by the result of research by Chevers and Chevers (2014) that to strengthen the implementation of corporate governance in SMEs, it must include information technology as an internal control. Hence, by the use of information technology, SMEs can run their business successfully.

Therefore, problem formulation in this research is: How is E-Corporate Governance model in SMEs that implement Financial Technology. The contribution of this article is to arrange a model of E-Corporate Governance for the development of SMEs in order to grow and survive in a long run through Fintech financing.

2   Literature Review

2.1   Fintech for Small Medium Enterprises

Fintech is a financial technology developed by information technology innovation which is able to provide financial solutions (Arner et al. 2015). One of Fintech model that is suitable to be applied to finance SMEs is Peer to Peer Lending (P2PL). This model is one of the forms of debt-based crowd-funding in the form of money lending practices where borrowers and investors are assembled through information technology. The P2P lending model is very attractive for SMEs to get financing because it does not need collateral, fast disbursement process, lower costs, competitive interest rates, paperless applications are easier and do not provide penalties for early repayment (Minerva 2016). Based on data from OJK, it shows that in 2018, the total number of Fintech operators that have been registered and obtained permission from OJK is 88 companies. Here are some examples of P2P lending in Indonesia: Modalku, Amartha, Crowdo, InvesTree, Mekar, and KoinWorks.

2.2   E-Corporate Governance for Small Medium Enterprises

There is no universal definition of corporate governance that can be applied to all situations, corporations or countries. Corporate governance appear as a solution from agency relationship if both parties to the relationship maximize the benefit, there is good reason to believe that the agent will not always act in the best interests of the principal (Jensen and Meckling 1967). This is strengthened by Panda and Leepsa (2017) with the statement that the conflict of interest and agency cost arises due to the separation of ownership from control, different risk preferences, asymmetric information and moral hazards. International Financial Corporation defines corporate governance as "the structures and processes for the direction and control of companies". Organization for Economic Cooperation and Development, "The internal means by which corporations are operated and controlled, which involve a set of relationships between a company's management, its board, its shareholders and other stakeholders, including creditors," (The Indonesian Corporate Governance Manual 2014). According to stakeholders theory, corporate governance model is a structure and process including the activities determined and operationalized in a company. Corporate governance in previous research is related to large companies and companies that go public where separation of ownership and control occurs within the company. Therefore the existence of corporate governance is intended to balance the importance and goals of owners and managers. However the corporate governance of SMEs in Indonesia has different characteristics from previous studies. First, there is no separation of ownership and control in SMEs, because mostly SMEs are family business where the owner, manager and employee are the same person (Anton, Muzakan, and Muhammad 2015). Second, agency conflict in SMEs is a conflict between creditors, Fintech Corporation and SMEs as debtors.

3   The Proposed Framework of E-Corporate Governance Model for SMEs

3.1   Mechanism of E-Corporate Governance Model for SMEs

There are two E-Corporate Governance mechanisms for SMEs, internal and external. The internal mechanism includes structures, systems and regulations that prevent moral hazard behavior digitally in SMEs. It includes SMEs' business process, compliance and risk management which are supported by the implementation of E-Corporate Governance. E-Corporate Governance includes transparency, accountability, responsibility, independency, and fairness as instruments to measure the implementation of E-Corporate Governance. The external mechanism means that there are supervision and monitor by external parties digitally, in this case by Fintech as creditors. Both mechanisms are interrelated. If the internal E-Corporate Governance works well, then it will increase the quality of external E-Corporate Governance (Fig. 2).

3.2   Measurement of E-Corporate Governance Model for SMEs

The measurement of E-Corporate Governance for SMEs is shown in several components: transparency, accountability, responsibility, independence and fairness in order to reduce the risk of bad credit, to improve performance and sustainability of SMEs (Kurniawati et al. 2018). *Transparency* is openness in decision making and SMEs in disclosing truthful information and providing financial or non-financial reports digitally in a determined time both to Fintech Corporation and other stakeholders. *Accountability* is clarity of functions, authority and accountability of SMEs' managers to Fintech Corporation and other stakeholders digitally, so the operations of SMEs can

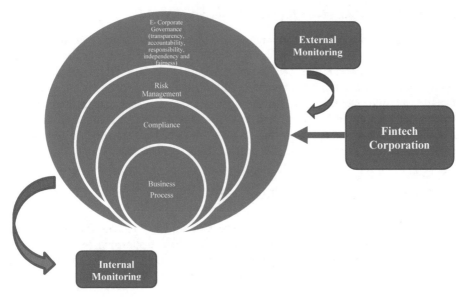

Fig. 2. Mechanism of E-Corporate Governance

run transparently, effectively and efficiently. *Responsibility* is compliance of SMEs' management with the legislation in financing field and ethical values as well as standards, principles and practices of healthy financing. *Independency* means that SMEs must be managed independently, professionally and free from conflicts of interest and influence or pressure from any party; and *Fairness* refers to equality, balance and justice in fulfilling the rights of SMEs' stakeholders which include: Fintech Corporation, managers and employees so that no party is harmed (Fig. 3).

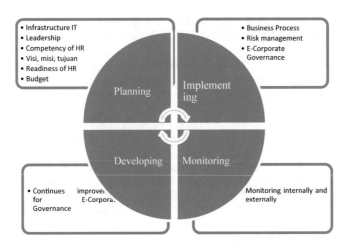

Fig. 3. Implementation of E-Corporate Governance

3.3   Implementation of E-Corporate Governance Model

For the implementation of E-Corporate Governance, it starts from planning, implementing, monitoring and developing. The Implementation has advantages which are faster, more efficient and cheaper. Planning requires adequate IT infrastructure, HR competencies, leadership commitment, HR readiness and budget. Implementing includes the implementation of business process and risk management. Monitoring is carried out both internally and externally. Developing includes continuous development to realize E-Corporate Governance goals which are to reduce moral hazard and credit risk as well as to improve and sustain SMEs' performance.

4   Conclusion and Future Research

In conclusion digital era is driving Fintech Corporation to grow rapidly. One of financing customers of Fintech Corporation is SME. The relationship between the two encourages the emergence of agency problem between creditors and debtors. SMEs are referred to as high risk borrowers, so to reduce the risk in today's digital era, it is necessary to apply E-Corporate Governance model.

For future research: (a) This article is still in the form of frame work on the Financial Technology and E-Corporate Governance Model for Small Medium Enterprises, so that in the future it is necessary to test the model by conducting empirical research related to the topic, (b) This article is a solution to agency problems between creditor and debtor, for future research the paper still needs to be developed in term of E-Corporate Governance Model as a solution for a broader agency conflict among other stakeholders, for example between employees, consumers, governments and communities.

References

Aaron, M., Rivadeneyra, F., Sohal, S.: Fintech: is this time different? A framework for assessing risks and opportunities for central banks. Bank of Canada Staff Discussion Paper, vol. 10 (2017)

Anton, S.A., Muzakan, I., Muhammad, W.F.: An assessment of SME competitiveness in Indonesia. J. Competitiveness **7**(2), 60–74 (2015). https://doi.org/10.7441/joc.2015.02.04

Arner, D.W., Barberis, J., Buckley, R.P.: The evolution of Fintech: a new post-crisis paradigm? SSRN Electron. J. **47**(4), 1271–1319 (2015)

Chevers, D.A., Chevers, J.E.: The impact of information technology material weakness on corporate governance changes in family-owned businesses. Sir Arthur Lewis Inst. Soc. Econ. Stud. **5**(6), 87–96 (2014)

Haider, H.: Constraints to business growth in low- and medium-income countries (2018)

Jensen, M.C., Meckling, W.H.: Theory of the firm : managerial behavior, agency costs and ownership structure theory of the firm : managerial behavior, agency costs and ownership structure. J. Financ. Econ. **3**(4), 305–360 (1976). http://www.sfu.ca/~wainwrig/Econ400/jensen-meckling.pdf

Karim Consulting Indonesia.: November 2 (Outlook Perbankan Syariah 2018) (2017)

Kurniawati, S.L., Sari, L.P., Kartika, T.P.D.: Development of good SME Governance in Indonesia: an empirical study of Surabaya. Int. J. Econ. Manage. **12**(April), 305–319 (2018)

Minerva, R.: The potential of the Fintech industry to support the growth of SMEs in Indonesia 35142338-8 Raras Minerva Management Strategy and Industry Evolution, pp. 1–60 (2016)

Mutamimah, S., Hendar, H.: Islamic Financial Inclusion: Supply Side Approach, pp. 1–9 (2017)

Panda, B., Leepsa, N.M.: Agency theory : review of theory and evidence on problems and perspectives (2017). https://doi.org/10.1177/0974686217701467

The Indonesia Corporate Governance Manual.: (First). International Finance Corporation (2014)

Yoshino, N., Taghizadeh-Hesary, F.: Asian Development Bank Institute (564) (2016)

Zairani, Z., Zaimah, Z.A.: Difficulties in securing funding from banks : success factors for small and medium enterprises (SMEs). J. Adv. Manage. Sci. **1**(4), 354–357. https://doi.org/10.12720/joams.1.4.354-357

Digital Knowledge Supply Chain for Creative Industry: A Conceptual Framework

Ardian Adhiatma[1]([⊠]), Olivia Fachrunnisa[1], and Mustafa[2]

[1] Department of Management, Faculty of Economics,
Sultan Agung Islamic University, Semarang, Indonesia
{ardian, olivia.fachrunnisa}@unissula.ac.id
[2] Department of Informatics Technology, Faculty of Industrial Technology,
Sultan Agung Islamic University, Semarang, Indonesia
mustafa@unissula.ac.id

Abstract. This paper aims to develop a framework for digital knowledge supply chain in creative industry. Creative industry is an industry that intensifies information and creativity by relying on ideas and stock of knowledge from its human resources (HR) as the main production factor in its economic activities. Industry stakeholders have committed to improve the quality of human resources involved in the creative industry. Commitments such as creating professional standardization, quality of education, competencies that must be possessed by each HR that are part of creating a competitive advantage of the nation through the creation of goods and services in the creative industry. The next problem that arises is that each institution or government body has an HR competency development program for creative industries, but until now it has not been integrated into the national standardization concept. Hence, we proposed a framework for digital knowledge supply chain to help this problem. Initial data is collected through in-depth interview with 13 informants. Results shows that creative industry has good understanding about knowledge management through talent development. Further research is discussed to validate the proposed framework.

Keywords: Digital knowledge supply chain · Creative industry · Competitive advantage

1 Introduction

Efforts to improve product quality and the competitiveness of products produced by the creative industry have been done in many ways. However, it is very limited or there has not even been an attempt to think that a good product comes from the quality of knowledge possessed by competent and standardized human resources and the collaboration model among them. Therefore, developing HR competency standardization in this industrial field is a very significant need. After efforts to develop HR standards were completed, the next step is to develop an information technology-based collaboration model among industry players both in terms of supply and demand.

The Indonesian government in this case through policies in several ministries such as the Ministry of Industry, Ministry of Cooperatives and MSMEs, the Ministry of Communication and Information, Ministry of research and higher education and BNSP

© Springer Nature Switzerland AG 2020
L. Barolli et al. (Eds.): CISIS 2019, AISC 993, pp. 914–924, 2020.
https://doi.org/10.1007/978-3-030-22354-0_85

(Professional Standardization National Agency) have sought to make general commitments regarding the quality of human resources involved in the creative industry. However, problem that arises is that each institution has competency development program for creative industries, but until now it has not been integrated into the national standardization concept. For example, the Ministry of Industry has a training program for the development of HR in creative industries, but some of the training has not been oriented to the standardization and professional certification of creative industries. Indeed, the competency standard has a single criterion that will be referred to by any institution. This has become a significant need because the achievement of competency standards involves a long process starting from the early age HR.

Therefore, this study aims to design a digital information technology-based collaboration network to meet the needs of industry and education as a competent source of human resources. This digital collaboration network is designed to create a business ecosystem for the Indonesian internet-based creative industry and web services. So, the focus of this research is to build a model and prototype of digital collaboration network software, an open platform as a virtual business ecosystem to facilitate collaboration between economic actors in an industry.

2 Literature Review

In the new economics and management literature, a knowledge-based theory of the firm is contributed to develop. It states that the main reason for firms' existence as being the creation, integration, and utilization of knowledge [1]. The knowledge-based view (KBV) has its roots in the resource-based view of the firm, which focuses on strategic assets as the main source of competitive advantages [2]. In contrast, under the KBV, knowledge is the main strategic resource, which, when properly managed, allows the firm to create value from its exploitation of production [4]. Therefore, through the combinative–dynamic capabilities, it results a knowledge bearing entity that manages its knowledge resources result the firm [3].

Consequently, by developing and implementing a series of activities or initiatives, it will deploy organizational capability and extract value; so that can be called as KM practices [4]. The main purpose of the use of KM organization is to get knowledge cognition, individually and collectively, and to shape itself in such a way as to make the use of firm knowledge to achieve effective and efficient performance. Alavi and Leidner [5] explained that the use of KM practices aims to give positive organizational outcomes such as enhanced communication and participation levels among staff members, efficiencies in finding solution and time-to-market, more energetic financial performance, better marketing practices, and improved project team performance, therefore, the widespread acknowledgement of KM contributed to succeed all organization. Nonetheless, in creative-intensive industries where the capability of firms to continually develop new products or processes, innovation seems to be the most important challenge for KM which is heavily depended by competitive advantage [6].

2.1   Digital Supply Chain

Digital Supply Chain (DSC) is defined as a customer-centric platform model that captures and maximizes the utilization of real-time data coming from a variety of sources. It enables demand stimulation, matching, sensing and management to optimize performance and minimize risk. Other definition of digital supply chain, argued that the digital supply chain is a process of networking between individuals and organizations involved in a business deal that is initiated in a paperless environment, using web-enabled capabilities. Whereas a supply chain is the simple networking between all individuals, organization, and activities involved in a business process from the manufacturer to the end user. It can be concluded that supply chain is a set of two or more individuals or organizations directly involved in selling or buying services, products, information from a source to a buyer. The whole process done with the aid of digital technology is termed as digital supply chain.

There are several key considerations in transforming a supply chain to a Digital Supply Chain. First and fore-most understands your customer and the end consumer in a more sophisticated way by utilizing data analytics and advanced technologies. Production will be driven by customer demand and not by manufacturing efficiencies. This will lead to improved inventory management while better meeting customer needs. Instead of being a function that focuses on the less visible aspects of business (everything from procurement through delivery), the supply chain will evolve into an integrated role alongside sales, marketing and product development. Data analytics and advanced technologies enable companies to have greater visibility into supply chain risks and prevent or mitigate negative impacts.

Digital supply chain integration is becoming increasingly dynamic. Access to customer demand needs to be shared effectively, and product and service deliveries must be tracked to provide visibility in the supply chain (SC). Business process integration is based on standards and reference architectures, which should offer end-to-end integration of product data. Companies operating in supply chains establish process and data integration through the specialized intermediate companies, whose role is to establish interoperability by mapping and integrating company specific data for various organizations and systems. Capabilities of core inter organizational processes, such as customer relationship management, supply chain management, and contract manufacturing, are suggested as critical to firm performance [7]. Their digitization across the extended enterprise is being enabled by Web technologies, workflow tools and portals for customers, suppliers, and employees, and information technology innovations targeted at supply chains and customer relationships. Firms are investing in these technologies and related partnerships to develop their extended enterprise capabilities.

Rubenstein [3] argued that many organizations are now engaging in Knowledge management (KM) in order to leverage both within their organization and externally to their customers and suppliers. KM is an important role in selecting the right information at the right time from several pertinent resources while converting it to useful insight. Effective knowledge management can help the enterprise to accumulate core knowledge, build corporate intelligence and obtain a competitive competence.

Kant [8] took a case study approach to develop a framework to guide KM implementation in supply chains. They introduced the construct of "value proposition" as the strategic knowledge sources identified across the SC. Li [9] aimed to outline the significance of SC knowledge sharing using the Prisoner's Dilemma Model of game theory. Mc Laughlin [10] mentioned that organizations can identified their core business processes as being responsive and flexible, or otherwise, less- responsive. The four KM process, knowledge creation, knowledge storage, knowledge transfer and knowledge application are associated with customer service management, decision-making, forecasting/demand planning and global SC.

In addition, Patil [11] determined the barriers and critical success factor for KM adoption within SC context. The knowledge-based view has been used as a theoretical fundamental to discuss the link between KM and SC performance. The main assumption of the knowledge-based view theory is the understanding of knowledge as a primary productive resource with a strategic connotation in the value adding process. Thus, from a knowledge-based view perspective, "knowledge can be viewed as a source of competitive advantages in supply chain and improved supply chain outcomes" [12]. The resource-based theory also has been noticed by the researchers to approach the KM discipline. From this theoretical perspective, as Halley [13] argue that the efficient use of resources as the notion of organizational learning and the efficient accumulation of resources will be useful to face future needs as the task corresponding to KM.

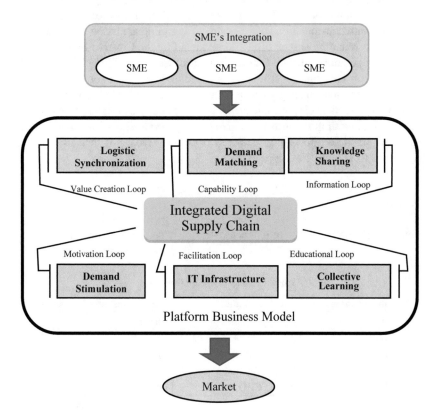

Fig. 1. Conceptual framework of digital knowledge supply chain

2.2   Conceptual Framework

The digital knowledge supply chain framework will have a design for mapping professional competencies needed by industry, designing an information technology-based digital collaboration network to facilitate the supply chain knowledge system by meeting the needs of the industrial and educational world as a competent source of human resources to produce high quality products. The framework can be pictorially described in Fig. 1.

3   Method

For initial validation of the proposed framework, we conduct explorative qualitative study to get clearer and accurate data. Interview and observation are done by creative industry actor who has been operating for at least two years. Based on the Presidential Regulation of Indonesia Number 72 Year 2015, which is an amendment of Presidential Regulation No. 6 of 2015 on Creative Economy Agency states that the creative industry sub-sector in Indonesia is divided into 16 categories which are: application sub-sectors and game developers; architecture; interior design; visual communication design; product design; fashion; movies, animations, and videos; photography; crafts; culinary; music; publishing; advertising; performing Arts; art; as well as television and radio [14]. Certain considerations of reference to sampling are based on the results of a special creative economy survey of 2017 conducted by the Creative Economy Agency (Bekraf) in cooperation with the Central Bureau of Statistics (CBS) and the leader or business owner or HR manager from each sub-sector creative economy as they directly involved with the talent management practices and policies. The two main questions are about their understanding on talent management and talent management practices. Those questions are important to get understanding about SMEs knowledge on talent management practice. Information gathered from this study will then use as input to design regard to Digital HR Supply Chain.

4   Result

We obtained 13 informants with characteristics provided in the Table 1.

Table 1. Informants' description

No	Informant code*	Gender	Position	Creative industry sub sector
1	Inf1	Female	Owner	Culinary
2	Inf2	Female	Owner/designer	Fashion
3	Inf3	Female	Owner	Craft
4	Inf4	Male	Owner	Art
5	Inf5	Male	CEO	IT

(*continued*)

Table 1. (*continued*)

No	Informant code*	Gender	Position	Creative industry sub sector
6	Inf6	Female	President Director	Music
7	Inf7	Male	HRD Manager	Publication
8	Inf8	Male	HRD Manager	Communication
9	Inf9	Male	Founder/manager	Arts
10	Inf10	Male	CEO Assistant	Advertisement
11	Inf11	Male	Solidarity Maker	Arts
12	Inf12	Male	Interior Designer	IT
13	Inf13	Male	Founder/owner	IT

Research Question 1 (RQ1): Understanding of Talent and Talent Management Definition

The results of research have found significant variations of meaning from the understanding of talent management phenomenon, which can be called as challenge of talent identification. Some of informants' statements are following (Table 2):

Table 2. Informants' statement on RQ1

Informants	Statements
Inf1, Inf2, Inf4, Inf6, Inf8	Talent is something that has existed since we were born or grace given by God since we were born. Talent cannot change over time. Talent needs to be developed. If it is not developed it will be useless. The point is in development and maintenance. Many persons still have not seen his talent, but if they try to practice more they can. Essentially a person has a talent, because he/she usually does. And of course it takes time, cannot be instant
Inf3	"Associated with talent management, it is necessary to be able to produce maximum creation. But it depends on every individual, if it's been enjoyed to keep working hard to make the craft, it needs perseverance"
Inf5, Inf9	"Talent management tends to manage talent's person, either personally or organization in order to be successful and beneficial, both for himself and others. Talent is really necessary in management. If there is no management, it will get messed up. Talent management is the goal to manage and direct employee to develop talent and get more successful. If we have successful employees automatically we also succeed"
Inf10, Inf12	"The talent management manages the talents of employees well for each individual or sub-division, and the most important thing is we can develop the talent they have. Talent needs to be managed by company management. If it is not managed then the talent development will be slow. By doing good management, it will be more focused on the better talent development"

(*continued*)

Table 2. (*continued*)

Informants	Statements
Inf12, Inf13	"The talent is already possessed since everybody was born. Talent is defined as art, so it needs to be developed, because otherwise it will fade. But if the person does not have talent, he/she will be directed to do anything difficult. So, it will be not maximal. Therefore, it needs to be developed and studied to be honed. Although talent can not be lost, but it could fade if he pursues a field that is not in accordance with the talent possessed"
Inf11	"Talent management is exploring the potential or prospective employees to be able to develop the unit that he did. The talent must be managed so that it really becomes something that produces, is completely unearthed. The talent management is more about how we can organize, and manage talent that exists within the organization, not only for us but also can be useful for others"

Based on the presentation of the talent's understanding by the 13 informants above, it can be concluded that talent is a gifted skill, attitude, character, knowledge, experience, and ability given by God to someone since he was born to continue to learn and grow; as well as a favorite that is done repeatedly. Talent depends on the developers, it does not depend on what talent it has. Someone who has a talent but he does not know and cannot be compared to people who does not have talent but he is diligent to learn, then eventually the work they create will give different results. While talent management is an effort to organize or manage the talent and potential possessed by a person or employees to be developed and dug deeper in order to produce work that is beneficial for himself, others, and his organization.

Research Question 2: Talent Management Practice

4.1   Primer Position

According to Collings and Mellahi [15] talent management begins with the identification of important positions that contribute differently to the competitive advantage of an organization. This is then followed by development of high performing potential performers to fill this role, and development of differentiated employees to manage the old players as well as to ensure their commitment to the organization.

> Inf4: "the difficult one is that to find freelance crews, because they are becoming the core of the cinematographic venture. For example, if they have already been taken or contracted by another production, we should look for another."

While the seventh informant (Inf7) argues that talent management practices are the same as employee management practices which only more focus on talented people. The practice of talent management is very important when recruiting people to occupy important positions within the organization. In addition, eighted informant (Inf8) argues that to apply talent management, first thing to do is to determine the important positions that require special talent, followed by employee's talent planning.

4.2 Hiring Talent

After identifying the key positions in the organization that is to be filled by talented employees, the next step is to plan for the provision of talented employees to fit the needs or how to hire talented employees. 12 out of 13 informants said that proper talent management practices within the organization should initially plan talent procurement or how to employ talented employees. This talent procurement is the first step in terms of getting talented employees for organization's sustainability (Table 3).

Table 3. Informants' statements on hiring practices

Informants	Statements
Inf1, Inf2, Inf3, Inf9	"The practice of talent management is more focused on finding early employees talents or by recruitment and selection. The point is to find employees that must be really - talented, intent, and honest. In essence we both need, employees need me, I also need them"
Inf5, Inf7	"Talent recruitment, for example when we want to find the talent, the company distributes questionnaires to find out the employment of candidates of employees, or tested psychologically. Recruitment of talent not only tends to important positions in the company, but more to all positions in the company"
Inf11	"In the recruitment of members, there are various ways, some of which we offer, we know, some are indeed holding recruitment, due internships, there is a fit together, a project like that we offered"
Inf12, Inf13	"Broadly, talent management practice is almost the same as employee management because it is only more likely to talent that each individual has especially in recruitment"

Informants 4, 6 and 10 specifically state that they need skilled employees for the talents they need in the workplace through casting and designs at recruitment time and selection.

4.3 Identifying Talent

Successful talent management practices must first be initiated by the identification of talents possessed by individuals or the organization. Five informants argue that one of the implementation practices of talent management is by identifying the talents of employees, by knowing the talents possessed then individuals or organizations will more easily do the development of the talent owned.

Inf3: "The main factor is the identification of talent. The craftsmen should be given such stimulus so they are interested in their creative work. If they are interested, then we facilitate them to start creating *batik* (such Indonesian creative work). Well, from there he can be caught his talent in what field. In addition, to know that talent, we need others to know, especially from their parents, or exactly mothers."

Inf2, Inf7: "Employees must start from the talent identification first so as not to misstep. If the first step is wrong, then the next step will be useless. So, I think the main factor is on the identification of talent, whether it is by the HR self individually or by the organization where he works."

Inf11, Inf12: "The most important factor is on talent detection. It can through psychological tests, such as steving, MBTI, temubakat.com, and so forth. So, individuals must know in advance what talents they have. But most of us know that the talents are not by ourselves but more likely by others. You tend to be ignorant and unaware."

4.4   Developing Talent

After an organization gets the talented employees they are looking for, the next step is to develop employees' talents. Four informants stated that, talent development can be adopted on how to develop employee in employees' management such as training, seminars, workshops, and so on.

Inf2: "… After we get the appropriate human resources with their talents, then we develop them with training, seminars, workshops, etc. It's just that the talent management that we apply is not up to the retirement."

Inf5: "… Also there is potential development. Honed and developed their talents, whether soft skills or hard skills. It is very necessary to be developed because to achieve the expected corporate or organizational goals."

Inf8: "There is no special activity, we know, digging, or assessing the talent of human resources at the time of monitoring or training during the first 3 months earlier. But it is not guaranteed, because of the monitoring process it will only be known 10–20% of talent owned by human resources. Not yet overall. For the next potential excavation, we put him in his new 10–20% initial talent just as he has."

4.5   Retain Talent

The last talent management practice is to retain talent. Once the organization identifies key positions that require talented human resources, then the organization has found the potential talent they are looking for, after the talent that has been acquired must be developed in order to align with the goals of the organization to be achieved, the last step is to retain talent. As we know, that the difficulty of maintaining potential employees is a problem experienced by many organizations in Indonesia. Therefore, it is important for us to know how to practice in maintaining potential employees in order to contribute to the organization and not to be taken by other organizations. Based on the results of the study, three informants said it was important for them to retain talent in their organizations.

Inf4: "Practice is more on how to retain talent. It will be more difficult than recruiting our way to maintain the talent of employees."

Inf9: "The practice is more about retaining members' talents."

Inf11: "It does not matter if they want to work elsewhere, we are collective art, so if they think they cannot collectively money they have to work elsewhere, but still have to sacrifice a lot. providing money, time, and energy to create projects. It is precisely in our way that gives them freedom that makes them comfortable and endures here."

Based on the results of research efforts to maintain the talent of employees in the organization in order to make them not run to other organizations, make them with not too pressed. Pressed necessary but not too forced, should also be approached in a personal and intensive, especially if it is very talented in the field of employees. In addition, we should give full confidence to the employees. Let them hone their own skills, but still be directed.

5 Conclusion and Future Research

The digital knowledge supply chain framework is expected to increase the connection of stakeholders in the creative industry, especially to standardize the quality of human resources. This is expected to make it easier for businesses to get information starting to get high-skill and certified human resources, places or skill grading institutions so they can create competitive advantages for the creative industry. The results of this study are expected to provide recommendations to related parties, especially in an effort to improve the quality of HR in creative industries that have an impact on improving product quality.

Result of initial observation in this study produce such information that talent management policies which will affect organizational performance in the creative economy are caused by five practices, namely the identification of important positions in the organization (pivotal position), procuring or employing talented employees (hiring of talent), identifying talent identification, and develop talents of employees (talent development), and retain talented employees to remain a member of the organization (talent retention).

Further research will be carried out to validate the framework design using mixed methods and triangulation approaches (quantitative and qualitative) and computer simulation. Data collection will be using a questionnaire survey approach and more in-depth interviews with creative industry stakeholders. Another in depth interviews are used to validate the findings of the findings obtained from the gap analysis survey. While the prototype will be designed, developed, validated and evaluated with a computer simulation approach using the Multi Agent System (MAS) method.

Acknowledgments. This research is funded by Directorate of Research and Higher Education, Indonesia, Applied Research Excellence for University (PTUPT) Scheme, 2019.

References

1. Melancon, J.P., Griffith, D.A., Noble, S.M., Chen, Q.: Synergistic effects of operant knowledge resources. J. Serv. Mark. **24**(5), 400–411 (2010)
2. Amit, R., Schoemaker, P.J.H.: Strategic assets and organizational rent. Strateg. Manage. J. **14**(1), 33–46 (1993)
3. Rubenstein-Montano, B., Liebowitz, J., Buchwalter, J., McCaw, D., Newman, B., Rebeck, K.: A systems thinking framework for knowledge management. Decis. Support Syst. **31**(1), 5–16 (2001). https://doi.org/10.1016/S0167-9236(00)00116-0

4. Ghazi, A., Edien, A.: Effects of job analysis on personnel innovation. Int. J. Bus. Manage. Inven. **4**(10), 9–18 (2015)
5. Alavi, M., Leidner, D.: Knowledge management and knowledge management systems: conceptual foundations and research issues. MIS Quart. **25**(1), 107–136 (2001)
6. Nonaka, I., Takeuchi, H.: The knowledge creating company: how Japanese companies create the dynamics of innovation. Oxford University Press, New York (1995)
7. Hagel, J., Singer, M.: Unbundling the corporation. Harvard Bus. Rev. **77**(2), 133–141 (1999)
8. Kant, R., Singh, M.D.: Knowledge management implementation in supply chains: a strategic plan. Int. J. Bus. Inf. Syst. **4**(6), 655 (2009). https://doi.org/10.1504/IJBIS.2009.026697
9. Li, X., Hu, J.: Business impact analysis based on supply chain's knowledge sharing ability. Procedia Environ. Sci. **12**, 1302–1307 (2012). https://doi.org/10.1016/j.proenv.2012.01.425
10. McLaughlin, S.: Six tenets for developing an effective knowledge transfer strategy. Vine **40** (2), 153182 (2010). https://doi.org/10.1108/03055721011050668
11. Patil, S.K., Kant, R.: A fuzzy DEMATEL method to identify critical success factors of knowledge management adoption in supply chain. J. Inf. Knowl. Manage. **12**(3), 1–15 (2013). https://doi.org/10.1142/S0219649213500196
12. Sangari, M.S., Hosnavi, R., Zahedi, M.R.: The impact of knowledge management processes on supply chain performance: An empirical study. Int. J. Logist. Manage. **26**(3), 603–626 (2015). https://doi.org/10.1108/IJLM-09-2012-0100
13. Halley, A., Nollet, J., Beaulieu, M., Roy, J., Bigras, Y.: The impact of the supply chain on core competencies and knowledge management: directions for future research. Int. J. Technol. Manage. **49**(4), 297 (2010). https://doi.org/10.1504/IJTM.2010.030160
14. Bekraf: Indonesia creative economy sub sector (2017). http://www.bekraf.go.id/subsektor/page. Accessed 2 Nov 2017
15. Collings, D.G., Mellahi, K.: Strategic talent management: a review and research agenda. Hum. Resour. Manage. Rev. **19**(4), 304–313 (2009)

Relational Selling Strategy
on SMEs Marketing Performance:
Role of Market Knowledge
and Brand Management Capabilities

Hendar[(⊠)], Ken Sudarti, and Intan Masfufah

Faculty of Economics, Universitas Islam Sultan Agung (UNISSULA),
Semarang, Indonesia
{hendar,kensudarti}@unissula.ac.id,
intanmasfufah03@gmail.com

Abstract. This conceptual paper explains the role of Relational Selling Strategy in the relationship between Market sensing capabilities and brand management capabilities with Marketing Performance in Small Business Enterprises (SMEs). Market sensing capabilities is part of marketing knowledge, specifically related to market scanning, while brand management capabilities is part of cross-functional capability related to brand knowledge. The linkages between market sensing capabilities and brand management capabilities with relational selling strategy and marketing performance are part of the development of Marketing Dynamic theory that had previously been developed by experts such as Day (1994) and Morgan (2012).

Keywords: Brand management capabilities · Market sensing capabilities · Relational selling strategy · Marketing performance

1 Introduction

This paper aims to explain the role of relational selling strategy in the relationship between market sensing capabilities and brand management capabilities with marketing performance. Market sensing capabilities as an important part of market knowledge and brand management capabilities as part of brand knowledge are two forms of marketing capabilities needed to develop marketing strategy. Marketing literature highlights market knowledge and cross-functional collaboration as the two fundamental resources and capabilities for innovation product performance (Luca and Atuahene-Gima 2007). Market knowledge concerns four dimensions such as; market scanning, market information transmission, market information interpretation, and market knowledge utilization (Lertputtarak 2011). Market scanning is the activity of seeking, obtaining and gathering information involving formal and informal about the current market and peripheral environment that is not emphasized by the company today (Veflen Olsen and Sallis 2006), such as examining customer needs and desires, competitor strategies, retailers, distributors, suppliers, technology, demographics, economics, and other environmental forces that can affect company performance (Kohli and Jaworski 1990;

L. Barolli et al. (Eds.): CISIS 2019, AISC 993, pp. 925–933, 2020.
https://doi.org/10.1007/978-3-030-22354-0_86

Day 1994; Slater and Narver 1994). To obtain valuable market information, each company needs to have market sensing capabilities, namely the ability to gather and use information needed to commercialize patented innovations from the market (Lin and Wang 2015). SMEs that have good market sensing capabilities will have the convenience of compiling their marketing strategies and then get better marketing performance.

Cross-functional collaboration involves the integration of several special marketing capabilities such as product management, pricing management, channel management, marketing communications, selling, and market research with input from special capabilities in other functions. Three of the marketing capabilities across the most important functions are brand management, customer relationship management and new product development (Morgan 2012). SMEs that have three types of capabilities will have a better ability to make decisions and implement marketing strategies, eventually leading to marketing performance. This paper focuses on the role of market sensing capabilities and brand management capabilities in improving SMEs marketing performance. Brand management involves combining various specialized marketing capabilities such as market research, product management, price, and marketing communication capabilities for certain brands. Likewise, the ability to design and monitor innovation, accounting, production, and operations to develop and implement a brand-level business plan (Morgan 2012).

Marketing literature shows that marketing capabilities are the initial source of the marketing process and are the key to successful marketing differentiation and performance (Morgan 2012). Capability marketing is an integrative process designed to apply the resources needed by the company to market-related needs of the company, enabling companies to add value and meet competitive demands (Day 1994; Martin et al. 2017). A capability is a process where companies choose value propositions intended to target customers and use resources to provide these value offerings in pursuit of desired goals (Vorhies and Morgan 2005; Martin and Javalgi 2016). Two types of marketing capabilities become a concern of this paper, they are, market sensing capabilities and brand management capabilities. Market sensing capabilities is the company's ability to see or learn from the market, find and obtain information about customers, competitors and existing market opportunities (Lindblom et al. 2008; Morgan et al. 2009). While brand management capabilities is the company's ability to improve, maintain, and manage brand image through the utilization of resources owned by a company, as well as the ability to inform the image of a company's brand to customers (Morgan et al. 2009; Morgan 2012). Companies that have both of these capabilities will have a better ability to make marketing strategy decision and implementation, and they will also have better ability to improve their marketing performance (Morgan et al. 2009).

Empirically, studies of market sensing capabilities and brand management capabilities, and their influence on marketing performance have received the attention of many researchers. Several previous studies have shown that market sensing capabilities has a direct role in increasing marketing performance (Osakwe et al. 2016), but other researchers have shown the opposite (Morgan et al. 2009; Ardyan 2016). This also applies the relationship between brand management capabilities and marketing performance. On the one hand, brand management capabilities is an important driver of marketing performance (Lee et al. 2008; Merrilees et al. 2011; Osakwe et al. 2016), but on the other hand, it is not a determinant of marketing performance (Morgan *et al.* 2009). The knowledge gap shows that market sensing capabilities and brand management capabilities are not always able to improve marketing performance or company performance. To overcome this gap, the marketing strategy literature suggests the need for a marketing strategy decision and marketing strategy implementation as variables that mediate the relationship between marketing capabilities and marketing performance. That means the marketing strategy can be a solution to the gap between market sensing capabilities and brand management capabilities with marketing performance. The marketing strategy meant involves market entry strategies, marketing mix strategies (product, price, distribution, and promotion), selling strategies, and customer care strategies. As part of the marketing strategy, relational selling strategy has the potential to mediate the relationship between market sensing capabilities and brand management capabilities with marketing performance. The assumption is that SMEs with market sensing capabilities and brand management capabilities that make the entrepreneurs easier to design relational selling strategy well will have the ability to increase marketing performance.

2   Literature Review

2.1   Relational Selling Strategy

Adoption of relational strategies is important for building customer relationships and gaining competitive advantage through the acquisition of intangible assets such as customer trust and commitment (Morgan and Hunt 1994). Meanwhile, a relational selling strategy can be built if there is interdependence between suppliers and customers, important information exchange, trust between partners and stable relationships, which allows each party to benefit from a fair return on investment (Guenzi *et al.* 2007). Therefore, the company's capacity to create superior customer value clearly depends on its ability to coordinate and integrate decisions and activities managed by sales and marketing personnel (Guenzi and Troilo 2007). Many researchers argue that sales team who adopt a relational approach provide the basis for gaining a competitive advantage by adding value to customers and influencing future purchase intentions (Boles *et al.* 2000). The relational selling strategy meant is a strategic approach developed by suppliers or sales team who are willing to build long-term and mutually beneficial relationships with several business partners and customers (Jolson 1997; Guenzi *et al.* 2007). Compared to traditional short-sighted selling strategies for the purpose of maximizing direct sales volume, a relationship-oriented approach requires leadership abilities and specific behaviors of sales team to change or adopt certain behaviors (Paparoidamis and Guenzi 2009). Relational selling strategy can work well in the long run if the sales teamhas several criteria in a relational selling strategy, such as Customer oriented selling, Adaptive selling, Sportsmanship, Conscientiousness, and Selling team (Guenzi et al. 2007).

Market sensing capabilities and relational selling strategy. A company is said to have good market sensing capabilities if it fulfills characteristics such as having the ability to learn customer needs and desires, find competitor strategies, gain insight into channel members, and learn about a broad market environment (Morgan *et al.* 2009). Some of these capabilities will make a seller or entrepreneur easier to take marketing strategy decision and implementation (Morgan 2012). Relational selling strategy is part of the marketing strategy, so companies that have good market sensing capabilities will be easier to run relational selling strategy. This is very possible because the knowledge gained from continuous market sensing activities, an entrepreneur or seller will continue to learn about the market that is happening, learn to find market opportunities, know the changing needs or desires of customers and take advantage of the current market opportunities, then develop strategies the right relational selling to customers. Therefore, proposition 1 is set:

P1: Market Sensing Capabilities has a positive relationship with Relational Selling Strategy.

BMC and relational selling strategy. Because product features are easily copied, brands are considered as the marketer's main tool for creating product differentiation (Kotler and Gertner 2002). Brands are becoming increasingly important as a foundation for marketing strategies. Assessing brand image and comparing it with the competitor's image are important steps in designing marketing strategies (Kotler and Gertner 2002). Therefore, BMC is an integral part of the success of marketing strategies. The brand management capability meant is the ability that companies use to build, develop, choose, and utilize company brands (Morgan *et al.* 2009). Companies with Strong BMC are characterized by several characteristics, such as the ability to use customer insight to identify valuable brand positions, develop brand image among target customers, develop brand awareness among target customers, establish desired brand associations in the minds of customers, achieve brand awareness which is high in the market, utilizing high brand equity in the market (Morgan *et al.* 2009). Some of these capabilities make the companies easier to make marketing strategy decision and implementation (Morgan 2012). Since the relational selling strategy is part of the marketing strategy, it is very possible that brand management capabilities will also lead to relational selling strategy. Therefore, proposition 2 is set:

P2: Brand Management Capabilities has a positive relationship with Relational Selling Strategy.

2.2 Marketing Performance

Marketing is a dynamic process, and marketing performance is a multidimensional process that involves adaptability, effectiveness, and efficiency (Morgan *et al.* 2002). Vorhies and Morgan (2003) describe marketing performance as marketing effectiveness and efficiency. Marketing effectiveness is measured by marketer perceptions regarding the extent to which the company achieves market share growth, sales growth, and achieving the market position. While marketing efficiency is measured by the ratio of marketing costs and sales proceeds to the company's gross operating income. In the perspective of small and medium enterprises, marketing performance relates to the combination of the results of marketing activities that are felt by company owners or managers about achieving sales revenue growth, increasing sales volume, customer growth, expanding marketing areas, increasing market share, increasing customer satisfaction and increasing profits (Soliman 2011; Hendar *et al.* 2017).

Market sensing capabilities and marketing management. Market information is a strategic resource as an initial source of increased marketing performance (Ferrell *et al.* 1999; Kotler and Armstrong 2011; Morgan 2012). Several studies have shown companies that have a tendency to seek information about markets, share information among interested parties and give positive responses based on market knowledge gained, have better ability to improve their marketing performance (Barnabas and Mekoth 2010; Beneke *et al.* 2016; Jogaratnam 2017). Market sensing capabilities as one of the important dimensions of the market learning orientation has positive relations with profitability (Day 1994; Osakwe *et al.* 2016), new product development performance (Mu 2015), and growth performance (Lindblom *et al.* 2008). Some of these findings illustrate how important market sensing capabilities is in supporting marketing performance. Therefore, proposition 3 is set:

P3: There is a positive relationship between market sensing capabilities and marketing performance.

Brand management capabilities and marketing performance. Brand management is a high-level marketing capability that is a potential determinant of marketing performance (Merrilees *et al.* 2011). Brand management capability concerns the systems and processes used to develop, grow, maintain, and enhance the company's brand assets (Morgan 2012). Brand management capabilities reflect the capabilities that are not only to create and maintain high levels of brand equity but also to use these resources in ways that are in line with the market environment (Morgan *et al.* 2009). This capability will lead to the company's marketing performance. Therefore, Proposition 4 is set:

P4: There is positive relationship between Brand Management Capability and Marketing Performance.

Relational selling strategy and marketing performance. Marketing literature explain that one of the important factors driving marketing performance is the selling strategy. Empirical findings also show that selling strategies have a positive effect on market performance (Terho *et al.* 2015), adaptive selling strategies have a positive effect on sales team performance (Kara *et al.* 2013; Singh and Das 2013; Chakrabarty *et al.* 2014), relational selling strategy influences customers satisfaction and customers loyalty (Lai *et al.* 2013), and relational selling strategy is an important factor in determining sales performance (Haghighi 2009). These significant relationships are very possible because to achieve the relationship selling strategy, sales team tends to be involved in ever-increasing interactions and interdependence with customers. As a result, they have access to a large amount of customer information (Haghighi 2009). Extensive knowledge of customers is the main capital in improving sales performance. Therefore, proposition 5 is proposed:

P5: Relational Selling Strategy affects Marketing Performance.

2.3 Role of Relational Selling Strategy Mediation in the Relationship Between Brand Management Capabilities and Market Sensing Capabilities with Marketing Performance

Marketing literature explains marketing capabilities, such as market sensing capabilities, brand management capabilities, specialized marketing, architectural marketing, and new product development become the important factors in the marketing strategy decision and marketing strategy implementation (Morgan 2012). That means market sensing capabilities and brand management capabilities lead to quality marketing strategies (including relational selling strategy). Furthermore, companies with good relational selling strategy will encourage an increase in marketing performance. That means relational selling strategy has the potential to be mediated in the relationship between market sensing capabilities and brand management capabilities with marketing performance. Therefore, propositions 6 and 7 are set:

P6: Relational Selling Strategy mediates the relationship between Brand Management Capabilities and Marketing Performance.

P7: Relational Selling Strategy mediates the relationship between Market Sensing Capabilities and Marketing Performance (Fig. 1).

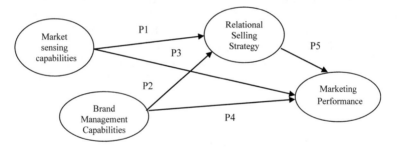

Fig. 1. Grand theoretical model of research

3 Conclusion

The strategic marketing literature has explained marketing capabilities such as market sensing capabilities and brand management capabilities, which are the foundations for the creation of a company's quality marketing strategy, including a relational sales strategy. Meanwhile, companies with good quality relational sales strategy have better opportunities to improve marketing performance. Thus, conceptually, relational selling strategy has a strategic role in mediating the relationship between market sensing capabilities and brand management capabilities with marketing performance.

4   Research Agenda

Empirical research is still needed to prove the mediation role of relational sales strategy in relation to market sensing capabilities and brand management capabilities with marketing performance. Research is more complex by adding other antecedent variables of marketing capabilities, such as customer relationship management capability, marketing innovativeness, marketing of specialized capabilities or architectural marketing capabilities will become the realm of interesting research in future.

References

Ardyan, E.: Market sensing capability and SMEs performance: the mediating role of product innovativeness success. DLSU Bus. Econ. Rev. **25**(2), 79–97 (2016)

Barnabas, N., Mekoth, N.: Autonomy, market orientation and performance in Indian retail banking. Asia Pac. J. Mark. Logistics **22**(3), 330–350 (2010)

Beneke, J., Blampied, S., Dewar, N., Soriano, L.: The impact of market orientation and learning orientation on organisational performance: a study of small to medium-sized enterprises in Cape Town, South Africa. J. Res. Mark. Entrepreneurship **18**(1), 90–108 (2016)

Boles, J.S., Johnson, J.T., Barksdale, J.H.C.: How salespeople build quality relationships: a replication and extension. J. Bus. Res. **48**, 75–81 (2000)

Chakrabarty, S., Widing, R.E., Brown, G.: Selling behaviours and sales performance: the moderating and mediating effects of interpersonal mentalizing. J. Pers. Selling Sales Manage. **34**(2), 112–122 (2014)

Day, G.S.: The capabilities of market-driven organizations. J. Mark. **58**(4), 37 (1994)

Ferrell, O.C., Hartline, M.D., Lucas, G.H., Luck, D.J.: Marketing Strategy. Dryden Press, Fort Worth (1999)

Guenzi, P., Pardo, C., Georges, L.: Relational selling strategy and key account managers' relational behaviors: an exploratory study. Ind. Mark. Manage. **36**(1), 121–133 (2007)

Guenzi, P., Troilo, G.: The joint contribution of marketing and sales to the creation of superior customer value. J. Bus. Res. **60**(2), 98–107 (2007)

Haghighi, G.M.A.a.M.: The effect of selling strategies on sales performance. Bus. Strategy Series **10**(5), 266–282 (2009)

Hendar, H., Ferdinand, A.T., Nurhayati, T.: Introducing the religio-centric positional advantage to Indonesian small businesses. Manage. Mark. **12**, 1 (2017)

Jogaratnam, G.: How organizational culture influences market orientation and business performance in the restaurant industry. J. Hospitality Tourism Manage. **31**, 211–219 (2017)

Jolson, M.A.: Broadening the scope of relationship selling. J. Pers. Selling Sales Manage. **17**(4), 75–88 (1997)

Kara, A., Andaleeb, S.S., Turan, M., Cabuk, S.: An examination of the effects of adaptive selling behavior and customer orientation on performance of pharmaceutical salespeople in an emerging market. J. Med. Mark.: Device, Diagn. Pharm. Mark. **13**, 102–114 (2013)

Kohli, A.K., Jaworski, B.J.: Market orientation: the construct, research propositions, and managerial implications. J. Mark. **54**, 1–18 (1990)

Kotler, P., Armstrong, G.: Principles of Marketing. Prentice Hall, Upper Saddle River (2011)

Kotler, P., Gertner, D.: Country as brand, product, and beyond: a place marketing and brand management perspective. Brand Manage. **9**(4), 249–261 (2002)

Lai, M.-C., Chou, F.-S., Cheung, Y.-J.: Investigating relational selling behaviors, relationship quality, and customer loyalty in the medical device industry in Taiwan. Int. J. Bus. Inf. **8**(1), 137–149 (2013)

Lee, J., Park, S.Y., Baek, I., Lee, C.-S.: The impact of the brand management system on brand performance in B-B and B–C environments. Ind. Mark. Manage. **37**(7), 848–855 (2008)

Lertputtarak, S.: The influence of HR, IT, and market knowledge competencies on the performance of HR managers in food exporting companies in Thailand. Int. Bus. Res. **5**, 1 (2011)

Lin, J.-H., Wang, M.-Y.: Complementary assets, appropriability, and patent commercialization: market sensing capability as a moderator. Asia Pac. Manage. Rev. **20**(3), 141–147 (2015)

Lindblom, A.T., Olkkonen, R.M., Mitronen, L., Kajalo, S.: Market-sensing capability and business performance of retail entrepreneurs. Contemp. Manage. Res. **4**(3), 17 (2008)

Luca, L.M.D., Atuahene-Gima, K.: Market knowledge dimensions and cross-functional collaboration: examining the different routes to product innovation performance. J. Mark. **71**(1), 95–112 (2007)

Martin, S.L., Javalgi, R.G.: Entrepreneurial orientation, marketing capabilities and performance: the moderating role of competitive intensity on Latin American International new ventures. J. Bus. Res. **69**(6), 2040–2051 (2016)

Martin, S.L., Javalgi, R.G., Cavusgil, E.: Marketing capabilities, positional advantage, and performance of born global firms: contingent effect of ambidextrous innovation. Int. Bus. Rev. **26**(3), 527–543 (2017)

Merrilees, B., Rundle-Thiele, S., Lye, A.: Marketing capabilities: antecedents and implications for B2B SME performance. Ind. Mark. Manage. **40**(3), 368–375 (2011)

Morgan, N.A.: Marketing and business performance. J. Acad. Mark. Sci. **40**(1), 102–119 (2012)

Morgan, N.A., Clark, B.H., Gooner, R.: Marketing productivity, marketing audits, and systems for marketing performance assessment Integrating multiple perspectives. J. Bus. Res. **55**, 363–375 (2002)

Morgan, N.A., Slotegraaf, R.J., Vorhies, D.W.: Linking marketing capabilities with profit growth. Int. J. Res. Mark. **26**(4), 284–293 (2009)

Morgan, R.M., Hunt, S.D.: The commitment-trust theory of relationship marketing. J. Mark. **58** (3), 20 (1994)

Mu, J.: Marketing capability, organizational adaptation and new product development performance. Ind. Mark. Manage. **49**, 151–166 (2015)

Osakwe, C.N., Chovancova, M., Ogbonna, B.U.: Linking SMEs profitability to brand orientation and market-sensing capability: a service sector evidence. Periodica Polytech. Soc. Manage. Sci. **24**(1), 34–40 (2016)

Paparoidamis, N.G., Guenzi, P.: An empirical investigation into the impact of relationship selling and LMX on salespeople's behaviours and sales effectiveness. Eur. J. Mark. **43**(7/8), 1053–1075 (2009)

Singh, R., Das, G.: The impact of job satisfaction, adaptive selling behaviors and customer orientation on salesperson's performance: exploring the moderating role of selling experience. J. Bus. Ind. Mark. **28**(7), 554–564 (2013)

Slater, S.F., Narver, J.C.: Does competitive environment moderate the market orientation-performance relationship? J. Mark. **58**, 46–55 (1994)

Soliman, H.S.: Customer relationship management and its relationship to the marketing performance. Int. J. Bus. Soc. Sci. **2**(10), 166–182 (2011)

Terho, H., Eggert, A., Haas, A., Ulaga, W.: How sales strategy translates into performance: the role of salesperson customer orientation and value-based selling. Ind. Mark. Manage. **45**, 12–21 (2015)

Veflen Olsen, N., Sallis, J.: Market scanning for new service development. Eur. J. Mark. **40**(5/6), 466–484 (2006)

Vorhies, D., Morgan, N.: Benchmarking marketing capabilities for sustainable competitive advantage. J. Mark. **69**, 80–94 (2005)

Vorhies, D.W., Morgan, N.A.: A configuration theory assessrnent of marketing organization fit with business strategy and its relationship with marketing performance. J. Mark. **67**, 100–115 (2003)

Communal Identity and Shared Value Toward Organizational Performance in the Context of Religious Knowledge Management

Nurhidayati[✉] and Olivia Fachrunnisa

Department of Management, Faculty of Economics,
Sultan Agung Islamic University, Semarang, Indonesia
{nurhidayati, olivia.fachrunnisa}@unissula.ac.id

Abstract. The purpose of this study is to develop a model of religious communal identity in strengthening organizational performance. Shared religious values as an organizational asset are able to play professional working behavior guidance among the group members and increase organizational performance. However, gender and culture community sometimes hamper and make more complex of the relationship. Thus, this study offers a conceptual model to maximize organizational performance, particularly, which running by women manager. In this article, the understanding of the concept and dimensions of shared religious values and religious communal identity and its relation to organizational performance will be discussed.

Keywords: Religious communal identity · Shared religious values · Collective engagement

1 Introduction

Community of practice (CoP) becomes a phenomenon in knowledge-based economy especially to support the lack of and dispersion of knowledge among the community. CoP defined as a group people who agree to develop expertise, which basis of knowledge or experience or expertise that they have. In knowledge-based economy, CoP is encouraged comprehensively use in identifying problems, evaluating, retrieving and sharing all the organizational information and practices to improve the performance of the organization among the community members.

Recently, the spiritual or religious paradigm gradually increases getting an attention from scholars to seek an understanding relationship between religious and organizational performance. Moreover, religion as a significant social force on organizational practices yet has been relatively under examine in the organizational theory. The problem arises from the study literature is the lack of studies on religious shared values and the impact to organizational performance. In the organization that is bonded with particular value such as religious shared, people will lead to act, perform and obey in the religious framework.

Prior studies of employee engagement is important to contribute in this study placed the concept on individual level, thus drawing up a description concept of

© Springer Nature Switzerland AG 2020
L. Barolli et al. (Eds.): CISIS 2019, AISC 993, pp. 934–938, 2020.
https://doi.org/10.1007/978-3-030-22354-0_87

collective engagement in a group or collective level is important to explore in this study. Therefore, the aim of this study is to offer a conceptual model for maximizing organizational performance, which is tied to religious values, in particular, by female managers. Subsequently validation the dimension and scale of measurement of the basic concepts will develop.

2 Literature Review

2.1 Shared Religious Value

Religion defined as "a particular institutionalized or personal system of beliefs, values, and practices relating to the divine – a level of reality or power that is regarded as the 'source' or 'ultimate', transcending yet immanent in the realm of human experience" [1]. The concept of religion in line with institutional logics principal, which is state that in social community there is a belief system or code of living and intensity in term of obedience, reverence, and worship towards a divine and imagined ultimate power that is considered superhuman [2].

Social communities are formed based on the specific aspects such as race, nationality, religion, and others. Culturally, in Indonesia, community that attracting to join is a community based on religion. Community-based religious values generates social processes that able to create social capital, empower resources, and provide arrangements for members and community change efforts to relate religious principles [3]. The benefits that the members are able to obtain when they participate in a community-based religious, they might create the organization as a place to share and collect religious knowledge (religious thought). Then, the members will apply shared religious knowledge into their business and daily life. Furthermore, a person's self-resolution in community-based religious releases with the group followed [4].

The members receive social support and transfer in term of information and knowledge among the members in community-based religious. This support leads to a positive impact on business performance and quality of life [5]. Social interactions among the members also improve quality relationships between members [6], so that a sense of mutual support and transfer knowledge in term of shared religious arises. Thus, it will strengthen religious communal identity.

2.2 Collective Engagement

Collective means the individual feels an attachment to the organization the individuals is participating in [7]. The strong feeling is felt by the individuals and they considers that the organization is a part of him. Engagement occurs when the individuals are able to express themselves actively within the organization [8]. The individuals regard the organizational environment is a very comfortable place so that they are able to be strongly involved in the organization. Collective engagement becomes a sign that the organization's environment motivates its members to do something such as doing a business.

2.3 Religious Communal Identity

Identity theory states that a person's sense of who they are based on their group membership which the groups are an important source of pride and self-esteem. Becomes the member of the group society, people try to interpret themselves in the society. Identity influences individuals' act and decision in making a policy [9]. There is a cause why individuals join a community, it is because they share similar identity to one another [10]. Identity is a longlife construction process [11]. It clings on the individuals, that is why it is important. Therefore, in joining a community, there will be religious organizational values or knowledge matching with individual values.

Religion and spirituality have positive effect on today's business environment [12]. Religion basic organization is able to create a social process which create social capital, empower sources, and give the settings to all members in linking the religious principal and the effort to change the society [3]. The values cover in the organization also create some supports from the members and the result is, they will perform well in the organizational performance.

Religion communal identity leads the individuals to be bounded with the organization followed [7]. Identity and values of the organization are being part of himself, in term of trust, support, profile, etc., that would be accepted automatically. When the individual is able to tie himself in the organization, then it can be said that the individuals own the strong organizational identification. In the program or activity which is held by the organization, the individuals would join with high enthusiastic as they feel pride. They will support each other and proud to be the part of the organization. From the explanation, it can be concluded that the identity bonding is able to create self-esteem and high confidence to perform better.

2.4 Task Sharing Household Responsibility

Women managers often experience double roles that lead to work-family conflict. Particular in patriarchy culture society, such in Indonesia, where social role dichotomy is evidence. Women's main role is not as a breadwinner of the family, but they have main responsibility in domestic roles, as a wife and a mother. Recently, women are doubled roles. Women managers are confident that they are able to run business as well as men managers. Particular in small business entrepreneur (SME), prior studies find that characteristic of profit is influenced by gender [13]. Particularly, women prefer to run small and simpler business than men as women try to avoid personal conflict which is usually emerge in the bigger business size [14].

As work-family conflict is the evidence for women with double roles [15], thus, sharing of household tasks are critical to achieve better performance and reduce such conflict. Spouse supports this term is able to reduce tension of work targets. There are some ways to measure the success of a business, there are resilience, profit, investment, sales growth, number of employees, happiness, reputation of cooperation, etc. [16]. Communal religious identity will lead women managers to be more confident in doing their job. They will do the job without guilty as they have to conflict with work-family responsibility as long as the relation moderates with task sharing from spouse and family members.

3 Conceptual Model

Based on literature review in the previous section, the conceptual models can be described in Fig. 1 and hypothesis proposed as follows:

H1: Shared religious value will increase religious communal identity
H2: Collective engagement will increase religious communal identity
H3: Religious communal identity will increase organizational performance
H4: Task sharing will moderates the relationship between communal identity and organizational performance

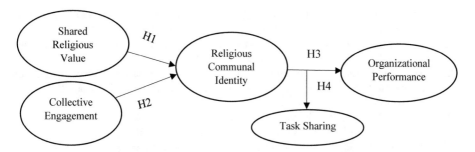

Fig. 1. Conceptual model of religious communal identity and organization performance

4 Conclusion and Further Research

The concept of religious communal identity and shared religious value toward organizational performance have not been widely discussed in the current literature. This article aims to build a conceptual model for shared religious value, collective engagement, religious communal identity, tasks sharing and organizational performance. In the future, authors plan to further validation the proposed model, in steps follow:

1. Develop concept of religious communal identity and shared religious values.
2. Develop a measurement of each variable in this study.
3. Conduct some pilot studies to improve variable measurement.
4. Conduct a survey to the respondents to test the validity of each instrument developed by structural and nomological.
5. Test the proposed model.

References

1. Worden, S.: Religion in strategic leadership: a positivistic, normative/theological, and strategic analysis. J. Bus. Ethics **57**(3), 221–239 (2005)
2. Gümüsay, A.A.: The potential for plurality and prevalence of the religious institutional logic. Bus. Soc. 1–36 (2017)
3. Todd, N.R.: Religious networking organizations and social justice: an ethnographic case study. Am. J. Community Psychol. **50**(1–2), 229–245 (2012)
4. Williams, J., Heiser, R., Chinn, S.J.: Social media posters and lurkers: The impact on team identification and game attendance in minor league baseball. J. Direct Data Digit. Mark. Pract. **13**(4), 295–310 (2012)
5. Ysseldyk, R., Matheson, K., Anisman, H.: Personality and social psychology review religiosity as identity: toward an a social identity perspective. Pers. Soc. Psychol. **14**(1), 60–71 (2010)
6. Pantano, E., Corvello, V.: The impact of experience on companies' reactions to negative comments on social networks. J. Direct. Data Digit. Mark. Pract. **14**(3), 214–223 (2013)
7. Gutierrez-garcia, J.O., Go, S.F.: Collective action in organizational structures. Comput. Math. Organ. Theory **24** (1), 1–33 (2018)
8. Cao, W., Xu, L., Liang, L., Chaudhry, S.S.: The impact of team task and job engagement on the transfer of tacit knowledge in e-business virtual teams. Inf. Technol. Manag. **13**(4), 333–340 (2012)
9. Utesheva, A., Jason, R., Fitzgerald, F.S.: Identity metamorphoses in digital disruption: a relational theory of identity. Eur. J. Inf. Syst. **25**, 344–363 (2016)
10. Wong, Kwok-Fu, Sam: Does investing in identity work? An examination of the ethnocentric social capital theory from an oriental perspective. Asia Eur. J. **3**(2), 209–227 (2005)
11. Chasserio, S., Pailot, P., Poroli, C.: When entrepreneurial identity meets multiple social identities: Interplays and identity work of women entrepreneurs. Int. J. Entrep. Behav. Res. **20**(2), 128–154 (2014)
12. Vasconcelos, A.F.: Spiritual development in organizations: a religious-based approach. J. Bus. Ethics. **93**, 607–622 (2010)
13. Arráiz, I.: Time to share the load: gender differences in household responsibilities and business profitability, pp. 57–84 (2018)
14. Mukhtar, S.: Differences in male and female management characteristics: a study of owner-manager Businesses, pp. 289–311 (2002)
15. Chang, X., Zhou, Y., Wang, C., Heredero, C.D.P.: How do work-family balance practices affect work-family conflict? The differential roles of work stress. Front. Bus. Res. China **11**, 8 (2017)
16. Radzi, K.M., Nazri, M., Nor, M.: The impact of internal factors on small business success: a case of small enterprises under the FELDA scheme. Asian Acad. Manag. J. **22**(1), 27–55 (2017)

Knowledge Management and Religiosity: A Conceptual Development of Islamic Vanguard Spirit

Ken Sudarti[1]([✉]) and Moch. Zulfa[2]

[1] Ph.D Student of Management Science, Faculty of Economics,
Universitas Islam Sultan Agung (UNISSULA), Semarang, Indonesia
ken.sudarti@unissula.ac.id
[2] Lecturer of Management, Faculty of Economics,
Universitas Islam Sultan Agung (UNISSULA), Semarang, Indonesia
zulfa@unissula.ac.id

Abstract. This paper tries to propose a new concept that combines the concepts of knowledge management and religiosity. A new concept called Islamic Vanguard Spirit (IVS) is the result of the synthesis of knowledge management, manifest needs theory, social capital, and Islamic value. This new concept is based on the phenomenon of individualist salespeople and the neglect of religious values in sales activities that ultimately have an impact on sales performance that is not optimal. This concept promotes team collaboration as a lever of sales performance. The resources possessed by each individual are integrated into a team strength that complements and strengthens through knowledge sharing and knowledge donating activities. Through the spirit of doing the best, continuous improvement, asking others and helping others as IVS dimensions are expected to be able to increase selling performance. Testing the validity of each dimension is needed, which can then be empirically tested by adding several variables such as customer-oriented team, team-oriented leadership and adaptive selling capability.

Keywords: Knowledge management · Religiosity · Islamic vanguard spirit

1 Introduction

Facing an increasingly dynamic market situation requires companies to improve their capabilities such as adaptive capabilities. Pehrsson (2014) defines adaptive capability as the company's ability to adapt in response to market changes including changes in the needs and desires of customers and then allocates its resources in order to provide value-added to overcome customer problems. The service condition is also experienced by the service industry. Service companies face some challenges in their marketing strategies, especially those related to service characteristics, namely inseparability, (Blut et al. 2014) had the consequences of the dominant role of "people" including sales team. As a representation of companies dealing directly with customers, they often face various "moments of truth" that require them to be able to make adaptive selling.

© Springer Nature Switzerland AG 2020
L. Barolli et al. (Eds.): CISIS 2019, AISC 993, pp. 939–945, 2020.
https://doi.org/10.1007/978-3-030-22354-0_88

To have the ability to make adaptive sales, sales team must have complete information about the needs and desires of customers. In other words, sales team must be customer-oriented. Customer orientation reflects the level of collection and application of customer knowledge and actions in response to customer desires (Tang 2014). Some of the results of previous studies concluded that a customer orientation culture that is always responsive to customer demands has an important role in improving company performance in the form of sales performance (Ali and Leifu 2016); and customer value (Guenzi and Troilo 2007). But some empirical findings produce different conclusions. In other words, the ability to respond to customers does not have a good impact on firm performance (Ismail *et al.* 2011).

The different findings of the relationship between customer orientation and performance are thought to be due to the motives of the sales team. One of the motivation forms is achievement motivation which refers to one of the manifest needs theory dimension (Khan et al. 2015). Someone who has achievement motivation is someone who wants to be the best. Unfortunately, the theory and results of the study still look from an individual perspective (Staman et al. 2017; Khan et al. 2015). It is still very rare to see from a team performance perspective and try to persuade other people to move forward together, help each other and strengthen each other.

Considering the weakness of the concept of manifest needs theory, this study proposes a new concept namely Islamic Vanguard Spirit (IVS). This concept is a synthesis of manifest needs theory, a social capital concept, and Islamic values. This concept is a unique and comprehensive concept. This uniqueness is reflected, *first*, from the time dimension, namely world life and hereafter. Someone who has IVS always wants to do the best with the intention to worship. They believe that what they get in the world is only as *wasilah* (intermediary) for their lives in the hereafter. In Qur'an (2: 148), it has been explicitly stated that every Muslim must be firm in doing good. Not just doing good but keep trying to do continuous improvement. *Second*, in order to carry out this continuous improvement, other parties are considered as partners and objects benchmark. Someone who has an Islamic Vanguard Spirit does not only think of themselves but also persuades and helps others in an effort to do well. *Third*, achievement motivation is measured at the team level.

In order to improve the performance of this team, someone gave the best example first, then persuade and help other team members to do the same. This persuasion and help will awaken each other's potential, arouse enthusiasm as a trigger for performance, both at the individual and team level. This persuasion activity contains knowledge sharing and knowledge donating activities. Team members who have more capability, are eager to share and donate their knowledge to other members who have less capable. Conversely, members who feel less capable, also have the spirit to make continuous improvements based on information obtained from the knowledge sharing and knowledge donation activities. With the two unique characteristics, the IVS concept is more powerful to increase sales performance. In addition, the results of this study are expected to be able to enrich and strengthen the Manifest Needs Theory especially in improving sales performance.

2   Literature Review

2.1   Manifest Needs Theory

Manifest need theory is often called as achievement motivation theory. This theory was first stated by David Mc. Clelland. This theory is based on the assumption that some people have a strong desire to achieve, while others do not. According to Mc Clelland, there are certain needs that are learned and obtained when they interact with the environment. These needs can be classified into three major groups, namely: the need for power, the need for affiliation and the need for achievement.

The study of achievement motivation has been carried out a lot and becomes an important factor in career success. Khan et al. (2015) had conducted study in nine business incubators in Austria, it is concluded that in general, the diversity of need for achievement had a negative impact on the effectiveness and efficiency of the entrepreneurial team. However, diversity in the need for achievement can increase team effectiveness when the existing (average) team is low. Even so, similarities in the need for achievement can help the team be more successful. Differences in the need for achievement will only cause conflict, overlap, and avoidance of tasks.

So, someone who has a need for achievement challenges himself to achieve better performance than before. Unfortunately, individuals tend to see from their own interests and very rarely think about the achievements of others. This is very different from the view of Islam. In Islam, achievement motivation is highly recommended (QS. Jumuah: 20). Motivation in Islam is not based on egoistic nature, but rather as worship which leads to devotion to God. Achievements must be beneficial for the society because the best of human beings are those who benefit the people. If conventional achievement motivation contains competing elements, then in Islam, other people are more regarded as partners.

2.2   Social Capital

The concept of social capital has been widely used in marketing studies, but there is no agreement in the measurement (Ferri et al. 2009). Yu *et al.* (2013) distinguishes social capital between level teams and individual levels. At the team level, social capital is measured by norm density, cognitive commonality, and cooperative norm. Cooperative norm in a team usually includes a willingness to respect and respond to diversity, openness to critical thinking, and hope for reciprocity and cooperation. This openness to critical thinking and reciprocal hope allows an individual to be willing to persuade one another and help his colleagues to do their best. Rosopa et al. (2013) state that employees who engage in altruistic behavior have greater potential for advancement, respect recommendations and are considered more emotionally stable, open, pleasant, and thorough.

2.3   Islamic Vanguard Spirit

Islamic Vanguard Spirit (IVS) is a person's spirit associated with religion, especially Islam. The concept of IVS contains 4 dimensions. *The first dimension* is the spirit to always try to do the best. Someone who has IVS believes that God is watching everything he does so he will do his job as best he can. Giving the best and efficient (Qur'an (25:67)). On the other hand, someone must do best in front of humans by carrying out their work obligations, achieving set targets, avoiding mistakes and having a high work ethic and providing total service.

In Islam, it is also known as an order to get achievement. Good and bad standards in Islam are based on Islamic principles (Al-Aidaros et al. 2013). The command to compete in goodness is in Qur'an (2: 148) and Qur'an (5: 48). This command is not interpreted as deadly competition with each other, not being "the best" but being "better". The indicator to measure the first dimension is (1) work as well as possible; (2) use resources as efficiently as possible; (3) completing tasks according to standards; and (4) minimizing errors.

The second dimension is continuous improvement. That is, someone will not feel enough if they have done their best, but this desire continues to be maintained and evaluated. They realize that the best size is dynamic so that adjustments and improvements must always be made. As stated in the hadith "seeking knowledge from swing to grave" so that they will take lessons from the events that occur around them. The indicators that can be used to measure this dimension are (1) doing the best continuously, (2) developing skills continuously, (3) taking good lessons from people's experiences around, and (4) making other people's failures as a very valuable lesson.

The third dimension is the motivation to get others to do what they do. A Muslim is obliged to persuade to the good and prevent *munkar* (badness) (QS, 3: 104), message one another with patience and affection (QS, 90: 17-18). This command contains elements of *da'wah* (missionary endeavor). *Da'wah* is a communication problem. For this reason, the use of the right language in the right place for the right person is very important. Giving examples first becomes an important factor so that the contents of *da'wah* can be accepted (QS. As-Shaff: 2-3). This activity involves knowledge sharing among team members. Important information is always shared as an effort to make improvements. Team members have no intention of hiding information for personal gain. Indicators used to measure these dimensions are: (1) enthusiastic in persuading others to do improvement, (2) using good language when persuading others, (3) persuading by giving examples first.

The fourth dimension is helping others in their best efforts. Someone who has IVS sees himself as part of a community that needs each other. He believes that he cannot live alone and he realized that he is imperfect, so that he must complement each other by helping one another (103: 1-3); QS (Al Maidah: 2). In western literature, the concept of helping others outside their role is called Organizational Citizenship Behavior (OCB) including altruistic as one dimension of OCB. Altruistic is the desire to help others. Altruistic begins with feelings of empathy for someone who then arouses helpful motivation (Quinn et al. 2010). People who wish to help usually have prosocial motivation. This activity is related to knowledge donating. Helping other team members is that not only in physical activity, but also donating their knowledge in completing their sales activities. If team members are helping each other, the sales activity will be resolved faster.

(Olowookere and Adekeye 2016) tried to relate a person's religious level (religiosity) to OCB and then concluded that someone who has high religiosity tends to do additional work outside his obligations. The indicators that can be used in measuring this dimension are: (1) eager to help others in terms of goodness, (2) eager to help others without being asked, (3) encouraging others to help without expecting a reward, (4) eager to help even there is no relation with assignments.

2.4   Sales Performance

Performance is a multidimensional concept that does not only show the measurement of results but also the process in achieving results and conditions that enable the achievement of results (da Gama 2011). This understanding allows the emergence of several performance measures, some of which measure financial and non-financial performance. Marketing performance can still be reduced to more specific measures such as sales performance (quality, quantity and time), customer satisfaction, customer loyalty, the addition of new customers. Román and Martín (2014) state that consumer satisfaction is considered to be one of the important performance indicators for sales team and not just the amount of sales. Sales performance can be divided into individual performance and team performance by using various indicators, such as the percentage of team achievements compared to company performance (Rapp *et al.* 2010); return on sales, return on investment and growth of sales (Liozu 2015).

Through an in-depth study, the study finally succeeded in synthesizing Islamic values, manifest needs theory and social capital concepts. The Islamic vanguard spirit propositions can be prepared as follows:

Islamic Vanguard Spirit is the spirit of the salespeople to always try to do the best and make continuous improvements while persuading and helping in an effort to do the best. This Islamic Vanguard Spirit has the potential to increase sales performance.

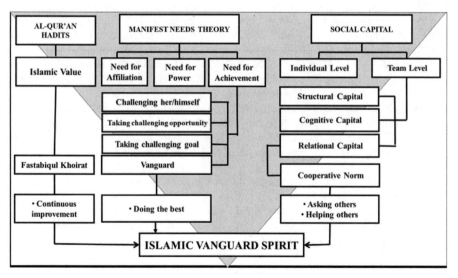

3 Conclusion

Islamic vanguard spirit (IVS) is an Islamic character that is thought to be able to increase selling performance. Someone who has IVS will always do the best according to their resources. Realizing that the best measure is dynamic, they always make continuous improvement. His desire to be better, not only for himself but accompanied by efforts to persuade and help others. If each individual in a team has IVS, then the strength of the congregation will be formed and very powerful in improving performance.

4 Future Research

For future research agenda, it is better to involve validating the measurement of each variable and testing the Islamic vanguard spirit concept at the empirical level. The antecedent variables offered are customer orientated team, team-oriented leadership, and adaptive selling capability. These four variables are expected to be able to increase selling performance. This model is very precisely tested in the insurance industry which has applied Islamic values in its operations with salespeople as its unit of analysis.

References

Al-Aidaros, A.-H., Shamsudin, F.M., Idris, K.M.: Ethics and ethical theories from an Islamic perspective. Int. J. Islamic Thought 4, 1 (2013). Available at http://search.ebscohost.com/login.aspx?direct=true&db=edb&AN=93517717&site=eds-live

Ali, R., Leifu, G.: The impact of technology orientation and customer orientation on firm performance: evidence form chinese firms. Int. J. Manag. Mark. Res. 9(1), 1–11 (2016)

Blut, M., et al.: The impact of service characteristics on the switching costs-customer loyalty link. J. Retail. 90(2), 275–290 (2014). https://doi.org/10.1016/j.jretai.2014.04.003

Ferri, P.J., Deakins, D., Whittam, G.: The measurement of social capital in the entrepreneurial context. J. Enterprising Communities: People Places Glob. Econ. 3(2), 138–151 (2009). https://doi.org/10.1108/17506200910960842

da Gama, A.P.: An expanded model of marketing performance. Mark. Intell. Plan. 29(7), 643–661 (2011). https://doi.org/10.1108/02634501111178677

Guenzi, P., Troilo, G.: The joint contribution of marketing and sales to the creation of superior customer value. J. Bus. Res. 60(2), 98–107 (2007). https://doi.org/10.1016/j.jbusres.2006.10.007

Ismail, M., et al. (2011) Market orientation adoption strategies for small restaurants : a study in the Eastern Sri Lanka. J. Manage.

Khan, M.S., Breitenecker, R.J., Schwarz, E.J.: Adding fuel to the fire: need for achievement diversity and relationship conflict in entrepreneurial teams. Manag. Decis. 53(1), 75–99 (2015). https://doi.org/10.1108/MD-02-2014-0066

Liozu, S.M.: Pricing superheroes: how a confident sales team can influence firm performance. Ind. Mark. Manage. 47, 26–38 (2015). https://doi.org/10.1016/j.indmarman.2015.02.003

Olowookere, E.I., Adekeye, O.A.: Relationship between religiosity and citizenship behaviours in organizations : empirical evidence from selected organisations in Lagos State. 7(4), 475–484 (2016). https://doi.org/10.5901/mjss.2016.v7n4p

Pehrsson, A.: Firms' customer responsiveness and performance: the moderating roles of dyadic competition and firm's age. J. Bus. Ind. Mark. **29**(1), 34–44 (2014). https://doi.org/10.1108/JBIM-01-2011-0004

Quinn, C., Clare, L., Woods, R.T.: The impact of motivations and meanings on the wellbeing of caregivers of people with dementia: a systematic review:43–55 (2010). https://doi.org/10.1017/s1041610209990810

Rapp, A. et al.: Managing sales teams in a virtual environment. Int. J. Res. Mark. **27**(3), 213–224 (2010). https://doi.org/10.1016/j.ijresmar.2010.02.003

Román, S., Martín, P.J.: Does the hierarchical position of the buyer make a difference? The influence of perceived adaptive selling on customer satisfaction and loyalty in a business-to-business context (December 2012) (2014). https://doi.org/10.1108/jbim-05-2012-0092

Rosopa, P.J., Schroeder, A.N., Hulett, A.L.: Helping yourself by helping others: examining personality perceptions. J. Manag. Psychol. **28**(2), 147–163 (2013). https://doi.org/10.1108/02683941311300676

Staman, L. (Laura), Timmermans, A. C. (Anneke), Visscher, A. J. (Adrie): Effects of a data-based decision making intervention on student achievement. Stud. Educ. Eval. **55**(July), 58–67 (2017). https://doi.org/10.1016/j.stueduc.2017.07.002

Tang, T.W.: Becoming an ambidextrous hotel: the role of customer orientation. Int. J. Hospitality Manage. **39**, 1–10 (2014). https://doi.org/10.1016/j.ijhm.2014.01.008

Yu, Y., et al.: A multilevel model for effects of social capital and knowledge sharing in knowledge-intensive work teams. Int. J. Inf. Manage. **33**(5), 780–790 (2013). https://doi.org/10.1016/j.ijinfomgt.2013.05.005

Effect of Islamic Leadership on Teacher Performance Through It Intervention Competency

Siti Sumiati[✉] and Erni Yuvitasari

Faculty of Economics, Universitas Islam Sultan Agung, Semarang, Indonesia
sitisumiati@unissula.ac.id,
erniyuvitasaril1419@gmail.com

Abstract. The purpose of this study is to analyze the effect of Islamic Leadership on Teacher Performance through Competency based on Information Technology (IT) Intervention and Quality of Work Life. This study used 100 teachers as a sample with Purposive Random Sampling. The result of this research showed that Islamic Leadership, Competency based on IT Intervention, and Quality of Work Life positively affect Teacher Performance. Based on data analysis, only Quality of Work Life on Teacher Performance which has no significant effect. However, Competency based on IT Intervention and Quality of Work Life cannot be considered as intervening variable in this research.

Keywords: Islamic leadership · Quality of work life ·
Competency based on IT intervention · Teacher performance

1 Introduction

Teachers play an important role in the development of education by building quality human resources (Leftwich 2009; Phillips 2011; Schweisfurth et al. 2018; Jones et al. 2014; Brautigam and Diolle 2009). The role of teachers in education is for human growth in all aspects: spiritual, intellectual, imaginative, physical, scientific, linguistic both individually and in groups (Aabed and Randall 2006; Dincer et al. 2015; Schweisfurth et al. 2018; Brannelly and Lewis 2011). Teacher performance is an achievement of a teacher in performing his duties and responsibilities that refers to teaching activities. Assessment of teacher performance is measured by individual teacher achievement, contribution to the learning process, student achievement, and graduates produced by the school (Clawson 2003).

Based on the Regulation of the Minister of National Education of the Republic of Indonesia No. 16 of 2007 concerning Teacher Competency Standards, it is stated that teacher competencies consist of pedagogical competence, personality competence, social competence, and professional competence. Assessment for the teacher performance can be said to be good if the teacher has all four competencies in carrying out his duties. Based on previous observations made by the authors in 3 Public Elementary Schools in Kudus, Central Java, Indonesia, the performance of elementary school teachers is still not optimal. The learning process conducted by teachers is still

© Springer Nature Switzerland AG 2020
L. Barolli et al. (Eds.): CISIS 2019, AISC 993, pp. 946–953, 2020.
https://doi.org/10.1007/978-3-030-22354-0_89

conventional, the sense of morale is still lacking and there is still a lack of Teacher Competency Test (UKG) in 2015 for 20% of teachers in Kudus, Central Java, Indonesia, declared not to have passed the UKG test.

Leadership has an important role in changing human resources (Sarachek 1968; Asrar-ul-Haq and Anwar 2018; Schweisfurth et al. 2018). The ability to manage and empower the Principal must be improved in order to work up teacher performance. Islamic leadership plays a very big role in improving teacher performance because it has the principle that an ideal worker makes the *Qur'an* and *Hadith* as a source of knowledge. In addition, the quality of work life must be considered more than the organization (school) because it is seen as being able to work up the participation of the teacher towards the school itself.

2 Literature Review

2.1 Islamic Leadership

Islamic leadership is a leadership that is always guided by the Qur'an and Hadith as a leader. The principle of Islamic leadership in an organization is that employers and employees in carrying out their daily activities are always based on and imbued with Islamic values and culture (Ahmad and Ogunsola 2011; Schweisfurth et al. 2018). There are several leadership characteristics of the Prophet Muhammad who can be exemplary, they are, *Siddiq* (Honesty), *Amanah* (Trust), *Fathonah* (Intelligence), and *Tabligh* (Openness).

Research conducted on BMT (Sharia Cooperatives) employees in Temanggung, Central Java, Indonesia, *PT. Bank Syariah Mandiri, Tbk*, and *Rabbani* show a positive and significant influence between leadership and quality of work life. These results raise the suspicion of the influence of Islamic leadership on the performance, competence and quality of the teacher's work life.

H1: Islamic Leadership has a positive and significant effect on Teacher Performance.
H2: Islamic Leadership has a positive and significant effect on the Quality of Work Life.
H3: Islamic Leadership has a positive and significant effect on Competency based on IT Intervention.

2.2 Quality of Work Life

The Quality of Work Life is to change and improve one's performance so that the end result is as expected (Nursalam et al. 2018; Punch et al. 2019; Boyle et al. 2018). The quality of work life is a sense of satisfaction that teachers feel in participating school organizations (Swamy et al. 2015; Zubair et al. 2017).

There are several indicators that represent the Quality of Work Life variable, namely, health and well-being; job security; job satisfaction; competency development; and the balance of work and life; (Greenhaus et al. 2003; Amstrong 2006; Sarmiento et al. 2004).

H4: Quality of Work Life has a positive and significant effect on Teacher Performance.

2.3   Competency Based on IT Intervention

Competency based on IT Intervention is an ability possessed by a teacher in carrying out his professional duties which will have an effect on the quality of education (Leijen et al. 2016; Nursalam et al. 2018). There are also two types of competence, namely visible competencies and hidden competencies (Sanchez and Lehnert 2019).Teachers must fulfill four basic competencies according to Minister of Education Regulation No. 16. 2007, they are, Pedagogic Competence, Personality Competence, Social Competence, and Professional Competence (Bersh 2018; Greene et al. 2012; Gläser-Zikuda and Fu$^\beta$ 2018).

H5: Competency based on IT Intervention has a positive and significant effect on Teacher Performance.

2.4   Teacher Performance

Teacher performance is a process of interaction carried out by the teacher in his work environment must be in accordance with predetermined criteria. Based on the Regulation of the Minister of National Education of the Republic of Indonesia No. 16 of 2007 concerning Academic Qualification Standards and Teacher Competencies, there are five indicators that are able to build Performance variable, namely, Quality of Work, Quantity of Work, Timeliness, Attendance, and Ability to Cooperate. In addition, there are several factors that can influence a person's performance, namely, individual factors related to expertise, leadership factors related to the quality of group factors/co-workers, system factors related to systems/work methods, situation factors related to environmental change.

3   Research Model

Based on the analysis of the literature review and the results of previous studies which gave rise to allegations between variables with one another, then it was described in the empirical model of the study as follows (Fig. 1):

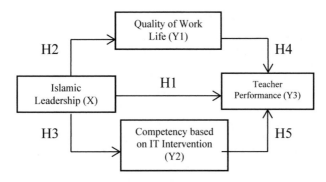

Fig. 1. Empirical Research Model

4 Research Method

The population of this study is the public elementary school teacher in Kudus, Central Java, Indonesia, which consists of 129 teachers. Respondents used in the study amounted to 100 teachers with the sampling technique used was Purposive Random Sampling. The sample characteristics that will be used in this study consist of the status of teacher positions taken by permanent employees, principals and teachers must be Muslim, work experience of at least 2 years, ages 22–60 years, and final education at least Diploma (Undergraduate).

5 Finding and Discussion

5.1 Findings

Based on the test results, the field research data shows that the research data in this study was valid and reliable. The data in this study also showed that there were no symptoms of multicollinearity which meant that there was no high correlation in the three regression models of this study. Linearity testing on the three research regression models also states that it meets linearity assumptions which means the regression model is considered correct.

The data of the study also carried out heteroscedasticity testing showed that the three regression models of this study has no heteroscedasticity and normal so that both models were used for forecasting (estimation). F Hypothesis test aims to test the significance of the regression coefficients of independent variables (X, Y1, and Y2) simultaneously on the dependent variable (Y3) that the three regression models of this study show significant results. That means there is a significant relationship among variables in each regression model (Table 1).

Table 1. Test of path analysis

Analysis of the Effect of Islamic Leadership (X) on Teacher Performance (Y3) through Quality of Work Life (Y1)		Analysis of the Effect of Islamic Leadership (X) on Teacher Performance (Y3) through Competency based on IT Intervention (Y2)	
P1	P2 × P4	P1	P3 × P5
0,228	0,369 × 0,097 = 0,036	0,228	0,247 × 0,292 = 0,072
Total influence: P1 + (P2 × P4) = 0,264		Total influence: P1 + (P3 × P5) = 0,300	
P1 > (P2 × P4) Quality of Work Life (Y1) cannot be an intervening variable		P1 > (P3 × P5) Teacher competency (Y2) cannot be an intervening variable	

Source: Primary Data processed, 2017.

Table 2. Summary of calculation results and data processing

Regression model	Stand. Coef.	F_{-count}	Sig. F	R	R^2	Status	Hypothesis	T_{count}	Sign. T	Status
Regression I:		8,052	0,000	0,448	0,176	Sign.	**H1**	2,261	0,026	Positive
Constant	0,228						(X → Y3)			and Sign
Islamic Leadership (X)										
Quality of Work life (Y1)	0,097									
Competency based on IT Intervention (Y2)	0,292									
The equation of the linear regression model I: Y3 = b1 X + b2 Y1 + b3 Y2 + e Y3 = 0.228 X + 0.097 Y1 + 0.292 Y2							**H2** (X → Y1)	3,929	0,000	Positive and Sign
Regression II:		15,439	0,000	0,369	0,136	Sign.	**H3**	2,519	0,013	Positive
Constant	0,369						(X → Y2)			and Sign
Islamic Leadership (X)										
The equation of the linear regression model II: Y1 = b1 X + e Y1 = 0.369 X + e							**H4** (Y1 → Y3)	0,987	0,326	Positive and not Sign
Regression III:		6,345	0,013	0,247	0,061	Sign.	**H5**	3,099	0,003	Positive
Constant	0,247						(Y2 → Y3)			and Sign
Islamic Leadership (X)										
The equation of the linear regression model III: Y2 = b1 X + e Y2 = 0.247 X + e										

Source: Primary Data processed, 2017.

5.2 Discussion

5.2.1 The Effect of Islamic Leadership on Teacher Performance

Table 2 explains that hypothesis 1 has a positive and significant effect between Islamic leadership (X) on teacher performance (Y3), meaning that teacher performance will increase if the implementation of Islamic leadership style carried out by the Principal is also improved. Indicators of Islamic leadership that include the Principal's honesty attitude, trust possessed by the Principal in leading, intelligence, and the Principal's openness toward teachers. Indicators of teacher performance include the quality of work, quantity of work, the timeliness of the teacher in carrying out his duties and responsibilities as an educator, the presence of the teacher, and also the ability to cooperate between coworkers and the Principal.

5.2.2 The Effect of Islamic Leadership on the Quality of Work Life

Table 2 shows that the second hypothesis is accepted, meaning that there is a positive and significant influence between Islamic leadership on the quality of work life. The implication is that if the intelligence of a Principal improved, it will work up the health and well-being of a teacher. If the Principal's attitude of openness is enhanced in the work environment, it will have a good effect on job security and the development of a teacher's competence. An increase in the Principal's sense of trust is also able to increase the sense of job satisfaction for the teachers. In addition, the high attitude of honesty of the Principal will increase the balance of time between work and life for a teacher.

5.2.3 The Effect of Islamic Leadership on Teacher Competency Based on IT Intervention

Based on the results of the t-test in Table 2 show that the third hypothesis is accepted. This means that this study found a positive and significant effect between Islamic leadership on competency based on IT Intervention. If a school principal has a high level of honesty, the teacher's personal competence is expected to increase due to the awe that arises from within the teacher to the principal. The high trust level of the Principal to the teachers is expected to be able to improve teacher professional competence related to the development of IT in work. The high intelligence possessed by the principal is expected to be able to help the teachers to improve pedagogical competence and IT development in work, namely, by sharing ideas and opinions. A high level of openness from the principal will be able to improve the teacher's social competence by establishing harmonious relationships between principal, teacher, students, and guardians of the student.

5.2.4 Effect of Quality of Work Life on Teacher Performance

The t-test results in Table 2 conclude that the fourth hypothesis in this study was rejected, which means that there is no significant effect between the Quality of Work Life and Teacher Performance because a teacher working in a Public Elementary School is required to always perform high under any conditions. These demands are considered to be the responsibility of Public Elementary School as institutions that operate in the field of public services, namely education services. These results have similarities with the results obtained by Pamungkas (2016) on the employees of the Central Bureau of Statistics in Yogyakarta, Indonesia. Although institutionally under the auspices of the Ministry of Education and Culture (*Kemendikbud*) continues to strive to improve the Quality of Work Life of a teacher, the teachers already have high self-awareness to remain and improve their performance as well as possible. The results of this study indicate that all public elementary school teachers are highly dedicated as educators to remain high-performing under any conditions.

5.2.5   The Effect of Competency Based on IT Intervention on Teacher Performance

The t-test results in Table 2 conclude that the fifth hypothesis is accepted, meaning that it has similarities with the real conditions in the field. The teacher professional competence which is supported by the ability of IT development must be applied as well as possible and improved to be able to have a positive impact on the quality and quantity of teacher performance in an organization. The demand to improve social competence in its environment aims to increase empathy and concern for others both in the work environment, family environment, and community environment. The well-established communication will be able to improve the quality of the teacher's work in communicating and having good relations with human beings on this earth.

6   Conclusion

The results of this study indicate a positive effect between Islamic leadership on the quality of work life, competency based on IT interventions, and teacher performance. Only the relationship between the Quality of Work Life on Teacher Performance that has insignificant results. This also provides a solution related to the still less optimal performance of Public Elementary School teachers in Indonesia, namely, the application of Islamic Leadership by Principal. This study suggested that the Principal should improve the application of Islamic Leadership styles in Public Elementary School. The form of leadership style must be based on honesty and intelligence. This study used only one independent variable, namely Islamic Leadership variable, resulting in a low effect percentage on the Quality of Work Life, teacher's competency based on IT intervention, and Teacher Performance. For future researchers, it is recommended to use other variables, such as inspirational motivation and knowledge management. In addition, data collection methods not only use the questionnaire method but also can use the interview method in order to obtain detailed and more accurate information as research material.

References

Aabed, A., Randall, V.: A study of Islamic leadership theory and practice in K–12 Islamic schools in Michigan. All Theses and Dissertations, 3206991, 220–220 (2006)

Ahmad, K., Ogunsola, O.K.: An empirical assessment of Islamic leadership principles. Int. J. Commer. and Manage. **21**(3), 291–318 (2011)

Amstrong. (n.d.). A Handbook of Human Resources Management Practice

Asrar-ul-Haq, M., Anwar, S.: The many faces of leadership: proposing research agenda through a review of literature. Future Bus. J. **4**(2), 179–188 (2018)

Bersh, L.C.: Writing autobiography to develop culturally responsive competency: teachers' intersections of personal histories, cultural backgrounds and biases. Int. J. Educ. Res. **88** (December 2017), 31–41 (2018)

Boyle, M.P., Milewski, K.M., Beita-Ell, C.: Disclosure of stuttering and quality of life in people who stutter. J. Fluency Disord. **58**, 1–10 (2018)

Brannelly, Lewis., N.: Higher education and the formation of developmental elites: a literature review and preliminary data analysis (2011)

Brautigam, D., Diolle, T.: Coalitions, capitalists and credibility: overcoming the crisis of confidence at independence in Mauritius. In V. T. T. M. S. G. Holden (ed.) Contemporary African Political Economy, pp. 17–67. Springer Nature Switzerland, Cham (2009)

Clawson, J.G.: Level three leadership : getting below the surface (2003). Retrieved from

Dincer, F.I., Dincer, M.Z., Yilmaz, S.: The economic contribution of Turkish tourism entrepreneurship on the development of tourism movements in Islamic countries. Procedia – Soc. Behav. Sci. **195**, 413–422 (2015)

Gläser-Zikuda, M., Fu$^\beta$, S.: Impact of teacher competencies on student emotions: A-multi method approach. Int. J. Educ. Res. **47**(2), 136–147 (2018)

Greene, J.A., Hutchinson, L.A., Costa, L.J., Crompton, H.: Investigating how collage students' task definitions and plans relate to self-regulated learning processing and understanding of a complex science topic. Contemp. Educ. Psychol. **37**, 307–320 (2012)

Greenhaus, J.H., Collins, K.M., Shaw, J.D.: The relation between work-family balance and quality of life. J. Vocat. Behav. **63**(3), 510–531 (2003)

Jones, A., Jones, C., Ndaruhutse, S.: Higher education and developmental leadership: the case of Ghana (26), 130 (2014)

Leftwich, A.: Bringing agency back in: politics and human agency in building institutions and states (June), 38 (2009)

Leijen, Ä. li, Slof, B., Malva, L., Hunt, P., Tartwijk, J. van, Schaaf, M. van der. Performance-based competency requirements for student teachers and how to assess them. Int. J. Inform. Educ. Technol. **7**(3):190–194 (2016)

Nursalam, N., Fibriansari, R.D., Yuwono, S.R., Hadi, M., Efendi, F., Bushy, A.: Development of an empowerment model for burnout syndrome and quality of nursing work life in Indonesia. Int. J. Nurs. Sci. **5**(4), 390–395 (2018)

Phillips, S.: Division in a crisis state (February 2011)

Punch, J.L., Hitt, R., Smith, S.W.: Hearing loss and quality of life. J. Commun. Disord. **78**, 33–45 (2019)

Sanchez, Carol M., Lehnert, K.: The Unbearable haviness of Leadership: The effects of competency, negatives, and experience on women's aspirations to leadership. J. Bus. Res. **95** (2), 182–194 (2019)

Sarachek, B.: Greek concepts of leadership. Acad. Manag. J. **11**(1), 39–48 (1968)

Sarmiento, T.P., Laschinger, H.K., Iwasiw, C.: Nurse educators' workplace empowerment, burnout, and job satisfaction: testing Kanter's theory. J. Adv. Nurs. **46**(2), 134–143 (2004)

Schweisfurth, M., Davies, L., Symaco, L. P., Valiente, O.: Higher education, bridging capital, and developmental leadership in the Philippines: learning to be a crossover reformer. Int. J. Educ. Dev. **59**(September 2017), 1–8 (2018)

Swamy, D.R., Nanjundeswaraswany, T.S., Rashmi, S.: Quality of work life: scale development and validation. Int. J. Caring Sci. **8**(2), 281–300 (2015)

Zubair, M.H., Hussain, L.R., Williams, K.N., Grannan, K.J.: Work-related quality of life of US general surgery residents: is it really so bad? J. Surg. Educ. **74**(6), 138–146 (2017)

Blue Accounting of the Marine Knowledge and Sustainable Seas: A Conceptual Model

Winarsih$^{(\boxtimes)}$, Khoirul Fuad, and Hendri Setyawan

Department of Accounting, Faculty of Economics,
Universitas Islam Sultan Agung, Semarang, Indonesia
{winarsih,khoirulfuad,hendri}@unissula.ac.id

Abstract. The purpose of this study is to develop a model in the framework of marine knowledge and blue accounting. The pollution and degradation of the marine environment due to human intervention which damage and harm marine life and produces strong risks band threats to this environment. To reinforce the marine knowledge, the blue accounting will provide to the citizen and the society valuable information based on accounting standards that identify, measure, value, and report with new opportunities for marine and maritime sustainability. Special target to be achieved is an understanding on the application of the concept of blue accounting as essential to support the ocean strategy, because sciences are interdependent and scarcity of marine resources demands knowledge and for the mitigate uncertainties and risk.

Keywords: Blue accounting · Marine knowledge · Sustainable sea

1 Introduction

Marine resource are a public good which is available to everybody without payment or compensations to this collective pressure of human activity. In order to reinforce the marine knowledge, the blue accounting will provide to the citizen, to the organization and the society valuable information based on accounting standards that identify, measure, value, and report this blue growth that is the ocean strategy with new opportunities for marine and maritime sustainability. Blue Accounting's mission is to improve decision making and increase efficiency of water management issues in the basin. This a conceptual model of the marine knowledge explores the framework from Appeltans et al. (2012) in general, and the blue accounting, in particular. Blue accounting will develop with an unprecedented urgency, because is a associated much more than financial accounting and management accounting (Boonstra et al. 2018). These information needs to be accountable to be reported and it demands innovations that can increase human well-being and a the same time enhance the capacity of ecosystems to produce services (Mober 2016).

Blue accounting bases the creation of marine knowledge that begins with the sea which is an asset of blue economy (UNDESA 2014). From blue accounting, it can be used to analyze the blue economy that the sea and ocean can emphasize wave energy, coastal protection, to be accountablefisheries, cultivation, and waste management. Blue accounting is expected to help stakeholders, especially in the marine sector, to always

© Springer Nature Switzerland AG 2020
L. Barolli et al. (Eds.): CISIS 2019, AISC 993, pp. 954–958, 2020.
https://doi.org/10.1007/978-3-030-22354-0_90

monitor the condition of the sea and its surroundings and the existence of transparency to minimize the negative impacts caused. Furthermore, stakeholders can make initial recognition, timely involvement and prudential principles in every decision making. In addition, the blue accounting will provide valuable information to the user, citizen, and the society based on the new accounting standard that identifies, measures, assesses and reports this blue growth that is the ocean strategy with new opportunities for marine and maritime sustainability. Thus an understanding of blue accounting will reduce the uncertainty of marine knowledge (FAO 2018). Recognition of the past, current and future estimates of marine resources make the strengths and weaknesses in accounting. The main question of the blue assets are no willing buyer neither willing seller. The gap between the demand of clean seawater and their availability is completely different when compare the dirty sea water.

The opportunity to explore maritime resources requires serious thinking to produce information on how to disclose assets, liabilities, and revenue from blue accounting. Besides that, whether blue accounting can reduce the negative impact in supporting the blue economy.

2 Literature Review

2.1 Blue Accounting

From the literature review, Gray et al. (1996) to promote the first insights of the blue accounting to explain the main objective and to refocus attention on the social accounting which involves the communication of information concering the impact of the maritime assets and its activities on the society. Hopwood (1992) defends that accounting is used from the active construction and transformation of organizational and social truths, associated economic truths and consequently, political truths.

The objective of the blue accounting standard is to prescribe the accounting treatment and the disclosure related with maritime activity. International Accounting Standard (EC 2008), the most important definitions are:

1. A gain or loss arising on initial recognition of a blue asset at fair value less costs to sell and from a change in fair value less costs to sell of a blue asset shall be included in profit or loss for the period in which it arise
2. An unconditional government grant related to a blue asset measured at its fair value less costs to sell shall be recognized in profit or loss when, and only when. the government grant becomes receivable
3. An understanding the data on the marine environment.

Blue accounting will provide information for users based on accounting standards, namely identifying, measuring, assessing, and reporting. In several functions, blue accounting is based on the creation of marine knowledge that begins with marine and maritime as an asset. Thus blue accounting must be used for public disclosure through accountability as a legitimacy tool, implementing accounting treatment, and disclosures related to maritime activities. In addition, the role of blue accounting minimizes negative impacts. The blue accounting process relies on agreed-upon, overarching the desired use of and values associated. The need of recognition and the measurement of the blue asset start with wasting water (Laughlin and Varangu 1991; Rodrigues et al. 2014; Santos et al. 2005). Indeed, wasting assets are defined by RICS (2017: 1), an asset with finite life which, when consumed, cannot be renewed in the existing physical location in which they occur. The absence of a specific international accounting standard weakens the reliability and the comparability of the data about the sea and the maritime resources of the blue accounting.

2.2 Recognition of the Blue Asset

The recognition of blue assets is established to recognize and measure the ocean and maritime items as assets based on the valuation process. Blue expenditure is considered an asset if there are activities to avoid or reduce damage, preserve resources, provide economic benefits in the future, expenditures used to reduce or avoid environmental contamination that may occur due to company activities in the future. From the literature review, Kovel (2002), pollution control or prevention facilities and machinery acquired to comply with environmental laws and regulation and rights or similar elements acquired for reasons associated with the impact of business activities on the environment, such as patents and pollution rights. When the book value of an asset is considered a loss for environmental reasons, costs can be considered as assets. Furthermore, the costs incurred are considered as assets that must be distributed systematically.

2.3 Recognition of the Blue Liability

Recognition of the Blue Liability is intended to recognize justified liabilities by the ocean and maritime resources. Recognition of the Blue liability when it is probable that the outflow of resources embodying economic benefits will result in the settlement of current environmental obligations as a result of past events that can be measured reliably. Thus the estimate of the costs arising from the obligation must be clear, and if the amount of the liability cannot be estimated then the contingent liability must be recognized even though contingent liabilities are not recognized in the balance sheet, but are recognized in the notes for annual accounts.

2.4 Recognition of the Blue Expenses

Recognition of the blue expenses is support on the adaptation of the recommendation 2001/453/EC (EC 2001) that presents several definitions that may help new accounting standard. Emerging costs consist of improving the environment caused by its operations, waste disposal and avoidance, the protection of clean water and climate, noise reduction, etc. Recognition of the blue expenses is a condition upon the period in which they are incurred. Blue expenses are related to losses that occurred in a previous year must be recorded in the year in which they are recognized and cannot be considered as adjustments of that previous year. The recognition of the expenses that an organization is required to bear in respect of the recovery of contaminated sites and the recognition foreseen for environmental liabilities at the estimated value for total liabilities, either totally or progressively.

2.5 Measurement of the Blue Liability

From the literature review, Rute Abreu et al. (2019), the blue liabilities are recognized whereby it is possible to make an estimate of the expenditure to meet the underlying obligation. The amount of the liability depends on the appropriate estimate of the expenses required to settle. If adequate estimates are difficult to measure, they are considered as a contingent liability. For blue liabilities that will not be settled in the near future, they should be measured at the present value. The method chosen to the measurement should be disclosed in the notes. The estimated amount of discounted cash is the amount expected to be paid on the date of completion and calculated using the cleaning or repair plan assumption.

3 Conclusion and Future Research

Blue accounting must be used for public disclosure through accountability as a tool to legitimize and ensure accounting and financial stability for the overall objective of maritime affairs. In addition, blue accounting can be used to determine the strengths and weaknesses of marine stakeholders and maritime resources that are continuously informed. This is expected to make a quick and precise decision based on financial statements.

Further researchers should develop the blue accounting issues that are based on marine and maritime issues, despite the strong limitation of literature because the marine and maritime resources are complex and there are a lot of data that are not in the marine scope. The use of this blue strategy demands the need for the blue accounting and fundamental science to empower the blue accounting as an important measurement and disclosure approach.

References

Appeltans, W., Ahyong, S.T., Anderson, G., Angel, M.V., Artois, T., Bailly, N., et al.: The magnitude of global marine species diversity. Curr. Biol. **22**, 2189–2202 (2012)

Boonstra, W.J., Valman, M., Bjorkvik, E.: A sea of many colours – How relevant is Blue Growth for capture fisheries in the Global North, and vice versa? Marine Policy **8**(January), 340–349 (2018)

European Commission: Recommendation 2001/453/EC, 21st May on the recognition, measurement and disclosure of environmental issues in the annual accounts and annual reports of companies. Brussels: Official Publication of The European Commission. L.156, on 2001, 13th June (2001)

European Commission (EC): Action plan for a Marine Strategy in the Atlantic Area. COM (2013) 279 final, Brussels: Official Publication of the European Commission (2013)

European Marine Observation and Data Network (EMODnet): EMODnet Geology, Finland (2018)

Food and Agriculture Organization (FAO): Blue Growth Initiative. FAO, Rome (2018)

Gray, R., Owen, D., Adams, C.: Acounting and Accountability. Prentice Hall, London (1996)

Hopwood, A.G.: Accounting calculation and the shifting sphere or the economic. Eur. Account. Rev. **1**(1), 125–143 (1992)

International Accounting Standards Board (IASB: International Accounting Standards. IASB, London (2018)

Kovel, J.: Thee enemy of nature: The end of capitalism or the end of the world?. Zed Books, London (2002)

Laughlin, B., Varangu, L.K.: Accounting for waste or garbage accounting: Some thoughts from non-accountants. Account. Audit. Account. J. (1991)

Rodrigues, D.B.B., Gupta, H.V., Mebdiondo, E.M.: A blue/green water-based accounting framework for assessment of water security. Water Resour. Res. **50**(9), 7187–7205 (2014)

Royal Institution of Chartered Surveyors (RICS): Consultation paper of The IVCS Standards Board Agenda Consultation London. RICS (2017)

Rute Abreu, F.D.: Blue Accounting: Looking for a New Standard (2019)

Santos, L.L., Mouga, T., Viana, A.S.: Proteccaoambiental e informacaofinanceira: Osimpactos, osriscos, e a suadivulgacaoemdireccaoa um desenvolvimentosustentavel. 1st GECAMB: Conference of Centre for Social and Environmental Accounting Research (CSEAR) (2005)

Seelbach, P.: Great Lakes Blue Accounting: Empowering Decisions to Realize Regional Water Values. A report to Council of Great Lakes Governors, in response to governor's 2013 resolution on water monitoring, 17 (2014)

United Nations Department of Economic and Social Affairs (UNDESA): Blue Economy concept paper. New York: UNDESA – United Nation (2014)

World Ocean Council (WOC): Ocean 2050: The Ocean business community and sustainable seas. Washington: WOC (2013)

World Ocean Council (WOC): Ocean 2030: Sustainable development goals and the ocean business community. Rotterdam :WOC (2016)

Fraud Prevention on Village Government: The Importance of Digital Infrastructure Supervision

Provita Wijayanti[✉], Nurhidayati, and Rustam Hanafi

Economics Faculty, Universitas Islam Sultan Agung, Semarang, Indonesia
{provita.w, nurhidayati, rustam}@unissula.ac.id

Abstract. This paper aims to develop a framework of digital infrastructure supervision to optimize fraud prevention in village government, Indonesia. Village is the smallest area in the country of Indonesia. The village government carries out the tasks of village management which is assisted by the facilitator and the village supervisor to account for village funds. Digital infrastructure supervision is used to detect and reduce the opportunity of fraud. The infrastructure of village government supervisors includes Regional Inspectorate, village community empowerment agency, village representative agency, village assistant, and digital supervision system. These infrastructures become a benchmark to increase accountability and transparency of village fund management. Future research will include validating the proposed framework using an empirical data.

Keywords: Fraud prevention · Digital infrastructure supervision

1 Introduction

Government promises to realize an equitable welfare of rural communities by the issuance of Law No. 6 of 2014 concerning Village which includes Village Funds which is started in 2015. Village funds are intended to help building infrastructure in villages that can facilitate economic activities in the countryside. Village funds can also be used to help rural poor communities to have a decent place or environment to live.

It is known that basically everything that is related to funds (money) should be intended to the public, but it is mostly misused for the fulfillment of the desires of certain individuals or groups. The misappropriation of village funds in the past three years has been widely reported that the funds were misused by the authorities mandated to allocate the funds. The village fund (fraud) cases began to be widely reported and occurred in almost every district in Indonesia. The report of (*Indonesian Corruption Watch* 2018) states that the number of village corruption cases has doubled and more from year to year. The total cases in 2015–2017 reached 154 cases. The number of losses donated reached Rp. 47.56 billion (*Indonesian Corruption Watch* 2018).

Fraud prevention is a necessity. One of the efforts taken is the appointment of village assistants. The main task of the village facilitator is to facilitate village

© Springer Nature Switzerland AG 2020
L. Barolli et al. (Eds.): CISIS 2019, AISC 993, pp. 959–965, 2020.
https://doi.org/10.1007/978-3-030-22354-0_91

development and financial planning, implementation of village development, village financial management, and evaluation of village development implementation.

Digital infrastructure can be interpreted as the use of ICT in government or commonly referred to *e-government* according to Siau; while according to Gupta and Debashish Jana, *E-government* is the application of IT in government which aims to simplify the work process in the government to be more accurate, responsive and to form a transparent government (qtd. in Tyas et al. 2016). As a result of rapid technological development and information growth the government is required to provide services to the community in accordance with developing technology and with e-government the government can reduce the possibility of fraud because the system that has been implemented in e-government must be carried out according to procedures and difficult to deceive.

In these tasks, the village facilitator is contained in the role of the village government supervisor and the internal control system. These two aspects are expected to minimize the possibility of fraud that will be carried out by irresponsible parties. The improvement of both village government supervisor and internal control system will be the most suitable method to prevent fraud (Omar and Bakar 2012; Othman *et al.* 2015).

In the process of supervising, village government supervisor and the internal control system, in today's era where everything can be facilitated by digital advancement, a digital system for supervision is proposed. The system of infrastructure can be altered to digital infrastructure. By having the digital infrastructure supervision, the access to detect and prevent fraud will be able to be done by protecting the software or application that is related to the infrastructure (Bierstaker et al. 2006; Rahman and Anwar 2014).

Based on the background, the purpose of this study is to find out whether digital infrastructure supervision will be able to moderate the internal control system and village management supervisor in order to prevent fraud.

2   Literature Review

According to the Legal Dictionary, legally, fraud is deliberate deception to secure unfair or unlawful profits or to eliminate victims from legal rights (Lawrence and Wells 2004).

Such problem must be fought by both the public and private sectors, especially in Indonesia. Internal auditors who are competent in evaluating financial statements and are able to make organizational operations effectively are needed to minimize fraud risk (Wilopo 2006). Strengthening the internal control structure, optimizing control activities, as well as effective internal audit functions, are fraud prevention strategies (Wuysang *et al.* 2016).

Ground theory to underline this research is the agency and GONE theory. Agency theory (Jensen and Meckling 1976) in Sudarma and Putra (2014) defines the relationship between principal and agent in detecting fraud. This theory aims to solve the problem in the agency relationship between principal and different agents called agency problems (Kusumastuti and Meiranto 2012). A GONE theory is described as follows: *Greed*, *Opportunity* to commit fraud, *Need* (the need to support life), and *Exposure* of actions or consequences for perpetrators of fraud if the perpetrator is proven to commit fraud. Exposure is related to the learning process of fraud because it is considered a sanction that is classified as mild. Greed and need are personal and difficult to eliminate, so they tend to violate regulations, while the Opportunity and Exposure factors are related to victims (communities, institutions, and organizations) who feel disadvantaged by fraud.

The components of the agency and GONE theory mentioned above can be minimized through the role of the village government supervisor and internal control systems including the management of the resources. The government's oversight function which is still minimal is also a factor causing fraud in village fund. Institutions such as Village Consultative Agency (BPD) have not been fully optimal in carrying out budget oversight in the village. BPD should be able to play an important role in preventing corruption of village funds, including encouraging other citizens to jointly monitor development in the village (*Indonesian Corruption Watch* 2018).

The functions of supervision on fraud prevention have been frequently published. Qualitative study (Wibisono and Purnomo 2017) reports that weak village fund management or the tendency of fraud in village fund is due to the non-functioning role of village facilitators, elements of coaching and supervision from the sub-district and also from the Government Guard and Development Team (TP4) are weak, lack of community participation in supervision of village funds, high non-budgetary costs, as well as the lack of ability of village fund managers and village heads. Qualitative studies conducted by (Wida et al. 2017) in Rogojampi, Banyuwangi Regency, reported the supervision phase of village fund allocations that had run well and was expected to prevent fraud.

Meanwhile the supervision includes internal control system and quality of village management supervisor especially the human resource. Atmadja et al. (2017) report that the internal control system influences the fraud prevention in Buleleng district in Bali and Widiyarta (2017) reports that the internal control system has a positive influence on the fraud prevention in village fund management in the village government in Buleleng district. (Wijayanti et al. 2018; Shanmugam et al. 2012) state that in order to have a better internal control system, internal parties including the village facilitators as the people who are mandated to account for the village funds, should be improved; because the fraud tends to be done by the internal people who have direct access to the funds. Henceforth, the improvement of effective internal control system will be able to combat fraud.

To ease the village government supervisor in performing supervision, digital advancement is launched to assist the program. According to Suryanto (2016), the involvement of information technology in the management of a community will help increasing the performance. Even though it may need many tasks to carry out the method, digital infrastructure supervision with the digital government technology is supposed to be able to mildly reduce the practice of fraud (Ojha et al. 2008).

3 Proposed Framework

In order to develop the framework of digital infrastructure supervision to optimize fraud prevention in village government through improving the quality of village government supervisor and the internal control system, here is the research model (Fig. 1).

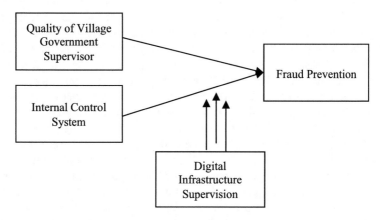

Fig. 1. Research model

Digital Infrastructure Supervision is the implementation of the grand design of village financial supervision strategies that are the priority of the Indonesian government which refers to Presidential Regulation number 54 of 2018 concerning the national strategy for preventing corruption in 2019–2020 (Merdeka.com 2018). According to Braa et al. (2007) and Tilson et al. (2010) the definition of digital infrastructure is a collection of combination of components between technology and human such as the networks, the systems and the processes which contributes to the information systems' functioning (qtd. in Sharma 2017). In addition, according to Ben et al. digital infrastructure consists of approach to ICT and level of internet use, the enthusiasm and skill in gaining access to the ICT and internet services. Digital development is affected by the industrial trends such as internet of things (IoT) and artificial intelligence (AI), development of consumption patterns and acceptance of new devices and applications such as smart phones, tablets and smart TVs (2017). Sharma supplements a concept of hierarchy of digital infrastructure which consists of three points, there are policy level, program level and project level. The policy level relates to the planning of policy and finding resources. The program level includes preparation in strategy and prioritizing as well as the structures and procedures. While the project level covers the arrangement, the conceptualization and the technology (2017).

Village government supervisor has the role to supervise the account of village funds and should be responsible in the accountability of the funds' usage. The quality of village government supervisor should be improved to have better governance and keep it safe from the misappropriation of the use of village funds. In order to do that, internal control system should be improved simultaneously.

There are two types of internal controls: preventive and detective controls. Preventive control is a control that can help prevent fraud happened. These control activities prevent reimbursement of costs without review or knowledge of leaders. Detective control is a control that can help reveal fraud schemes. This control is generally less popular than preventive control because this control is known to work after fraud occurs. Those two types of internal control also apply in two phases of the different fraud cycle (Ramamoorti and Dupree 2010). There is a proactive stage, which uses preventive controls to minimize opportunities for fraud. There is also a reactive stage, which relies on the detective's control to find fraud after that happens. After fraud occurs or control deficiencies have been discovered, compensation control can be applied to reduce the effects of adverse financial reporting (Gramling et al. 2010). Compensation control can be a preventive or detective control applied to businesses when there is a deficiency of control and work as a support.

For example, previous preventive controls that require lead reviews on reimbursement of costs before being processed are useless because management has never taken the time to authorize daily reimbursement, so strong compensation controls will require a receipt to be attached to the replacement form. In this way, the person responsible for reimbursement can verify that the expenditure is legal and available. As seen in this example, compensation control is generally used when actual controls are too expensive or time-consuming to set, but they are less desirable than preventive controls because they usually occur at the reactive stage of the fraud cycle.

The internal control system in this research can also be included in detective internal control where internal control is carried out after fraud has occurred. Related to the agency theory, internal control should be carried out from the beginning of the village government apparatus to compile a budget plan for the use of village funds to the end result of the realization of the village funds usage, so that the implementation of the internal control system carries out in all these stages can prevent self-interest and cause individuals to not be bounded rationality and strive for risk-averse. Regarding GONE theory, the internal control system at each stage of village fund allocation will limit a person to being greed, limiting opportunities for cheating (opportunity), not dare to act to enrich themselves (needs) and fear that fraud is known so that they will face sanctions or penalties obtained (exposure).

This is also in accordance with previous researches that the weak village fund management or the tendency of fraud in village fund was due to the non-functioning role of village facilitators, elements of coaching and supervision from the sub-district and also from the Government Guard and Development Team (TP4) are weak, lack of community participation in supervision of village funds, high non-budgetary costs, as well as the lack of ability of village fund managers and village heads (Wibisono and Purnomo 2017). In addition, the qualitative study conducted by Wida et al. (2017) in Rogojampi, Banyuwangi Regency also reported that the supervision phase of village fund allocation that had run well could prevent fraud.

When the accountability and responsibility of the use of village funds are in the right path after the improvement in internal control system and the quality of village government supervisor, digital infrastructure supervision has moderating role to support and to facilitate the supervision process with an integrated system in digital technology. By using this digital infrastructure supervision, facilitators of the village can oversight the condition in the village and take immediate action to prevent any kinds of fraud.

4 Conclusion

Thus in implementing the concept of the research, the mechanism is to improve the internal control system and village management supervisor by fixing the facilitators of the village i.e. the human resource. So, digital infrastructure supervision acts as moderating factor to support the effort in preventing fraud.

For future research this paper is still in the form of concept on fraud prevention and digital infrastructure supervision in village government, therefore it is necessary to test the concept's model by conducting empirical research related to the topic. This paper is a solution to fraud prevention and still needs to be developed in term of digital infrastructure supervision model.

References

Atmadja, A.T., Adi, K., Saputra, K.: Fraud prevention in village fund management. J. Sci. Account. Bus. **12**(1), 7–16 (2017)

Shenglin, B., et al.: Digital infrastructure overcoming the digital divide in China and the European Union. Emerging Market Sustainability Dialogues (2017)

Bierstaker, J.L., Brody, R.G., Pacini, C.: Accountants' perceptions regarding fraud detection and prevention method. Manag. Audit. J. **21**, 520–535 (2006)

Gramling, A.A. et al.: Audit partner evaluation of compensating controls: a focus on design effectiveness and extent of auditor testing. Audit.: J. Pract. Theory **2**(29), 175–187 (2010)

Indonesian Corruption Watch: The surge of corruption in village (2018)

Jensen, M.C., Meckling, H.: J. Financ. Econ. **3**, 305–360 (1976). 115.248.176.49

Kusumastuti, N.R., Meiranto, W.: Analysis of factors that influence unethical behavior as intervening variables. Diponegoro J. Account. **1**(1), 1–15 (2012)

Lawrence, G.M., dan Wells, J.T.: Basic Legal Concepts (2004)

Merdeka.com: Prevent fraud, KPK and government creates Timnas PK. https://m.merdeka.com/peristiwa/cegah-korupsi-kpk-dan-pemerintah-bentuk-timnas-pk.html?utm_source=GoogleAMP&utm_medium=CrossLink&utm_campaign=Mdk-AMP-Crosslink&utm_content=Artikel-4. (2018)

Ojha, A., Palvia, S., Gupta, M.P.: A model for impact of e-government on corruption: exploring theoretical foundations. *Foundations*, 160–170 (2008). www.csi-sigegov.org

Omar, N., Bakar, K.M.A.: J. Mod. Account. Audit. **8**, 15–30 (2012)

Othman, R. et al.: Procedia Econ. Financ. **28**, 59–67 (2015)

Rahman, R.A., Anwar, I.S.K.: Procedia Soc. Behav. Sci. **145**, 97–102 (2014)

Ramamoorti, S., Dupree, J.: Continuous monitoring can help deter and prevent fraud. Financ. Exec. 66–67 (2010)

Shanmugam, J.K., Haat, M.H.C., Ali, A.: Int. J. Bus. Res. Manag. (IJBRM) **3** (2012)

Sharma, R.: Digital infrastructure in India. In: Symposium on Collaborative Regulation for Digital Societies (2017)

Sudarma, M., Putra, I.W.: Effect of good corporate governance on agency costs. E-J. Account. Udayana Univ. **9**(3), 591–607 (2014)

Tyas, D.L., Budiyanto, A.D., Santoso, A.J.: Measurement of digital community gap in Pekalongan. In: SENTIKA (2016)

Wibisono, N., Purnomo, H.: Revealing public supervision phenomenon on village funds in district. AKSI J. (Account. Inform. Syst.) **2**, 8–19 (2017)

Wida, S.A., Supatmoko, D., Kurrohman, T.: Akuntabilitas Pengelolaan Alokasi Dana Desa (ADD) di Desa – Desa Kecamatan Rogojampi Kabupaten Banyuwangi (The accountability in the management of the village fund allocation in villages at Rogojampi District, Banyuwangi Regency) **IV**(2), 148–152 (2017)

Widiyarta, K., et al.: Fraud prevention in management of village funds (Empirical study in village government in district). E-J. Bachelor Account. Univ. of Pendidikan Ganesha **1** (2017)

Wilopo: Analysis of factors influencing accounting fraud trend: a study of public companies and BUMN in Indonesia. In: SNA 9. Padang (2006)

Wuysang, R.V.O., Nangoi, G., Pontoh, W.: Analysis of the implementation of forensic accounting and investigative audits on prevention and disclosure of frauds in regional financial management on representative of BPKP Sulawesi Utara Province. J. Acc. Forensic **31–53**, 2 (2016)

The Role of the Human Capital and Network in Maintaining the Sustainability of IMFI in the Digital Era: An Islamic Perspective

Widiyanto bin Mislan Cokrohadisumarto[(⊠)]

Department of Management, Faculty of Economics,
Sultan Agung Islamic University,
Islamic Economic Studies and Thought Centre (IESTC), Semarang, Indonesia
widiyantopunt@hotmail.com

Abstract. The issue of poverty still needs to get a long-term solution. This requires a sustainable microfinance institution. In the digital era, the sustainability of microfinance institutions will face many challenges due to technological changes. This paper offers a model for increasing the sustainability of Islamic microfinance institutions in the digital era on the basis of Islamic human capital and information technology-based network that are imbued with Islamic values. Islamic human capital has characteristics that are very different from the concept of human capital in general. Likewise, network is imbued with Islamic values.

Keywords: Islamic microfinance · Islamic human capital · Network · Sustainability

1 Introduction

The spread of poverty in various parts of the world is still a problem that needs to get a sustainable solution, given that the solution to poverty takes a long time. Microfinance still seems to be considered important and strategic in solving this problem. This is evidenced by the many microfinance institutions that have emerged and developed in various countries (such as in Georgia, Bolivia United States, United Kingdom, Sudan, Pakistan, Bangladesh, Malaysia, Indonesia) that provide capital assistance and development to poor people with the intention of making them work actively and able to improve their lives. Al-Qardawi (2006) states that work is a way to distance themselves from poverty. In addition to the establishment of thousands of conventional microfinance institutions, in the past few decades Islamic MicroFinance Institutions (IMFI) have also emerged to participate in poverty reduction efforts as mentioned (especially in countries that are predominantly Muslim). Kamel and Jalel (2015) stated that microfinance is potential for poverty reduction. Furthermore, the results of research conducted by Adnan and Ajija (2015) show that IMFI is effective in reducing poverty. The broader impact of this is Islamic microfinance provides opportunities for participants to have an important role in economic development (Samer et al. 2015). Good results as mentioned above need to be continued so that the implementation of poverty alleviation can be achieved as expected.

© Springer Nature Switzerland AG 2020
L. Barolli et al. (Eds.): CISIS 2019, AISC 993, pp. 966–971, 2020.
https://doi.org/10.1007/978-3-030-22354-0_92

The digital era will be a challenge for efforts to maintain the survival of IMFI, which in that situation technology will change very rapidly. According to Jens et al. (2018) that in the digital era every organization will only be able to become a technology taker (there is no choice but to use technology that is always updated and interconnected). This technology has the potential to weaken processes, people and organizations while creating new more efficient ways of working. Technology also demands a certain price from users: to play in the digital era, one must change behavior at work. Furthermore, organizations must use sustainable management of change to control how modern technology, created externally, affects their organizations. The above situation shows that in the digital age IMFI will be faced with threats and opportunities. Readiness to face threats and challenges will be the key to maintaining the sustainability of IMFI, and it lies in the readiness of the human resources and network which are owned by the organization.

Therefore, IMFI is an organization based on Islamic values, so related to IMF sustainability in the digital era, this paper will highlight the important role of human capital (in an Islamic perspective which will be called Islamic human capital) and the network which is imbued by Islamic values.

2 Literature Review

2.1 Islamic Microfinance Institutions Sustainability in the Digital Era

IMFI sustainability seems to be a necessity for achieving long-term poverty alleviation programs. It must be realized that eradicating poverty is not like giving a panacea once given will cure. In general, Zubair (2016) stressed sustainability on the ability of organizations to operate the adopted systems in order to achieve long-term sustainability. More detailed, sustainability, according to Bouljelbene and Fersi (2016), refers to long term on-going programs of microfinance institutions that continuously generate income for its client and able to keep combination from the three performance; managerial, social and financial. However, Doshi (2015) mentioned that the most important is the sustainability of benefits available to the clients in the form of income generation. It implies that the potential to continue as a self-generating system in a closed reinforcing loop. It helps microfinance institutions balance social value with financial goals and reach the scales necessary to achieve the goal of poverty alleviation.

The discussion above shows that achieving IMFI sustainability is not only related to one variable (which is generally only seen from the financial aspect), but also includes three things that must be achieved, namely those relating to managerial, social and financial aspects. Thus, in the digital era - where technology will continue to grow and potentially weaken processes, people, organizations and also create new ways of working more efficiently in the discussion above - IMFI in achieving its survival in serving the lower classes of society is demanded to have adequate capabilities to face any changes that occur. it means that IMFI has good management, provides the widest possible benefits for the community, especially the poor (such as creating jobs, increasing the incomes of the poor), and IMFI must get the advantages that will be used for its operational sustainability without having to get subsidies. To be able to do this, IMFI must have quality human resources (from the moral, spiritual and physical dimensions).

2.2 The Role of Islamic Human Capital

According to Ugbaja (2005), human capital is both men and women who work in order to achieve company goals. Becker (1964) defines Human capital as a knowledge, idea, and skill that is owned and obtained by someone through various activities such as formal education and through other job training. Roshayani et al. (2015) mention human capital with attitudes, competencies, skills, knowledge and innovation. Islam views that human capital not only as an intellectual and emotional emphasis, but also the most important is the spiritual aspect of divinity that must be owned by someone (Thaib, 2013). Therefore, Bilal et al. (2010) explain that Islam emphasizes dual quality: moral and professional, which then development must include knowledge of skills and Islamic knowledge. It must be realized that human existence on earth is as the representative of God for the prosperity of life on earth (QS: 2:30, 6; 165) and the human obligation to do organizational work is part of his duty as God's representative on earth. The following are some important values (elements) that must be possessed by each individual as God's representative on earth (including in carrying out organizational duties) in the Qur'an and the Sunnah of the Prophet. The values that can be guided are that every Muslim must have faith and knowledge (QS: 58; 11), faith, the practice of virtue, mutual counsel, and patience (QS: 103; 1–3), faith in Allah and practice virtue (QS: 98; 7), obedient to worship (QS: 51; 56), Praying, willing to sacrifice, and practice virtue (QS: 108; 2), knowledge (HR. Ibn Majah. No. 224), experts his field (HR. Bukhari No. 6015), pioneer and innovative (SDM. Muslim No. 1107), improve themselves: want to change to be better (creative) (QS: 13; 11), behave well (QS: 99; 6–7), dare to invite goodness and prevent evil (*munkar*) (QS: 3; 110), fulfil promises, and be patient (QS: 2; 177).

From the table above, it can be seen that Islamic human capital is not only related to attitudes, competencies, skills, knowledge and innovation, but also a wider scope, with its main base being a factor of faith. With this faith, it will be able to mobilize other things, such as working on virtue, studying, becoming a pioneer, innovative, creative and so on. Thus, the organization (IMFI) that has human capital with the characteristics as mentioned in the table above will be able to deal with changes that occur in the digital era. Rafiki et al (2014) state that Islamic human capital can influence the performance and growth of an organization.

2.3 The Role of Network

Ranjay (n.d.) describes the organizational network as a group of organizations that care each other, where there is a close relationship (because of common interest or values) among the members of the organization. Business networks have an important role, Kariv et al. (2009) found that business networks have a significant role in the measurement of business performance, especially in sales and survival. It is in line with the finding of Nzuve and Bundi (2014); that human capital would boost the company's performance in the presence of a strong network. Kalm (2012) found a positive relationship between networking and company growth. Relating to network quality; Mutaqin and Cokrohadisumarto (2018) found that network quality (i.e., help each other, learn from each other, commitment to move forward together, tell each other) is able to strengthen the impact of human capital on the IMFI performance. The above discussion shows that that network has an important role on the improvement of organizational performance, and then it would improve the organizational sustainability.

To face the digital era, IMFI must improve the quality of its network including sharing knowledge, sharing technology, sharing values, and increasing attitudes of help. In implementing information technology networking, it is necessary to have awareness and involvement of all parties in the organization. The existence of information technology networking encourages the capabilities and competencies of human resources owned by the organization to be better. Networks built based on technology and information will support the creation of long-term organizational growth (especially the Islamic microfinance institution). Given that technology and information are important aspects to face the digital era development, especially in determining organizational strategy.

3 Proposed Sustainability Model

This paper offers a sustainability model of IMFI in the digital era with basing on Islamic human capital and information technology-based network that is imbued of Islamic values.

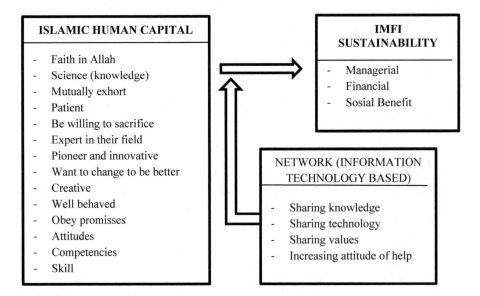

Fig. 1. Sustainability Model of Islamic Microfinance Institutions (IMFI). Improvement of sustainability will be built through Islamic values-based human resources supported by the strength of information technology-based networks inspired by Islamic values.

4 Conclusion and Future Research

The existence of IMFI is still very much needed to be able to help solve the problem of poverty. Therefore, the sustainability of these institutions needs to be maintained. To maintain the sustainability of IMFI in the digital era, human capital with Islamic characteristics and network quality based on technology is needed. This matter is very important to get attention related to the sustainability of IMFI in the digital era. For future research, exploring deeper the important role of Islamic human capital and information technology-based networks on the sustainability of IMFI is very necessary.

References

Adnan, M., Ajija, S.: The effectiveness of Baitul Maal wat Tammwil in reducing poverty: the case of Indonesia Islamic Microfinance Institution. Humanomics **31**(2), 160–182 (2015)

Al-Qardawi, A.Y.: Economic Security in Islam. Abdul Naeem for Islamic Book Service, New Delhi (2006)

Becker, G.S.: Human Capital: A Theoretical and Empirical Analysis, with Special Reference to Education. National Bureau of Economic Research, London (1964)

Bilal, K., Farooq, A., Hussain, Z.: Human resource management: an Islamic perspective. Asia-Pacific J. Bus. Adm. **2**(1), 17–34 (2010). https://doi.org/10.1108/17574321011037558

Bouljelbene, M., Fersi, M.: The determinants of the performance and the sustainability of confentional and Islamic microfinance institions. Econ. World **4**(5), 197–215 (2016)

Doshi, K.: Sustainability and impact of microfinance institutions: a case study of ACCION San Diego. In: Positive Design and Appreciative Construction: From Sustainable Development to Sustainable Value, pp. 275–295 (2015). Published online 09 Mar 2015

Jens P. Flanding, Grabman, G.M., Cox, J.D.S.Q.: Leading change in the digital era. In: The Technology Takers, pp. 159–172. https://doi.org/10.1108/978-1-78769-463-720181005 (2018). Published online 07 Nov 2018

Kamel, B.H.M., Jalel-Eddine, B.R.: Microfinance and poverty reduction: a review and synthesis of empirical evidence. Procedia. Soc. Behav. Sci. **195**, 705–712 (2015)

Kalm, M.: The impact of networking on firm performance. Evidence from small and medium-sized firms in emerging technology areas. ETLA – The Research Institute of the Finnish Economy Discussion Paper No. 1278. (2012)

Kariv, D., Menzeis, T.V., Brenner, G.A., Filion, L.: Transnational networking and business performance?: ethnic entrepreneurs in Canada. Entrepreneurship Reg. Dev. **21**(3), 239–264 (2009)

Mutaqin, M., Cokrohadisumarto, W.b.M.: The role of network quality as moderating variable on the performance enhancement of Islamic microfinance institutions (Cases of Baitul Mal Wat Tamwil). Iqtishadia **11**(2) (2018). P-ISSN: 1979-0724 E-ISSN: 2502-3993. https://doi.org/10.21043/iqtishadia.v11i2.3198

Nzuve, S.N.M., Bundi, E.G.: Human capital management practices and firms performance: a survey of commercial banks in Kenya. J. Econ. Lit. **39**(4), 1–13 (2014)

Rafiki, A., Alkhalifah, K., Buchori, I.: Islamic human capital and firm performance: an evidence of small and medium enterprises in Bahrain. Int. J. Bus. Manage. **9**(4), 173 (2014)

Roshayani, A., Noorb, A.H.M., Yahya, A.: Human capital and islamic-based social impact model: small enterprise perspective. Procedia Eco. Finance **31**, 510–519 (2015)

Samer, S., Majid, I., Rizal, S., Muhamad, M., Halim, S.: The impact of microfinance on poverty reduction: empirical evidence from Malaysian perspective. Procedia Soc. Behav. Sci. **195**, 721–728 (2015)

Thaib, L.: Human capital development from Islamic perspective: Malaysia's experience. Eur. J. Manag. Sci. Econ. **1**(1), 11–23 (2013)

Ugbaja, C.O.: A Modern Approach to Management. Pacific Publishers, Obosi (2005)

Zubair, M.K.: Analisis Faktor-Faktor Sustainabilitas Lembaga Keuangan Mikro Syariah. Iqtishadia: Jurnal Kajian Ekonomi dan Bisnis Islam STAIN Kudus **9**(2), 201–226 (2016)

Collaborative Agility Capital: A Conceptual Novelty to Support Knowledge Management

Tri Wikaningrum[1,2(✉)], Heru Sulistyo[1], Imam Ghozali[2],
and Ahyar Yuniawan[2]

[1] Department of Management, Faculty of Economics,
Sultan Agung Islamic University, Semarang, Indonesia
{wika,heru}@unissula.ac.id
[2] Diponegoro University, Semarang, Indonesia
ghozali_imam@yahoo.com, ahyar_yuniawan@yahoo.com

Abstract. Research that examines the correlation of human resource management practices and knowledge management showed inconsistencies in findings,
and differences in perspective in understanding those two variables. Previous
studies confirmed the effect of HRM practices on knowledge management in
organizations. However, a number of studies did not show a significant effect
between the two. Therefore, a study of the development of socio-cultural factors
is needed to strengthen this practice. The authors propose the concept of collaborative agility capital developed from a synthesis of social exchange theory
and resource-based theory of organization. This competency is expected to
contribute meaningfully to improve the quality of knowledge management
process in the organization. This article also provides direction on research to
test this conceptual novelty.

Keywords: Collaborative agility capital · Knowledge management ·
HRM practices · Social exchange theory ·
Resource-based theory of organization

1 Introduction

The study that examined the correlation between Human Resource Management
(HRM) practices and knowledge management showed inconsistencies and different
perspectives in understanding those two variables. A number of studies confirm the
effect of HRM practices on knowledge management in organizations. However, some
studies do not show significant effect between the two. The research of Lin [21] showed
that when knowledge management moves to a higher position, the company should
develop HR-oriented practices to motivate and encourage their interaction. Similarly,
other studies have placed HR management as part of the infrastructure used by companies to support their knowledge management activities [13]. This is in line with the
research of Cabrera and Cabrera [6], confirming that companies can create a conducive
environment for the creation or knowledge sharing through HRM practices such as
team work, promotion, socialization programs, performance appraisal, and compensation. Furthermore, Chen and Huang [9] also analyzed the role of mediation of

© Springer Nature Switzerland AG 2020
L. Barolli et al. (Eds.): CISIS 2019, AISC 993, pp. 972–980, 2020.
https://doi.org/10.1007/978-3-030-22354-0_93

knowledge management capacity (acquisition, transfer, and application) in the correlation between strategic HR practices (training, compensation, performance assessment, selection and participation) and innovation performance (technical, administrative). Likewise, Lopez-Cabrales et al. [23] also found the intervening effect of employee knowledge uniqueness in the correlation of the number of collaborative HRM practices and innovation performance.

Several studies showed different findings. The research of Afacan Fındıklı et al. [2] showed a difference in the effect of HR practices on knowledge management capacity. The researcher examined the HRM practices variable which included training, compensation, performance appraisal, staffing, and employee recruitment. The study showed only performance and compensation assessments have a significant effect on knowledge acquisition, knowledge application and knowledge sharing. Then, training practices only have a significant effect on the application of knowledge. So, training practices do not have a significant effect on the acquisition of knowledge as well as knowledge sharing. Moreover, in relation to the application of knowledge, training practices have the least effect of training compared to the practice of performance appraisal and compensation. Meanwhile, staffing practices and employee recruitment have no significant effect on the three knowledge management dimensions (knowledge acquisition, knowledge sharing, and knowledge application).

The study of Chen and Huang [9] also showed the inconsistency results. This study aims to examine the role of knowledge management capacity in the correlation of strategic human resource practices and innovation performance from a knowledge standpoint. This is based on the idea that innovation depends on the organization's human capital, including the skills, knowledge and attitudes of employees, which can be affected and sharpened through existing HR management practices. However, knowledge in the company that is attached to the individual in it is not easily transferred to other employees. Transfer of knowledge requires the capacity to manage knowledge and motivation as well as the willingness of employees to share and apply it. The variable of HRM Practices examined included staffing, participation, compensation, training, and performance appraisal functions, while the variable of Knowledge Management Capacity included dimensions of knowledge acquisition, knowledge application, and knowledge sharing. The results showed that staffing and participation have a positive and significant effect on knowledge acquisition, application and sharing. However, training only has a positive and significant effect on knowledge acquisition and application. Likewise, compensation only has a positive and significant effect on application and knowledge sharing. Meanwhile, the practice of performance appraisal has no significant effect on the three dimensions of knowledge management.

Research Prieto Pastor et al. [25] which examined the correlation of HR management and knowledge management also provided different findings. HRM practices are expected to have an effect on the ability, motivation, and opportunities of employees to participate in knowledge management, making it possible to share, maintain and create knowledge within the organization. HRM practices that are expected to have an effect on employee capabilities include training and development practices, while motivation includes valuation practices and rewards, and employee opportunities include providing support for trustworthy collaborative relationships. The results showed that HRM practices aimed at motivating and giving employees the opportunity to behave as the company's expectation, significantly affecting knowledge sharing and maintenance of knowledge. Then, both mediate the correlation between motivation and opportunities and knowledge creation. However, HRM practices aimed at increasing employee capabilities do not have a significant effect on knowledge sharing, knowledge maintenance, and knowledge creation. So, knowledge sharing and knowledge maintenance are not also variables that mediate training and development practices with knowledge creation.

Knowledge Sharing depends on the capability of employees to share knowledge. This capability can be improved through appropriate HRM practices. Therefore, it is important to take it into account when designing HRM practices that are capable of supporting knowledge sharing capabilities among employees. Abdul-Jalal et al. [1] examined whether employee perceptions of the capability of knowledge sharing have an effect on the success of knowledge sharing. It resulted that a combination of ability, motivation and opportunity is a key mechanism to support the flow of knowledge within the company. The researcher stated that the opportunity to share knowledge depends on HRM practices that enable social exchange to support the development of formal and informal employment relationships.

Companies engaged in the service industry whose competitive advantage is gained through the creation of new knowledge show that their knowledge process does not only transfer knowledge from people to documents [16], however, more focus on personalization strategies i.e. to improve social processes to facilitate sharing of tacit knowledge among members of the organization. The knowledge personalization strategy emphasizes human factors, not technology. That is why socio-cultural factors are needed to increase knowledge management activities.

2 Concept Development

The presentation in the previous paragraph underlies the idea of the need to examine concepts related to socio-cultural factors in the correlation of HRM practices and knowledge management. Some efforts to encourage socio-cultural factors are needed to increase the willingness of employees to participate in knowledge acquisition. As Hislop [19] stated, HRM practices can affect socio-cultural factors that have an effect on the desire of employees to support the company's knowledge management activities.

This study develops the concept of collaborative agility capital developed from a synthesis of social exchange theory and resource-based theory of organization. This concept is related to collaborative interpersonal interaction behaviour in the context of learning process. Collaborative interactions between individuals in organizations will facilitate the creation of new knowledge to support the organization's ability to face external challenges. Such behaviour not only minimizes the company's dependence on certain potential individuals, but also increases the capability of knowledge creation through the flow of knowledge at the level of individuals, groups, and organizations. Therefore, it is clear that there is a connection between social network and learning process.

2.1 Collaborative Capital in the Perspective of Social Exchange Theory

Collaborative capital is a concept derived from the theory of social exchange which was coined by George C. Homans in the mid 20th century or the 1950s, and later developed by, one of them, Peter M Blau. Homans's exchange theory rests on the assumption that people engage in behaviour to get reward or avoid punishment, in other words they minimize costs and increase profits. Exchange of behaviour to obtain rewards is a basic principle in simple economic transactions. Homans saw all social behaviour and economic behaviour as a form of exchange to get rewards. Rewards referred here include extrinsic rewards such as salary/wages, facilities, and intrinsic rewards such as satisfaction with work results, friendship relationships, and job prestige.

Clear articulation of exchange orientation for the first time was given by Peter Blau [15]. He stated that individuals are basically driven by hedonic motivations, in which all actions are directed towards seeking pleasure and reducing painful things. To get rewards and reduce penalties, they must carry out various behaviours. Thus, social life is based on a series of transactions in which compensation and costs depend on the exchange of behaviour with other individuals.

Social exchange theory is a supporter in the synthesis of theories that produce new concepts in this study. Social exchange theory is the basis for developing social capital in the context of the organization. Social relation both in the form of formal and informal among individuals is the capital to implement collaboration that supports the effectiveness of an interactive learning process. The existence of trust in these relationships has the potential to increase collaboration and encourage information sharing, both among employees and managers and among organizational units.

Social exchange theory and the norms of reciprocity on the basis of trust lead to the quality of resource exchange among members of the organization, contributing to collaborative capital [27]. The concept of collaborative capital is actually social capital according to Putnam version. In contrast to social capital in Burt's view, which focuses on private goods, Putnam views social capital as public property. Thus, the social exchange that occurs will avoid opportunism and increase collective action between the two parties. In relation to knowledge management activities, those attitudes and actions are important for the organization.

2.2 Agility Learning in the Perspective of Resource-Based Theory of Organization

Resource-based theory of organization states that competitive advantage can be achieved through the empowerment of human resources. In one hand, organizations that achieve their goals to compete using assets and resources that are valuable, rare, and difficult to imitate will be able to achieve competitive advantage [3, 4]. The ability of organizations to attract potential human resources consistently and effectively, develop and renew their capabilities, increase their commitment in achieving organizational goals, generate ideas for continuous improvement, and maintain their existence, will perpetuate them. On the other hand, organizations which ignore those all which do not consider their human talents as valuable assets will lose in competition.

Human resources are considered valuable assets because they are a source of knowledge while being able to access knowledge and conduct learning in a sustainable manner. Such individuals are said to be high potential talents that must be managed by the organization. High-Po talent is not just a learning ability and high competence, but also agility and consistency to continue learning that is able to increase the knowledge needed and valuable for the organization in responding to the challenges of change. That is why resource-based theory of organization underlies talent management.

Talent management is a continuous process that covers all processes of human resource activities leading to selection, outreach, maintenance and development [14, 26]. The management of these employee talents is expected to support organizational agility in dealing with market complexity and the dynamics of change. Employee talents are a source of knowledge because of their ability to access external knowledge, generate ideas/knowledge, transfer and apply the knowledge within their organization [11, 32]. So, in order for companies to remain competitive, the company may as well, must consistently manage the acquisition and application of knowledge among its employees; also ensure the transfer of knowledge among them through increasing social capital. Jones [20] stated that to build social capital, management of employee talent becomes an important thing that must be done by the company.

In its later development, talent is defined as high potential and high performance [31]. High potential is someone who has the ability, engagement and aspiration to develop, while high performance is associated with individuals who have a high level of expertise, leadership behaviour, creativity, and initiatives based on self-confidence. Some authors recommend that a critical component of talent management is the development of a structured process to identify high potential [8, 18]. At present and in the future, organizations need high potential that has an open character, a desire to learn and experience something new, and tolerance for high ambiguity, innovation, and flexibility in carrying out complex strategies. Such individuals are a figure of learning agility which is currently seen as a high potential key indicator [11, 29].

Human resources as a source of competitive advantage must support their organizations with the willingness and ability to learn something new, tolerance for ambiguity, flexibility, innovation, mobile, and education [5, 12]. The key words are not just "ability to learn" and "increased competence". This is because both of them only improve individual abilities, openness to experience, motivation to learn, and seek self-development opportunities. However, the way to have continuously and skillfully "ability to learn" to support the organization is able to deal with change and seize opportunities. Paauwe and Boselie [24] and Lombardo and Eichinger [22] assert that action is learning agility.

2.3 Collaborative Agility Capital

In the context of developing human resources, organizations become potential workplaces to support the learning process. The learning process is expected to provide opportunities for potential individuals to develop their competencies (knowledge, expertise, and abilities) through existing social networks. This social network can be considered as a supporter of the learning process, because learning is not an individual activity but there are elements of interaction in it. Therefore, there is a clear link between talent management, learning, and social networks.

Collaborative capital is a concept derived from social exchange theory. This theory explains that social life is colored by a series of exchange transactions between one individual with another. There is a phenomenon of cooperation, mutual help behaviour, the formation and acceptance of norms, and reciprocal actions. This explains the development of social capital theory which is a network of relationships that allows the flow of knowledge resources to be exchanged.

Social capital both internally and externally increases opportunities for employees to capture knowledge from inside and outside the organization. Although the employee network ties create opportunities for sharing knowledge with colleagues, adequate norms and trust are needed to exploit these opportunities [34]. This shows that trust is a fundamental dimension of Social Capital. The research also explores the mechanisms underlying the effect of Social Capital on Knowledge Sharing (interactive condition).

In an organizational context, social exchange theory is used as a basis for understanding the feelings of obligation and pro-organizational members of the organization. The greater the diversity of exchange relations between employees, the lower the feeling of being obliged to reciprocate the actions of coworkers, and the lower the identification of themselves with groups or organizations. As stated by Wikaningrum [33], the diversity of exchange relationships often causes negative impacts on communication and interpersonal attraction. Communication problems can reduce group cohesiveness, and low personal attractiveness has an effect on organizational outcomes through the level of desire to maintain membership in the organization.

Based on the previous description, it can be concluded that the study of interpersonal relations is not sufficiently discussed at the individual level, therefore, it is important to examine at the level of relations. At this level, there is a two-way interaction between the receiver and sender in resource exchange activities. This is what is meant by collaboration, which is an interactive and continuous exchange relationship among coworkers. Meanwhile, collaborative capital shows the quality of the exchange relationship.

The quality of the exchange relationship between individuals in the organization will support the effectiveness of an interactive learning process. The results of learning increase the capacity of individuals who have four aspects of dexterity as coined by Lombardo and Eichinger [22]. First is agility in human relations that describes a person's ability to recognize themselves, learns from experience, treats others constructively, and calmly faces the pressure of change. Second, dexterity results which describe someone who achieves under the harsh conditions, inspires others to perform more, and their existence which is able to build the trust of others. Third is mental agility which is related to one's ability to face complexity, ambiguity, and explains thoughts to others. Finally, dexterity changes reflect to someone who has curiosity, enthusiasm to come up with ideas, and engage in activity building skills. This shows that the agility of learning will have an effect and benefit for other individuals. If all individuals in the organization undergo the learning, and collaboration between them occurs, then beneficial resource exchanges will be more effective.

3 Research Direction

The implementation of human resource management practices has a role in encouraging employee behaviour and positive attitudes towards learning activities. As a statement of Hatch and Dyer [17] and Streumer [30] that HRM practices also has a positive effect on learning behaviour. So that, it has an impact on the ability of employees to respond to business changes quickly and flexibly. The dynamics of change require exploration of knowledge. However, knowledge creation requires social interaction and continuous learning among individuals. Both will facilitate the movement of knowledge from the individual level to the level of groups and organizations, and potentially transfer explicit and tacit knowledge. This becomes the background of the correlation between learning competencies and knowledge creation in organizations.

Collaborative agility capital as a conceptual novelty in this study is defined as an agile learning competency. It is supported by a balanced and sustainable quality of resource exchange. This competency is characterized by respect, trust, willingness to learn, inspiring, flexible, partnership, and learning speed. They have the potential to improve knowledge management process, especially knowledge acquisition and knowledge transfer. Furthermore, it is necessary to test the contribution of collaborative agility capital in strengthening the relationship of HRM practices and knowledge management.

Fig. 1. Proposed empirical model

Further research needs to be implemented to test empirically the concept of collaborative agility capital, as the research model in Fig. 1. Testing this research model can be applied to the settings of the banking industry which can be based on these reasons. First, banking is a knowledge-intensive organization where the knowledge workers interact highly with external stakeholders. Second, the principle of operationalizing the work has been standardized and must be obeyed by all employees. Third, the level of potential employee turnover is quite high in the banking industry. Fourth, the dynamics of change and intense competition appear in the era of financial technology (fintech). The four of them increase banking interests to transfer knowledge at the individual level to organizational knowledge. This is not only to minimize organization dependence on certain individuals, but also to encourage the learning process among employees.

The data analysis will be conducted using path analysis to test the regression equation mediating variable. HRM practices variable uses the concept developed by Chuang [10], i.e. with the three dimensions which include motivation, opportunity, and competency-enhancing HR practices. The variable of collaborative agility capital uses the indicators of flexibility, ability to inspire, willingness to learn, respect, trust, partnership, and learning speed. Then, the variable of knowledge management uses the concept of Nonaka developed by Song and Kolb [28] and Chang [7], i.e. with the four dimensions which include socialization, externalization, combination, and internalization.

4 Conclusion

Social exchange theory is very important in explaining the support of social networks in individual learning processes that encourage the transfer/exchange of knowledge resources. The process is supported by interactions based on a relationship of mutual trust and mutual effort. The synthesis of the social exchange theory with resource-based theory of organization creates a novelty concept; collaborative agility capital. This learning competency is not only at the individual level. However, ongoing collaborative learning has an impact on other members of the organization and supports the ability of organizations to follow the challenges of change. This can be a reference for further research to consider collaborative agility capital as a variable that plays a role in mediating the correlation between HRM practices and knowledge management.

References

1. Abdul-Jalal, H., Toulson, P., Tweed, D.: Knowledge sharing success for sustaining organizational competitive advantage. Procedia Econ. Finance 7, 150–157 (2013)
2. Afacan Fındıklı, M., Yozgat, U., Rofcanin, Y.: Examining organizational innovation and knowledge management capacity the central role of strategic human resources practices (SHRPs). Procedia – Soc. Behav. Sci. 181, 377–387 (2015)
3. Barney, J.: Firm resources and sustained competitive advantage. J. Manag. 17(1), 99–120 (1991)
4. Barney, J.B.: Looking inside for competitive advantage. Acad. Manage. Executive 9(4), 49–61 (1995)
5. Briscoe, D., Schuler, R., Claus, E.: International Human Resource Management. Routledge, London (2009)
6. Cabrera, E.F., Cabrera, A.: Fostering knowledge sharing through people management practices. Int. J. Hum. Resour. Manage. 16(5), 720–735 (2005)
7. Chang, C.M., Hsu, M.H., Yen, C.H.: Factors affecting knowledge management success: the fit perspective. J. Knowl. Manage. 16(6), 847–861 (2012)
8. Charan, R.: Ending the CEO succession crisis. Harvard Bus. Rev. 83(2), 72–81 (2005)
9. Chen, C.-J., Huang, J.-W.: Strategic human resource practices and innovation performance - the mediating role of knowledge management capacity. J. Bus. Res. 62(1), 104–114 (2009)
10. Chuang, C.H., Jackson, S.E., Jiang, Y.: Can knowledge-intensive teamwork be managed? examining the roles of HRM systems, leadership, and tacit knowledge. J. Manag. 42(2), 524–554 (2013)
11. De Meuse, K.P., Dai, G., Hallenbeck, G.S.: Using learning agility to identify high potentials around the world. Korn/Ferry Institute, pp. 1–22 (2008)
12. Dries, N., Vantilborgh, T., Pepermans, R.: The role of learning agility and career variety in the identification and development of high potential employees. Pers. Rev. 41(3), 340–358 (2012)

13. Gold, A.H., Malhotra, A., Segars, A.H.: Knowledge management: an organizational capabilities perspective. J. Manage. Inf. Syst., Summer **18**, 1 (2001)
14. Goldsmith, M., Carter, L.: Best Practices in Talent Management: How the World's Leading Corporations Manage, Develop, and Retain Top Talent. Wiley, San Francisco (2009)
15. Greenberg, M.S.: Social Exchange: Advances in Theory and Research. Plenum Press, New York (1980)
16. Hansen, M.T., Nohria, N., Tierney, T.: What's your strategy for managing knowledge? In: The Knowledge Management Yearbook 2000–2001, pp. 55–69. Butterworth–Heinemann, Woburn (1999)
17. Hatch, N.W., Dyer, J.H.: Human capital and learning as a source of sustainable competitive advantage. Strateg. Manag. J. **25**(12), 1155–1178 (2004)
18. Hewitt Associates: The top companies for leaders. J. Hum. Resour. Plann. Soc. **28**(3), 18–23 (2005)
19. Hislop, D.: Knowledge Management in Organizations: A Critical Introduction, 3rd edn. Oxford University Press, Oxford, United Kingdom (2013)
20. Jones, R.: Social capital: bridging the link between talent management and knowledge management. In: Smart Talent Management: Building Knowledge Assets for Competitive Advantage. Edward Elgar Publishing, Cheltenham (2010)
21. Lin, H.-F.: The effects of employee motivation, social interaction, and knowledge management strategy on KM implementation level. Knowl. Manage. Res. Pract. **9**(3), 263–275 (2011)
22. Lombardo, M.M., Eichinger, R.W.: High potentials as high learners. Hum. Resour. Manage. **39**(4), 321–330 (2000)
23. Lopez-Cabrales, A., Pérez-Luño, A., Cabrera, R.V.: Knowledge as a mediator between HRM practices and innovative activity. Hum. Resour. Manage. **48**(4), 485–503 (2009)
24. Paauwe, J., Boselie, J.P.P.E.F.: HRM and performance: what's next? Hum. Resour. Manage. J. **15**(4), 68–83 (2005)
25. Prieto Pastor, I.M., Pérez Santana, M.P., Martín Sierra, C.: Managing knowledge through human resource practices: empirical examination on the Spanish automotive industry. Int. J. Hum. Resour. Manage. **21**(13), 2452–2467 (2010)
26. Schweyer, A.: Talent Management Systems: Best practices in Technology Solutions for Recruitment, Retention and Workforce Planning. Wiley, Hoboken (2004)
27. Smith, M.L.: Team-member exchange and individual contributions to collaborative capital in organizations. Adv. Interdisc. Stud. Work Teams **11**, 161–181 (2005)
28. Song, J.H., Kolb, J.A.: The influence of learning culture on perceived knowledge conversion: an empirical approach using structural equation modelling. Hum. Resour. Dev. Int. **12**(5), 529–550 (2009)
29. Spreitzer, G.M., McCall, M.W., Mahoney, J.D.: Early identification of international executive potential. J. Appl. Psychol. **82**(1), 6 (1997)
30. Streumer, J.N.: Work-Related Learning, pp. 103–106. Springer, Dordrecht (2006)
31. Tansley, C.: What do we mean by the term "talent" in talent management? Ind. Commercial Training **43**(5), 266–274 (2011)
32. Vaiman, V., Vance, C.: Smart Talent Management: Building Knowledge Assets for Competitive Advantage. Edward Elgar Publishing, Cheltenham (2010)
33. Wikaningrum, T.: Coworker exchange, leader-member exchange, and work attitudes. Gadjah Mada Int. J. Bus. **9**(2), 187–215 (2007)
34. Yen, Y.-F., Tseng, J.-F., Wang, H.-K.: The effect of internal social capital on knowledge sharing. Knowl. Manage. Res. Pract. **13**(2), 214–224 (2013)

Fostering Absorptive Capacity and Self-efficacy on Knowledge Sharing Behavior and Innovation Capability: An Empirical Research

Heru Sulistyo[(⊠)] and Tri Wikaningrum

Faculty of Economic, Department of Management, Universitas Islam Sultan
Agung, Semarang, Indonesia
{heru,wika}@unissula.ac.id

Abstract. This study aims to examine the effect of absorption capacity and self-efficacy on knowledge sharing behavior in improving the SMEs' innovation capability. The sample of this study was 106 SMEs in Central Java. Primary data collection is done by using questionnaire and interview instruments. The results showed that absorption capacity had a significant effect on knowledge sharing behavior and innovation capability. In addition, self-efficacy has a significant effect on knowledge sharing behavior and innovation capability, while Knowledge sharing behavior has a significant effect on innovation capability.

Keywords: Absorption capacity · Self-efficacy ·
Knowledge sharing behavior · Innovation capability · SMEs

1 Introduction

Knowledge sharing is one of the knowledge management activities in SMEs that play a role in increasing knowledge resources, especially owners of SMEs. Knowledge sharing is an important process in modern organizations because the success of knowledge sharing will result in shared intellectual capital and increasingly quality resources. Thus, it is necessary to strengthen knowledge sharing behavior in SMEs in order to be able to increase intellectual capital in encouraging increased innovation in increasingly dynamic global competition. Many SMEs have knowledge, especially related to the creative fashion industry, but are rarely shared among fellow entrepreneurs, both through formal and informal meetings, so that each of them innovates according to their mastery of knowledge. Low knowledge sharing behavior will lead to low innovation capabilities in products and processes, marketing innovation and innovation in the field of financial management. Some studies have examined the factors that influence the knowledge sharing behavior in various manufacturing and service industries. The current study examines knowledge sharing behavior as a mediation of self-efficacy and capacity to absorb knowledge of innovation capability in Fashion SMEs. By increasing knowledge capacity and fostering self-efficacy, it will encourage an increase in knowledge sharing behavior for employees in organizations that impact on the innovation of small and medium enterprises (SMEs).

© Springer Nature Switzerland AG 2020
L. Barolli et al. (Eds.): CISIS 2019, AISC 993, pp. 981–990, 2020.
https://doi.org/10.1007/978-3-030-22354-0_94

Innovation is an important factor in improving organizational performance especially small and medium enterprises (SMEs). Liao et al. [1] and Daniel and Raquel [2] stated that innovation activities mostly focus on improving technology and product development. The importance of innovation capability in achieving organizational competitive advantage encourages many researchers to test several antecedent factors of innovation capability. Innovation capability is built through intangibles assets, especially the concept of knowledge sharing in organization. Knowledge is important for human resources. If knowledge in the form of tacit knowledge possessed by each member of the organization is not shared with the organization, it will be difficult for employees to improve their performance, due to difficulties in innovating. Some of the weaknesses of SMEs in Indonesia is that the low innovation capability [3] The low level of innovation is due to the lack of knowledge sharing among employees in the organization and with other organizations.

The knowledge absorption capacity is related to efforts to find, acquire, change, and transfer expertise from knowledge sources to knowledge-based systems. Knowledge absorption is the first activity in the form of receiving knowledge from the external environment and changing so as to be used in organization. Buckley et al. [4] stated that the knowledge absorption is the transfer of knowledge resources of the company with the aim of gaining knowledge to learn. The higher the knowledge absorption capacity possessed by each member of the organization, it will facilitate the occurrence of knowledge sharing behavior in SMEs. Yang and Farn [5] found that the knowledge absorption had a significant effect on knowledge sharing activities. The knowledge absorption capacity is a major success factor in enhancing knowledge sharing behavior and innovation capability of SMEs. In addition, Sulistyo and Ayuni [6] found that the absorption capacity of knowledge was able to increase the innovation.

Some empirical studies show a positive relationship between self-efficacy and a person's tendency to be involved in sharing information. Kankanhalli et al. [7] argue that perceived expertise will increase one's confidence in what they can do. In the context of knowledge, when a person has confidence in the ability of his knowledge, he will tend to be brave enough to share the knowledge to fellow colleagues in the organization. Self-efficacy is one component of the Theory of Planned Behavior (TPB). According to Rahab et al. [8], self-efficacy is an individual assessment of their ability to manage and implement the actions needed to achieve performance. Rachna and Cevahir [9] showed that self-efficacy has a significant effect on innovation. In addition, Irene et al. [10] and Xi Zhang [11] found that there was significant influence between self-efficacy and knowledge sharing activities. This study aims to examine the effect of self-efficacy and knowledge absorption capacity in enhancing knowledge sharing behavior and innovation capability in SMEs. Several studies on the factors that influence the knowledge sharing behavior and innovation capability have been carried out by several researchers, but most of them rarely use self-efficacy as one component of the Theory of Planned Behavior (TPB) as an antecedent of knowledge sharing behavior in SMEs. Many researches have been done on established manufacturing and service organizations. Therefore, this study contributes the role of knowledge sharing behavior as a mediator between self-efficacy and knowledge absorption capacity on innovation capability.

2 Literature Review

2.1 Knowledge Absorption Capacity as a Determinant of Knowledge Sharing Behavior

Some researchers have defined the absorption capacity of knowledge varies. The knowledge absorption capacity is a way to acquire, change, and transfer expertise from a knowledge source to a knowledge-based system. Another definition of knowledge absorption capacity is an activity related to finding, obtaining, changing, and transferring expertise from knowledge sources to knowledge-based systems. Knowledge absorption is the first activity in the form of receiving knowledge from the external environment and changing so as to be used in organization. Buckley et al. [4] stated that the knowledge absorption is the transfer of knowledge resources of the company with the aim of gaining knowledge to learn. Yang and Farn [5] found that the knowledge absorption affects knowledge sharing activities. Agarwal et al. [12] concluded that knowledge absorption has a significant effect on knowledge sharing. The study results of Li-fen Liao [13] and Monica Hu et al. [14] concluded that knowledge sharing has a positive effect on increasing organizational innovation.

H1: Knowledge absorption capacity has a significant effect on knowledge sharing behavior.

2.2 Self-efficacy as a Determinant of Knowledge Sharing Behavior

Several empirical studies have reported a positive relationship between information self-efficacy and a person's tendency to be involved in sharing information. Kankanhalli et al. [7] argue that perceived expertise increases a person's confidence in what they can do. This perception, in turn, inspires individuals to share knowledge in organization. Self-efficacy is an important factor developed from social cognitive theory and represents an individual's assessment of his ability to perform certain actions or behaviors [15]. High efficacy individuals will tend to be stronger in facing obstacles and more active in acquiring and sharing knowledge [16]. Ahearne et al. [17] describe self-efficacy as a belief in an individual's ability to organize and carry out an action to achieve the necessary goals. To facilitate knowledge sharing environments, institutions need to create, disseminate and adopt knowledge. In the perspective of knowledge sharing, people will have a higher tendency to share his knowledge if he feels how important social norms are. Xi Zhang [11] also concluded that self-efficacy had a significant effect on knowledge sharing activities explicitly and tacit. Based on TPB's theory, if in organizations especially SMEs, the more self-efficacy employees have, it will increase knowledge sharing behavior. Chen et al. [10] found that there was a significant effect between self-efficacy and knowledge sharing activities.

H2: Self-efficacy has a significant effect on knowledge sharing behavior.

2.3   Knowledge Absorption Capacity as a Determinant of Innovation Capability

Knowledge absorption capacity possessed by employees in the organization will facilitate interaction and knowledge sharing in an organization which then has an impact on increasing innovation capability. Liao [18] found that knowledge absorption significantly affected innovation. Studies conducted by Muscat and Deery [19] found that knowledge absorption capacity is very important in predicting organizational capabilities. The study conducted by Cohen and Levinthal [20] found that increasing the knowledge absorption capacity will have an impact on increasing innovation.

H3: Knowledge absorption capacity has a significant effect on innovation capability.

2.4   Self-efficacy as a Determinant of Innovation Capability

Innovation capability is the implementation and creation of technology that is applied to new systems, policies, programs, products, processes and services to the organization. Innovation capability is also the ability to absorb and use external information to be transferred to new knowledge [20]. Innovation capability is a comprehensive set of characteristics from organizations that facilitate and encourage innovation strategies. Innovation is an important organizational capability because the success of new products is an engine of growth and has an impact on increasing sales, profits, and competitive power for many organizations [21]. The study conducted by Jen et al. [22] found that high self-efficacy will produce innovation, due to direct involvement in interacting to produce creative ideas.

H4: Self-efficacy has a significant effect on innovation capability.

2.5   Knowledge Sharing Behavior as a Determinant of Innovation Capability

Researchers have defined knowledge sharing based on their perspectives on knowledge sharing and research contexts [23]. Bartol and Srivastava [24] define knowledge sharing as individuals who share information, ideas, suggestions, and expertise that are organizationally relevant to each other. Knowledge sharing is also interpreted as an act of sharing experiences, events, thoughts or understanding of anything in the hope of gaining more insight and understanding about something for curiosity [25]. Hooff et al. [26] stated that knowledge sharing is a process between two individuals to share knowledge by bringing knowledge and gaining knowledge, so the knowledge sharing behavior involves sharing information between two or more individuals. Liao [13] also explained that knowledge sharing activities have a positive effect on improving organizational innovation. In addition, Monica Hu et al. [14] found a positive relationship between knowledge sharing and innovation activities. Furthermore,

H5: Knowledge sharing behavior has a significant effect on the innovation capability.

3 Methods

The sample of this study was Batik SMEs in Central Java with 106 respondents. Batik is a craft in the field of clothing that has high artistic value and has been part of Indonesian culture for a long time and has been recognized by UNESCO as an Intangible Cultural Heritage. Batik is one of the creative fashion industries in Indonesia that has contributed greatly to improve the Indonesian economy. The majority of batik fashion SMEs is in the Central Java, Indonesia. Sampling uses a purposive sampling method based on the consideration of Batik SMEs that have been operating for at least five years and still exist today. Primary data collection is done by using questionnaire and interview instruments. Questionnaires were given to SME owners. Knowledge absorption capacity measured by five items adapted from Liao et al. [18]. A Sample item is "I actively use various sources of information to support the completion of work. Measurement of self-efficacy was measured by four items adapted from [10]. A Sample item is "I feel confident I can express ideas". Knowledge sharing of items adapted from Liao et al. [18]. A Sample item is "I often share the new information to my colleagues". Innovation capability was measured by four items adapted from Andreeva and Kianto [27]. A sample item is "I often use new ideas to get things done" (Table 1).

Table 1. Result of outer loading

	Original sample	T Statistics (|O/STDEV|)	P Values
X1.1 ⇐ X1	0,927	51.243	0.000
X1.2 ⇐ X1	0,943	65.201	0.000
X1.3 ⇐ X1	0,895	35.655	0.000
X1.4 ⇐ X1	0,925	35.848	0.000
X1.5 ⇐ X1	0,922	71.503	0.000
X2.1 ⇐ X2	0,922	76.666	0.000
X2.2 ⇐ X2	0,934	85.317	0.000
X2.3 ⇐ X2	0,929	67.708	0.000
X2.4 ⇐ X2	0,910	42.558	0.000
Y1.1 ⇐ Y1	0,870	24.625	0.000
Y1.2 ⇐ Y1	0,897	34.341	0.000
Y1.3 ⇐ Y1	0,794	14.483	0.000
Y2.1 ⇐ Y2	0,926	58.870	0.000
Y2.2 ⇐ Y2	0,910	10.978	0.000
Y2.3 ⇐ Y2	0,791	13.840	0.000
Y2.4 ⇐ Y2	0,770	51.243	0.000

P-values $\leq 0,05$

Note: X1: Knowledge Absorption Capacity
        X2: Self-efficacy
        Y1: Knowledge Sharing Behavior
        Y2: Innovation Capability

4 Result

Based on the results of outer loading using the PLS program, all variable indicators of Self-efficacy, absorption capacity, knowledge sharing behavior, and innovation capability are all valid with a p-value ≤ 0.05 (Fig. 1, Table 2).

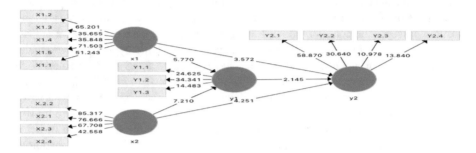

Fig. 1. The path coefficient results using PLS

Table 2. Effects among variables

	Original Sample (O)	Sample Mean (M)	Standard Deviation	T Statistics	P Values
X1 ⇒ Y1	0.413	0.411	0.072	5.770	0.000
X1 ⇒ Y2	0.325	0.324	0.091	3.572	0.000
X2 ⇒ Y1	0.508	0.510	0.071	7.210	0.000
X2 ⇒ Y2	0.397	0.401	0.093	4.251	0.000
Y1 ⇒ Y2	0.197	0.196	0.092	2.145	0.032

P-Values ≤ 0.05

Based on the PLS analysis, the original sample values obtained by the effect of absorption capacity on knowledge sharing behavior were 0.413 with p-values of 0,000 ≤ 0.05. Absorption capacity has a significant effect on knowledge sharing behavior, so H1 is supported. Absorption capacity also has a significant effect on the innovation capability, with an original sample value of 0.325 and p-values of 0.000 ≤ 0.05, so that H2 is supported. Self-efficacy has a significant effect on knowledge sharing behavior, p-values 0.000 ≤ 0.05, meaning that H3 is supported. Self-efficacy also has a significant effect on innovation capability with p values of 0.000 ≤ 0.05, so that H4 is supported. The results also showed that knowledge sharing behavior had a significant effect on innovation capability with p-values of 0.032 ≤ 0.05, so H5 is supported.

The results of the study show that the higher the absorption capacity of knowledge in organizations, especially SMEs, the higher individuals in an organization is willing to share knowledge in enhancing innovation capability. Self-efficacy also plays an important role in enhancing knowledge sharing behavior and innovation capability of SMEs. If the owners of SMEs have a perception of ease in gaining knowledge, it will be easier to share knowledge.

5   Discussion

Knowledge absorption capacity has a significant and positive effect on knowledge sharing behavior. The results of this study support previous studies by Yang and Farn [5] and Agarwal et al. [12]. In the context of knowledge sharing, strong self-efficacy becomes the main determinant in shaping optimism towards knowledge sharing [28]. For SME owners, efforts must be made to rely on the knowledge, skills, and resources, so that among SME owners can share information in encouraging increased innovation capabilities [15, 29]. Absorption capacity in obtaining high knowledge will facilitate the process of knowledge sharing in organizations. This study supports previous research by Li and Tao [30] who found that self-efficacy has an important effect in knowledge sharing. The results of this study also showed that self-efficacy had a significant effect on knowledge sharing. This is in line with the findings of Xi Zhang [11] who also concluded that self-efficacy had a significant effect on knowledge sharing activities explicitly and tacit. In addition, this also supports the findings of Chen et al. [10] and Xi Zhang [11] who stated that there is significant effect between self-efficacy and knowledge sharing activities. However, the research findings contradict the findings of Rahab et al. [8], Kwakye et al. [31] who found an insignificant relationship between self-efficacy towards knowledge sharing activities. Absorption capacity is very important in improving innovation capability. The higher a person in the organization, he will get a lot of knowledge, both explicit and tacit and even it can be absorbed and stored, it will increase the innovation capability in an organization. The results of this study support the findings of Liao et al. [18]; Muscat and Deery [19] and Cohen and Levinthal [20] that knowledge absorption capacity is very important in predicting organizational capabilities. The results of this study also show that high self-efficacy in the organization will improve the innovation capability. SME owners who have strong self-confidence and believe in their ability to carry out actions that will be adopted will be easier to innovate. This supports the findings of Jen et al. [23] who found that high self-efficacy will result in innovation, due to direct involvement in interaction resulting in creative ideas. Bandura [15] stated that perceived self-efficacy affects individual decisions about what behaviors should be done, the level of effort they will take in pursuit of the goals adopted, and the level of goal behavior that they set for themselves. The stronger the perceived efficacy, it will be the higher the individual's efforts to carry out activities. Knowledge sharing behavior has a significant effect on innovation capability. The results of this study support the findings of Liao [13] that knowledge sharing activities have a positive effect on improving organizational innovation. In addition, the results of this study also support the results of Monica Hu et al. [14] who found that knowledge sharing activities had a significant effect on innovation.

6 Conclusion

This study discusses the importance of the factors of absorption capacity and self-efficacy in improving the knowledge sharing behavior and innovation capabilities of SMEs. To overcome the obstacles faced by SMEs in improving innovation is by improving knowledge sharing behavior both formally and informally among SME owners and business associations. Knowledge is the main source for developing organizational innovation. Thus, it is necessary to strengthen the self-efficacy of each SME actor in order to have the trust and confidence to share knowledge with other parties among SME owners. A large number of studies also argue that one's self-efficacy can encourage them to share knowledge because they are more willing to share knowledge that will be valuable to others. This kind of thing needs to get the attention of SME players in order to encourage increased innovation capability. In addition to the factors of self-efficacy, this study also states the importance of knowledge absorption capacity in improving knowledge sharing behavior and innovation capability. The higher knowledge sharing among SME owners has the ability to absorb high knowledge that will facilitate to share knowledge with other parties so that it will encourage innovation capability.

7 Managerial Implication

SMEs need to improve the innovation capability through the creation of a good atmosphere related to the willingness of each individual to share knowledge, especially tacit knowledge. Managers cannot impose the willing to share tacit knowledge, but efforts are needed to improve the self-efficacy of employees in the organization combined with the knowledge absorption capacity. This study examines the importance of increasing knowledge sharing behavior among batik fashion SME owners in improving innovation capability through the ability to absorb knowledge and self-efficacy. This study only used one component of the theory of planned behavior in improving knowledge sharing and innovation capability. Therefore, future research needs to add subjective attitude and norm variables, so that Theory of Planned Behavior implementation is more holistic on SMEs with knowledge management. The sample of this study is still limited to the owners of the batik fashion SMEs whose numbers are still relatively limited, so that future research needs to be studied broader and uses wider range of SMEs.

References

1. Liao, S., Fei, W., Chen, C.: Knowledge sharing, absorptive capacity and innovation capability: an empirical study of taiwan's knowledge-intensive industries. J. Info. Sci. **33**(3), 160–167 (2009)
2. Daniel Jiménez, J., Raquel Sanz, V.: Could HRM supporting organizational innovation? Int. J. Hum. Resour. Manag. **19**(7), 1208–1221 (2008). Routledge Taylor & Francis

3. Siyamtinah, S.H.: Innovation capability of SMEs through entrepreneurship, marketing capability, relational capital and empowerment. Asia Pac. Manag. Rev. **21**(4), 196–203 (2016)
4. Buckley, P.J., Glaister, K.W., Klijn, E., Tan, H.: Knowledge accession and knowledge acquisition in strategic alliances: the impact of supplementary and complementary dimensions. Br. J. Manag. **20**, 598–609 (2009)
5. Yang, S.C., Farn, C.K.: Investigating tacit acquisition and sharing from the perspective of social relationship—a multilevel model. Asia Pac. Manag. Rev. **15**(2), 167–185 (2010)
6. Sulistyo, H., Sri, A.: How does knowledge absorption foster performance ? the mediating effect of innovation capability. Bagaimana Knowledge Absorption Dapat Meningkatkan Kinerja ? Efek Mediasi Dari Innovation Capability **9**(36), 114–125 (2018)
7. Kankanhalli, A., Tan, B.C.Y., Wei, K.K.: Contributing knowledge to electronic knowledge repositories: an empirical investigation. MIS Quarterly **29**(1), 113–143 (2005)
8. Rahab, Sulistyandari, Sudjono: The development of innovation capability of small medium enterprises through knowledge sharing process: an empirical study of Indonesian creative industry. Int. J. Bus. Soc. Sci. **2**(21), 112–123 (2011)
9. Kumar, R., Uzkurt, C.: Investigating the effect of self efficacy on innovativeness and the moderating impact of cultural dimensions. J. Bus. Cult. Stud. **4**(1), 1–15 (2010)
10. Chen, I.Y.L., Chen, N.S., Kinshuk: Examining the factor influencing participants' knowledge sharing behavior in virtual learning communities. Educ. Tech. Soc. **12**(1), 134–148 (2009)
11. Zhang, X.: Cultural influences on explicit and implicit knowledge sharing behaviour in virtual teams. Int. J. Comput. Sci. Info. Tech. **3**(4), 29–44 (2011)
12. Agarwal, P.D., Kiran, R., Verma, A.K.: Knowledge sharing for stimulating learning environment in institutions of higher technical education. Afr. J. Bus. Manage. **6**(16), 5533–5542 (2012)
13. Liao, L.-F.: A learning organization perspective on knowledge-sharing behavior and firm innovation. Hum. Syst. Manage. IOSS Press **25**, 227–236 (2006)
14. Monica Hu, M.L., OU, T.L., Chiou, H.J., Lin, L.C.: Effects social exchange and trust on knowledge sharing and service innovation. Soc. Behav. Pers. **40**(5), 783–800 (2012)
15. Bandura, A.: Self-Efficacy: The Exercise of Control. Freeman, New York, NY (1997)
16. Cabrera, A., Collins, W.C., Salgado, J.F.: Determinants of individual engagement in knowledge sharing. Int. J. Hum. Res. Manag. **17**(2), 245–264 (2006)
17. Ahearne, M., Mathieu, J., Rapp, A.: To empower or not to empower your sales force? An empirical examination of the influence of leadership empowerment behavior on customer satisfaction and performance. J. Appl. Psychol. **90**(5), 945–995 (2005)
18. Liao, S.H., Wu, C.F., Chih, C.C.: Knowledge sharing, absorptive capacity and innovation capability: an empirical study of taiwan's knowledge intensive industries. J. Info. Sci. **33**(3), 1–20 (2007)
19. Muskat, B., Deery, M.: Knowledge transfer and organizational memory: an events Perspective. Event Manage. **21**(4), 431–447 (2017)
20. Cohen, W.M., Levinthal, D.A.: Absorptive capacity: a new perspective on learning and innovation. Adm. Sci. Q. **35**(1), 128–152 (1990)
21. Sivadas, E., Dwyer, F.R.: An Examination of organizational factors influencing new product success in internal and alliance based processes. J. Mark. **64**(1), 31–49 (2000)
22. Tangaraja, G., Mohd Rasdi, R., Abu Samah, B., Ismail, M.: Fostering knowledge sharing behaviour among public sector managers: a proposed model for the malaysian public service. J. Knowl. Manag. **19**(1), 121–140 (2015)
23. Chang, J.C., Hsiao, H.C., Tu, Y.L., Chen, S.C.: The influence of teachers' self-efficacy on innovative work behavior. Int. Conf. Soc. Sci. Humanity **5**(6), 233–237 (2011)

24. Bartol, M.K., Srivastava, A.: Encouraging knowledge sharing: the role of organizational reward systems. J. Leadersh. Organ. Stud. **9**(1), 64 (2002)
25. Sohail, M.S., Daud, S.: Knowledge sharing in higher education institutions: perspectives from Malaysia. J. Info. Knowl. Manag. Syst. **39**(2), 125–142 (2009)
26. Van Den Hooff, B., De Ridder, J.A.: Knowledge sharing in context: the influence of organizational commitment, communication climate and CMC use on knowledge sharing. J. Knowl. Manag. **8**(6), 117–130 (2004)
27. Andreeva, T., Kianto, A.: Knowledge processes, knowledge-intensity and innovation: a moderated mediation analysis. J. Knowl. Manag. **15**(3), 1016–1034 (2011)
28. Ye, S., Chen, H., Jin, X.: An empirical study of what drives users to share knowledge in virtual communities. In: Lang, J., Lin, F., Wang, J. (eds.) Knowledge Science, Engineering and Management, pp. 563–575. Springer, Berlin (2006)
29. Huang, J.W., Li, Y.H.: Managing knowledge resource practices and innovation performance. Academy of Management Annual Meeting, pp. 1–13 (2008)
30. Li, Z., Zhu, T., Wang, H.: A study on the influencing factors of the intention to share tacit knowledge in the university research team. J. Softw. **5**(5), 538–545 (2010)
31. Kwakye Okyere, E., Nor, K.Md., Ziaei, S.: The influence of altruism, self efficacy and trust on knowledge sharing. J. Knowl. Eco. Knowl. Manage. **6**, 31–39 (2011)

Author Index

© Springer Nature Switzerland AG 2020
L. Barolli et al. (Eds.): CISIS 2019, AISC 993, pp. 991–994, 2020.
https://doi.org/10.1007/978-3-030-22354-0

Printed in the United States
By Bookmasters